Teacher's Edition

PRACTICAL MATHEMATICS

CONSUMER APPLICATIONS

Holt, Rinehart and Winston, Inc.

Austin New York San Diego Chicago Toronto Montreal

Authors

Marguerite M. Fredrick
Computer Consultant, K–12
Pleasantville Union Free School District
Pleasantville, New York
Formerly Math Chairperson, 7–12
The Ursuline School
New Rochelle, New York

Steven J. Leinwand
Mathematics Consultant, K–12
Connecticut State Department of Education
Hartford, Connecticut

Robert D. Postman
Professor and Chairperson
Mathematics Education
Mercy College
Westchester, New York

Laurence R. Wantuck
Mathematics Supervisor, K–12
Broward County Schools
Fort Lauderdale, Florida

For Photo and Art Credits, see page 486.

We would like to thank the following teachers and administrators who reviewed this text or who field-tested material from this text in their classrooms.

Pam Campbell
Mathematics Teacher
La Porte High School
La Porte, Texas

Donna M. Carter
Mathematics Teacher
Point Loma High School
San Diego, California

Carol Denicole
Mathematics Department
Trinity Preparatory of Florida
Winter Park, Florida

Dan Hall
Mathematics Instructor
Yerba Buena High School
San Jose, California

James Hill
Chairman, Mathematics Department
Lew Wallace High School
Gary, Indiana

E. Gillette Irby
Independent Mathematics Consultant
Corpus Christi, Texas

Donald Langlee
Mathematics Specialist
Duluth Public Schools
Duluth, Minnesota

Judy Needham
Mathematics Teacher
La Porte High School
La Porte, Texas

Mae Pierson
Mathematics Teacher
Washington Junior High School
Duluth, Minnesota

Diane Pors
Mathematics, Subject Area Coordinator
East Side Union High School District
San Jose, California

Patricia A. Yarbrough
Division Chairperson, Mathematics and Science
Rich East High School
Park Forest, Illinois

Printed in the United States of America

ISBN: 0-03-012769-6

7890 039 987654321

CONTENTS

CHAPTER 1 PAPER-AND-PENCIL COMPUTATIONS

CHAPTER 2 THE CALCULATOR AND ESTIMATION

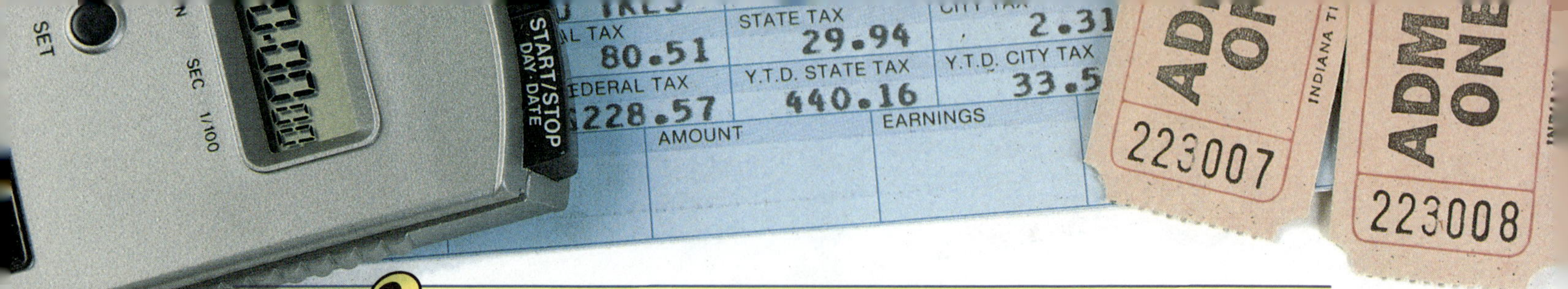

CHAPTER 3 PART-TIME AND SUMMER JOBS

CHAPTER 4 FULL-TIME WORK

CHAPTER 5 RECREATION AND SPORTS

CHAPTER 6 BASIC PURCHASES

CHAPTER 7 CHECKING AND SAVINGS ACCOUNTS

CHAPTER 8 CREDIT

CHAPTER 9 AUTOMOBILE EXPENSES

CHAPTER 10 TRANSPORTATION

CHAPTER 11 TAXES

CHAPTER 12 HOUSING

CHAPTER 13 TRADE INDUSTRIES

CHAPTER 14 OTHER PROFESSIONS

CHAPTER 15 PERSONAL FINANCE

CHAPTER 16 INVESTMENTS

CHAPTER 17 PROBABILITY

CHAPTER 18 EQUATIONS

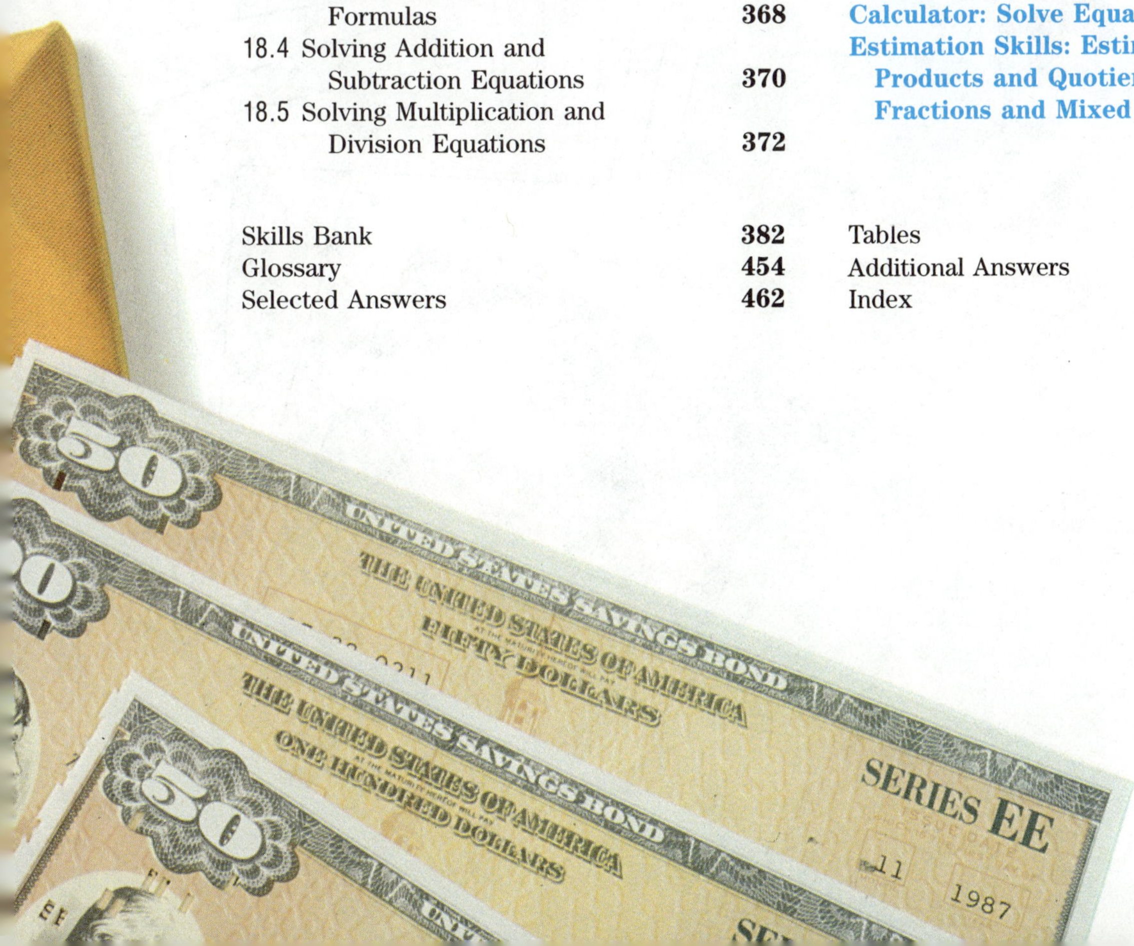

USING PRACTICAL MATHEMATICS
CONSUMER APPLICATIONS

***Practical Mathematics* is organized to help students overcome initial skill deficiencies before proceding to the essential consumer content.**

The fourteen applications chapters that follow focus on basic survival skills and essential consumer topics such as checking and savings accounts, housing, and personal finance.

Chapters One and Two review basic computation, estimation techniques, and the basic operations of a calculator. (This equips students to deal with the more complex problems found in later chapters.)

The book concludes with chapters on probability and equations.

The Skills Bank, located at the back of the student's book, is an abridged version of the first-year general math course and provides a resource for students who need to review basic skills as they learn consumer applications.

(The Skills Bank is extensive—more than 70 pages—and provides instruction, models, and numerous exercises for each skill.)

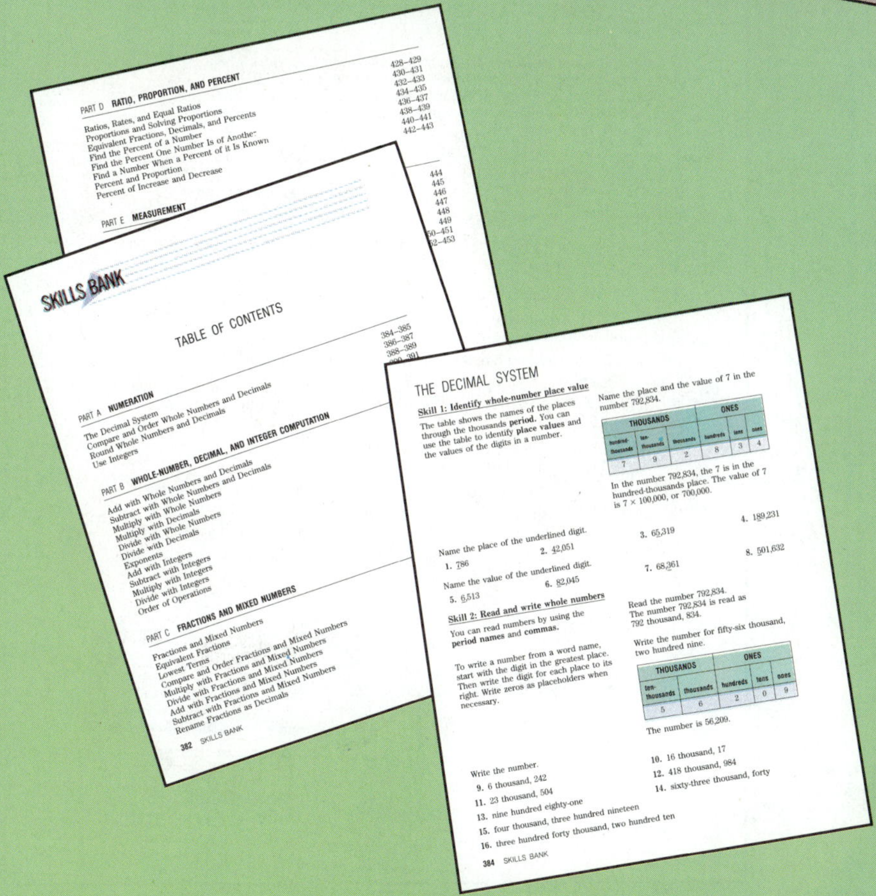

PART D RATIO, PROPORTION, AND PERCENT

PART E MEASUREMENT

SKILLS BANK

TABLE OF CONTENTS

PART A NUMERATION

PART B WHOLE-NUMBER, DECIMAL, AND INTEGER COMPUTATION

PART C FRACTIONS AND MIXED NUMBERS

382 SKILLS BANK

THE DECIMAL SYSTEM

Skill 1: Identify whole-number place value

The table shows the names of the places through the thousands **period.** You can use the table to identify **place values** and the values of the digits in a number.

Name the place and the value of 7 in the number 792,834.

THOUSANDS			ONES		
hundred-thousands	ten-thousands	thousands	hundreds	tens	ones
7	9	2	8	3	4

In the number 792,834, the 7 is in the hundred-thousands place. The value of 7 is $7 \times 100{,}000$, or 700,000.

Name the place of the underlined digit.

1. 786
2. 42,051
3. 65,319
4. 189,231

Name the value of the underlined digit.

5. 6,513
6. 82,045
7. 68,361
8. 501,632

Skill 2: Read and write whole numbers

You can read numbers by using the **period names** and **commas.**

Read the number 792,834.
The number 792,834 is read as 792 thousand, 834.

To write a number from a word name, start with the digit in the greatest place. Then write the digit for each place to its right. Write zeros as placeholders when necessary.

Write the number for fifty-six thousand, two hundred nine.

THOUSANDS		ONES		
ten-thousands	thousands	hundreds	tens	ones
5	6	2	0	9

The number is 56,209.

Write the number.

9. 6 thousand, 242
10. 16 thousand, 17
11. 23 thousand, 504
12. 418 thousand, 984
13. nine hundred eighty-one
14. sixty-three thousand, forty
15. four thousand, three hundred nineteen
16. three hundred forty thousand, two hundred ten

384 SKILLS BANK

TEACHING THE LESSON

***Practical Mathematics* presents mathematical skills and concepts in meaningful, real-life situations. The varied examples, exercises, and applications help students become proficient, confident problem solvers and consumers.**

CLEAR INSTRUCTION AND EXAMPLES

Each lesson is introduced with a problem of interest to students.

Examples illustrate the skill and relevant sub-skills in a step-by-step manner.

Think offers helpful hints for solving problems and emphasizes the mental processes students should be using.

For Discussion provides an additional opportunity to assess students' understanding of the skills and concepts taught.

15.1 PURCHASING POWER

It seems that you pay more for things each year. Indeed, most prices have risen steadily during the past decades. The **Consumer Price Index (CPI)** has gone up over 200% since it was first used in 1967. This steady increase in prices is called **inflation.** The **inflation rate** is the percent that prices increase.

EXAMPLE 1 Suppose that, in January, the average cost of food for your family was \$143 per wk. At the end of the year, the average cost of food was \$159 per wk. What was the inflation rate for your food costs?

INFLATION RATE = PRICE INCREASE ÷ ORIGINAL PRICE

1. Subtract to find the price increase. $\$159 - \$143 = \$16$
2. Divide to find the inflation rate. $\$16 \div \$143 = 0.1118881$
3. Rename as a percent. 11.2%

The inflation rate was about 11.2%.

As the inflation rate goes up, the **purchasing power** goes down.

EXAMPLE 2 Your net income was \$17,000 per y. During the year, the inflation rate was 8%. What was the purchasing power of your income after inflation?

1. Multiply to find the decrease in purchasing power.
   THINK: 8% = 0.08 $0.08 \times \$17{,}000 = \$1{,}360$
2. Subtract to find the purchasing power after inflation. $\$17{,}000 - \$1{,}360 = \$15{,}640$

Your \$17,000 salary had the purchasing power of \$15,640 after inflation.

FOR DISCUSSION

1. How could you be sure that your income was keeping up with inflation?
2. Why is inflation particularly difficult for people with fixed incomes?

302 CHAPTER 15

AMPLE AND VARIED PRACTICE

The abundant exercises are patterned on the examples in the lesson.

Exercises are structured and sequenced to build students' competence—and confidence. The first set of exercises reinforces the first example; the second group supports the second example.

Extensions offer a more challenging problem related to the lesson's theme.

PRACTICE EXERCISES Remember to estimate whenever you use your calculator.

Complete the table to find the inflation rate to the nearest tenth of a percent.

Item	Sports tickets	Weekly food	Monthly rent	Car rental	Gallon of gasoline	Slacks	Paper clips
Current price	\$12.00	\$183.00	\$262.00	\$104.00	\$1.24	\$33.89	\$0.89
Original price	\$10.00	\$167.00	\$257.00	\$89.00	\$0.98	\$31.95	\$0.63
Price increase	1. ■	2. ■	3. ■	4. ■	5. ■	6. ■	7. ■
Inflation rate	8. ■	9. ■	10. ■	11. ■	12. ■	13. ■	14. ■

Complete the table to find the purchasing power after inflation.

Net income	\$19,000	\$22,500	\$24,600	\$28,750	\$30,325	\$34,875
Inflation rate	10%	8%	4%	11%	6.5%	7.5%
Decrease in purchasing power	15. ■	16. ■	17. ■	18. ■	19. ■	20. ■
Purchasing power after inflation	21. ■	22. ■	23. ■	24. ■	25. ■	26. ■

Solve.

27. Originally, a box of your favorite cereal cost \$1.29. Now the price is \$1.47. What was the inflation rate to the nearest tenth of a percent?

28. During the year when Andy's income was \$26,480, the inflation rate was 7.5%. What was the purchasing power of his income after inflation?

EXTENSION Deflation

A steady decrease in prices is called **deflation.** The prices of some goods and services have deflated during the past decades. Originally, a certain computer cost \$1,900. A few years later, the same computer cost \$1,100. What was the rate of deflation?

1. Subtract to find the price decrease. \$1,900 − \$1,100 = \$800
2. Divide to find the deflation rate. \$800 ÷ \$1,900 = 0.4210526

The deflation rate was about 42.1%.

Find the deflation rate to the nearest tenth of a percent.

1. Current price: \$350
Original price: \$420

2. Current price: \$1,165
Original price: \$1,830

3. Current price: \$0.99
Original price: \$1.29

PERSONAL FINANCE **303**

PROBLEM-SOLVING STRATEGY LESSONS

The Problem-Solving Strategy lessons provide an intensive, structured approach to teaching varied methods of solving problems.

Real-world situations show how math fills a need in everyday life.

The strategy is always identified.

The lesson shows how to apply the strategy in a step-by-step manner.

Practice Exercises reinforce students' ability to use the problem-solving strategy.

PROBLEM *Solving* STRATEGY

13.6 DRAWING A DIAGRAM

Situation:

Mike owns a landscape-contracting company. Mike's crew built a fence around a square piece of land that measured 4 yd on each side. The fence posts were placed 1 yd apart. How many posts did Mike need?

Strategy:

Drawing a diagram can help you to solve a problem.

Applying the Strategy:

1. Draw a square.
2. Mark the 4 corner posts.
3. Mark posts 1 yd apart.
4. Count the number of posts needed.

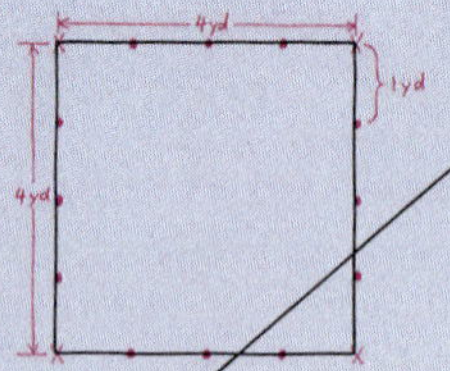

Mike needed 16 posts.

FOR DISCUSSION

Mike designed a flower bed for a customer. The first row of flowers was red. The second row of flowers was yellow. This pattern was repeated.

1. What would you draw to help you see the flower bed?
2. What color were the flowers in the ninth row?

PRACTICE EXERCISES

Copy the diagram. Then fill in the missing numbers to solve the problem.

1. Mike's crew planted a row of pine trees on the 60-ft side of Mr. Wesley's yard. A pine was planted every 10 ft. How many pines were planted?

2. A 12-ft-long flower bed was cut into a rectangular lawn. The lawn is 32 ft long. The flower bed is in the center of the lawn. How many feet of lawn are on each side of the flower bed?

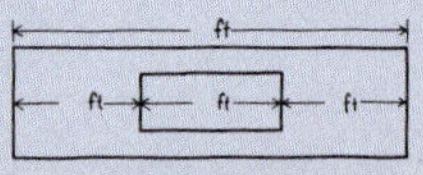

Solve the problem by drawing a diagram.

3. Mike's crew built a fence around a square piece of land. The land was 9 yd on each side. The fence posts were placed 3 yd apart. How many posts were used?
4. Mike planned a garden with 15 rows of flowers. The first row was blue, the second was white, the third was pink. How many times was this pattern repeated?
5. An azalea garden contains 49 azaleas, 7 in each row. The center and end azaleas in each row are pink. How many pink azaleas are there in all?
6. A row of trees was planted along Maple Drive. The length of the row was 148 ft. A tree was planted every $18\frac{1}{2}$ ft. How many trees were planted?
7. Mike built a fence around a rectangular garden that was 16 yd long and 12 yd wide. The posts were placed 2 yd apart. How many posts did he need?
8. A rectangular lawn contains three 7-ft-square flower beds. There is a border of 3 ft between each flower bed and the perimeter of the plot. How long is the plot?

DECISION-MAKING LESSONS

The open-ended *Decision-Making* lessons reinforce logical reasoning and move beyond computation to a consideration of all factors involved in making sound decisions.

Problems are placed in a realistic situation.

After considering the **Decision-Making Factors**, students practice simple, efficient ways to organize the information necessary for **Decision-Making Comparisons.**

The open-ended problems promote decision making based on logical reasons and choices.

DECISION MAKING

6.7 COMPARING FOOD COSTS

After a busy day it is often easiest to just take out food or to go out to a nearby restaurant for dinner. But not preparing dinner at home can become very costly.

PROBLEM

Les and Wanda have been ordering take-out food and eating in restaurants frequently. Because they are trying to save money for a new car, they decide to examine a sample of their meal costs (excluding tax) to see if they can cut expenses. They made the following notes.

Chicken dinner at home:	4 pieces of chicken	$3.60	Time: 50 min
	Salad ingredients (serves two)	$2.00	Not very convenient
	1 package potatoes	$1.59	
	Rolls	$0.40	
	Juice (two glasses)	$0.40	
Chicken dinner from Barney's Take-Out:	4 pieces of chicken	$2.00 each	Time: 10 min
	Salads	$1.50 each	Very convenient
	Potatoes	$0.90 per order	
	Rolls	Included	
	Juice	$0.75 per person	
Chicken dinner at the Capital Diner:	Complete dinner (includes chicken, salad, potatoes, roll)	$9.95 each	Time: 35 min
			Convenient
	Juice	$1.00 per glass	
	Tip	15%	

DECISION-MAKING FACTORS

Cost Time Convenience Quality of food

DECISION-MAKING COMPARISONS

Compare the 3 options by completing the table.

Factor	Home	Barney's Take-Out	Capital Diner
Cost for two: Chicken	1. ■	2. ■	Included
Salad	3. ■	$3.00	4. ■
Potatoes	$1.59	5. ■	6. ■
Rolls	7. ■	Included	8. ■
Juice	9. ■	10. ■	$2.00
Total cost	11. ■	$14.30	12. ■
Time	50 min	13. ■	14. ■
Convenience	15. ■	16. ■	Average

MAKING THE DECISIONS

Where should Les and Wanda get their dinner:

17. If cost were the only factor?

18. If time were the only factor?

19. If convenience were the only factor?

20. How much can Les and Wanda save by eating at home instead of using Barney's Take-Out?

21. If they have been using Barney's Take-Out an average of twice a week, how much could they save in a year by preparing dinner at home?

22. How much could Les and Wanda save by preparing dinner at home instead of going to the Capital Diner?

23. If they have been going to the diner about once a week, how much could they save in a year by preparing dinner at home?

24. Where do you think the quality of the food is the highest? Why?

25. Is the cost of the food alone an accurate measure of the cost of preparing dinner at home? What other costs might be included?

26. Is the time for food preparation alone an accurate measure of the time involved in preparing a meal at home? Why?

REVIEW, TEST, AND MAINTENANCE FEATURES

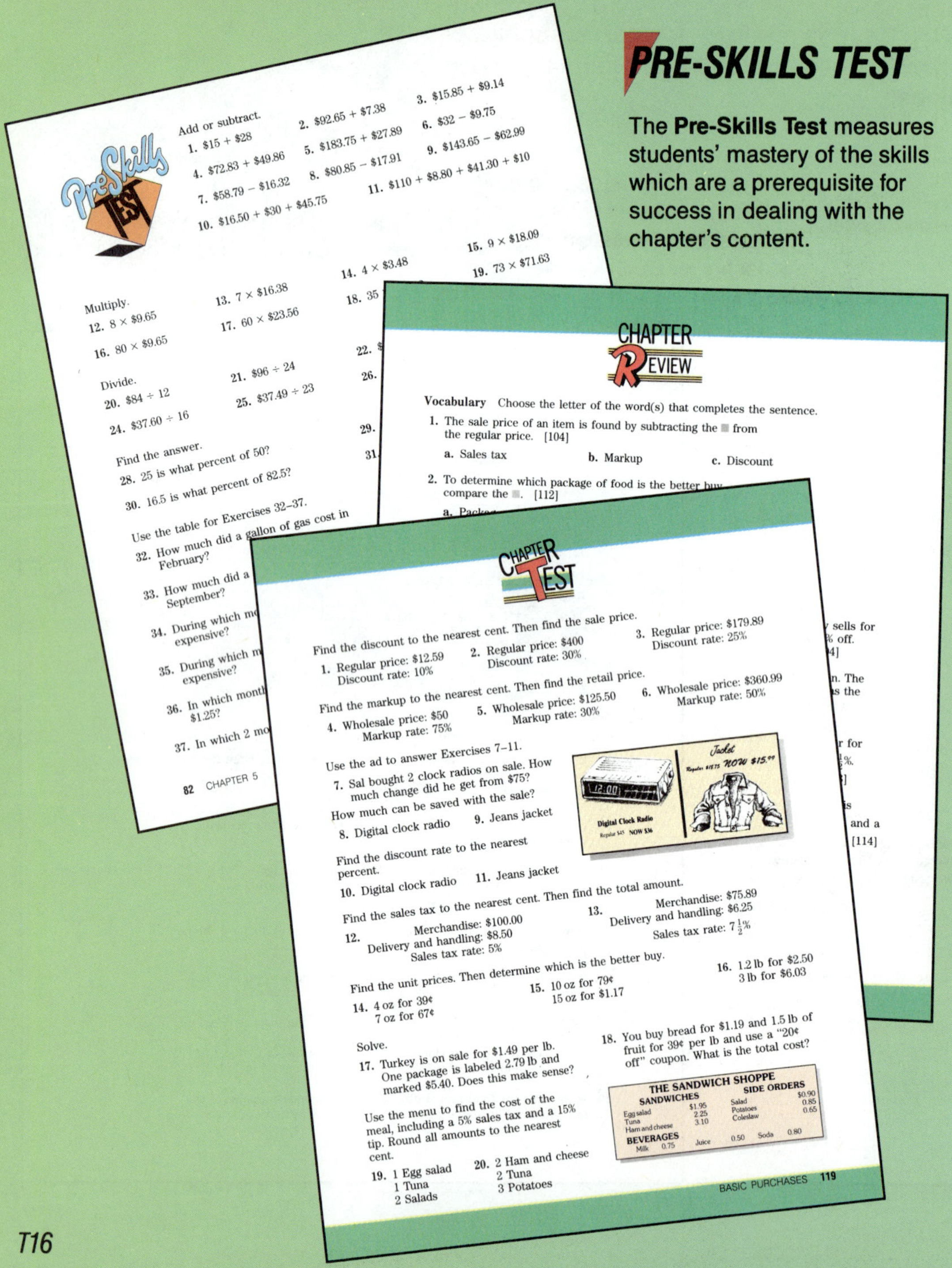
Pre-Skills Test

Add or subtract.
1. $15 + $28 2. $92.65 + $7.38 3. $15.85 + $9.14
4. $72.83 + $49.86 5. $183.75 + $27.89 6. $32 − $9.75
7. $58.79 − $16.32 8. $80.85 − $17.91 9. $143.65 − $62.99
10. $16.50 + $30 + $45.75 11. $110 + $8.80 + $41.30 + $10

Multiply.
12. 8 × $9.65 13. 7 × $16.38 14. 4 × $3.48 15. 9 × $18.09
16. 80 × $9.65 17. 60 × $23.56 18. 35 19. 73 × $71.63

Divide.
20. $84 ÷ 12 21. $96 ÷ 24 22.
24. $37.60 ÷ 16 25. $37.49 ÷ 23 26.

Find the answer.
28. 25 is what percent of 50? 29.
30. 16.5 is what percent of 82.5? 31.

Use the table for Exercises 32–37.
32. How much did a gallon of gas cost in February?
33. How much did a September?
34. During which m expensive?
35. During which m expensive?
36. In which month $1.25?
37. In which 2 mo

82 CHAPTER 5

Chapter Review

Vocabulary Choose the letter of the word(s) that completes the sentence.
1. The sale price of an item is found by subtracting the ■ from the regular price. [104]
a. Sales tax b. Markup c. Discount
2. To determine which package of food is the better buy compare the ■. [112]

Chapter Test

Find the discount to the nearest cent. Then find the sale price.
1. Regular price: $12.59 Discount rate: 10%
2. Regular price: $400 Discount rate: 30%
3. Regular price: $179.89 Discount rate: 25%

Find the markup to the nearest cent. Then find the retail price.
4. Wholesale price: $50 Markup rate: 75%
5. Wholesale price: $125.50 Markup rate: 30%
6. Wholesale price: $360.99 Markup rate: 50%

Use the ad to answer Exercises 7–11.
7. Sal bought 2 clock radios on sale. How much change did he get from $75?

How much can be saved with the sale?
8. Digital clock radio 9. Jeans jacket

Find the discount rate to the nearest percent.
10. Digital clock radio 11. Jeans jacket

Find the sales tax to the nearest cent. Then find the total amount.
12. Merchandise: $100.00 Delivery and handling: $8.50 Sales tax rate: 5%
13. Merchandise: $75.89 Delivery and handling: $6.25 Sales tax rate: $7\frac{1}{2}$%

Find the unit prices. Then determine which is the better buy.
14. 4 oz for 39¢ 7 oz for 67¢
15. 10 oz for 79¢ 15 oz for $1.17
16. 1.2 lb for $2.50 3 lb for $6.03

Solve.
17. Turkey is on sale for $1.49 per lb. One package is labeled 2.79 lb and marked $5.40. Does this make sense?
18. You buy bread for $1.19 and 1.5 lb of fruit for 39¢ per lb and use a "20¢ off" coupon. What is the total cost?

Use the menu to find the cost of the meal, including a 5% sales tax and a 15% tip. Round all amounts to the nearest cent.
19. 1 Egg salad 1 Tuna 2 Salads
20. 2 Ham and cheese 2 Tuna 3 Potatoes

THE SANDWICH SHOPPE			
SANDWICHES		SIDE ORDERS	
Egg salad	$1.95	Salad	$0.90
Tuna	2.25	Potatoes	0.85
Ham and cheese	3.10	Coleslaw	0.65
BEVERAGES			
Milk 0.75	Juice 0.50	Soda 0.80	

BASIC PURCHASES 119

PRE-SKILLS TEST

The **Pre-Skills Test** measures students' mastery of the skills which are a prerequisite for success in dealing with the chapter's content.

CHAPTER REVIEW

Each full-page **Chapter Review** reinforces key vocabulary, skills, and concepts.

CHAPTER TEST

The **Chapter Test** evaluates students' mastery of basic skills and concepts as well as applications and strategies. Each test item is keyed to a chapter objective in the *Teacher's Edition.*

CUMULATIVE-REVIEW

The **Cumulative Review** provides abundant practice for reviewing key concepts, skills, applications, and strategies. The reviews are carefully spaced throughout the book (after Chapters 4, 8, 12 and 16).

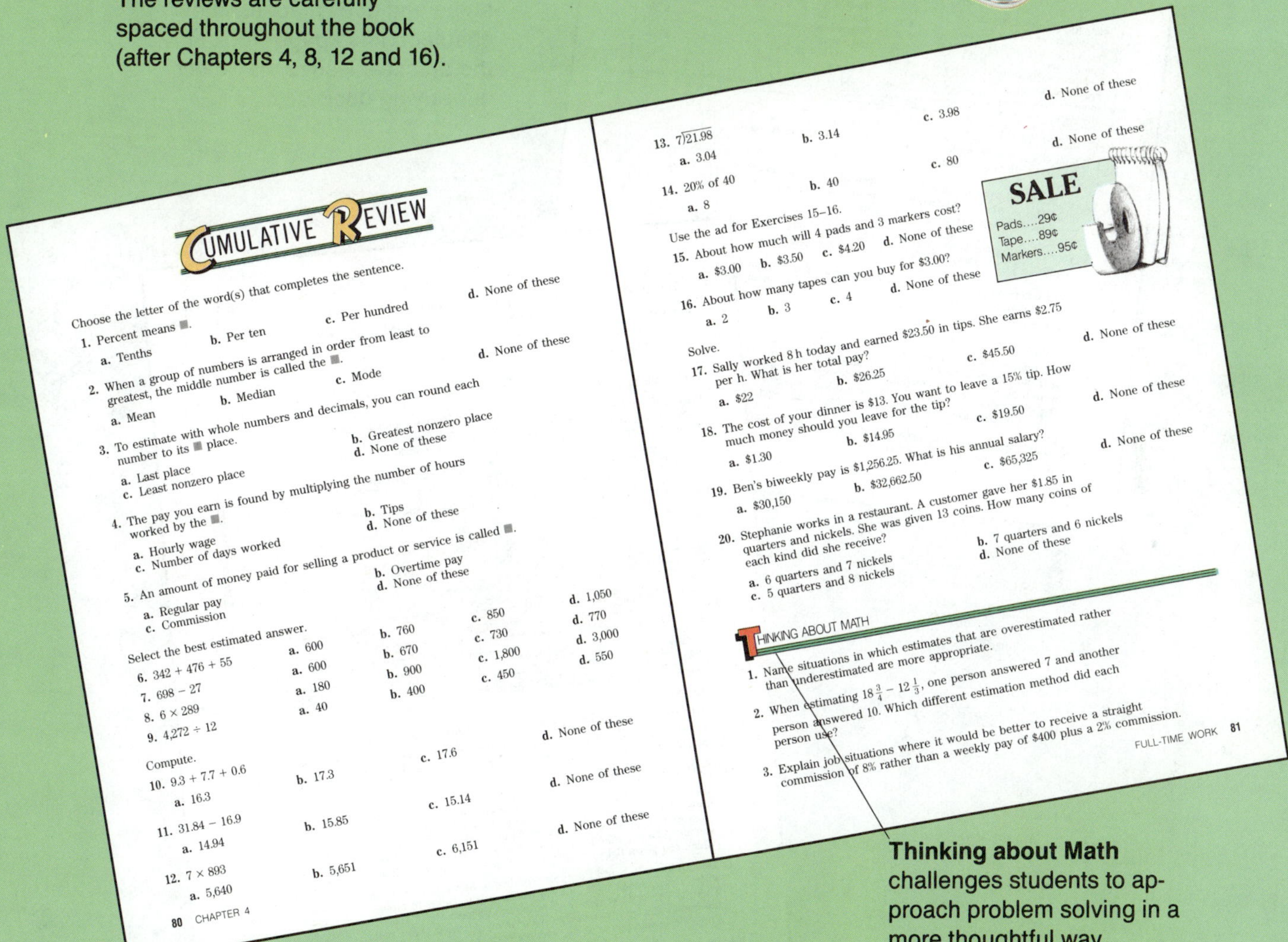

CUMULATIVE REVIEW

Choose the letter of the word(s) that completes the sentence.

1. Percent means ■.
   a. Tenths b. Per ten c. Per hundred d. None of these
2. When a group of numbers is arranged in order from least to greatest, the middle number is called the ■.
   a. Mean b. Median c. Mode d. None of these
3. To estimate with whole numbers and decimals, you can round each number to its ■ place.
   a. Last place b. Greatest nonzero place c. Least nonzero place d. None of these
4. The pay you earn is found by multiplying the number of hours worked by the ■.
   a. Hourly wage b. Tips c. Number of days worked d. None of these
5. An amount of money paid for selling a product or service is called ■.
   a. Regular pay b. Overtime pay c. Commission d. None of these

Select the best estimated answer.

6. 342 + 476 + 55 a. 600 b. 760 c. 850 d. 1,050
7. 698 − 27 a. 600 b. 670 c. 730 d. 770
8. 6 × 289 a. 180 b. 900 c. 1,800 d. 3,000
9. 4,272 ÷ 12 a. 40 b. 400 c. 450 d. 550

Compute.

10. 9.3 + 7.7 + 0.6 a. 16.3 b. 17.3 c. 17.6 d. None of these
11. 31.84 − 16.9 a. 14.94 b. 15.85 c. 15.14 d. None of these
12. 7 × 893 a. 5,640 b. 5,651 c. 6,151 d. None of these

80 CHAPTER 4

13. $7\overline{)21.98}$ a. 3.04 b. 3.14 c. 3.98 d. None of these
14. 20% of 40 a. 8 b. 40 c. 80 d. None of these

Use the ad for Exercises 15–16.

SALE
Pads....29¢
Tape....89¢
Markers....95¢

15. About how much will 4 pads and 3 markers cost?
    a. $3.00 b. $3.50 c. $4.20 d. None of these
16. About how many tapes can you buy for $3.00?
    a. 2 b. 3 c. 4 d. None of these

Solve.

17. Sally worked 8 h today and earned $23.50 in tips. She earns $2.75 per h. What is her total pay?
    a. $22 b. $26.25 c. $45.50 d. None of these
18. The cost of your dinner is $13. You want to leave a 15% tip. How much money should you leave for the tip?
    a. $1.30 b. $14.95 c. $19.50 d. None of these
19. Ben's biweekly pay is $1,256.25. What is his annual salary?
    a. $30,150 b. $32,662.50 c. $65,325 d. None of these
20. Stephanie works in a restaurant. A customer gave her $1.85 in quarters and nickels. She was given 13 coins. How many coins of each kind did she receive?
    a. 6 quarters and 7 nickels b. 7 quarters and 6 nickels c. 5 quarters and 8 nickels d. None of these

THINKING ABOUT MATH

1. Name situations in which estimates that are overestimated rather than underestimated are more appropriate.
2. When estimating $18\frac{3}{4} - 12\frac{1}{3}$, one person answered 7 and another person answered 10. Which different estimation method did each person use?
3. Explain job situations where it would be better to receive a straight commission of 8% rather than a weekly pay of $400 plus a 2% commission.

FULL-TIME WORK 81

Thinking about Math challenges students to approach problem solving in a more thoughtful way.

LEARNING AIDS AND EXTENSIONS

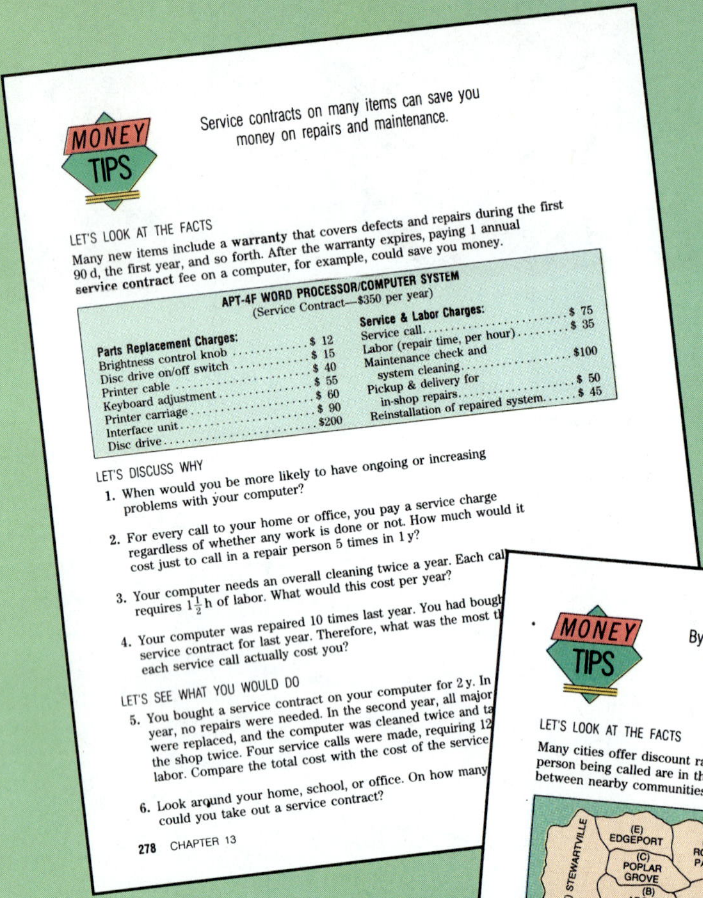

MONEY TIPS

Service contracts on many items can save you money on repairs and maintenance.

LET'S LOOK AT THE FACTS

Many new items include a **warranty** that covers defects and repairs during the first 90 d, the first year, and so forth. After the warranty expires, paying 1 annual **service contract** fee on a computer, for example, could save you money.

APT-4F WORD PROCESSOR/COMPUTER SYSTEM
(Service Contract—$350 per year)

Parts Replacement Charges:		Service & Labor Charges:	
Brightness control knob	$ 12	Service call	$ 75
Disc drive on/off switch	$ 15	Labor (repair time, per hour)	$ 35
Printer cable	$ 40	Maintenance check and system cleaning	$100
Keyboard adjustment	$ 55	Pickup & delivery for in-shop repairs	$ 50
Printer carriage	$ 60	Reinstallation of repaired system	$ 45
Interface unit	$ 90		
Disc drive	$200		

LET'S DISCUSS WHY

1. When would you be more likely to have ongoing or increasing problems with your computer?
2. For every call to your home or office, you pay a service charge regardless of whether any work is done or not. How much would it cost just to call in a repair person 5 times in 1 y?
3. Your computer needs an overall cleaning twice a year. Each cal requires $1\frac{1}{2}$ h of labor. What would this cost per year?
4. Your computer was repaired 10 times last year. You had bough service contract for last year. Therefore, what was the most t each service call actually cost you?

LET'S SEE WHAT YOU WOULD DO

5. You bought a service contract on your computer for 2 y. In year, no repairs were needed. In the second year, all major were replaced, and the computer was cleaned twice and ta the shop twice. Four service calls were made, requiring 12 labor. Compare the total cost with the cost of the service
6. Look around your home, school, or office. On how many could you take out a service contract?

278 CHAPTER 13

MONEY TIPS

The **Money Tips** pages appear at the end of every consumer chapter. These pages look at the practical issues that affect our buying decisions.

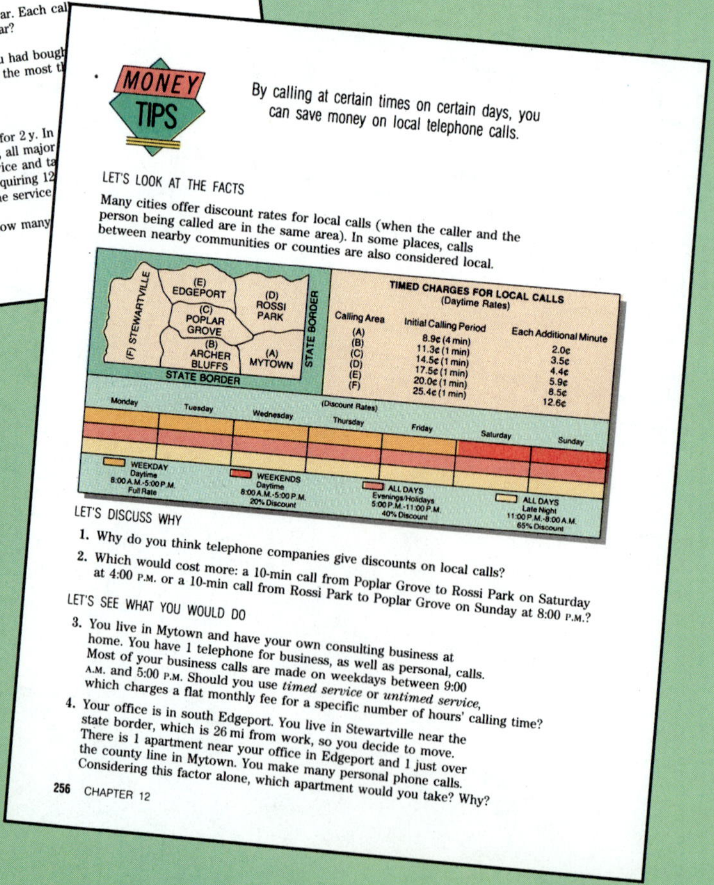

MONEY TIPS

By calling at certain times on certain days, you can save money on local telephone calls.

LET'S LOOK AT THE FACTS

Many cities offer discount rates for local calls (when the caller and the person being called are in the same area). In some places, calls between nearby communities or counties are also considered local.

TIMED CHARGES FOR LOCAL CALLS
(Daytime Rates)

Calling Area	Initial Calling Period	Each Additional Minute
(A)	8.9¢ (4 min)	2.0¢
(B)	11.3¢ (1 min)	3.5¢
(C)	14.5¢ (1 min)	4.4¢
(D)	17.5¢ (1 min)	5.9¢
(E)	20.0¢ (1 min)	8.5¢
(F)	25.4¢ (1 min)	12.6¢

LET'S DISCUSS WHY

1. Why do you think telephone companies give discounts on local calls?
2. Which would cost more: a 10-min call from Poplar Grove to Rossi Park on Saturday at 4:00 P.M. or a 10-min call from Rossi Park to Poplar Grove on Sunday at 8:00 P.M.?

LET'S SEE WHAT YOU WOULD DO

3. You live in Mytown and have your own consulting business at home. You have 1 telephone for business, as well as personal, calls. Most of your business calls are made on weekdays between 9:00 A.M. and 5:00 P.M. Should you use *timed service* or *untimed service*, which charges a flat monthly fee for a specific number of hours' calling time?
4. Your office is in south Edgeport. You live in Stewartville near the state border, which is 26 mi from work, so you decide to move. There is 1 apartment near your office in Edgeport and 1 just over the county line in Mytown. You make many personal phone calls. Considering this factor alone, which apartment would you take? Why?

256 CHAPTER 12

ESTIMATION SKILLS

The **Estimation Skills** pages extend students' understanding of estimation techniques.

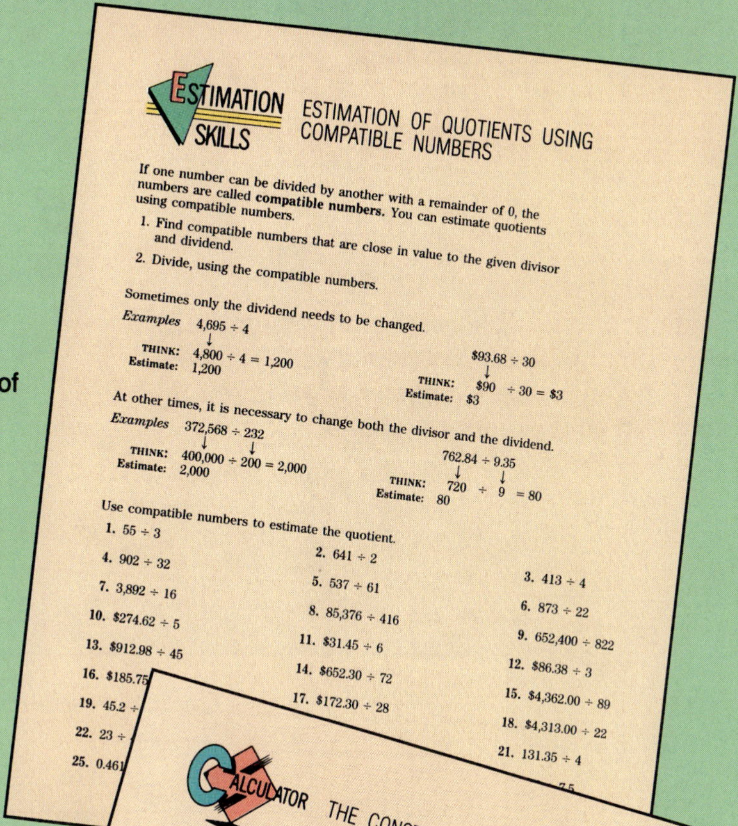

ESTIMATION SKILLS — ESTIMATION OF QUOTIENTS USING COMPATIBLE NUMBERS

If one number can be divided by another with a remainder of 0, the numbers are called **compatible numbers.** You can estimate quotients using compatible numbers.

1. Find compatible numbers that are close in value to the given divisor and dividend.
2. Divide, using the compatible numbers.

Sometimes only the dividend needs to be changed.

Examples 4,695 ÷ 4 — THINK: 4,800 ÷ 4 = 1,200 Estimate: 1,200 — \$93.68 ÷ 30 — THINK: \$90 ÷ 30 = \$3 Estimate: \$3

At other times, it is necessary to change both the divisor and the dividend.

Examples 372,568 ÷ 232 — THINK: 400,000 ÷ 200 = 2,000 Estimate: 2,000 — 762.84 ÷ 9.35 — THINK: 720 ÷ 9 = 80 Estimate: 80

Use compatible numbers to estimate the quotient.

1. 55 ÷ 3
2. 641 ÷ 2
3. 413 ÷ 4
4. 902 ÷ 32
5. 537 ÷ 61
6. 873 ÷ 22
7. 3,892 ÷ 16
8. 85,376 ÷ 416
9. 652,400 ÷ 822
10. \$274.62 ÷ 5
11. \$31.45 ÷ 6
12. \$86.38 ÷ 3
13. \$912.98 ÷ 45
14. \$652.30 ÷ 72
15. \$4,362.00 ÷ 89
16. \$185.75
17. \$172.30 ÷ 28
18. \$4,313.00 ÷ 22
19. 45.2 ÷
21. 131.35 ÷ 4
22. 23 ÷
25. 0.461

CALCULATOR

The **Calculator** pages teach the keys and functions commonly available on calculators. These pages emphasize the time-saving benefits of using these tools correctly.

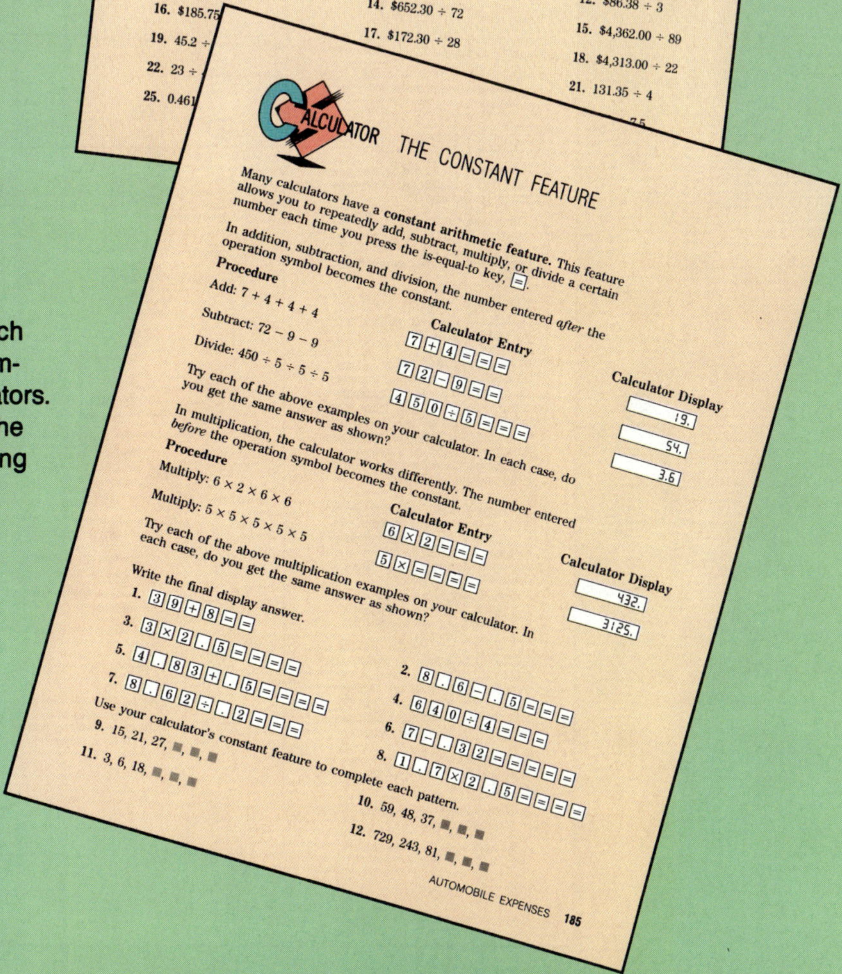

CALCULATOR — THE CONSTANT FEATURE

Many calculators have a **constant arithmetic feature.** This feature allows you to repeatedly add, subtract, multiply, or divide a certain number each time you press the is-equal-to key, [=].

In addition, subtraction, and division, the number entered *after* the operation symbol becomes the constant.

Procedure	Calculator Entry	Calculator Display
Add: 7 + 4 + 4 + 4	7 + 4 = = =	19.
Subtract: 72 − 9 − 9	7 2 − 9 = =	54.
Divide: 450 ÷ 5 ÷ 5 ÷ 5	4 5 0 ÷ 5 = = =	3.6

Try each of the above examples on your calculator. In each case, do you get the same answer as shown?

In multiplication, the calculator works differently. The number entered *before* the operation symbol becomes the constant.

Procedure	Calculator Entry	Calculator Display
Multiply: 6 × 2 × 6 × 6	6 × 2 = = =	432.
Multiply: 5 × 5 × 5 × 5 × 5	5 × = = = =	3125.

Try each of the above multiplication examples on your calculator. In each case, do you get the same answer as shown?

Write the final display answer.

1. 3 9 + 8 = =
2. 8 . 6 − . 5 = = =
3. 3 × 2 . 5 = = = =
4. 6 4 0 ÷ 4 = = =
5. 4 . 8 3 + . 5 = = = =
6. 7 − . 3 2 = = = = =
7. 8 . 6 2 ÷ . 2 = = =
8. 1 . 7 × 2 . 5 = = = =

Use your calculator's constant feature to complete each pattern.

9. 15, 21, 27, ■, ■, ■
10. 59, 48, 37, ■, ■, ■
11. 3, 6, 18, ■, ■, ■
12. 729, 243, 81, ■, ■, ■

AUTOMOBILE EXPENSES 185

TEACHER'S EDITION

The *Teacher's Edition* is practical and flexible, with all the materials needed for a lesson presented in one place. The variety of suggestions minimizes preparation time and makes it easier to adapt each lesson to your teaching style and your students' needs.

Objectives—clearly stated goals that students should attain by the end of the lesson.

Warmup—a brief activity for reviewing the lesson's prerequisite skills.

Introduction—material you can use to capture students' interest at the beginning of the lesson.

Instruction—useful teaching notes that help you derive maximum benefit from every page and every feature.

OBJECTIVES

- Compute the inflation rate.
- Compute purchasing power after inflation.

TEACHING THE LESSON

WARM-UP Write the following exercises on the chalkboard or an overhead transparency.

Rename as a percent.

1. 0.52 **2.** 0.716 **3.** 0.608

Divide. Round to the nearest thousandth.

4. \$23 ÷ \$213 **5.** \$0.39 ÷ \$1.40

(**1.** 52% **2.** 71.6% **3.** 60.8% **4.** 0.108 **5.** 0.279)

INTRODUCTION Use the introductory situation to review the meaning of *inflation.* Explain to students that the Consumer Price Index (CPI) is 1 way of measuring how expensive things are now compared to the past. Write a recent CPI on the chalkboard and compare it to the CPI in 1967 (100) and the CPI in 1985 (322.2).

INSTRUCTION Use Example 1 to show students how to find the inflation rate. In Step 2, remind students that they should divide by the original price, not the current price. Show students that this sequence of keystrokes would yield the inflation rate.

[1] [5] [9] [−] [1] [4] [3] [÷] [1] [4] [3] [=]

Use Example 2 to show students that money has less purchasing power as the inflation rate goes up. Explain to students that a lower salary in the past may have had more purchasing power than a higher salary today.

SKILLS BANK You may want to refer students to the following skills in the Skills Bank:

Subtract With Whole Numbers and Decimals, Skill 6

Multiply With Decimals, Skill 5

302

Chapter 15 Personal Finance

15.1 PURCHASING POWER

It seems that you pay more for things each year. Indeed, most prices have risen steadily during the past decades. The **Consumer Price Index (CPI)** has gone up over 200% since it was first used in 1967. This steady increase in prices is called **inflation.** The **inflation rate** is the percent that prices increase.

EXAMPLE 1 Suppose that, in January, the average cost of food for your family was \$143 per wk. At the end of the year, the average cost of food was \$159 per wk. What was the inflation rate for your food costs?

INFLATION RATE = PRICE INCREASE ÷ ORIGINAL PRICE

1. Subtract to find the price increase. \$159 − \$143 = \$16
2. Divide to find the inflation rate. \$16 ÷ \$143 = 0.1118881
3. Rename as a percent. 11.2%

The inflation rate was about 11.2%.

As the inflation rate goes up, the **purchasing power** goes down.

EXAMPLE 2 Your net income was \$17,000 per y. During the year, the inflation rate was 8%. What was the purchasing power of your income after inflation?

1. Multiply to find the decrease in purchasing power.
   THINK: 8% = 0.08 0.08 × \$17,000 = \$1,360
2. Subtract to find the purchasing power after inflation. \$17,000 − \$1,360 = \$15,640

Your \$17,000 salary had the purchasing power of \$15,640 after inflation.

FOR DISCUSSION See TE side column.

1. How could you be sure that your income was keeping up with inflation?
2. Why is inflation particularly difficult for people with fixed incomes?

302 CHAPTER 15

Reduced Student Pages—clear, readable student pages. Bullets next to some exercises tell the teacher which answers appear in the back of the student's book.

Assignments—guidelines for gearing lessons to student's varying abilities, making lessons challenging and interesting, yet attainable for each student.

Follow-up Activity—an additional activity for reinforcing or extending the skills taught in the lesson. The nature of the activity is identified in boldfaced type.

Supplementary Materials—correlations to the extra support materials in the *Teacher's Resource Package.*

PRACTICE EXERCISES

Remember to estimate whenever you use your calculator.

Complete the table to find the inflation rate to the nearest tenth of a percent.

Item	Sports tickets	Weekly food	Monthly rent	Car rental	Gallon of gasoline	Slacks	Paper clips
Current price	$12.00	$183.00	$262.00	$104.00	$1.24	$33.89	$0.89
Original price	$10.00	$167.00	$257.00	$89.00	$0.98	$31.95	$0.63
Price increase	• 1. ■	• 2. ■	3. ■	4. ■	5. ■	6. ■	7. ■
Inflation rate	8. ■	9. ■	10. ■	11. ■	12. ■	13. ■	14. ■

1. $2.00 2. $16.00 3. $5.00 4. $15.00 5. $0.26 6. $1.94 7. $0.26 8. 20% 9. 9.6% 10. 1.9% 11. 16.9% 12. 26.5% 13. 6.1% 14. 41.3%

Complete the table to find the purchasing power after inflation.

Net income	$19,000	$22,500	$24,600	$28,750	$30,325	$34,875
Inflation rate	10%	8%	4%	11%	6.5%	7.5%
Decrease in purchasing power	$1,900 • 15. ■	$1,800 • 16. ■	$984 17. ■	$3,162.50 18. ■	$1,971.13 19. ■	$2,615.63 20. ■
Purchasing power after inflation	$17,100 21. ■	$20,700 22. ■	$23,616 23. ■	24. ■	25. ■	26. ■

24. $25,587.50 25. $28,353.87 26. $32,259.37

Solve.

27. Originally, a box of your favorite cereal cost $1.29. Now the price is $1.47. What was the inflation rate to the nearest tenth of a percent? 14.0%

28. During the year when Andy's income was $26,480, the inflation rate was 7.5%. What was the purchasing power of his income after inflation? $24,494

EXTENSION Deflation

A steady decrease in prices is called **deflation.** The prices of some goods and services have deflated during the past decades. Originally, a certain computer cost $1,900. A few years later, the same computer cost $1,100. What was the rate of deflation?

1. Subtract to find the price decrease. $1,900 − $1,100 = $800
2. Divide to find the deflation rate. $800 ÷ $1,900 = 0.4210526

The deflation rate was about 42.1%.

Find the deflation rate to the nearest tenth of a percent.

1. Current price: $350 Original price: $420 16.7%
2. Current price: $1,165 Original price: $1,830 36.3%
3. Current price: $0.99 Original price: $1.29 23.3%

PERSONAL FINANCE **303**

FOR DISCUSSION

1. Students should explain that their income is keeping up with inflation if their percent of pay increase matches the inflation rate. They are moving behind inflation if their percent increase is less than the inflation rate.
2. Students should explain that people with fixed incomes often cannot keep up with inflation. Even retired people with inflation protection in their pensions usually fall behind the inflation rate.

EXTENSION Explain that the usual trend is for prices to move upward. However, there may be periods of *deflation* when prices move downward. Periods of deflation are not necessarily better because during these periods many people are out of work.

ASSIGNMENTS

Level	Exercises
Level 1	1–26
Level 2	1–28, EX
Level 3	5–7, 12–14, 18–20, 24–28, EX

FOLLOW-UP ACTIVITY

ENRICHMENT Write these CPIs on the chalkboard: Year 1—323.8; Year 2—341.7. Have students use the steps from Example 1 to find the inflation rate. Encourage students to use calculators. Students should subtract to find the difference (17.9) and then divide the difference by the Year 1 CPI to find the inflation rate (5.5%). Ask students to find the purchasing power of these salaries after 5.5% inflation:
$19,000; $23,500; $37,860. ($17,955; $22,207.50; $35,777.70)

SUPPLEMENTARY MATERIALS

TRP Practice, p. 101
TRP Reteaching, p. 82
TRP Applications, p. 17
TRP Transparency 51

303

TEACHER'S RESOURCE PACKAGE

One source provides
- **Maximum reaching flexibility.**
- **Materials for students of all ability levels.**
- **Reinforcement and extension of skills.**
- **All items in blackline-master format.**

PRACTICE, RETEACHING AND ACTIVITY BOOK

- **Practice Sheets**—one page of practice exercises per lesson.
- **Reteaching Sheets**—one page of practice with instruction and worked examples for every skills lesson in the text.
- **Applications Sheets**—extensions of the consumer applications in the text.
- **Group Projects**—cooperative-learning projects for each chapter.
- **Lesson Aids**—useful forms, such as job applications, tax forms, and bank loan applications.

TRANSPARENCIES

- Includes a transparency for every lesson in the text, enabling you to work through problems as a class activity.
- Features two lessons on each transparency.

TEST BOOK

- **Skills Inventory**—a multiple-choice inventory test that evaluates students initial skill proficiencies.
- **Post-Tests**—a two-page test for each chapter in the book—available in alternate A and B forms.
- **Cumulative Tests**—multiple-choice tests for every Cumulative Review in the book.
- **Final Test**—a multiple-choice end-of-year exam.
- **Management Forms**—tools for recording and tracking student progress.

THE ANSWER BOOK

- Contains the answers to every worksheet and test.

PACING CHART

This pacing chart is a general guide for using Holt Practical Mathematics during the school year. It offers a suggested number of days to spend on each chapter for 3 levels of student ability. Your own individual needs and those of your students may require alternate plans.

		Suggested Number of Days		
	Chapter	Level 1 Below Average Students	Level 2 Average Students	Level 3 Above Average Students
1	Paper-and-Pencil Computations	12	10	6
2	The Calculator and Estimation	11	9	9
3	Part-Time and Summer Jobs	9	8	8
4	Full-Time Work	12	12	10
5	Recreation and Sports	9	9	8
6	Basic Purchases	9	9	7
7	Checking and Savings Accounts	10	8	8
8	Credit	10	11	12
9	Automobile Expenses	8	8	8
10	Transportation	10	10	10
11	Taxes	8	9	10
12	Housing	12	12	12
13	Trade Industries	10	9	9
14	Other Professions	8	9	9
15	Personal Finance	9	9	10
16	Investments	11	11	12
17	Probability	6	8	10
18	Equations	6	9	12
	Total	170	170	170

Chapter Objectives

Chapter 1 PAPER-AND-PENCIL COMPUTATIONS
A. Add whole numbers and decimals.
B. Subtract whole numbers and decimals.
C. Multiply whole numbers and decimals.
D. Divide whole numbers and decimals.
E. Rename fractions, decimals, and percents.
F. Find the percent of a number.
G. Find the mean, the median, and the mode.
H. Interpret data in tables and bar graphs.

Chapter 2 THE CALCULATOR AND ESTIMATION
A. Use a calculator to compute sums, differences, products, and quotients.
B. Mentally compute sums, differences, products, and quotients.
C. Estimate sums and differences.
D. Estimate products and quotients.
E. Estimate reasonable answers.
F. Solve problems by determining the best way to find an answer.
G. Estimate to solve problems.

Chapter 3 PART-TIME AND SUMMER JOBS
A. Compute total pay.
B. Compute hourly wages.
C. Compute the amount of tips.
D. Compute total pay based on hourly wages and tips.
E. Compute total pay and amounts withheld for Social Security.
F. Solve problems by using guessing and checking.

Chapter 4 FULL-TIME WORK
A. Compute regular pay and overtime pay.
B. Compute pay from time sheets and time cards.
C. Compute piecework pay.
D. Compute commissions.
E. Compute weekly, biweekly, semimonthly, and monthly pay, and annual salary.
F. Compute net pay and payroll deductions.
G. Compute health insurance costs.
H. Compute life insurance costs.

Chapter 5 RECREATION AND SPORTS
A. Compute the costs of hobby material.
B. Compute the costs of tickets and the amount of change.
C. Compute the total cost of sports activities.
D. Compute the cost per visit of a sports activity.
E. Solve problems by interpreting remainders.

Chapter 6 BASIC PURCHASES
A. Compute discount, sale price and discount rate.
B. Compute markup and retail price.
C. Compute sales tax and total cost.
D. Compute and compare unit prices.
E. Determine if purchase prices make sense.
F. Compute total purchase cost when using coupons.
G. Use a menu to find the total cost of a meal, including tax and tip.

Chapter 7 CHECKING AND SAVINGS ACCOUNTS
A. Write a check.
B. Compute a running balance in a check register.
C. Reconcile a checking account.
D. Compute a savings account passbook balance.
E. Compute simple interest.
F. Use a table to compute compound interest.

Chapter 8 CREDIT
A. Compute a credit unpaid balance, finance charge, and new balance (fixed finance rate).
B. Compute a credit finance charge and new balance (variable finance rate).
C. Compute overdraft checking interest using the sum of the daily balances and a daily interest rate.
D. Compute loan interest, total to be repaid, and monthly payment using monthly payment per $100.
E. Compute an installment price and finance charge.
F. Solve problems by looking for a pattern.

Chapter 9 AUTOMOBILE EXPENSES
A. Compute the amount owed for a car.
B. Estimate mpg.
C. Compute a yearly base car insurance premium.
D. Compute a rated yearly car insurance premium.
E. Compute a car rental cost.
F. Estimate the cost of gasoline for a trip.
G. Compute the finance charge for an automobile loan.

Chapter 10 TRANSPORTATION
A. Estimate and compute travel time.
B. Solve problems by using a map.
C. Use bus and train schedules.
D. Use subway and airline schedules.
E. Compute taxi fares.
F. Compute the cost of a trip.

Chapter 11 TAXES
A. Read and interpret W-2 Forms.
B. Find AGI, taxable income, and Federal income tax.
C. Determine Federal income tax and the amount owed or refund due.
D. Determine state income tax and the amount owed or refund due.
E. Determine city income tax and the amount owed or refund due.
F. Solve problems by projecting estimates.

Chapter 12 HOUSING
A. Compute the maximum amount to be spent on rent.
B. Compute the maximum amount to borrow to buy a house.
C. Compute the appreciation and the appreciated value of a house.
D. Compute the total monthly cost of an apartment or a condominium.
E. Compute the monthly payments for a mortgage and the closing costs.
F. Compute the assessed valuation and the annual real estate tax of a property.
G. Compute homeowner's insurance coverages.
H. Compute the costs of utilities.
I. Compute decorating and remodeling costs.

Chapter 13 TRADE INDUSTRIES
A. Compute construction and renovation costs.
B. Compute the costs of carpentry, plumbing, electrical, painting, or masonry supplies.
C. Compute the inside and outside diameter of pipes.
D. Compute the quantities of electrical supplies needed.
E. Compute labor charges.
F. Find the area of walls to be painted.
G. Solve problems by drawing diagrams.

Chapter 14 OTHER PROFESSIONS
A. Compute the costs of advertising.
B. Convert money amounts in other currencies to dollars.
C. Make quality-control predictions.
D. Compute medication dosages.

Chapter 15 PERSONAL FINANCE
A. Compute the inflation rate.
B. Compute purchasing power after inflation.
C. Compute the budget for living expenses.
D. Find the monthly share of an annual expense.
E. Compute the total monthly expenses and the monthly balance.
F. Compute the costs of raising children.
G. Compute total assets, total liabilities, and net worth.

Chapter 16 INVESTMENTS
A. Compute the yield on an investment.
B. Compute the value of a U.S. Savings Bond.
C. Compute the interest earned on a CD.
D. Compute the value of shares of stock.
E. Compute the profit or loss on a stock transaction.
F. Compute the interest earned on a bond.
G. Compute pension fund benefits.
H. Solve problems by working backward.
I. Compare dividends and interest on various investments.

Chapter 17 PROBABILITY
A. Find the probabilities of simple events.
B. Find the probability of 2 independent events.
C. Find the probability of 2 dependent events.
D. Draw tree diagrams or sample spaces.
E. Use the counting principle to find possible outcomes.
F. Use probability to predict outcomes.
G. Solve problems by making organized lists.

Chapter 18 EQUATIONS
A. Simplify numerical expressions.
B. Translate words into algebraic expressions or equations.
C. Evaluate algebraic expressions.
D. Write and solve 1-step equations.
E. Write and solve 2-step equations.
F. Solve problems by using rounded numbers in simpler problems.

To: The Teachers

From: The Authors

Re: **Philosophy for Practical Mathematics: Consumer Applications**

As consumers, we all live in a complicated world of buying and selling, saving and investing, borrowing and renting, and insuring and taxing. We are bombarded with come-ons and gimmicks, sales pitches, and deals. And we are frequently forced to answer "Which one should I choose?" and "Is it worth it?"

To chart a successful and prudent course as a consumer, we need knowledge, experience, and the ability to make good decisions.

That is what **Practical Mathematics: Consumer Applications** is about. It was written to and for teenagers who are beginning to cope with the complex options of being a consumer at the mall, on the job, in the bank, or even at play. We have written the book to provide students with the following:

—The fundamental *knowledge* they need to understand their options and to make reasonable decisions;

—Numerous opportunities to *experience* how to apply this knowledge in real situations; and

—Realistic *decision-making* activities that draw on a range of skills and understandings.

Beyond the consumer issues, this book also shows students how valuable mathematical skills and understandings are for wise consumerism. However, rather than stressing isolated computational skills, we have assumed that, like most consumers, they will have and use a calculator for most arithmetic calculations. This shift allows us to focus on when and why to use mathematics and on its applications for consumers.

Overall, in our lessons and exercises, we have tried to show students the usefulness and importance of mathematical skills and logical reasoning in surviving in the real world. We hope, therefore, that this book will help them better bridge the gap between what they learn in school and what they do outside of school now and in the future.

Marguerite M. Fredrick

Steven J. Leinwand

Laurence R. Wantuck

Robert P. Postman

Chapter 1 Paper- and Pencil-Computations

OBJECTIVE

- Test the prerequisite skills needed to learn the concepts developed in Chapter 1.

USING THE TEST

The Pre-Skills Test is designed to diagnose students' strengths and weaknesses on prerequisite skills necessary to study the mathematics in Chapter 1.

Assign the Pre-Skills Test. Allow students to work together in pairs or in small groups. Group members should help those who demonstrate a misunderstanding of a concept or a weakness in a skill.

The table below correlates the items on the Pre-Skills Test with the prerequisite skill and with the lesson(s) in the chapter for which it is needed.

Item(s)	Prerequisite skill	Lesson(s)
1–8	Regroup numbers.	1.1, 1.2
9–23	Add and subtract with basic facts.	1.1, 1.2
24–38	Multiply and divide with basic facts.	1.3, 1.4, 1.5, 1.6, 1.8
39–42	Add three numbers.	1.1, 1.8
43–46	Multiply and then add.	1.3
47–50	Write ratios as fractions.	1.5, 1.6
51–56	Write ratios as percents.	1.5, 1.6

References to specific skills that are reviewed in the Skills Bank are given on the individual lessons themselves.

You may wish to use the Skills Inventory Test available in the Teacher's Resource Package to establish groups for 3 levels of instruction or to determine students' readiness to begin the text.

SUPPLEMENTARY MATERIALS

TRP Skills Inventory Test, pp. 1–8

Pre-Skills Test

Copy and complete.

1. 87 = 8 tens ■ ones = 7 tens ■ ones 7; 17
2. 92 = ■ tens 2 ones = ■ tens 12 ones 9; 8
3. 326 = 3 hundreds ■ tens 6 ones = 2 hundreds ■ tens 6 ones 2; 12
4. 583 = 5 hundreds 8 tens ■ ones = 5 hundreds 7 tens ■ ones 3; 13

Copy and complete.

5. 2.8 = 2 ones and ■ tenths = 1 one and ■ tenths 8; 18
6. 4.1 = ■ ones and 1 tenth = ■ ones and 11 tenths 4; 3
7. 1.74 = 1 one and ■ tenths 4 hundredths = 1 one and ■ tenths 14 hundredths 7; 6
8. 9.23 = 9 ones and 2 tenths ■ hundredths = 9 ones and 1 tenth ■ hundredths 3; 13

Add or subtract.

9. $5 + 3$ 8
10. $6 + 2$ 8
11. $8 + 4$ 12
12. $5 + 8$ 13
13. $8 + 7$ 15
14. $9 + 6$ 15
15. $7 + 7$ 14
16. $9 - 5$ 4
17. $8 - 4$ 4
18. $12 - 3$ 9
19. $14 - 6$ 8
20. $11 - 9$ 2
21. $16 - 7$ 9
22. $16 - 9$ 7
23. $18 - 9$ 9

Multiply or divide.

24. 3×6 18
25. 4×4 16
26. 6×5 30
27. 5×9 45
28. 8×6 48
29. 4×9 36
30. 7×9 63
31. $15 \div 5$ 3
32. $28 \div 7$ 4
33. $32 \div 4$ 8
34. $8\overline{)64}$ 8
35. $6\overline{)42}$ 7
36. $7\overline{)56}$ 8
37. $8\overline{)72}$ 9
38. $9\overline{)81}$ 9

Compute.

39. $2 + 5 + 3$ 10
40. $3 + 7 + 8$ 18
41. $12 + 0 + 4$ 16
42. $9 + 6 + 7$ 22
43. $5 \times 3 + 6$ 21
44. $2 \times 9 + 3$ 21
45. $8 \times 0 + 3$ 3
46. $8 \times 3 + 0$ 24

Write the ratio in fraction form.

47. squares to all shapes $\frac{4}{13}$
48. circles to all shapes $\frac{7}{13}$
49. triangles to all shapes $\frac{2}{13}$
50. circles and squares to all shapes $\frac{11}{13}$

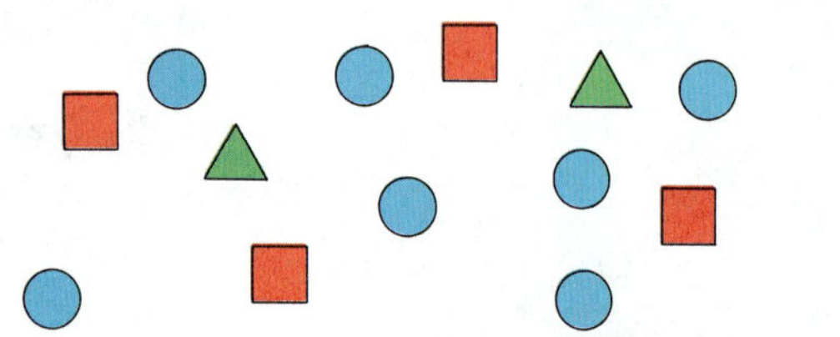

Write the ratio as a percent.

51. 6 per 100 6%
52. 17 per 100 17%
53. 80 per 100 80%
54. 8 out of 100 8%
55. 21 out of 100 21%
56. 75 out of 100 75%

Chapter 1
PAPER -AND- PENCIL COMPUTATIONS

Many people enjoy participating in or watching a baseball game. How are numbers used in baseball, both during the game and in keeping performance records?

OBJECTIVE

- Formulate problems, using text and data.

USING THE CHAPTER OPENER

Discuss students' answers to the question on the page. Most of them should have a fairly good understanding of baseball, but may not have thought about the game in terms of numbers. Some of the uses are as follows: The bases have the ordinal numbers first, second, and third; each field position is numbered for ease in keeping score; players wear numbers on their uniforms; the field is designed with specific measurements; 3 strikes, you're out, and 4 balls, you walk; players' statistics are percentages of total times at bat or total innings pitched. There are many other examples that students should be able to come up with.

You may want to discuss briefly if it would be possible to play baseball without using numbers.

COOPERATIVE LEARNING

Assign heterogeneous groups of students the task of finding all the ways numbers are used in various sports. Assign 1 sport to each group. These might include football, tennis, hockey, soccer, bowling, or basketball. Some uses of numbers are obvious, but research will turn up others. Measurements may be involved, including pounds of air in an inflated ball, and time is used, including how long a team can possess the basketball before they must shoot or the length of a penalty in hockey.

Have groups present their findings to the class, allowing them to be the judge of each group's thoroughness and accuracy.

SUPPLEMENTARY MATERIALS

TRP Group Projects, p. 1

OBJECTIVE

- Add whole numbers and decimals.

TEACHING THE LESSON

WARM-UP Write the following exercises on the chalkboard or an overhead transparency.

Name the place value of the underlined digit.

1. 3<u>4</u>5 **2.** <u>4</u>,092 **3.** 5.<u>7</u>9 **4.** 82.0<u>7</u>

Add.

5.	**6.**	**7.**	**8.**
22	436	3.4	0.4
+35	+533	+2.5	+0.4
(57)	(969)	(5.9)	(0.8)

(**1.** 4 tens; **2.** 4 thousands; **3.** 7 tenths; **4.** 7 hundredths)

INTRODUCTION Begin the discussion by asking students to describe situations in which they would need to add. Examples might include: joining 2 or more groups, or finding a total number of objects, measures, or money amounts.

Remind students that adding decimals is similar to adding amounts of money. Encourage them to name some reasons for adding money.

INSTRUCTION Review the addition vocabulary at the top of the page. After discussing each skill, have students name the addends and the sum.

Emphasize the steps used for each paper-and-pencil skill. You may want to write each of the examples on the chalkboard and ask students to describe the steps. If students have difficulty with a given example, work through a similar example on the chalkboard. One similar example is given below for each skill.

For Skill 1, emphasize adding from right to left. Explain that in Step 1 we are regrouping 12 ones as 1 ten 2 ones and that in Step 2 we are regrouping 16 tens as 1 hundred 6 tens. Similar example: 6,038 + 2,897

1.1 ADDING WHOLE NUMBERS AND DECIMALS

Addition Vocabulary: ADDEND + ADDEND = SUM

ADDEND
+ADDEND
SUM

SKILL 1 *Adding 2 whole numbers*

Add: 4,068 + 5,794

1. Add ones. Regroup.	2. Add tens. Regroup.	3. Add hundreds.	4. Add thousands.
1	1 1	1 1	1 1
4,068	4,068	4,068	4,068
+5,794	+5,794	+5,794	+5,794
2	62	862	9,862

SKILL 2 *Column addition with whole numbers*

Add: 453 + 4,802 + 327

Remember to line up the places when you write the example in vertical form.

1. Add ones. Regroup.	2. Add tens.	3. Add hundreds. Regroup.	4. Add thousands.
1	1	1 1	1 1
453	453	453	453
4,802	4,802	4,802	4,802
+ 327	+ 327	+ 327	+ 327
2	82	582	5,582

SKILL 3 *Adding 2 decimals*

Add: 4.6 + 3.7

1. Line up the decimal points and the decimal places.	2. Write the decimal point for the sum.	3. Add as you would whole numbers.
		1
4.6	4.6	4.6
+3.7	+3.7	+3.7
		8.3

SKILL 4 *Column addition with decimals*

Add: 5.16 + 8.7 + 4.02

1. Line up the decimal points and the decimal places. **THINK:** 8.7 = 8.70	2. Write the decimal point for the sum.	3. Add as you would whole numbers.
5.16 8.70 +4.02	5.16 8.70 +4.02	5.16 8.70 +4.02 17.88

PRACTICE EXERCISES

Add.

• **1.** 521 +236 757	**2.** 837 +185 1,022	**3.** 908 + 39 947	**4.** 2,346 +7,385 9,731	**5.** 7,368 + 612 7,980	**6.** 568 +2,394 2,962
• **7.** 18.5 +11.3 29.8	**8.** 0.86 +1.62 2.48	**9.** 9.47 +6.8 16.27	**10.** 0.73 +0.64 1.37	**11.** 2.5 +29.63 32.13	**12.** 16.2 +35.97 52.17
•**13.** 802 311 + 89 1,202	**14.** 875 918 +6,213 8,006	**15.** 4,275 5,728 + 982 10,985	**16.** 1,805 4,633 +5,065 11,503	**17.** 5,024 810 + 265 6,099	**18.** 411 6,807 + 253 7,471
•**19.** 7.3 4.2 +3.5 15.0	**20.** 9.6 8.7 +0.4 18.7	**21.** 38.6 19.8 +24.2 82.6	**22.** 2.94 3.65 +13.08 19.67	**23.** 13.2 14.57 + 0.7 28.47	**24.** 0.5 17.83 + 0.27 18.60

25. 4,892 + 3,605 8,497 **26.** 567 + 9,028 9,595 **27.** 3,841 + 187 4,028

28. 19.3 + 48.6 67.9 **29.** 9 + 8.5 17.5 **30.** 18.34 + 9.7 28.04

31. 4,387 + 8,062 + 964 13,413 **32.** 4,361 + 904 + 627 5,892 **33.** 2,053 + 87 + 763 2,903

34. 12.3 + 8.9 + 17.1 38.3 **35.** 9.6 + 13.05 + 10 32.65 **36.** 2.5 + 0.18 + 12.08 14.76

Solve.

•**37.** Ron spent $28.36 at the sports store and $16.09 at the clothing store. How much did he spend all together? $44.45

38. Margaret sold 189 ice cream cones on Friday, 308 on Saturday, and 67 on Sunday. How many cones did she sell all together? 564 cones

39. It is 789 mi from Houston, Texas, to Atlanta, Georgia, and 2,496 mi from Atlanta to San Francisco, California. How many miles is it all together from Houston to San Francisco? 3,285 mi

40. During the first week, 10.8 in. of snow fell. During the second week 18 in. of snow fell, and 0.8 in. fell during the third week. What was the total snowfall for these 3 weeks? 29.6 in.

For Skill 2, emphasize lining up the place values. Do not hesitate to have students write column addition examples vertically on graph paper to help them align the columns properly. Point out that addition can sometimes be made easier if they can find numbers that "make ten." When adding the ones, it is easy to first add 3 + 7 to get 10 and then add 10 + 2 to get 12. Review the regrouping in Steps 2 and 3.
Similar example: 320 + 5,896 + 484

For Skill 3, emphasize aligning the decimal points and the decimal places. Point out that 13 tenths are regrouped as 1 one and 3 tenths.
Similar example: 2.8 + 5.6

For Skill 4, emphasize writing zero as a placeholder since 8.7 and 8.70 are equivalent decimals.
Similar example: 7.18 + 9.7 + 3.03

SKILLS BANK You may want to refer students to the following skills in the Skills Bank:

The Decimal System, Skills 1, 3

Add With Whole Numbers and Decimals, Skills 1–6

ASSIGNMENTS

Level 1	Odd 1–23, 37–38
Level 2	1–39
Level 3	25–40

FOLLOW-UP ACTIVITY

APPLICATION Have groups of students plan trips they would like to make between United States cities by car. Ask students to look up distances between these cities using distance charts on road maps or in an almanac. Then ask students to add the distances to find the total lengths of the trips.

SUPPLEMENTARY MATERIALS

TRP Practice, p. 1
TRP Reteaching, p. 1
TRP Transparency 1

OBJECTIVE

- Subtract whole numbers and decimals.

TEACHING THE LESSON

WARM-UP Write the following exercises on the chalkboard or an overhead transparency.

Complete.

1. 7 tens 4 ones = ■ tens 14 ones(6)
2. 9 ones and 2 tenths = ■ ones and 12 tenths (8)

Subtract.

3.	4.	5.	6.
83	592	6.8	7.19
−62	−250	−1.3	−3.08
(21)	(342)	(5.5)	(4.11)

INTRODUCTION Begin the discussion by asking students to describe situations in which they would need to subtract. Examples might include: separating a group, finding how much is left, finding how many more are needed, or comparing numbers of objects, measures, or money amounts.

Remind students that subtracting decimals is similar to subtracting amounts of money. Ask them to name reasons for subtracting money.

INSTRUCTION Review the vocabulary at the top of the page. After discussing each skill, have students name the parts of each subtraction example.

Emphasize the steps used for each paper-and-pencil skill. You may want to write each of the examples on the chalkboard and ask students to describe the steps. If students have difficulty with a given example, work through a similar example on the chalkboard. One similar example is given below for each skill.

For Skill 1, emphasize the regrouping in Step 1 of 7 tens 4 ones as 6 tens 14 ones so that 9 can be subtracted from 14.
Similar example: 7,864 − 5,628

1.2 SUBTRACTING WHOLE NUMBERS AND DECIMALS

Subtraction Vocabulary:
Addition and subtraction are related operations.

ADDEND + ADDEND = SUM
SUM − ADDEND = ADDEND (DIFFERENCE)

```
    SUM
 −ADDEND
  ADDEND (DIFFERENCE)
```

SKILL 1 *Subtracting whole numbers*

Subtract: 8,674 − 6,319

1. Regroup tens. Subtract ones.	2. Subtract tens.	3. Subtract hundreds.	4. Subtract thousands.

<pre>   6 14
 8,674
−6,319
     5</pre>
	<pre> 6 14
	 8,674
	−6,319
	    55</pre>
		<pre> 6 14
		 8,674
		−6,319
		   355</pre>
			<pre> 6 14
			 8,674
			−6,319
			 2,355</pre>

SKILL 2 *Subtracting whole numbers across zeros*

Subtract: 8,012 − 7,635

1. Regroup tens. Subtract ones.	2. No hundreds to regroup. Regroup thousands.	3. Regroup hundreds.	4. Complete the subtraction.

<pre>   0 12
 8,012
−7,635
     7</pre>
	<pre>7 10 0 12
	 8,012
	−7,635
	     7</pre>
		<pre> 9 10
		7 10 0 12
		 8,012
		−7,635
		     7</pre>
			<pre> 9 10
			7 10 0 12
			 8,012
			−7,635
			   377</pre>

SKILL 3 *Subtracting decimals*

Subtract: 9.2 − 0.7

1. Line up the decimal points and the decimal places.	2. Write the decimal point for the difference.	3. Regroup ones. Subtract tenths.	4. Subtract ones.
<pre> 9.2 −0.7</pre>	<pre> 9.2 −0.7</pre>	<pre> 8 12 9.2 −0.7 .5</pre>	<pre> 8 12 9.2 −0.7 8.5</pre>

SKILL 4 *Subtracting decimals by writing additional zeros*

Subtract: 6.3 − 2.15

1. Line up the decimal points and the decimal places. **THINK: 6.3 = 6.30**	2. Write the decimal point for the difference.	3. Regroup.	4. Subtract.
6.30 −2.15	6.30 −2.15 .	6.3̸0̸ (2 10) −2.15 .	6.3̸0̸ (2 10) −2.15 4.15

PRACTICE EXERCISES

Subtract.

• 1. 683 − 251 = 432	2. 872 − 839 = 33	3. 964 − 87 = 877	4. 4,829 − 1,237 = 3,592	5. 8,658 − 2,672 = 5,986	6. 5,216 − 397 = 4,819
• 7. 5,007 − 3,129 = 1,878	8. 8,730 − 5,372 = 3,358	9. 4,900 − 634 = 4,266	• 10. 7.2 − 3.6 = 3.6	11. 7.34 − 4.92 = 2.42	• 12. 78.5 − 39.58 = 38.92
13. 0.83 − 0.57 = 0.26	14. 68.21 − 46.3 = 21.91	15. 81.1 − 79.3 = 1.8	16. 90.7 − 39.8 = 50.9	17. 30.06 − 4.87 = 25.19	18. 20.08 − 17.3 = 2.78

19. 937 − 425 512
20. 602 − 418 184
21. 537 − 189 348
22. 4,862 − 3,794 1,068
23. 6,238 − 429 5,809
24. 3,072 − 1,691 1,381
25. 8,006 − 5,237 2,769
26. 6,521 − 875 5,646
27. 5,307 − 87 5,220
28. 43.6 − 18.7 24.9
29. 58.32 − 29.54 28.78
30. 53.1 − 40.2 12.9
31. 90.08 − 69.79 20.29
32. 36.2 − 14.17 22.03
33. 50.3 − 15.82 34.48
34. 9 − 6.3 2.7
35. 27 − 12.89 14.11
36. 60 − 7.35 52.65

Solve.

• 37. It is 2,078 mi from Dallas, Texas, to Seattle, Washington. You have already traveled 1,693 mi. How many more miles do you need to travel? 385 mi

38. Your class needs to raise $2,004 for a class trip. So far the class has $796. How much more money does the class need to raise? $1,208

39. You need to buy some school supplies. The total cost of the supplies is $9.06. You have $7.98. How much more money do you need? $1.08

40. In March, there were 3.69 in. of rain. In April, 4 in. of rain fell. How much more rain fell in April than in March? 0.31 in.

For Skill 2, emphasize how to subtract while regrouping across zeros. Point out that Steps 2 and 3 show that, in order to get enough tens to subtract, the regrouping process must start with the thousands and then move from left to right.
Similar example: 5,023 − 4,357

For Skill 3, emphasize aligning the decimal points and the decimal places. Point out that 9 ones and 2 tenths are regrouped as 8 ones and 12 tenths so that 7 can be subtracted from 12 in the tenths column.
Similar example: 6.3 − 0.8

For Skill 4, emphasize writing zero as a placeholder since 6.3 and 6.30 are equivalent decimals.
Similar example: 8.4 − 3.28

SKILLS BANK You may want to refer students to the following skills in the Skills Bank:

The Decimal System, Skills 1, 3

Subtract With Whole Numbers and Decimals, Skills 1–6

ASSIGNMENTS

Level	Exercises
Level 1	Odd 1–17, 37–38
Level 2	1–39
Level 3	25–40

FOLLOW-UP ACTIVITY

COOPERATIVE LEARNING
Divide the class into small heterogeneous groups. Ask each group to look through a catalog and/or newspaper advertisements to find the same or similar items advertised at different prices. Then ask students to subtract to find the amount they would save if they bought the items at the lower prices. Have the groups compare their findings orally.

SUPPLEMENTARY MATERIALS

TRP Practice, p. 2
TRP Reteaching, p. 2
TRP Transparency 1

OBJECTIVE

- Multiply whole numbers and decimals.

TEACHING THE LESSON

WARM-UP Write the following exercises on the chalkboard or an overhead transparency.

Compute.

1. $6 \times 3 + 4$ (22) **2.** $9 \times 0 + 6$ (6)

3. $7 \times 3 + 2$ (23) **4.** $8 \times 1 + 5$ (13)

Multiply.

5. 43×2 (86) **6.** 123×3 (369) **7.** 0.4×2 (0.8) **8.** 0.04×2 (0.08)

INTRODUCTION Present this situation: You want to buy 5 cassette tapes. Each tape costs $6.00. How could you add to find the total cost of the 5 tapes? How could you multiply?

Point out that, since all of the addends are the same, multiplying is faster than adding. Lead students in a discussion about other situations in which they could use multiplication.

INSTRUCTION Review the multiplication vocabulary at the top of the page. After discussing each skill, have students name the factors and the product.

Emphasize the steps used for each paper-and-pencil skill. You may want to write each of the examples on the chalkboard and ask students to describe the steps. If students have difficulty with a given example, work through a similar example on the chalkboard. One similar example is given below for each skill.

For Skill 1, emphasize that, in Step 1, $6 \times 8 = 48$ and that the 8 ones is placed in the product and the 4 tens above the tens column. In Step 2, point out that we multiply 6 and 3 (tens) to get 18 (tens) and then add 4 (tens) to get 22 (tens), or 2 (hundreds) 2 (tens).

1.3 MULTIPLYING WHOLE NUMBERS AND DECIMALS

Multiplication Vocabulary: FACTOR × FACTOR = PRODUCT

$$\begin{array}{r} \text{FACTOR} \\ \times\text{FACTOR} \\ \hline \text{PRODUCT} \end{array}$$

SKILL 1 *Multiplying whole numbers*

Multiply: 6×38

1. Multiply ones. Regroup.

$$\begin{array}{r} {}^{4} \\ 38 \\ \times\ 6 \\ \hline 8 \end{array}$$

2. Multiply tens. Then add 4 tens.

$$\begin{array}{r} {}^{4} \\ 38 \\ \times\ 6 \\ \hline 228 \end{array}$$

SKILL 2 *Multiplying whole numbers with zeros*

Multiply: 9×407

1. Multiply ones. Regroup.

$$\begin{array}{r} {}^{6} \\ 407 \\ \times\ 9 \\ \hline 3 \end{array}$$

2. Multiply tens. Then add 6 tens.
   THINK: 0 tens + 6 tens = 6 tens

$$\begin{array}{r} {}^{6} \\ 407 \\ \times\ 9 \\ \hline 63 \end{array}$$

3. Multiply hundreds.

$$\begin{array}{r} {}^{6} \\ 407 \\ \times\ 9 \\ \hline 3{,}663 \end{array}$$

SKILL 3 *Multiplying whole numbers and decimals in tenths*

Multiply: 7×3.4

1. Multiply as you would whole numbers.

$$\begin{array}{r} {}^{2} \\ 3.4 \\ \times\ 7 \\ \hline 238 \end{array}$$

2. Add the numbers of decimal places in the factors. There are that many decimal places in the product.

$$\begin{array}{rl} 3.4 & \leftarrow 1 \text{ decimal place} \\ \times\ 7 & \leftarrow 0 \text{ decimal place} \\ \hline 23.8 & \leftarrow 1 \text{ decimal place} \end{array}$$

SKILL 4 *Multiplying whole numbers and decimals in hundredths*

Multiply: 8 × 1.73

1. Multiply as you would whole numbers.

```
  5 2
  1.73
×    8
  1384
```

2. Add the numbers of decimal places in the factors. There are that many decimal places in the product.

```
   1.73 ← 2 decimal places
×     8 ← 0 decimal place
  13 84 ← 2 decimal places
```

PRACTICE EXERCISES

Multiply.

• 1. 24 × 2 48
2. 12 × 4 48
3. 34 × 3 102
4. 39 × 3 117
5. 26 × 9 234
6. 89 × 7 623
7. 213 × 2 426
• 8. 401 × 3 1,203
9. 418 × 3 1,254
10. 372 × 9 3,348
11. 609 × 5 3,045
12. 307 × 8 2,456
•13. 3.2 × 3 9.6
14. 2.4 × 2 4.8
15. 0.9 × 2 1.8
16. 0.5 × 6 3.0
17. 6.8 × 3 20.4
18. 8.2 × 9 73.8
•19. 4.32 × 2 8.64
20. 1.31 × 4 5.24
21. 0.32 × 6 1.92
22. 0.65 × 7 4.55
23. 7.07 × 4 28.28
24. 1.35 × 8 10.8

25. 4 × 52 208
26. 3 × 81 243
27. 8 × 37 296
28. 9 × 16 144
29. 3 × 221 663
30. 6 × 167 1,002
31. 9 × 618 5,562
32. 7 × 442 3,094
33. 8 × 3.6 28.8
34. 5 × 7.3 36.5
35. 3 × 1.3 3.9
36. 6 × 1.9 11.4
37. 8 × 0.65 5.2
38. 2 × 7.12 14.24
39. 4 × 3.75 15
40. 9 × 3.39 30.51
41. 3 × 80 240
42. 7 × 104 728
43. 2 × 907 1,814
44. 8 × 408 3,264
45. 5 × 1.02 5.1
46. 8 × 6.07 48.56
47. 4 × 0.09 0.36
48. 9 × 0.03 0.27

Solve.

•49. It is 379 mi from Los Angeles to San Francisco. You made this trip 7 times. How many miles did you travel? 2,653 mi

50. You bought 6 lb of vegetables that cost $0.49 per lb. How much money did you spend? $2.94

51. Your class is selling school shirts for $2.39 each. The class sold 8 shirts during the first hour of the sale. How much money was collected during that hour? $19.12

52. During one year, the annual consumption of fresh citrus fruits was about 94.8 lb per person. About how many pounds did a family of 4 consume that year? 379.2 lb

Some students may mistakenly multiply 6 × 4. Some may add 4 + 3 and then multiply 6 × 7, instead of 6 × 3. Similar example: 7 × 49

For Skill 2, emphasize Step 2. Show that 9 × 0 (tens) = 0 (tens) and that 0 (tens) + 6 (tens) = 6 (tens). Similar example: 8 × 609

For Skills 3 and 4, emphasize the rule for determining the number of decimal places in a product. Mention that the rule also works when there are decimal places in both factors. Similar examples: 6 × 5.7; 7 × 2.38

SKILLS BANK You may want to refer students to the following skills in the Skills Bank:

Multiply With Whole Numbers, Skills 2–3

Multiply With Decimals, Skill 2

ASSIGNMENTS

Level 1	Odd 1–23, 49–50
Level 2	1–51
Level 3	29–52

FOLLOW-UP ACTIVITY

ENRICHMENT Write these multiplication examples on the chalkboard or on an overhead transparency. Ask students to discover a pattern.

1	2	3	4	5	6	7	8	9
×9	×9	×9	×9	×9	×9	×9	×9	×9
9	18	27	36	45	54	63	72	81

Lead students in a discussion that reveals that the sum of the digits in each product is always 9 and that the tens digit in each product is always 1 less than the "top" factor. Point out to students how these patterns can help them multiply by 9.

SUPPLEMENTARY MATERIALS

TRP Practice, p. 3
TRP Reteaching, p. 3
TRP Transparency 2

OBJECTIVE

- Divide whole numbers and decimals.

TEACHING THE LESSON

WARM-UP Write the following exercises on the chalkboard or an overhead transparency.

Divide.

1. $3\overline{)600}$ (200) **2.** $4\overline{)800}$ (200) **3.** $3\overline{)240}$ (80) **4.** $4\overline{)280}$ (70)

Round to the nearest tenth.

5. 0.24 (0.2) **6.** 9.68 (9.7)

INTRODUCTION Present this situation: You see a sign that advertises 6 records for $24. How could you find the cost of each record?

Point out that division can be used in 2 ways: (1) to find how many equal groups or (2) to find how many in each equal group. The situation above is an example of the second use of division. There are 6 equal groups in $24. How much is there in each equal group? ($4)

INSTRUCTION Review the division vocabulary at the top of the page. After discussing each skill, have students name the dividend, the divisor, the quotient, and the remainder.

Emphasize the steps used for each paper-and-pencil skill. You may want to write each of the examples on the chalkboard and ask students to describe the steps. If students have difficulty with a given example, work through a similar example on the chalkboard. One similar example is given below for each skill.

For Skill 1, make sure that students remember the series of steps used to complete a division example: Divide, Multiply, Subtract, and Bring Down. Since 6 > 3, there are enough hundreds in the dividend to divide. Therefore, the first digit in the quotient is hundreds.
Similar example: 957 ÷ 3

1.4 DIVIDING WHOLE NUMBERS AND DECIMALS

Division Vocabulary: DIVIDEND ÷ DIVISOR = QUOTIENT AND (R)REMAINDER

$$\begin{array}{r} \text{QUOTIENT (R)REMAINDER} \\ \text{DIVISOR}\overline{)\text{DIVIDEND}} \end{array}$$

SKILL 1 *Dividing 3-digit numbers by 1-digit numbers*

Divide: 648 ÷ 3

1. Divide hundreds.
   - $3\overline{)648}$, quotient so far: 2
   - 6

2. Divide tens.
   - $3\overline{)648}$, quotient so far: 21
   - 6↓
   - 4
   - 3
   - 1

3. Divide ones.
   - $3\overline{)648}$, quotient: 216
   - 6↓
   - 4
   - 3↓
   - 18
   - 18
   - 0

SKILL 2 *Dividing 4-digit numbers by 1-digit numbers*

Divide: 2,417 ÷ 3

1. Divide hundreds.
   - $3\overline{)2,417}$, quotient so far: 8
   - 2 4

2. Divide tens.
   - $3\overline{)2,417}$, quotient so far: 80
   - 2 4↓
   - 1
   - 0
   - 1

3. Divide ones.
   - $3\overline{)2,417}$, quotient: 805 R2
   - 2 4↓
   - 1
   - 0↓
   - 17
   - 15
   - 2

SKILL 3 *Dividing decimals by whole numbers*

Divide: 18.3 ÷ 6

1. Place the decimal point in the quotient directly above the decimal point in the dividend.
2. Divide as you would whole numbers. Write additional zeros in the dividend as needed.

Worked example:
- $6\overline{)18.30}$, quotient: 3.05
- 18↓
- 3
- 0↓
- 30
- 30
- 0

For Skill 2, there are not enough thousands in the dividend to divide since $2 < 3$. Have students look to the next digit, 4. Since $24 > 3$, there are enough hundreds in the dividend to divide. Therefore, the first digit in the quotient is hundreds. Point out the placement of the remainder.
Similar example: 3,631 ÷ 3

For Skill 3, emphasize placing the decimal point in the quotient before dividing. Point out that an additional zero is needed in the dividend so that you can continue dividing until a zero remainder is found.
Similar example: 24.4 ÷ 8

For Skill 4, emphasize that the quotient must be found to thousandths, the place to the right of the place you will be rounding to. In order to do this, 2 additional zeros are needed in the dividend.
Similar example: 4.3 ÷ 6

SKILLS BANK You may want to refer students to the following skills in the Skills Bank:

Round Whole Numbers and Decimals, Skill 3

Divide With Whole Numbers, Skills 2–5

Divide With Decimals, Skill 2

SKILL 4 *Rounding decimal quotients*

Divide and round to the nearest hundredth: 3.4 ÷ 7

1. Place the decimal point in the quotient directly above the decimal point in the dividend.
2. Divide one place beyond the place you are rounding to.
   THINK: Divide to thousandths, round to hundredths.
3. Round the quotient.
   Remember: ≈ means *is approximately equal to.*

$$\begin{array}{r} 0.485 \\ 7\overline{)3.400} \\ \underline{2\,8} \\ 60 \\ \underline{56} \\ 40 \\ \underline{35} \\ 5 \end{array}$$

0.485 ≈ 0.49

PRACTICE EXERCISES

Divide.

• **1.** $2\overline{)84}$ 42 **2.** $3\overline{)39}$ 13 **3.** $2\overline{)37}$ 18 R1 • **4.** $6\overline{)282}$ 47 **5.** $8\overline{)751}$ 93 R7

6. $4\overline{)872}$ 218 **7.** $5\overline{)530}$ 106 **8.** $7\overline{)723}$ 103 R2 • **9.** $6\overline{)8{,}790}$ 1,465 **10.** $2\overline{)6{,}210}$ 3,105

11. $8\overline{)6{,}616}$ 827 **12.** $3\overline{)2{,}249}$ 749 R2 **13.** $6\overline{)15.3}$ 2.55 **14.** $8\overline{)6.8}$ 0.85 **15.** $7\overline{)6.3}$ 0.9

16. $4\overline{)8.12}$ 2.03 **17.** $5\overline{)0.52}$ 0.104 **18.** $6\overline{)11.22}$ 1.87 **19.** $7\overline{)30.87}$ 4.41 **20.** $9\overline{)8.37}$ 0.93

21. 83 ÷ 4 20 R3 **22.** 71 ÷ 6 11 R5 **23.** 842 ÷ 2 421 **24.** 3,748 ÷ 4 937 **25.** 6,790 ÷ 9 754 R4

26. 7.2 ÷ 5 1.44 **27.** 5.6 ÷ 4 1.4 **28.** 8.19 ÷ 9 0.91 **29.** 12.34 ÷ 4 3.085 **30.** 23.5 ÷ 8 2.9375

Divide and round to the nearest tenth.

• **31.** 9.92 ÷ 4 2.5 **32.** 7.41 ÷ 3 2.5 **33.** 5.48 ÷ 8 0.7 **34.** 7.63 ÷ 9 0.8 **35.** 20.3 ÷ 6 3.4

Divide and round to the nearest hundredth.

• **36.** 8.32 ÷ 6 1.39 **37.** 8.59 ÷ 4 2.15 **38.** 6.93 ÷ 8 0.87 **39.** 6.87 ÷ 7 0.98 **40.** 39.7 ÷ 3 13.23

Solve.

• **41.** An 8-oz box of chocolates costs \$3.92. How much is the cost per ounce?
\$0.49

42. The total cost of a restaurant dinner for 6 people was \$73.80. How much was each person's equal share?
\$12.30

43. It is about 2,628 mi from Los Angeles to Washington, DC. You made the trip in 6 d, traveling the same distance each day. How many miles did you travel each day?
438 mi

44. During one year, a family of 5 consumed about 750.2 lb of fresh vegetables. About how much was that per family member?
150.04 lb

ASSIGNMENTS

Level 1	1–20, Odd 31–35, 36–37
Level 2	1–43
Level 3	21–44

FOLLOW-UP ACTIVITY

APPLICATION Show students how to check answers to division examples. Ask students to check the answers to Practice Exercises 1–10.

SUPPLEMENTARY MATERIALS

TRP Practice, p. 4
TRP Reteaching, p. 4
TRP Transparency 2

OBJECTIVE

- Rename fractions, decimals, and percents.

TEACHING THE LESSON

WARM-UP Write the following exercises on the chalkboard or an overhead transparency.

Multiply. Identify the patterns.

1. 10×9 (90) **2.** 10×0.9 (9.0)

100×9 (900) 100×0.9 (90.0)

(Multiply by 10: Move the decimal point 1 place to the right. Multiply by 100: Move it 2 places to the right.)

Divide. Identify the patterns.

3. $400 \div 10$ (40) **4.** $62.3 \div 10$ (6.23)

$400 \div 100$ (4) $62.3 \div 100$ (0.623)

(Divide by 100: Move the decimal point 1 place to the left. Divide by 100: Move it 2 places to the left.)

INTRODUCTION Write a fraction, a decimal, and a percent on the chalkboard: $\frac{1}{4}$; 0.65; 43%

Ask students to read these different numbers and to tell where and how they have used numbers like these or seen them used.

INSTRUCTION Review the percent vocabulary at the top of the page. Reinforce the idea that percent means "per hundred, or hundredths." Have students write 12%, 37%, 3%, and 9% as decimals and as fractions. (0.12, $\frac{12}{100}$; 0.37, $\frac{37}{100}$; 0.03, $\frac{3}{100}$; 0.09, $\frac{9}{100}$)

For Skill 1, point out what happens when a number is multiplied by 100. Note that an additional zero is needed as a placeholder.

For Skill 2, emphasize that the fraction $\frac{1}{5}$ means "1 out of 5" or "1 divided by 5." Point out that this decimal quotient, when written in hundredths, can easily be renamed as a percent.

1.5 FRACTIONS, DECIMALS, AND PERCENTS

Percent Vocabulary: **Percent (%)** means **per hundred.**
18% means 18 per hundred, 0.18, or $\frac{18}{100}$.

SKILL 1 *Renaming decimals as percents*

Rename 0.9 as a percent.

1. Multiply by 100 by moving the decimal point 2 places to the right. Write additional zeros if necessary. — 0.90
2. Write the percent sign. — 90%

Other Examples

0.89 → 0.89 → 89% 0.034 → 0.034 → 3.4% 8.4 → 8.40 → 840%

SKILL 2 *Renaming fractions as percents*

Rename $\frac{1}{5}$ as a percent.

1. Write the fraction as a decimal. Divide the numerator by the denominator. Write additional zeros if necessary. — $\frac{1}{5} \rightarrow 5\overline{)1.0}$ = 0.2
2. Write the decimal as a percent. — 0.20 → 20%

Other Examples

$2\frac{1}{5} = 2 + \frac{1}{5} \rightarrow 200\% + 20\% = 220\%$ $\frac{5}{8} \rightarrow 8\overline{)5.000}$ = 0.625 = 62.5%

SKILL 3 *Renaming percents as decimals*

Rename 3% as a decimal.

1. Divide by 100 by moving the decimal point 2 places to the left. Write additional zeros if necessary. — 3% → 003%
2. Remove the percent sign. — 0.03

Other Examples

43% → 043% → 0.43 5.7% → 005.7% → 0.057 287% → 287% → 2.87

10 CHAPTER 1

SKILL 4 *Renaming percents as fractions*

Rename 80% as a fraction.

1. Write the percent as a fraction with a denominator of 100. $80\% = \frac{80}{100}$
2. Write the fraction in lowest terms. $\frac{80}{100} = \frac{80 \div 20}{100 \div 20} = \frac{4}{5}$

Other Examples

$75\% = \frac{75}{100} = \frac{75 \div 25}{100 \div 25} = \frac{3}{4}$ $150\% = \frac{150}{100} = \frac{150 \div 50}{100 \div 50} = \frac{3}{2} = 1\frac{1}{2}$

PRACTICE EXERCISES

Rename as a percent.

• 1. $\frac{23}{100}$ 23% 2. $\frac{12}{100}$ 12% 3. $\frac{7}{100}$ 7% 4. $\frac{3}{100}$ 3% 5. $\frac{186}{100}$ 186% 6. $\frac{600}{100}$ 600%

• 7. 0.63 63% 8. 0.07 7% 9. 0.058 5.8% 10. 0.2 20% 11. 9 900% 12. 6.21 621%

Rename as a decimal and as a percent.

• 13. $\frac{1}{4}$ 0.25; 25% 14. $\frac{4}{5}$ 0.8; 80% 15. $\frac{7}{8}$ 0.875; 87.5% 16. $\frac{3}{8}$ 0.375; 37.5% 17. $\frac{2}{5}$ 0.4; 40% 18. $\frac{7}{10}$ 0.7; 70%

• 19. $3\frac{1}{2}$ 3.5; 350% 20. $7\frac{3}{4}$ 7.75; 775% 21. $9\frac{1}{8}$ 9.125; 912.5% 22. $4\frac{5}{8}$ 4.625; 462.5% 23. $5\frac{3}{5}$ 5.6; 560% 24. $8\frac{3}{10}$ 83; 830%

Rename as a decimal.

• 25. 24% 0.24 26. 56% 0.56 27. 18% 0.18 28. 4% 0.04 29. 2% 0.02 30. 9% 0.09

• 31. 2.8% 0.028 32. 0.8% 0.008 33. 0.05% 0.0005 34. 300% 3 35. 438% 4.38 36. 703% 7.03

Rename as a fraction.

• 37. 70% $\frac{7}{10}$ 38. 60% $\frac{3}{5}$ 39. 85% $\frac{17}{20}$ 40. 140% $1\frac{2}{5}$ 41. 110% $1\frac{1}{10}$ 42. 180% $1\frac{4}{5}$

Complete the table of estimated job opening changes from 1980 to 1990.

Occupation	Percent	Fraction	Decimal
Welder	up 25%	up $\frac{1}{4}$	• 43. up ■ 0.25
Librarian	up 5%	• 44. up ■ $\frac{1}{20}$	up 0.05
Secretary	45. up ■ 32%	46. up ■ $\frac{8}{25}$	up 0.32
Computer technician	up 112%	47. up ■ $1\frac{3}{25}$	48. up ■ 1.12
High school teacher	49. down ■	down $\frac{7}{50}$	50. down ■

49. 14% 50. 0.14

For Skill 3, point out what happens when a number is divided by 100. Note that 2 additional zeros are needed as placeholders.

For Skill 4, remind students that they can multiply or divide the numerator and the denominator of a fraction by the same number to find an equivalent fraction.

SKILLS BANK You may want to refer students to the following skills in the Skills Bank:

- Multiply With Decimals, Skill 1
- Divide With Decimals, Skill 1
- Write Fractions as Decimals, Skills 1–2
- Equivalent Fractions, Decimals, and Percents, Skills 1–5

ASSIGNMENTS

Level 1	Odd 1–41
Level 2	Even 2–42, 43–50
Level 3	Odd 19–41, 43–50

FOLLOW-UP ACTIVITY

APPLICATION Newspapers often contain stories that use decimals, percents, or fractions. Ask students to find these stories and circle the decimals, percents, and fractions. Have them rename each circled amount in the other 2 forms.

SUPPLEMENTARY MATERIALS

TRP Practice, p. 5
TRP Reteaching, p. 5
TRP Transparency 3

OBJECTIVE

- Find the percent of a number.

TEACHING THE LESSON

WARM-UP Write the following exercises on the chalkboard or an overhead transparency.

Write as a decimal and as a fraction.

1. 50% $(0.50, \frac{1}{2})$ **2.** 40% $(0.40, \frac{2}{5})$

3. 5% $(0.05, \frac{1}{20})$ **4.** 9% $(0.09, \frac{9}{100})$

Multiply.

5. 0.5×60 (30) **6.** 0.4×16 (6.4)

7. 0.05×45 (2.25)

INTRODUCTION Present this situation: A used book is selling for 40% of its original $19 cost. How could you find the cost of the used book?

Lead students to the realization that they cannot multiply with a percent and so they will need to rename 40% either as a decimal or as a fraction.

INSTRUCTION Begin by illustrating on the chalkboard how the process of finding a given percent of a number can be translated into a number sentence. Point out that the word *of* in mathematics often means to multiply to find a part of a greater number.

For Skill 1, emphasize the order of the steps. Although 40% technically is equal to 0.40, 0.4 is easier to use when multiplying. Remind students that the product will have 1 decimal place.

For Skill 2, refer students to the table. Review the percent–fraction equivalencies. Encourage them to memorize the equivalencies.

Emphasize that instead of multiplying 16 by 0.75, they can multiply 16 by $\frac{3}{4}$ because $75\% = 0.75 = \frac{3}{4}$. Step 3 requires students to multiply a whole number by a fraction informally.

1.6 FINDING THE PERCENT OF A NUMBER

SKILL 1 *Expressing percents as decimals*

You can express any percent as a decimal.

Find 40% of 19.

1. Write the problem as a number sentence. 40% of 19 is ■
   ↓ ↓
   $40\% \times 19 = ■$
2. Rename the percent as a decimal. $0.4 \times 19 = ■$
   THINK: 40% = 0.40 = 0.4
3. Solve. $0.4 \times 19 = 7.6$

So 40% of 19 is 7.6.

SKILL 2 *Expressing percents as fractions*

Sometimes it is easier to express a percent as a fraction.

This table shows the equivalent fractions for some percents.

$25\% = \frac{1}{4}$	$50\% = \frac{1}{2}$	$75\% = \frac{3}{4}$	
$20\% = \frac{1}{5}$	$40\% = \frac{2}{5}$	$60\% = \frac{3}{5}$	$80\% = \frac{4}{5}$
$16\frac{2}{3}\% = \frac{1}{6}$	$33\frac{1}{3}\% = \frac{1}{3}$	$66\frac{2}{3}\% = \frac{2}{3}$	$83\frac{1}{3}\% = \frac{5}{6}$
$12\frac{1}{2}\% = \frac{1}{8}$	$37\frac{1}{2}\% = \frac{3}{8}$	$62\frac{1}{2}\% = \frac{5}{8}$	$87\frac{1}{2}\% = \frac{7}{8}$

Find 75% of 16.

1. Write the problem as a number sentence. 75% of 16 is ■
   ↓ ↓
2. Rename the percent as a fraction. $75\% \times 16 = ■$
   THINK: $75\% = \frac{3}{4}$ $\frac{3}{4} \times 16 = ■$
3. Solve.
   THINK: $\frac{1}{4} \times 16 \rightarrow 16 \div 4 = 4$
   $\frac{3}{4} \times 16 \rightarrow 3 \times 4 = 12$
   $\frac{3}{4} \times 16 = 12$ So 75% of 16 is 12.

PRACTICE EXERCISES

Find the answer. Express the percent as a decimal.

• **1.** 10% of 40 4 • **2.** 20% of 30 6 **3.** 20% of 600 120 **4.** 60% of 50 30

5. 40% of 15 6 **6.** 30% of 320 96 **7.** 90% of 36 32.4 **8.** 80% of 52 41.6

9. 50% of 73 36.5 **10.** 5% of 65 3.25 **11.** 2% of 95 1.9 **12.** 8% of 57 4.56

Find the answer. Express the percent as a fraction.

• **13.** 25% of 24 6 • **14.** 20% of 25 5 **15.** 50% of 36 18 **16.** 60% of 45 27

17. 75% of 56 42 **18.** 40% of 85 34 **19.** 80% of 75 60 **20.** 75% of 24 18

21. $16\frac{2}{3}\%$ of 36 6 **22.** $12\frac{1}{2}\%$ of 16 2 **23.** $33\frac{1}{3}\%$ of 39 13 **24.** $66\frac{2}{3}\%$ of 54 36

Find the answer. Decide whether to express the percent as a decimal or as a fraction.

• **25.** 60% of 35 21 • **26.** 30% of 36 10.8 **27.** 80% of 50 40 **28.** 40% of 35 14

29. $33\frac{1}{3}\%$ of 36 12 **30.** 75% of 60 45 **31.** $87\frac{1}{2}\%$ of 72 63 **32.** 20% of 62 12.4

33. $37\frac{1}{2}\%$ of 64 24 **34.** 25% of 400 100 **35.** $62\frac{1}{2}\%$ of 120 75 **36.** $16\frac{2}{3}\%$ of 39 6.5

Solve.

• **37.** The trip between 2 towns is exactly 90 mi. You have gone 40% of this distance. How far have you gone? 36 mi

• **38.** Roy is reading a book that has 320 pages. He has already read 90% of the book. How many pages has he already read? 288 pages

39. It takes Cara 35 min to walk to school. It takes Sue 80% of Cara's time. How long does it take Sue? 28 min

40. Larry's math test score was 90. Andrew's test score was 20% less. How much less was Andrew's score? 18 pts.

41. A sports watch originally cost $56. It is now being sold for $87\frac{1}{2}\%$ of its original price. How much does the watch cost now? $49

42. A ski outfit originally cost $42.60. It is now being sold at 30% off. How much has been deducted from the original cost of the outfit? $12.78

43. After a taste test, 40% of the 30 people interviewed preferred a new vegetable drink over orange juice. The rest preferred orange juice. How many people preferred the vegetable drink? How many people preferred orange juice? 12 people; 18 people

44. At a store promotion, 150 free televisions were to be given away to persons who arrived during the first 15 min after the store opened. On the first day, 30% of the televisions were given away. How many televisions was that? How many were left for the following days? 45 televisions; 105 televisions

Carefully review the THINK. If you prefer students to use the formal multiplication algorithm, refer them to the appropriate Skills Bank pages.

Students will sometimes have difficulty deciding whether to rename a percent as a decimal or as a fraction. If the percent ends in a zero, you usually rename it as a decimal. (Example: 70% = 0.7) A fractional percent can usually be renamed as a fraction. (Example: $66\frac{2}{3}\% = \frac{2}{3}$) Sometimes a percent can be easily renamed as a decimal or as a fraction. In that case, use this general rule: If the denominator of the fraction divides evenly into the number you are finding a percent of, use the fraction. Otherwise, use the decimal. (Examples: $40\% \times 25 = \frac{2}{5} \times 25 = 10$; $40\% \times 63 = 0.4 \times 63 = 25.2$)

SKILLS BANK You may want to refer students to the following skills in the Skills Bank:

Multiply With Decimals, Skill 2

Multiply With Fractions and Mixed Numbers, Skills 1–3

ASSIGNMENTS

Level	Exercises
Level 1	Odd 1–23, 37–38
Level 2	Even 2–24, 25–42
Level 3	25–44

FOLLOW-UP ACTIVITY

APPLICATION Ask students to look for advertisements or other announcements that involve a percent of a number. Ask students to cut out these newspaper advertisements or to copy announcements they find in stores. Have them compute the answers to the problems implied.

SUPPLEMENTARY MATERIALS

TRP Practice, p. 6
TRP Reteaching, p. 6
TRP Transparency 3

Objectives

- Solve problems by interpreting data in tables and bar graphs.
- Determine the placements of bars that fall halfway between the given numbers on a bar graph scale.

Teaching the Lesson

WARM-UP Write the following exercises on the chalkboard or an overhead transparency.

Compute.

1. 75 − 45 (30) **2.** 80 − 35 (45)

3. 60 + 75 + 46 (181)

4. 50 + 27 + 80 + 39 (196)

INTRODUCTION Ask students where they have seen information given in table form or in bar graph form. Responses might include newspaper articles, business reports, department store advertisements, or reference books in the library. Discuss the fact that tables and graphs allow a reader to see differences easily and to make comparisons quickly.

INSTRUCTION Present the problem situation to the class. Point out that Situations A and B will both be used to answer the initial question.

Use Situation A to show how to interpret data given in a table. Emphasize the importance of carefully reading table headings. Point out that, in Step 1, students are comparing 5 numbers in order to determine the greatest.

Use Situation B to show how to interpret data given in a bar graph. Have students identify what information is given on the vertical and horizontal scales. Emphasize the importance of paying attention to the intervals used on the vertical scale (intervals of 50 cars and trucks). Encourage students to use a straightedge to line up the tops of the bars with the vertical scale.

PROBLEM *Solving* STRATEGY

1.7 INTERPRETING DATA FROM TABLES AND GRAPHS

Situation:

The sales staff at Donney Motors keeps records of their car and truck sales. Contests are sometimes held to encourage special efforts to sell various cars and trucks. How can these records be used to identify a salesperson's performance?

Strategy:

You can use information in a **table** or a **bar graph** to solve a problem.

Applying the Strategy:

A. The salesperson who sold the greatest number of trucks in October won a videocassette recorder. Who was it?

THINK: Look at the column labeled "Number of Trucks Sold."

OCTOBER SALES

Salesperson	Number of trucks sold
Ruth	22
Art	15
John	9
Eric	12
Mindy	4

1. Which number is the greatest? 22
2. Which name is on the same line as 22? Ruth

Ruth sold the greatest number of trucks in October and won the VCR.

B. Eric sold the greatest number of cars and trucks last year and won a free trip. How many cars and trucks did he sell?

THINK: Look at the bar above Eric's name.

1. Between which 2 numbers does the bar lie? Between 250 and 300
2. Is the bar nearer to 250 or 300? It is halfway between 250 and 300.
3. What number is halfway between 250 and 300? 250 + 300 = 550; 550 ÷ 2 = 275

Eric sold 275 cars and trucks last year.

PRACTICE EXERCISES

Use the table of October Sales for Exercises 1–3.

1. How many trucks did Art sell? 15 trucks
2. How many trucks did Mindy sell? 4 trucks
3. How many more trucks did Art sell than Mindy? 11 trucks

Use the bar graph of Car and Truck Sales Last Year for Exercises 4–7.

• 4. How many did Ruth sell? 150 cars and trucks
5. How many did Mindy sell? 125 cars and trucks
6. What was the total number sold by Ruth and Mindy together? 275 cars and trucks
7. Who sold the same number last year as Ruth and Mindy together? Eric

The sales staff posted this bar graph to show the numbers of cars and trucks Donney Motors leased last year.

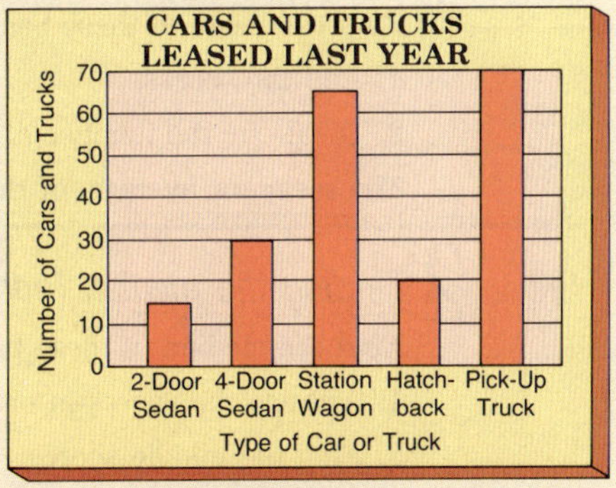

• 8. Which type was leased the most? pickup truck
9. How many station wagons were leased? 65 station wagons
10. How many more hatchbacks were leased than 2-door sedans? 5 hatchbacks
11. What fraction of the vehicles leased were 4-door sedans? Write this as a percent. $\frac{3}{20}$; 15%

12. Use the information in the table at the right to construct a bar graph. Use this vertical scale: 0, 5, 10, 15, 20, 25, 30, 35. Check students' graphs.

DECEMBER SALES

Salesperson	Ruth	Art	John	Eric	Mindy
Number of cars sold	10	25	30	15	20

You may wish to explain the solution to Step 3 using this alternate method:

$300 - 250 = 50$
$50 \div 2 = 25$
$250 + 25 = 275$

Formal instruction in finding an average (or mean) will be presented in Lesson 1.8.

SKILLS BANK You may want to refer students to the following skills in the Skills Bank:

Compare and Order Whole Numbers and Decimals, Skills 1–2

Divide With Whole Numbers, Skills 2–5

ASSIGNMENTS

Level 1	1–10, 12–13
Level 2	1–14
Level 3	8–16

FOLLOW-UP ACTIVITY

SITUATIONAL PROBLEM SOLVING Present students with the following problem: You are interested in researching the numbers and kinds of vehicles owned by your classmates' families.

Some questions to ask include: How will you go about gathering the data? How will you record your findings?

Some approaches to solving the problem include: Develop a plan. Show the data in both a table and a bar graph.

SUPPLEMENTARY MATERIALS

TRP Practice, p. 7
TRP Applications, p. 1
TRP Transparency 4

OBJECTIVE

- Find the mean, the median, and the mode of a set of data.

TEACHING THE LESSON

WARM-UP Write the following exercises on the chalkboard or an overhead transparency.

Add.

1. 234 + 76 + 905 **2.** 5.6 + 3.8 + 9.7

Divide.

3. 161 ÷ 7 **4.** 26.6 ÷ 7 **5.** 93 ÷ 5

Write in order from least to greatest.

6. 23, 58, 46, 72, 31

7. 3.6, 2.9, 3.2, 4.1, 3.0, 1.8
(**1.** 1,215 **2.** 19.1 **3.** 23 **4.** 3.8
5. 18.6 **6.** 23, 31, 46, 58, 72
7. 1.8, 2.9, 3.0, 3.2, 3.6, 4.1)

INTRODUCTION Encourage students to describe the averages they see or use. These averages may include average height, average weight, the average of test scores, and batting averages.

INSTRUCTION Ask students to read aloud the definitions of *mean (average), median,* and *mode.* Explain that they are all measures of *central tendency,* or ways to describe data.

Use Skill 1 to show students how to find the mean, or average. Explain that the mean does not have to be one of the given scores. Emphasize that a score that appears more than once is counted each time it appears. (The mean of 5, 6, 5, 8 is 24 ÷ 4 = 6. The 5 is counted twice.)

Use Skill 2 to show students how to find the median of an odd number of scores. Point out that this median is always one of the given scores. Students may make mistakes when trying to identify the middle score before arranging the scores in order. You should emphasize the importance of Step 1.

1.8 MEAN, MEDIAN, AND MODE

Vocabulary: **Mean (average)**—The sum of a group of numbers divided by number of addends.
Median—The middle number when a group of numbers is arranged in order from least to greatest.
Mode—The number that occurs most frequently in a group of numbers.

SKILL 1 *Finding the mean*

Find the mean of these basketball player's scores: 48, 36, 51, 72, 58.

1. Add the scores. 48 + 36 + 51 + 72 + 58 = 265
2. Divide by the number of scores. 265 ÷ 5 = 53

The mean, or average, of these scores is 53.

SKILL 2 *Finding the median (odd number of scores)*

Find the median of these bowling scores: 126, 108, 145, 108, 117.

1. Arrange the scores in order. 108 108 117 126 145
2. Find the middle score. 117

The median of these scores is 117.

SKILL 3 *Finding the median (even number of scores)*

Find the median of these bowling scores: 139, 106, 145, 113, 128, 109.

1. Arrange the scores in order. 106 109 113 128 139 145
2. Find the middle score.
   THINK: There is no one middle number. 113 128
3. Find the mean of the two middle scores. 113 + 128 = 241
   241 ÷ 2 = 120.5

The median of these scores is 120.5.

SKILL 4 *Finding the mode*

Find the mode of these race times: 9.3, 9.6, 9.2, 10.2, 9.6, 10.1, 9.5.

Find the time that occurs most often. 9.6 occurs twice.

The mode of these scores is 9.6.

PRACTICE EXERCISES

Find the mean.

• **1.** 86, 90, 94 90
2. 420, 185, 874 493
3. 4.2, 8.5, 2.6 5.1
4. 15, 34, 19, 52 30
5. 172, 550, 293, 413 357
6. 1.2, 6.3, 9.5, 10.2 6.8
7. 20, 18, 6, 10, 16 14
8. 2.7, 1.8, 4.5, 5.1, 3.4 3.5

Find the median.

• **9.** 7, 13, 25, 46, 8 13
10. 9, 12, 7, 24, 3, 18 10.5
11. 1.3, 6.2, 8.9, 2.4, 5.6 5.6
12. 2.3, 5.4, 0.7, 1.6, 4.6, 3.7 3.0

Find the mode.

• **13.** 1, 3, 5, 9, 3, 6, 7, 8 3
14. 2.6, 3.5, 0.8, 4.7, 6.2, 0.8, 1.2 0.8

Find the mean, the median, and the mode. Round the mean and median to nearest tenth.

• **15.** 56, 78, 92 75.3; 78; none
16. 108, 135, 87 110; 108; none
17. 6.3, 2.5, 10.1 6.3; 6.3; none
18. 31, 78, 56, 31 49; 43.5; 31
19. 302, 220, 220, 208 237.5; 220; 220
20. 14.8, 16.2, 4.8, 12.1 12.0; 13.5; none
21. 75, 93, 89, 75, 84 83.2; 84; 75
22. 158, 213, 107, 213, 213 180.8; 213; 213
23. 6.1, 5.9, 4.2, 5.9, 10.7 6.6; 5.9; 5.9

Solve.

24. At noon each day last week, Fran took temperature readings. They were 16°F, 23°F, 10°F, 9°F, and 14°F. What are the mean and median of these temperatures? 14.4°F; 14°F

25. Six judges scored a diving contest. On 1 dive, 2 judges gave a 7.5. The other scores were 8.0, 7.9, 6.8, and 7.3. What are the mean, median, and mode of these scores? 7.5; 7.5; 7.5

EXTENSION Using a Tally

Each time Ed played miniature golf, he made a tally mark next to his score.

1. How many games did Ed play? 21 games
2. What are his two mode scores? 70; 69
3. What is his median score? 69

What is the total of the scores of:

4. 69? 483 **5.** 68? 204 **6.** 67? 134

7. What is his mean score? 69.2

ED'S GOLF SCORES

Score	Total	Score	Total
72 I	72	69 𝍸 II	
71 I	71	68 III	
70 𝍸 II	490	67 II	

Use Skill 3 to show students how to find the median of an even number of scores. Point out that this median need not be one of the given scores. Emphasize the importance of Steps 2 and 3.

Use Skill 4 to show students how to find the mode. Point out that there may be no mode or there may be more than one mode.

SKILLS BANK You may want to refer students to the following skills in the Skills Bank:

Compare and Order Whole Numbers and Decimals, Skills 2, 4

Add With Whole Numbers and Decimals, Skill 5

Divide With Whole Numbers, Skills 2–5

Divide With Decimals, Skill 2

EXTENSION Explain that the information about Ed's golf scores has been grouped. Grouped data is usually easier to work with. Assign the exercises after having students read aloud the number of times Ed got each score.

ASSIGNMENTS

Level 1	1–17
Level 2	1–25
Level 3	15–25, EX

FOLLOW-UP ACTIVITY

ENRICHMENT Ask students to find their ages in months. Tell them to multiply their age by 12 and add the months since their birthday. Have each student write his or her age in months on the chalkboard. Ask the students to find the mean, the median, and the mode of this data.

SUPPLEMENTARY MATERIALS

TRP Practice, p. 8
TRP Reteaching, p. 7
TRP Transparency 4

OBJECTIVES

- Review vocabulary.
- Practice key chapter concepts and skills.

USING THE REVIEW

The Chapter Review is designed to help students prepare for taking the Chapter Test. The first section focuses on vocabulary. It requires students to select words to complete statements. The second section presents practice of key chapter skills.

Each item on the Chapter Review is referenced to the page on which the topic is taught in the Pupil's Edition. You may wish to have students refer to these pages to help review any concepts or skills they have not yet mastered.

It is suggested that students work in small-sized heterogeneous cooperative learning groups. Some cooperative learning methods that may be used are as follows:

1. After each student has independently completed the Chapter Review, a discussion should follow within each group about the solutions to the problems.
2. The group can complete the entire Chapter Review by working together to solve the problems.

End the lesson with an entire class discussion in which any questions brought up in group discussions are presented and answered.

CHAPTER REVIEW

Vocabulary Choose the letter of the word(s) that completes the sentence.

1. Percent means ■. [10] c
   a. Per ten **b.** Tenths **c.** Per hundred
2. The sum of a group of numbers divided by the number of addends is called the ■. [16] a
   a. Mean **b.** Median **c.** Mode
3. The middle number when a group of numbers is arranged in order from least to greatest is called the ■. [16] b
   a. Mean **b.** Median **c.** Mode

Skills Add. [2]

4. 682 + 327 1,009 **5.** 6,981 + 2,217 9,198 **6.** 741 + 2,037 + 667 3,445
7. 4.7 + 3.6 8.3 **8.** 5.81 + 7.99 13.8 **9.** 3.16 + 4.2 + 3.55 10.91

Subtract. [4]

10. 753 − 541 212 **11.** 9,832 − 6,754 3,078 **12.** 5,003 − 2,324 2,679
13. 9.6 − 6.4 3.2 **14.** 11.07 − 9.21 1.86 **15.** 10.6 − 3.58 7.02

Multiply. [6]

16. 2 × 43 86 **17.** 6 × 725 4,350 **18.** 7 × 209 1,463
19. 3 × 6.5 19.5 **20.** 5 × 3.23 16.15 **21.** 4 × 8.02 32.08

Divide. [8]

22. 468 ÷ 2 234 **23.** 644 ÷ 7 92 **24.** 6,072 ÷ 8 759
25. 7.8 ÷ 3 2.6 **26.** 0.6 ÷ 4 0.15 **27.** 40.02 ÷ 5 8.004

Rename as a percent. [10]

28. $\frac{6}{100}$ 6% **29.** $\frac{57}{100}$ 57% **30.** $\frac{400}{100}$ 400% **31.** 0.87 87% **32.** 0.09 9% **33.** 7 700%

Rename as a decimal. [10]

34. 43% 0.43 **35.** 98% 0.98 **36.** 4% 0.04 **37.** 0.6% 0.006 **38.** 600% 6 **39.** 153% 1.53

Find the answer. [10, 12]

40. 40% of 25 10 **41.** 70% of 96 67.2 **42.** $12\frac{1}{2}$% of 48 6

CHAPTER TEST

Add.

1. 682 + 527 1,209 **2.** 8,732 + 1,487 10,219 **3.** 23.6 + 7.5 31.1 **4.** 16.7 + 0.89 + 3.65 21.24

Subtract.

5. 739 − 486 253 **6.** 4,376 − 2,923 1,453 **7.** 48.7 − 32.81 15.89 **8.** 50.06 − 17.39 32.67

Multiply.

9. 4 × 321 1,284 **10.** 7 × 803 5,621 **11.** 4 × 8.2 32.8 **12.** 7 × 6.08 42.56

Divide.

13. 652 ÷ 4 163 **14.** 2,464 ÷ 3 821 R1 **15.** 7.2 ÷ 4 1.8 **16.** 4.68 ÷ 9 0.52

Divide and round to the nearest hundredth.

17. 9.27 ÷ 4 2.32 **18.** 3.86 ÷ 7 0.55 **19.** 7.31 ÷ 6 1.22 **20.** 5.08 ÷ 6 0.85

Rename as a percent.

21. 0.75 75% **22.** 0.062 6.2% **23.** 4.87 487% **24.** $\frac{2}{5}$ 40% **25.** $\frac{3}{4}$ 75%

Rename as a decimal and as a fraction.

26. 60% 0.6; $\frac{3}{5}$ **27.** 95% 0.95; $\frac{19}{20}$ **28.** 5% 0.05; $\frac{1}{20}$ **29.** 160% 1.6; $1\frac{3}{5}$ **30.** 250% 2.5; $2\frac{1}{2}$

Find the answer. Express the percent as a decimal or as a fraction.

31. 30% of 70 21 **32.** 40% of 45 18 **33.** $33\frac{1}{3}$% of 42 14 **34.** 75% of 64 48

Group of numbers: 15, 12, 10, 15

Find:

35. the mean 13 **36.** the median 13.5 **37.** the mode 15

Use the bar graph for Exercises 38–40.

38. Who sold the greatest number of cars and trucks?
John

39. How many cars and trucks did Ruth sell? 25 cars and trucks

40. How many more cars and trucks did Mindy sell than Art?
5 cars and trucks

OBJECTIVE

- Evaluate the achievement of chapter objectives.

USING THE TEST

The Chapter Test may be used as a post-test to evaluate student achievement. However, you may wish to use the Chapter Post-Test offered in the Teacher's Resource Package or to design your own chapter test. If this page is not used as a test, you may want to assign it as additional review or practice.

The test items are correlated to the chapter objectives in the table below.

Chapter Objectives	Test Items
A. Add whole numbers and decimals.	1–4
B. Subtract whole numbers and decimals.	5–8
C. Multiply whole numbers and decimals.	9–12
D. Divide whole numbers and decimals.	13–20
E. Rename fractions, decimals, and percents.	21–30
F. Find the percent of a number.	31–34
G. Find the mean, the median, and the mode.	35–37
H. Solve problems by interpreting data in tables and bar graphs.	38–40

SUPPLEMENTARY MATERIALS

TRP Ch. 1 Post-Test Form A, pp. 1–2
TRP Ch. 1 Post-Test Form B, pp. 1–2

OBJECTIVE

- Test the prerequisite skills needed to learn the concepts developed in Chapter 2.

USING THE TEST

The Pre-Skills Test is designed to diagnose students' strengths and weaknesses on prerequisite skills necessary to study the mathematics in Chapter 2.

Assign the Pre-Skills Test. Allow students to work together in pairs or in small groups. Group members should help those who demonstrate a misunderstanding of a concept or a weakness in a skill.

The table below correlates the items on the Pre-Skills Test with the prerequisite skill and the lesson(s) in the chapter for which it is needed.

Item(s)	Prerequisite skill	Lesson(s)
1–25	Round whole numbers and decimals.	2.3, 2.4, 2.6
26–30	Round fractions.	2.5
31–40	Add with multiples of 10, 100, 1,000, or 10,000.	2.2, 2.3, 2.6
41–50	Subtract with multiples of 10, 100, 1,000, or 10,000.	2.2, 2.3, 2.6
51–58	Multiply with multiples of 10 or 100.	2.2, 2.4, 2.6
59–66	Divide with multiples of 10 or 100.	2.2, 2.4, 2.6

References to specific skills that are reviewed in the Skills Bank are given on the individual lessons themselves.

Pre-Skills Test

Round to the nearest ten.

1. 58 60 **2.** 62 60 **3.** 27 30 **4.** 34 30 **5.** 95 100

Round to the nearest hundred.

6. 138 100 **7.** 594 600 **8.** 737 700 **9.** 461 500 **10.** 975 1,000

Round to the nearest thousand.

11. 8,659 9,000 **12.** 9,823 10,000 **13.** 1,844 2,000 **14.** 2,425 2,000 **15.** 6,432 6,000

Round to the nearest tenth.

16. 0.86 0.9 **17.** 0.68 0.7 **18.** 0.128 0.1 **19.** 0.268 0.3 **20.** 0.985 1.0

Round to the nearest hundredth.

21. 0.067 0.07 **22.** 0.084 0.08 **23.** 0.0524 0.05 **24.** 0.037 0.04 **25.** 0.098 0.10

Round to the nearest whole number.

26. $2\frac{3}{4}$ 3 **27.** $8\frac{1}{5}$ 8 **28.** $9\frac{1}{2}$ 10 **29.** $36\frac{2}{3}$ 37 **30.** $57\frac{3}{16}$ 57

Add or subtract.

31. 20 + 17 = 37 **32.** 50 + 38 = 88 **33.** 700 + 125 = 825 **34.** 400 + 369 = 769 **35.** 900 + 873 = 1,773

36. 2,000 + 3,000 = 5,000 **37.** 6,000 + 4,000 = 10,000 **38.** 10,000 + 40,000 = 50,000 **39.** 30,000 + 50,000 = 80,000 **40.** 70,000 + 20,000 = 90,000

41. 65 − 30 = 35 **42.** 88 − 40 = 48 **43.** 592 − 300 = 292 **44.** 910 − 500 = 410 **45.** 798 − 700 = 98

46. 5,000 − 3,000 = 2,000 **47.** 8,000 − 4,000 = 4,000 **48.** 60,000 − 50,000 = 10,000 **49.** 40,000 − 20,000 = 20,000 **50.** 90,000 − 60,000 = 30,000

Multiply or divide.

51. 6 × 30 180 **52.** 8 × 10 80 **53.** 54 × 500 27,000 **54.** 8 × 900 7,200

55. 30 × 100 3,000 **56.** 50 × 500 25,000 **57.** 70 × 800 56,000 **58.** 90 × 400 36,000

59. 40 ÷ 4 10 **60.** 20 ÷ 4 5 **61.** 700 ÷ 7 100 **62.** 400 ÷ 5 80

63. 60 ÷ 30 2 **64.** 80 ÷ 40 2 **65.** 100 ÷ 20 5 **66.** 200 ÷ 50 4

Chapter 2 THE CALCULATOR AND ESTIMATION

You will probably be using a calculator to work the exercises in this book. Why have people always been interested in inventing devices for computing numbers?

OBJECTIVE

- Formulate problems, using text and data.

USING THE CHAPTER OPENER

Discuss students' answers to the question on the page. Some possible answers include: to do calculations faster and more accurately; to make work with greater numbers easier, or even possible; to enable non-mathematicians to solve complex problems.

Some students may believe calculators are faster for all computations. If they don't believe that this is true, challenge them to a race. Have 1 group of students use paper and pencil and 1 group use a calculator. Read the following to them rather quickly:
$3 + 5$; 7×7; $9 - 5$; $10 + 20$; $50 \div 2$; $2{,}000 + 4{,}000$; and $50{,}000 \div 2$.

The students using paper and pencil should be able to keep up with you while the students using calculators should fall behind.

COOPERATIVE LEARNING

Write the following on the chalkboard.

True or *False?*

1. Most people use a calculator at work.
2. Most people use a calculator at home.

Ask students to answer the questions and record the class results. Assign heterogeneous groups to research the question in the community. They are to organize their data into a format of their choosing and present the information to the class.

Compare the results to the class survey. Conclude by asking if they think calculator use will increase in the future.

SUPPLEMENTARY MATERIALS

TRP Group Projects, p. 2

OBJECTIVES

- Add, subtract, multiply, and divide on a calculator.
- Use the CLEAR ENTRY key.

TEACHING THE LESSON

WARM-UP Write the following numbers on the chalkboard or an overhead transparency. Ask students to explain how each would be entered on a calculator.

1. 8,432 **2.** 4.03 **3.** $82.15 **4.** 0.125
(**1.** without the comma **2.** with the decimal point **3.** without the dollar sign but with the cents point **4.** without the zero)

INTRODUCTION Begin the discussion by asking students when they would use a calculator. Point out that calculators are used for most routine calculations when speed and accuracy are important. Remind students that calculators have their own "personalities" and that although they may have the same buttons, the buttons each work a little differently.

INSTRUCTION Use the sample 1-step computations at the top of the page to review how the 4 operation keys on a calculator can be used to compute sums, differences, products, and quotients. Remind students of the importance of pushing the Equals key to complete a calculation.

Use Example 1 to show how a series of operations can be computed without using the Equals key after each operation. Ask students when they might do a series of operations. Examples might include averaging, balancing a checkbook, and so on.

Use Example 2 to show how the Clear Entry key erases the display when you make a mistake and allows you to reenter a number without going back to the beginning of the computation. Explain that many calculators have a CE/C key. Pushing this key once clears only what is currently in the display, while pushing it twice clears the current calculation.

2.1 BASIC OPERATIONS ON A CALCULATOR

When you want to compute quickly and accurately with greater numbers, you can use a calculator.

The four basic operations can be performed easily.

Operation	Calculator Entry	Display
Add: 49,567 + 78,078	[4] [9] [5] [6] [7] [+] [7] [8] [0] [7] [8] [=]	127,645
Subtract: 34.014 − 5.708	[3] [4] [.] [0] [1] [4] [−] [5] [.] [7] [0] [8] [=]	28.306
Multiply: 908 × 0.045	[9] [0] [8] [×] [.] [0] [4] [5] [=]	40.86
Divide: $0.078\overline{)4.9452}$	[4] [.] [9] [4] [5] [2] [÷] [.] [0] [7] [8] [=]	63.4

You can use a calculator to do a series of operations without using the [=] **key (is equal to)** after each operation.

EXAMPLE 1 Compute: 45.09 − 6 + 4.7 − 18

Operation	Calculator Entry	Display
1. Subtract.	[4] [5] [.] [0] [9] [−] [6] [+] ← **Get ready to add.**	39.09
2. Add.	[4] [.] [7] [−] ← **Get ready to subtract.**	43.79
3. Subtract.	[1] [8] [=]	25.79

So 45.09 − 6 + 4.7 − 18 = 25.79.

The [CE] **key (Clear Entry)** can help you when you have entered a wrong number into the calculator.

EXAMPLE 2 Correct, then complete the following computation on a calculator.

Multiply: 4.5 × 78 × 0.09 — **Error**: 87 entered instead of 78

	Calculator Display
Calculator Entry: [4] [.] [5] [×] [8] [7]	87
1. Press [CE] to clear 87 from the calculator.	0
2. Complete the computation.	
Enter: [7] [8] [×]	351
Enter: [.] [0] [9] [=]	31.59

So 4.5 × 78 × 0.09 = 31.59.

FOR DISCUSSION See TE side column.

1. For which operations will the order in which you enter two numbers not affect the answer? Why?

2. The entry below was made on two different calculators. One calculator displayed the answer 6.15. The other calculator displayed the answer 16.4. Explain the different answers.
[4] [.] [5] [×] [3] [+] [7] [−] [8] [.] [2] [÷] [2] [=]

PRACTICE EXERCISES

Use a calculator to compute.

• 1. 234,634 + 33,568 + 995,400 1,263,602
• 2. 1,294,588 − 346,183 948,405
3. 4,983 × 8,834 44,019,822
4. 27,587 ÷ 49 563
5. 256.3 + 89.375 + 5,832.6 6,178.275
6. 735.8 − 35.635 700.165
7. 93.53 × 7.394 691.56082
8. 425.36 × 18.3 7,784.088
9. 39.447 + 874.336 − 175.68 738.103
10. 46.84 × 7.03 ÷ 2 164.6426
11. 114,289 divided by 1,126 101.5
12. 36.5 divided into 65,700 1,800
13. 678.68 divided by 7.6 89.3
14. 0.4 divided into 0.028 0.07

Divide and round to the nearest tenth.

•15. 737 ÷ 15 49.1
16. 18.2)375 20.6
17. 13,275 ÷ 1,083 12.3
18. 3.79)463 122.2

Divide and round to the nearest hundredth.

•19. 14.83 ÷ 0.21 70.62
20. 2.9)0.37 0.13
21. 72.6 ÷ 0.046 1,578.26
22. 53.8)2.847 0.05

Compute from left to right.

•23. 687.3 + 39.4 − 398.46 − 4.236 324.004
24. 933.6 − 3.498 + 364.9 − 483.78 811.222
25. 76.2 × 3.9 ÷ 7.5 × 35 1,386.84
26. 28.64 ÷ 0.2 × 0.45 ÷ 15 4.296

Solve.

•27. George spent $27.80 for a pair of pants and $14.50 for a shirt. He paid a sales tax of $2.54. How much did he spend in all? $44.84

28. Joe's car odometer read 27,314 mi when he left for vacation. Three weeks later, the odometer read 35,278 mi. How far had Joe traveled? 7,964 mi

29. Susan is reimbursed 21.5¢ per mi when she drives for business. She drove 85 mi last week. How much was she reimbursed? $18.28

30. Gary ordered 500 sheets of paper. When they arrived, they formed a pile 4.5 cm high. How thick was each sheet? 0.009 cm

SKILLS BANK You may want to refer students to the following skill in the Skills Bank:

Order of Operations, Skill 1

FOR DISCUSSION

1. Students should explain that order does not affect the answer when you add or multiply because these operations are commutative.
2. Students should explain that some calculators compute in the order that the numbers are entered, in this case getting 6.15. However, other calculators are programmed to use algebraic reasoning and compute products and quotients first and then sums and differences. This kind of calculator, using the conventional order of operations, would get an answer of 16.4.

ASSIGNMENTS

Level 1	Odd 1–25, 27–28
Level 2	1–30
Level 3	9–30

FOLLOW-UP ACTIVITY

ENRICHMENT Write a 3- or 4-digit number on the chalkboard. Ask students to use guessing and checking and their calculators to find the number that when multiplied by the number on the chalkboard results in an answer closest to 1 million. Repeat for a 5- or 6-digit dividend and find the divisor that results in a quotient closest to 100.

SUPPLEMENTARY MATERIALS

TRP Practice, p. 9
TRP Reteaching, p. 8
TRP Transparency 5

OBJECTIVES

- Add and subtract mentally.
- Multiply and divide by 10, 100, and 1,000 mentally.

TEACHING THE LESSON

WARM-UP Orally present the following exercises on mental computation from left to right.

1. 2 + 7 − 3 × 5 + 6 (36)

2. 14 + 6 ÷ 2 × 7 − 15 (55)

3. 36 ÷ 4 + 9 ÷ 3 − 5 × 100 (100)

INTRODUCTION Ask students how they might determine the change for a $5.95 record that is paid for with a $10 bill. Encourage them to share their methods. Focus on the ease of adjusting the $5.95 to $6 and model how change is given: 5¢ more is $6 and $4 more is $10, and so the change is $4.05.

INSTRUCTION Use Example 1 to show how mental computation with adjusting (compensation) is often far easier than using pencil and paper. Explain how adjusting numbers to multiples of 10 or 100 makes the computation simpler.

Use Example 2 to help students understand that when one of the numbers is adjusted, the answer must also be adjusted. Help students articulate their reasoning verbally. For Example 2b: "Since I took away 10 more than needed, the answer must be 10 more."

Use Examples 3 and 4 to reinforce the understanding that when multiplying or dividing by 10, 100, or 1,000, you move the decimal point as many places as there are zeros in the 10, 100, or 1,000.

SKILLS BANK You may want to refer students to the following skills in the Skills Bank:

Multiply With Decimals, Skill 1

Divide With Decimals, Skill 1

2.2 COMPUTING MENTALLY

You may often find it easier to compute mentally than to use pencil and paper or even a calculator. You can add mentally by using numbers that are **multiples of 10** and then adjusting.

EXAMPLE 1 You are buying a shirt for $19 and a jacket for $44. Mentally compute the cost of the shirt and the jacket.

Add: $19 + $44

THINK: $19 is $1 less than $20. $20 + $44 = $64

So $19 + $44 is $1 less than $64, or $63.

The shirt and the jacket will cost $63.

You can subtract mentally in the same way.

EXAMPLE 2 Subtract:

a. 94 − 18

THINK: 18 is 2 less than 20.

94 − 20 = 74
So 94 − 18 is 2 more than 74, or 76.

b. 465 − 190

THINK: 190 is 10 less than 200.

465 − 200 = 265
So 465 − 190 is 10 more than 265, or 275.

Mental computation is also commonly used when you multiply or divide by **powers of 10,** such as 10, 100, or 1,000.

EXAMPLE 3 Multiply: 100 × 87.3

THINK: The product must be greater than 87.3, so move the decimal point to the right.

1 0 0 × 8 7.3 0 = 8,730

2 zeros — 2 places right

EXAMPLE 4 Divide: 38,430 ÷ 1,000

THINK: The quotient must be less than 38,430, so move the decimal point to the left.

3 8,4 3 0 ÷ 1,0 0 0 = 38.43

3 places left — 3 zeros

FOR DISCUSSION See TE side column.

1. Ricky argues that mental computation is a waste of time, since he has a calculator. How would you convince Ricky that he is wrong?

2. Enter 245.8 into your calculator. Press [÷] [10] [=]. Read the display. Press [=]. Read the display. Press [=] and read. What is happening?

PRACTICE EXERCISES

Use mental computation to add or subtract.

• **1.** 56 + 19 75 **2.** 25 + 48 73 **3.** 263 + 390 653 **4.** 467 + 310 777

5. 343 + 590 933 **6.** 36¢ + 29¢ 65¢ **7.** $638 + $206 $844 **8.** $6.55 + $3.39 $9.94

9. 1.20 + 0.62 1.82 **10.** 6.57 + 2.99 9.56 • **11.** 80 − 38 42 **12.** 47 − 19 28

13. 586 − 420 166 **14.** 376 − 102 274 **15.** 80¢ − 29¢ 51¢ **16.** $875 − $280 $595

17. $200 − $61 $139 **18.** $1.70 − $0.61 $1.09 **19.** 4.63 − 1.98 2.65 **20.** 7.80 − 4.38 3.42

Use mental computation to multiply or divide.

• **21.** 10 × 127 1,270 **22.** 100 × 4,806 480,600 **23.** 10 × 37.24 372.4 **24.** 100 × 6.8 680

25. 1,000 × 2.8 2,800 **26.** 100 × 759.5 75,950 **27.** 1,000 × 0.43 430 **28.** 10 × 0.532 5.32

• **29.** 3,200 ÷ 100 32 **30.** 6.5 ÷ 10 0.65 **31.** 0.27 ÷ 100 0.0027 **32.** 19.4 ÷ 10 1.94

33. 459 ÷ 1,000 0.459 **34.** 28.5 ÷ 100 0.285 **35.** 4.3 ÷ 1,000 0.0043 **36.** 2,045 ÷ 100 20.45

Solve using mental computation.

• **37.** The tax on a $17.60 blouse is $0.98. What is the cost of the blouse including tax?
$18.58

• **38.** A $550 refrigerator is marked down $59. How much does the refrigerator now sell for?
$491

39. An average person can lift 90 lb. If a forklift can lift 1,000 times as much, how much can the forklift lift?
90,000 lb

40. A box contains 425 g of cereal. How much will a bowl containing one-tenth of the cereal weigh?
42.5 g

41. Chin bought groceries that cost $1.46, 80¢, and 69¢. How much did he spend in all?
$2.95

42. The 100 members of a band agree to split the $7,340.00 cost of their trip. What is each member's share?
$73.40

EXTENSION Multiplying Mentally by 50 and by 25

Multiply: 50 × 2.8

THINK: 100 × 2.8 = 280

Since 50 = 100 ÷ 2,
then 50 × 2.8 = 280 ÷ 2 = 140.

Multiply: 25 × 16.4

THINK: 100 × 16.4 = 1,640

Since 25 = 100 ÷ 4,
then 25 × 16.4 = 1,640 ÷ 4 = 410.

Use mental computation to multiply.

1. 50 × 4.6 230 **2.** 50 × 84 4,200 **3.** 25 × 4.8 120 **4.** 25 × 240 6,000

5. 50 × 0.9 45 **6.** 50 × 120 6,000 **7.** 25 × 0.72 18 **8.** 25 × 360 9,000

9. 0.24 × 50 12 **10.** 1.08 × 50 54 **11.** 404 × 25 10,100 **12.** 82.8 × 25 2,070

FOR DISCUSSION

1. Students should explain that many times the numbers being used are simple enough to do the computations in your head. They might also argue that you might not always have a calculator and that for some computations, it's much quicker to do it mentally than to use a calculator.
2. Students should explain that the decimal point moves 1 place to the left each time the Equals key is pushed.

EXTENSION Using their knowledge of mentally multiplying a number by 100, students can learn to multiply mentally by 50 and by 25. Point out that $50 = \frac{1}{2}$ of 100, or 100 ÷ 2, and that $25 = \frac{1}{4}$ of 100, or 100 ÷ 4.

ASSIGNMENTS

Level 1	Odd 1–39
Level 2	1–42
Level 3	Even 1–36, 37–42, EX

FOLLOW-UP ACTIVITY

COOPERATIVE LEARNING
Provide students with a list of 6 grocery items and their prices. For example:

eggs 89¢	soup 41¢	mustard 49¢
tuna 99¢	soda $1.19	cheese $1.99

Have students, working in pairs, select 2 or 4 items from the list. One student should use mental addition to find the total cost of the items. Then the other student should use mental subtraction to find the change from $5.00. Students should explain their reasoning to each other and then switch roles.

SUPPLEMENTARY MATERIALS

TRP Practice, p. 10
TRP Reteaching, p. 9
TRP Transparency 5

OBJECTIVE

- Estimate sums and differences of whole numbers, money amounts, and decimals.

TEACHING THE LESSON

WARM-UP Write the following exercises on the chalkboard or an overhead transparency.

Round to the nearest 1,000.

1. 8,720 (9,000) **2.** 3,489 (3,000)

Round to the nearest dollar.

3. $1.95 ($2.00) **4.** $6.20 ($6.00)

Round to the greatest nonzero place.

5. 27.4 (30) **6.** 0.659 (0.7)

INTRODUCTION Ask students about how much a record album costs, how much a shirt costs, or how much a television costs. Elicit from students "about $9," "about $400," and so on. Remind students that we use estimates all the time and that these are often as sufficient for our needs as exact figures.

INSTRUCTION Use Example 1 to demonstrate a common situation in which an estimate would be used. Each number is rounded to its greatest place and then added to get an estimate. Have students name the greatest place for each addend. (ten thousands for 21,284, 38,372, and 10,480; thousands for 3,487) Emphasize that in the process of rounding each number to its greatest place, some numbers are rounded up and some are rounded down, more or less balancing each other out.

Use Example 2 to show how the same process works with decimals and with money amounts. Emphasize that with decimals it is necessary to overlook any initial zeros and to identify the greatest nonzero place. Make sure that students round *each* number separately to its greatest nonzero place. Have them name the place for each number.

2.3 ESTIMATING SUMS AND DIFFERENCES

A common way to **estimate** sums is to round each number to its **greatest place** and then add mentally.

EXAMPLE 1 About how much is the total population of Fairview County?

Town	Greenfield	Salem	Goshen	Wells
Population	21,284	3,487	38,372	10,480

1. Round.
   21,284 → 20,000
   3,487 → 3,000
   38,372 → 40,000
   +10,480 → +10,000
2. Add. 73,000

73,000 is a good estimate for the total population.

When adding decimals or money amounts, estimate by rounding each number to its **greatest nonzero place.**

EXAMPLE 2 Estimate:

a. 31.07 + 0.6 + 3.87

1. Round.
   31.07 → 30
   0.6 → 0.6
   + 3.87 → + 4
2. Add. 34.6

b. Estimate: 95¢ + $2.09 + $5.75

1. Round.
   $0.95 → $1
   $2.09 → $2
   +$5.75 → +$6
2. Add. $9

The same estimating rules are used for subtraction.

EXAMPLE 3 Estimate:

a. 27,387 − 2,163

1. Round.
   27,387 → 30,000
   − 2,163 → − 2,000
2. Subtract. 28,000

b. 0.37 − 0.097

1. Round.
   0.37 → 0.4
   −0.097 → −0.1
2. Subtract. 0.3

FOR DISCUSSION

See TE side column.

1. How is mental computation different from estimation?
2. To estimate $4.80 + $2.25 + $3.40, Gail used $5 + $2 + $3 = $10. Gail's mother rounded up and used $5 + $3 + $4 = $12. What are some advantages of doing estimation the second way?

PRACTICE EXERCISES

Estimate the sum or difference.

• 1. 47 + 29 80

2. 378 + 128 500

3. 4,572 + 8,752 14,000

4. 572 + 628 + 84 1,280

5. 27,480 + 4,749 + 32,564 65,000

• 6. \$6.84 + 3.14 \$10.00

7. \$7.85 + 0.98 \$9.00

8. \$2.18 + 0.95 + 1.32 \$4.00

9. 4.98 + 5.1 10

10. 893.5 + 37.75 940

11. 4.78 + 0.7 + 1.26 6.7

12. 0.099 + 0.43 0.5

• 13. 778 − 259 500

14. 432 − 19 380

15. 286 − 63 240

16. 88,271 − 27,557 60,000

17. 72,800 − 4,290 66,000

• 18. \$0.82 − 0.17 \$0.60

19. \$28.14 − 14.80 \$20

20. \$13.72 − 0.84 \$9.20

21. 9.3 − 2.87 6

22. 248.2 − 24.9 180

23. 1.37 − 0.0294 0.97

24. 4.89 − 0.75 4.2

25. 0.56 − 0.096 0.5

26. 47,564 + 6,891 57,000

27. \$795.66 + \$24.13 \$820

28. 0.028 + 0.0095 0.04

• 29. 61,285 − 8,721 51,000

30. \$795.77 − \$9.56 \$790

31. 0.082 − 0.0099 0.07

Solve.

• 32. Ben bought lunch for \$1.88. About how much change did he get from \$5.00?
\$3

33. Robin bought a radio for \$46.30. About how much change did she get from \$100.00?
\$50

34. A television is advertised at Bob's Appliances for \$379. The same television is on sale at Acme Audio for \$285. About how much can be saved by buying it at Acme?
\$100

35. One store sells microwave ovens for \$229.50. Another sells them for \$309.88. About how much can be saved by buying the less expensive microwave oven?
\$100

Use the Fair Attendance sign for Exercises 36–37.

FAIR ATTENDANCE	
Friday	18,275
Saturday	37,587
Sunday	22,812

36. About how many people attended in all?
80,000 people

37. About how many more people attended the fair on Saturday than on Friday?
20,000 people

Example 3 shows how the same rounding process is used to estimate differences. Again, have students name the greatest nonzero place for each number. Remind students that estimates like these are most valuable when you need a ballpark figure or to check calculator computations.

SKILLS BANK You may want to refer students to the following skills in the Skills Bank:

Round Whole Numbers and Decimals, Skills 1–4

Add With Whole Numbers and Decimals, Skill 4

Subtract With Whole Numbers and Decimals, Skill 4

FOR DISCUSSION

1. Students should explain that mental computation is a process for getting an exact answer, whereas estimation is a process for getting an approximate answer.
2. Students should explain that when buying things we often round prices up to be sure our estimate is an overestimate so that we can make sure we have enough money.

ASSIGNMENTS

Level	Exercises
Level 1	Odd 1–25, 32–33
Level 2	1–37
Level 3	Even 2–30, 32–37

FOLLOW-UP ACTIVITY

ENRICHMENT Provide students with a copy of the "Most Active Stocks" table from a recent newspaper. Develop oral exercises or a worksheet that reinforces estimating sums and differences.

SUPPLEMENTARY MATERIALS

TRP Practice, p. 11
TRP Reteaching, p. 10
TRP Transparency 6

OBJECTIVE

- Estimate products and quotients of whole numbers, money amounts, and decimals.

TEACHING THE LESSON

WARM-UP Write the following exercises on the chalkboard or an overhead transparency.

Round to the greatest nonzero place.

1. \$3.90 (\$4.00) **2.** 0.367 (0.4)

Multiply or divide mentally.

3. 6×20 (120) **4.** 60×200 (12,000)

5. $400 \div 5$ (80) **6.** $400 \div 50$ (8)

INTRODUCTION Ask students, "About how much will 4 pairs of jeans cost?" Ask how they arrived at their estimates. List on the chalkboard the various estimates for 1 pair and how students estimated the cost of 4 pairs. (For example: Multiplied by 4, doubled and then doubled again, etc.)

INSTRUCTION Use Example 1 to demonstrate a common situation in which an estimate would be sufficient. Explain that 6 need not be rounded in order to estimate this quotient. Reinforce the idea that rounding a number to its greatest place allows you to compute mentally to make an estimate.

Use Example 2 to show how rounding is used similarly to estimate both products and quotients of whole numbers. Have students name the greatest place for each number. In Example 2a, show students how $3{,}000 \times 50$ is the same as 3×5, or 15, and 4 zeros. In Example 2b, show students that $20{,}000 \div 40$ is the same as $2{,}000 \div 4$, or $20 \div 4$, and 2 zeros.

Use Example 3 to show how a decimal or money amount is rounded to the greatest nonzero place in order to make an estimate. Instruct students to overlook the zero in \$0.29 and round it to the nearest tenth of a dollar, or \$0.30.

2.4 ESTIMATING PRODUCTS AND QUOTIENTS

A common way to estimate products and quotients is to round each number to its **greatest place** and then compute mentally.

EXAMPLE 1 There were 285 wildlife pamphlets left to be distributed. Six friends shared the task. About how many pamphlets must each person hand out if they share the job?

1. Round. 285 rounds to 300. Since 6 is a 1-digit number, it does not need to be rounded.
2. Divide. $300 \div 6 = 50$.

So each person will hand out about 50 pamphlets.

EXAMPLE 2

a. Estimate: $2{,}789 \times 48$
1. Round. $3{,}000 \times 50$
2. Multiply. 150,000

b. Estimate: $22{,}270 \div 39$
1. Round. $20{,}000 \div 40$
2. Divide. 500

When multiplying or dividing decimals or money amounts, estimate by rounding each number to its **greatest nonzero place.**

EXAMPLE 3

a. Estimate: $78 \times \$0.29$
1. Round. $80 \times \$0.30$
2. Multiply. \$24

b. Estimate: $324.8 \div 4.87$
1. Round. $300 \div 5$
2. Divide. 60

FOR DISCUSSION See TE side column.

1. Chicken costs 89¢ per lb. When Rob purchased 3.7 lb, the price sticker read \$5.24. How did Rob know immediately that he was being overcharged?
2. Laura calculated that 3.2×16.8 is 5.376. Estimate and explain why Laura's answer cannot be correct.

PRACTICE EXERCISES

Estimate the product or quotient.

• **1.** 7 × 23 140	**2.** 2 × 192 400	**3.** 612 × 289 180,000	**4.** 4,812 × 44 200,000
• **5.** 7 × $1.98 $14	**6.** 52 × $12.75 $500	**7.** 480 × $4.90 $2,500	**8.** 360 × $19.95 $8,000
9. 8 × 9.7 80	**10.** 82 × 5.8 480	**11.** 713 × 3.87 2,800	**12.** 8.9 × 3.13 27
13. 4.845 × 84.1 400	**14.** 27.19 × 8.47 240	**15.** 67 × $0.39 $28	**16.** 385 × $0.82 $320
17. 32 × 0.638 18	**18.** 185 × 0.89 180	**19.** 0.43 × 0.18 0.08	**20.** 0.65 × 0.34 0.21
• **21.** 87 ÷ 3 30	**22.** 413 ÷ 8 50	**23.** 182 ÷ 21 10	**24.** 216 ÷ 39 5
25. 3,485 ÷ 11 300	**26.** 875 ÷ 182 4.5	**27.** 7,742 ÷ 390 20	• **28.** $11.80 ÷ 5 $2
29. $625.50 ÷ 3 $200	**30.** $43.60 ÷ 22 $2	**31.** $198.75 ÷ 38 $5	**32.** $293 ÷ 53 $6
33. 39.21 ÷ 23 2	**34.** 87 ÷ 2.87 30	**35.** 292 ÷ 6.14 50	**36.** 957.5 ÷ 46 20
37. 7.8 ÷ 4.3 2	**38.** 35.7 ÷ 5.4 8	**39.** 286.4 ÷ 2.88 100	**40.** 962.1 ÷ 19.3 50

Use the menu for Exercises 41–44.

• **41.** About how much will 6 hot dogs cost?
$6

42. About how much will 5 juices cost?
$2.50

43. About how many hot dogs can be bought for $6.00?
6 hot dogs

44. About how many hamburgers can be bought for $18.00?
9 hamburgers

MENU

Hog dogs	$1.15
Hamburgers	$1.89
Juice	$0.49

Solve.

• **45.** Denim cloth sells for $7.79 per yd. About how much will 8 yd of denim cloth cost?
$64

46. Cassette tapes cost $4.75 each. About how much will 12 tapes cost?
$60

47. School sweaters cost $18.50 each. About how many sweaters can be bought for $57.00?
3 sweaters

48. Men's suits cost $175.50 each. About how many suits can be bought for $820.00?
4 suits

SKILLS BANK You may want to refer students to the following skills in the Skills Bank:

Round Whole Numbers and Decimals, Skills 1–4

Multiply With Whole Numbers, Skill 1

Multiply With Decimals, Skills 2, 4

Divide With Whole Numbers, Skill 1

FOR DISCUSSION

1. Students should explain that if both numbers are rounded up, Rob bought less than 4 lb at less than 90¢ per lb. An overestimate would be $3.60, far less than the $5.27 sticker price.
2. Students should explain that Laura could round this to 3 × 20, or 60. The 5.376 is not even close. (In fact, Laura pressed 1.68 instead of 16.8. Her answer is off by a ×10.)

ASSIGNMENTS

Level 1	Odd 1–43
Level 2	1–48
Level 3	Even 2–40, 41–48

FOLLOW-UP ACTIVITY

APPLICATION Provide students with the following sale sign.

SALE!!!	
Erasers 19¢	Pads $1.95
Pencils 29¢	Pens $2.95

Develop oral exercises that reinforce estimating products and quotients. Include questions such as: About how much will 4 erasers and 2 pencils cost? About how much will 3 pens cost? About how many pads can you buy for $10? Name 3 ways you could spend about $5.00.

SUPPLEMENTARY MATERIALS

TRP Practice, p. 12
TRP Reteaching, p. 11
TRP Transparency 6

OBJECTIVE

- Estimate sums, differences, products, and quotients of mixed numbers.

TEACHING THE LESSON

WARM-UP Write the following exercises on the chalkboard or an overhead transparency.

Round to the nearest whole number.

1. $3\frac{3}{4}$ (4) **2.** $5\frac{2}{10}$ (5)

3. $6\frac{5}{6}$ (7) **4.** $8\frac{4}{9}$ (8)

5. $16\frac{11}{16}$ (17) **6.** $1\frac{1}{5}$ (1)

In checking students' answers, use a number line to show how each mixed number lies between 2 whole numbers.

INTRODUCTION Ask students to name situations in which they might have to estimate with fractions or mixed numbers. Discuss situations such as computing time (quarter hours), measuring lumber (eighths of an inch), measuring distance (quarter miles), and so on.

INSTRUCTION Use the introductory situation and Example 1 to show how to estimate sums of mixed numbers by rounding. Reinforce the idea that all mixed numbers can be rounded up or down to a whole number. If the fraction part is less than $\frac{1}{2}$, round down. If it is $\frac{1}{2}$ or greater, round up.

Use Example 2 to show how the same rounding rules are used to estimate differences. Explain, as necessary, that when the numerator of the fraction is less than half of the denominator, the fraction is less than $\frac{1}{2}$ and you round down. Otherwise, you round up.

Use Examples 3 and 4 to demonstrate how the same rounding rules change mixed numbers to whole numbers that can be easily multiplied or divided to arrive at an estimate.

2.5 ESTIMATING WITH MIXED NUMBERS

Sonya sometimes baby-sits for a neighbor after school. Since she is interested in an approximation of how much time she has worked, an estimate will be sufficient. Estimating sums and differences is often used with mixed numbers because they are easily rounded up or down to the **nearest whole number.**

EXAMPLE 1 About how many hours did Sonya baby-sit last week?

1. Round.
$7\frac{1}{2} \rightarrow 8$
$3\frac{1}{4} \rightarrow 3$
$+5\frac{3}{4} \rightarrow +6$
2. Add. 17

Sonya baby-sat about 17 h.

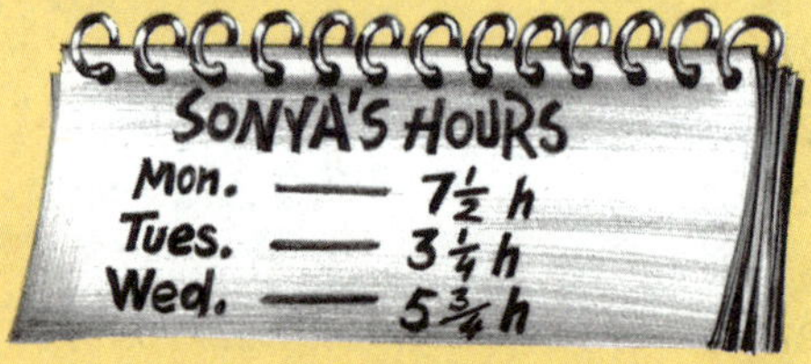

EXAMPLE 2

a. Estimate: $67\frac{7}{8} - 41\frac{3}{4}$
1. Round. $68 - 42$
2. Subtract. 26

b. Estimate: $12\frac{13}{16} - 7\frac{3}{8}$
1. Round. $13 - 7$
2. Subtract. 6

Rounding mixed numbers is also helpful when estimating products and quotients.

EXAMPLE 3

a. Estimate: $4 \times 7\frac{3}{4}$
1. Round. 4×8
2. Multiply. 32

b. Estimate: $8\frac{7}{8} \times 3\frac{1}{8}$
1. Round. 9×3
2. Multiply. 27

EXAMPLE 4

a. Estimate: $1{,}000 \div 19\frac{7}{8}$
1. Round. $1{,}000 \div 20$
2. Divide. 50

b. Estimate: $14\frac{3}{4} \div 3\frac{1}{16}$
1. Round. $15 \div 3$
2. Divide. 5

FOR DISCUSSION

See TE side column.

When estimating $48\frac{3}{4} - 26\frac{1}{9}$, Nancy rounded to the nearest 10 and subtracted 50 − 30. Jill rounded to the nearest whole number and subtracted 49 − 26. Compare and contrast these two methods.

PRACTICE EXERCISES

Estimate the answer.

• **1.** $7\frac{3}{4} + 10\frac{3}{4}$ 19
2. $2\frac{1}{4} + 8\frac{7}{8}$ 11
3. $27\frac{1}{6} + 3\frac{7}{8}$ 31
4. $43\frac{3}{8} + 22\frac{5}{6}$ 66

5. $2\frac{2}{3} + 5\frac{1}{4} + 6$ 14
6. $4\frac{4}{5} + 6\frac{5}{6} + 10\frac{1}{3}$ 22
7. $2\frac{1}{2} + 8\frac{1}{4} + 17\frac{7}{8}$ 29
8. $15\frac{1}{4} + 8\frac{4}{5} + 14\frac{9}{10}$ 39

• **9.** $12\frac{2}{3} - 7\frac{3}{4}$ 5
10. $6\frac{3}{4} - 2\frac{5}{8}$ 4
11. $3\frac{3}{4} - 3\frac{1}{8}$ 1
12. $19\frac{7}{8} - 10\frac{1}{3}$ 10

13. $100 - 17\frac{1}{8}$ 83
14. $37\frac{4}{5} - 18\frac{1}{5}$ 20
15. $97\frac{1}{3} - 26\frac{5}{8}$ 70
16. $77\frac{3}{5} - 52\frac{5}{6}$ 25

• **17.** $6\frac{3}{4} \times 5$ 35
18. $8\frac{1}{3} \times 4\frac{7}{8}$ 40
19. $7\frac{4}{5} \times 9\frac{1}{6}$ 72
20. $100 \times 8\frac{5}{8}$ 900

21. $200 \times 3\frac{7}{8}$ 800
22. $\frac{7}{8} \times 30\frac{1}{4}$ 30
23. $29\frac{1}{2} \times 2\frac{1}{16}$ 60
24. $39\frac{1}{2} \times 29\frac{11}{16}$ 1,200

• **25.** $16\frac{1}{4} \div 1\frac{7}{8}$ 8
26. $55\frac{3}{4} \div 7\frac{7}{8}$ 7
27. $59\frac{3}{5} \div 6\frac{1}{5}$ 10
28. $1{,}000 \div 4\frac{7}{10}$ 200

29. $2{,}000 \div 9\frac{3}{4}$ 200
30. $27\frac{5}{8} \div 1\frac{7}{8}$ 14
31. $72\frac{3}{8} \div 35\frac{7}{8}$ 2
32. $89\frac{4}{5} \div 44\frac{15}{16}$ 2

Use the chart for Exercises 33–36.

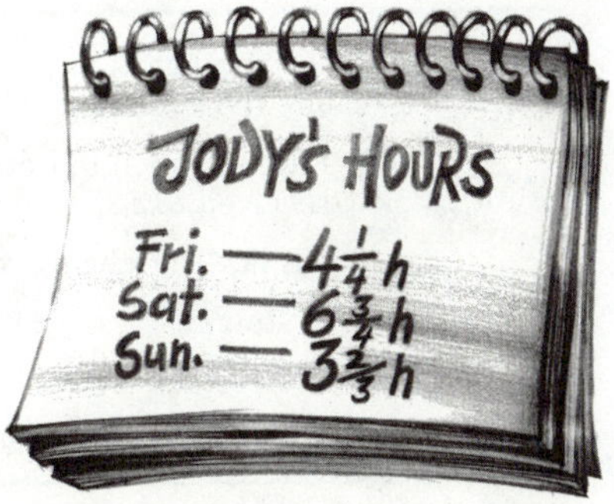

• **33.** About how many hours did Jody work altogether?
15 h

34. About how many more hours did Jody work on Saturday than on Sunday?
3 h

35. If Jody works twice as long next Saturday, about how long will she work?
14 h

36. At this rate, about how many 3-d work periods will Jody need to work before she has worked 75 h?
5 3-day work periods

Solve.

• **37.** A piece of pipe $42\frac{3}{4}$ in. long is cut from a piece $75\frac{1}{4}$ in. long. About how much pipe is left?
32 in.

38. A hiker covered $24\frac{8}{10}$ mi in $4\frac{3}{4}$ h. About how fast was she traveling?
5 mi per h

39. Aaron worked $45\frac{1}{2}$ h last week and $39\frac{3}{4}$ h this week. About how many hours did he work during the 2 wk?
86 h

40. Sue is cutting pieces of ribbon $8\frac{1}{8}$ in. long. About how many pieces can she cut from a 6-ft length of ribbon?
9 pieces

SKILLS BANK You may want to refer students to the following skills in the Skills Bank:

Fractions and Mixed Numbers, Skill 2

Multiply With Whole Numbers, Skills 2–4

Divide With Whole Numbers, Skills 2–5

FOR DISCUSSION Students should explain that the closer the rounded number is to the original number, the closer the estimate is to the actual answer. Therefore, by rounding to the nearest whole number rather than to the nearest 10, Jill's rounded numbers are closer to the original figures and her estimate of 49 − 26, or 23, is closer to the actual answer than Nancy's 50 − 30, or 20. Both are good estimates, but one is closer to the actual answer than the other.

ASSIGNMENTS

Level	Exercises
Level 1	Odd 1–35
Level 2	1–39
Level 3	Even 2–32, 33–40

FOLLOW-UP ACTIVITY

ENRICHMENT Provide students with a copy of the "Most Active Stocks" table from a recent newspaper as was suggested in the Follow-Up Activity for Lesson 2.3. Develop oral exercises or a worksheet that reinforces estimating with mixed numbers. Include questions such as: About how much did ____ increase yesterday? If you owned 200 shares of ____, about how much would you have earned or lost? If you had $800, about how many shares of ____ could you buy?

SUPPLEMENTARY MATERIALS

TRP Practice, p. 13
TRP Reteaching, p. 12
TRP Transparency 7

OBJECTIVES

- Use estimation when only approximate answers are needed.
- Use estimation to determine the reasonableness of answers.
- Use estimation to determine if you have enough money to make purchases.

TEACHING THE LESSON

WARM-UP Write the following exercises on the chalkboard or an overhead transparency.

Estimate the answer.

1. 47.9 + 31.8 **2.** 389 − 121

3. 5 × $7.98 **4.** $88.75 ÷ 9

1. 300 **2.** 80 **3.** $10 **4.** $40

INTRODUCTION Ask students what time they got up this morning. Discuss how many estimates they made in order to decide what was the best time to get up in order to get to school on time. For example, about how long to shower, about how long to eat breakfast, about how long to walk to the bus, and so on. Use this exercise to remind students of how commonplace estimation is.

INSTRUCTION Use the introductory situation and Example 1 to demonstrate how estimation is often needed in situations in which an exact answer is not necessary. Explain that with many variable measures, such as miles per gallon, not only is an estimate useful but often an exact amount is meaningless.

Use Example 2 to show how to use estimation to quickly check that you are not being overcharged or shortchanged in a store.

Use Example 3 to show how rounding up is a good practice in order to make sure you have enough money to make purchases. Explain how rounding up also usually leaves enough room for the sales tax. Discuss the difference between an underestimate and an overestimate.

2.6 USES OF ESTIMATION

Matt has just filled the gas tank in his car with 9.8 gal of gasoline. He has driven 213 mi since his last fill-up and wants to check his gas mileage.

Matt could use a calculator and actually divide 213 by 9.8, getting 21.734693, or 21.7 mi per gal. But of what practical value is an answer that exact? An estimate would do here.

EXAMPLE 1 About how many miles per gallon did Matt's car get?

THINK: Miles per gallon means 213 mi divided by 9.8 gal.

1. Round. $213 \div 9.8 \approx 200 \div 10$
2. Divide. 20 mi per gal

Matt's car got about 20 mi per gal.

Another use for estimation is in judging the reasonableness of answers, especially those computed on calculators or cash registers.

EXAMPLE 2 Karen's lunch bill came to $5.75. She gave the cashier $20.00 and received $12.25 in change. Was this a reasonable amount of change?

THINK: The change will be $20.00 − $5.75.

1. Round. $20 − $5.75 ≈ $20 − $6
2. Subtract. $14 in change

No, $12.25 was not a reasonable amount of change.

Estimation is also used when we want to be sure we have enough money to make purchases. When estimating the cost of a purchase, round all money amounts up, getting an **overestimate.**

EXAMPLE 3 Willie selected 3 ties for $8.40 each and 2 sweaters for $19.50 each. Will $60 be enough to buy these items?

1. Estimate the cost of the ties. 3 × $8.40 ≈ 3 × $9 = $27
2. Estimate the cost of the sweaters. 2 × $19.50 ≈ 2 × $20 = $40
3. Add the estimates. $27 + $40 = $67

Since $60 < $67, $60 will not be enough to buy these items.

FOR DISCUSSION See TE side column.

1. Name other situations in which you would round up all amounts in order to estimate. Tell why.
2. Name situations in which estimates would not be appropriate. Tell why.

PRACTICE EXERCISES

Estimate to choose the reasonable answer.

		a.	b.	c.
• 1.	487 + 357 + 621 b	a. 148	b. 1,483	c. 2,483
2.	6.2 + 15.9 + 0.3 a	a. 22.4	b. 2.2	c. 224
3.	5,946 − 2,873 c	a. 307	b. 8,819	c. 3,073
4.	15.87 − 0.6 c	a. 9.87	b. 5.81	c. 15.27
• 5.	27 × $18.50 c	a. $49.95	b. $4,995	c. $499.50
6.	7.81 × 0.32 a	a. 2.4992	b. 24.992	c. 0.24992
7.	58.24 ÷ 3.2 a	a. 18.2	b. 1.82	c. 182
8.	$18,540 ÷ 45 b	a. $41.20	b. $412	c. $4,120

Solve.

• 9. Beth needs $48 for a new dress. If her allowance is $5 per wk, about how many weeks will it take her to save the $48?
10 wk

10. Ethan spent $7.85 on books. About how much change did he receive when he gave the clerk $20.00?
$12

11. A car went 487 mi on 19.6 gal of gasoline. About how many miles per gallon is that?
25 mi per gal

12. About how far can a car go on 15 gal of gasoline if it averages 29.3 mi per gal?
600 mi

Use the ad and estimate for Exercises 13–16.

•13. Will $15.00 be enough to buy 3 pairs of socks?
yes

14. Will $15.00 be enough to buy a scarf and a belt?
no

15. Will $15.00 be enough to buy 2 pairs of socks and a tie?
no

16. Is $30.00 enough to buy 1 of each item?
yes

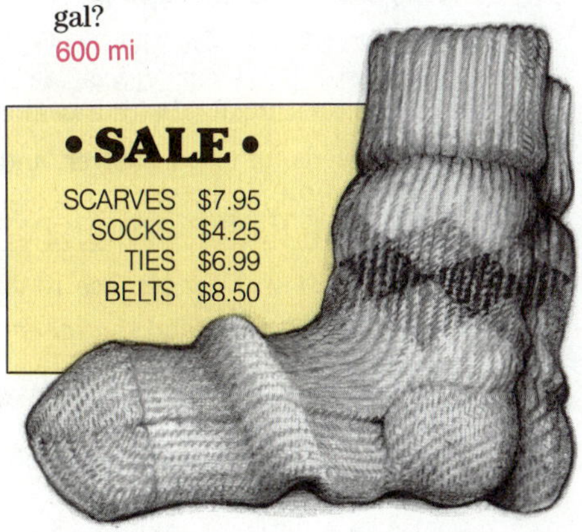

SKILLS BANK You may want to refer students to the following skills in the Skills Bank:

Round Whole Numbers and Decimals, Skills 1–4

FOR DISCUSSION

1. Answers will vary but might include when buying wood or material for a project; when estimating how much wallpaper, tile, or carpeting you need; or other situations in which you would rather not be left short with an underestimate.
2. Answers will vary but might include any situation in which an exact answer is needed. Some examples are bank deposits, figuring exact costs, completing an order form, and paying a bill.

ASSIGNMENTS

Level	Exercises
Level 1	1–10, 13–14
Level 2	1–16
Level 3	Odd 1–7, 9–16

FOLLOW-UP ACTIVITY

COOPERATIVE LEARNING
Have students work in pairs with a calculator and a worksheet with computation problems. The first student uses the calculator to compute the answer to the first example and states either a reasonable answer or bluffs one that is not reasonable. The second student must then estimate and decide whether the stated answer is reasonable or not. The second student gets a point if the reasonable answer is indeed reasonable or if the unreasonable answer is indeed unreasonable. Students then switch roles and play until one gets 5 points.

SUPPLEMENTARY MATERIALS

TRP Practice, p. 14
TRP Reteaching, p. 13
TRP Transparency 7

OBJECTIVE

- Solve problems by determining whether solutions are best attained by using paper and pencil, a calculator, or mental computation.

TEACHING THE LESSON

INTRODUCTION Students often jump immediately to paper and pencil to solve problems and do not stop to ask themselves whether the problem could be more easily solved with mental computation. Similarly, after some experience with a calculator, students immediately rush to their calculators to solve all problems before asking themselves whether they really need the calculator. Explain to students that mathematics is usually done with a combination of machine (calculator or computer), paper (to record), and brain (mental computation and reasoning). Ask students to stop and think about a problem and to plan a solution before rushing to a single solution method.

INSTRUCTION Present the problem situation to the class and ask them to solve the problem the way they are most comfortable solving it. Discuss the different methods used by the class and show how the problem would be done with paper and pencil, how it would be done with a calculator, and how it would be done mentally. Discuss the relative advantages and disadvantages of each method.

Use Situations A, B, and C to reinforce the idea that different situations require different strategies. While different methods can obviously be used to solve a problem, one is likely to be easier or more efficient than others. On the other hand, a student's personal preference may not always be the easiest or most efficient method. Have students make some generalizations about which method seems to be the best under each condition.

PROBLEM Solving STRATEGY

2.7 WHICH WAY TO COMPUTE?

Situation:

Suppose you are asked to find the cost of 5 pairs of socks at \$1.95 a pair plus a sales tax of \$0.40. Which way would you use to compute the answer?

Strategy:

Use paper and pencil skills, a calculator, or mental computation skills depending on the situation, the numbers involved, or your own personal preference.

Applying the Strategy:

Joan took out a pencil and computed:

$$\begin{array}{r} \$1.95 \\ \times \quad 5 \\ \hline \$9.75 \\ +\ 0.40 \\ \hline \$10.15 \end{array}$$

Jill took out a calculator and computed:

[5] [×] [1] [.] [9] [5] [+] [.] [4] [=] [10.15]

Jackie thought:
\$1.95 is \$2 less 5¢.
So 5 pairs are \$10 less 25¢, or \$9.75, plus 40¢.

40¢ is 25¢ plus 15¢.
So \$9.75 plus 25¢ is \$10, plus 15¢ is \$10.15.

Notice that Joan, Jill, and Jackie all got the same answer.

Other Situations

A. What is the best way to compute the amount of change that Ray received if he paid \$5.00 for a \$2.97 meal?

Ray can easily compute this mentally.
He thinks: \$2.97 is 3¢ less than \$3.00.
\$5.00 − \$3.00 is \$2.00.

So \$5.00 − \$2.97 is \$2.00 plus 3¢, or \$2.03.

B. What is the best way for Alvin to compute the new balance in the class treasury? The balance was \$357.82. He made a deposit of \$182.14 and then made a withdrawal of \$78.50.

Alvin needs an exact answer, and the numbers are too great to use mental computation. So he uses his calculator or paper and pencil to get \$461.46.

[3] [5] [7] [.] [8] [2] [+] [1] [8] [2] [.] [1] [4] [−] [7] [8] [.] [5] [=] [461.46]

C. What is the best way for Mr. Lee to compute the total length of pipe? One piece is $4\frac{3}{4}$ ft long, and the other is $2\frac{7}{8}$ ft long.

If Mr. Lee wants an exact answer, it is unlikely that he will use mental computation or convert to decimals and use a calculator. He will probably use paper and pencil.

$$\begin{array}{rcr} 4\frac{3}{4} & = & 4\frac{6}{8} \\ +2\frac{7}{8} & = & +2\frac{7}{8} \\ \hline & & 6\frac{13}{8} = 7\frac{5}{8}\text{ ft} \end{array}$$

FOR DISCUSSION See TE side column.

1. Show how Alvin could have solved his problem with paper and pencil instead of a calculator. Discuss which method you prefer and why.

2. Show how Mr. Lee could have solved his problem using a calculator. Discuss why this answer is different from what Mr. Lee found.

PRACTICE EXERCISES

Identify whether you would most likely use *paper and pencil, a calculator,* or *mental computation* to compute. Answers may vary.

• 1. Find the change from $10.00.
mental computation

2. Find the tip for a waiter.
mental computation

• 3. Balance a checkbook.
paper and pencil or calculator

4. Find the walking time to school.
mental computation

5. Find the total cost of 2 items that cost $127.95 and $264.85.
paper and pencil or calculator

6. Find the cost of 17 items in the grocery store.
calculator

7. Find how much gasoline you need for a trip across the country.
calculator

8. Find the average weight of the students in your class.
calculator

Use 2 different methods to compute. Identify the most efficient method. Answers may vary.

• 9. The change from $10.00 for an $8.95 record.
$1.05; mental computation

10. The actual distance between 2 cities that are $2\frac{1}{4}$ in. apart on a map with a scale of 1 in. = 40 mi.
90 mi; mental computation or paper and pencil

11. The cost per person if 30 people spent a total of $150.
$5; mental computation

12. The cost per person if 27 people spent a total of $200.
$7.41; calculator

35

SKILLS BANK You may want to refer students to the following skills in the Skills Bank:

Add With Fractions and Mixed Numbers, Skill 2

Multiply With Fractions and Mixed Numbers, Skill 3

FOR DISCUSSION

1. Have students compute the answer using paper and pencil. Students should explain that the calculator is faster and probably more accurate.
2. Students should explain that he would need to convert $4\frac{3}{4}$ to 4.75 and $2\frac{7}{8}$ to 2.875. His answer of 7.625 is equivalent to $7\frac{5}{8}$. It is probably not as useful for ordering pipe.

ASSIGNMENTS

Level 1	1–12
Level 2	1–12
Level 3	1–12

FOLLOW-UP ACTIVITY

SITUATIONAL PROBLEM SOLVING Provide students with a department store advertisement and tell them that they can go on a shopping spree and spend as close to $500 as possible without going over it. They cannot buy more than 3 of any one item.

Some questions to ask include: How much will 3 of each cost? What combinations will cost close to $500?

Some approaches to solving the problem include: Use estimation, mental computation, or a calculator. Make a list of possible combinations of items that cost close to $500.

SUPPLEMENTARY MATERIALS

TRP Practice, p. 15
TRP Applications, p. 2
TRP Transparency 8

OBJECTIVES

- Review vocabulary.
- Practice key chapter concepts and skills.

USING THE REVIEW

The Chapter Review is designed to help students prepare for taking the Chapter Test. The first section focuses on vocabulary. It requires students to select words to complete statements. The second section presents practice of key chapter skills.

Each item on the Chapter Review is referenced to the page on which the topic is taught in the Pupil's Edition. You may wish to have students refer to these pages to help review any concepts or skills they have not yet mastered.

It is suggested that students work in small-sized heterogeneous cooperative learning groups. Some cooperative learning methods that may be used are as follows:

1. After each student has independently completed the Chapter Review, a discussion should follow within each group about the solutions to the problems.
2. The group can complete the entire Chapter Review by working together to solve the problems.

End the lesson with an entire class discussion in which any questions brought up in group discussions are presented and answered.

CHAPTER REVIEW

Vocabulary Choose the letter of the word(s) that completes the sentence.

1. The ■ calculator key can be used when you have entered the wrong number into the calculator. [22] b

   a. Is equal to b. Clear entry c. Clear

2. You can estimate with whole numbers and decimals by rounding each number to its ■ place. [26, 28] b

   a. Last place b. Greatest nonzero place c. Least nonzero place

3. You can estimate with mixed numbers by rounding each number to the ■. [30] c

   a. Greatest nonzero place b. Nearest tenth c. Nearest whole number

Skills Use a calculator to compute. [22]

4. 4,689 + 8,873 + 472 14,034
5. 36,748 − 7,974 28,774
6. 58.93 × 39.58 2,332.4494
7. $38.5\overline{)938.63}$ 24.38
8. 7 + 39.88 − 15.938 30.942
9. 5.02 × 21.25 ÷ 5 21.335

Compute mentally. [24]

10. 357 + 298 655
11. \$394 − \$290 \$104
12. 10 × 8.9 89
13. 100 × 14.7 1,470
14. 866 ÷ 100 8.66
15. 37,480 ÷ 1,000 37.48

Estimate. [26, 28]

16. 6,835 + 3,128 10,000
17. 14.87 + 7.299 17
18. 13.89 − 4.935 5
19. 29 × 63 1,800
20. 67 × \$7.88 \$560
21. 387.5 ÷ 3.74 100

Estimate. [30]

22. $3\frac{2}{3} + 22\frac{1}{4}$ 26
23. $46\frac{7}{8} - 9\frac{3}{4}$ 37
24. $63\frac{1}{3} - 12\frac{1}{4}$ 51
25. $300 \times 3\frac{7}{8}$ 1,200
26. $9\frac{4}{5} - 1\frac{1}{3}$ 9
27. $26\frac{5}{8} - 2\frac{15}{16}$ 24

Write *yes* or *no*. [32]

28. Records are on sale for \$6.79 each. Will \$25 be enough to buy 4 records? no
29. Will \$30.00 be enough to buy 6 books that cost \$3.98 each? yes
30. You multiply 8.73 × 2.57. Is a calculator display of 11.30 reasonable? no

Chapter Test

Use a calculator to compute.

1. 498.38 + 98.47 596.85
2. 42,827 − 28,894 13,933
3. 489 × 3,983 1,947,687
4. 8.72)780.44 89.5
5. 596.35 − 56.217 + 482.1 1,022.233
6. 141.82 − 15.6 + 302.88 429.1
7. 56.2 × 6.25 ÷ 0.05 7,025

Use mental computation to add or subtract.

8. 453 + 210 663
9. 563 − 98 465
10. \$45 + \$39 \$84
11. \$5.00 − \$2.98 \$2.02

Use mental computation to multiply or divide.

12. 100 × 23 2,300
13. 1,000 × 8.27 8,270
14. 78.4 ÷ 10 7.84
15. 6.4 ÷ 100 0.064

Estimate the sum or difference.

16. 875 + 786 1,700
17. 687 − 312 400
18. 5.24 + 18.9 25
19. 53.2 − 18.66 30
20. \$4.86 + \$8.52 \$14
21. \$0.83 − \$0.38 \$0.40
22. $32\frac{1}{6} + 9\frac{3}{4}$ 42
23. $97\frac{2}{3} - 20\frac{1}{4}$ 78

Estimate the product or quotient.

24. 62 × 78 4,800
25. 217 ÷ 18 10
26. 6.385 × 9.82 60
27. 62.38 ÷ 2.8 20
28. 29 × \$7.85 \$240
29. \$839.50 ÷ 18 \$40
30. $14\frac{7}{8} \times 1\frac{7}{8}$ 30
31. $35\frac{1}{16} \div 6\frac{15}{16}$ 5

Estimate to choose the reasonable answer.

32. 86 + 74.9 + 365.4 b — **a.** 5.263 **b.** 526.3 **c.** 5,263
33. 18.26 − 5.8 a — **a.** 12.46 **b.** 15.46 **c.** 23.46
34. 2.3 × 12.8 b — **a.** 2.944 **b.** 29.44 **c.** 294.4
35. 6.34 × 0.86 a — **a.** 5.4524 **b.** 54.524 **c.** 545.2
36. 5,860 ÷ 15 c — **a.** 3.907 **b.** 39.07 **c.** 390.7

Use the ad for Exercises 37–40.

37. How much will 2 pens and 4 erasers cost? \$2
38. How much will 4 pencils and 5 notebooks cost? \$5.80
39. Will \$4.00 be enough to buy 10 erasers? yes
40. About how many pens can you buy for \$2.00? 5 pens

OBJECTIVE

- Evaluate the achievement of chapter objectives.

USING THE TEST

The Chapter Test may be used as a post-test to evaluate student achievement. However, you may wish to use the Chapter Post-Test offered in the Teacher's Resource Package or to design your own chapter test. If this page is not used as a test, you may wish to assign it as additional review or practice.

The test items are correlated to the chapter objectives in the table below.

Chapter objectives	Test items
A. Use a calculator to compute sums, differences, products, and quotients.	1–7
B. Mentally compute sums, differences, products, and quotients.	8–15
C. Estimate sums and differences.	16–23
D. Estimate products and quotients.	24–31
E. Estimate reasonable answers.	32–36
F. Solve problems by determining the best way to find answers.	37–38
G. Estimate to solve problems.	39–40

SUPPLEMENTARY MATERIALS

TRP Ch. 2 Post-Test Form A, pp. 1–2
TRP Ch. 2 Post-Test Form B, pp. 1–2

OBJECTIVE

- Test the prerequisite skills needed to learn the concepts developed in Chapter 3.

USING THE TEST

The Pre-Skills Test is designed to diagnose students' strengths and weaknesses on prerequisite skills necessary to study the mathematics in Chapter 3.

Assign the Pre-Skills Test. Allow students to work together in pairs or in small groups. Group members should help those who demonstrate a misunderstanding of a concept or a weakness in a skill.

The table below correlates the items on the Pre-Skills Test with the prerequisite skill and the lesson(s) in the chapter for which it is needed.

Item(s)	Prerequisite skill	Lesson(s)
1–8	Add money amounts.	3.1, 3.2, 3.3, 3.6
9–16	Subtract money amounts.	3.1, 3.2, 3.3, 3.4, 3.6
17–24	Multiply with money amounts.	3.2, 3.3, 3.6
25–32	Divide with money amounts.	3.2, 3.3, 3.6
33–40	Rename percents as decimals.	3.3, 3.4
41–46	Find the percent of a number.	3.3, 3.4

References to specific skills that are reviewed in the Skills Bank are given on the individual lessons themselves.

Pre-Skills Test

Add.

1. $23 + $28 $51
2. $56 + $87 $143
3. $138 + $76 $214
4. $207 + $906 $1,113
5. $18.27 + $16.30 $34.57
6. $84.26 + $13.97 $98.23
7. $125.62 + $9.87 $135.49
8. $379.05 + $298 $677.05

Subtract.

9. $78 − $26 $52
10. $92 − $38 $54
11. $416 − $47 $369
12. $801 − $136 $665
13. $39.48 − $13.17 $26.31
14. $86.27 − $54.38 $31.89
15. $316.32 − $58.26 $258.06
16. $410.09 − $89 $321.09

Multiply.

17. 9 × $8 $72
18. 7 × $1.40 $9.80
19. 6 × $8.90 $53.40
20. 2 × $4.08 $8.16
21. 16 × $3.86 $61.76
22. 39 × $5.17 $201.63
23. 12 × $17.36 $208.32
24. 28 × $11.07 $309.96

Divide.

25. $32 ÷ 5 $6.40
26. $66 ÷ 8 $8.25
27. $23 ÷ 4 $5.75
28. $270 ÷ 3 $90
29. $146 ÷ 8 $18.25
30. $24.48 ÷ 6 $4.08
31. $85.25 ÷ 5 $17.05
32. $696.40 ÷ 4 $174.10

Rename as a decimal.

33. 45% 0.45
34. 78% 0.78
35. 19.4% 0.194
36. 72.41% 0.7241
37. 4.56% 0.0456
38. 0.08% 0.0008
39. 7.51% 0.0751
40. 7.15% 0.0715

Find the answer to the nearest cent.

41. 25% of $100 $25
42. 30% of $50 $15
43. 8% of $55 $4.40
44. 6.5% of $94 $6.11
45. 7.15% of $78.50 $5.61
46. 7.51% of $149.75 $11.25

Many teenagers hold part-time or summer jobs. How can these jobs help prepare them for full-time work when they are older?

OBJECTIVE

- Formulate problems, using text and data.

USING THE CHAPTER OPENER

Begin the discussion of the question on the page by having students relate their personal job experiences and how these experiences helped prepare them for future work. Then move on to the more general discussion covering jobs with which students do not have first-hand experience. Some conclusions might be that part-time jobs help you learn (1) responsibility; (2) how an office is run; (3) how to use a cash register, copy machine, or word processor; and (4) numerous other skills.

COOPERATIVE LEARNING

Bring several editions of local newspapers to class. Assign heterogeneous groups and tell students that they are to work together to help each other find jobs in the want ads. The jobs they choose to "pursue" should provide some experience that will be of value to them in the future as they see it. Explain that many of them will not find exactly the kind of work for which they are looking and that this often happens in reality. Therefore, they are to make the best choice from what is available at this time.

Some students may actually want to pursue one of the jobs and they should be encouraged to do so. Ask them privately to keep you informed of their progress, so that if they are successful you can ask them to tell the class about the job.

Have the rest of the students present their job choices; tell how it will help prepare them for future work; and, given the pay, how long it will take them to earn $1,000.

SUPPLEMENTARY MATERIALS

TRP Group Projects, p. 3

OBJECTIVES

- Understand the practical aspects of job hunting.
- Compute total pay, given daily earnings.
- Compare earnings.

TEACHING THE LESSON

WARM-UP Write the following exercises on the chalkboard or an overhead transparency.

Add or subtract.

1. $18.50 + $27.80 ($46.30)

2. $19.23 + $327.78 ($347.01)

3. $46.70 − $10.15 ($36.55)

4. $438.97 − $18.98 ($419.99)

INTRODUCTION Ask students to describe any jobs they have held. Also ask students with jobs to describe how they got them. Encourage students to discuss the things that a potential employer expects of a job applicant.

INSTRUCTION Familiarize students with the process of job hunting by reading the introductory story. Ask students to name some reasons why they themselves would want to earn money.

Review the job application and ask students to supply this information on a separate sheet of paper or to use the complete job application offered in the Teacher's Resource Package. Emphasize the importance of neatness and correct spelling on the application. Ask students to name some reasons why an employer would want to know about other jobs they have held.

Discuss the job-interviewing process. Encourage students to think of questions that an applicant would probably ask. Ask students to describe how they would prepare for an interview.

Use the Example to demonstrate how part-time earnings depend on the total number of days you work and, sometimes, on the specific days.

3.1 FINDING A JOB

You need a part-time job. You want to earn some money to buy things and go places with your friends. You begin your job search by visiting stores and restaurants near your home.

You apply for several different jobs. For each job you have to fill out a **job application** like the one below. Later you will have to come back for a personal interview.

Copy the top of this job application.
Supply all this information about yourself.

Name ______________________ Phone ____________

Address ______________________________________
Street Town State Zip

Social Security Number ______________________

Position applied for ______________________

Circle last grade completed 5 6 7 8 9 10 11 12

Other Jobs You Have Held

Employer	Job	Dates	Supervisor

You went on an **interview** for a part-time job at a drugstore and you were hired. The manager told you that you would earn $12.50 for each day you worked after school during the week. You would also earn $26.75 for working all day Saturday.

EXAMPLE You decide to work 2 d after school and all day Saturday. How much money will you earn each week?

THINK: You make $12.50 each weekday and $26.75 on Saturday.

Add. $12.50 + $12.50 + $26.75 = $51.75

You will earn $51.75 per wk.

FOR DISCUSSION See TE side column.

1. What information should you find out about a job? Why?

2. What kinds of questions might you be asked during a job interview?

PRACTICE EXERCISES

Remember to estimate whenever you use your calculator.

Solve.

• 1. Sue earns $12 working after school in a department store. She earns $18.50 working in the store on Saturday. She worked 1 d after school and 1 Saturday. How much did she earn that week?
$30.50

• 2. Ron has a part-time job in a bakery. He earns $13.25 when he helps the baker. He earns $10.95 when he works at the sales counter. He helped the baker for 1 d and worked at the counter 2 d. How much did he earn?
$35.15

3. Fred works part time at a cleaners. Last week, he worked extra hours and earned $14.50 more than he usually does. He usually earns $70.50. How much did he earn last week?
$85.00

4. Marita usually earns $50 per wk working weekdays at a movie theater. Last week, she earned $74.50 because she worked all day Saturday. How much did she earn last Saturday?
$24.50

5. Sarah usually earns $80.25 per wk working in a pizza parlor. This week, she took some time off. She earned $9.50 less than she usually does. How much did she earn?
$70.75

6. Morey works part time at a fast food restaurant. Last week, he earned $40.45. That was $19.75 less than he usually earns. How much does he usually earn?
$60.20

This chart shows the wages of some people who work Saturdays in the Recreation Program.

Assistant Director	$23
Arts and Crafts Instructor	$18
Sports Coach	$18
Junior Counselor	$11.50

• 7. How much more does an Assistant Director earn each Saturday than a Junior Counselor?
$11.50

8. Eirin worked 2 Saturdays as a Sports Coach and 1 Saturday as a Junior Counselor. How much did she earn?
$47.50

9. Derek worked 1 Saturday as an Assistant Director and 2 Saturdays as an Arts and Crafts Instructor. How much did he earn?
$59

10. Fran usually earns $87 per mo. Last month, she missed 1 Saturday as an Assistant Director. How much did she earn last month?
$64

11. Last Saturday, there were 2 Assistant Directors, 1 Arts and Crafts Instructor, and 3 Junior Counselors working. What was the payroll that day?
$98.50

Remind students to use estimation to check the reasonableness of their answers. Point out that when using calculators, estimation will point out any errors made when pressing the keys.

SKILLS BANK You may want to refer students to the following skills in the Skills Bank:

Add With Whole Numbers and Decimals, Skills 3, 5–6

Subtract With Whole Numbers and Decimals, Skills 5–6

FOR DISCUSSION

1. Write a list on the chalkboard, eventually leading students to a discussion of these considerations: Type of job—Can I do it? Hours—Can I be there? Salary—How much will I earn? Working conditions—Will I like them?
2. Students should explain that employers might be interested in knowing about an applicant's school grades, available work time, work reasons, out-of-school activities, and hobbies, at-home responsibilities, etc.

ASSIGNMENTS

Level 1	1–6
Level 2	1–10
Level 3	1–11

FOLLOW-UP ACTIVITY

ENRICHMENT Ask students to make up fantastic imaginary jobs that pay great amounts of money. Have them decide on the daily pay for each job. Ask students to use their calculators to compute their earnings for 1 wk, 1 mo, 6 mo, and 1 y.

SUPPLEMENTARY MATERIALS

TRP Practice, p. 16
TRP Reteaching, p. 14
TRP Lesson Aids, p. 1
TRP Transparency 8

OBJECTIVES

- Compute pay, given hours worked, and hourly wage or weekly pay.
- Compute weekly pay, daily pay, and hourly wage, given a season rate.

TEACHING THE LESSON

WARM-UP Write the following exercises on the chalkboard or an overhead transparency.

Multiply.

1. 8×7 **2.** $7 \times \$4.50$ **3.** $15 \times \$5.48$
(**1.** 56 **2.** \$31.50 **3.** \$82.20)

Divide.

4. $63 \div 7$ (9) **5.** $\$42.40 \div 8$ (\$5.30)

6. $\$105.40 \div 5$ (\$21.08)

INTRODUCTION Discuss the minimum wage and tell students what the current minimum wage is. Ask students with jobs to talk about the hourly wages they receive and whether their wages are more or less than the minimum wage. Point out that there are some teenage jobs and others that involve tips, where the minimum wage does not apply.

INSTRUCTION Ask students to read and interpret the want ad. Ask them to name the things they can tell about the cashier job from the ad.

Use Example 1 to emphasize that hourly wage and weekly pay are 2 important pay factors to consider when looking for a part-time job. Discuss the formula in terms of the units used. Point out that weekly pay can be found if you know the number of hours worked that week and the hourly wage.

Use Example 2 to demonstrate how a season rate can be broken down into weekly and daily pay. Point out that many summer-camp jobs offer a season rate that usually applies to a 7- or 8-wk season.

3.2 COMPUTING PAY

You want to try to find a job that pays more than the **minimum wage.** The minimum wage is set by federal law. It is the least amount of money that a person can be paid per hour. You are interested in this want ad.

> ★★ **CASHIER** ★★
> DOWNTOWN RECORD STORE
> M – Th 6:00 – 9:00 P.M.
> **\$4.85/h**

EXAMPLE 1 If you got this job, how much would you earn each week?

THINK: Monday—Thursday: 4 d 6:00—9:00 P.M.: 3 h

PAY = HOURS WORKED × HOURLY WAGE

1. Multiply to find how many work hours. $4 \times 3 = 12$ h
2. Multiply to find your weekly pay. $12 \times \$4.85 = \$58.20$

You would earn \$58.20 per wk.

Your friend Brian is interested in this want ad for a summer job.

> ***CAMP VANGUARD***
> Counselor in Training
> JULY–AUGUST
> 8 weeks 6 days a week
> \$432 Season—Room and Board

EXAMPLE 2 If Brian took this job, how much money would he earn each week? Each day?

THINK: Season rate: \$432 Weeks: 8 Days per week: 6

1. Divide to find the weekly pay. $\$432 \div 8 = \$54$

Brian would earn \$54 per wk.

2. Divide to find the daily pay. $\$54 \div 6 = \$9$

Brian would earn \$9 per d.

FOR DISCUSSION See TE side column.

Why do you think a camp counselor can usually earn more by working at a day camp than at a sleep-away camp?

PRACTICE EXERCISES

Remember to estimate whenever you use your calculator.

Use these want ads for Exercises 1–6.

P/T Fred's Franks
Afternoons
12 hours weekly
$55.80 weekly

P/T Counter Help
★ *Jan's Cleaners* ★
T, W, Th 4:00 – 8:00 P.M.
$4.05/h

WILLOW LAKE DAY CAMP
Counselor 7 weeks
JULY – AUGUST
$945 season

• **1.** What is the hourly wage for working at Fred's Franks?
$4.65

• **2.** What is the weekly pay for working at Jan's Cleaners?
$48.60

3. What is the weekly pay for working at Willow Lake Day Camp?
$135

4. Which of these jobs has the highest weekly pay?
Willow Lake Day Camp

5. Which job has the higher hourly wage, working for Jan's Cleaners or for Fred's Franks?
Fred's Franks

6. Assuming an 8-h work day, about what is the hourly wage for working at Willow Lake Day Camp?
$2.81

Use these want ads for Exercises 7–14.

Temporary Help
Holiday Sale Days
RAY'S DEPT. STORE
15 hours a week
$65.55 weekly

Hamburger Barn
Counter Person
After School
M-F 6:00 – 9:00 P.M.
$4.18/h

CAMP CRESTMONT
JUNIOR COUNSELOR
July – August
Room & Board
$576 Season

• **7.** Which of these is a summer job? A permanent part-time job? A temporary part-time job? Camp Crestmont; Hamburger Barn; Ray's Dept. Store

• **8.** As a junior counselor at Camp Crestmont, you would work for 8 wk. What is the weekly pay?
$72

9. What is the hourly wage at Ray's Department Store?
$4.37

10. How much will you earn each week working at the Hamburger Barn?
$62.70

11. How much will you earn working for 50 wk at the Hamburger Barn?
$3,135

12. You would work for 48 d at Camp Crestmont. What is the daily pay?
$12

13. Which job has the highest weekly pay?
Camp Crestmont

14. Which job has the highest hourly wage? Assume an 8-h work day and 6 d per wk at Camp Crestmont.
Ray's Dept. Store

Some students may have difficulty deciding whether to multiply or divide when converting between pay periods. Advise them to follow the following guidelines. If they are going from a lesser unit of time to a greater unit of time, they multiply. (For example: hourly to weekly in Example 1.) If they are going from a greater unit of time to a lesser unit of time, they divide. (For example: weekly to daily in Example 2.)

SKILLS BANK You may want to refer students to the following skills in the Skills Bank:

Multiply With Decimals, Skills 2, 5

Divide With Decimals, Skills 2, 8

FOR DISCUSSION Students should explain that sleep-away camp counselors are paid less money because minimal room and board (meals) is taken out of their pay. Since a day camp does not need to provide for its counselors in this way, the counselors receive full pay.

ASSIGNMENTS

Level 1	1–5, 7–10
Level 2	1–13
Level 3	7–14

FOLLOW-UP ACTIVITY

COOPERATIVE LEARNING
Have students work in pairs to write want ads. Each pair should write 2 ads for part-times jobs, using a variety of pay descriptions, and 2 ads for summer-season jobs. Have pairs switch their ads with each other and, for each ad, use a calculator to determine monthly, weekly, daily, and hourly pay.

SUPPLEMENTARY MATERIALS

TRP Practice, p. 17
TRP Reteaching, p. 15
TRP Transparency 9

OBJECTIVES

- Compute the amount of tips.
- Compute total earnings from wages and tips.

TEACHING THE LESSON

WARM-UP Write the following exercises on the chalkboard or an overhead transparency.

Rename as a decimal.

1. 10% (0.1) **2.** 5% (0.05) **3.** 15% (0.15)

Multiply.

4. 0.1×35 **5.** 0.2×40 **6.** 0.15×47

(**4.** 3.5 **5.** 8 **6.** 7.05)

INTRODUCTION Begin discussion by asking students to describe any jobs that they have had or know of that involve tips. Encourage students to discuss the advantages and disadvantages of this type of job.

INSTRUCTION Read the introductory material. Direct students' attention to the tipping standards. Ask them to describe what they consider to be poor, average–good, and excellent service.

Use Example 1 to teach the procedure for computing a 15% tip. Remind students that this procedure represents finding a percent of a given number. Therefore, the percent needs to be renamed as either a decimal or a fraction.

Emphasize that mental computation is what many people use to determine the amount of a tip. Carefully review the thinking steps involved. Step 3 reminds students to add 10% of $18 to 5% of $18 to get 15% of $18, or $2.70. Explain to students that, when leaving tips, many people would, for the sake of convenience, round up and leave $3.00 instead of $2.70.

Use Example 2 to illustrate how some people work for a combination of a low hourly wage and tips. Carefully review the 3-step solution with students.

3.3 TIPS

Many students your age have jobs that involve tips. A **tip** is the amount of money a customer leaves as a gift for a person who has provided service. The word TIPS comes from these four words—To Insure Prompt Service. Usually, the better the service, the greater the tip.

If a job involves tips, there may not be any hourly wage. If there is an hourly wage, it will probably be less than the minimum wage.

People usually leave a tip as a percent of the bill according to these standards.

Poor Service	10% or less
Average–Good Service	15%
Excellent Service	20%

EXAMPLE 1 You eat at a restaurant and the meal costs $18.00. You want to leave a 15% tip. How much should you leave?

Multiply to find the tip.

THINK: 15% = 0.15 $\quad 0.15 \times \$18.00 = \$2.70$

You should leave a $2.70 tip.

You can also compute the tip mentally.

1. Find 10% of $18.00.
   THINK: 10% = 0.1 $\quad 0.1 \times \$18.00 \rightarrow \$18.00 \rightarrow \$1.80$
2. Find 5% of $18.00. ($5\% = \frac{1}{2}$ of 10%) $\quad$ 5% of $\$18 = \frac{1}{2}$ of $\$1.80 = \$0.90$
3. Add. $\quad \$1.80 + \$0.90 = \$2.70$

EXAMPLE 2 Your wage at a resort is $1.75 per h. The resort adds 15% to each bill. You receive that amount as a tip. One day, you worked 5 h and served meals costing $325. How much did you earn that day?

1. Multiply to find your earnings from wages. $\quad 5 \times \$1.75 = \$8.75$
2. Multiply to find your earnings from tips. $\quad 0.15 \times \$325 = \$48.75$
3. Add to find your total earnings. $\quad \$8.75 + \$48.75 = \$57.50$

You earned $57.50.

FOR DISCUSSION See TE side column.

Discuss the importance of hourly wage, food prices and the number of customers when choosing a restaurant job that involves tips.

PRACTICE EXERCISES

Remember to estimate whenever you use your calculator.

Solve.

• **1.** You ate at a restaurant and the meal cost $14.00. You want to leave a 15% tip. How much should you leave? $2.10

2. You served a meal that cost your customers $47.25. You are expecting a 20% tip from them. How much are you expecting? $9.45

Six students worked at a resort on Saturdays. The table below shows the hourly wages, hours worked, and the costs of the meals served on 1 Saturday. The resort adds 15% to the cost of the meals. Each person gets that amount as a tip. Find each person's earnings.

Employee	Hourly wage	Hours worked	Cost of meals	Earnings from wages	Tips	Total earnings
Janice	$1.75	6	$180	• **3.** ■	• **4.** ■	**5.** ■
Rob	$1.75	5	$160	**6.** ■	**7.** ■	**8.** ■
Sven	$1.75	7	$240	**9.** ■	**10.** ■	**11.** ■
Susan	$1.95	8	$520	**12.** ■	**13.** ■	**14.** ■
Ted	$2.05	6.5	$380	**15.** ■	**16.** ■	**17.** ■
Arlene	$2.15	6	$439	**18.** ■	**19.** ■	**20.** ■

3. $10.50 **4.** $27 **5.** $37.50 **6.** $8.75 **7.** $24 **8.** $32.75 **9.** $12.25 **10.** $36 **11.** $48.25 **12.** $15.60 **13.** $78 **14.** $93.60 **15.** $13.33 **16.** $57 **17.** $70.33 **18.** $12.90 **19.** $65.85 **20.** $78.75

EXTENSION Finding What Percent Tip is Given

You served a meal that cost your customers $25.75. They gave you a $4.50 tip. You are curious to know what percent $4.50 is of $25.75.

1. Divide. $\$4.50 \div \$25.75 = 0.1747572$
2. Rename the decimal as a percent. $0.1747572 = 17.47572\%$
3. Round to the nearest tenth. $17.47572\% \approx 17.5\%$

Your customers left you a 17.5% tip.

Find the percent of tip to the nearest tenth of a percent.

1. Cost of meal: $18.50 Tip: $2.80 15.1%

2. Cost of meal: $28.00 Tip: $3.60 12.9%

3. Cost of meal: $58.25 Tip: $10.50 18%

SKILLS BANK You may want to refer students to the following skills in the Skills Bank:

Equivalent Fractions, Decimals, and Percents, Skill 4

Find the Percent of a Number, Skill 2

Find the Percent One Number Is of Another, Skill 2

FOR DISCUSSION Students should explain that (1) an hourly wage would benefit them during slow business times; (2) the greater the prices of the meals, the greater their share as tips; and (3) the more customers you can wait on, the more tips you can earn.

EXTENSION Emphasize that finding what percent a tip is of the cost of a meal is the same as finding what percent one number is of another.

ASSIGNMENTS

Level 1	1–11
Level 2	1–20
Level 3	3–20, EX

FOLLOW-UP ACTIVITY

APPLICATION Write the table below on the chalkboard. Ask students to compare Jobs A and B and find each amount of tips.

	Job A	Job B
Hourly wage (8 h)	$2/h	$2.75/h
Average meal cost	$28/ meal	$21/ meal
Average number of meals per day	16 meals	19 meals
Percent of tip	15%	15%
Average daily amount of tips	($83.20)	($81.85)

SUPPLEMENTARY MATERIALS

TRP Practice, p. 18
TRP Reteaching, p. 16
TRP Transparency 9

OBJECTIVE

- Compute FICA contributions.

TEACHING THE LESSON

WARM-UP Write the following exercises on the chalkboard or an overhead transparency.

Rename as a decimal.

1. 7.15% (0.0715) **2.** 7.51% (0.0751)

Find the answer.

3. 7.15% of \$200 **4.** 7.51% of \$200

(**3.** \$14.30 **4.** \$15.02)

INTRODUCTION Ask how many students have a Social Security number. Ask those with one to explain why and how long they have had it. Point out that most students their age probably have a Social Security number since they are needed in order to be employed; they were also needed for custodial investment accounts that may have been opened for them by their parents. If possible, give students the address and phone number of the local Social Security office where they can visit to get their number.

INSTRUCTION Read and discuss the introductory material about the Social Security system and its benefits. Emphasize that a certain amount is withheld from each paycheck for Social Security and that, right now, that amount is 7.51% of their pay.

Use Example 1 to show how to compute the amount of Social Security Tax that would be withheld from a paycheck. Remind students to write 7.51% as 0.0751 in order to multiply to find a percent of a number. Emphasize using estimation to check for reasonableness or to get an approximation.

Use Example 2 to demonstrate how you could determine the amount left in your paycheck after the Social Security deduction.

3.4 SOCIAL SECURITY

You got your first paycheck. When you took it out of the envelope, you noticed that it was for less than you had earned. You headed into the bookkeeper's office to find out why.

The bookkeeper explained: "We all have amounts withheld from our paychecks. There are taxes, including FICA, withheld. **FICA** stands for **Federal Insurance Contributions Act,** or **Social Security.** Remember when you got your job here? You had to get a Social Security card with your Social Security number. You are the only person alive with that number."

"The money you contribute to Social Security is used to provide the following benefits:

1. retirement income
2. income for disabled persons who can no longer work
3. survivor's benefits for the dependents of a worker who dies
4. medical costs for persons covered by Medicare

The money you contribute is not yours, but the government knows how much you have paid. Each month, a percent of a paycheck is withheld for Social Security. Right now, that amount is **7.51%** of your pay."

EXAMPLE 1 You earned \$300 last month. How much was withheld for Social Security?

THINK: The withholding rate is 7.51%. 7.51% = 0.0751

Multiply to find the amount withheld. 0.0751 × \$300 = \$22.53

The amount withheld was \$22.53.

You could use estimation to check the reasonableness of your answer.

THINK: 7.51% is between 7% and 8%. 0.07 × \$300 = \$21

7% = 0.07 8% = 0.08 0.08 × \$300 = \$24

The amount withheld was between \$21 and \$24.

So, \$22.53 is reasonable.

EXAMPLE 2 Your friend Alicia earned $475 last month.
How much was withheld for Social Security?
How much was left over after paying Social Security?

1. Multiply to find the amount withheld. $0.0751 \times \$475 = \$35.6725$

The amount withheld was $35.67.

2. Subtract to find the amount left over. $\$475 - \$35.67 = \$439.33$

Alicia had $439.33 left over after paying Social Security.

FOR DISCUSSION

See TE side column.

1. In what type of situations do you think you would need to give your Social Security number?

2. Many people believe that Social Security benefits are only for senior citizens. How can Social Security benefit young families?

PRACTICE EXERCISES

Remember to estimate whenever you use your calculator.

Use the rate of 7.51% to find the Social Security withholding for the earnings. Round the amount to the nearest cent.

• 1. $200 $15.02 • 2. $700 $52.57 3. $4,200 $315.42 4. $6,200 $465.62 5. $945 $70.97

6. $382 $28.69 7. $837 $62.86 8. $62.78 $4.71 9. $489.28 $36.74 10. $183.47 $13.78

Remember that the people named in the table worked in a resort. The resort added 15% to the cost of meals and gave it to each person as a tip.

Find the earnings from wages, tips, total pay and the FICA withheld for each person. Round each amount to the nearest cent.
See page 476 for answers to Exercises 11–34.

Employee	Hourly wage	Hours worked	Cost of meals	Earnings from wages	Tips	Total pay	FICA withheld
Janice	$1.75	15	$900	• 11. ■	• 12. ■	13. ■	14. ■
Rob	$1.75	10	$750	15. ■	16. ■	17. ■	18. ■
Sven	$1.75	12	$800	19. ■	20. ■	21. ■	22. ■
Susan	$1.95	20	$1,100	23. ■	24. ■	25. ■	26. ■
Ted	$2.05	20.5	$1,320	27. ■	28. ■	29. ■	30. ■
Arlene	$2.15	15	$1,845.96	31. ■	32. ■	33. ■	34. ■

SKILLS BANK You may want to refer students to the following skills in the Skills Bank:

Equivalent Fractions, Decimals, and Percents, Skill 4

Find the Percent of a Number, Skill 2

FOR DISCUSSION

1. Encourage students to speculate freely about when they would need to give their Social Security number. Uses include as a company, group, or organization identification number; when applying for a job; when opening a bank account.

2. Students should explain that, in case of the disability or death of a working person, Social Security pays benefits to the person's survivors to make up for lost income.

ASSIGNMENTS

Level 1	1–18
Level 2	6–34
Level 3	11–34

FOLLOW-UP ACTIVITY

ENRICHMENT Inform students that, as of 1990, the FICA rate changes to 7.56%. Ask students to predict the impact of the rate change. Then write some sample weekly paycheck amounts on the chalkboard. For each amount, have students answer these questions.

1. How much is withheld using an FICA rate of 7.51%?

2. How much is withheld using an FICA rate of 7.56%?

3. How much more *per year* is deducted when using 7.56% than when using 7.51%?

SUPPLEMENTARY MATERIALS

TRP Practice, p. 19
TRP Reteaching, p. 17
TRP Transparency 10

OBJECTIVE

- Solve problems by using guessing and checking.

TEACHING THE LESSON

WARM-UP Write the following exercises on the chalkboard or an overhead transparency.

Multiply.

1. 10 × $0.15 **2.** 8 × $0.10

3. 6 × $0.25 **4.** 12 × $0.05

5. 9 × $0.05 **6.** 14 × $0.25

(**1.** $1.50 **2.** $0.80 **3.** $1.50
4. $0.60 **5.** $0.45 **6.** $3.50)

INTRODUCTION Pose this situation to the class: The sum of 2 numbers is 24. Their difference is 6. What are the 2 numbers?

Ask if the 2 numbers guessed satisfy both conditions of the problem. Have students continue guessing and checking until the correct numbers, 9 and 15, are found.

INSTRUCTION Present the problem situation to the class. Explain that sometimes there is no obvious way to compute a solution to a problem. A guess is a quick and easy way to explore the problem further.

Use the information in Guesses 1 and 2 to show that the guessing-and-checking strategy requires students to interpret information obtained on the first guesses and to use that information as a basis for making subsequent guesses.

You may wish to have 6 quarters and 15 nickels available for demonstration purposes. Encourage students to check their work using a calculator.

SKILLS BANK You may want to refer students to the following skills in the Skills Bank:

Multiply With Decimals, Skills 2, 5

PROBLEM *Solving* STRATEGY

3.5 GUESSING AND CHECKING

Situation:

Carol works part time at the Golden Arch Diner. A customer left Carol a tip of $1.90 in quarters and nickels. She was given 18 coins in all. How many coins of each kind did she receive?

Strategy:

Guessing and checking can help you arrive at a solution.

Applying the Strategy:

Guess 1

Carol received
3 quarters and 15 nickels.

Check

3 × $0.25 = $0.75
15 × $0.05 = + 0.75
$1.50

THINK: $1.50 < $1.90
Make another guess. Try more quarters to raise the total.

Guess 2

Carol received
6 quarters and 12 nickels.

Check

6 × $0.25 = $1.50
12 × $0.05 = + 0.60
$2.10

THINK: $2.10 < $1.90
Make another guess. Try fewer quarters to raise the total.

Guess 3

Carol received
5 quarters and 13 nickels.

Check

4 × $0.25 = $1.25
13 × $0.05 = + 0.65
$1.90

Carol received 5 quarters and 13 nickels.

FOR DISCUSSION See TE side column.

Ernie has a part-time job at a bakery. He worked a total of 35 h in 2 wk. During the first week, he worked 5 h more than the second week.

1. You want to know how many hours Ernie worked in each of the 2 wk. What facts are you given?
2. Explain why 25 h the first week and 10 h the second week cannot be correct.

PRACTICE EXERCISES

Remember to estimate whenever you use your calculator.

Check each guess. Choose the correct guess.

1. Rosa has a part-time job at a supermarket. She worked a total of 44 h in 2 wk. During the first week, she worked 8 h more than the second week. How many hours did Rosa work in each of the 2 wk?
   c
   a. Guess: 16 h, 24 h
   b. Guess: 18 h, 26 h
   c. Guess: 26 h, 18 h

2. Ted works part time at the Star Diner. A customer gave him an $0.85 tip in quarters and nickels. He was given 9 coins. How many coins of each kind did he receive?
   b
   a. Guess: 3 quarters, 6 nickels
   b. Guess: 2 quarters, 7 nickels
   c. Guess: 1 quarter, 8 nickels

Solve the problem by using guessing and checking.

• 3. Jacob worked a total of 170 h in 2 mo. During the first month, he worked 20 h more than the second month. How many hours did Jacob work in each of the 2 mo?
95 h, 75 h

• 4. A customer left Adrian a tip of $3.25 in quarters and dimes. He was given 16 coins in all. How many coins of each kind did he receive?
11 quarters, 5 dimes

5. Diana has a part-time job at a gas station. She earned a total of $70 in 2 wk. During the first week, she earned $20 less than the second week. How much did Diana earn in each of the 2 wk?
$25, $45

6. Sarah works part-time at Bill's Deli. She worked 72 h in April and May. She worked one-half as many hours in April as she did in May. How many hours did Sarah work in each month?
24 h, 48 h

7. Christine works at the Mountain View Restaurant on weekends. A customer left her a tip of $2.55 in quarters, dimes, and nickels. She was given 18 coins in all. How many coins of each kind was Christine given?
6 quarters, 9 dimes, 3 nickels

8. Isaac works at the General Store on Mondays and Tuesdays. Last week, he worked a total of $9\frac{3}{4}$ h for both days. He worked 135 min longer on Monday than on Tuesday. How long did Isaac work each day? 6 h, $3\frac{3}{4}$ h

FOR DISCUSSION

1. Students should explain that they are given 2 facts: (1) Ernie worked a total of 35 h in 2 wk and (2) he worked 5 h more during the first week than the second week.
2. Students should explain that 25 h the first week and 10 h the second week cannot be correct because both conditions of the problem are not met. Even though the total is 35 h, there should be 5 h more the first week instead of 15 h more.

ASSIGNMENTS

Level 1	1–5
Level 2	1–7
Level 3	3–8

FOLLOW-UP ACTIVITY

SITUATIONAL PROBLEM SOLVING Present students with the following problem: You have just accepted a part-time job at the local store. You need to determine which hours you want to work, how many hours you want to work, and how much you want to earn each week.

Some questions to ask include: What other activities do you participate in each week? At what times? Is there enough time available for work so that you can earn the amount of money you want?

Some approaches to solving the problem include: making a weekly schedule, and guessing and checking to determine the number of hours you can work versus the pay you want.

SUPPLEMENTARY MATERIALS

TRP Practice, p. 20
TRP Reteaching, p. 18
TRP Applications, p. 3
TRP Transparency 10

OBJECTIVE

- Decide, given relevant information, which help-wanted ad to pursue.

TEACHING THE LESSON

WARM-UP Write the following exercises on the chalkboard or an overhead transparency.

Complete the table.

	Hours worked	Hourly wage	Total pay
1.	9	$3.50	■
2.	■	$5.50	$49.50
3.	12	■	$54.00

(**1.** $31.50 **2.** 9 **3.** $4.50)

INTRODUCTION Present the problem situation to the class. Ask students to recall some of the factors they previously listed as important when looking for a part-time or summer job. Write on the chalkboard any factors that are given.

INSTRUCTION Review the want ads in Problem A. Ask students to interpret each want ad, identifying the information given in each. Have them point out any special features of each job. Discuss the decision-making factors.

Have small groups of students complete Exercises 1–10 in the table in order to compare the 3 help-wanted ads. Ask them to write statements that describe the data now in the table. Then ask them to write conclusions that can be drawn from the data. Discuss the appropriateness of these statements and conclusions and the possible misinterpretations of the data.

DECISION MAKING

3.6 INVESTIGATING WANT ADS

(Open-ended problem solving)

The first step in selecting a job is to locate and compare the want ads you are interested in.

PROBLEM A

Rebecca is looking for a part-time job. She cut out these want ads.

COUNTER HELP
Big Time Burger
Mon. – Fri. 5:00 – 7:00 P.M.
Saturday 8:00 – 11:00 P.M.
$4.15/h

HIGH SCHOOL STUDENT
Light Stock Work
Morse's Pharmacy
T, W 4:00–8:00 P.M.
$34.00/wk

SALES CLERK
Dime Department Store
Experience Necessary
Fri. 4–10 Sat. 9–4
$4.10/h

DECISION-MAKING FACTORS

Weekly schedule Hours per week Hourly wage Weekly pay Job responsibilities

DECISION-MAKING COMPARISONS

Complete the table to compare the 3 jobs.

	Big Time Burger	Morse's Pharmacy	Dime Department Store
Weekly schedule	M–F 5–7, S 8–11	• **1.** ■	**2.** ■
Hours per week	• **3.** ■	8 h	**4.** ■
Hourly wage	$4.15	**5.** ■	**6.** ■
Weekly pay	**7.** ■	$34	**8.** ■
Job responsibilities	**9.** ■	**10.** ■	Sales clerk

1. T, W 4–8 **2.** F 4–10, S 9–4 **3.** 13 h **4.** 13 h **5.** $4.25 **6.** $4.10 **7.** $53.95 **8.** $53.30 **9.** Counter help **10.** Light stock work

MAKING THE DECISIONS

•**11.** If Rebecca's main interest were weekly pay, which job would be best? Big Time Burger

•**12.** If Rebecca's main interest were hourly wage, which job would be best? Morse's Pharmacy

13. If Rebecca likes to work around food, which job would be best? Big Time Burger

14. If Rebecca has never worked before, which job could she not get? Dime Department Store

15. If Rebecca cannot work more than 12 h a week, which job would be best? Morse's Pharmacy

16. If Rebecca likes to work with people, which job would be the worst? Morse's Pharmacy

PROBLEM B

You are looking for a summer job in a day camp (5 d per wk). You marked these want ads A, B, and C. Now you need to decide which job is best for you.

CAMP COUNSELOR
$4.50/h
8 hours per day
Supervise 15 8-year-olds
HS junior/senior
Some overnight trips

APPLE DAY CAMP
Nature Counselor
Work with 8 to 12-year-olds
Planned program of activities
9:00–4:30 $35/day

ARTS & CRAFTS COUNSELOR
$1,000 for 7 wk, M-F
35 h per wk
Plan, organize,
direct activities

DECISION-MAKING FACTORS

Length of work day
Job responsibilities
Weekly pay
Other information
Hourly wage

DECISION-MAKING COMPARISONS

Complete the table to compare the 3 jobs.

	Camp A	Camp B	Camp C
Length of day	8 h	•17. ■	18. ■ 7 h
Weekly pay	•19. ■	$175	20. ■ $142.86
Hourly wage	$4.50	21. ■	22. ■ $4.08
Job responsibilities	23. ■	24. ■	Plan, organize, direct arts and crafts activities
Other information	25. ■	26. ■	Plan *own* activities

17. $7\frac{1}{2}$ h
19. $180
21. $4.67
23. supervise
24. nature work

25. 15 8-year-olds; some overnight trips **26.** 8 to 12-year-olds; planned program

MAKING THE DECISIONS

•**27.** If your main interest were weekly pay, which camp would be best?
Camp A

•**28.** If your main interest were hourly wage, which camp would be best?
Camp B

29. If your main interest were nature, which camp would be best?
Camp B

30. If you like to work with children under 10, which camp would be best?
Camp A; information not given for Camp C

31. If your main interest were arts and crafts, which camp would be best?
Camp C

32. If you like to plan your own activities, which camp would be best?
Camp C

33. Only you can decide which job has the best combination of pay and other factors. Which of these camp jobs would be best for you? Why?
Answers may vary.

Have students complete Exercises 11–16 in small groups, having students explain and defend their answers and decisions. Continually reinforce the fact that, in job considerations, best answers depend upon individual situations and preferences. Encourage all students to participate and to express their opinions.

Repeat the same instructional procedure for Problem B. Point out that Exercise 33 provides an opportunity for students to use their own personal preferences for a job. Lead a discussion in which each student indicates his or her choice of job and provides a clear rationale for that choice.

ASSIGNMENTS

Level 1	1–33
Level 2	1–33
Level 3	1–33

FOLLOW-UP ACTIVITY

APPLICATION Ask students to collect want ads or announcements about part-time or summer jobs. You may want to post these ads on a bulletin board or on a table in the room. Encourage students to contact these potential employers directly to answer any questions the students may have about the advertisements.

Once enough ads have been collected, ask students to choose 3 ads that interest them and to make their own list of factors to help decide which ad to pursue. Finally have them choose an ad and give the reasons for their decision.

SUPPLEMENTARY MATERIALS

TRP Practice, p. 21
TRP Transparency 11

OBJECTIVES

- Review vocabulary.
- Practice key chapter concepts and skills.

USING THE REVIEW

The Chapter Review is designed to help students prepare for taking the Chapter Test. The first section focuses on vocabulary. It requires students to select words to complete statements. The second section presents practice of key chapter skills.

Each item on the Chapter Review is referenced to the page on which the topic is taught in the Pupil's Edition. You may wish to have students refer to these pages to help review any concepts or skills they have not yet mastered.

It is suggested that students work in small-sized heterogeneous cooperative learning groups. Some cooperative learning methods that may be used are as follows:

1. After each student has independently completed the Chapter Review, a discussion should follow within each group about the solutions to the problems.
2. The group can complete the entire Chapter Review by working together to solve the problems.

End the lesson with an entire class discussion in which any questions brought up in group discussions are presented and answered.

CHAPTER REVIEW

Vocabulary Choose the letter of the word(s) that completes the sentence.

1. The pay you earn is found by multiplying the number of hours worked by the ■. [42] b

   a. Number of days worked **b.** Hourly wage **c.** Tips

2. The amount of money a customer leaves for a person who has provided service is called a ■. [44] c

   a. Hourly wage **b.** Total pay **c.** Tip

3. Another name for the Federal Insurance Contributions Act is ■. [46] b

   a. Payroll deductions **b.** Social Security **c.** Paycheck

Skills Find the answer.

4. The hourly wage is $4.10. John worked 7 h. How much did he earn? [42]
   $28.70
5. The hourly wage is $4.85. May worked 9 h. How much did she earn? [42]
   $43.65
6. Serita worked 6 h and earned $23.40. What is her hourly wage? [42]
   $3.90
7. Sal worked 9 h and earned $39.15. What is his hourly wage? [42]
   $4.35
8. Andrew received a 15% tip for serving a $25 meal. How much was his tip? [44]
   $3.75
9. Opray received a 15% tip for serving a $28.65 meal. How much was her tip? [44]
   $4.30
10. A meal you ordered cost $24.00. You left a 15% tip. How much did you leave? [44]
    $3.60
11. Jacqueline earned $1.75 per h for 6 h plus 15% in tips on $220 of meals. What was her total pay? [42, 44]
    $43.50
12. Enrico earned $1.95 per h for 6 h. He also earned tips of 15% on $275 of meals. What was his total pay? [42, 44]
    $52.95
13. Ann earned $2.35 per h for 5 h plus 15% in tips on $318.65 of meals. What was her total pay? [42, 44]
    $59.55
14. Mark's total pay was $400. How much was withheld for Social Security? [46]
    $30.04
15. Manuel earned $8.75 per h for 25 h. How much was withheld for Social Security? [46]
    $16.43
16. Regina earned $2.85 per h for 7 h plus 15% in tips on $417 worth of served meals. What was her total pay? How much was withheld for Social Security? [46]
    $82.50; $6.20
17. Ira earned $1.95 per h for 5 h plus 15% in tips on $225 worth of served meals. What was his total pay? How much was withheld for Social Security? [46]
    $43.50; $3.27

Chapter Test

Find the total pay.

1. Monday: $18
Tuesday: $16
$34

2. Monday: $21
Tuesday: $16
Friday: $37
$74

3. Monday: $10.95
Tuesday: $10.95
Saturday: $28.49
$50.39

4. Hours worked: 6
Hourly wage: $4.30
$25.80

5. Hours worked: 11
Hourly wage: $3.85
$42.35

6. Hours worked: 8.5
Hourly wage: $4.65
$39.53

Find the hourly wage.

7. Amount earned: $27.65
Hours worked: 7
$3.95

8. Amount earned: $29.10
Hours worked: 6
$4.85

9. Amount earned: $38.52
Hours worked: 9
$4.28

How much was the tip?

10. Cost of meal: $27
Tip: 10%
$2.70

11. Cost of meal: $34.50
Tip: 15%
$5.18

12. Cost of meal: $64.54
Tip: 15%
$9.68

Find the total pay.

13. Hours worked: 6
Hourly wage: $3.85
Tips: $15.50
$38.60

14. Hours worked: 7
Hourly wage: $1.90
Tips: $28.50
$41.80

15. Hours worked: 4.5
Hourly wage: $2.35
Tips: $38.65
$49.23

Find the total pay and the amount withheld for Social Security.
Use 7.51% as the withholding rate.
Round each amount to the nearest cent.

16. Hours worked: 5
Hourly wage: $4.26
$21.30; $1.60

17. Hours worked: 6
Hourly wage: $1.85
Cost of meals: $235
Tip: 15%
$46.35; $3.48

18. Hours worked: 8
Hourly wage: $2.05
Cost of meals: $485
Tip: 18%
$103.70; $7.79

Solve the problem using guessing and checking.

19. Anna worked a total of 150 h in 2 mo. During the first month, she worked 20 h more than the second month. How many hours did Anna work in each of the 2 mo?
85 h, 65 h

20. A customer left Bob a tip of $2.30 in quarters and dimes. He was given 14 coins in all. How many coins of each kind did he receive?
6 quarters, 8 dimes

Objective

- Evaluate the achievement of chapter objectives

Using the Test

The Chapter Test may be used as a post-test to evaluate student achievement. However, you may wish to use the Chapter Post-Test offered in the Teacher's Resource Package or to design your own chapter test. If this page is not used as a test, you may want to assign it as additional review or practice.

The test items are correlated to the chapter objectives in the table below.

Chapter objectives	Test items
A. Compute total pay.	1–6
B. Compute hourly wages.	7–9
C. Compute the amount tips.	10–12
D. Compute total pay based on hourly wages and tips.	13–15
E. Compute total pay and amounts withheld for Social Security.	16–18
F. Solve problems by using guessing and checking.	19–20

Supplementary Materials

TRP Ch. 3 Post-Test Form A, pp. 1–2
TRP Ch. 3 Post-Test Form B, pp. 1–2

OBJECTIVE

- Understand that transportation costs can affect actual take-home pay.

USING THE PAGE

Begin discussion by asking students to name some modes of transportation that could be used to go to and from a part-time job. Ask them to approximate the weekly cost of each mode in their area.

Read the information given in the table. Have students work individually or in pairs to answer Exercises 1–4. Summarize the answers as follows: A job that originally appears to be the highest paying may not be when transportation costs are considered.

Have students work individually to answer Exercises 5–6. Then have them share their answers and reasons in a whole-class discussion.

ANSWERS/EXPLANATIONS

3. To the Bank: bicycle or walk
   To the Bake Shop: take 2 buses 1 way and walk the other way; or take 1 bus each way and walk the difference; or bicycle both ways.

MONEY TIPS

Transportation costs should be considered when you want to know your actual take-home pay.

LET'S LOOK AT THE FACTS

If you use public transportation, it will cost you money to go to and from your job. Look at the comparison table below for 3 different part-time jobs.

	Diana's Health & Fitness Center	Pattycake Bake Shop	Winston National Bank
Weekly pay	$200	$190	$150
Distance from your home	3 mi	4 mi	2 mi
Transportation (1 way)	1 express bus	2 local buses	1 local bus

LET'S DISCUSS WHY See TE side column for answer to Exercise 3.

1. Compute how much transportation would cost weekly (5 d) for each of the 3 jobs if a local bus ride costs $1.00 and an express bus ride costs $4.00.
   $40; $20, $10
2. Subtract the cost of weekly transportation from each weekly pay. Which job would leave you with the greatest actual take-home pay?
   Pattycake Bake Shop—$170
3. How might you reduce transportation costs to the Bank? To the Bake Shop?
4. For each job, how much money could you save if you walked to work every morning and took public transportation only at night?
   $20; $10; $5

LET'S SEE WHAT YOU WOULD DO

5. You and 3 friends all work at the Bake Shop on the same days and during the same hours. How could the 4 of you get to work besides taking the bus?
   walk, bicycle, skate, take turns driving
6. You are eager to save as much money as possible each week from a part-time job in order to buy a stereo cassette player and some new clothes. If you would not mind walking up to 4 mi per d, which of the jobs could you take to get an actual take-home pay of $180?
   Diana's Health & Fitness Center or Pattycake Bake Shop

CALCULATOR STANDARD AND SCIENTIFIC CALCULATORS

There are 2 kinds of calculators, standard and scientific. They perform a series of calculations differently.

A **standard calculator** performs operations in the order in which they are entered.

Evaluate $6 + 4 \times 3$, using a standard calculator.

If you enter [6] [+] [4] [×] [3] [=], a standard calculator will display an incorrect answer of 30. First, it adds 6 and 4 to give 10. Then it multiplies 10 by 3.

To obtain the correct answer on a standard calculator, you must follow the correct order of operations.

Procedure	Calculator Entry	Calculator Display
1. Perform the multiplication.	[4] [×] [3] [=]	12.
2. Perform the addition.	[+] [6] [=]	18.

So $6 + 4 \times 3 = 18$.

A **scientific calculator** performs operations according to the correct order of operations. If you enter [6] [+] [4] [×] [3] [=] on a scientific calculator, the display will show the correct answer, 18.

1. Using $6 + 4 \times 3$, determine whether your calculator is standard or scientific. Answers may vary.

Use a calculator to evaluate.

2. $12 + 3 \times 4$ 24
3. $18 + 5 \times 3$ 33
4. $2 \times 6 - 3$ 9
5. $12 + 9 \div 3$ 15
6. $135 + 10 \div 5$ 137
7. $16.9 + 8.6 \div 2$ 21.2
8. $432 + 15.6 \times 3$ 478.8
9. $152 + 8.1 \times 9$ 224.9
10. $200 + 49.6 \div 12.4$ 204
11. $36 + 46 + 1.8 \times 5$ 91
12. $8.6 - 4.5 + 2.3 \times 7$ 20.2
13. $16 \div 2 \times 4 + 8.5$ 40.5

There is a 1-digit number missing in the number sentence. First estimate, then use your calculator to find the missing number.

14. $12 + 4 \times ■ = 24$ 3
15. $6.8 + 3 \times ■ = 30.8$ 8
16. $41 + 7 \times ■ = 76$ 5
17. $60 - 35 + 5 \times ■ = 45$ 4

OBJECTIVES

- Perform a series of operations on a standard calculator.
- Identify differences between standard and scientific calculators.

USING THE PAGE

Tell students that the 2 major types of calculators are *scientific* and *standard (nonscientific).* They will probably be working with a standard calculator.

If possible, display a scientific calculator for students to examine. Have a volunteer compare and contrast the keys on the standard and scientific calculators. Scientific calculators have keys for parentheses, exponents, pi (π), reciprocals, and so on.

Tell students that in this lesson they will study 1 important difference between the 2 types of calculators. A scientific calculator evaluates an expression according to the correct algebraic order of operations; a standard calculator evaluates an expression by performing operations in the order in which they are entered.

At this point, you may wish to review the algebraic order of operations with students: parentheses, multiplication and division, addition and subtraction. (Order of operations is covered more fully in Chapter 18.)

Tell students they must think carefully before they enter numbers in their standard calculators. The order in which numbers and operation signs are entered is very important.

Have students use their calculators to work through the example in the text. You may wish to point out that it may not be necessary to press the is-equal-to key in the first step.

Objective

- Test the preprequisite skills needed to learn the concepts developed in Chapter 4.

Using the Test

The Pre-Skills Test is designed to diagnose students' strengths and weaknesses on prerequisite skills necessary to study the mathematics in Chapter 4.

Assign the Pre-Skills Test. Allow students to work together in pairs or in small groups. Group members should help those who demonstrate a misunderstanding of a concept or a weakness in a skill.

The table below correlates the items on the Pre-Skills Test with the prerequisite skill and the lesson(s) in the chapter for which it is needed.

Item(s)	Prerequisite skill	Lesson(s)
1–6	Rename mixed numbers and percents as decimals.	4.1, 4.2, 4.5, 4.7
7–10	Identify the correct operations in problem situations.	all
11–24	Subtract, multiply, and divide money amounts.	all
25–28	Compute elapsed times.	4.2
29–32	Add and subtract with mixed numbers.	4.2
33–36	Find the percent of a number.	4.5, 4.7

References to specific skills that are reviewed in the Skills Bank are given on the individual lessons themselves.

Chapter 4 Full-Time Work

Pre-Skills Test

Rename as a decimal.

1. $35\frac{1}{2}$ 35.5
2. $17\frac{3}{4}$ 17.75
3. $35\frac{1}{4}$ 35.25
4. 27% 0.27
5. 4.5% 0.045
6. 2.7% 0.027

Identify the correct operation.

7. To find how much greater one number is than another
subtraction
8. To find the total amount of 3 unequal measures
addition
9. To find the total amount of 24 equal measures
multiplication
10. To find the size of each equal portion of an amount
division

Subtract.

11. \$258.99 − \$37.56 − \$74.57 \$146.86
12. \$5,389 − \$368.90 − \$35 \$4,985.10

Multiply.

13. 8 × \$4.67 \$37.36
14. 5 × \$2.74 \$13.70
15. 52 × \$3.58 \$186.16
16. 30 × \$472 \$14,160
17. 1.5 × \$4.50 \$6.75
18. 1.5 × \$8.90 \$13.35
19. 25.5 × \$6.75 \$172.13
20. 35.75 × \$12.50 \$446.88

Divide. Round to the nearest cent.

21. \$18,900 ÷ 12 \$1,575
22. \$673.40 ÷ 40 \$16.84
23. 26)\$8,940 \$343.85
24. 24)\$23,800 \$991.67

Find the elapsed time.

25. Find 7:15 A.M. to 3:45 P.M. $8\frac{1}{2}$ h
26. From 8:10 A.M. to 5:15 P.M. 9 h 5 min
27. From 9:20 A.M. to 2:10 P.M. 4 h 50 min
28. From 8:45 A.M. to 3:30 P.M. $6\frac{3}{4}$ h

Add or subtract.

29. $2\frac{1}{2} + 5\frac{3}{4} + 6$ $14\frac{1}{4}$
30. $5 + 7\frac{1}{2} + 5\frac{3}{4}$ $18\frac{1}{4}$
31. $37\frac{1}{2} - 5\frac{3}{4}$ $31\frac{3}{4}$
32. $20 - 6\frac{1}{4}$ $13\frac{3}{4}$

Find the answer.

33. 80% of \$385 \$308
34. 40% of \$36.50 \$14.60
35. $6\frac{1}{2}$% of \$4,500 \$292.50
36. 3.5% of \$53,860 \$1,885.10

56 CHAPTER 4

In many factories, workers must punch in and punch out on a time clock. What factors, other than regular hours worked, could determine how much a factory worker is paid for a week's work?

OBJECTIVE

- Formulate problems, using text and data.

USING THE CHAPTER OPENER

Discuss students' answers to the question on the page. Some possible answers are as follows. *Overtime,* which could be more than the regular hourly wage, increases a worker's pay. This could be time and a half or even double time for working on a holiday. *Night bonuses* are occasionally paid to people as an incentive for working the night shift. In many cases, this would be a percentage of the base pay. *Piecework* involves paying a worker by the piece rather than by the hour. *Bonuses* may be paid to workers who exceed their quota.

COOPERATIVE LEARNING

Write the following on the chalkboard.

FACTORY WORKERS WANTED

L.M.N., Inc.
40 h per wk $8.00 per h
Day shift: 7:00 A.M.–4:00 P.M.
Night shift: 3:00 P.M.–12:00 Mid.
(10% bonus)
No overtime

O.P.Q., Inc.
35 h per wk $8.25 per h
Day shift: 6:00 A.M.–3:00 P.M.
Night shift: 3:00 P.M.–12:00 Mid.
Saturday overtime available twice a month (time and a half).

Assign heterogeneous groups and have them study the 2 want ads. Tell them to find the weekly pay for each job, excluding overtime, and to write some ways that overtime and bonuses could affect that pay. Then have them list the reasons why they would or would not choose to work at each job.

SUPPLEMENTARY MATERIALS

TRP Group Projects, p. 4

OBJECTIVES

- Compute regular pay, given hours worked and hourly wage.
- Compute overtime pay at $1\frac{1}{2}$ times regular hourly wage.

TEACHING THE LESSON

WARM-UP Write the following exercises on the chalkboard or an overhead transparency.

Find the earnings.

1. 5 h at \$2 per h (\$10)

2. 2 h at \$3.75 per h (\$7.50)

3. 8 h at \$4.50 per h (\$36.00)

INTRODUCTION Remind students that regular pay is derived from a regular hourly wage. Ask various students who have jobs how much they earn per hour and about how many hours they work per week.

INSTRUCTION Use the introductory situation and Example 1 to review the procedure for computing regular pay. Discuss how the formula works in terms of the units used.

Explain that after working 40 h during a week workers are eligible for overtime pay. Students sometimes forget that the overtime rate applies only to the overtime hours. Explain that earnings for the first 40 h do not change, even if you work more than 40 h.

Use Example 2 to demonstrate the step-by-step procedure for computing total pay including overtime. Explain how to find the overtime rate. This is a good opportunity to review the use of calculator memory to develop a single key sequence to find total pay.

SKILLS BANK You may want to refer students to the following skills in the Skills Bank:

Write Fractions as Decimals, Skill 1

Add With Whole Numbers and Decimals, Skill 6

Multiply With Decimals, Skills 2, 4

4.1 HOURLY WAGES AND OVERTIME PAY

You work as a mechanic in a large garage and earn \$8.40 per h. For any hours you work above 40 h per wk, you are eligible for **overtime pay** at a rate of $1\frac{1}{2}$ times your regular hourly wage.

EXAMPLE 1 What is your regular weekly pay if you work $37\frac{1}{2}$ h each week?

THINK: Hours worked: $37\frac{1}{2}$, or 37.5   Hourly wage: \$8.40

HOURS WORKED	×	HOURLY WAGE	=	PAY
37.5	×	\$8.40	=	\$315.00

Your regular weekly pay is \$315.00.

EXAMPLE 2 During a very busy week, you were asked to work 46 h. What was your total pay for that week?

1. Multiply to find the regular pay. — 40 × \$8.40 = \$336.00
2. Find the overtime pay.
   - Subtract to find the overtime hours. — 46 − 40 = 6 h
   - Multiply to find the overtime rate. — 1.5 × \$8.40 = \$12.60 per h
   - Multiply to find the overtime pay. — 6 × \$12.60 = \$75.60
3. Add to find the total pay. — \$336.00 + \$75.60 = \$411.60

Your total pay for the 46 h was \$411.60.

FOR DISCUSSION See TE side column.

1. Why do you think workers have asked for overtime pay at a higher rate than their regular hourly wage?

2. Why does the government set a minimum wage? How does this affect you?

PRACTICE EXERCISES

Remember to estimate whenever you use your calculator.

Find each employee's regular pay. Round the amount to the nearest cent.

Employee	Hours worked	Hourly wage	Regular pay
Henderson	40	$4.85	• 1. ■
Rodriquez	25	$7.39	• 2. ■
Martin	$37\frac{3}{4}$	$5.40	3. ■

Employee	Hours worked	Hourly wage	Regular pay
Lieberman	$16\frac{1}{4}$	$6.12	4. ■
Washington	$32\frac{1}{2}$	$12.80	5. ■
Jones	$34\frac{3}{4}$	$18.42	6. ■

1. $194 **2.** $184.75 **3.** $203.85 **4.** $99.45 **5.** $416 **6.** $640.10

Complete the table. Use an overtime rate of $1\frac{1}{2}$ times the hourly wage for work over 40 h. Round each amount to the nearest cent.

Employee	Hours worked	Hourly wage	Regular pay	Overtime pay	Total pay
Carter	38	$6.48	• 7. ■	• 8. ■	9. ■
Honesdale	43	$5.00	10. ■	11. ■	12. ■
Chang	$46\frac{3}{4}$	$8.20	13. ■	14. ■	15. ■
Hernandez	$52\frac{1}{2}$	$7.48	16. ■	17. ■	18. ■
Santini	$49\frac{1}{4}$	$18.42	19. ■	20. ■	21. ■

7. $246.24 **8.** $0 **9.** $246.24 **10.** $200 **11.** $22.50
12. $222.50 **13.** $328 **14.** $83.03 **15.** $411.03 **16.** $299.20
17. $140.25 **18.** $439.45 **19.** $736.80 **20.** $255.58 **21.** $992.38

Solve.

• **22.** Sarah earned $230 for 40 h of work. What was her hourly wage? $5.75

23. Jim earned $313.50 for 32 h of work. What was his hourly wage? $9.80

24. Cheryl earns $7.58 per h. If her regular pay last week was $272.88, how many hours did she work? 36 h

25. Bart earns $6.30 per h. If his regular pay last week was $236.25, how many hours did he work? $37\frac{1}{2}$ h

EXTENSION

Overtime Hours

Matt earns $9.00 per h. His total weekly pay, including overtime, was $414.00.

1. For how many hours did he earn $9.00 per h? 40 h

2. What was his regular pay? $360

3. What was his overtime pay? $54

4. What was his overtime rate of pay? $13.50

5. How many overtime hours did he work? 4 h

FOR DISCUSSION

1. Students should explain that working extra hours beyond a normal work week should be rewarded in some way. Also, since workers must be at their job and away from their families or other activities more, a higher overtime rate is an incentive.

2. Students should explain that the minimum wage is set to protect workers from being forced to work for unfairly low wages and to prevent abuses during times of high unemployment when workers might bid the hourly wage down just to have a job.

EXTENSION This sequential problem allows students to work backward from a given total pay to determine the number of overtime hours worked. Questions 1–4 are all subproblems that lead to the answer of Question 5.

ASSIGNMENTS

Level 1	1–15
Level 2	4–21, 22–24, EX
Level 3	7–25, EX

FOLLOW-UP ACTIVITY

APPLICATION Ask students to select 3 jobs from newspaper help-wanted ads that advertise an hourly wage (or provide students with 3 representative jobs and the average hourly wage for each). Construct a table that shows the total pay for each job for a 25-, 35-, $37\frac{1}{2}$-, 40-, and 50-h work week.

SUPPLEMENTARY MATERIALS

TRP Practice, p. 22
TRP Reteaching, p. 19
TRP Transparency 11

OBJECTIVES

- Complete time sheets and compute hours worked.
- Use time cards to compute hours worked.
- Use time sheets or cards to compute pay.

TEACHING THE LESSON

WARM-UP Write the following exercises on the chalkboard or an overhead transparency.

Add.

1. $2\frac{1}{2} + 5 + 3\frac{1}{4}$ $(10\frac{3}{4})$

2. $4\frac{3}{4} + 3\frac{1}{2} + 5\frac{1}{4}$ $(13\frac{1}{2})$

3. $7\frac{1}{4} + 4 + 3\frac{1}{2} + 5$ $(19\frac{3}{4})$

INTRODUCTION Begin discussion by directing students to the time sheet and the time card on page 60. Ask them to describe how each is used to record work hours.

INSTRUCTION Use the introductory situation and Example 1 to demonstrate how to compute weekly earnings from a time sheet. Step 2 represents the formula:

$$\text{HOURS WORKED} \times \text{HOURLY WAGE} = \text{PAY}$$

Point out that, in order to use a calculator, $22\frac{1}{2}$ needs to be renamed as 22.5. Remind students always to estimate to check the reasonableness of their answers. These calculations are important to check the accuracy of your paycheck. In order to complete Exercises 16–17, students may use the time sheet forms offered in the Teacher's Resource Package.

Use Example 2 to demonstrate how to compute weekly pay from a time card. First, review the process for rounding times down to the nearest quarter hour. Have students explain how each IN and OUT time is rounded in Step 1. Now review how elapsed times can be mentally computed in Step 1. For example: 8 to 4 is 4 h to noon plus 4 h more, or 8 h.

4.2 TIME SHEETS AND TIME CARDS

You are a freelance proofreader for a large publisher. You work at home and earn $8.50 per h. You keep track of the hours you work on a **time sheet** that you submit weekly to the publisher.

TIME SHEET

DATE	HOURS
9/6	6 3/4
9/17	5
9/19	2 1/2
9/20	8 1/4
TOTAL	

EXAMPLE 1 According to your time sheet, how many hours did you work during the week? How much did you earn?

1. Add to find the total hours. $6\frac{3}{4} + 5 + 2\frac{1}{2} + 8\frac{1}{4} = 22\frac{1}{2}$ h

You worked a total of $22\frac{1}{2}$ h.

2. Multiply to find the earnings. $22\frac{1}{2} \times \$8.50 = 22.5 \times \$8.50 = \$191.25$

You earned $191.25.

Many workers "punch in" and "punch out" on a **time clock** that records their arrival and departure times on a weekly **time card.** The time card is then used to determine how many hours were worked. Times from time cards are usually rounded *down* to the nearest quarter hour.

Employee Sarah Williams
Employee # 2733
Week of 10/7

DATE	IN	OUT
10/7	8:02	4:05
10/8	7:55	4:52
10/10	9:01	3:34
10/11	8:12	4:17
TOTAL HOURS		

EXAMPLE 2 Sarah works as a cashier in a food store and earns $5.18 per h. Use her time card to determine how many hours she worked. Then find her regular pay for the week.

1. Round to find the hours worked.
   - 10/7: 8 to 4 is 8 h
   - 10/8: 7:45 to 4:45 is 9 h
   - 10/10: 9 to 3:30 is $6\frac{1}{2}$ h
   - 10/11: 8 to 4:15 is $8\frac{1}{4}$ h
2. Add to find the total hours. $31\frac{3}{4}$ h

According to Sarah's time card, she worked $31\frac{3}{4}$ h.

3. Multiply to find the regular pay. $31\frac{3}{4} \times \$5.18 = 31.75 \times \$5.18 = \$164.465$

Sarah's regular pay for this week was $164.47.

FOR DISCUSSION See TE side column.

1. What are the advantages of using a time clock in a factory?
2. Why are hours worked usually rounded down to the nearest quarter hour?

PRACTICE EXERCISES

Remember to estimate whenever you use your calculator.

Complete the table to find the hours worked. Round down to the nearest quarter hour. **3.** $8\frac{1}{4}$ h **6.** 9 h **9.** $7\frac{1}{4}$ h **12.** $5\frac{1}{2}$ h **15.** $8\frac{1}{2}$ h

Employee	Time in	Time out	Rounded time in	Rounded time out	Hours worked
Meyer	7:10 A.M.	3:16 P.M.	• 1. ■ 7:00	• 2. ■ 3:15	3. ■
Labeau	7:54 A.M.	4:57 P.M.	4. ■ 7:45	5. ■ 4:45	6. ■
Modica	8:19 A.M.	3:43 P.M.	7. ■ 8:15	8. ■ 3:30	9. ■
Jefferson	7:32 A.M.	1:13 P.M.	10. ■ 7:30	11. ■ 1:00	12. ■
Anderson	8:41 A.M.	5:14 P.M.	13. ■ 8:30	14. ■ 5:00	15. ■

Write a time sheet for the person, including the total. Use the time sheet in Example 1 as your model. Then answer the question. Check students' time sheets.

16. Rob worked 6 h on Mon., $5\frac{1}{2}$ on Tues., $4\frac{1}{4}$ on Wed., $7\frac{1}{2}$ on Thurs., and $4\frac{3}{4}$ on Fri. If Rob is paid $4.75 per h, how much did he earn? Total: 28 h; $133

17. Julie worked $5\frac{1}{4}$ h on Mon., $3\frac{1}{2}$ on Tues., $8\frac{3}{4}$ on Thurs., and $7\frac{1}{4}$ on Fri. If Julie earns $6.86 per h, how much did she earn?
Total: $24\frac{3}{4}$ h; $169.79

Use the time cards below for Exercises 18–19.

Employee Sam Longo
Employee # 25830
Week of 12/15

DATE	IN	OUT
12/15	8:10	4:08
12/16	7:55	2:58
12/17	8:16	4:29
12/18	7:57	4:45
12/19	8:57	12:31
TOTAL HOURS		

Employee Tina Baker
Employee # 4755
Week of 11/20

DATE	IN	OUT
11/20	7:05	3:58
11/21	7:30	5:01
11/22	8:01	5:32
11/23	7:22	4:46
11/24	6:55	4:31
TOTAL HOURS		

18. How many hours did Sam work? If he earns $7.55 per h, what was his pay?
$35\frac{3}{4}$ h; $269.91

19. How many hours did Tina work? If she earns $5.35 per h, what was her pay?
47 h; $251.45

Point out in Step 3 that, although a calculator will display 164.465 as the answer, students will need to round to the nearest cent to record the answer.

SKILLS BANK You may want to refer students to the following skills in the Skills Bank:

Write Fractions as Decimals, Skill 1

Add With Fractions and Mixed Numbers, Skills 4–5

Multiply With Decimals, Skills 4–5

FOR DISCUSSION

1. Students should explain that a time clock assures accuracy and provides a record of arrival and departure times, and when computerized, they can be easily used to find the total hours worked and the total pay received.
2. Students should explain that quarter hours are a reasonable unit to work with rather than pay workers for each minute.

ASSIGNMENTS

Level 1	1–9, 16
Level 2	1–19
Level 3	10–19

FOLLOW-UP ACTIVITY

APPLICATION Prepare a worksheet with 4 columns: Starting time, Ending time, Hourly wages, and Earnings. Fill in times and hourly wages for the first 3 columns for 5 or 10 rows. Ask students to estimate the daily earnings for each row and to add their estimates to find weekly earnings. Have students then compute the actual earnings.

SUPPLEMENTARY MATERIALS

TRP Practice, p. 23
TRP Reteaching, p. 20
TRP Lesson Aids, p. 2
TRP Applications, p. 4
TRP Transparency 12

OBJECTIVES

- Compute weekly, monthly, semimonthly, or biweekly pay, given annual salaries.
- Compute annual salaries, given weekly, monthly, semimonthly, or biweekly pay.

TEACHING THE LESSON

WARM-UP Make a list of bi-words in one column (biannual, bifocals, bipartisan) and a list of semi-words in another column (semicircle, semiannual, semicolon). Use the lists to differentiate between bi-, meaning 2, and semi-, meaning half. Explain why biweekly means every other week and results in 26 periods per y, while semimonthly means half-monthly, or twice a month, and results in 24 periods per y.

INTRODUCTION Use the help-wanted ads from a local newspaper and make a list on the chalkboard of the various wages and salaries listed: hourly, weekly, and annual. Ask students how they could determine which job pays the most and which pays the least.

INSTRUCTION Use Example 1 to demonstrate how to convert annual salary to weekly pay. Discuss how the formula works in terms of the units used. Point out that the computed answer must be recorded to the nearest cent.

Use Example 2 to demonstrate the conversion of annual salary to semimonthly pay. Redo Example 2 to find biweekly pay ($838.46 per paycheck).

Use Example 3 to show how to compare 2 jobs on the basis of pay. Explain that this can be done by converting either the annual salary to weekly pay or the weekly pay to an annual salary. Point out to students that an alternate form of the formula in Example 1 can be used to find yearly salary:

NUMBER OF PERIODS × PAY PER PERIOD = YEARLY SALARY

4.3 SALARY

You take a job as a computer-service technician that pays a **salary** of $18,500 per y. You will be paid weekly.

EXAMPLE 1 What is your pay per week?

THINK: Your salary is divided into 52 equal pay periods.

YEARLY SALARY	÷	NUMBER OF PAY PERIODS	=	PAY PER PERIOD
$18,500	÷	52	=	$355.76923

Your pay per week is $355.77.

Many people are paid **biweekly,** or every other week. This means they will receive 26 paychecks per y. Other workers are paid **semimonthly.** These workers receive 2 paychecks per mo, or 24 paychecks per y.

EXAMPLE 2 Kathleen takes a job that pays $21,800 per y. If she is paid semimonthly, what is the amount of each paycheck?

THINK: Semimonthly means 24 equal pay periods per year.

Divide to find the pay per period. $21,800 ÷ 24 = $908.33333

Kathleen's semimonthly paycheck is $908.33.

Some jobs are advertised with a weekly pay or an hourly wage. It is often necessary to convert these to annual salaries to compare jobs.

EXAMPLE 3 Richard needs to choose between a job advertised at $16,250 per y and another advertised at $350 per wk. Which job has the greater annual salary?

1. Multiply to find the annual salary. 52 × $350 = $18,200
2. Compare the annual salaries. $18,200 > $16,250

The $350-per-wk job has the greater annual salary.

FOR DISCUSSION See TE side column.

What are some advantages to being paid biweekly instead of monthly?

PRACTICE EXERCISES

Remember to estimate whenever you use your calculator.

Complete the table.

Employee	Annual salary	Pay period	Pay per period	
Parker	$8,750	weekly	• 1. ■	$168.27
Smith	$12,300	biweekly	2. ■	$473.08
Oliviero	$14,890	semimonthly	3. ■	$620.42
Yeh	$24,700	monthly	4. ■	$2,058.33

Complete the table.

Employee	Pay period	Pay per period	Annual salary	
Rogers	weekly	$247.55	• 5. ■	$12,872.60
Silverbear	biweekly	$638.85	6. ■	$16,610.10
Halbacher	semimonthly	$487.32	7. ■	$11,695.68
Cohen	monthly	$1,587.95	8. ■	$19,055.40

Which job pays more? How much more?

	Job A	Job B
• 9.	$450 per wk	$25,800 per y Job B; $2,400
10.	$8,450 per y	$775 per mo Job B; $850
11.	$439 semimonthly	$420 biweekly Job B; $384
12.	$8.95 per h (40 h per wk)	$16,400 per y Job A; $2,216
13.	$12.25 per h (40 h per wk)	$28,600 per y Job B; $3,120
14.	$10.95 per h (35 h per wk)	$18,500 per y Job A; $1,429

Solve.

•15. Kelly has been offered one job that pays $740 per mo and another that pays $9,200 per y. Which job pays more per year? How much more?
$9,200-per-y job; $320

16. Jeff presently earns $230 per wk. He is offered another job that pays $13,200 per y. How much will his salary increase per year? How much more will he earn per week?
$1,240; $23.85

17. Roger earns a salary of $22,800 per y. He works a 40-h week. What is his hourly wage?
$10.96

18. Nicole earns $7.65 per h for a 35-h week. How much more per year will she earn if she is promoted to a job that pays an annual salary of $16,750?
$2,827

Help students understand that in going from shorter periods of time to longer periods of time, the amount of a paycheck increases and you must multiply. Conversely, when going from longer periods of time to shorter periods of time, the amount of a paycheck decreases and you must divide.

SKILLS BANK You may want to refer students to the following skills in the Skills Bank:

Multiply With Decimals, Skill 5

Divide With Decimals, Skill 8

FOR DISCUSSION Students should explain that the major advantage is that you get paid more than twice as often and are less likely to run out of money just before your next paycheck.

ASSIGNMENTS

Level 1	1–8, Odd 9–15
Level 2	1–16
Level 3	Even 2–8, 9–18

FOLLOW-UP ACTIVITY

ENRICHMENT Have students use their calculators to solve.

1. If you earned $100,000 per y, how much would you earn per 1 mo, wk, d, h, min, and s?
2. If you earned 3¢ per s for every second during the day, how much would you earn in a year?

SUPPLEMENTARY MATERIALS

TRP Practice, p. 24
TRP Reteaching, p. 21
TRP Transparency 12

OBJECTIVES

- Compute piecework pays, given piece rates and the numbers of items produced.
- Find the numbers of items that need to be produced, given piece rates and the pays desired.

TEACHING THE LESSON

WARM-UP Write the following exercises on the chalkboard or an overhead transparency.

Multiply mentally.

1. $2 × 500 **2.** $0.02 × 500

3. $20 × 50 **4.** $0.20 × 50

(**1.** $1,000 **2.** $10 **3.** $1,000 **4.** $10)

Review with students how to mentally compute by multiplying the greatest nonzero places, 2 and 5, and then adding the correct number of zeros.

INTRODUCTION Begin discussion by asking students if they have ever been paid by the number of jobs they completed or the number of items they sold rather than by the hour. (For example: selling magazine subscriptions or delivering newspapers.) Use students' examples to present the idea of piecework as differentiated from hourly wage.

INSTRUCTION Use Example 1 to demonstrate how piecework pay is computed. Discuss how the formula works in terms of the units used. Since this is a 2-step solution, point out that Step 1 shows how to find the number of items produced and Step 2 represents the use of the formula.

Use Example 2 to show how you can work backward to determine the number of items you need to complete in order to earn a desired amount of money. Emphasize the step-by-step reasoning used to find the number of pages Alex needs to type. Point out that the answer to Step 2, 15.873015 h, can be rounded to 16 h before it is used in Step 3.

4.4 PIECEWORK

In some industries, workers are paid based on the number of items they produce or jobs they complete. This is called **piecework** since workers' wages depend on the number of pieces or jobs completed.

EXAMPLE 1 Suppose you are paid $4.67 for each garment you complete while working in a clothing factory. If you complete an average of 2 garments per h, what is your average pay for a 40-h work week?

THINK: Work week: 40 h Items produced: 2 per h Piece rate: $4.67

PIECE RATE × ITEMS PRODUCED = PAY

1. Multiply to find the average number of items produced. 40 × 2 = 80 items
2. Multiply to find the pay. $4.67 × 80 = $373.60

You earn about $373.60 for a 40-h week.

EXAMPLE 2 Alex works part time at home, typing papers for others on his word processor. Alex charges $1.75 per page and averages 9 pages per h. Alex needs to earn $250 per wk. About how many hours must he work each week? About how many pages will he need to type each week?

THINK: Piece rate: $1.75 Items produced: 9 per h
Pay needed: $250 per wk

1. Multiply to find the average pay per hour. $1.75 × 9 = $15.75 per h
2. Divide to find the number of hours needed. $250 ÷ $15.75 = 15.873015 h

Alex must work about 16 h to earn $250.

3. Multiply to find the number of typed pages needed. 16 × 9 = 144 pages

Alex will need to type about 144 pages to earn $250.

FOR DISCUSSION See TE side column.

1. How does the piecework system motivate workers?
2. What are some advantages and disadvantages of the piecework system?

PRACTICE EXERCISES

Remember to estimate whenever you use your calculator.

Find the pay for the job.

1. $12.25 2. $161.20 3. $180 4. $500 5. $87.50 6. $30 7. $36 8. $5,400

	Piece rate	Items produced
• 1.	$1.75	7 garments made
• 2.	$12.40	13 machine parts made
3.	$20	9 cleaned carpets
4.	$1.25	400 proofread pages
5.	$3.50	25 typed pages
6.	$2.00	15 grocery deliveries

	Piece rate	Items produced
7.	$0.08	450 delivered newspapers
8.	$45.00	120 illustrations
9.	$4.75	300 washed cars
10.	$6.35	55 packed cartons
11.	$63.50	18 car tune-ups
12.	$85.75	25 repaired televisions

9. $1,425 10. $349.25 11. $1,143 12. $2,143.75

Solve.

• 13. Becky is paid $1.25 per shirt. She averages 5 shirts per h. What is her average pay for a 35-h week?
$218.75

• 14. Juan earns 4.2¢ for each circuit. He averages 120 circuits per h. What is his average pay for a 30-h week?
$151.20

15. Leon assembles an average of 12 machines per d. He is paid $15.60 per machine. What is his average pay for a 5-d week?
$936

16. Jennifer averages 5 deliveries per h and earns $2.40 per delivery. What is her average pay for a 35-h week?
$420

17. Blair is paid 16¢ for each telephone call. She earned $148.16 last week. How many calls did she make?
926 calls

18. Jake earns $5.25 for each lawn he mows. About how many lawns must he mow in order to earn $200 per wk?
38 lawns

19. Roberta earns $35 per illustration. It takes her about $1\frac{1}{2}$ h to complete each illustration. How many illustrations must she complete to earn $525? How many hours will this take her?
15 illustrations; $22\frac{1}{2}$ h

20. Guido is paid 85¢ per page to proofread manuscripts. He proofreads about 15 pages per h. About how many hours must he work in order to earn over $400 per week? About how many pages will he need to proofread to earn this much?
32 h; 480 pages

EXTENSION Mixed Rates of Pay

Eric is paid $265 per wk plus 80¢ for each part he machines.

1. What is Eric's piecework pay for machining 97 parts?
$77.60
2. What is Eric's total pay for a week in which he machines 97 parts?
$342.60
3. What is his average hourly wage if he works a 35-h week?
$9.79

SKILLS BANK You may want to refer students to the following skills in the Skills Bank:

Multiply With Decimals, Skills 2, 5

Divide With Decimals, Skills 4, 8

FOR DISCUSSION Students should explain that an advantage of the piecework system is the way it rewards workers for the amount of work they complete. However, piecework is also often very high-pressure work and can sometimes lead to problems of quantity over quality.

EXTENSION Use this problem to show how some workers receive mixed rates of pay with an hourly wage or weekly pay component as well as a piecework component. Discuss why this arrangement might be advantageous to workers.

ASSIGNMENTS

Level 1	1–14
Level 2	1–12, Odd 13–19
Level 3	13–20, EX

FOLLOW-UP ACTIVITY

ENRICHMENT Ask students to suppose that they are paid $2 per product produced, that they can produce 15 per h, and that they work 40 h per wk. Ask students to use their calculators to estimate how many years they would need to work to earn 1 million dollars (about 83 y).

SUPPLEMENTARY MATERIALS

TRP Practice, p. 25
TRP Reteaching, p. 22
TRP Transparency 13

OBJECTIVES

- Compute straight commissions.
- Compute pay composed of salary plus commission.

TEACHING THE LESSON

WARM-UP Write the following exercises on the chalkboard or an overhead transparency.

Rename as a decimal.

1. 5% **2.** 8.5% **3.** $16\frac{1}{4}$%

Find the answer.

4. 10% of $350 **5.** 50% of $240

6. 2% of $800 **7.** 20% of $400
(**1.** 0.05 **2.** 0.085 **3.** 0.1625 **4.** $35 **5.** $120 **6.** $16 **7.** $80)

INTRODUCTION Ask students if they know how most automobile salespeople and real-estate agents are paid. Explain how the earnings of many salespeople are based on the value of the sales they make. Introduce the word *commission* in reference to what an automobile salesperson is paid. You may want to discuss how the $10,000 you pay for a car is divided up among the auto company, the salesperson, and the dealership.

INSTRUCTION Use Example 1 to demonstrate the meaning of commission and to compute a commission based on the *number* of sales.

Use Example 2 to show how to find commission based on the *value* of sales. In this case, the commission rate is expressed as a percent. Remind students to rename percents as decimals and to estimate first before computing.

Use Example 3 to introduce students to situations in which employees receive a combination of salary and commission. Carefully review the 2-step solution.

4.5 COMMISSION

You have taken a job selling magazine subscriptions. It offers a commission of $8.25 for each subscription sold. A **commission** is an amount of money paid for selling a product or a service.

EXAMPLE 1 What would your commission be if you sold 60 subscriptions?

THINK: Commission rate: $8.25 per subscription
Total sales: 60 subscriptions

COMMISSION RATE × TOTAL SALES = COMMISSION

$8.25 × 60 = $495.00

Your commission would be $495.00.

The **commission rate** is often a percent of the total value of sales. Remember to change the percent to a decimal or a fraction before multiplying.

EXAMPLE 2 A sign in a carpet-store window advertises a sales position that earns a commission of $5\frac{1}{4}$%. How much commission would you earn on sales of $6,000?

THINK: $5\frac{1}{4}\% = 0.0525$

COMMISSION RATE × TOTAL SALES = COMMISSION

0.0525 × $6,000 = $315

You would earn a commission of $315.

In the above situations, the only pay was a commission, known as **straight commission.** Some jobs offer a combination of **salary and commission.**

EXAMPLE 3 A sales position at an appliance store offers a base pay of $250 per wk plus a 12% commission on sales. How much would you earn on weekly sales of $1,750?

1. Multiply to find the commission.
   THINK: 12% = 0.12   0.12 × $1,750 = $210
2. Add to find the total earnings. $210 + $250 = $460

You would earn $460.

FOR DISCUSSION See TE side column.

If you could earn $300 per wk plus a 3% commission or a straight commission of 7%, which job would you choose? Why?

PRACTICE EXERCISES

Remember to estimate whenever you use your calculator.

Find the commission on the sale. 1. $133.75 2. $162.50 3. $655.20 4. $33.10 5. $790.77 6. $80.29 7. $756.77 8. $63.14

	Commission rate	Total sales
• 1.	$5.35 per order	25 orders
• 2.	25¢ per box	650 boxes
3.	9%	$7,280
4.	40%	$82.75

	Commission rate	Total sales
5.	3%	$26,359
6.	$12\frac{1}{2}\%$	$642.30
7.	19%	$3,983
8.	$6\frac{3}{4}\%$	$935.40

Solve.

• 9. Dwight is paid a commission of $3.75 for each class ring he sells. How much did he earn by selling 74 class rings? $277.50

• 10. Sylvia is in automobile sales. She earns a commission of 3.2% on the value of her auto sales. How much did she earn on sales of $46,280? $1,480.96

11. Ellen sells cosmetics and earns a weekly pay of $230 plus a commission of 12% on sales. How much did she earn last week when she sold $1,200 worth of cosmetics? $374

12. Will earns a monthly pay of $825 plus a commission of $4\frac{1}{2}\%$ on sales. How much did he earn last month when his sales were $13,850? $1,448.25

13. Luis is paid $245 per wk plus a 12.5% commission on all sales over $500. His sales last week totaled $1,340. What were his total earnings? $350

14. Emily earns $775 per mo plus a $3\frac{3}{4}\%$ commission on sales over $5,000. What were her total earnings last month when she had sales of $14,250? $1,121.88

EXTENSION Graduated Commission

In some situations, a different rate of commission is paid for different levels of sales. This is called **graduated commission.**

Each month, Lance receives a graduated commission of 8% on his first $5,000 of sales, 10% on his next $5,000, and 12% on sales over $10,000. Last month, Lance's sales totaled $14,250.

1. What was Lance's commission on the first $5,000? $400
2. What was Lance's commission on the next $5,000? $500
3. How much in sales over $10,000 did Lance have last month? $4,250
4. How much commission did Lance receive on the sales over $10,000? $510
5. What was Lance's total commission for the month? $1,410

SKILLS BANK You may want to refer students to the following skills in the Skills Bank:

Equivalent Fractions, Decimals, and Percents, Skill 4

Find the Percent of a Number, Skill 2

FOR DISCUSSION Students should explain that it could depend on expected sales. If sales were high, you would be better off with the 7% straight commission. However, if sales were questionable or seasonal, you may be better off with an assured base salary.

EXTENSION Explain how graduated commission provides an incentive to produce greater sales and how it is computed.

ASSIGNMENTS

Level	Exercises
Level 1	1–9, 11
Level 2	1–13
Level 3	9–14, EX

FOLLOW-UP ACTIVITY

COOPERATIVE LEARNING
Have students work in small heterogeneous groups to gather information from members of the community about various commission rates for various positions. Have each group estimate the value of an average sale for each position. Ask them to compute the average number of sales needed in each position to earn a commission of $1,000, of $10,000, and of $100,000.

Ask each group to prepare a table that summarizes their research and their answers. Display the tables and ask the groups to compare findings.

SUPPLEMENTARY MATERIALS

TRP Practice, p. 26
TRP Reteaching, p. 23
TRP Transparency 13

OBJECTIVES

- Compute net pay, given gross pay and payroll deductions.
- Compute annual payroll deductions.

TEACHING THE LESSON

WARM-UP Write the following exercises on the chalkboard or an overhead transparency.

Multiply.

1. 12 × \$3.75 **2.** 24 × \$18.50

3. 52 × \$42.00 **4.** 52 × \$25.60

Find the answer to the nearest percent.

5. 20 is what percent of 80?

6. \$250 is what percent of \$380?

(**1.** \$45 **2.** \$444 **3.** \$2,184
4. \$1,331.20 **5.** 25% **6.** 66%)

INTRODUCTION Using the earnings statement on page 68, discuss how the \$376 earned is reduced by payroll deductions. Discuss how payroll deductions can take away a significant amount of a paycheck and discuss payroll deductions over which you have some control. Ask students with part-time jobs to describe what happens to their earnings on their paychecks.

INSTRUCTION Use the earnings statement and Example 1 to show how gross earnings are reduced by payroll deductions to arrive at net pay, or take-home pay. Explain the purpose of each deduction on the earnings statement and whether or not it is an automatic deduction. Discuss how the formula works and explain the difference between gross pay and net pay.

Use Example 2 to show how much is deducted for each deduction over the period of a year. Discuss the psychology of weekly deductions of \$54.40 that you never really see as opposed to getting an annual bill of \$2,828.80 for Federal Withholding Tax.

4.6 PAYROLL DEDUCTIONS AND NET PAY

For a 40-h work week, you are paid a **gross pay** of \$376.00. Your "take-home pay," or **net pay,** was only \$253.44 because of **payroll deductions.**

City of Hampton
Employee *Your Name*

Employee Earnings Statement
Social Security # 000-00-0000

Earnings		*Taxes/Deductions*	
Regular Earnings:	\$376.00	Federal Taxes:	\$54.40
		FICA:	28.24
Overtime Earnings:	0.00	State Tax:	9.52
		Insurance:	17.60
Gross Earnings:	\$376.00	Retirement:	8.30
		Union Dues:	4.50
Week ending 10/25		NET PAY:	\$253.44

EXAMPLE 1 How much was deducted from your gross pay?

THINK: Gross pay: \$376.00 Net pay: \$253.44

GROSS PAY − NET PAY = TOTAL DEDUCTIONS

\$376.00 − \$253.44 = \$122.56

A total of \$122.56 was deducted from your gross pay.

The two highest deductions are for **Federal Withholding Tax** and for **Social Security Tax (FICA).**

EXAMPLE 2 How much will be deducted from your pay over the course of a year (52 weekly paychecks) for each tax?

THINK: Weekly Federal Withholding: \$54.40
Weekly Social Security: \$28.24

Multiply to find each annual deduction.

Federal Withholding Tax → 52 × \$54.40 = \$2,828.80
Social Security Tax → 52 × \$28.24 = \$1,468.48

One way to judge the size of your payroll deductions is to compute the percentage of your gross pay that you take home.

EXAMPLE 3 What percent of your gross pay is your net pay?

THINK: \$253.44 is what percent of \$376.00?

1. Divide to find a decimal. \$253.44 ÷ \$376.00 = 0.6740425
2. Rename the decimal as a percent. 0.6740425 = 67.40425%

Your net pay is about 67%, or about $\frac{2}{3}$ of your gross pay.

FOR DISCUSSION See TE side column.

1. Why is Federal Withholding Tax deducted from your pay?
2. Why is Social Security Tax deducted from your pay?

PRACTICE EXERCISES

Remember to estimate whenever you use your calculator.

1. $138.60 2. $109.98 3. $314 4. $209.52 5. $332.50

Complete the table. 6. $215.86 7. $115.81 8. $90.99 9. $83 10. $74.77

Employee	Hours worked	Hourly wage	Gross pay	Federal tax	Social Security	Other deductions	Net pay
Burbridge	36	$3.85	• 1. ■	$14.71	$10.41	$3.50	• 2. ■
Samuelson	40	$7.85	3. ■	$56.37	$23.58	$24.53	4. ■
Alloca	35	$9.50	5. ■	$60.42	$24.97	$31.25	6. ■
O'Donnell	$27\frac{1}{4}$	$4.25	7. ■	$9.75	$8.70	$6.37	8. ■
Garcia	$15\frac{3}{4}$	$5.27	9. ■	$0	$6.23	$2.00	10. ■

Josh works $37\frac{1}{2}$ h per wk at $10.73 per h. His Federal Withholding Tax is $75.40 and his Social Security Tax is $30.22. He also has $8.50 deducted each week for union dues, and he pays $15.30 per wk for insurance. Complete the earnings statement below for Josh.

Employee Josh Evans | Social Security # 000-00-0000

Earnings		*Taxes/Deductions*	
Regular Earnings:	• 11. ■ $402.38	Federal Taxes:	• 13. ■ $75.40
Overtime Earnings:	0.00	FICA:	14. ■ $30.22
Gross Earnings:	12. ■ $402.38	Insurance:	15. ■ $15.30
		Union Dues:	16. ■ $8.50
Week ending 11/18		NET PAY:	17. ■ $272.96

What is Josh's annual deduction for:

•18. Federal Withholding Tax? $3,920.80

19. Social Security Tax? $1,571.44

20. Insurance? $795.60

21. What percent of his gross pay does Josh take home? 68%

Solve.

22. Ken has $4.30 deducted twice a month for union dues. How much is deducted per year? $103.20

23. Robin's net weekly pay is $248.55. If her gross pay is $325.00, what percent of her gross pay is her net pay? 76%

Use Example 3 to show how to compute what percent net pay is of gross pay—in other words, what percent of gross pay is actually taken home.

SKILLS BANK You may want to refer students to the following skills in the Skills Bank:

Subtract With Whole Numbers and Decimals, Skill 6

Multiply With Decimals, Skills 2, 5

Find the Percent One Number Is of Another, Skill 2

Percent and Proportion, Skill 2

FOR DISCUSSION

1. Students should explain that Federal Withholding Tax deductions are prepayments on your annual federal income tax and that this money helps fund the activities of the federal government.
2. Students should explain that Social Security Taxes are worker contributions to the Social Security fund that pays benefits to older, retired, or disabled workers.

ASSIGNMENTS

Level	Exercises
Level 1	1–19, 21
Level 2	1–23
Level 3	11–23

FOLLOW-UP ACTIVITY

ENRICHMENT Create a simulated earnings statement including gross earnings, several deductions, and net pay. Delete the numbers and provide students with a copy of the numberless earnings statement and a list of the missing numbers. Ask students to use logic and their understanding of deductions to correctly place the numbers on the earnings statement.

SUPPLEMENTARY MATERIALS

TRP Practice, p. 27
TRP Reteaching, p. 24
TRP Transparency 14

OBJECTIVES

- Compute insurer's and insured's shares of medical bills.
- Compute insured's share of health insurance costs.

TEACHING THE LESSON

WARM-UP Write the following exercises on the chalkboard or an overhead transparency.

Find the answer.

1. a. 20% of $450 **b.** 80% of $450

2. a. 30% of $870 **b.** 70% of $870

3. a. 25% of $2,280 **b.** 75% of $2,280

(**1a.** $90 **1b.** $360 **2a.** $261
2b. $609 **3a.** $570 **3b.** $1,710)

Explain that the **b.** exercise in each pair could be done by subtracting the **a.** answer from the total. Discuss why this alternative approach works.

INTRODUCTION Ask the class how their medical bills get paid if they go to the doctor or to the hospital. Answers should vary from "we pay cash" to "insurance covers it all" to many in-between possibilities. Use these answers to explain that health insurance is a common fringe benefit of employment and that health insurance policies vary widely. Have students relate their family experiences with large medical bills as a lead-in to a discussion of health insurance.

INSTRUCTION Use Example 1 to explain the concept of *deductible* and to find both the insurance company's and the insured's share of a medical bill. Students are sometimes confused over when to subtract the deductible. Remind them that the deductible amount represents the first dollars in any bill or claim and must be paid by you regardless of the size of the bill or claim. So the deductible amount is part of your share and should be subtracted first. Remind students that since the insurance company pays 80%, your share is 20% of $1,170 plus the $75 deductible.

4.7 HEALTH INSURANCE

Most employees receive some form of **health insurance** as a "fringe benefit" of their employment. Their employer pays all or part of the cost of this insurance, which usually covers the cost of doctors, hospitals, and medicine, and sometimes of dental and eye care.

A common feature of health-insurance policies is a **deductible**—the amount that the person insured must pay. The insurance company then pays all or part of the remaining balance.

EXAMPLE 1 Your employer provides you with health insurance that has a $75 deductible. The insurance company will then pay 80% of the remaining medical expenses up to $20,000. After minor surgery, you incur medical expenses of $1,245. How much will your insurance carrier pay, and how much will you pay?

THINK: The insurance company will pay 80%, or 0.8, of the costs after subtracting the deductible.

1. Find the insurance company's share.
   - Subtract the deductible. $1,245 − $75 = $1,170
   - Multiply by the rate. 0.8 × $1,170 = $936

The insurance company will pay $936 of the bill.

2. Subtract to find your share. $1,245 − $936 = $309

Your share of the bill will be $309.

Sometimes the cost of health insurance is split between an employee and his or her employer. The employee's share is taken out as a **payroll deduction.**

EXAMPLE 2 Norma's company pays 60% of the cost of her family-coverage health insurance. The monthly premium for the insurance is $147.20. What is Norma's monthly payroll deduction for health insurance? How much does she pay per year for health insurance?

THINK: Her employer pays 60%.
Norma's share is 40%, or 0.4, of the premium.

1. Multiply to find Norma's share. 0.4 × $147.20 = $58.88

Norma's monthly deduction is $58.88.

2. Multiply to find the annual cost. 12 × $58.88 = $706.56

Norma pays $706.56 per y for health insurance for her family.

FOR DISCUSSION See TE side column.

1. Why do health-insurance policies usually have deductibles?

2. Why is health insurance usually less expensive when purchased for a large group than when purchased for an individual?

PRACTICE EXERCISES

Remember to estimate whenever you use your calculator.

Complete the table. 1. $300 2. $175 3. $0 4. $88 5. $136.50 6. $108.50 7. $2,146.50 8. $338.50 9. $442.65 10. $303.35

Medical bill	Deductible amount	% of Coverage over deductible	Amount paid by insurer	Amount paid by you
$475	$100	80%	• 1. ■	• 2. ■
$88	$100	80%	3. ■	4. ■
$245	$50	70%	5. ■	6. ■
$2,485	$100	90%	7. ■	8. ■
$746	$65	65%	9. ■	10. ■

Complete the table.

Monthly premium for insurance	% Paid by employer	Monthly deduction for insurance	Annual cost	
$38.70	50%	•11. ■ $19.35	•12. ■	$232.20
$46.00	80%	13. ■ $9.20	14. ■	$110.40
$128.50	75%	15. ■ $32.13	16. ■	$385.56
$72.80	90%	17. ■ $7.28	18. ■	$87.36

Solve.

•19. Terry's insurance pays 80% of all costs. Last year, Terry ran up doctor bills of $47.50, $73.00, and $26.75. How much did Terry pay her doctor? $29.45

20. Dan's bills were $275 for a doctor, $897 for the hospital, and $92.30 for medicine. If his policy has a $50 deductible and pays 80% of all costs over $50, how much did Dan pay? $292.86

21. From each of Janet's weekly paychecks, $4.50 is deducted for health insurance. If her employer is paying 80% of the cost of this insurance, what is the total annual premium for Janet's health insurance? $1,170

22. Ted's health insurance costs a total of $87.50 per mo, but his employer pays 70% of the premium. If Ted paid $283.50 for medical expenses last year, how much did he pay for health care, including insurance? $598.50

Use Example 2 to give students some sense of how expensive health insurance is. Explain that, while many individuals must purchase health insurance on their own, many others share the cost with their employer.

SKILLS BANK You may want to refer students to the following skill in the Skills Bank:

Find the Percent of a Number, Skill 2

FOR DISCUSSION

1. Students should explain that because there are many routine and relatively inexpensive medical expenses, the deductible allows these to be paid by you and leaves the insurance for greater expenses. Since with a deductible there will always be some cost to you for health care, people will be less inclined to rush to the doctor.
2. Students should explain that when pricing policies for large groups, insurance companies can use probabilities and prior experience to predict the kinds of medical expenses to expect. The risk can be more evenly shared.

ASSIGNMENTS

Level 1	1–18
Level 2	1–21
Level 3	5–10, 15–22

FOLLOW-UP ACTIVITY

APPLICATION Provide students with copies of several real hospital bills and copies of several health-insurance contracts. Apply the provisions of the different contracts to the bills and discuss the characteristics of the best health-insurance policies.

SUPPLEMENTARY MATERIALS

TRP Practice, p. 28
TRP Reteaching, p. 25
TRP Transparency 14

OBJECTIVES

- Identify the differences between term and straight life insurance.
- Compute life-insurance premiums using rate tables.

TEACHING THE LESSON

WARM-UP Write the following exercises on the chalkboard or an overhead transparency.

How many 1,000s are there in:

1. 20,000? (20) **2.** 25,000? (25)

3. 30,000? (30) **4.** 35,000? (35)

5. 75,000? (75) **6.** 100,000? (100)

INTRODUCTION Ask students to give reasons for buying life insurance. Discuss the desire to provide financial protection for one's family in case of the death of the income producer(s). Life insurance is a common fringe benefit of employment.

INSTRUCTION You may want to provide students with a brochure from a local bank explaining the provisions of savings bank life insurance. Discuss the similarities and differences between term and whole, or straight, insurance and define the key insurance terms on page 72.

Discuss the insurance-rate table and point out that rates increase with age. Also discuss why insurance rates for females are lower than for males the same age. (In general, females tend to live longer than males.)

Use the Example to show how to use the rate table to compute insurance premiums. Remind students that there are 30 $1,000s in $30,000. Have students compute the premiums for other policies to illustrate the difference in costs between term and straight life policies, males and females, and older and younger people.

SKILLS BANK You may want to refer students to the following skill in the Skills Bank:

Multiply With Decimals, Skill 5

4.8 LIFE INSURANCE

People buy **life insurance** for themselves or for others to provide financial protection against loss of income in the event that the person insured dies.

Key life insurance terms are:

Beneficiary—the person who receives the insurance money
Face value—the amount of money received from the policy

There are 2 major types of life insurance:

Term insurance policies provide protection for a limited period of time—usually 5, 10, 15, or 20 y—and can then be renewed at higher premiums. Term insurance is the simplest and least expensive form of life insurance.

Straight life insurance policies provide permanent lifetime protection for as long as you pay your premiums.

Life insurance rates vary depending on the age and gender of the insured person, and the face value of the policy.

SOME ANNUAL PREMIUMS FOR $1,000

Age at issue		Term Insurance		Straight life
Male	Female	10-Year	20-Year	
20	23	$2.86	$3.90	$13.67
25	28	$3.27	$5.74	$15.78
30	33	$3.72	$7.83	$18.87
35	38	$4.42	$10.22	$23.37
40	43	$5.87	$14.58	$28.51
45	48	$7.42	$18.46	$32.17

EXAMPLE Ronald is 25 y old and wants to buy $30,000 of insurance. What would his annual premium be for 10-y term insurance? For straight life?

1. Multiply to find the annual premium for 10-y term. $30 \times \$3.27 = \$98.10$

Ronald's annual premium would be $98.10 for 10-y term insurance.

2. Multiply to find the annual premium for straight life. $30 \times \$15.78 = \$473.40$

Ronald's annual premium would be $473.40 for straight life insurance.

FOR DISCUSSION See TE side column.

1. Why do term insurance rates increase as you get older?
2. Why are the premiums for straight life insurance higher than for term insurance?

PRACTICE EXERCISES

Remember to estimate whenever you use your calculator.

Use the premium table to compute the annual premium for each policy.

	Type of policy	Age	Gender	Face value	
• 1.	10-y term	25	Male	$35,000	$114.45
• 2.	Straight life	28	Female	$50,000	$789
3.	Straight life	20	Male	$25,000	$341.75
4.	20-y term	23	Female	$40,000	$156
5.	10-y term	30	Male	$75,000	$279
6.	Straight life	40	Male	$75,000	$2,138.25
7.	20-y term	38	Female	$50,000	$511
8.	Straight life	48	Female	$50,000	$1,608.50

Solve.

9. How much less per year would a 35-year-old male pay for $25,000 of 20-y term insurance than for straight life? $328.75

10. Jack purchased $150,000 of 20-y term insurance at age 25. What will his total premiums be over the 20 y? $17,220

EXTENSION Decreasing Term Insurance

As people get older, their need for financial protection often decreases. For this reason, **decreasing term insurance** is also available for less cost than level term insurance. As each year passes, the amount of insurance coverage decreases.

1. Sam bought $40,000 of 10-y decreasing term insurance. How much insurance coverage will he have in the 4th year of the policy? $29,200
2. Rachel bought $60,000 of 10-y decreasing term insurance. How much insurance coverage will she have in the 7th year of the policy? $27,600

COVERAGE PER $1,000 INITIAL VALUE 10-Y DECREASING TERM INSURANCE

Year	Coverage	Year	Coverage
1	$1,000	6	$555
2	$940	7	$460
3	$860	8	$340
4	$730	9	$250
5	$660	10	$140

FOR DISCUSSION

1. Students should explain that because the probability of your dying increases each year as you get older, the likelihood of the insurance company having to pay death benefits also increases. Therefore, your rates and your annual premium increase.
2. Students should explain that, unlike term insurance, your straight life premiums are covering more than just death benefits. Your premiums build up like a savings account and eventually you can decide to cash the policy in and receive this money. This savings feature adds to the cost of the policy relative to term insurance.

EXTENSION Describe decreasing term insurance. Review with students the possible advantages of purchasing this type of policy. Use the table to compute the face value of a $50,000 policy in each of the 10 y the policy is in force.

ASSIGNMENTS

Level 1	Odd 1–9
Level 2	1–10
Level 3	5–10, EX

FOLLOW-UP ACTIVITY

ENRICHMENT Provide students with a copy of an actual life insurance contract to read. Have each student prepare a list of questions on parts of the contract that he or she does not understand. Then invite a local life-insurance agent to class to answer the questions and to further describe how life insurance works and how to be a wise insurance consumer.

SUPPLEMENTARY MATERIALS

TRP Practice, p. 29
TRP Reteaching, p. 26
TRP Transparency 15

OBJECTIVE

- Decide, given relevant information, which type of life insurance to purchase.

TEACHING THE LESSON

WARM-UP Write the following exercises on the chalkboard or an overhead transparency.

Multiply.

1. 50 × \$2.60 **2.** 50 × \$3.40

3. 50 × \$9.50 (**1.** \$130 **2.** \$170 **3.** \$475)

INTRODUCTION Present the problem situation to the class. Ask what key factors would influence the decision. Discuss what kind of information would be needed to make an informed decision and where this information is available. Encourage students to name some factors that might be used to determine how much life-insurance coverage a person would need. (Examples include whether one or both spouses work; whether there are any children; and whether there are prospects of future children, raises, and investments.)

INSTRUCTION Discuss the decision-making factors. Review the terms *cash value* and *renewability* as well as the differences between term and straight life insurance. Point out that nonrenewable policies require a physical examination before a new policy can be issued. Ask students how this would affect some people's decisions to choose renewable or nonrenewable policies. Read the list of options that Jeff and Susan have prepared. Ask if there is any other information that they should have.

DECISION MAKING

4.9 BUYING INSURANCE

(Open-ended problem solving)

Two of the hardest decisions young people must make are deciding what type of life insurance to purchase and how much coverage to get.

PROBLEM

Jeff and Susan are both 20 y old, and they were just married. Susan has life insurance through her company; Jeff does not. They believe that while they are young and healthy, and before they have children, it would be wise to purchase life insurance for Jeff on their own. They need to decide what type of insurance they can afford at this time.

DECISION-MAKING FACTORS

Type of policy   Cost of policy   Term of policy

Cash value is the monetary value of your policy if, at some time, you choose to cancel it.

Renewable policies can be automatically renewed for additional periods of time without requiring a physical examination.

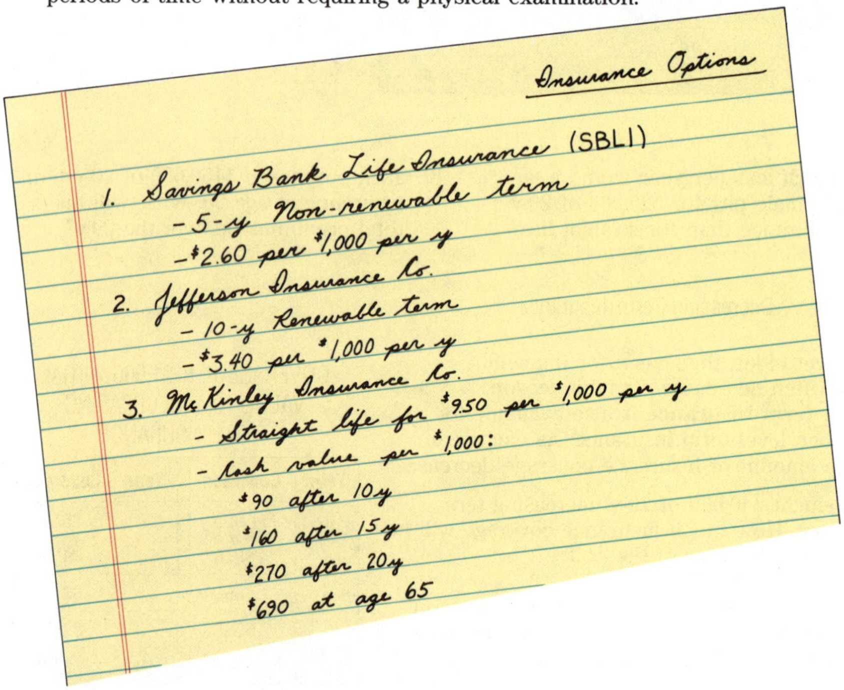

DECISION-MAKING COMPARISONS

Compare the 3 options by completing the table.

Factor	SBLI	Jefferson	McKinley	
Type of policy	• 1. ■	term	2. ■	term; straight life
Cost per $1,000	$2.60	• 3. ■	4. ■	$3.40; $9.50
Term of insurance	5. ■	10 y	6. ■	5 y; life
Cash value—yes or no	7. ■	8. ■	yes	no; no
Renewable—yes, no, or not applicable	no	9. ■	10. ■	yes; not applicable

MAKING THE DECISIONS

•11. Jeff and Susan decide that Jeff should have $50,000 worth of insurance. What is the annual premium for a $50,000 policy under each option?
$130; $170; $475

•12. Which policy should they purchase if cost is the only factor?
SBLI

13. What is the cash value of a $50,000 straight life policy from McKinley after 10 y? Remember: Cash values are per $1,000.
$4,500

14. What is the net cost of this McKinley policy if Jeff and Susan deduct the cash value from the total premiums paid over 10 y?
$250

15. Is the net cost of the McKinley policy more or less expensive than 10 y of the Jefferson policy? less expensive

16. Jeff learns that when he is 5 y older, the cost of a new SBLI policy will rise to $4.90 per $1,000 per y. Compare the total cost of the SBLI and the Jefferson policies over the 10 y. Which is the better buy? Remember: For 10 y, Jeff will need 2 consecutive SBLI policies of 5 y each.
Jefferson policy

17. If you were Jeff, which policy would you purchase? Why? Answers may vary.

Have small groups of students complete Exercises 1–10 in the table in order to compare the 3 policies. Ask them to write statements that describe the data now in the table. Then ask them to write conclusions that can be drawn from the data. Discuss the appropriateness of these statements and conclusions and the possible misinterpretations of the data.

Have students complete Exercises 11–17 in small groups, having students explain and defend their answers and their decisions. Some students are very wary of defending their point of view among their classmates. Continually reinforce the fact that, in many consumer issues, best answers depend upon individual situations and preferences. Encourage all students to participate and to express their opinions.

ASSIGNMENTS

Level	Exercises
Level 1	1–17
Level 2	1–17
Level 3	1–17

FOLLOW-UP ACTIVITY

COOPERATIVE LEARNING
Provide students with insurance brochures from banks, insurance agents, or the mail. Working in small heterogeneous groups, have students recreate Jeff and Susan's type of research and, after making a comparison table, come to a written group consensus on which policy to purchase and at what face value.

SUPPLEMENTARY MATERIALS

TRP Practice, p. 30
TRP Transparency 15

OBJECTIVES

- Review vocabulary.
- Practice key chapter concepts and skills.

USING THE REVIEW

The Chapter Review is designed to help students prepare for taking the Chapter Test. The first section focuses on vocabulary. It requires students to select words to complete statements. The second section presents practice of key chapter skills.

Each item on the Chapter Review is referenced to the page on which the topic is taught in the Pupil's Edition. You may wish to have students refer to these pages to help review any concepts or skills they have not yet mastered.

It is suggested that students work in small-sized heterogeneous cooperative learning groups. Some cooperative learning methods that may be used are as follows:

1. After each student has independently completed the Chapter Review, a discussion should follow within each group about the solutions to the problems.
2. The group can complete the entire Chapter Review by working together to solve the problems.

End the lesson with an entire class discussion in which any questions brought up in group discussions are presented and answered.

CHAPTER REVIEW

Vocabulary Choose the letter of the word(s) that completes the sentence.

1. For work above 40 h per wk, many employees often receive $1\frac{1}{2}$ times their regular wage, or ■. [58] a

   a. Overtime pay **b.** Hourly wage **c.** Piecework

2. An amount of money paid for selling a product or service is called ■. [66] b

   a. Regular pay **b.** Commission **c.** Annual salary

3. After payroll deductions, most people's ■ is much less than their gross pay. [68] c

   a. Biweekly pay **b.** Social Security deduction **c.** Net pay

Skills Find the answer.

4. Celia worked 40 h at $5.40 per h. What was her regular pay? [58] $216.00
5. Joe worked $37\frac{1}{2}$ h at $7.50 per h. What was his regular pay? [58] $281.25
6. Ellen worked $44\frac{1}{2}$ h at $6.20 per h. Her overtime rate is $1\frac{1}{2}$. What was her total pay? [58] $289.85
7. Bob earns an annual salary of $14,500. What is his weekly pay? [62] $278.85
8. Ted earns an annual salary of $23,500. What is his biweekly pay? [62] $903.85
9. Sharon's weekly pay is $385. What is her annual salary? [62] $20,020
10. Sarah earns a piece rate of $12.50 per delivery. How much did she earn by making 7 deliveries? [64] $87.50
11. Eli earns a 15% commission on sales. How much did he earn on sales of $3,680? [66] $552.00
12. David earns a $6\frac{1}{2}$% commission on sales. How much did he earn on sales of $159,000? [66] $10,335
13. John's gross pay was $548.96. His net pay was $458.75. How much was deducted from his paycheck? [68] $90.21
14. Kelly has $89.30 deducted each week for Social Security. How much is deducted in a year? [68] $4,643.60
15. Calvin earns $8.25 per h for a 40-h week. If his deductions total $71.90, what is his net pay? [68] $258.10
16. Your health insurance has a $50 deductible, after which it pays 80% of all costs. How much will your insurance pay on a $265 bill? [70] $172
17. At $12.84 per $1,000, what is the annual premium for a $75,000 straight life insurance policy? [72] $963

CHAPTER TEST

Find the earnings.

1. Hours worked: 38
   Hourly wage: $5.35
   $203.30
2. Hours worked: 42
   Hourly wage: $7.45
   Overtime rate: $1\frac{1}{2}$
   $320.35
3. Hours worked: $47\frac{1}{2}$
   Hourly wage: $6.60
   Overtime rate: $1\frac{1}{2}$
   $338.25
4. Hours worked: 5, $4\frac{1}{2}$, $8\frac{3}{4}$
   Hourly wage: $18.75
   $342.19
5. Hours worked: 3, $6\frac{1}{2}$, $5\frac{3}{4}$, $8\frac{1}{4}$
   Hourly wage: $20.50
   $481.75
6. Hours worked: 7:15 A.M. to 3:30 P.M.
   Hourly wage: $4.58
   $37.79
7. Hours worked: 8:00 A.M. to 5:30 P.M.
   Hourly wage: $7.25
   $68.88
8. Piece rate: $7.50 Items produced: 73 interviews $547.50
9. Commission rate: 30¢ Total sales: 850 boxes $255
10. Commission rate: $4\frac{1}{2}$% Total sales: $4,325 $194.63

Solve.

11. June packs an average of 12 cartons per h. She is paid 60¢ per carton. About how much does she earn for a 35-h work week?
    $252
12. Miguel worked 10 h and earned $175 constructing fencing. If he averages 8 yd per h, how much is he paid for each yard of fencing?
    $2.19
13. Grace's annual salary is $12,680. What is her weekly pay?
    $243.85
14. Dave's biweekly pay is $793.68. What is his annual salary?
    $20,635.68
15. Dawn works a 40-h week at $6.34 per h. If her deductions total $86.47, what is her net pay?
    $167.13
16. Lee's gross pay last week was $493.54. If his net pay was $386.47, how much was deducted from his pay?
    $107.07
17. Hal's health insurance pays 80% of his medical bills with a $100 deductible. How much did he pay on a $2,850 bill?
    $650
18. Tracy's employer pays 75% of the cost of her health insurance. How much does she pay per year if the total biweekly premium is $35.70?
    $232.05
19. Rita has $80,000 straight life insurance. It costs $3.87 per $1,000. What is her annual premium?
    $309.60
20. Diane has $50,000 10-y term life insurance. It costs $5.32 per $1,000. How much will she pay over the 10-y period?
    $2,660

OBJECTIVE

- Evaluate the achievement of chapter objectives.

USING THE TEST

The Chapter Test may be used as a post-test to evaluate student achievement. However, you may wish to use the Chapter Post-Test offered in the Teacher's Resource Package or to design your own chapter test. If this page is not used as a test, you may want to assign it as additional review or practice.

The test items are correlated to the chapter objectives in the table below.

Chapter objectives	Test items
A. Compute regular pay and overtime pay.	1–3
B. Compute pay from time sheets and time cards.	4–7
C. Compute piecework pay.	8, 11–12
D. Compute commissions.	9–10
E. Compute weekly, biweekly, semimonthly, and monthly pay, and annual salary.	13–14
F. Compute net pay and payroll deductions.	15–16
G. Compute health insurance costs.	17–18
H. Compute life insurance costs.	19–20

SUPPLEMENTARY MATERIALS

TRP Ch. 4 Post-Test Form A, pp. 1–2
TRP Ch. 4 Post-Test Form B, pp. 1–2

OBJECTIVE

- Understand that there are some financial benefits that company employees receive but self-employed people do not.

USING THE PAGE

Have students read the introductory material and information given in the table. Point out to students that self-employment tax is only paid in certain areas. Also tell students to estimate answers to Exercises 1 and 4 by doubling the Social Security company employees pay to find the amount self-employed people pay.

Have students work individually or in pairs to answer Exercises 1–3. Summarize the answers as follows: Even at the same salary, company employees could take home more pay because their companies pay many expenses for them.

Have students work individually to answer Exercises 4–5. Then have them share their answers and reasons in a whole-class discussion.

ANSWERS/EXPLANATIONS

3. Self-employed people pay other expenses that company employees do not. However, self-employed people have expenses (telephone costs, office supplies, and so forth) that they can deduct from income taxes.
4. Increased by $498.56. By staying with the agency, you would have paid $961.28 (7.51% of $12,800). Otherwise, you would pay $1,459.84 ($480.64 company + $979.20 self-employed).
5. Company employees are paid for a full day, even though they do not work every minute of the day. They are also paid for sick days and vacations, and receive disability payments and pension plans. Self-employed people are paid only for actual working time and must plan for their own retirement.

MONEY TIPS

Company-employed people pay less, or nothing at all, for certain benefits that self-employed people must pay for in full.

LET'S LOOK AT THE FACTS

As a company employee, you pay only 7.51% of your wages for Social Security. Your company pays the rest. As a self-employed person, you must pay 15.3%. There are also other financial advantages to working for a company. The table below shows some common deductions and expenses, not including basic income taxes.

INCOME DEDUCTIONS AND EXPENSES

	Company-employed	Self-employed
Gross salary	$25,000	$30,000
Social Security	$1,878	$4,590
Self-Employment Tax	None	$800
Health/Dental plan	Company paid	$1,800
Amount left over	$23,122	$22,810

LET'S DISCUSS WHY See TE side column for answers to Exercises 3–5.

1. If a self-employed person earned $25,000, how much more Social Security would he or she have to pay than the company-employed person?
$1,947.50
2. You are a company employee earning $40,000 per y. A self-employed friend also earns $40,000, but pays $1,200 self-employment tax and $1,600 for a health plan. How much will each of you have left over?
Company employee: $36,996 Self employed: $31,080
3. A self-employed person could take home less pay than a company-employed person even though he or she earns a higher gross salary. Why?

LET'S SEE WHAT YOU WOULD DO

4. June 30 is your last day with the Murray Hill Advertising Agency. Your salary up to that point totals $6,400. You become a freelance writer and earn another $6,400 by December 31. Has your annual Social Security deduction been increased or decreased? By how much?
5. Besides the benefits discussed above, can you think of any other situations in which a company-employed person might come out ahead financially?

ESTIMATION SKILLS

FRONT-END ESTIMATION OF SUMS

One way to estimate sums is by using **front-end estimation.**

1. Find the greatest number(s) to be added.
2. Identify the place of its leading nonzero digit.
3. Add only the digits in that place.

Examples

②1,284	$①.37	0.⑥32
2,487	④.25	0.043
+ ③8,372	+ 0.63	+ 0.②1
⑤	⑤	⑧
Estimate: 50,000	**Estimate:** $5.00	**Estimate:** 0.8

To get a more precise estimate, front-end estimation can be adjusted by *also* using the next greatest place.

Examples

②[1],284	$①.[3]7	0.⑥[3]2
[2],487	④.[2]5	0.0[4]3
+ ③[8],372	+ 0.[6]3	+ 0.②[1]
First estimate: ⑤0,000	**First estimate:** $⑤.00	**First estimate:** 0.⑧
+ about [1][0],000	+ about [1].[0]0	+ about 0.[1][0]
Final estimate: 60,000	**Final estimate:** $6.00	**Final estimate:** 0.9

Use front-end estimation with adjusting to estimate the sum.

1. 206 + 317 + 182	**2.** 3,605 + 294 + 7,413	**3.** 40,056 + 23,214 + 10,800	**4.** 7,265 + 1,207 + 23,666	**5.** 25,397 + 4,113 + 62,909
Answer: 700	Answer: 11,000	Answer: 70,000	Answer: 30,000	Answer: 90,000
6. $2.45 + 3.72 + 1.05	**7.** $1.10 + 8.57 + 8.39	**8.** $7.55 + 0.09 + 3.52	**9.** $0.42 + 0.08 + 0.32	**10.** $0.01 + 0.33 + 0.91
Answer: $7.00	Answer: $18.00	Answer: $11.00	Answer: $0.80	Answer: $1.20
11. 0.513 + 0.259 + 0.157	**12.** 0.724 + 0.08 + 0.238	**13.** 0.02 + 0.632 + 0.24	**14.** 0.0212 + 0.0337 + 0.0043	**15.** 0.006 + 0.04 + 0.0348
Answer: 0.9	Answer: 1.0	Answer: 0.9	Answer: 0.06	Answer: 0.08
16. 113 + 206 + 545 + 147	**17.** 31,873 + 4,071 + 21,343 + 34,902	**18.** $7.10 + 5.05 + 2.14 + 3.07	**19.** $0.43 + 0.05 + 0.20 + 0.21	**20.** 0.0327 + 0.011 + 0.0052 + 0.011
Answer: 1,000	Answer: 90,000	Answer: $17.00	Answer: $0.90	Answer: 0.06

OBJECTIVE

- Estimate sums, using front-end estimation.

USING THE PAGE

A simple first estimate made by the front-end method is particularly useful when a very quick approximation is required. It should be kept in mind that such an estimate is never greater than the actual sum. The greater the number of addends, the greater the chance that a simple front-end estimate will be considerably less than the actual sum.

When a closer estimate is required, it is worthwhile to adjust the first estimate by using the digits in the next greatest place. You might have students compare the actual sums for the 3 examples on the page with the first estimates and with the final estimates.

	21,284
	2,487
	+38,372
Actual sum:	62,143
First estimate:	50,000
Final estimate:	60,000

	$1.37
	4.25
	+ 0.63
Actual sum:	$6.25
First estimate:	$5.00
Final estimate:	$6.00

	0.632
	0.043
	+0.21
Actual sum:	0.885
First estimate:	0.8
Final estimate:	0.9

OBJECTIVES

- Review and maintain previously taught concepts and skills.
- Practice taking tests in multiple-choice format.

USING THE REVIEW

Assign the Cumulative Review to all students. Provide students with an answer sheet to record their answers.

Each Cumulative Review gives students an opportunity to practice taking tests that are written in multiple-choice format.

Provide appropriate remedial help for students having difficulty with any of the skills and concepts on this page.

CUMULATIVE REVIEW

Choose the letter of the word(s) that completes the sentence.

1. Percent means ■. c

   a. Tenths **b.** Per ten **c.** Per hundred **d.** None of these

2. When a group of numbers is arranged in order from least to greatest, the middle number is called the ■. b

   a. Mean **b.** Median **c.** Mode **d.** None of these

3. To estimate with whole numbers and decimals, you can round each number to its ■ place. b

   a. Last place **b.** Greatest nonzero place
   c. Least nonzero place **d.** None of these

4. The pay you earn is found by multiplying the number of hours worked by the ■. a

   a. Hourly wage **b.** Tips
   c. Number of days worked **d.** None of these

5. An amount of money paid for selling a product or service is called ■. c

   a. Regular pay **b.** Overtime pay
   c. Commission **d.** None of these

Select the best estimated answer.

6. 342 + 476 + 55 c	**a.** 600	**b.** 760	**c.** 850	**d.** 1,050
7. 698 − 27 b	**a.** 600	**b.** 670	**c.** 730	**d.** 770
8. 6 × 289 c	**a.** 180	**b.** 900	**c.** 1,800	**d.** 3,000
9. 4,272 ÷ 12 b	**a.** 40	**b.** 400	**c.** 450	**d.** 550

Compute.

10. 9.3 + 7.7 + 0.6 c

    a. 16.3 **b.** 17.3 **c.** 17.6 **d.** None of these

11. 31.84 − 16.9 a

    a. 14.94 **b.** 15.85 **c.** 15.14 **d.** None of these

12. 7 × 893 d

    a. 5,640 **b.** 5,651 **c.** 6,151 **d.** None of these

13. $7\overline{)21.98}$ b

a. 3.04 b. 3.14 c. 3.98 d. None of these

14. 20% of 40 a

a. 8 b. 40 c. 80 d. None of these

Use the ad for Exercises 15–16.

15. About how much will 4 pads and 3 markers cost? c

a. $3.00 b. $3.50 c. $4.20 d. None of these

16. About how many tapes can you buy for $3.00? b

a. 2 b. 3 c. 4 d. None of these

Solve.

17. Sally worked 8 h today and earned $23.50 in tips. She earns $2.75 per h. What is her total pay? c

a. $22 b. $26.25 c. $45.50 d. None of these

18. The cost of your dinner is $13. You want to leave a 15% tip. How much money should you leave for the tip? d

a. $1.30 b. $14.95 c. $19.50 d. None of these

19. Ben's biweekly pay is $1,256.25. What is his annual salary? b

a. $30,150 b. $32,662.50 c. $65,325 d. None of these

20. Stephanie works in a restaurant. A customer gave her $1.85 in quarters and nickels. She was given 13 coins. How many coins of each kind did she receive? a

a. 6 quarters and 7 nickels b. 7 quarters and 6 nickels
c. 5 quarters and 8 nickels d. None of these

THINKING ABOUT MATH See TE side column for answers.

1. Name situations in which estimates that are overestimated rather than underestimated are more appropriate.

2. When estimating $18\frac{3}{4} - 12\frac{1}{3}$, one person answered 7 and another person answered 10. Which different estimation method did each person use?

3. Explain job situations where it would be better to receive a straight commission of 8% rather than a weekly pay of $400 plus a 2% commission.

THINKING ABOUT MATH The questions in this section can be used in discussions with individual students or with small groups. Guide students in conveying their ideas clearly and precisely.

Listed below are expected student answers. However, accept any reasonable answers.

1. Estimates that are overestimated should be used in situations where you must know if you have enough. Some examples are when estimating
   –the cost of a purchase
   –gasoline
   –material needed for a particular job
   –time

2. The estimate of 7 is found by rounding $18\frac{3}{4}$ and $12\frac{1}{3}$ to the nearest whole numbers, $19 - 12 = 7$. The estimate of 10 is found by rounding $18\frac{3}{4}$ and $12\frac{1}{3}$ to the greatest nonzero places, $20 - 10 = 10$. The first method gives an estimated answer that is closer to the exact answer. However, the latter method is faster when a general estimate is required.

3. It would be more desirable to receive a straight commission of 6% rather than a weekly pay of $400 plus a 2% commission in the following situations:
   –when selling very expensive merchandise—such as houses, cars, and luxury items
   –when selling large quantities of moderately priced items—such as televisions, stereos, cameras, and so on

SUPPLEMENTARY MATERIALS

TRP Cum. Test, Ch. 1–3, pp. 1–4

Objective

- Test the prerequisite skills needed to learn the concepts developed in Chapter 5.

Using the Test

The Pre-Skills Test is designed to diagnose students' strengths and weaknesses on prerequisite skills necessary to study the mathematics in Chapter 5.

Assign the Pre-Skills Test. Allow students to work together in pairs or in small groups. Group members should help those who demonstrate a misunderstanding of a concept or a weakness in a skill.

The table below correlates the items on the Pre-Skills Test with the prerequisite skill and the lesson(s) in the chapter for which it is needed.

Item(s)	Prerequisite skill	Lesson(s)
1–11	Add and subtract money amounts.	5.1, 5.2, 5.3, 5.4, 5.5
12–19	Multiply with money amounts.	5.1, 5.2, 5.3, 5.4, 5.5, 5.7
20–27	Divide with money amounts.	5.3, 5.5, 5.7
28–31	Find what percent one number is of another.	5.5
32–37	Read a table.	5.1, 5.2, 5.3

References to specific skills that are reviewed in the Skills Bank are given on the individual lessons themselves.

Chapter 5 Recreation and Sports

Pre-Skills Test

Add or subtract.

1. $15 + $28 $43
2. $92.65 + $7.38 $100.03
3. $15.85 + $9.14 $24.99
4. $72.83 + $49.86 $122.69
5. $183.75 + $27.89 $211.64
6. $32 − $9.75 $22.25
7. $58.79 − $16.32 $42.47
8. $80.85 − $17.91 $62.94
9. $143.65 − $62.99 $80.66
10. $16.50 + $30 + $45.75 $92.25
11. $110 + $8.80 + $41.30 + $10 $170.10

Multiply.

12. 8 × $9.65 $77.20
13. 7 × $16.38 $114.66
14. 4 × $3.48 $13.92
15. 9 × $18.09 $162.81
16. 80 × $9.65 $772
17. 60 × $23.56 $1,413.60
18. 35 × $48.02 $1,680.70
19. 73 × $71.63 $5,228.99

Divide.

20. $84 ÷ 12 $7
21. $96 ÷ 24 $4
22. $527 ÷ $31 17
23. $972 ÷ $27 36
24. $37.60 ÷ 16 $2.35
25. $37.49 ÷ 23 $1.63
26. $71.06 ÷ 34 $2.09
27. $56.96 ÷ 64 $0.89

Find the answer.

28. 25 is what percent of 50? 50%
29. 15 is what percent of 60? 25%
30. 16.5 is what percent of 82.5? 20%
31. 56 is what percent of 14? 400%

Use the table for Exercises 32–37.

COST OF GAS PER GALLON

Month	Price
January	$1.21
February	$1.27
March	$1.19
April	$1.16
May	$1.23
June	$1.28
July	$1.31
August	$1.26
September	$1.17
October	$1.15
November	$1.19
December	$1.14

32. How much did a gallon of gas cost in February? $1.27
33. How much did a gallon of gas cost in September? $1.17
34. During which month was gas most expensive? July
35. During which month was gas least expensive? December
36. In which months did gas cost more than $1.25? February, June, July, August
37. In which 2 months did gas cost the same? March, November

Objective

- Formulate problems, using text and data.

Using the Chapter Opener

Discuss students' answers to the questions on the page. The factors that determine how students spend their free time might include schedule (after-school jobs or activities), finances (a limited budget is usually the case), availability of recreation facilities, availability of friends, and so on. A limited budget clearly limits choices. Some students will recognize that creating their own forms of recreation is feasible, as is seeking out free activities that might be available. Ask students to share their ideas about free or low-cost recreation.

Cooperative Learning

Assign heterogeneous groups of students the task of finding free or inexpensive recreational activities in their community. Tell them that each group will be responsible for entertaining an equal number of students from another part of the country for 1 wk. They should assume that these students share their own interests. Tell them that, since their budget is $10.00 each for the week, they will have to find free activities if they are to entertain their guests each day. (Note that they will not be responsible for food.) Have them use the local newspapers as aids for finding activities (Usually Friday editions carry an extended coverage of weekend and upcoming events.)

When each group presents its schedule for the week, have the students discuss each in terms of the number of activities planned, the use of free activities, the use of the $10.00, and whether they think the guests will enjoy the week.

Supplementary Materials

TRP Group Projects, p. 5

OBJECTIVES

- Compute the cost of attending a movie or show with fixed or variable ticket prices.
- Compute change.

TEACHING THE LESSON

WARM-UP Write the following exercises on the chalkboard or an overhead transparency.

Add.

1. \$9.85 + \$13.45 **2.** \$21.45 + \$8.95

Subtract.

3. \$40 − \$16.95 **4.** \$100 − \$89.80

Multiply.

5. 9 × \$4.50 **6.** 8 × \$19.75

(**1.** \$23.30 **2.** \$30.40 **3.** \$23.05
4. \$10.20 **5.** \$40.50 **6.** \$158.00)

INTRODUCTION Encourage students to describe the types of movies and shows they enjoy and to talk about how much they would be willing to pay to attend a movie or show.

INSTRUCTION Using the introductory situation, ask students to read aloud the different prices for attending the Quadplex movies and to compare these prices with their local movie theaters.

Use Example 1 to emphasize the importance of using the correct costs for attending a movie. Read the THINK aloud and ask students to tell how much each person would pay. Point out that estimation is a good way to find out whether or not they have enough money. Illustrate on the chalkboard how multiplication could have been used to find the exact answer:

(2 × \$6.25) + (2 × \$4.75) + \$3.50 = \$25.50

Use Example 2 to remind students that they can use multiplication when each ticket costs the same amount. Emphasize the importance of computing and counting their change.

5.1 GOING TO MOVIES AND SHOWS

One summer night, you went to the movies with your parents, your brother, and your sister. You and your sister are students over age 12. Your brother is 10 y old.

QUADPLEX MOVIES

Adults	\$6.25
Students (over 12)	\$4.75
Students (12 and under)	\$3.50
Children (under 5)	FREE

All prices include tax.

EXAMPLE 1 How much did it cost for your family to go to the movies? How much change did you get from \$30?

THINK: 2 adults, 2 students over 12, 1 student 12 and under

1. First estimate the total cost. \$6 + \$6 + \$5 + \$5 + \$4 = \$26
2. Add to find the total cost.
   \$6.25 + \$6.25 + \$4.75 + \$4.75 + \$3.50 = \$25.50

   Compare \$25.50 to the estimate. The answer is reasonable.
   It cost \$25.50 for your family to go to the movies.
3. Subtract to find the amount of change. \$30.00 − \$25.50 = \$4.50

   You got \$4.50 in change.

Sometimes you will buy tickets that each cost the same amount.

EXAMPLE 2 You and 6 of your friends are going to attend a concert. Each ticket, including tax, costs \$18.75. What is the total cost for all 7 people? How much change will you get from \$140?

1. First estimate the total cost. 7 × \$20 = \$140
2. Multiply to find the total cost. 7 × \$18.75 = \$131.25

   Compare \$131.25 to the estimate. The answer is reasonable.
3. Subtract to find the amount of change. \$140.00 − \$131.25 = \$8.75

   You will get \$8.75 in change.

FOR DISCUSSION See TE side column.

Many people get tickets for concerts, plays, and musicals from large ticket brokers. Why do you think brokers charge more per ticket than theaters?

PRACTICE EXERCISES

Remember to estimate whenever you use your calculator.

Use these movie prices for Exercises 1–6.

	Theater A	Theater B	Theater C
Adults	$6.25	$7.25	$5.75
Students (over 12)	$4.75	$3.75	$5.75
Students (12 and under)	$3.50	$3.75	$4.25
Children (under 5)	$2.00	FREE	$1.25

Jones family: 2 adults, 1 student over 12, 1 student 12 or under, 1 child under 5

• **1.** How much would it cost for the Jones family to go to Theater A? How much change would they get from $25?
$22.75; $2.25

• **2.** How much would it cost for the Jones family to go to Theater B? How much change would they get from $30?
$22.00; $8.00

3. How much would it cost for the Jones family to go to Theater C? How much change would they get from $40?
$22.75; $17.25

Wong family: 1 adult, 3 students over 12, 1 child under 5

4. How much would it cost for the Wong family to go to Theater A? How much change would they get from $40?
$22.50; $17.50

5. How much would it cost for the Wong family to go to Theater B? How much change would they get from $30?
$18.50; $11.50

6. How much would it cost for the Wong family to go to Theater C? How much change would they get from $25?
$24.25; $0.75

Solve.

• **7.** Tickets to the school play cost $7. How much will it cost for 8 people to go?
$56

8. Tickets to the laser-light show cost $9.50. How much will it cost for 6 people to go?
$57

• **9.** Sondra bought 6 tickets to a concert. Tickets cost $13.75 each. How much did she pay? How much change did she get from $100?
$82.50; $17.50

10. The Planetarium sells tickets in books of 12. Each ticket costs $4.35. How much did Ray spend for a book of tickets? How much change did he get from $60?
$52.20; $7.80

11. Lorraine wants to buy 11 tickets to an ice-skating exhibition. Tickets cost $9.85 each. She plans to buy them from a ticket broker who charges $2.50 extra for each ticket. She has $88. How much more does she need?
$47.85

12. The regular price of a theater ticket is $19.80. Ann has **twofer coupons.** Each one lets her buy 2 tickets for the price of 1. How much would 9 tickets cost if Ann uses her twofer coupons?
$99

SKILLS BANK You may want to refer students to the following skills in the Skills Bank:

Add With Whole Numbers and Decimals, Skill 6

Subtract With Whole Numbers and Decimals, Skill 6

Multiply With Decimals, Skill 5

FOR DISCUSSION Students should explain that brokers earn their money by making tickets available to the public and that they add a fee as payment for their services.

ASSIGNMENTS

Level 1	1–9
Level 2	1–12
Level 3	4–6, 10–12

FOLLOW-UP ACTIVITY

APPLICATION Ask students to gather information about the movies playing in their neighborhoods. Students should find out the prices the theaters charge and whether there are different prices for Saturday matinees, and so on. Have students draw signs like the one on page 84. Then ask students to determine how much it would cost for their families to attend a movie together.

SUPPLEMENTARY MATERIALS

TRP Practice, p. 31
TRP Reteaching, p. 27
TRP Transparency 16

OBJECTIVES

- Compute the total cost of tickets for parks and sports events.
- Decide if it is less expensive to pay for individual rides or to pay one price for an all-day pass.
- Compute change.

TEACHING THE LESSON

WARM-UP Write the following exercises on the chalkboard or an overhead transparency.

Compute.

1. \$2.75 + \$6.00 + \$9.50 (\$18.25)

2. \$1.35 + \$7.60 + \$4.05 (\$13.00)

3. (4 × \$1.50) + \$9.00 (\$15.00)

4. (3 × \$1.75) + (4 × \$2.35) (\$14.65)

INTRODUCTION Ask students to talk about the parks and sports events they have attended. Encourage students to describe the parks, rides, and events they enjoy and why they enjoy them.

INSTRUCTION Read the introductory situation and ask students if they know of any parks where they could pay for individual rides or get an all-day pass. Ask students how they would decide whether to pay for individual rides or get a pass.

Use Example 1 to show students how to decide whether they should pay for individual rides or an all-day pass and how to compute the amount they would save. Remind students that in Step 3 the cost of the all-day pass is subtracted from the total cost of the rides to find the amount saved. Encourage students to describe combinations of rides which would be less expensive than an all-day pass. Challenge students to find combinations of rides that would cost the same.

Use Example 2 to reinforce the importance of computing the amount of money left over after purchasing tickets. Since Sharon could have given all of the \$30 to the cashier, the \$6.30 could have been her change.

5.2 GOING TO PARKS AND SPORTS EVENTS

You are at Future Land Park. You can either pay one price for an **all-day pass** or pay separately for each ride.

FUTURE LAND PARK

All-Day Pass	\$14.25
Tilt Slide	\$ 1.75
Looper	\$ 2.50
Water Trough	\$ 1.75
Coaster-to-Coaster	\$ 3.50

EXAMPLE 1 You want to go twice each on the Water Trough and the Coaster-to-Coaster. You want to go on the other rides once. How much would you save by buying an all-day pass instead of paying for each ride?

1. Multiply to find the costs of the Trough and the Coaster rides. 2 × \$1.75 = \$3.50; 2 × \$3.50 = \$7.00
2. Add to find the total cost of all of the rides. \$1.75 + \$2.50 + \$3.50 + \$7.00 = \$14.75
3. Subtract to find the amount saved. \$14.75 − \$14.25 = \$0.50

You would save \$0.50 by buying an all-day pass.

Sharon went to the ballpark to pick up some baseball tickets for future games. Tickets for the better seats are more expensive.

TICKET PRICES

Bleachers	\$4.35
Grandstand	\$6.25
Reserved Seat	\$8.75
Box Seat	\$10.50

Tax included.

EXAMPLE 2 Sharon had \$30. She bought 2 Bleacher tickets, 1 Grandstand ticket, and 1 Reserved Seat ticket. How much money did she have left?

1. Multiply to find the cost of 2 Bleacher tickets. 2 × \$4.35 = \$8.70
2. Add to find the total cost of all of the tickets. \$8.70 + \$6.25 + \$8.75 = \$23.70
3. Subtract to find the amount left. \$30.00 − \$23.70 = \$6.30

She had \$6.30 left.

Sometimes you may be able to calculate the cost of tickets mentally.

EXAMPLE 3 Tad bought tickets for all 3 days of the Women's Tennis Tournament. Tickets cost $9.95 per d. How much did he spend?

THINK: $10.00 is $0.05 more than $9.95.

1. Multiply. $3 \times \$10 = \$30$ $3 \times \$0.05 = \$0.15$
2. Subtract to find the total cost. $\$30.00 - \$0.15 = \$29.85$

Tad spent $29.85 on tickets.

FOR DISCUSSION

See TE side column.

Amusement parks sometimes charge large groups, like a whole class, a **group rate.** Do you think that the cost per person with a group rate is more or less than the regular cost per person? Why?

PRACTICE EXERCISES

Remember to estimate whenever you use your calculator.

Use the Safari Park ticket prices.
You can ride all the Safari Park rides as many times as you wish for $18.50. Find the total cost of the given ride tickets. Decide if you should pay for each ride separately or pay the $18.50.

SAFARI PARK TICKETS

Tilt Slide	$1.75
Looper	$2.50
Water Trough	$1.75
Coaster-to-Coaster	$3.50
Giant Swing	$2.25
Puddle Jumper	$2.75
Merry-Go-Round	$1.25

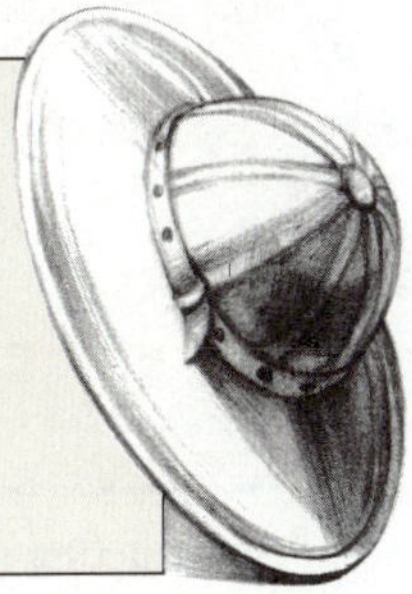

• **1.** 2 Tilt Slides
3 Water Troughs
1 Coaster
4 Merry-Go-Rounds
$17.25; pay separately

2. 3 Coasters
2 Jumpers
3 Swings
$22.75;
pay $18.50

3. You have $15. How many times could you ride the Puddle Jumper?
5 times

4. You have $14. How many times could you ride the Tilt Slide?
8 times

Use the baseball ticket prices on page 86. Find the total cost of the given tickets and the amount of change.

5. 1 Bleacher
2 Grandstands
2 Box Seats
Change from $50?
$37.85; $12.15

6. 3 Bleachers
2 Reserved Seats
1 Box Seat
Change from $45?
$41.05; $3.95

7. 2 Grandstands
2 Reserved Seats
1 Box Seat
Change from $60?
$40.50; $19.50

Solve by computing mentally.

8. You bought 7 baseball tickets for $6.95 each. How much change did you get from $50?
$1.35

9. You want to buy 6 soccer tickets for $8.75 each. You have $50. How much more money do you need?
$2.50

Use Example 3 to stress the importance of mental computation in real-life situations. Remind students that in Step 2, $0.15 is subtracted from $30.00 since the $30.00 is an overestimate and must be adjusted down. Point out to students that much of the computation they do in real life will be mental computation.

SKILLS BANK You may want to refer students to the following skills in the Skills Bank:

Add With Whole Numbers and Decimals, Skill 6

Multiply With Decimals, Skill 5

FOR DISCUSSION Students should explain that the cost per person will be less for a group rate because the owners of the park want to attract large groups of people and it is less expensive to process one large group than many individuals.

ASSIGNMENTS

Level	Exercises
Level 1	1–3, 5–7
Level 2	1–9
Level 3	3–9

FOLLOW-UP ACTIVITY

COOPERATIVE LEARNING
Have students work in small heterogeneous groups to gather information about various amusement parks or sports events that offer group rates. (Sports events as well as amusement parks offer group rates.) Have each group use 1 park or sports event and compute how much it would cost for the whole class to attend the event at the regular price and at the group price. Ask the students to find out how much each person would save with the group rate.

SUPPLEMENTARY MATERIALS

TRP Practice, p. 32
TRP Reteaching, p. 28
TRP Transparency 16

OBJECTIVES

- Compute the costs of hobby materials.
- Compute change.

TEACHING THE LESSON

WARM-UP Write the following exercises on the chalkboard or an overhead transparency.

Multiply or divide.

1. 7 × \$1.30 **2.** 16 × \$2.28

3. \$2.75 ÷ \$0.25 **4.** \$8.16 ÷ \$0.85

(**1.** \$9.10 **2.** \$36.48 **3.** 11 **4.** 9.6)

INTRODUCTION Read the introductory situation and ask students to name some reasons why many people enjoy stamp collecting. Tell them that there are only about 80 of the upside-down "inverted gennys" left. Remind them that one of these stamps is worth about \$80,000, and ask them to calculate how much all 80 stamps are worth. (about \$6,400,000) Encourage students to name their favorite hobbies and to explain why they enjoy them.

INSTRUCTION Use Example 1 to show students how to divide money amounts. Point out the estimation in Step 1. Explain that since \$0.35 has been rounded up to \$0.40, there is a good chance that the number of packets will be less than 15. Remind students that \$5.00 divided by \$0.35 means "How many 35 cents are there in 5 dollars?" Encourage students to discuss why the answer to this type of problem would not be a money amount. For Step 2, explain to students that a part of a packet of stamps cannot be bought, and so there is enough money for 14 packets but not enough money for 15 packets.

Use Example 2 to show students how to read and use information from a price list. Ask students to identify the cost of each item on the Materials Needed list. Point out how multiplication and addition are used to arrive at the total cost.

5.3 HOBBIES

Stamp collecting is a popular hobby. The rarest United States stamp is the "inverted genny" 24-cent stamp printed in 1918. One hundred of these stamps were printed upside-down by mistake. This stamp is now worth over \$80,000.

You do not need to buy expensive stamps to be a collector. You could start with just a few stamps.

EXAMPLE 1 You have \$5.00 to start your collection. The stamp shop has packets of stamps on sale for \$0.35 each. How many packets can you buy? How much change will you get?

1. First estimate the number of packets.

   THINK: $\$0.35 \approx \$0.40$

   3 packets cost 3 times the cost of 1. $3 \times \$0.40 = \$1.20 \approx \$1.00$
   15 packets cost 5 times the cost of 3. $5 \times \$1.00 = \$5.00$

   You can buy about 15 packets with \$5.00.

2. Divide to find exactly how many packets. $\$5.00 \div \$0.35 = 14.285714$

   You can actually buy 14, not 15, packets with \$5.00.

3. Multiply to find the cost. $14 \times \$0.35 = \$4.90$
4. Subtract to find the change. $\$5.00 - \$4.90 = \$0.10$

   You will get \$0.10 in change.

Constructing models is another popular hobby. Some people construct miniature doll houses. Others construct model cars or planes.

EXAMPLE 2 You need these materials to finish your model. What is the total cost of these materials? How much change will you get from \$15?

Materials Needed

$3\ \frac{1}{2}$-in. dowels
1 bag balsa wood
2 bottles paint
1 paint brush
1 package decals

HOBBY SHOP
PRICE LIST

¼-in. dowel	\$0.49
½-in. dowel	\$1.89
1 bag balsa wood	\$2.95
1 jar paste	\$0.75
1 bottle paint	\$1.05
1 paint brush	\$0.59
1 pkg. decals	\$1.50

Prices include tax.

1. Multiply to find the cost of the dowels. $3 \times \$1.89 = \$5.67$
2. Multiply to find the cost of the paint. $2 \times \$1.05 = \$2.10$
3. Add to find the total cost.
$\$5.67 + \$2.95 + \$2.10 + \$0.59 + \$1.50 = \$12.81$

The total cost of the materials is $12.81.

4. Subtract to find the change. $\$15.00 - \$12.81 = \$2.19$

You will get $2.19 in change.

FOR DISCUSSION

See TE side column.

What types of costs might be involved for photography? Your favorite hobby?

PRACTICE EXERCISES

Remember to estimate whenever you use your calculator.

Find the total cost of the purchases.

• 1. *Hobby: Stamp collecting*
Stamps: $7.20
Tweezers: $1.29
Display book: $12.95
$21.44

2. *Hobby: Painting*
Canvas: $6.00
Brushes: $7.50
Paints: $21.85
$35.35

3. *Hobby: Sewing*
Pattern: $4.45
Fabric: $18.70
Thread: $1.69
Zipper: $2.89
$27.73

Solve.

• 4. Juan has $18.00 to spend on stamp packets that sell for $1.85 each. How many packets can he buy? How much change will he get?
9 packets; $1.35

5. Debra has $21.00 to spend on clay that costs $3.75 per lb. How many pounds can she buy? How much change will she get?
5 lb; $2.25

6. Michael belongs to a book club. He orders 4 special books for $0.99 each and spends an additional $32.50 for other books. The shipping and handling costs are $15.25. What is the total cost of his order?
$51.71

7. Lois wants to buy 4 packets of stamps that cost $2.85 each, 1 stamp that costs $2.89, and a stamp catalog that costs $15.95. She has $24.35. How much more money does she need?
$5.89

Use the Hobby Shop price list on page 88. Find the total cost of the materials. Then find the amount of change.

8. $1\ \frac{1}{2}$-in. dowel
3 bags balsa wood
1 bottle paint
2 pkgs. decals
Change from $15?
$14.79; $0.21

9. $5\ \frac{1}{4}$-in. dowels
$1\ \frac{1}{2}$-in. dowel
3 jars paste
2 bottles paint
Change from $10?
$8.69; $1.31

10. 3 bags balsa wood
$1\ \frac{1}{2}$-in. dowel
3 bottles paint
4 paint brushes
Change from $20?
$16.25; $3.75

SKILLS BANK You may want to refer students to the following skills in the Skills Bank:

Multiply With Decimals, Skill 5

Divide With Decimals, Skill 8

FOR DISCUSSION Students should explain that the main costs in photography are the purchase of a camera and film and the expense of film processing. Students should also explain that the cost of a camera ranges from under $25 to thousands of dollars. Since many photography hobbyists do their own processing, students may also discuss these expenses, which include the costs of a darkroom, an enlarger, and other developing equipment and chemicals.

Encourage students to discuss their hobbies in detail along with the costs involved. If the opportunity presents itself, use this discussion to create sample related problems for all students to solve.

ASSIGNMENTS

Level 1	1–5, 8–9
Level 2	1–10
Level 3	3, 7–10

FOLLOW-UP ACTIVITY

ENRICHMENT Conduct a survey in your class to determine the favorite hobby of each student. Encourage students to survey other students in the school. Help students make a tally on the chalkboard to show how many prefer a particular hobby. Ask each student to construct a bar graph to show the information from the tally.

SUPPLEMENTARY MATERIALS

TRP Practice, p. 33
TRP Reteaching, p. 29
TRP Transparency 17

OBJECTIVES

- Compute the equipment fees, use fees, and transportation costs of a sports activity to find the total cost.
- Decide whether to buy or rent sports equipment.

TEACHING THE LESSON

WARM-UP Write the following exercises on the chalkboard or an overhead transparency.

Write > or <.

1. \$8.95 ● \$9.04 **2.** \$93.45 ● \$89.74

3. \$9.45 + \$8.65 ● \$11.36 + \$6.65

4. 4 × \$8.32 ● 9 × \$3.72

(**1.** < **2.** > **3.** > **4.** <)

INTRODUCTION Ask students to talk about their favorite sports. Encourage them to discuss the kinds of equipment they need to participate in the sport and if they need to go to a special place, such as a bowling alley, in order to participate.

INSTRUCTION Review the formula for TOTAL COST. Explain that if someone went bowling, the equipment cost would be the amount the person paid to buy or rent shoes and to buy a ball. Explain that the use fee would be the cost of bowling a game and that the transportation fee would be the cost of getting to and from the bowling alley. Ask students to give other examples.

Use Example 1 to show students how to find the total cost of a sports activity. Read through the problem, and ask students to identify whether each amount is an equipment cost, a use fee, or a transportation cost. Point out that only the underlined answers in Steps 1, 3, and 4 are used to find the total cost in Step 5.

Use Example 2 to help students understand the difference between buying or renting equipment. Students may also be curious to know how much is saved by renting the equipment. (\$255.45 − \$210 = \$45.45)

5.4 SPORTS EQUIPMENT AND FEES

Some sports cost almost nothing while others can be very expensive.

TOTAL COST = EQUIPMENT COSTS + USE FEES + TRANSPORTATION COSTS

EXAMPLE 1 You decide to go swimming once a week for the next 28 wk. You pay \$28.95 for a swimsuit and \$4.29 for goggles. You pay \$25 to join the swim club, and it costs \$2.75 each time you go swimming. The bus you take to and from the swim club costs \$1.30 per round trip. What is the total cost of this sports activity?

1. Add to find the cost of the equipment. \$28.95 + \$4.29 = <u>\$33.24</u>
2. Multiply to find the cost of the swim sessions. 28 × \$2.75 = \$77.00
3. Add to find the total cost of using the swim club. \$25 + \$77 = <u>\$102.00</u>
4. Multiply to find the cost of the bus transportation. 28 × \$1.30 = <u>\$36.40</u>
5. Add to find the total cost. <u>\$33.24</u> + <u>\$102</u> + <u>\$36.40</u> = \$171.64

The total cost of this sports activity is \$171.64.

For some sports, you may need to decide whether to buy or to rent equipment.

EXAMPLE 2 Sarah's ski club dues include transportation to and from the ski slopes and the lift tickets. It costs $20 to rent skis and $15 to rent poles and boots. It would cost her $169.50 to buy skis, $67.95 to buy boots, and $18 to buy poles. She hopes to go skiing with the club 6 times. Would it be less expensive to rent or to buy the equipment?

1. Add to find the cost of 1 day's rental. $20 + $15 = $35
2. Multiply to find the cost of 6 days' rentals. 6 × $35 = $210
3. Add to find the cost of buying equipment. $169.50 + $67.95 + $18 = $255.45
4. Compare the rental and buying costs. $210 < $255.45

It is less expensive to rent the ski equipment.

FOR DISCUSSION

See TE side column.

Name some sports that you think are inexpensive and some that are expensive. Tell why.

PRACTICE EXERCISES

Remember to estimate whenever you use your calculator.

Complete the table to find the total cost of the sports activity.

Sports activity	Equipment cost	Use fees	Round-trip transportation cost	Total cost
Ice skating	$59.50	• 1. 10 visits at $3.50/visit = ■ $35	$0.85	• 2. ■ $103
Tennis	$83.95	3. $45 membership + 25 h at $9.50/h = ■	$0.00	4. ■ $366.45
Swimming	$28.95	5. $35 membership + 18 visits at $4.50/visit = ■	$1.35	6. ■ $169.25
Racquetball	$76.85	7. $75 membership + 30 h at $6.50/h = ■	8. 25 trips at $0.90/trip = ■	9. ■ $369.35
Bowling	$48.79	10. 39 games at $2.75/game = ■	11. 14 trips at $2.60/trip = ■	12. ■ $192.44

3. $282.50 **5.** $116 **7.** $270 **8.** $22.50 **10.** $107.25 **11.** $36.40

Solve.

13. Julio goes ice skating. He can either rent or buy skates. Skates sell for $45.90 and rent for $1.75 per h. Each week, Julio skates for about 28 h. Is it less expensive for him to rent or to buy skates?
buy skates

SKILLS BANK You may want to refer students to the following skills in the Skills Bank:

Compare and Order Whole Numbers and Decimals, Skill 5

Multiply With Decimals, Skill 5

FOR DISCUSSION Students should explain that sports such as jogging are less expensive than sports such as skiing. Students should also explain that any one or combination of the 3 cost components (equipment, use fees, and transportation) can make a sport expensive.

ASSIGNMENTS

Level	Exercises
Level 1	1–6, 13
Level 2	1–13
Level 3	5–13

FOLLOW-UP ACTIVITY

APPLICATION Present the following problem: Jean budgeted 8% of her take-home pay for sports. Her weekly take-home pay is $218. How much did Jean budget for sports?

Explain that this sequence of calculator keystrokes will result in an answer of $17.44 per wk.

[.] [0] [8] [×] [2] [1] [8] [=] 17.44

Ask students to find each amount budgeted for sports.

1. Sports budget: 7%
Take home: $198

2. Sports budget: 6.5%
Take home: $238

(**1.** $13.86 **2.** $15.47)

SUPPLEMENTARY MATERIALS

TRP Practice, p. 34
TRP Reteaching, p. 30
TRP Transparency 17

OBJECTIVE

- Compute the cost per visit of a health club or fitness class membership.

TEACHING THE LESSON

WARM-UP Write the following exercises on the chalkboard or an overhead transparency.

Divide.

1. $288 ÷ 6 **2.** $13.50 ÷ 18

Divide. Round to the nearest cent.

3. $128.36 ÷ 17 **4.** $205.76 ÷ 23

(**1.** $48 **2.** $0.75 **3.** $7.55 **4.** $8.95)

INTRODUCTION Ask students to describe fitness classes or health clubs they have gone to or have seen advertised on TV or in newspapers. Have them discuss facilities, the costs of classes and clubs, and the special discounts offered in advertisements.

INSTRUCTION Read the introductory situation, and emphasize the importance of proper supervision. Review the formula for COST PER VISIT and explain to students that this may be the best way of determining the cost of an activity. Point out that the cost per visit goes down as the number of visits goes up.

Use Example 1 to show students how to find the cost per visit. In Step 2, remind students to look at the digit 3 places to the right of the cents point, or 9, to decide whether to round up or to round down to the nearest cent. (round up)

Use Example 2 to show students that more expensive, longer-term classes can cost less per class. Explain to students that this sequence of keystrokes on a calculator with memory keys will result in the difference in cost of the classes:

[1] [3] [8] [.] [7] [5] [÷] [7] [5] [M+]
[1] [0] [1] [.] [2] [5] [÷] [4] [5] [M−] [RM]

5.5 HEALTH CLUBS AND FITNESS CLASSES

You have decided to join a health club and to take a fitness class. As a member of a health club, you can usually go whenever you wish to use the exercise machines and other facilities. As you inquire about the costs of clubs and classes, you should also make sure that the activities are properly supervised.

You pay a **membership fee** to belong to a health club. The **cost per visit** is the membership fee divided by the number of visits.

COST PER VISIT = MEMBERSHIP FEE ÷ NUMBER OF VISITS

EXAMPLE 1 You want to join the All-American Health Club. The annual membership fee is $418. You expect to go about 3 times per wk for a year. What would be your cost per visit to the nearest cent?

1. Multiply to find the number of visits.
   THINK: 1 y = 52 wk 3 × 52 = 156
2. Divide to find the cost per visit. $418 ÷ 156 = $2.6794871

It would cost you about $2.68 per visit.

As a member of the health club, you still need to pay for any fitness classes you join. For many fitness classes, the cost per class decreases as the term of time for which you sign up increases.

EXAMPLE 2 You want to sign up for an aerobics class that meets 3 times per wk. The class costs $101.25 for 15 wk (45 classes) or $138.75 for 25 wk (75 classes). How much less would the cost per class be if you signed up for 25 wk instead of 15 wk?

1. Divide to find the cost of a 15-wk class. $101.25 ÷ 45 = $2.25
2. Divide to find the cost of a 25-wk class. $138.75 ÷ 75 = $1.85
3. Subtract to find the difference. $2.25 − $1.85 = $0.40

It would cost $0.40 less per class if you signed up for the longer term of 25 wk.

FOR DISCUSSION See TE side column.

Is it more important to know the total club membership fee or to compute your own cost per visit? Why?

PRACTICE EXERCISES

Remember to estimate whenever you use your calculator.

Find the cost per visit to the nearest cent. Then find the cost per visit including the transportation cost. 1. $2.38 2. $4.18 3. $2.38 4. $3.28 5. $1.73

Health club	Membership fee	Expected number of visits	Cost per visit	Transportation cost per visit	Cost per visit including transportation
USA Health	$233/y	98/y	• 1. ■	$1.80	• 2. ■
Slim Studio	$9.50/wk	4/wk	3. ■	$0.90	4. ■
Aerobics Unlimited	$19/mo	11/mo	5. ■	$2.80	6. ■
The Fitness Center	$462 for 3 y	63/y	7. ■	$0.00	8. ■
Health City	$193 for 6 mo	3/mo	9. ■	$3.15	10. ■

6. $4.53 7. $2.44 8. $2.44 9. $10.72 10. $13.87

Find how much can be saved, to the nearest cent, by signing up for the longer-term class.

• **11. Judo**
$210 for 15 wk (30 classes)
$340 for 26 wk (52 classes)
$0.46

12. Jazz Dance Workout
$98 for 5 wk (15 classes)
$125 for 8 wk (24 classes)
$1.32

13. Weight Training
$680 for 24 wk (72 classes)
$1,200 for 48 wk (144 classes)
$1.11

14. Floor Exercises
$220 for 12 wk (36 classes)
$420 for 25 wk (100 classes)
$1.91

EXTENSION

Percents of the Total Cost

What percent of the total cost did you spend on equipment, on fees, and on transportation for this sports activity?

Total cost: $230 Equipment: $83 Fees: $118 Transportation: $29

Divide each cost by the total cost.

$83 ÷ 230 = 0.3608695, or 36% Equipment was 36% of the total cost.

$118 ÷ 230 = 0.5130434, or 51% Fees were 51% of the total cost.

$29 ÷ 230 = 0.1260869, or 13% Transportation was 13% of the total cost.

Find the percent of the total cost spent on equipment, on fees, and on transportation. Round each percent to the nearest percent.

1. Total cost: $234.50 Equipment: $54.00 Fees: $153.50 Transportation: $27.00
2. Total cost: $256.20 Equipment: $62.00 Fees: $169.50 Transportation: $24.70

1. 23%; 65%; 12% 2. 24%; 66%; 10%

SKILLS BANK You may want to refer students to the following skills in the Skills Bank:

Divide With Decimals, Skill 8

Round Whole Numbers and Decimals, Skill 4

Find the Percent One Number Is Of Another, Skill 2

FOR DISCUSSION Students should explain that it is usually best to know their own cost per visit. They should also explain that it is important to be realistic about the number of times they plan to participate in an activity.

EXTENSION Point out to students, for example, that finding the percent of the total cost spent on equipment means finding the percent one number is of another. They should divide the amount spent on equipment by the total cost.

ASSIGNMENTS

Level 1	1–6, 11–12
Level 2	1–14
Level 3	5–14, EX

FOLLOW-UP ACTIVITY

ENRICHMENT Ask students to construct a circle graph from the situation explained in the Extension using the following instructions: (1) draw a circle and mark the center; (2) multiply each percent for the equipment, fees, and transportation by 360°, and round to the nearest degree; (3) measure these angles around the center of the circle; and (4) complete and label the graph. (36% × 360° ≈ 130°; 51% × 360° ≈ 184°; 13% × 360° ≈ 49°; Check students' graphs.)

SUPPLEMENTARY MATERIALS

TRP Practice, p. 35
TRP Reteaching, p. 31
TRP Applications, p. 5
TRP Transparency 18

OBJECTIVE

- Solve problems by interpreting quotients and remainders.

TEACHING THE LESSON

WARM-UP Write the following exercises on the chalkboard or an overhead transparency.

Divide.

1. 57 ÷ 6 (9 R3) **2.** 95 ÷ 8 (11 R7)

3. 79 ÷ 12 (6 R7) **4.** 139 ÷ 18 (7 R13)

INTRODUCTION Pose this situation to the class: The Bainbridge Hotel offers its guests a free van ride to the airport. The hotel van can hold 8 passengers, and 43 guests are waiting to go to the airport.

Ask the following questions: (1) If the van only goes to the airport when filled, how many trips will it make? (5 trips) (2) If all 43 guests must be taken, how many trips will the van make? (6 trips)

Discuss the fact that in the first question, the remainder was not important to finding the solution. In the second question, the fact that there was a remainder was important.

INSTRUCTION Present the problem situation to the class. Explain that 4 different situations requiring the given division computation will be presented. As you discuss Situation A, stress that because the only concern is with classes filled to the maximum, the remainder is not important.

For Situation B, stress that if all 140 people registered, the last class would have been partially filled with 14 students. If calculators are being used, review the procedure for finding a whole-number remainder.

PROBLEM *Solving* STRATEGY

5.6 INTERPRETING A REMAINDER

Situation:

Last week, the Brookdale Fitness Club held registration for its exercise classes. Each class can hold a maximum of 18 students. If 140 people registered, how could the classes have been arranged?

Strategy:

When you divide to solve a problem, be sure that you answer the question that was asked.

Applying the Strategy:

Divide 140 by 18.

$$\begin{array}{r} 7 \text{ R14} \\ 18\overline{)140} \\ \underline{126} \\ 14 \end{array}$$

Notice the different answers to each of the questions below.

A. How many exercise classes would have been filled to the maximum by 140 students?

Use only the quotient. 7 classes would have been filled to the maximum.

B. How many students would have been in the last class?

Use only the remainder. 14 students would have been in the last class.

C. How many exercise classes would have been needed to hold all 140 students?

Raise the quotient by 1. THINK: **7 full classes + 1 class with 14 students**
8 classes would have been needed in all.

D. What fraction of the last class would have been filled?

Use a fractional remainder. THINK: $\frac{\text{students in last class} \rightarrow 14}{\text{students in full class} \rightarrow 18} = \frac{7}{9}$

$\frac{7}{9}$ of the last class would have been filled.

PRACTICE EXERCISES

Write the letter of the correct answer.

1. Each weight-lifting class can hold a maximum of 8 students. If 38 students want to register, how many students will be in the last, or smallest, class? b
   - **a.** Use only the quotient.
   - **b.** Use only the remainder.
   - **c.** Raise the quotient by 1.

2. How many advanced exercise classes can be formed from 52 students if exactly 12 students can be in each class? a
   - **a.** Use only the quotient.
   - **b.** Use only the remainder.
   - **c.** Raise the quotient by 1.

The Brookdale Fitness club gives each student a tee shirt to wear during classes. Dan packed 12 tee shirts in each carton. He had 285 tee shirts to pack.

• **3.** How many cartons did Dan fill? 23 cartons

4. How many tee shirts did he put in the last partially filled carton? 9 tee shirts

5. How many cartons did Dan use to pack 285 tee shirts? 24 cartons

6. What fraction of the last carton was filled? $\frac{3}{4}$

Solve by interpreting a remainder.

• **7.** Sample classes are held during registration week. Each sample class costs $3. Tim has $10 to spend on sample classes. How many classes can he attend? 3 classes

8. There are 78 floor mats in a shipment that arrived at the club. Each exercise room fits 8 mats. After fitting 9 rooms, what fraction of the 10th room can be fitted with mats? $\frac{3}{4}$

9. One towel is needed for each of the 185 new students at the club. If a package contains 12 towels, how many packages are needed? 16 packages

10. The 40 students in a calisthenics class were divided into groups of 4 pairs for a floor exercise. How many groups were set up? 5 groups

For Situation C, point out that the size of the remainder is not important. If there were a remainder of any size at all, the 8th class would still be needed.

For Situation D, point out that the size of the remainder is important because writing a fraction requires knowing the number of students in 7 full classes as well as the number of students in a partially filled 8th class.

SKILLS BANK You may want to refer students to the following skills in the Skills Bank:

Divide With Whole Numbers, Skills 2–7

ASSIGNMENTS

Level	Exercises
Level 1	1–7
Level 2	1–9
Level 3	3–10

FOLLOW-UP ACTIVITY

SITUATIONAL PROBLEM SOLVING Present students with the following problem: Your school is going to have an after-school fitness program. Your class is in charge of determining what kinds of activities to include. Your class will need to present its ideas to the principal.

Some questions to ask include: What kinds of activities would interest the students in your school? How should you present your ideas to the principal?

Some approaches to solving the problem include making a student survey, such as using a questionnaire or providing accessible sign-up sheets, tallying results, or making a bar graph.

SUPPLEMENTARY MATERIALS

TRP Practice, p. 36
TRP Transparency 18

OBJECTIVE

- Decide, given relevant information, which sport to participate in.

TEACHING THE LESSON

WARM-UP Write the following table on the chalkboard or an overhead transparency.

Complete the table.

Equip. costs	Use fees	Trans. cost	Total cost	Cost per visit
$38.59	20 visits at $18 each	$0.45 per visit	1. ■ ($407.59)	2. ■ ($20.38)
$58.00	15 visits at $6.35 each	$2.35 per visit	3. ■ ($188.50)	4. ■ ($12.57)

INTRODUCTION Present the problem situation to the class. Encourage students to name some of the factors they would use to make an informed decision about which sport to choose and where this information is available. List the factors that are named on the chalkboard.

INSTRUCTION Read the list of options that Derek has prepared. Ask students to fully discuss the features and to mention those features that they find most appealing. Ask if there is any other information that Derek should have. Discuss the decision-making factors. Remind students that a class or club must be properly supervised for them to even consider it.

Have small groups of students complete Exercises 1–15 in the table in order to compare the 4 sports. Students will have the most difficulty computing the cost per hour (Exercises 4–6), and it may be necessary for them to refer back to the computations done in Lesson 5.5. Ask students to write statements that describe the data now in the table. Then ask them to write conclusions that can be drawn from the data.

DECISION MAKING

5.7 SELECTING A SPORT

(Open-ended problem solving)

You do not need to be a good athlete to participate in and enjoy a sport. In fact, there is probably a sport for everyone.

PROBLEM

Derek is trying to decide in which sport to participate. He is interested in tennis, racquetball, skiing, and bowling. He listed the features of each sport to help him decide which one to choose.

Tennis
The equipment costs about $90.
I'll have to pay $5.00 per h to play outside and $16.00 per h to play inside.
I'll need to travel about 15 min to the tennis courts at a cost of about $1.25 for a round trip.
Of the 4 sports, tennis ranks 3rd in the amount of exercise I'll get.

Racquetball
The equipment costs about $45.
I'll have to pay $18 per h to play.
The indoor racquetball courts are a short walk from my home.
Racquetball ranks 1st in the amount of exercise I'll get.

Skiing
The equipment costs about $275.
I'll have to pay $35 for a lift ticket every day I ski.
I will probably want to ski about 5 h each day.
It takes me about 2 h to get to the ski slopes at a cost of about $7.50 for a round trip.
Skiing ranks between tennis and racquetball in the amount of exercise I'll get.

Bowling
The equipment costs about $55.
I'll have to pay $2.75 for each game I play.
I can bowl about 3 games per h for 2 h.
It takes me about 1 h to get to the bowling alley at a round trip cost of $3.00.
Bowling trails well behind tennis in the amount of exercise I'll get.

DECISION-MAKING FACTORS

Indoor or outdoor   Cost per hour   Travel time   Travel cost   Amount of exercise

DECISION-MAKING COMPARISONS

Compare the 4 sports, based on 40 h of participation, by completing the table. Remember to include equipment cost in the cost per hour.

Factor	Tennis	Racquetball	Skiing	Bowling
Indoor or outdoor	Both	• 1. ■	2. ■	3. ■
Cost per hour	$7.25 outside $18.25 inside	• 4. ■	5. ■	6. ■
Round-trip travel time	7. ■	Few min	8. ■	9. ■
Travel cost	10. ■	11. ■	12. ■	$60.00
Amount of exercise	13. ■	14. ■	Ranks 2nd	15. ■

1. indoor **2.** outdoor **3.** indoor **4.** $19.13 **5.** $13.88 **6.** $9.63
7. 30 min **8.** 4 h **9.** 2 h **10.** $50 **11.** $0 **12.** $60
13. ranks 3rd **14.** ranks 1st **15.** ranks 4th

MAKING THE DECISIONS

Which sport would Derek choose if:

•**16.** He wanted to play indoors and outdoors?
tennis

17. Cost per hour were the only factor?
tennis outside

18. Being close to home were the only factor?
racquetball

19. Travel costs were the only factor?
racquetball

•**20.** If Derek did not want to spend more than $12 per h, which sports would he eliminate?
tennis inside, racquetball, skiing

21. If Derek wanted the sport that would give him the most exercise, which sport would he choose?
racquetball

22. If Derek wanted to exert himself as little as possible, which sport would he choose?
bowling

23. If Derek decided against racquet sports, which sports could he eliminate?
tennis, racquetball

24. Which one of these sports would you choose? Name at least 3 reasons why.
Answers may vary.

Discuss the appropriateness of these statements and conclusions. Then lead a discussion about interpretations and possible misinterpretations of the data.

Have students complete Exercises 16–24 in small groups, and ask them to explain and defend their answers and their decisions. Remind students that in Exercise 24 there is no one right answer. Explain that choices will vary and that even those with the same choice may have different reasons for their choice. Some students are very wary of defending their point of view among their classmates. Continually reinforce the fact that, in many consumer issues, best answers depend on individual situations and preferences. Encourage all students to participate and to express their opinions.

ASSIGNMENTS

Level 1	1–24
Level 2	1–24
Level 3	1–24

FOLLOW-UP ACTIVITY

COOPERATIVE LEARNING

Have students work in small heterogeneous groups to find out about a particular fitness class or health club. After each group has gathered and summarized all the relevant information, have each group report it's findings to the class. Then ask the class to use the decision-making factors and organize the information about each class or club under these factors. Finally, ask students to make individual decisions about which class or club they would choose and explain the basis for their decisions.

SUPPLEMENTARY MATERIALS

TRP Practice, p. 37
TRP Transparency 19

OBJECTIVES

- Review vocabulary.
- Practice key chapter concepts and skills.

USING THE REVIEW

The Chapter Review is designed to help students prepare for taking the Chapter Test. The first section focuses on vocabulary. It requires students to select words to complete statements. The second section presents practice of key chapter skills.

Each item on the Chapter Review is referenced to the page on which the topic is taught in the Pupil's Edition. You may wish to have students refer to these pages to help review any concepts or skills they have not yet mastered.

It is suggested that students work in small-sized heterogeneous cooperative-learning groups. Some cooperative learning methods that may be used are as follows:

1. After each student has independently completed the Chapter Review, a discussion should follow within each group about the solutions to the problems.
2. The group can complete the entire Chapter Review by working together to solve the problems.

End the lesson with an entire class discussion in which any questions brought up in group discussions are presented and answered.

CHAPTER REVIEW

Vocabulary Choose the letter of the word(s) that completes the sentence.

1. If you want to pay only 1 price to go on all of the rides in an amusement park, you can buy ■. [86] a
   a. An all-day pass **b.** Individual tickets **c.** A season pass
2. Amusement parks often charge large groups of people a ■. [86] c
   a. Season pass **b.** Use fee **c.** Group rate
3. Membership fees and costs for a sports activity are called ■. [90] b
   a. Equipment costs **b.** Use fees **c.** Transportation costs

Skills Find the answer.

4. Use the movie-price list on page 84. An adult, two 14-year-olds, and a 3-year-old went to the movies. What was the total cost? [84]
   $15.75
5. You and 3 friends went to a concert. Each ticket cost $23.50, including tax. What was the total cost? [84]
   $94.00
6. Use the Future Land Park ticket prices on page 86. Ana wants to go twice on each of the rides. How much would she save by buying an all-day pass? [86]
   $4.75
7. Use the baseball-ticket prices on page 86. Robert bought 2 Box Seat tickets, 2 Grandstand tickets, and 2 Bleachers tickets. How much money did he have left from $50? [86]
   $7.80
8. Tony bought stamps for $5.50, a display book for $9.99, and tweezers for $0.89. How much did Tony spend? [88]
   $16.38
9. Anita's hobby is photography. She bought a camera for $75.59, film for $7.20, and a camera case for $21.10. How much did Anita spend? [88]
   $103.89
10. Lisa went bowling and spent $2.75 a game for 13 games. She spent $4.75 on shoe rental and $4.70 on transportation. How much did Lisa spend? [90]
    $45.20
11. Ron joined a racquetball club for $55 and played for 28 h at $7.75 per h. The equipment cost $32, and Ron paid $13 for transportation. How much did Ron spend? [90]
    $317
12. Fran joined a health club for $144 per y. She expects to visit the club 7 times per mo. What is the cost per visit? [92]
    $1.71
13. A 10-class karate course costs $65. The 25-class course costs $149. How much can you save per class by signing up for 25 classes? [92]
    $0.54

CHAPTER TEST

Find the total cost of the purchases.

1. Wood: $5.75
   Paste: $0.65
   Paint: $2.99
   Paintbrush: $1.59
   $10.98

2. Paper: $ 6.20
   Charcoal: $ 2.25
   Pencils: $ 1.89
   Easel: $15.50
   $25.84

3. Seeds: $ 9.95
   Shovel: $12.00
   Fertilizer: $ 6.50
   Wheelbarrow: $25.99
   $54.44

Find the total cost of the tickets and the amount of change.

TICKET PRICES	
Bleachers	**$ 4.35**
Grandstand	**$ 6.25**
Reserved Seat	**$ 8.75**
Box Seat	**$10.50**

Tax included.

4. 6 Grandstands
   3 Reserved Seats
   2 Box Seats
   Change from $100?
   $84.75; $15.25

5. 3 Bleachers
   1 Box Seat
   2 Grandstands
   4 Reserved Seats
   Change from $80.00?
   $71.05; $8.95

Complete the table to find the total cost of the sports activity.

Sports activity	Equipment cost	Use fees	Round-trip transportation cost	Total cost
Bowling	$42.50	**6.** 18 games at $3.25/game = ■	$5.10	**7.** ■
Tennis	$92.39	**8.** $75 membership + 25 h at $10.75/h = ■	$30.55	**9.** ■
Swimming	$58.99	**10.** $99 membership + 12 visits at $2.75/visit = ■	**11.** 12 trips at $1.60/trip = ■	**12.** ■

6. $58.50 **7.** $106.10 **8.** $343.75 **9.** $466.69 **10.** $132 **11.** $19.20 **12.** $210.19

Complete the table to find the cost per visit to the nearest cent.

Health club	Membership fee	Expected number of visits	Cost per visit	Transportation cost per visit	Cost per visit including transportation
Health Club	$327/y	109/y	**13.** ■	$2.35	**14.** ■
Gym Center	$218 for 6 mo	16/mo	**15.** ■	$0.80	**16.** ■
Slim Salon	$402 for 2 y	9/mo	**17.** ■	$1.40	**18.** ■

13. $3.00 **14.** $5.35 **15.** $2.27 **16.** $3.07 **17.** $1.86 **18.** $3.26

Solve.

19. How much can be saved per class by signing up for 45 classes for $180 instead of 24 classes for $100?
    $0.17 per class

20. How many 12-person exercise classes can be filled by 40 people?
    3 classes

OBJECTIVE

- Evaluate the achievement of chapter objectives.

USING THE TEST

The Chapter Test may be used as a post-test to evaluate student achievement. However, you may wish to use the Chapter Post-Test offered in the Teacher's Resource Package or to design your own chapter test. If this page is not used as a test, you may want to assign it as additional review or practice.

The test items are correlated to the chapter objectives in the table below.

Chapter objectives	Test items
A. Compute the costs of hobby materials.	1–3
B. Compute the costs of tickets and the amount of change.	4–5
C. Compute the total costs of sports activities.	6–12
D. Compute the cost per visit of a sports activity.	13–19
E. Solve problems by interpreting remainders.	20

SUPPLEMENTARY MATERIALS

TRP Ch. 5 Post-Test Form A, pp. 1–2
TRP Ch. 5 Post-Test Form B, pp. 1–2

OBJECTIVE

- Understand that buying tickets to sports events for each game can cost much more than if you buy a season ticket.

USING THE PAGE

Read the information given in the table. Encourage students to pay special attention to the days of play and the number of games included in each of the 4 packages.

Have students work individually or in pairs to answer Exercises 1–6. Summarize the answers as follows: People who enjoy sports can save money by buying season tickets and attend as many games as they like on days that fit into their schedules.

Have students work individually to answer Exercises 7–8. Then have them share their answers and reasons in a whole-class discussion.

ANSWERS/EXPLANATIONS

2. Season tickets guarantee admission to games; season ticket holders have first preference for playoff tickets; no need to wait or stand on line for tickets for each game
3. Weekday games—not as crowded, better seats, preferable for people who work odd hours or on weekends; weekend games—better for people who work during the week, more fun to take friends or family along
7. (a) $728 ($408, 47 weekday + $320, 32 weekend box tickets)
   (b) $680 ($408, 47 weekday + $272, 32 weekend reserved tickets)
   (c) $750 (all games, box seat)
   (d) $600 (all games, reserved seat)
   (e) $790 (all games, box seats by buying single tickets)
   (f) $671.50 (all games, reserved seats by buying single tickets)

You save the most money with (d).

MONEY TIPS

Season tickets to sports events can save you money if you attend many games.

LET'S LOOK AT THE FACTS

Some people buy season tickets instead of buying tickets for 1 game at a time. This will cost less if you plan to attend many or all games. Read the ticket prices shown here carefully.

BASEBALL TICKET PRICES

TICKET	SEAT	PRICE	NUMBER OF GAMES
Single	Box	$10.00	1 only
Single	Reserved	$ 8.50	1 only
Season	Box	$750	79 - all home games
Season	Reserved	$600	79 - all home games
Season	Box	$408	47 - all weekday games
Season	Box	$187	30 - Mon/Wed/Thu games

IMPORTANT NOTE: Season price covers the cost of 79 tickets. Two doubleheaders are played at home, so that the season price covers the cost of 81 games.

LET'S DISCUSS WHY See TE side column for answers to Exercises 2–3 and 7.

1. Why do you think baseball teams offer season tickets?
   to encourage attendance; to make more tickets available to more people at a lower price
2. Besides saving money, why do you think some people prefer buying season tickets?
3. Some people attend more weekday games. Others attend more weekend games. What might be some reasons for this?
4. What would be the price of 1 box ticket at the season ticket price? Is it more or less than a single box ticket? Why?
   $9.49; less than a $10.00 single ticket because you're buying all 79 tickets
5. What is the price of 1 ticket of a season ticket for all 47 weekday games? For 30 weekday games? In each case, how much would you save over single box ticket prices?
   47 games: $8.68; save $1.32   30 games: $6.23; save $3.77
6. How much would single box tickets cost for all weekday games? What percent discount would you get with a season ticket?
   $470; 13%

LET'S SEE WHAT YOU WOULD DO

7. You want to attend all weekday games. Your friend wants to attend only weekend games. In how many ways could you both buy the tickets you want? Which way would save both of you the most money?
8. You bought a full-season ticket for $750. Your boss transfers you to an out-of-town branch office for the month of August. You will miss some games, and the ticket is nonrefundable. What can you do to avoid losing money?
   ask boss to reimburse you for money you'd lose; sell tickets to friends or relatives

CALCULATOR PERCENT ON A CALCULATOR

Some calculators have a **percent key**, [%]. When you use the percent key on most calculators, do not press the is-equal-to key, [=]. You can use a calculator to find a percent of a number.

Find 30% of 60.

Procedure	Calculator Entry	Calculator Display
1. Enter the base number for which you must find the percent.	[6] [0]	60.
2. Enter the [×] key.	[×]	60.
3. Enter the percent, including the [%] key.	[3] [0] [%]	18.

So 30% of 60 = 18.

Some calculators do not have a [%] key. If your calculator does not, first rename the percent as a decimal. Then multiply.

Find 125% of 40. Do not use the percent key, [%].

Procedure	Calculator Entry	Calculator Display
1. Enter 1.25. THINK: 125% = 1.25	[1] [.] [2] [5]	1.25
2. Multiply by 40.	[×] [4] [0] [=]	50.

So 125% of 40 = 50.

Use a calculator to find the answer.

1. 45% of 100 45
2. 15% of 400 60
3. 125% of 500 625
4. 28% of 66 18.48
5. 8% of 90 7.2
6. 175% of 1,750 3,062.5
7. 325% of 0.6 1.95
8. 2.5% of 420 10.5
9. 0.05% of 18 0.009
10. 13.2% of 650 85.8
11. 450% of 9.8 44.1
12. 0.03% of 5 0.0015

Use a calculator to solve.

13. For Sunday's game, 92% of the 52,000 tickets were sold. How many were sold? 47,840
14. Alex put down a 15% deposit on a $480 gym membership fee. How much did he pay? $72

OBJECTIVE

- Find a percent of a number, using a calculator.

USING THE PAGE

After students read the first example, remind them that *of* usually indicates multiplication. Then point out that in this example, the order in which the numbers are entered does make a difference. The base number, 60, must be entered first.

Point out that once the percent key is pressed, the computation is completed. On most calculators, students should not press the is-equal-to key after pressing the percent key. If they do press the is-equal-to key, the number 1,080 may appear in the display. You may wish to have students explain what happened. (18 was multiplied by the original factor of 60, which became a multiplication constant.)

In the second example, a percent must be renamed as a decimal (125% = 1.25) before the calculator is used. Review with students the rule for renaming percents as decimals. Point out that once the percent has been renamed as a decimal, the order in which the factors are entered does not matter.

Have students use their calculators to work through the 2 examples in the text.

OBJECTIVE

- Test the prerequisite skills needed to learn the concepts developed in Chapter 6.

USING THE TEST

The Pre-Skills Test is designed to diagnose students' strengths and weaknesses on prerequisite skills necessary to study the mathematics in Chapter 6.

Assign the Pre-Skills Test. Allow students to work together in pairs or in small groups. Group members should help those who demonstrate a misunderstanding of a concept or a weakness in a skill.

The table below correlates the items on the Pre-Skills Test with the prerequisite skill and the lesson(s) in the chapter for which it is needed.

Item(s)	Prerequisite skill	Lesson(s)
1–6	Round money amounts to the nearest cent.	6.1, 6.3, 6.5
7–12	Rename decimals as percents.	6.1
13–24	Rename percents as decimals.	6.1, 6.3, 6.6
25–34	Add and subtract money amounts.	All
35–40	Divide money amounts.	6.1, 6.2, 6.5
41–44	Find the percent of a number.	6.1, 6.3, 6.6

References to specific skills that are reviewed in the Skills Bank are given on the individual lessons themselves.

Pre-Skills TEST

Write the money amount to the nearest cent.

1. \$1.072 \$1.07
2. \$4.638 \$4.64
3. \$8.5552 \$8.56
4. \$12.0318 \$12.03
5. \$34.7966 \$34.80
6. \$83.24913 \$83.25

Rename as a percent. Round to the nearest percent.

7. 0.194 19%
8. 0.327 33%
9. 0.2318 23%
10. 0.7551 76%
11. 0.5983 60%
12. 0.2914 29%

Rename as a decimal.

13. 4% 0.04
14. 5% 0.05
15. 6% 0.06
16. 8% 0.08
17. 15% 0.15
18. 18% 0.18
19. $5\frac{1}{2}\%$ 0.055
20. $7\frac{1}{2}\%$ 0.075
21. $8\frac{1}{2}\%$ 0.085
22. 104% 1.04
23. 108% 1.08
24. 110% 1.10

Add.

25. \$10.50 + \$18.25 + \$13.89 \$42.64
26. \$9.35 + \$31.65 + \$42.69 \$83.69
27. \$110.63 + \$1.85 + \$6.25 \$118.73
28. \$216.40 + \$5.36 + \$11.75 \$233.51

Subtract.

29. 69¢ − 20¢ 49¢
30. 83¢ − 25¢ 58¢
31. 59¢ − 15¢ 44¢
32. \$35.29 − \$16.50 \$18.79
33. \$115.50 − \$20.75 \$94.75
34. \$249 − \$25.80 \$223.20

Divide. Round to the nearest tenth of a cent.

35. 79¢ ÷ 3 26.3¢
36. 55¢ ÷ 4 13.8¢
37. \$1.75 ÷ 4 \$0.438
38. \$2.25 ÷ 8 \$0.281
39. \$5.45 ÷ 12 \$0.454
40. \$9.95 ÷ 20 \$0.498

Find the answer to the nearest cent.

41. 4% of \$35 \$1.40
42. 6% of \$15.50 \$0.93
43. $5\frac{1}{2}\%$ of \$98 \$5.39
44. $7\frac{1}{2}\%$ of \$20.89 \$1.57

Chapter 6 BASIC PURCHASES

Most people like to buy new clothes. What factors would determine your choices if you had a limited amount of money to spend?

OBJECTIVE

- Formulate problems, using text and data.

USING THE CHAPTER OPENER

Discuss students' answers to the question on the page. Some factors that might determine choices include need (Does the student really need the item or simply want it?), use (Will the student get a lot of wear out of an item?), existing wardrobe (Does an item go with the student's other clothes, or will additional purchases have to be made?), and style (If a bargain item is totally out of style or unflattering to the student, should it be bought?).

Note that student opinions may vary on this topic. This could lead to an interesting debate about the need for fashionable clothes.

COOPERATIVE LEARNING

Have heterogeneous groups of students collect newspaper ads and catalogs from various stores, or have them go out to the actual stores to do the project. Explain that they are to put together 2 complete outfits. The budget for the first outfit is $100; the budget for the second is $1,000. They should not forget shoes and accessories they would normally wear. They may want to use the members of their group as fashion consultants.

After they select the clothing and the stores at which they would purchase the items, have them present their 2 outfits. This would include a description of each item and its cost. Then have them discuss the differences between the 2 outfits (other than the cost). These might include quality, status labels, style, and so on. Have the class decide if each budget was spent wisely.

SUPPLEMENTARY MATERIALS

TRP Group Projects, p. 6

OBJECTIVES

- Compute the sale price of a discounted item.
- Compute the discount rate.
- Compute the marked-up price of an item.

TEACHING THE LESSON

WARM-UP Write the following exercises on the chalkboard or an overhead transparency.

Rename as a decimal.

1. 32% **2.** 8.5% **3.** $12\frac{1}{2}$%

Find the answer.

4. 10% of \$56.80 **5.** 20% of \$240

(**1.** 0.32 **2.** 0.085 **3.** 0.125 **4.** \$5.68 **5.** \$48)

INTRODUCTION Explain that wise consumers often wait for sales to make large purchases. Discuss why many stores use sales to clear out old inventory. Ask students to name some ways that stores let us know about their upcoming sales.

INSTRUCTION Familiarize students with the term discount by discussing what a "20% Off" sale in a store means. Use the introductory situation to distinguish the amount of money saved, or the discount, from the percent off, or the discount rate.

Use Example 1 to show students how to find the sale price of an item. Discuss the formulas and the fact that finding the sale price is usually a 2-step procedure: (1) first finding the amount of the discount and (2) then finding the sale price. Explain that a 20% discount leaves 80% of the price. Show that the sale price can also be determined in 1 step and that this procedure results in the same answer.

Use Example 2 to show how the formula used in Example 1 can be transformed to find the discount rate. Review with students why the discount is divided by the regular price to find the discount rate.

6.1 BUYING AUDIO AND VIDEO EQUIPMENT

You have been saving for a new stereo system. The system you want regularly sells for \$329.99, but is now on sale for 20% off.

The amount that an item on sale has been reduced is called the **markdown,** or **discount.** When this amount is expressed as a percent, it is called the **discount rate.**

EXAMPLE 1 How much will the stereo cost on sale?

THINK: Discount rate: 20% or 0.2 **Regular price: \$329.99**

DISCOUNT RATE × REGULAR PRICE = DISCOUNT

REGULAR PRICE − DISCOUNT = SALE PRICE

1. Multiply to find the discount. $0.2 \times \$329.99 = \$65.998 \approx \$66.00$
2. Subtract to find the sale price. $\$329.99 - \$66.00 = \$263.99$

The stereo will cost \$263.99 on sale.

Sometimes the discount rate needs to be computed.

EXAMPLE 2 An AM/FM stereo cassette player that regularly sells for \$84.99 is on sale for \$59.98. What is the discount rate for this sale?

THINK: Regular price: \$84.99 **Sale price: \$59.98**

DISCOUNT ÷ REGULAR PRICE = DISCOUNT RATE

1. Subtract to find the discount. $\$84.99 - \$59.98 = \$25.01$
2. Divide to find the discount rate. $\$25.01 \div \$84.99 = 0.2942699$
3. Rename the decimal as a percent. $0.2942699 = 29.42699\%$

The discount rate is about 29%.

Most **retail** stores buy their merchandise **wholesale** and increase the price before selling it to their retail customers. This increase in price is called the **markup,** and is often expressed as a percent of the wholesale price.

EXAMPLE 3 A store buys a VCR for \$175 and marks the price up 40%. For what will the store sell the VCR?

THINK: Markup rate: 40% or 0.4 **Wholesale price: \$175**

1. Multiply to find the markup. $0.4 \times \$175 = \$70$
2. Add to find the retail price. $\$175 + \$70 = \$245$

The store will sell the VCR for \$245.

FOR DISCUSSION See TE side column.

The same color TV is advertised at one store for $200 less 20%, and at another store for $230 less 25%. Which store offers the greater discount? Which offers the better buy?

PRACTICE EXERCISES

Remember to estimate whenever you use your calculator.

Find the sale price to the nearest cent.

Item	Regular price	Discount rate	Sale price	
Cassette recorder	$25.99	30%	• 1. ■	$18.19
Turntable	$178.50	25%	2. ■	$133.87
19″ color TV	$339.99	20%	3. ■	$271.99
4-head VCR	$399.50	33%	4. ■	$267.66
Digital receiver	$429.50	15%	5. ■	$365.07

Find the discount rate to the nearest percent.

Item	Regular price	Sale price	Discount rate	
Video-cassette tape	$3.99	$3.29	• 6. ■	18%
B/W TV	$79.99	$59.99	7. ■	25%
Stereo cabinet	$69.90	$55.00	8. ■	21%
AM/FM car stereo	$89.99	$79.99	9. ■	11%
Large-screen TV	$1,299.00	$949.99	10. ■	27%

Find the retail price to the nearest cent.

Item	Wholesale price	Markup rate	Retail price	
Compact-disc player	$75.00	80%	• 11. ■	$135.00
VCR	$129.95	50%	12. ■	$194.93
Video camera	$760.00	45%	13. ■	$1,102.00
Color TV	$249.99	24%	14. ■	$309.99

Solve.

15. A stereo regularly sells for $229.00. The regular discount is 10%. During a sale, it is reduced an additional 25%. What is the final price?
$154.57

16. A store buys a VCR for $120. It marks the item up 60% and then sells it at a 10% discount. What is the final retail price?
$172.80

Remind students that if they are using a calculator, the display answer must be rounded and then the decimal changed to a percent.

Use Example 3 to demonstrate how to find the marked-up price of an item. Carefully review the terms retail and wholesale.

SKILLS BANK You may want to refer students to the following skills in the Skills Bank:

Subtract With Whole Numbers and Decimals, Skill 6

Multiply With Decimals, Skill 5

Find the Percent of a Number, Skill 2

Find the Percent One Number Is of Another, Skill 2

FOR DISCUSSION Students should explain that the $200 TV at 20% off would sell for $160, while the $230 TV at 25% off would sell for $172.50. Even though the second store offers the greater discount rate (25%), the first store offers the lower price.

ASSIGNMENTS

Level 1	1–3, 6–8, 11–13
Level 2	1–15
Level 3	3–5, 8–10, 13–16

FOLLOW-UP ACTIVITY

ENRICHMENT Provide students with the wholesale price of a particular piece of audio or video equipment. Select a markup rate and a discount rate and then determine the percent the sale price is of the original wholesale price. Investigate the difference between a 40% markup followed by a 25% discount and a 25% markup followed by a 40% discount.

SUPPLEMENTARY MATERIALS

TRP Practice, p. 38
TRP Reteaching, p. 32
TRP Transparency 19

OBJECTIVES

- Estimate and compute the cost of a purchase.
- Compute the change from a purchase.

TEACHING THE LESSON

WARM-UP Write the following exercises on the chalkboard or an overhead transparency.

Estimate the product.

1. 5 × $8.95 **2.** 4 × $19.79

3. 8 × $14.89 **4.** 6 × $28.99

(**1.** $45 **2.** $80 **3.** $80 **4.** $180)

INTRODUCTION Ask students to make a list of all the articles of clothing they are wearing. Then ask students to estimate whether $25 would be enough to replace everything they are wearing. Discuss with students how they arrived at an estimate for the cost of their clothing, paying particular attention to the highest and the lowest estimates.

INSTRUCTION Use the introductory situation and Example 1 to demonstrate how you use an advertisement to decide whether you can afford a particular purchase. Use the Clothes Sale ad to have students mentally estimate the cost of other purchases.

Use Example 2 to review how to compute the exact cost of a purchase and the amount of change you should receive. Explain that, although the store clerk will probably be using an electronic cash register, it is still important to estimate the cost of each purchase yourself to be sure you are not accidentally overcharged.

Use Example 3 to show how to determine how much is saved by buying clothes on sale. Discuss with students the advantages and disadvantages of waiting until particular items go on sale.

6.2 BUYING CLOTHES

You have been given $50 for new clothes. You see in the newspaper that jeans are on sale for $19.49 per pair at the local department store.

EXAMPLE 1 Do you have enough money to buy 2 pairs of jeans?

THINK: The jeans are about $20 each.

Estimate. 2 × $20 = $40

Yes, you have enough money for 2 pairs, even with sales tax.

EXAMPLE 2 Excluding sales tax, exactly how much will it cost to buy 2 flannel shirts and a pair of jeans? How much money will you have left?

1. Add to find the total cost. $12.99 + $12.99 + $19.49 = $45.47

The items will cost you $45.47.

2. Subtract to find how much is left. $50.00 − $45.47 = $4.53

You will have $4.53 left.

EXAMPLE 3 How much would you save by buying 2 pairs of jeans on sale?

THINK: **Sale price: $19.49** **Regular price: $25.99**

1. Subtract to find the savings on each pair. $25.99 − $19.49 = $6.50
2. Multiply to find the total savings. 2 × $6.50 = $13.00

You would save $13.00 by taking advantage of the sale.

FOR DISCUSSION See TE side column.

1. When items are on sale, why do you usually save more money on the more expensive items?
2. When estimating the cost of several items, why is it best to round each price up?

PRACTICE EXERCISES

Remember to estimate whenever you use your calculator.

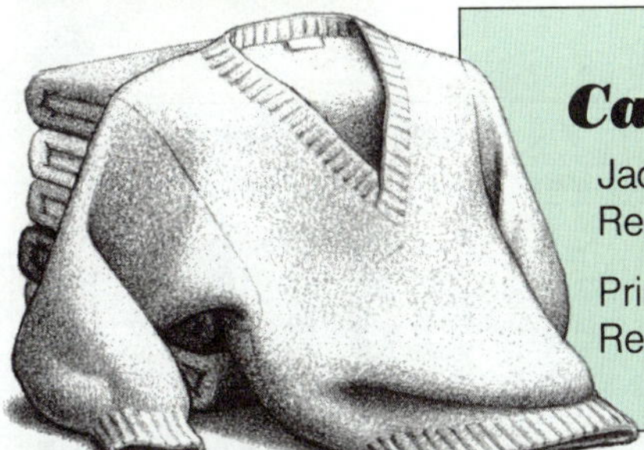

Casual Coordinates 25% off

Jacquard Sweaters Reg. $17.99 . . $13.49	Corduroy Skirts Reg. $15.99 . . $11.99
Print Shirts Reg. $13.99 . . $10.49	Corduroy Pants Reg. $16.99 . . $12.74

Use the ad to complete the table, excluding sales tax.

Purchase made	Estimated cost	Exact cost	Amount given	Change	
2 Sweaters	• 1. ■	• 2. ■	$30.00	3. ■	**3.** $3.02
1 Shirt and 1 skirt	4. ■	5. ■	$25.00	6. ■	**6.** $2.52
2 Shirts and 2 pants	7. ■	8. ■	$50.00	9. ■	**9.** $3.54
2 Skirts and 1 sweater	10. ■	11. ■	$50.00	12. ■	**12.** $12.53
3 Pants and 2 sweaters	13. ■	14. ■	$100.00	15. ■	**15.** $34.80

1. $28.00 **2.** $26.98 **4.** $23.00 **5.** $22.48 **7.** $48.00 **8.** $46.46 **10.** $38.00 **11.** $37.47 **13.** $67.00 **14.** $65.20

Use the ad above to solve.

16. Robin bought 1 pants. How much was saved by buying them on sale? $4.25

17. Joan bought 2 sweaters. How much was saved by buying them on sale? $9.00

	Tee Shirts	Shorts	Bathing Suits
Summer sale	$9.99	$15.99	$13.49
Regular price	$12	$22	$18

Use the Summer Sale ad to complete the table.

Purchase made	Regular price	Sale price	Amount saved	
1 Shirt and 1 shorts	•18. ■	•19. ■	20. ■	**20.** $8.02
3 Bathing suits	21. ■	22. ■	23. ■	**23.** $13.53
2 Shirts and 1 bathing suit	24. ■	25. ■	26. ■	**26.** $8.53
3 Shorts and 2 bathing suits	27. ■	28. ■	29. ■	**29.** $27.05

18. $34 **19.** $25.98 **21.** $54 **22.** $40.47 **24.** $42 **25.** $33.47 **27.** $102 **28.** $74.95

SKILLS BANK You may want to refer students to the following skills in the Skills Bank:

Add With Whole Numbers and Decimals, Skill 6

Subtract With Whole Numbers and Decimals, Skill 6

Multiply With Decimals, Skills 2, 5

FOR DISCUSSION

1. Students should explain that the same discount rate will produce greater savings on a more expensive item. Also, since the markup is usually greater on more expensive items, the discount can be greater too, and the store will still make a profit.
2. Students should explain that by rounding each price up, you generate an overestimate. This should assure that you have enough money to cover the purchase and the sales tax.

ASSIGNMENTS

Level 1	1–9, 16, 18–23
Level 2	1–29
Level 3	7–17, 24–29

FOLLOW-UP ACTIVITY

APPLICATION Provide students with a copy of a department-store or a discount-store advertising circular and tell them that they have $100 to spend. Ask students to select at least 4 items and spend as close to the $100 as possible. Try more difficult examples like buying exactly 5 items and spending as close to $80 as possible.

SUPPLEMENTARY MATERIALS

TRP Practice, p. 39
TRP Reteaching, p. 33
TRP Applications, p. 6
TRP Transparency 20

OBJECTIVES

- Compute the sales tax on a purchase.
- Use a table to find the sales tax on a purchase.

TEACHING THE LESSON

WARM-UP Write the following exercises on the chalkboard or an overhead transparency.

Rename as a decimal.

1. 3% **2.** $4\frac{1}{2}$% **3.** $8\frac{1}{4}$%

Compute mentally.

4. 1% of \$45 **5.** 2% of \$36

6. 10% of \$87.40 **7.** 5% of \$24.80

(**1.** 0.03 **2.** 0.045 **3.** 0.0825
4. \$0.45 **5.** \$0.72 **6.** \$8.74
7. \$1.24)

INTRODUCTION Use the introductory material to discuss what a sales tax is and why it is used. Ask students to name some of the many local or state services provided by revenue from a sales tax. Provide a sales tax table for the area in which they live. Discuss why many politicians consider the sales tax to be "painless tax."

INSTRUCTION Use Example 1 to demonstrate how to find the sales tax by using a sales tax table or by multiplying. Remind students that percents are easily converted to decimals and can then be entered into a calculator. Use a calculator to find the sales tax on several successive amounts to show how the sales tax table intervals are determined and how the sales tax is rounded up or down.

Use Example 2 to demonstrate how the sales tax on a purchase can often be computed mentally. Remind students that 1% of an amount can be found easily by moving the decimal point 2 places to the left. Students can verify the mental computation by using their calculators to multiply: 0.03 × \$42.75 = \$1.2825 ≈ \$1.28

6.3 SALES TAX

The most common form of tax we pay is the **sales tax.** Sales tax is computed as a percent of the total sales of goods and services. The money raised from the sales tax is a major source of revenue in many states and municipalities. So the sales tax rate varies from state to state and from city to city.

$7\frac{1}{2}$% Sales Tax Table

Amount		of sales tax	Amount		of sales tax	Amount		of sales tax
0.01 -	0.06	0.00	8.40 -	8.73	0.65	17.27 -	17.30	1.30
.07 -	.19	.01	8.74 -	8.86	.66	17.40 -	17.63	1.31
.28 -	.33	.02	8.87 -	8.97	.67	17.54 -	17.66	1.32
.34 -	.46	.03	9.00 -	9.13	.68	17.67 -	17.79	1.33
.47 -	.53	.04	9.14 -	9.26	.69	17.80 -	17.93	1.34
.60 -	.73	.05	9.27 -	9.39	.70	17.94 -	18.06	1.35
.74 -	.85	.06	9.40 -	9.53	.71	18.07 -	18.19	1.36
.87 -	.90	.07	9.54 -	9.66	.72	18.20 -	18.33	1.37
1.00 -	1.13	.08	9.67 -	9.79	.73	18.34 -	18.46	1.38
1.14 -	1.26	.09	9.80 -	9.93	.74	18.47 -	18.59	1.39
1.27 -	1.30	.10	9.94 -	10.06	.75	18.60 -	18.73	1.40
1.40 -	1.53	.11	10.07 -	10.19	.76	18.74 -	18.86	1.41
1.54 -	1.66	.12	10.20 -	10.33	.77	18.87 -	18.99	1.42
1.67 -	1.79	.13	10.34 -	10.46	.78	19.00 -	19.13	1.43
1.80 -	1.93	.14	10.47 -	10.59	.79	19.14 -	19.26	1.44
1.94 -	2.06	.15	10.60 -	10.73	.80	19.27 -	19.39	1.45
2.07 -	2.19	.16	10.74 -	10.86	.81	19.40 -	19.53	1.46
2.20 -	2.33	.17	10.87 -	10.99	.82	19.54 -	19.66	1.47

EXAMPLE 1 How much sales tax will you pay on a \$10.80 book if the sales tax rate is $7\frac{1}{2}$?

THINK: $7\frac{1}{2}\% = 0.075$

Method 1

Use the partial table. Since \$10.80 is between \$10.74 and \$10.86, the sales tax will be \$0.81.

Method 2

Multiply to find the tax.
0.075 × \$10.80 = \$0.81
The sales tax will be \$0.81.

EXAMPLE 2 About how much sales tax will you pay on a purchase of \$42.75 worth of records if the sales tax rate is 3%?

THINK: $1\% = \frac{1}{100}$ $\quad \frac{1}{100}$ **of \$42.75 = \$42.75 ÷ 100 = \$0.4275, or about \$0.43.**

Multiply mentally: 3% is about 3 × \$0.43, or \$1.29.

The sales tax will be about \$1.29.

EXAMPLE 3 If the sales tax rate is 4%, what is the total cost of a \$19.50 pair of sneakers?

THINK: 4% = 0.04 104% = 1.04

Method 1

1. Multiply to find the sales tax.
   0.04 × \$19.50 = \$0.78
2. Add to find the total cost.
   \$19.50 + \$0.78 = \$20.28

Method 2

THINK: Adding 4% makes the total cost 104% of the original price.

Multiply to find the total cost.

1.04 × \$19.50 = \$20.28

The total cost of the sneakers will be \$20.28.

FOR DISCUSSION See TE side column.

Some people travel to other states or municipalities to do their shopping. Why do you think they do this? What hidden expenses are involved?

PRACTICE EXERCISES

Remember to estimate whenever you use your calculator.

Use the partial sales tax table on page 108 to find the $7\frac{1}{2}$% sales tax on the purchase price.

• 1. $2.22 $0.17 • 2. $18.32 $1.37 3. $10.85 $0.81 4. $17.75 $1.33 5. $1.95 $0.15 6. $19.10 $1.43

Find the sales tax to the nearest cent.

Purchase price	Sales tax rate	Sales tax
$23.50	4%	• 7. ■
$98.99	5%	• 8. ■
$375.00	6%	9. ■

Purchase price	Sales tax rate	Sales tax
$165.95	$2\frac{1}{2}$%	10. ■
$46.85	$5\frac{1}{2}$%	11. ■
$3,575.00	$8\frac{1}{4}$%	12. ■

7. $0.94 8. $4.95 9. $22.50 10. $4.15 11. $2.58 12. $294.94

Complete the table to find the estimated sales tax and exact sales tax. Round each amount to the nearest cent.

Purchase price	Sales tax rate	Estimated sales tax	Exact sales tax
$21.30	3%	• 13. ■	• 14. ■
$31.80	4%	15. ■	16. ■
$80.98	5%	17. ■	18. ■

13. $0.63 14. $0.64 15. $1.28 16. $1.27 17. $4.05 18. $4.05

Complete the table to find the sales tax and total cost. Round each amount to the nearest cent.

Purchase price	Sales tax rate	Sales tax	Total cost	
$17.50	5%	• 19. ■	• 20. ■	$0.88; $18.38
$85.99	$4\frac{1}{2}$%	21. ■	22. ■	$3.87; $89.86
$257.00	6%	23. ■	24. ■	$15.42; $272.42
$4,786.50	$8\frac{1}{2}$%	25. ■	26. ■	$406.85; $5,193.35
$10,612.75	$7\frac{1}{4}$%	27. ■	28. ■	$769.42; $11,382.17

Use Example 3 to show how the total cost, including sales tax, can be found in 2 steps or, more easily, in 1 step.

SKILLS BANK You may want to refer students to the following skills in the Skills Bank:

Round Whole Numbers and Decimals, Skill 4

Equivalent Fractions, Decimals, and Percents, Skills 2, 4

Multiply With Decimals, Skills 4, 5

Find the Percent of a Number, Skill 2

FOR DISCUSSION Students should explain that the sales tax varies from city to city and from state to state. For example, the same $10,000 automobile will cost less in a state with a 4% sales tax than in one with a 7% sales tax. However, the costs of traveling to the other state to buy, return, or repair the item are hidden costs that must be considered.

ASSIGNMENTS

Level 1	1–12, 13–14, 19–22
Level 2	1–28
Level 3	4–6, 13–18, 23–28

FOLLOW-UP ACTIVITY

ENRICHMENT Provide students with a blank mock-up of a sales tax table. Ask students to create their own sales tax table for a $5\frac{1}{2}$% sales tax on purchases from $0.01 to $5.00. Remind students that all money amounts are rounded up or down to form the intervals on the table and suggest that they use the constant multiplier function on their calculators.

SUPPLEMENTARY MATERIALS

TRP Practice, p. 40
TRP Reteaching, p. 34
TRP Applications, p. 7
TRP Transparency 20

OBJECTIVE

- Complete a catalog purchase order form.

TEACHING THE LESSON

WARM-UP Write the following exercises on the chalkboard or an overhead transparency.

Compute and simplify.

1. 2 lb 4 oz + 3 lb 8 oz

2. 4 lb 6 oz + 6 lb 10 oz

3. 10 lb 9 oz + 8 lb 12 oz

4. 3 × 5 lb 4 oz **5.** 4 × 6 lb 5 oz

(**1.** 5 lb 12 oz **2.** 11 lb **3.** 19 lb 5 oz **4.** 15 lb 12 oz **5.** 25 lb 4 oz)

INTRODUCTION Explain that when buying from a catalog, you must specify exactly what you want to order. Provide students with various catalogs. Have them use the catalogs to make a list of the variables that must be considered when making purchases (type of item, price, size, color, style, number). Ask students what other information must be included on a catalog form in order to ensure that you get what you want (name, address, correct calculation of cost).

INSTRUCTION Use the introductory situation and the Example to demonstrate how you complete a catalog order form. You may wish to use the Catalog Order Form offered in the Teacher's Resource Package and have each student complete the order form along with you. The Catalog Order Forms can also be used for Exercises 7–10.

Explain that the total price is computed by multiplying the item price by the quantity. Review how the shipping weight of 1 lb 12 oz is doubled to get 3 lb 8 oz and how the shipping weight is totaled to get 7 lb 4 oz. Point out how the delivery charge is found in the table and that the 5% sales tax is computed *only* on the $140.00 total for merchandise.

6.4 CATALOG SHOPPING

Many people shop at home by buying items through **catalogs.**

You find some winterwear that you really like in a catalog and decide to order 1 medium blue coat and 2 pairs of size 10 white stirrup pants.

EXAMPLE Complete the order form for this purchase. Find the total cost.

1. Enter the information needed on the top 2 lines.
2. Total the price and shipping weight.
3. Use the partial delivery charges table to find the delivery charge for Zone 3.
4. Add the handling charge and the 5% sales tax to find the total amount.

WINTERWEAR

TOGGLE COAT plays soft, brushed cotton flannel against rugged wide wale cotton corduroy. Polyester fiberfill; cotton flannel lining. Machine wash.
Misses: S(4-6), M(8-10), L(12-14).
B9157 1291 - Peach.
B9157 1289 - Blue.
B9157 1192 - Pink.
B9157 1290 - White.
State color and size letter.
(3 lb 12 oz) 64.00

STIRRUP PANTS with V yoke below the waist. Leg tapers to the ankle, elastic stirrups. Wide wale cotton corduroy. Machine wash.
Misses: 4, 6, 8, 10, 12, 14.
B9157 2230 - Peach.
B9157 2228 - Blue.
B9157 2231 - Pink.
B9157 2229 - White.
State color and size.
(1 lb 12 oz) 38.00

BOOTS finish the long line. Woven cotton outside with warm acrylic pile lining. Rubber soles.
B(Medium) full sizes: 5, 6, 7, 8, 9, 10.
For ½ sizes, order next largest full size.
W9157 1479 - Peach.
W9157 1476 - Blue.
W9157 1480 - Pink.
W9157 1478 - White.
State color, size and B width.
(2 lb 4 oz) 50.00

name of item	item number	color	size	qty.	price each	total price	ship. wt.
TOGGLE COAT	B9157 1289	BLUE	M	1	64 00	64 00	3 12
STIRRUP PANTS	B9157 2229	WHITE	10	2	38 00	76 00	3 8

Find your Zone at right ▶ **DELIVERY CHARGES**

Shipping Weight ▼	Zone 1	Zone 2	Zone 3	Zone 4
Minimum charge	$ 3.75	$ 4.00	$ 4.25	$ 4.50
2 lb 1 oz to 5 lb	5.25	5.50	6.00	6.50
5 lb 1 oz to 10 lb	7.00	7.50	8.25	9.00
10 lb 1 oz to 15 lb	8.75	9.50	10.25	11.50
15 lb 1 oz to 20 lb	10.00	10.75	12.00	14.75
20 lb 1 oz to 25 lb	11.50	12.50	13.75	16.50

total for merchandise	$ 140 00	7 lb 4 oz
total delivery charge	8 25	On charge orders, we do this part for you. Credit orders are subject to credit approval.
handling charge	$ 1 00	
add sales tax of 5%	7 00	
TOTAL AMOUNT	$ 156 25	

FOR DISCUSSION See TE side column.

1. What are some advantages and disadvantages of catalog shopping?
2. Why do you often pay more for an item when you purchase it through a catalog?

PRACTICE EXERCISES

Remember to estimate whenever you use your calculator.

Use the partial delivery charges table on page 110 to find the delivery charge for the order.

1. 6 lb 4 oz to Zone 2 $7.50
2. 22 lb to Zone 4 $16.50
3. 1 lb 15 oz to Zone 2 $4.00

Use the catalog information on page 110 to find the total cost to Zone 1.

• 4. 1 pair of boots $58.75
5. 1 pair of stirrup pants $44.65
6. 2 toggle coats $142.40

Use the catalog information at the right and the delivery charges table on page 110 to complete an order form for the order. Then find the total cost. Check students' order forms.

• 7. 1 large blue-denim jacket to Zone 3 $58.45
8. Size 31 pleated cords, 1 pair gray and 1 pair blue, to Zone 2 $66.35
9. 1 medium navy sweater and 2 large blue oxford shirts to Zone 4 $99.89
10. 1 small blue-denim jacket, 2 size 29 black pleated cords, and 2 small green sweaters to Zone 1 $203.28

D. THE JACKET IS 30% OFF. Denim on the outside, flannel on the inside, and corduroy on the collar. Zip front, snap cuffs, corduroy elbow patches. Two pockets. Pure cotton; dry clean. Blue.
Chests: S(34-36); M(38-40); L(42-44); XL(46).
State size letter. (3 lb)
A4656 1146T..............Reg. $70 NOW 49.00

THE SWEATER IS $15 OFF. It's loaded with paisley. Pure wool; long sleeves. Hand wash.
Chests: S(34-36); M(38-40); L(42-44); XL(46).
B4258 4855 - Navy. B4258 4856 - Green.
State color, size letter. (1 lb 6 oz)
Reg............................$55 NOW 39.99

OXFORD CLOTH SHIRT IS 20% OFF. Strong on stripes, with button-down collar; chest pocket; long sleeves. In yarn-dyed cotton/polyester. Machine wash.
Neck sizes: S(14-14 1/2); M(15-15 1/2); L(16 - 16 1/2); XL(17-17 1/2).
State color, size letter.
B4258 1002 - Blue/Green Stripe (shown)
B4258 1001 - Green/Burgundy Stripe
(12 oz).........................Reg. $30 NOW 24.00

PLEATED CORDS ARE 25% LESS than in many stores. In colors as versatile as they are uncommon. Pure cotton, with straight legs; back pockets. 35" finished bottoms. Machine wash.
Waists: 29, 30, 31, 32, 33, 34, 36, 38.
B4756 6200 - Taupe
B4756 6198 - Mallard Blue
B4756 6199 - Willow Gray
B4756 6788 - Black
State color, size. (1 lb 10 oz)
Elsewhere..................$38. NOW 28.50

Solve. Remember to include the delivery charge, the handling charge, and the sales tax.

11. How much is saved by ordering a denim jacket at the sale price? $22.05
12. How much is saved by buying a pair of pleated cords in a store with a 5% sales tax instead of ordering them from the catalog to Zone 4? $5.50

Discuss with students the purpose of a handling charge and why the delivery charge varies by zone.

SKILLS BANK You may want to refer students to the following skills in the Skills Bank:

Multiply With Decimals, Skill 5

Customary Units of Capacity and Weight, Skill 2

Find the Percent of a Number, Skill 2

FOR DISCUSSION

1. Students should explain that the advantages of catalog buying are convenience and a wide choice of merchandise. However, the disadvantages are not being able to see the item before buying it and the time it takes to receive an item.
2. Students should explain that delivery charges and handling charges often result in higher costs.

ASSIGNMENTS

Level 1	1–8
Level 2	1–11
Level 3	4–12

FOLLOW-UP ACTIVITY

ENRICHMENT Distribute pages from a catalog and blank order forms to all students. Have them place an order for merchandise of their choice. Ask students to use estimation skills and fill out the order form with the goal being to make the order come as close to $150 as possible, including delivery and handling charges and sales tax. After students have completed the task, discuss estimation methods used to get close to $150.

SUPPLEMENTARY MATERIALS

TRP Practice, p. 41
TRP Reteaching, p. 35
TRP Lesson Aids, p. 3
TRP Transparency 21

OBJECTIVES

- Determine whether a labeled price makes sense.
- Use unit prices to determine the better buy.
- Compute the total cost of a purchase when using coupons.

TEACHING THE LESSON

WARM-UP Write the following exercises on the chalkboard or an overhead transparency.

Estimate the product or quotient.

1. 5 × 89¢ **2.** 2.1 × $4.89

3. $1.99 ÷ 4 **4.** $2.39 ÷ 7.8

(**1.** $4.50 **2.** $10 **3.** $0.50 **4.** $0.25)

INTRODUCTION Ask students to bring in food ads from newspapers and to estimate their family's weekly food bill. Determine a rough average for the class and multiply by 52 to give the class a sense for how much an average family spends in a year.

INSTRUCTION Use Example 1 to demonstrate how to use estimation to check the accuracy of food labels, especially on meats, fish, deli items, and produce. Remind students to round up to get an overestimate.

Use Example 2 to show how unit pricing allows a shopper to compare prices and determine the better or best buy when an item comes in 2 or more package sizes. Provide a comparison involving a conversion of units (for example, 15 oz vs. 2 lb 10 oz). Explain why unit prices must be computed to the same unit.

Use Example 3 to review briefly how coupons and double-coupon promotions reduce costs.

SKILLS BANK You may want to refer students to the following skills in the Skills Bank:

Divide With Decimals, Skill 8

Compare and Order Whole Numbers and Decimals, Skill 5

6.5 BUYING FOOD

The weekly trip to the supermarket provides an excellent opportunity to shop wisely and save money. Since there are so many opportunities for errors in labeling or at the checkout line, it is very useful to estimate food costs.

EXAMPLE 1 Cut-up frying chicken is on sale for $1.99 per lb. One package is labeled 2.86 lb and marked $7.42. Does this price make sense?

THINK: Almost 3 lb at about $2.00 per lb

Multiply to estimate the cost. 3 × $2.00 = $6.00

So $7.42 does *not* make sense. It is too much.

Another way to shop wisely is to purchase the most economically-sized package of a certain food. A good way to determine which package or container is the better buy is to compare the **unit prices.**

EXAMPLE 2 Raisin bran is available in 15-oz boxes for $1.89 or in 20-oz boxes for $2.29. Which is the better buy? Divide to find the cost per ounce for each package.

15-oz box	**20-oz box**
$1.89 ÷ 15 = $0.126, or 12.6¢ per oz	$2.29 ÷ 20 = $0.1145, or 11.5¢ per oz

The 20-oz box is the better buy.

Another way to save money on food is to use **coupons.**

EXAMPLE 3 The 6-oz container of frozen orange juice sells for 79¢. How much will you pay for the juice if you use your "15¢ off" coupon during the double-coupon promotion at the store?

THINK: During the promotion, the coupon doubles in value to 30¢.

Subtract to find the sale price. 79¢ − 30¢ = 49¢

You will pay only 49¢ for the juice.

FOR DISCUSSION See TE side column.

1. Why do people sometimes buy frozen dinners? Why are these dinners usually more expensive than food prepared at home?
2. Why do people sometimes buy items that are not on their shopping lists? What do supermarkets do to encourage this type of buying?

PRACTICE EXERCISES

Remember to estimate whenever you use your calculator.

Estimate to determine whether the labeled price *does* or *does not* make sense.

- • **1.** 2 lb of apples at 89¢ per lb — $1.78
- • **2.** 3.4 lb of hamburger at $1.49 per lb — $3.29
- **3.** 0.76 lb of salami at $2.79 per lb — $2.69
- **4.** a 10.6-lb turkey at 89¢ per lb — $9.43
- **5.** 1.24 lb of cole slaw at 49¢ per lb — 89¢
- **6.** a 3.89-lb steak at $2.79 per lb — $10.85

1. does **2.** does not **3.** does not **4.** does **5.** does not **6.** does

Find the unit prices to the nearest tenth of a cent. Identify which is the better buy.

See page 476 for answers to Exercises 9, 12, 15, 18.

• **7.** 6 oz for 49¢ 8.2¢ per oz
10 oz for 69¢ 6.9¢ per oz
10 oz for 69¢

• **8.** 9 oz for 89¢ 9.9¢ per oz
14 oz for $1.49 10.6¢ per oz
9 oz for 89¢

9. 5 oz for 75¢
10 oz for $1.25

10. 20 oz for $1.79 9.0¢ per oz
32 oz for $2.99 9.3¢ per oz
20 oz for $1.79

11. 24 oz for $2.69 11.2¢ per oz
35 oz for $3.49 10.0¢ per oz
35 oz for $3.49

12. 12 oz for $3.36
18 oz for $4.70

13. 15 oz for $2.69 17.9¢ per oz
1 lb 9 oz for $4.29 17.2¢ per oz
1 lb 9 oz for $4.29

14. 24 oz for $1.59 6.6¢ per oz
2 lb 5 oz for $2.39 6.5¢ per oz
2 lb 5 oz for $2.39

15. 6.5 lb for $5.85
10 lb for $8.75

16. 1.75 lb for $3.79 $2.17 per lb
2.5 lb for $4.99 $2.00 per lb
2.5 lb for $4.99

17. 1 L for $1.19 $1.19 per L
2.5 L for $2.49 $1.00 per L
2.5 L for $2.49

18. 0.5 L for 79¢
2 L for $2.79

Find the total cost of the shopping list.

19.

2 cans tomato sauce at 39¢ each
4 lb hamburger at $1.59 per lb
2.3 lb grapes at $1.19 per lb
1 box muffins at $1.79
1 can coffee at $3.89

You have a "50¢ off" coupon on the coffee during a double-coupon sale.
$14.56

20.

1 box detergent at $3.89
2 boxes cereal at $1.79 each
5 cans beans at 2 for 89¢
1.45 lb sliced ham at $2.79 per lb
3 cans tuna at $1.09 each

You have a "25¢ off" coupon on the cereal and a "50¢ off" coupon on the detergent during a double-coupon sale.
$15.52

EXTENSION Yearly Savings

Small savings each week can add up quickly. Large bottles of apple juice can be purchased for $1.69 for the name brand and $1.09 for the store brand. If you consume 2 bottles of apple juice each week, how much can you save in a year by buying the store brand?
$62.40

FOR DISCUSSION

1. Students should explain that frozen dinners are convenience foods that can be prepared quickly and easily in either a conventional oven or a microwave oven. They are usually more expensive than preparing the meal from scratch at home because, in addition to buying the food, the producer must also cook and package the food, adding to its final cost.
2. Students should explain that stores often create large displays of certain food items, place items that are popular and often more expensive at eye level, and display items that are not really needed at the checkout lines. Each of these strategies encourages customers to purchase extra items.

EXTENSION Show students how easy it is to save significant amounts of money by purchasing store brands of certain items.

ASSIGNMENTS

Level	Exercises
Level 1	1–3, 7–12, 19
Level 2	1–20, EX
Level 3	4–6, 13–20, EX

FOLLOW-UP ACTIVITY

APPLICATION Distribute a weekly supermarket sales flyer to each student and ask students to develop a 10-item shopping list. Remind students to note how many of each item or the weight of each item. Ask students to first estimate the total cost of the shopping list. Then have them compute the exact cost of the items on the list and compare it to their estimate.

SUPPLEMENTARY MATERIALS

TRP Practice, p. 42
TRP Reteaching, p. 36
TRP Transparency 21

OBJECTIVES

- Use a menu to estimate the cost of a meal.
- Compute the cost of meals, including tax and tip.

TEACHING THE LESSON

WARM-UP Write the following exercises on the chalkboard or an overhead transparency.

Estimate the sum.

1. $4.95 + $3.25 + $1.85 ($10)
2. $9.90 + $2.89 + $0.90 ($14)
3. $0.39 + $0.79 + $1.10 ($2)
4. $5.90 + $2.35 + $3.20 ($11)

INTRODUCTION Discuss with students that, as lifestyles change and more and more families have 2 working parents, more and more meals are eaten outside of the home. Explain that restaurant meals are often much more convenient than meals at home, but that they are also often much more expensive. For this reason, it is important to be able to estimate the cost of meals in restaurants.

INSTRUCTION Use the Country Sandwich Shop menu and Example 1 to show how to estimate the cost of the 2 lunches. Remind students to round up in order to get an overestimate. Ask students to use estimation to create orders that would be about $5, $15, and $25.

Use Example 2 to show how to calculate the actual cost of a meal, including tax and tip. Remind students how to calculate the 5% sales tax and the 15% tip on a calculator. Explain that the tip is usually calculated on the total cost of the meal, excluding sales tax. Point out that many people compute a 15% tip mentally by taking 10% of the cost and then adding half of that (5%) to arrive at 15%. Give students several samples of meal costs and ask them to estimate the tip.

6.6 EATING IN A RESTAURANT

SANDWICHES

Tuna $2.75
Roast Beef 3.75
Ham 3.45
Egg Salad 2.50
Steak 3.95

SPECIALS

Tuna Sandwich
Lunch $4.20
(includes potatoes, medium juice)
Steak Sandwich
Lunch 4.90
(includes salad or cole slaw, medium juice)

SIDE ORDERS

Potatoes . . . . $ .95
Onions 1.25
Salad 65
Cole Slaw 75

BEVERAGES

Juice . . . . small .50
medium .70 large .90
Milk .60 Coffee .45
Tea .45

You and a friend shop for lunch at the Country Sandwich Shop. You order a roast beef sandwich, potatoes, and a medium juice. Your friend orders an egg salad sandwich and a large juice.

EXAMPLE 1 Will $12 be enough to cover both lunches? Estimate the total cost.

Your lunch:	roast beef sandwich	$3.75	
	potatoes	0.95	About $6.00
	medium juice	0.70	
Friend's lunch:	egg salad sandwich	2.50	
	large juice	0.90	+ About $4.00
			About $10.00

So $12 should be enough for the lunches.

EXAMPLE 2 Including a 5% sales tax and a 15% tip, exactly how much will the lunches cost?

1. Add to find the total cost.

$3.75 + $0.95 + $0.70 + $2.50 + $0.90 = $8.80

2. Multiply to find the sales tax.

THINK: 5% = 0.05 0.05 × $8.80 = $0.44

3. Multiply to find the tip.

THINK: 15% = 0.15 0.15 × $8.80 = $1.32

4. Add the total cost, the sales tax, and the tip. $8.80 + $0.44 + $1.32 = $10.56

The entire lunch with sales tax and tip will cost $10.56.

FOR DISCUSSION See TE side column.

1. Why do you think the Tuna Sandwich Special is less expensive than ordering each of the items separately?
2. In some cities the sales tax rate is about 8%. Why do many people just double the sales tax to compute a 15% or 16% tip?

PRACTICE EXERCISES

Remember to estimate whenever you use your calculator.

Use the Country Sandwich Shop menu to estimate whether the meal should cost *more than $5.00* or *less than $5.00.*

• 1. Tuna sandwich, coffee less than $5.00
• 2. Steak sandwich, onions, milk 2. more than $5.00
3. Tuna Sandwich Special, tea less than $5.00
4. Steak Sandwich Special, small juice 4. more than $5.00
5. Ham sandwich, cole slaw, coffee less than $5.00
6. Egg salad sandwich, potatoes, milk less than $5.00

Use the menu to compute the cost of the meal, including a 6% sales tax and a 15% tip. Then find the amount of change from $20.

• 7. Roast beef sandwich
Salad
Large juice $6.42; $13.58

• 8. Ham sandwich
Potatoes
Coffee $5.87; $14.13

9. 2 Tuna sandwiches
Onions
2 Medium juices
9. $9.86; $10.14

10. Ham sandwich
Roast beef sandwich
2 Cole slaws
2 Coffees
$11.62; $8.38

11. Tuna Sandwich Special
2 Egg salad sandwiches
1 Milk
2 Medium juices
$13.55; $6.45

12. 2 Steak Sandwich Specials
2 Salads
2 Large juices
$15.61; $4.39

Solve.

• 13. A meal costs $21.50. The sales tax rate is 8%. Find the sales tax. Mentally compute a 16% tip.
$1.72; $3.44

14. A meal costs $28.00. The 16% tip is $4.48. Mentally compute the 8% sales tax.
$2.24

15. A complete lunch consists of 1 sandwich, 1 side order, and 1 beverage. What is the difference between the cost of the most expensive lunch and the least expensive lunch?
$2.50

16. Suppose a family of 4 eats lunch at the Country Sandwich Shop. Make up an order that comes closest to $20.00, including a 5% sales tax and a 15% tip.
Answers may vary. Check students' answers.

SKILLS BANK You may want to refer students to the following skills in the Skills Bank:

Add With Whole Numbers and Decimals, Skill 6

Find the Percent of a Number, Skill 2

FOR DISCUSSION

1. Students should explain that "specials" often come on 1 plate, while ordering the items separately means 2 or more plates. Also, the portions served on specials are sometimes smaller than when the items are ordered separately.
2. Students should explain that, when the sales tax is about 8%, doubling the tax is the same as finding about 16% of the meal cost. Since a tip is generally about 15%, doubling the 8% tax is a simple way to figure out the tip.

ASSIGNMENTS

Level	Exercises
Level 1	Odd 1–11, 13–14
Level 2	1–15
Level 3	4–6, 10–16

FOLLOW-UP ACTIVITY

COOPERATIVE LEARNING
Distribute to students copies of a restaurant menu. Pair students and ask them to take turns placing an order that comes as close as possible to an amount named by the other student. The students then calculate the bill, including tax and tip, to see how close the first student came.

SUPPLEMENTARY MATERIALS

TRP Practice, p. 43
TRP Reteaching, p. 37
TRP Transparency 22

OBJECTIVE

- Decide, given relevant information, where and how to eat dinner.

TEACHING THE LESSON

WARM-UP Write the following exercises on the chalkboard or an overhead transparency.

Multiply.

1. 2 × $1.50 **2.** 2 × $6.75

3. 2 × $0.50 **4.** 2 × $0.30

(**1.** $3.00 **2.** $13.50 **3.** $1.00 **4.** $0.60)

INTRODUCTION Present the problem situation to the class. Encourage students to name some of the factors they would use to make an informed decision about where to eat or to prepare a meal at home. List the factors that are named on the chalkboard and discuss how to find out information on each of these factors. Encourage students to name some factors they would use to determine to which restaurant to go.

INSTRUCTION Read the list of options that Les and Wanda have prepared. Ask students to discuss fully the features of each. Ask if there is any other information that Les and Wanda should have. Discuss the decision-making factors.

Have small groups of students complete Exercises 1–16 in the table in order to compare the 3 options. Ask students to write statements that describe the data now in the table. Then ask them to write conclusions that can be drawn from the data. Discuss the appropriateness of these statements and conclusions and lead a discussion about interpretations and possible misinterpretations of the data.

DECISION MAKING

6.7 COMPARING FOOD COSTS

(Open-ended problem solving)

After a busy day it is often easiest to just take out food or to go out to a nearby restaurant for dinner. But not preparing dinner at home can become very costly.

PROBLEM

Les and Wanda have been ordering take-out food and eating in restaurants frequently. Because they are trying to save money for a new car, they decide to examine a sample of their meal costs (excluding tax) to see if they can cut expenses. They made the following notes.

Chicken dinner at home:	4 pieces of chicken	$3.60	Time: 50 min
	Salad ingredients (serves two)	$2.00	Not very convenient
	1 package potatoes	$1.59	
	Rolls	$0.40	
	Juice (two glasses)	$0.40	
Chicken dinner from Barney's Take-Out:	4 pieces of chicken	$2.00 each	Time: 10 min
	Salads	$1.50 each	Very convenient
	Potatoes	$0.90 per order	
	Rolls	Included	
	Juice	$0.75 per person	
Chicken dinner at the Capital Diner:	Complete dinner (includes chicken, salad, potatoes, roll)	$9.95 each	Time: 35 min
			Convenient
	Juice	$1.00 per glass	
	Tip	15%	

DECISION-MAKING FACTORS

Cost Time Convenience Quality of food

DECISION-MAKING COMPARISONS

Compare the 3 options by completing the table.

Factor	Home	Barney's Take-Out	Capital Diner	
Cost for two: Chicken	• 1. ■	• 2. ■	Included	$3.60; $8.00
Salad	3. ■	$3.00	4. ■	$2.00; Included
Potatoes	$1.59	5. ■	6. ■	$1.80; Included
Rolls	7. ■	Included	8. ■	$0.40; Included
Juice	9. ■	10. ■	$2.00	$0.40; $1.50
Total cost	11. ■	$14.30	12. ■	$7.99; $25.19
Time	50 min	13. ■	14. ■	10 min; 35 min
Convenience	15. ■	16. ■	Average	Not very; very

MAKING THE DECISIONS

Where should Les and Wanda get their dinner:

•17. If cost were the only factor?
Home

18. If time were the only factor?
Barney's Take-Out

19. If convenience were the only factor?
Barney's Take-Out

•20. How much can Les and Wanda save by eating at home instead of using Barney's Take-Out?
$6.31

21. If they have been using Barney's Take-Out an average of twice a week, how much could they save in a year by preparing dinner at home?
$656.24

22. How much could Les and Wanda save by preparing dinner at home instead of going to the Capital Diner?
$17.20

23. If they have been going to the diner about once a week, how much could they save in a year by preparing dinner at home?
$894.40

24. Where do you think the quality of the food is the highest? Why?
Answers may vary.

25. Is the cost of the food alone an accurate measure of the cost of preparing dinner at home? What other costs might be included?
Answers may vary.

26. Is the time for food preparation alone an accurate measure of the time involved in preparing a meal at home? Why?
Answers may vary.

Have students complete Exercises 17–26 in small groups, asking students to explain and defend their answers and decisions. Many answers involve subjective decisions and will therefore vary from student to student. Some students are very wary of defending their point of view among their classmates. Continually reinforce the fact that, in many consumer issues, best answers depend upon individual situations and preferences. Encourage all students to participate and express their opinions.

ASSIGNMENTS

Level 1	1–26
Level 2	1–26
Level 3	1–26

FOLLOW-UP ACTIVITY

COOPERATIVE LEARNING

Provide students with grocery-store advertisements and restaurant menus in order to compare the costs of a spaghetti dinner. Working in small heterogeneous groups, have students recreate Les and Wanda's type of research and, after making a comparison table, come to a written group consensus on which restaurant to patronize or whether to cook dinner at home.

SUPPLEMENTARY MATERIALS

TRP Practice, p. 44
TRP Transparency 22

OBJECTIVES

- Review vocabulary.
- Practice key chapter concepts and skills.

USING THE REVIEW

The Chapter Review is designed to help students prepare for taking the Chapter Test. The first section focuses on vocabulary. It requires students to select words to complete statements. The second section presents practice of key chapter skills.

Each item on the Chapter Review is referenced to the page on which the topic is taught in the Pupil's Edition. You may wish to have students refer to these pages to help review any concepts or skills they have not yet mastered.

It is suggested that students work in small-sized heterogeneous cooperative learning groups. Some cooperative learning methods that may be used are as follows:

1. After each student has independently completed the Chapter Review, a discussion should follow within each group about the solutions to the problems.
2. The group can complete the entire Chapter Review by working together to solve the problems.

End the lesson with an entire class discussion in which any questions brought up in group discussions are presented and answered.

CHAPTER REVIEW

Vocabulary Choose the letter of the word(s) that completes the sentence.

1. The sale price of an item is found by subtracting the ■ from the regular price. [104] c

   a. Sales tax **b.** Markup **c.** Discount

2. To determine which package of food is the better buy, compare the ■. [112] b

   a. Package prices **b.** Unit prices **c.** Package sizes

3. To find the total cost of a meal in a restaurant, include the sales tax and the ■. [114] a

   a. Tip **b.** Markup **c.** Coupon value

Skills Find the answer.

4. A knit shirt that regularly sells for \$11.99 is now on sale at 10% off. What is the sale price? [104] \$10.79
5. A ski sweater that regularly sells for \$35.00 is now on sale at 20% off. What is the sale price? [104] \$28.00
6. A cardigan sweater that regularly sells for \$25.00 is now on sale for \$22.50. How much can be saved by buying 2 sweaters on sale? [106] \$5.00
7. Lana bought a \$300 television. The sales tax rate is 6%. What was the total cost? [108] \$318
8. Fred bought a \$25 book. The sales tax rate is $7\frac{1}{2}$%. What was the total cost? [108] \$26.88
9. Karen bought a cassette player for \$75.50. The sales tax rate is $5\frac{1}{2}$%. What was the total cost? [108] \$79.65
10. A chef's-salad lunch special is \$4.50. Including a 5% sales tax and a 15% tip, what is the total cost? [114] \$5.40
11. A tuna-sandwich lunch special is \$3.75. Including a $7\frac{1}{2}$% sales tax and a 15% tip, what is the total cost? [114] \$4.59

Find the sales tax to the nearest cent. Then find the total amount. [110]

12. Merchandise: \$50.00
    Delivery and handling: \$4.10
    Sales tax rate: 6%
    \$3.00; \$57.10
13. Merchandise: \$150.00
    Delivery and handling: \$16.25
    Sales tax rate: $5\frac{1}{2}$%
    \$8.25; \$174.50

Which is the better buy? [112]

14. 3 oz for 95¢ or 8 oz for \$2.40
    8 oz for \$2.40
15. 1.56 lb for \$1.34 or 3 lb for \$2.64
    1.56 lb for \$1.34

CHAPTER TEST

Find the discount to the nearest cent. Then find the sale price.

1. Regular price: $12.59
   Discount rate: 10%
   $1.26; $11.33
2. Regular price: $400
   Discount rate: 30%
   $120; $280
3. Regular price: $179.89
   Discount rate: 25%
   $44.97; $134.92

Find the markup to the nearest cent. Then find the retail price.

4. Wholesale price: $50
   Markup rate: 75%
   $37.50; $87.50
5. Wholesale price: $125.50
   Markup rate: 30%
   $37.65; $163.15
6. Wholesale price: $360.99
   Markup rate: 50%
   $180.50; $541.49

Use the ad to answer Exercises 7–11.

7. Sal bought 2 clock radios on sale. How much change did he get from $75? $3

How much can be saved with the sale?

8. Digital clock radio
   $9
9. Jeans jacket
   $2.76

Find the discount rate to the nearest percent.

10. Digital clock radio
    20%
11. Jeans jacket
    15%

Find the sales tax to the nearest cent. Then find the total amount.

12. Merchandise: $100.00
    Delivery and handling: $8.50
    Sales tax rate: 5%
    $5.00; $113.50
13. Merchandise: $75.89
    Delivery and handling: $6.25
    Sales tax rate: $7\frac{1}{2}\%$
    $5.69; $87.83

Find the unit prices. Then determine which is the better buy.

14. 4 oz for 39¢ 9.75¢ per oz
    7 oz for 67¢ 9.57¢ per oz
    7 oz for 67¢
15. 10 oz for 79¢ 7.9¢ per oz
    15 oz for $1.17 7.8¢ per oz
    15 oz for $1.17
16. 1.2 lb for $2.50
    3 lb for $6.03

16. $2.08 per lb; $2.01 per lb; 3 lb for $6.03

Solve.

17. Turkey is on sale for $1.49 per lb. One package is labeled 2.79 lb and marked $5.40. Does this make sense?
    No
18. You buy bread for $1.19 and 1.5 lb of fruit for 39¢ per lb and use a "20¢ off" coupon. What is the total cost?
    $1.58

Use the menu to find the cost of the meal, including a 5% sales tax and a 15% tip. Round all amounts to the nearest cent.

THE SANDWICH SHOPPE

SANDWICHES		SIDE ORDERS	
Egg salad	$1.95	Salad	$0.90
Tuna	2.25	Potatoes	0.85
Ham and cheese	3.10	Coleslaw	0.65

BEVERAGES

Milk	0.75	Juice	0.50	Soda	0.80

19. 1 Egg salad
    1 Tuna
    2 Salads
    $7.20
20. 2 Ham and cheese
    2 Tuna
    3 Potatoes
    $15.90

OBJECTIVE

- Evaluate achievement of chapter objectives.

USING THE TEST

The Chapter Test may be used as a post-test to evaluate student achievement. However, you may wish to use the Chapter Post-Test offered in the Teacher's Resource Package or to design your own chapter test. If this page is not used as a test, you may want to assign it as additional review or practice.

The test items are correlated to the chapter objectives in the table below.

Chapter objectives	Test items
A. Compute discount, sale price, and discount rate.	1–3, 7–11
B. Compute markup and retail price.	4–6
C. Compute sales tax and total cost.	12–13
D. Compute and compare unit prices.	14–16
E. Determine if purchase prices make sense.	17
F. Compute total purchase cost when using coupons.	18
G. Use a menu to find the total cost of a meal, including tax and tip.	19–20

SUPPLEMENTARY MATERIALS

TRP Ch. 6 Post-Test Form A, pp. 1–2
TRP Ch. 6 Post-Test Form B, pp. 1–2

OBJECTIVE

- Understand that if you know when items are on sale, you can save money.

USING THE PAGE

Begin discussion by asking students what items they would like to buy, and when and where they plan to buy these items.

Read the names of the items given in the monthly lists. Have students work individually or in pairs to answer Exercises 1–11. Summarize the answers as follows: You can save money and buy the items you want or need during off-months when prices are at their lowest. Remind students there are also discount stores that sell many items at even lower than regular store sale prices.

Have students work individually to answer Exercise 12. Then have them share their answers and reasons in a whole-class discussion.

ANSWERS/EXPLANATIONS

4. To make room for spring/summer inventory; to make a profit on left-over inventory before styles and seasons change
5. September, October; to clear out remaining items and make room for stock for the new winter season
10. To get rid of older models; to make room for new inventory
11. September, October; to sell off summer inventory (which also takes up a lot of storage space)
12. March—garden supplies (prepare to plant after the last frost); April—painting equipment and spring cleaning supplies (start to spruce up homes for warm weather); April—lawn/garden furniture (for more outdoor fun and entertaining); sales on all these items so stores can make room for spring/summer inventory

MONEY TIPS

You can get the best buys on some items if you buy them at certain times of the year.

LET'S LOOK AT THE FACTS

February, March, and April are considered some of the best months to buy some items. Read the list of sale items for each month.

FEBRUARY
- Air conditioners
- Curtains
- Humidifiers
- Housewares
- Lamps
- Stereo equipment
- Storm windows
- Used cars

MARCH
- Children's shoes
- Garden supplies
- Hosiery
- Infants' clothing
- Luggage
- Winter coats
- Winter sports equipment

APRIL
- Painting equipment
- Spring cleaning supplies
- Fabrics
- Lawn and garden furniture

LET'S DISCUSS WHY

See TE side column for answers to Exercises 4–5 and 10–12.

Winter Coats

1. Which month is a sale month? March
2. During which months are they worn? Nov., Dec., Jan., Feb., March
3. When do you think stores would like to empty out their inventory of them? What would they replace them with in the store? Feb. and March; spring coats and jackets
4. Why are winter coat sales held in March?
5. During which other month(s) of the year might winter coats also be on sale? Why?

Air Conditioners

6. Which month is a sale month? February
7. During which months are they used most often? July and August
8. By which time do you think some stores would like to make room for newer models? April or May
9. By which time do you think other stores would like to use these sales to bring in customers after the holiday sales? January and February
10. Why are air conditioner sales held in February?
11. During which other month(s) of the year might air conditioners also be on sale? Why?

LET'S SEE WHAT YOU WOULD DO

12. Next summer, you plan to repaint your house, to plant a garden, and to furnish your patio. Which items could you buy on sale in February, March, or April? Tell why each would be on sale at that time.

ESTIMATION SKILLS DECIMAL SUMS APPROXIMATELY EQUAL TO 1

Front-end estimation of some sums can be adjusted by examining decimals for sums that are approximately equal to 1.

1. Add the whole numbers for a first estimate.
2. Examine the decimals.
3. Adjust the first estimate.

Examples

	②. 6	
	②. 5	THINK: 0.6 + 0.5 is about [1].
	①. 7	
	+ 0 . 2 9	THINK: 0.7 + 0.29 is about [1].
First estimate:	⑤ +	about [2] = 7 **Final estimate**

	⑧. 5 2	
	①. 1 3	THINK: 0.52 + 0.5 is about [1].
	⑥. 5	THINK: 0.13 + 0.79 is about [1].
	+ ⑤. 7 9	
First estimate:	⑳ +	about [2] = 22 **Final estimate**

Use front-end estimation with adjusting to estimate the sum.

1.	2.	3.	4.	5.
1.9	10.8	1.2	4.9	8.52
2.1	5.3	2.3	10.4	6.5
3.3	2.4	5.3	6.4	1.13
+4.7	+ 5.5	+6.2	+ 3.2	+5.79
12	24	15	25	22

6.	7.	8.	9.	10.
6.31	2.22	20.71	21.62	30.54
7.8	10.59	30.18	40.2	50.12
2.27	7.15	9.78	10.39	18.8
+6.59	+ 4.29	+40.24	+ 5.75	+21.5
23	24	101	77	121

11.	12.	13.	14.	15.
7.1	3.2	10.1	8.32	4.72
4.9	5.1	4.4	2.71	2.09
3.1	2.4	20.5	1.49	2.29
2.4	1.1	5.5	4.5	5.88
1.1	2.2	15.6	5.16	7.42
+6.6	+4.3	+10.2	+1.8	+1.8
25	18	66	24	24

OBJECTIVE

- Adjust front-end estimated sums of decimals.

USING THE PAGE

You might remind students that an unadjusted front-end estimate of a sum is never an overestimate. The adjustment method described in this lesson provides a closer estimate, which may be either slightly greater than or slightly less than the actual sum.

The procedure involves examining the decimals for single decimals that are approximately equal to 1 or for groups of 2 or more decimals that have a sum equal to or approximately equal to 1. This can often be done in more than 1 way. In Exercise 4, for example, one student might think: $0.9 \approx 1$ and $0.4 + 0.4 + 0.2 = 1$. Another student might think: $0.9 + 0.2 \approx 1$ and $0.4 + 0.4 \approx 1$. In either case, the sum is about 23 + 2, or 25.

After students master this skill, they may wish to see how closer estimates can be obtained in some cases by looking for decimals with a sum of about 1 *and* for decimals with a sum of about 0.5. The following example illustrates the procedure.

3.942	
5.136	$0.942 + 0.136 \approx 1$
7.127	
1.384	$0.127 + 0.384 \approx 0.5$
+2.025	

First estimate: 18
\+ about: 1.5
Final estimate: 19.5

OBJECTIVE

- Test the prerequisite skills needed to learn the concepts developed in Chapter 7.

USING THE TEST

The Pre-Skills Test is designed to diagnose students' strengths and weaknesses on prerequisite skills necessary to study the mathematics in Chapter 7.

Assign the Pre-Skills Test. Allow students to work together in pairs or in small groups. Group members should help those who demonstrate a misunderstanding of a concept or a weakness in a skill.

The table below correlates the items on the Pre-Skills Test with the prerequisite skill and the lesson(s) in the chapter for which it is needed.

Item(s)	Prerequisite skill	Lesson(s)
1–4	Write money amounts in words.	7.1, 7.3
5–10	Add and subtract money amounts.	7.1, 7.2, 7.3
11–14	Find the percent of a number.	7.4
15–20	Multiply money amounts, decimals, and fractions.	7.4
21–24	Convert units of time.	7.4
25–28	Multiply and add money amounts.	7.5

References to specific skills that are reviewed in the Skills Bank are given on the individual lessons themselves.

Chapter 7 Checking and Savings Accounts

Pre-Skills Test

Write the money amount in words.

1. $48.75
2. $104.50
3. $279.30
4. $1,483.89

Compute.

5. $63.80 + $58.25 + $124.50 $246.55
6. $293.84 + $53.32 + $127.03 $474.19
7. $93.53 − $4.24 + $83.18 $172.47
8. $530.49 − $214.96 + $430.51 $746.04
9. $27.39 + $192.80 − $67.31 − $63.81 − $3.92 $85.15
10. $1,943.84 − $294.73 − $97.59 − $326.16 − $27.27 + $639.20 $1,837.29

Find the answer to the nearest cent.

11. 5% of $293.00 $14.65
12. $4\frac{1}{2}$% of $374.22 $16.84
13. $6\frac{1}{2}$% of $93.48 $6.08
14. $5\frac{3}{4}$% of $1,200 $69

Multiply.

15. $800 × 0.03 × $\frac{1}{4}$ $6
16. $1,500 × 0.09 × $2\frac{1}{2}$ $337.50
17. $2,200 × 0.055 × 3 $363
18. $840 × 0.0675 × $4\frac{1}{2}$ $255.15

Multiply. Round to the nearest cent.

19. $280 × 1.3728 $384.38
20. $1,400 × 2.5182 $3,525.48

Solve.

21. How many quarter years are there in 4 y? 16
22. How many half years are there in $6\frac{1}{2}$ y? 13
23. How many months are there in $2\frac{1}{4}$ y? 27
24. How many quarter years are there in $7\frac{1}{2}$ y? 30

Find the answer.

25. 13 × 25¢ + $4.00 $7.25
26. 7 × 20¢ + $5.00 $6.40
27. 19 × 15¢ + $6.00 $8.85
28. 9 × 30¢ + $7.50 $10.20

1. forty-eight dollars and seventy-five cents
2. one hundred four dollars and fifty cents
3. two hundred seventy-nine dollars and thirty cents
4. one thousand, four hundred eighty-three dollars and eighty-nine cents

OBJECTIVE

- Formulate problems, using text and data.

USING THE CHAPTER OPENER

Discuss students' answers to the question on the page. Banks offer a variety of types of accounts to meet different customer needs, but also to compete with other banks. Also, the banks in a community will often engage in a "war," each trying to offer better gifts for opening an account; more services such as automatic teller machines, drive-up windows, or evening and Saturday hours; and more kinds of accounts. Interest rates vary based on the type of account, the life of a certificate, the minimum balance required, and so on.

COOPERATIVE LEARNING

Write the following on the chalkboard.

MONTHLY BUDGET	
Food	$320
Housing	$750
Telephone	$ 50
Transportation	$125
Clothing	$
Miscellaneous	$
Savings	$
	$1,600

Assign heterogeneous groups and explain that the budget is for a family of 4 that includes a couple and their 2 children ages 16 and 12. Tell the groups that they are to fill in the 3 blanks in the budget based on their own best judgment, but they must include something for savings. Then they are to visit local banks and gather information about available savings plans. They are to use this information to develop a savings plan for the family that will provide the greatest possible interest.

SUPPLEMENTARY MATERIALS

TRP Group Projects, p. 7

OBJECTIVES

- Write checks and complete checking account deposit slips.
- Compute a running balance in a check register.

TEACHING THE LESSON

WARM-UP Write the following exercises on the chalkboard or an overhead transparency.

Compute.

1. $28.39 + $75.89 − $60.43 − $13.38
2. $249.53 − $79.42 − $124.86 − $8.29
3. $527.83 − $8.20 − $83.57 + $135.00

(**1.** $30.47 **2.** $36.96 **3.** $571.06)

INTRODUCTION Begin the discussion of checking accounts by asking students why someone would want to have a checking account. Discuss the advantages of a checking account, including the convenience, the need to carry less cash, and the ease of paying bills and keeping records. Remind students that, since it is unwise to send cash through the mail, some bills can be paid conveniently only by check. You may want to get the forms necessary to open a checking account from a local bank and discuss how you open a checking account.

INSTRUCTION Use Example 1 and a large facsimile of a check on the chalkboard to show how to write a check. Blank check forms are offered in the Teacher's Resource Package. Point out the various features of the check and the importance of the computer numbers on the bottom that tell from which account to take the money. Explain that the amount of money the check is for is written both in numbers and in words to protect you from anyone tampering with your check.

7.1 CHECKING ACCOUNTS

The most commonly used bank account is a **checking account.** **Checks** are a convenient and safe way to make purchases and to pay bills. Checking accounts also help you to keep records on when and where you spend your money.

EXAMPLE 1 Suppose you wrote a check for $35.99 to pay the College Bookstore for your textbooks. Review the parts of the check below and describe each of your entries.

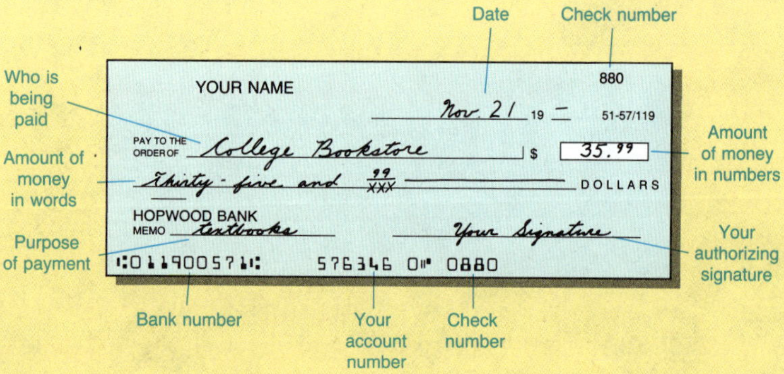

It is important that you have enough money in your checking account to cover each check that you write.

EXAMPLE 2 You earned $136.00 waiting on tables at the Student Union. You also had $20.00 in cash from tips. You deposited your paycheck and the cash in your checking account. Review the **deposit slip** below and describe each of your entries.

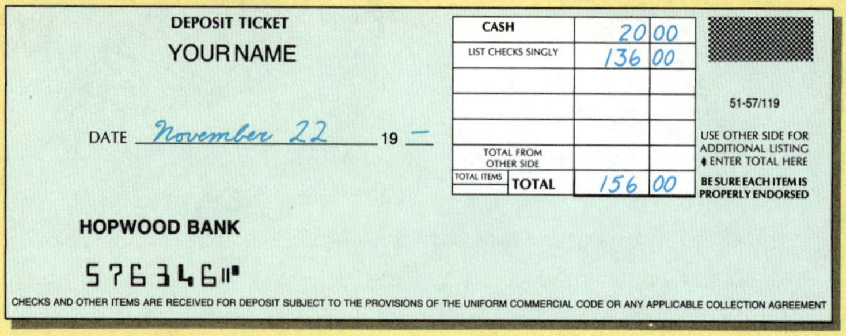

All checks and deposits should be recorded in your **check register**, which comes with your checks.

EXAMPLE 3 You completed the check register for your check to the bookstore and for your deposit. What is your new checking account balance?

THINK: The amount of each check must be subtracted from the balance. The amount of each deposit must be added to the balance.

Review the check register below.

RECORD ALL CHARGES OR CREDITS THAT AFFECT YOUR ACCOUNT

NUMBER	DATE	DESCRIPTION OF TRANSACTION	PAYMENT/DEBIT (−)	✓ T	FEE (IF ANY) (−)	DEPOSIT/CREDIT (+)	BALANCE	
							$ 679	09
877	11/20	Dr. Sandra Stone	$ 15.00		$	$	− 15	00
		dog's medicine					664	09
878	11/20	Mike's Sports Shop	42.00				− 42	00
		bowling equipment					622	09
879	11/21	Atlantic Electric Co.	87.42				− 87	42
		electric bill					534	67
880	11/21	College Bookstore	35.99				− 35	99
		textbooks					498	68
	11/22	Deposit				156.00	+ 156	00
							654	68

Your new balance is $654.68.

EXAMPLE 4 Susan's checking account balance was $214.72. She then wrote checks for $45.77, $79.35, and $105.60. She made a deposit of $90.50. Does she have enough money in her account to pay her $78.85 electric bill?

1. Add the deposit to the old balance. $214.72 + $90.50 = $305.22
2. Subtract the checks.
$305.22 − $45.77 − $79.35 − $105.60 = $74.50

Susan's new balance is $74.50. Therefore, she does not have enough money to cover the $78.85 electric bill.

FOR DISCUSSION See TE side column.

1. What does it mean to **bounce** a check? How can you prevent bounced checks?

2. Why should you start writing the amount of the check in numbers and in words as close to the left as possible and add a wavy line after writing the amount in words?

Point out these other facts about checks. (1) Sign a check only after completing all other parts of the check. (2) When a discrepancy occurs between the amount written in numbers and the amount written in words, the bank will use the amount written in words. (3) When you want to have cash taken from your account, you can issue a check made out either to yourself or to "Cash." (4) If you postdate a check, it will not be cashed by the bank until that date.

Use Example 2 and a large facsimile of a deposit slip on the chalkboard to show how to complete a deposit slip. Blank checking account deposit slip forms are offered in the Teacher's Resource Package. Explain that each deposited check is entered separately and that you should always get a receipt for your deposits. Also explain why and how to endorse checks before depositing them. Point out that, when you deposit checks into your account, you cannot immediately withdraw the amount deposited. Depending on individual banks, there is a clearance period.

Explain that, when depositing their paychecks, many people take a part of the check in cash and deposit the rest. Show how to complete a deposit slip for a situation involving a $225.00 paycheck with $50 cash received, leaving a net deposit of $175.00.

Use Example 3 to show how to maintain a check register and keep a running balance for a checking account. Blank check register forms are offered in the Teacher's Resource Package. Point out how the check and deposit from Examples 1 and 2 have been entered in the register.

Use Example 4 to show why it is important to keep an accurate and up-to-date record of your checking account. Remind students that keeping track of your checking account balance is just like keeping track of how much money you have in your purse or wallet.

SKILLS BANK You may want to refer students to the following skills in the Skills Bank:

Add With Whole Numbers and Decimals, Skill 6

Subtract With Whole Numbers and Decimals, Skill 6

FOR DISCUSSION

1. Students should explain that a check is "bounced" when there is not enough money in the account to cover it. Banks often charge you a $5 or $10 fine when checks are returned because of "insufficient funds." Students should also explain that this can be prevented by keeping a careful and accurate record of checks and deposits.
2. Students should explain that this prevents anyone from tampering with your check and changing the amount for which it is written. For example, if you leave room when writing $42.85, someone could easily make it $142.85.

ASSIGNMENTS

Level 1	1–13
Level 2	1–14
Level 3	1–14

FOLLOW-UP ACTIVITY

APPLICATION Provide each student with a page from a check register and a beginning balance of $300.00. Ask students to simulate an average family's checking account by making weekly paycheck deposits and entering checks to cover standard expenses. Tell students to keep a running balance and to be sure not to overdraw their account.

SUPPLEMENTARY MATERIALS

TRP Practice, p. 45
TRP Reteaching, p. 38
TRP Lesson Aids, pp. 4–6
TRP Transparency 23

PRACTICE EXERCISES

Remember to estimate whenever you use your calculator.

Make up a check for the payment. Use today's date and your own name and signature. Check students' work.

1. Check #256 to Robert's Department Store for $38.75
2. Check #257 to The Record Outlet for $27.30
3. Check #258 to the Telephone Company for $75.42
4. Check #259 to American National Bank for $189.50
5. Check #260 to 800 Equities for $489.75

Make up a deposit slip for the deposit. Check students' work.

• 6. A check for $145.00 and a check for $15.86 total deposit: $160.86
7. A check for $85.16 and $50.00 in cash total deposit: $135.16
8. Checks for $48.30, $194.74, and $83.90 total deposit: $326.94
9. Checks for $239.70 and $80.53 and $120.00 in cash total deposit: $440.23
10. Checks for $82.89, $130.15, and $480.25 and $250.00 in cash total deposit: $943.29

Make up a check register for the checking account transactions. Find the new balance. Check students' work.

•11. Old balance: $574.83
Check: #179 to Regal Gifts for $25.85 on December 15
Check: #180 to Monitor Leasing for $325.00 on December 18 for rent
Deposit: $288.92 on December 20
Check: #181 to The Toy Shoppe for $81.74 on December 22
Check: #182 to Dr. Louis Amara for $45.00 on December 23
$386.16

12. Old balance: $237.80
Check: #54 to The Boutique for $72.58 on March 3
Check: #55 to Home Finance Co. for $124.77 on March 3
Service charge: $6.40 on March 5
Deposit: $253.15 on March 8
Check: #56 to The Vision Place for $119.75 on March 10
Check: #57 to Pam's Fruit Stand for $25.39 on March 11
Deposit: $125.00 on March 12
$267.06

Solve.

13. Alan had $342.09 in his checking account. He then wrote checks for $142.50, $53.80, and $29.85. What is the least amount he must transfer from his savings account in order to cover his car payment of $185.00? $69.06

14. Rosa began the month with a checking account balance of $74.30. She then deposited $275.88 and wrote checks for $113.70, $65.36, and $97.25. How much must she deposit in order to pay her $295.00 rent? $221.13

OBJECTIVE

- Reconcile a bank statement balance with a check register balance.

TEACHING THE LESSON

WARM-UP Write the following exercises on the chalkboard or an overhead transparency.

Compute.

1. \$34.58 + \$28.95 + \$93.37

2. \$934.75 + \$35.90 − \$5.25

3. \$734.72 + \$147.83 + \$85.43 − \$6.89

4. \$84.86 − \$6.14 + \$36.85 − \$27.54

(**1.** \$156.90 **2.** \$965.40 **3.** \$961.09 **4.** \$88.03)

INTRODUCTION Discuss the various features of the checking account statement, including the account summary, the listing of checks, the deposits, the outstanding checks, and the service charge.

Ask students why it is likely that the statement ending balance will be different from the balance in their check register. Discuss the possibilities that all the checks you have written have not yet been received by the bank or that a recent deposit does not appear on the statement. Also remind students that the monthly service charge on the statement must be deducted in the register, or that an error might have been made in either the statement or the register.

INSTRUCTION Use the Example to demonstrate how to reconcile a bank statement with a check register. Explain to students that the process of reconciling balances involves adjusting each balance to account for missing information. It is often easiest to think in terms of What does the bank know that my register does not include? and What do I know that the bank has not included?

7.2 RECONCILING A CHECKING ACCOUNT STATEMENT

When you have a checking account, your bank returns your **canceled checks** each month and sends you a computerized **bank statement** that includes a listing of your deposits and the amount paid by each check you have written. To check for any errors, it is important to reconcile your **check register** with the monthly bank statement.

EXAMPLE Your 12/10 bank statement shows your deposits and checks for the past month and shows an ending balance of \$430.36. But, your check register shows a balance of \$350.19 on that date. Reconcile this account statement given the facts that it shows a \$7.25 service charge and that check #879 for \$87.42 is still outstanding.

ACCOUNT SUMMARY

BEGINNING BALANCE	TOTAL DEPOSITS & TRANSFERS AMOUNT	NUMBER	TOTAL CHECKS & PAYMENTS AMOUNT	NUMBER	SERVICE CHARGE	ENDING BALANCE
414 09	468 00	2	444 48	13	7 25	430 36

TRANSACTION DETAIL

CHECKING ACCOUNT 5763460

CHECK NUMBER	DATE POSTED	AMOUNT	DESCRIPTION OF TRANSACTION
			LAST STATEMENT BALANCE
	11-14	312 00	DEPOSIT
	11-22	156 00	DEPOSIT
874	11-18	2 00	PERSONAL CHECK
875	11-19	3 00	PERSONAL CHECK
876	11-19	42 00	PERSONAL CHECK
877	11-20	15 00	PERSONAL CHECK
878	11-20	42 00	PERSONAL CHECK
* 880	11-21	35 99	PERSONAL CHECK
881	11-23	46 00	PERSONAL CHECK
882	11-23	25 00	PERSONAL CHECK
883	11-25	23 00	PERSONAL CHECK
884	11-25	23 92	PERSONAL CHECK
885	11-25	56 04	PERSONAL CHECK
886	11-28	90 00	PERSONAL CHECK
887	12-10	40 53	PERSONAL CHECK
	12-10	7 25	SERVICE CHARGE

DAILY BALANCE SUMMARY

DATE	BALANCE
11-12	414 09
11-14	726 09
11-18	724 09
11-19	679 09
11-20	622 09
11-21	586 10
11-22	742 10
11-23	671 10
11-25	568 14
11-28	478 14
12-10	430 36

Check # 879 is outstanding.

Ending balance

Service charge deducted from account

THINK: The bank statement balance must be adjusted upward to account for outstanding deposits and downward to account for outstanding checks. Your register balance must be adjusted downward to account for the service charge.

Bank balance:	\$430.36	Register balance:	\$350.19
Plus outstanding deposits:	+\$0	Minus service charges:	−\$7.25
Minus outstanding checks:	−\$87.42	Adjusted balance:	\$342.94
Adjusted balance:	\$342.94		

Reconciled balances

SKILLS BANK You may want to refer students to the following skills in the Skills Bank:

Add With Whole Numbers and Decimals, Skill 6

Subtract With Whole Numbers and Decimals, Skill 6

FOR DISCUSSION

1. Students should explain that their canceled checks serve as records of their checking account transactions. These checks should be saved and compared to your checkbook register to ensure that you have not noted a check that was written. Many canceled checks are also needed for tax purposes.
2. Students should explain that many banks charge a fee per check, depending on your average balance. The fee appears as a service charge for keeping your account.

ASSIGNMENTS

Level 1	1–8
Level 2	1–8
Level 3	1–8

FOLLOW-UP ACTIVITY

ENRICHMENT Point out to students that another reason it is important to reconcile your checking account each month is to find arithmetic errors that you might have made in your check register. Present the following problem. One month, your check register balance was $65.80. Your bank statement balance was $151.20. There were outstanding checks of $32.40, $5.85, and $48.35, and a service charge of $1.10. Ask students to reconcile this account. By how much are the 2 balances off? What is the most likely reason?

SUPPLEMENTARY MATERIALS

TRP Practice, p. 46
TRP Reteaching, p. 39
TRP Transparency 23

FOR DISCUSSION

See TE side column.

1. Why is it important to save all of your canceled checks?
2. Why do banks often charge customers a monthly service fee on a checking account?

PRACTICE EXERCISES

Remember to estimate whenever you use your calculator.

Use the bank statement and the check register below for Exercises 1–6.

CHECK NUMBER	DATE POSTED	AMOUNT	DESCRIPTION OF TRANSACTION	DATE	BALANCE
CHECKING		ACCOUNT	5763460		
			LAST STATEMENT BALANCE	10-10	243 24
	10-24	312 00	DEPOSIT	10-24	555 24
683	10-27	18 00	PERSONAL CHECK	10-27	537 24
684	10-29	80 40	PERSONAL CHECK	10-29	456 84
* 686	11-12	38 00	PERSONAL CHECK	11-12	414 09
	11-12	4 75	SERVICE CHARGE		

NUMBER	DATE	DESCRIPTION OF TRANSACTION	PAYMENT/DEBIT (−)	✓ T	FEE (IF ANY) (−)	DEPOSIT/CREDIT (+)	BALANCE
							$ 243 24
	10/24	Deposit	$		$	$312 00	+ 312 00
							555 24
683	10/25	Norma's Yarn Shop Sweater yarn	18 00				− 18 00
							537 24
684	10/26	Southwestern Phone Co. telephone bill	80 40				− 80 40
							456 84
685	11/3	Worth's Catalog Sales jeans	19 85				− 19 85
							436 99
686	11/10	Wesley School	38 00				− 38 00
							398 99

What is the:

• 1. Statement ending balance? $414.09

• 2. Check register balance? $398.99

3. Outstanding check number? 685

4. Amount of the outstanding check? $19.85

5. Service charge? $4.75

6. Actual reconciled statement balance? $394.24

Use the given information to reconcile the bank statement balance with the check register balance. Find the adjusted balance for each.

7. Check register balance: $479.70
Statement ending balance: $582.43
Outstanding checks: $76.40, $29.83
Service charge: $3.50
$476.20

8. Check register balance: $551.38
Statement ending balance: $556.63
Outstanding deposit: $75.00
Outstanding checks: $62.45, $18.75
Service charge: $0.95
$550.43

7.3 SAVINGS ACCOUNTS

You earn about $80.00 per wk in a part-time job. You decide to open a **savings account** in order to save enough money for a down payment on a used car.

EXAMPLE 1 You deposited $50.00 in cash from your earnings plus a $25 birthday check in your savings account. Review the **deposit slip** below and describe each of your entries.

SAVINGS DEPOSIT
CITY SAVINGS BANK

OFFICE OF ACCOUNT: *Main St.*
DATE: *1/15/–*
PRINT NAME(S): *YOUR NAME*
ACCOUNT NUMBER: *073-51203*

	DOLLARS	CENTS
BILLS	50	00
COIN		
CHECK NO.	25	00
TOTAL ▶	$ 75	00

1-0622 2/89[76616] DO NOT WRITE BELOW THIS LINE

⑆5040⑈0000⑆

EXAMPLE 2 After several months, you have saved over $900. You withdraw $650.00 to make the down payment on the used car. Review the **withdrawal slip** below and describe each of your entries.

SAVINGS WITHDRAWAL
CITY SAVINGS BANK

DATE: *6/1/–*

OFFICE OF ACCOUNT: *Main St.*

PAY TO MYSELF OR BEARER	DOLLARS	CENTS
$	650	00

Six hundred fifty and 00/xx ———— DOLLARS

PRINT ACCOUNT NAME: *Your Name*
ACCOUNT NUMBER: *073-51203*
SIGNATURE: *Your Signature*

1-0622 2/89[76616] DO NOT WRITE BELOW THIS LINE

⑆5050⑈0000⑆

OBJECTIVES

- Complete savings account deposit and withdrawal slips.
- Compute a running savings account passbook balance.

TEACHING THE LESSON

WARM-UP Write the following exercises on the chalkboard or an overhead transparency.

Compute.

1. $87.53 + $174.75 + $13.69

2. $2,187.59 − $483.62 − $294.26

3. $492.36 + $238.42 + $4.33 + $39.52

4. $5,839.53 − $847.00 + $3,829.79

(**1.** $275.97 **2.** $1,409.71 **3.** $774.63 **4.** $8,822.32)

INTRODUCTION Begin the discussion of savings accounts by asking students why someone would want to open a savings account. Discuss the safety factors of keeping money in a bank, the benefit of earning interest, and the value of saving money for large purchases.

INSTRUCTION Use the introductory situation and Example 1 to demonstrate how money is deposited into a savings account. Blank savings account deposit slip forms are offered in the Teacher's Resource Package. Show students how the savings account deposit is like a checking account deposit, and how bills, coins, and checks are each entered separately.

Use Example 2 to demonstrate how money is withdrawn from a savings account. Blank sample savings account withdrawal slips are offered in the Teacher's Resource Package. Remind students that withdrawal slips are very similar to checks written to yourself for your own money. Review the importance of writing the amount of money both in numbers and in words to avoid errors or tampering.

Use Example 3 to show how a savings account record is kept in a passbook. Compare the passbook record with a checking account register. Remind students that they should check the calculations and the accuracy of each transaction every time they use their passbook.

Give students a beginning balance and simulate a series of savings account transactions, keeping a running balance and being sure not to overdraw the account.

SKILLS BANK You may want to refer students to the following skills in the Skills Bank:

Add With Whole Numbers and Decimals, Skill 6

Subtract With Whole Numbers and Decimals, Skill 6

FOR DISCUSSION

1. Students should explain that savings accounts are generally used to accumulate money while checking accounts are used to spend money conveniently. Also, interest is always paid on savings accounts, but only sometimes on checking accounts. Additionally, withdrawals from savings accounts are usually cash withdrawals made with a passbook, while withdrawals from checking accounts are made with checks.
2. Students should explain that interest is the cost of using someone's money. Since the bank is using your money by lending it to others, it pays you for this privilege. The bank makes its money by lending your money out at higher interest rates than it pays you.

Savings account records are kept in a **passbook.** You need to bring your passbook to the bank each time you make a deposit or a withdrawal. The bank's computer enters each transaction in your passbook.

EXAMPLE 3 You had a balance of $935.00 on June 2. You made a deposit of $62.50 on June 5, a withdrawal of $650.00 on June 11, another deposit of $52.00 on June 18, and had interest of $2.82 posted on July 1. What is your new savings account balance?

THINK: Add deposits and interest. Subtract withdrawals.

OLD BALANCE + DEPOSITS + INTEREST − WITHDRAWALS = NEW BALANCE
$935.00 + $114.50 + $2.82 − $650.00 = $402.32

Your new balance is $402.32. Review the passbook page below.

Name Your Name   Account # 073-51203

Date	Deposit	Withdrawal	Interest	Balance
6/2	$65.00			$935.00
6/5	$62.50			$997.50
6/11		$650.00		$347.50
6/18	$52.00			$399.50
7/1			$2.82	$402.32

FOR DISCUSSION See TE side column.

1. How are savings accounts different from checking accounts?
2. Why do banks pay interest on savings accounts?

PRACTICE EXERCISES

Remember to estimate whenever you use your calculator.

Make up a deposit slip for the deposit. Use today's date and your own name. Check students' work.

• 1. $85.00 in cash and a check for $43.29
total deposit: $128.29

2. $32.00 in cash, $2.75 in coins, and checks for $134.80 and $32.53
total deposit: $202.08

3. Checks for $45.89, $187.40, and $14.50 and $25.00 in cash
total deposit: $272.79

4. $10.50 in coins, $297.00 in cash, and checks for $135.00 and $273.65
total deposit: $716.15

Make up a withdrawal slip for the withdrawal. Use today's date and your own name and signature. Check students' work.

5. $280.00   6. $437.88   7. $47.00   8. $1,350.65

Complete the table.

Old balance	Transaction	New balance
$143.78	Deposit: $44.79	• 9. ■
$76.95	Withdrawal: $35.00	• 10. ■
$388.39	Deposit: $128.73	11. ■
$37.51	Interest payment: $5.83	12. ■
$1,836.68	Deposit: $387.24	13. ■
$502.31	Withdrawal: $280.00	14. ■

9. $188.57 10. $41.95 11. $517.12 12. $43.34 13. $2,223.92 14. $222.31

Compute the running balances for this savings account passbook.

Date	Deposit	Withdrawal	Interest	Balance	
10/3				$ 73.98	
10/5	$54.87			• 15. ■	$128.85
10/9		$45.00		• 16. ■	$83.85
10/12		$27.50		17. ■	$56.35
10/19	$86.36			18. ■	$142.71
10/20			$2.56	19. ■	$145.27
10/27		$64.00		20. ■	$81.27
10/30	$15.84			21. ■	$97.11

ASSIGNMENTS

Level 1	1–12, 15–21
Level 2	1–21
Level 3	1–21

FOLLOW-UP ACTIVITY

COOPERATIVE LEARNING
Provide each pair of students with a facsimile of a passbook. Ask students to decide who will play the banker and who will play the customer. Tell each pair to start with a balance of $100.00 and have the customer make 5 deposits and 2 withdrawals, each less than $100, without overdrawing the account. For each transaction, the student playing the banker should enter the amount in the passbook and use a calculator to keep a running balance. Students should then switch roles and check each other's work.

SUPPLEMENTARY MATERIALS

TRP Practice, p. 47
TRP Reteaching, p. 40
TRP Lesson Aids, pp. 7–8
TRP Applications, p. 8
TRP Transparency 24

OBJECTIVES

- Compute simple interest, using the interest formula.
- Compute compound interest with and without a table.

TEACHING THE LESSON

WARM-UP Write the following exercises on the chalkboard or an overhead transparency.

Find the percent of the number.

1. 5% of \$3,650 **2.** $4\frac{1}{2}$% of \$525

3. $6\frac{3}{4}$% of \$900 **4.** $7\frac{1}{4}$% of \$6,500

(**1.** \$182.50 **2.** \$23.63 **3.** \$60.75 **4.** \$471.25)

INTRODUCTION Ask students why banks pay interest on savings accounts. Explain that interest is really the cost of the bank using your money. Ask students what rates of interest the banks in their area usually pay on savings accounts.

INSTRUCTION Use the introductory material and Example 1 to demonstrate how to compute simple interest. Explain the meaning of the components of the interest formula and how they are used to calculate interest. Point out that the time is always calculated in years and discuss how to convert months to years. Remind students that interest is added to an account.

Use Example 2 to demonstrate how compounding of interest works in the depositor's favor. Review the terms *day-of-deposit* and *day-of-withdrawal* as they apply to the interest earned on savings accounts. Explain that care needs to be taken when you withdraw money so that you do not withdraw everything and thus lose earned interest. Point out that some banks require you to maintain a minimum balance of \$100. Explain how the annual interest rate must be divided by 2 for semiannual compounding and by 4 for quarterly compounding of interest.

7.4 SIMPLE AND COMPOUND INTEREST

The **interest** you earn on your checking or savings accounts is determined by:
the interest **rate** expressed as an annual percent,
the **principal,** or amount of money in your account, and
the length of **time** (in years) the money is used.

Simple interest is computed by using this formula.

INTEREST = PRINCIPAL × RATE × TIME, or $I = p \times r \times t$

EXAMPLE 1 You deposit \$400 in a savings account that pays 6% interest per y. How much interest will you earn in 3 mo? How much will then be in your account?

THINK: Principal: \$400 Rate: 6% = 0.06 Time: 3 mo = $\frac{1}{4}$ y = 0.25 y

1. Multiply to find the interest earned. $I = p \times r \times t$
   $I = \$400 \times 0.06 \times 0.25$
   $I = \$6$

You will earn \$6 in interest.

2. Add to find the new balance. \$400 + \$6 = \$406

At the end of 3 mo, you will have \$406 in your account.

For longer periods of time, banks **compound** your interest. They periodically add the interest to your account and compute interest in successive periods on each new higher principal.

EXAMPLE 2 Jeff deposits \$750.00 in a savings account that pays 5% interest compounded quarterly. How much will be in the account at the end of 1 y? How much interest will he earn in 1 y?

THINK: Principal: \$750 Rate: 5% = 0.05 Time: each $\frac{1}{4}$ y = 0.25 y

1. Compute 1st-quarter interest and the new balance (principal). \$750.00 × 0.05 × 0.25 = \$9.38; \$750.00 + \$9.38 = \$759.38
2. Compute 2nd-quarter interest and the new balance (principal). \$759.38 × 0.05 × 0.25 = \$9.49; \$759.38 + \$9.49 = \$768.87
3. Compute 3rd-quarter interest and the new balance (principal). \$768.87 × 0.05 × 0.25 = \$9.61; \$768.87 + \$9.61 = \$778.48
4. Compute 4th-quarter interest and the new balance (principal). \$778.48 × 0.05 × 0.25 = \$9.73; \$778.48 + \$9.73 = \$788.21

At the end of 1 y, Jeff's new savings account balance will be \$788.21.

5. Subtract to find the interest. \$788.21 − \$750.00 = \$38.21

Jeff will earn \$38.21 in interest.

Banks use a compound interest table to compute compound interest.

EXAMPLE 3 Delia deposits $2,000 in a savings account that pays 8% interest compounded quarterly. How much will she have at the end of 3 y?

THINK: 4 quarters per y means 12 quarters in 3 y.
8% annual interest is equal to 2% interest per quarter.

1. Use the table below to find the compounded value of $1.00. $1.00 at 2% for 12 periods will grow to $1.2682.
2. Multiply to find the new balance (principal). $2,000 × $1.2682 = $2,536.40

At the end of 3 y, Delia will have $2,536.40.

COMPOUND INTEREST TABLE

No. of Periods	1.5%	2%	2.5%	3%	3.5%	4%	5%
1	1.0150	1.0200	1.0250	1.0300	1.0350	1.0400	1.0500
2	1.0302	1.0404	1.0506	1.0609	1.0712	1.0816	1.1025
3	1.0457	1.0612	1.0769	1.0927	1.1087	1.1248	1.1576
4	1.0614	1.0824	1.1038	1.1255	1.1475	1.1699	1.2155
5	1.0773	1.1041	1.1314	1.1593	1.1877	1.2167	1.2763
6	1.0934	1.1262	1.1597	1.1941	1.2293	1.2653	1.3401
7	1.1098	1.1487	1.1887	1.2299	1.2723	1.3159	1.4071
8	1.1265	1.1717	1.2184	1.2668	1.3168	1.3686	1.4775
9	1.1434	1.1951	1.2489	1.3048	1.3629	1.4233	1.5513
10	1.1605	1.2190	1.2801	1.3439	1.4106	1.4802	1.6289
11	1.1779	1.2434	1.3121	1.3842	1.4600	1.5395	1.7103
12	1.1956	1.2682	1.3449	1.4258	1.5111	1.6010	1.7959
13	1.2136	1.2936	1.3785	1.4685	1.5640	1.6651	1.8856
14	1.2318	1.3195	1.4130	1.5126	1.6187	1.7317	1.9799
15	1.2502	1.3459	1.4483	1.5580	1.6753	1.8009	2.0789
16	1.2690	1.3728	1.4845	1.6047	1.7340	1.8730	2.1829
17	1.2880	1.4002	1.5216	1.6528	1.7947	1.9479	2.2920
18	1.3073	1.4282	1.5597	1.7024	1.8575	2.0258	2.4066
19	1.3270	1.4568	1.5987	1.7535	1.9225	2.1068	2.5270
20	1.3469	1.4859	1.6386	1.8061	1.9898	2.1911	2.6533
21	1.3671	1.5157	1.6796	1.8603	2.0594	2.2788	2.7860
22	1.3876	1.5460	1.7216	1.9161	2.1315	2.3699	2.9253
23	1.4084	1.5769	1.7646	1.9736	2.2061	2.4647	3.0715
24	1.4295	1.6084	1.8087	2.0328	2.2833	2.5633	3.2251
25	1.4509	1.6407	1.8539	2.0938	2.3673	2.6658	3.3864

FOR DISCUSSION See TE side column.

1. Which earns more in a year, an account compounded quarterly or the same account compounded daily? Why?
2. Why do banks pay you less interest than they earn on your money?

Explain to students that most accounts have interest compounded quarterly and the annual interest rate must be divided by 360 to calculate daily interest on your account. Remind students that large computers can calculate this interest very quickly.

Use Example 3 to show how a compound interest table makes the work of finding interest much easier. Explain that the numbers in the table refer to the value of $1.00 compounded at a particular interest rate for a specific period of time. Demonstrate how to use the table by matching the appropriate interest rate for the period (year, half year, or quarter year) with the appropriate number of periods.

Provide students with several examples of situations involving the use of the table, especially those involving the conversion of rates and periods. An example is converting 5% semiannually for 4 y to $2\frac{1}{2}$% for 8 periods.

SKILLS BANK You may want to refer students to the following skills in the Skills Bank:

Equivalent Fractions, Decimals, and Percents, Skill 4

Find the Percent of a Number, Skill 2

FOR DISCUSSION

1. Students should explain that the account that is compounded daily earns more than the account that is compounded quarterly. This is because the principal grows each day, rather than only once each quarter, in an account that is compounded daily.
2. Students should explain that banks take your money and lend this money to others at interest rates higher than those they pay their depositors so that they can earn a profit on the use of your money.

EXTENSION Explain the differences between regular checking accounts and new NOW and money market accounts. Discuss the various advantages and disadvantages of these accounts and why such interest-bearing accounts are not for everyone; for example, there are minimum balance requirements and heavy penalties for dropping below that minimum balance.

ASSIGNMENTS

Level 1	Odd 1–15
Level 2	1–16, EX
Level 3	9–16, EX

FOLLOW-UP ACTIVITY

APPLICATION To illustrate the impact of more frequent compounding periods, tell students that they have $1,000 to deposit into a savings account that pays 6% interest. Compute the value of the account at the end of 1 y if the interest is compounded (a) annually ($1,060), (b) semiannually ($1,060.90), and (c) quarterly ($1,061.40).

To illustrate the impact of small changes in interest rates, tell students that they have $50,000 to deposit into a savings account that compounds annually. Compute the value of the account at the end of 1 y if the interest rate is (a) 4% ($52,000), (b) 6% ($53,000), (c) 8% ($54,000), and (d) 10% ($55,000).

SUPPLEMENTARY MATERIALS

TRP Practice, p. 48
TRP Reteaching, p. 41
TRP Transparency 24

PRACTICE EXERCISES

Remember to estimate whenever you use your calculator.

Compute the simple interest earned and the new balance (principal).

	Principal	Rate	Time	
• 1.	$250	6%	4 y	$60; $310
• 2.	$500	5%	3 y	$75; $575
3.	$1,250	7%	2 y	$175; $1,425
4.	$8,020	8%	8 y	$5,132.80; $13,152.80

	Principal	Rate	Time	
5.	$3,500	$5\frac{1}{2}$%	7 y	$1,347.50; $4,847.50
6.	$850	$8\frac{1}{2}$%	10 y	$722.50; $1,572.50
7.	$2,630	$4\frac{3}{4}$%	3 y	$374.78; $3,004.78
8.	$5,875	$5\frac{3}{4}$%	5 y	$1,689.06; $7,564.06

Compute the new balance (principal) and the compound interest earned.

	Principal	Rate	Time	
• 9.	$1,000	5% compounded semiannually	1 y	$1,050.63; $50.63
10.	$5,000	$6\frac{1}{4}$% compounded semiannually	2 y	$5,654.91; $654.91
11.	$400	4% compounded quarterly	1 y	$416.24; $16.24
12.	$7,500	$7\frac{1}{2}$% compounded quarterly	18 mo	$8,384.30; $884.30

Use the compound interest table to find the new balance (principal).

	Principal	Rate	Time	
• 13.	$4,000	6% compounded semiannually	5 y	$5,375.60
14.	$6,500	8% compounded semiannually	7 y	$11,256.05
15.	$10,000	10% compounded quarterly	4 y	$14,845.00
16.	$3,200	6% compounded quarterly	6 y	$4,574.40

EXTENSION Money Market and NOW Accounts

Banks offer special checking accounts that allow you to earn interest. Most NOW and Money Market accounts require that you keep a minimum balance of $1,000 or $2,000. Interest is usually compounded daily and paid monthly on the basis of your average daily balance.

1. How do NOW and Money Market accounts differ from regular checking accounts?
   They pay interest. They require a minimum balance.
2. If your $5\frac{1}{2}$% NOW account has an average daily balance of $2,500 over the course of a year, about how much interest would you earn?
   $137.50

7.5 WHICH BANK ACCOUNT IS BEST?

(Open-ended problem solving)

DECISION MAKING

When you decide to open a checking account or a savings account, you are often confronted by many different types of accounts. You must examine the features of each and decide which is best for your own particular needs.

PROBLEM A

Victoria has just started a new job and needs to open a checking account in order to deposit her weekly paychecks. So that she can choose which type of account is best for her, the bank's customer service representative gave Victoria a brochure describing the 4 types of checking accounts.

REGULAR CHECKING

Each month, your canceled checks are returned to you with a statement describing your account activity for the month. There is a $5.00 monthly service charge plus 25¢ per check.

NOW CHECKING

This account pays you 5% interest on your checking account balance which is compounded daily and paid monthly. Your NOW account is free if you keep the required minimum daily balance of $1,000 in your account. Below the minimum there is a $7.50 monthly service charge plus 25¢ per check.

VALUE CHECKING

This checking account is ideal for people who do not write many checks. There is no minimum balance and you can write up to 5 checks per month for free. If you write more than 5 checks, there is a monthly service charge of $6.00 plus 20¢ per check.

MONEY MARKET CHECKING

This account pays you about 7% on your checking account balance. Your money market account is free if you keep a minimum daily balance of $2,000 in your account. Below the minimum there is a $7.50 monthly service charge plus 25¢ per check.

DECISION-MAKING FACTORS

Minimum balance requirement	Monthly service charge
Cost per check	Interest paid

OBJECTIVE

- Decide, given relevant information, what type of checking account or savings account to open.

TEACHING THE LESSON

WARM-UP Write the following exercises on the chalkboard or an overhead transparency.

Compute.

1. 12 × $0.25 + $3.00

2. 15 × $0.20 + $5.00

3. 18 × $0.30 + $7.50

4. 7 × $0.15 + $4.00

(**1.** $6.00 **2.** $8.00 **3.** $12.90 **4.** $5.05)

INTRODUCTION Present Problem A to the class. Encourage students to name some of the factors they would use to make an informed decision about which checking account to open. List the factors that are named on the chalkboard and discuss how to find out information on each factor. Encourage students to name the factors that are most important to them.

INSTRUCTION Read the descriptions of the 4 types of checking accounts. Ask students to discuss fully the features of each. Ask if there is any other information that Victoria should have. Discuss the decision-making factors.

Have small groups of students complete Exercises 1–12 in the table in order to compare the 4 accounts. Ask students to write statements that describe the data now in the table. Ask them to write conclusions that can be drawn from the data. Discuss the appropriateness of these statements and conclusions and lead a discussion about interpretations and possible misinterpretations of the data.

Have students complete Exercises 13–22 in small groups, having students explain and defend their answers and decisions. Remind students that there is no one right answer to either Exercise 21 or Exercise 22. Explain that answers will vary and even students with the same answer may have different reasons for choosing that answer. Continually reinforce the fact that, in many consumer issues, best answers depend on individual situations and preferences. Encourage students to participate and express their opinions.

DECISION-MAKING COMPARISONS

Compare the 4 options by completing the table.

Factor	Regular	NOW	Value	Money Market
Minimum balance	None	• 1. ■	• 2. ■	3. ■
Monthly service charge (if conditions not met)	4. ■	5. ■	$6.00	6. ■
Cost per check (if conditions not met)	7. ■	25¢	8. ■	9. ■
Interest paid	10. ■	11. ■	12. ■	Yes; about 7%

1. $1,000 **2.** None **3.** $2,000 **4.** $5.00 **5.** $7.50 **6.** $7.50 **7.** 25¢

MAKING THE DECISIONS **8.** 20¢ **9.** 25¢ **10.** No **11.** Yes; 5% **12.** No

Which account should Victoria open:

• **13.** If the amount of interest earned were the only factor?
Money Market

14. If she expects to write about 10 checks per mo and does not expect to have more than $1,000 in her account?
Regular

15. If she expects to maintain a low balance and to write only 3 or 4 checks per mo?
Value

16. If she has no problem maintaining a minimum balance of $1,000?
NOW

• **17.** What is the monthly service charge if she opens a Regular checking account and writes 12 checks per mo? How much is this per year?
$8.00; $96.00

18. What is the monthly service charge if she opens a NOW account and writes 12 checks per mo and maintains a minimum balance of $1,000?
$0.

19. What is the monthly service charge on a NOW account if she writes 15 checks per mo and does not maintain a minimum balance of $1,000?
$11.25

20. Suppose Victoria can earn an average of $5.00 interest per mo on a Money Market account and can maintain a minimum balance of $2,000. What would be her net annual savings over the Regular account in Exercise 17?
Interest + fee = $156

21. Do you think that the free checking accounts are worth the minimum balance requirements and the higher service charge penalties?
Answers may vary.

22. Which account would you choose? Why?
Answers may vary.

PROBLEM B

After Victoria has worked for several months, her checking account balance has been consistently high enough that she decides to open a savings account. She gets information on the types of savings accounts available so that she can choose which one is best for her.

REGULAR SAVINGS	MONEY MARKET SAVINGS
★ Can be opened with as little as $5 ★ Unlimited deposits and withdrawals ★ 5½% interest ★ No service charge	★ Can be opened with $1,500 or more ★ Limit of 5 withdrawals per month ★ Variable interest from 6½% up, adjusted weekly ★ $3.00 per month service charge if balance falls below $1,500

DECISION-MAKING FACTORS

Initial deposit Access to money Interest rate Service charge

DECISION-MAKING COMPARISONS

Compare the 2 options by completing the table.

Factor	Regular savings	Money Market savings
Initial deposit	•23. ■ $5	•24. ■ $1,500
Access to money	Any day	25. ■
Interest rate	26. ■	Variable
Service charge	27. ■ None	28. ■

25. 5 withdrawals per mo **26.** $5\frac{1}{2}\%$ **28.** $3 per mo if balance falls below $1,500

MAKING THE DECISIONS

Which account should Victoria open:

•**29.** If she only has a small amount of money?
Regular savings

•**30.** If she wants to earn the most interest and she does have $3,000 to deposit?
Money Market savings

31. If she expects to make frequent deposits and withdrawals?
Regular savings

32. If she would like to earn 8% interest?
Money Market savings

33. If the Money Market savings account averaged $7\frac{1}{2}\%$ interest, how much more could Victoria earn per year over a Regular savings account on a $10,000 deposit?
$200

34. Which account would you open? Why?
Answers may vary.

Repeat this same process for Problem B, which involves savings accounts.

ASSIGNMENTS

Level 1	1–34
Level 2	1–34
Level 3	1–34

FOLLOW-UP ACTIVITY

ENRICHMENT Invite the customer-service representative from a local bank to your class. Ask him or her to bring brochures and other materials describing the various checking and savings accounts offered by the bank. Have students prepare questions for the service representative ahead of time.

SUPPLEMENTARY MATERIALS

TRP Practice, p. 49
TRP Transparency 25

OBJECTIVES

- Review vocabulary
- Practice key chapter concepts and skills.

USING THE REVIEW

The Chapter Review is designed to help students prepare for taking the Chapter Test. The first section focuses on vocabulary. It requires students to select words to complete statements. The second section presents practice of key chapter skills.

Each item on the Chapter Review is referenced to the page on which the topic is taught in the Pupil's Edition. You may wish to have students refer to these pages to help review any concepts or skills they have not yet mastered.

It is suggested that students work in small-sized heterogeneous cooperative learning groups. Some cooperative learning methods that may be used are as follows:

1. After each student has independently completed the Chapter Review, a discussion should follow within each group about the solutions to the problems.
2. The group can complete the entire Chapter Review by working together to solve the problems.

End the lesson with an entire class discussion in which any questions brought up in group discussions are presented and answered.

CHAPTER REVIEW

Vocabulary Choose the letter of the word(s) that completes the sentence.

1. Money placed by you into a savings or a checking account is called ■. [124, 129] b

a. A withdrawal **b.** A deposit **c.** Interest

2. Money that is added to your account by the bank is called ■. [129] c

a. A withdrawal **b.** A deposit **c.** Interest

3. When interest is compounded quarterly, it is paid to you ■ times per year. [132] c

a. One **b.** Two **c.** Four

Skills Find the new bank account balance.

4. Old balance: $387.87
Check: $37.75
Deposit: $123.30
Check: $23.08
Check: $76.57 [124]
$373.77

5. Old balance: $1,835.94
Check: $274.33
Check: $88.27
Service charge: $3.50
Deposit: $385.89 [124]
$1,855.73

6. Old balance: $500.00
Withdrawal: $175.00
Withdrawal: $60.00
Deposit: $37.50 [129]
$302.50

7. Old balance: $675.30
Deposit: $50.00
Interest: $2.87
Withdrawal: $185.00 [129]
$543.17

Use the given information to reconcile the bank statement balance with the check register balance. Find the adjusted balance for each. [127]

8. Check register balance: $217.84
Statement ending balance: $279.50
Outstanding checks: $15.80, $49.71
Service charge: $3.85
$213.99

9. Check register balance: $87.50
Statement ending balance: $137.90
Outstanding check: $75.00
Outstanding deposit: $19.85
Service charge: $4.75
$82.75

Find the answer. [132]

10. How much simple interest will be earned in 3 y on $2,500 in an account that pays 4.5%?
$337.50

11. What will be the new balance on $800 in an account that earns $6\frac{1}{2}$% simple interest for 5 y?
$1,060

12. Use the compound interest table on page 133. How much interest will be paid on $400 earning 6% compounded quarterly after 4 y? $107.60

13. Use the compound interest table on page 133. What will be the new balance on $5,000 earning 4% compounded semiannually after 10 y?
$7,429.50

CHAPTER TEST

Make up a check for the payment. Use today's date and your own name and signature. Check students' work.

1. Check #136 to Dr. Robin Yeats for $65.00

2. Check #137 to Security Insurance Co. for $258.34

Find the new checking account balance.

3. Old balance: $783.21
   Check: $73.90
   Check: $182.70
   Check: $57.24
   Service charge: $8.30
   $461.07

4. Old balance: $1,284.89
   Deposit: $231.48
   Check: $83.90
   Check: $374.50
   Check: $214.38
   $843.59

Use the given information to reconcile the bank statement balance with the check register balance. Find the adjusted balance for each.

5. Check register balance: $86.30
   Statement ending balance: $171.54
   Outstanding checks: $3.90, $87.74
   Service charge: $6.40
   $79.90

6. Check register balance: $486.35
   Statement ending balance: $500.75
   Outstanding checks: $29.40, $165.00
   Outstanding deposit: $176.50
   Service charge: $3.50
   $482.85

Solve.

7. Jim opened a savings account with a deposit of $250.00. He later deposited $45.00 and then withdrew $120. The bank posted interest of $3.28. What is his new balance?
   $178.28

8. Sarah's savings account balance was $218.00. She withdrew $50 one month and deposited $40 the next. She then withdrew $75. The bank posted interest of $6.83. What is her new balance?
   $139.83

Find the simple interest.

9. $600 at 5% for 10 y $300

10. $2,300 at $6\frac{1}{4}$% for 4 y $575

Use the compound interest table on page 133 to find the new balance.

11. $2,500 at 6% compounded semiannually for 4 y $3,167

12. $6,000 at 8% compounded quarterly for 5 y $8,915.40

OBJECTIVE

- Evaluate the achievement of chapter objectives.

USING THE TEST

The Chapter Test may be used as a post-test to evaluate student achievement. However, you may wish to use the Chapter Post-Test offered in the Teacher's Resource Package or to design your own chapter test. If this page is not used as a test, you may want to design it as additional review or practice.

The test items are correlated to the chapter objectives in the table below.

Chapter objectives	Test items
A. Write a check.	1–2
B. Compute a running balance in a check register.	3–4
C. Reconcile a checking account.	5–6
D. Compute a savings account passbook balance.	7–8
E. Compute simple interest.	9–10
F. Use a table to compute compound interest.	11–12

SUPPLEMENTARY MATERIALS

TRP Ch. 7 Post-Test Form A, pp. 1–2
TRP Ch. 7 Post-Test Form B, pp. 1–2
TRP Lesson Aids, p. 4

OBJECTIVE

- Understand that ATMs are an easy way to conduct bank transactions that will save you time and money.

USING THE PAGE

Begin discussion by asking students if their family members use ATMs. If so, have students ask if they can go along to the bank to see how an ATM is used. Encourage students to write down what appears on the ATM screen for a bank deposit.

Have students read the introductory material. Point out that a receipt is issued for every ATM transaction and it should be checked on the spot for accuracy. Also mention that ATM transactions should be checked against monthly bank statements to be sure that no computer errors were made.

Have students work individually or in pairs to answer Exercises 1–7. Summarize the answers as follows: By comparing the ATM policies of a few banks, you could save money in transaction fees and cut down the time you have to spend at the bank.

Have students work individually to answer Exercises 8–9. Then have them share their answers and reasons in a whole-class discussion.

ANSWERS/EXPLANATIONS

8. Pros: charges no fee; you may only need 1 ATM location (if your office is also near home); also has an airport ATM, convenient if you do a lot of traveling

Cons: neither of the 2 locations may be convenient; a bank with only 2 ATM locations is probably small and would not offer other banking services you may want

9. They could use it to get at your money; criminals often know how to use a PIN with a phony ATM card or can manipulate the machine as long as they have a valid PIN with which to work.

MONEY TIPS

You can save time at the bank by using ATMs (Automated Teller Machines) and save money on transaction fees if the ATMs are owned by the bank.

LET'S LOOK AT THE FACTS

Many banks have **ATMs,** which allow customers to use a special bank card to conduct certain transactions *at any time*, even if the bank itself is closed. Some of these banks do not own their ATMs and may charge up to $1.00 or more for every transaction. With bank-owned ATMs, there is either a much lower transaction fee or no fee at all.

LET'S DISCUSS WHY

1. For what types of transactions might an ATM be used?
   cash withdrawals, deposits, transfer funds, get account information
2. What are the advantages of dealing with a bank that has ATMs?
   no waiting on long lines; open 24 h per d, 7 d per wk
3. Why do you think banks which do not own their ATMs charge higher transaction fees?
   to pay for the ATM service; to cover costs of operation
4. More banks charge an ATM transaction fee for withdrawals than for deposits. What might be the reason for this?
   it costs more to give cash; it depletes the bank's funds
5. If the ABC Bank charges $1.00 per transaction and you use its ATMs 12 times per mo, how much would you pay per year for this service?
   $144
6. If you use the XYZ Bank's ATMs and pay only 20¢ per transaction, how much would you save in a year over what you would pay to use the ABC Bank's ATMs in Exercise 5?
   $115.20
7. How could you find out if you are paying ATM transaction fees?
   read your ATM agreement; check your monthly statement; call the bank

LET'S SEE WHAT YOU WOULD DO See TE side column for answers to Exercises 8–9.

8. Last year, it cost you $90 (75¢ per transaction) to use ATMs at the Capital City Bank, which has 16 locations around the city. The Warshaw County Bank charges no fee but has only 2 ATM locations, one near your office and one at the airport. Discuss the pros and cons of switching your accounts to the Warshaw Bank.

9. The Olympia Federal Bank issues you a special bank card and assigns you a **Personal Identification Number (PIN)** so you can use its ATMs. This is your number exclusively and must be entered into an ATM before any transaction can be conducted. Why should you never give your PIN to someone else?

CALCULATOR FRACTIONS ON A CALCULATOR

You have learned that you can rename fractions and mixed numbers as decimals by dividing the numerator by the denominator. You can also use a calculator to rename a fraction or mixed number as a decimal.

Rename $2\frac{4}{5}$ as a decimal.

Procedure	Calculator Entry	Calculator Display
1. Divide the numerator of the fraction by the denominator.	[4] [÷] [5] [=]	0.8
2. Add the whole number.	[+] [2] [=]	2.8

So $2\frac{4}{5} = 2.8$.

To rename a fraction or a mixed number as a percent, first use the calculator to rename it as a decimal. Then multiply by 100.

Rename $\frac{3}{8}$ as a percent.

Procedure	Calculator Entry	Calculator Display
1. Divide the numerator by the denominator.	[3] [÷] [8] [=]	0.375
2. Multiply by 100.	[×] [1] [0] [0] [=]	37.5

So $\frac{3}{8} = 37.5\%$.

Use a calculator to rename the fraction or mixed number as a decimal. When necessary, round your answer to the nearest thousandth.

1. $\frac{5}{8}$ 0.625
2. $\frac{7}{20}$ 0.35
3. $\frac{3}{7}$ 0.429
4. $4\frac{3}{5}$ 4.6
5. $6\frac{1}{6}$ 6.167

Use a calculator to rename the fraction or mixed number as a percent. When necessary, round your answer to the nearest tenth of a percent.

6. $\frac{3}{4}$ 75%
7. $\frac{9}{20}$ 45%
8. $\frac{11}{25}$ 44%
9. $\frac{5}{7}$ 71.4%
10. $\frac{1}{6}$ 16.7%
11. $3\frac{3}{5}$ 360%
12. $9\frac{12}{25}$ 948%
13. $6\frac{9}{16}$ 656.3%
14. $7\frac{5}{6}$ 783.3%
15. $8\frac{1}{3}$ 833.3%
16. If your calculator has a percent key, [%], discover how to rename $\frac{3}{8}$ as a percent.

   [3] [÷] [8] [%] 37.5

OBJECTIVE

- Rename a fraction as a percent, using a calculator.

USING THE PAGE

Review with students the meaning of *percent* and the relationship between fractions, decimals, and percents.

Remind students that to rename a fraction or mixed number as a percent, they should first rename it as a decimal. The first example shows how to rename a mixed number as a decimal on a calculator. The order of operations is important here. The fraction, $\frac{4}{5}$, should be renamed as a decimal first, and then the whole number, 2, should be added to it.

In the second example, students complete the entire process. The fraction is renamed as a decimal. Then the decimal is renamed as a percent. The process shown will work with calculators that have and do not have percent keys. At this time, you may want to discuss Exercise 16.

After students have successfully completed both examples on their calculators, you may wish to point out that pressing the is-equal-to key at the end of the first step may not be necessary.

Before assigning the exercises, you may wish to work through an example with students in which they rename a mixed number as a percent, such as $3\frac{1}{5}$.

OBJECTIVE

- Test the prerequisite skills needed to learn the concepts developed in Chapter 8.

USING THE TEST

The Pre-Skills Test is designed to diagnose students' strengths and weaknesses on prerequisite skills necessary to study the mathematics in Chapter 8.

Assign the Pre-Skills Test. Allow students to work together in pairs or in small groups. Group members should help those who demonstrate a misunderstanding of a concept or a weakness in a skill.

The table below correlates the items on the Pre-Skills Test with the prerequisite skill and the lesson(s) in the chapter for which it is needed.

Item(s)	Prerequisite skill	Lesson(s)
1–6	Round money amounts to the nearest cent.	8.2, 8.3, 8.4, 8.7
7–18	Rename percents as decimals.	8.2, 8.3
19–38	Multiply money amounts.	8.2, 8.3, 8.4, 8.5, 8.7
39–46	Find the percent of a money amount to the nearest cent.	8.2, 8.3

References to specific skills that are reviewed in the Skills Bank are given on the individual lessons themselves.

Pre-Skills Test

Round to the nearest cent.

1. $27.868 $27.87
2. $4.923 $4.92
3. $30.905 $30.91
4. $86.2073 $86.21
5. $0.0753 $0.08
6. $0.0523 $0.05

Rename as a decimal.

7. 15% 0.15
8. 53% 0.53
9. 7% 0.07
10. 3% 0.03
11. 105% 1.05
12. 156% 1.56

Rename as a decimal.

13. 1% 0.01
14. 1.5% 0.015
15. $1\frac{1}{4}$% 0.0125
16. 0.0405% 0.000405
17. 0.06308% 0.0006308
18. 0.05506% 0.0005506

Multiply.

19. 0.9 × $40 $36
20. 0.8 × $35 $28
21. 0.7 × $78 $54.60
22. 0.6 × $835 $501
23. 0.35 × $700 $245
24. 0.46 × $950 $437
25. 0.79 × $748 $590.92
26. 0.86 × $739 $635.54

Multiply. Round to the nearest cent.

27. 6 × $5.128 $30.77
28. 4 × $3.752 $15.01
29. 40 × $1.499 $59.96
30. 0.73 × $678.20 $495.09
31. 0.43 × $809.49 $348.08
32. 0.74 × $287.51 $212.76
33. 0.67 × $526.43 $352.71
34. 0.83 × $1,012.56 $840.42
35. 0.015 × $2,023.63 $30.35
36. 0.075 × $59.50 $4.46
37. 0.0013 × $319.26 $0.42
38. 0.00046 × $607.43 $0.28

Find the answer to the nearest cent.

39. What is 13% of $528.20? $68.67
40. What is 17% of $832.78? $141.57
41. What is 18.6% of $409.86? $76.23
42. What is 19.65% of $689.53? $135.49
43. What is $1\frac{1}{2}$% of $25.50? $0.38
44. What is 1.3% of $62.82? $0.82
45. What is 0.04657% of $33? $0.02
46. What is 0.02342% of $18.75? $0.00

Many people pay for major purchases with a credit card. What factors should you consider before choosing and applying for a credit card?

OBJECTIVE

- Formulate problems, using text and data.

USING THE CHAPTER OPENER

Discuss students' answers to the question on the page. Their answers should include interest rates, whether payment is due monthly or can be extended, annual membership fees, and any penalties for missed payments. Hidden interest rates should also be discussed. For example, some department stores do not charge interest if a bill is paid in full; however, if only a partial payment is made, interest is charged (18% is not unusual). While these rates are detailed on monthly statements, many consumers ignore them or don't understand them.

One other important factor to discuss is a person's ability to exercise self-control with credit cards. People have been known to put themselves into serious debt by charging far more than they could pay for.

COOPERATIVE LEARNING

Have heterogeneous groups of students collect credit-card application forms and other data from banks and stores. Ask them to compare the advantages and disadvantages of each card. They should also provide examples of how credit can increase the original price of a charged item.

As a summary, each group should recommend how credit can be used by people of average means and people of very limited means. (Generally speaking, people of average means should attempt to pay their bills in full each month. People of limited means are often forced to extend payments so that the payments are smaller.)

SUPPLEMENTARY MATERIALS

TRP Group Projects, p. 8

OBJECTIVES

- Read a credit card receipt.
- Read a credit card statement.

TEACHING THE LESSON

WARM-UP Write the following exercises on the chalkboard or an overhead transparency.

Add.

1. \$156.48 + \$38.91 (\$195.39)

2. \$207.56 + \$18.49 (\$226.05)

Subtract.

3. \$118.45 − \$68.32 (\$50.13)

4. \$107.43 − \$62.93 (\$44.50)

INTRODUCTION Pass around a credit card and ask students to identify the features of the card. Emphasize that a credit card carries with it a great deal of responsibility. Encourage students to discuss how they would use a credit card wisely.

Ask students to discuss a time when they, or someone they were with, used a credit card. Encourage students to describe the procedures the salesperson, gas station attendant, or so forth followed.

INSTRUCTION Read the introductory situation about charging a sweater at The Clothing Mart. Point out to students that they would first need to make sure that the store accepted their credit card. Stores usually have signs near the entrance or at the cash registers that show which credit cards they accept.

Use Example 1 to help students interpret a credit card receipt. Explain that salespeople in stores often check with the credit card company to be sure that the charge is authorized (that the card is not a stolen or lost one; that the customer is not over his or her credit limit). Usually an automatic-dialing machine reads the credit card number from the black strip on the card and sends the number to the credit company's computer.

8.1 USING CREDIT CARDS

Only you can decide whether or not you should get a **credit card.** Once you get a card, you must accept the responsibility for using it.

You used your credit card to charge a sweater at The Clothing Mart. The sweater cost \$34.97. In the store, you looked over the **charge receipt** to make sure that all the entries were correct.

EXAMPLE 1 Identify the parts of the charge receipt.

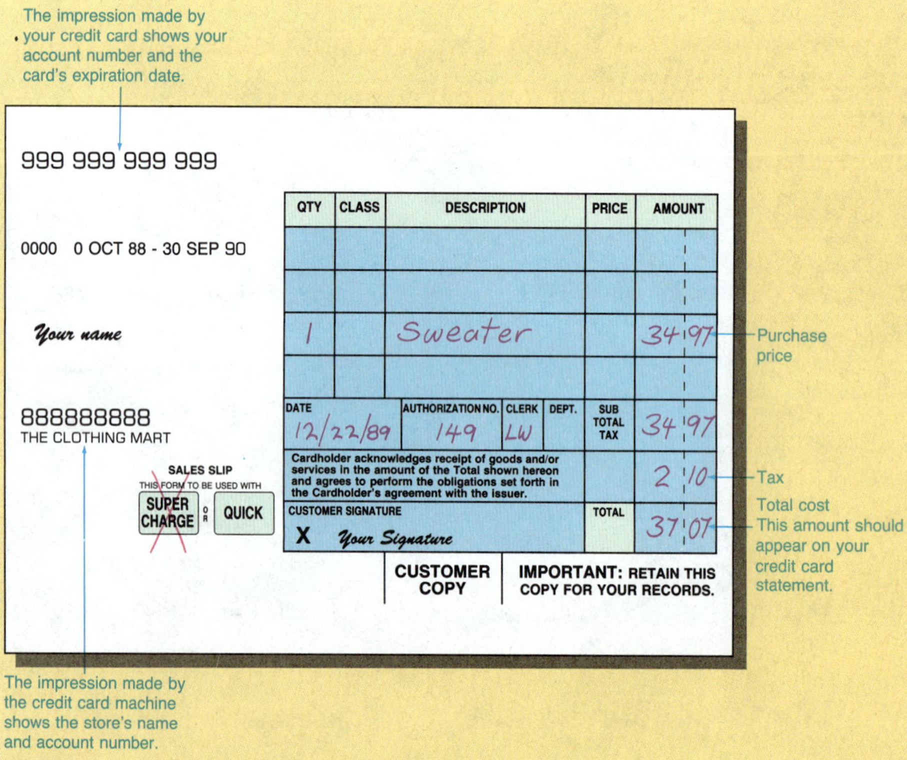

At the end of the month, you got a statement that showed your charges, payments, the amount you owe, and other important information. You compared the statement with your charge receipts.

EXAMPLE 2 Identify the parts of the monthly **credit card statement.**

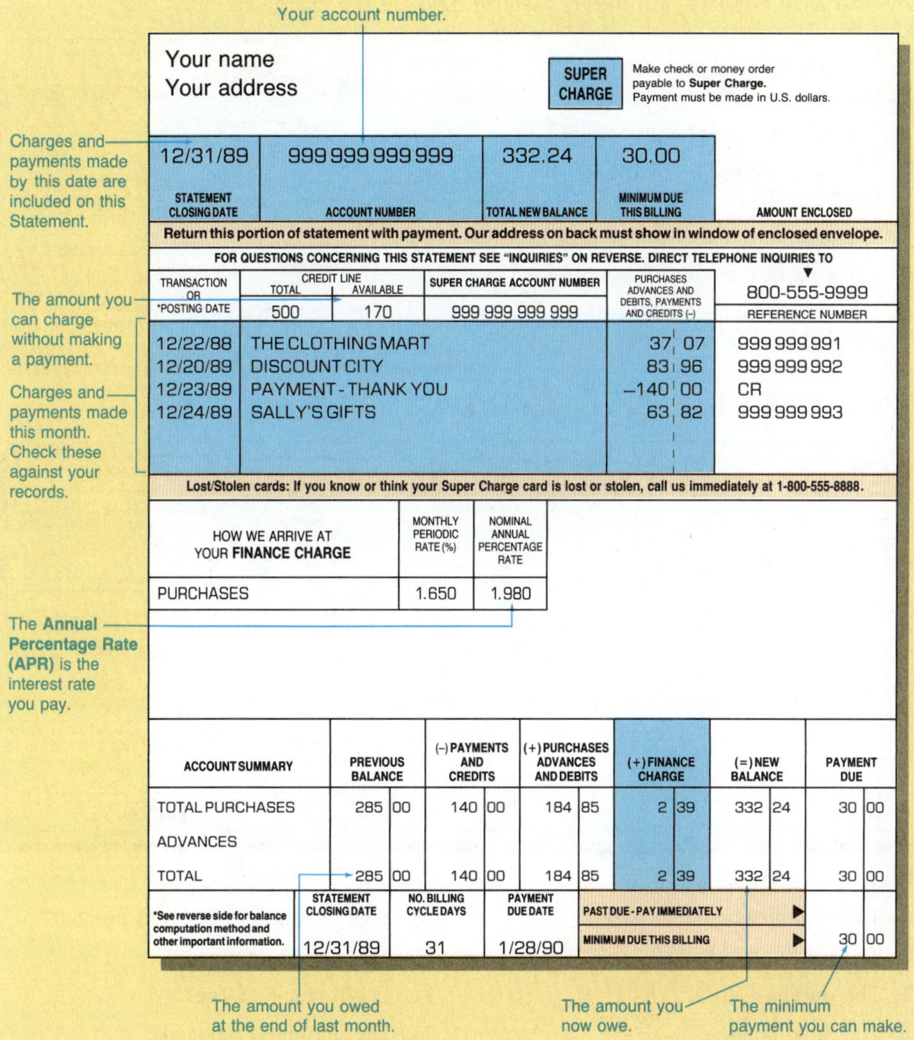

FOR DISCUSSION See TE side column.

How would you check your credit card receipts against a credit card statement?

The computer checks the credit card file to be sure that the charge is authorized. Then the computer sends an authorization number, which the clerk writes on the receipt. Ask students to identify the authorization number. Ask students to distinguish between (1) the impression made by the credit card, (2) the impression made by the credit card machine, (3) the information written by the salesperson, and (4) the information written by the customer. Ask students to discuss why it is necessary to destroy the carbon paper with which duplicate copies of charge receipts are made.

Use Example 2 to help students interpret a monthly credit card statement. Explain that the statement is produced by a computer, but that people make many of the entries into the computer. Just because a statement is produced by a computer does not mean that there will be no mistakes.

Actively involve the students in identifying and discussing the various entries. Point out that they should have a receipt for each purchase on the statement and explain that the statement should be checked against the receipts. Emphasize the summary at the bottom of the statement, which shows how the New Balance was computed.

SKILLS BANK You may want to refer students to the following skills in the Skills Bank:

Add With Whole Numbers and Decimals, Skill 6

Subtract With Whole Numbers and Decimals, Skill 6

FOR DISCUSSION Students should explain that they should check to be sure that the total amount on each receipt is the same as the amount on the statement. They should also check to be sure that there are no charges on the statement that they did not make and that any payments are recorded as credits. Point out that students should keep all of their credit card receipts.

ASSIGNMENTS

Level 1	1–16
Level 2	1–18
Level 3	1–18

FOLLOW-UP ACTIVITY

APPLICATION Ask students to speak to store owners or employees they know or take the class to visit a local store that accepts credit cards. Ask the manager to arrange a demonstration of the procedure for processing charge purchases. Engage the students in a discussion about the hidden costs of credit. Most stores pay the credit card company a fee that is usually a percent of the amount charged. These fees are added to the costs of merchandise and are paid for by the customer.

SUPPLEMENTARY MATERIALS

TRP Practice, p. 50
TRP Reteaching, p. 42
TRP Transparency 25

PRACTICE EXERCISES

Remember to estimate whenever you use your calculator.

You also have a credit card to buy gasoline. Look at the charge receipt. What is:

123 789 456 2 — $11.10 TOTAL CHARGE
00-K00 — 30 JUNE 92
Your name — TRU VALUE
Midville Tru Value, Closter NJ — DATE 2/22/89 — SALES TICKET NUMBER 189453

LICENSE	STATE	TRU VALUE PRODUCTS	QUANTITY	PRICE INCLUD. TAX	AMOUNT
AFT-702	NJ	GASOLINE ☑ DIESEL ☐	11.3	98.2	11 10
COMPANY ISSUING CARD	AUTHORIZATION RB				
BUYER'S SIGNATURE X Your Name		TOTAL MUST AGREE WITH TOTAL CHARGE ON TOP ▶		TOTAL	11 10

Thank you • Please come again! — CUSTOMER'S COPY

- • 1. Your credit card account number? 123 789 456 2
- • 2. The date of this charge receipt? 2/22/89
- 3. The expiration date of the credit card? June 30, 1992
- 4. The total amount of this charge? $11.10
- 5. The cost of a gallon of gasoline? 98.2¢
- 6. The license plate number of the car that got the gasoline? AFT-702

Look at the gasoline credit card statement below.

TRU VALUE — 123 789 456 2 CREDIT CARD NUMBER — MINIMUM PAYMENT 75.00 — NEW BALANCE 142.85

FEB. 28, 1987 STATEMENT CLOSING DATE — Your name, Your address

TRANSACTION DATE	SALES TICKET NUMBER	CARD NUMBER	TRANSACTION DESCRIPTION & PURCHASE LOCATION (See Reverse Side For Description of Codes)					AMOUNT
1 03	4368221	001	01	210	KINDRKMCK	WESTWOOD	NJ	10.85
1 12	1527363	001	01	681	PIERMONT	CLOSTER	NJ	9.60
2 19				PAYMENT - THANK YOU				79.39CR
2 20	1523642	001	01	681	PIERMONT	CLOSTER	NJ	9.00
2 22	1196543	001	01	1	MIDVALE	CLOSTER	NJ	10.50
2 22	1894583	001	01	1	MIDVALE	CLOSTER	NJ	11.10
2 27	4365454	001	01	210	KINDRKMCK	WESTWOOD	NJ	11.00

Previous Balance		Payments & Credits		Purchases & Adjustments		FINANCE CHARGES		New Balance
158.88	−	79.39	+	62.05	+	1.39	=	142.93

FINANCE CHARGE INFORMATION	PERIODIC RATE	ANNUAL PERCENTAGE RATE	Statement Closing Date	Credit Card Number	Minimum Payment
To $ 500	1.75%	21%	FEB 28 89	123 789 456 2	75.00
Over $ 500	1.50%	18%	To Avoid Additional FINANCE CHARGE New Balance Must Be Received By MAR 20 89		TRU VALUE

What is:

- • 7. The statement closing date? Feb. 28, 1989
- 8. Your credit card account number? 123 789 456 2
- • 9. The previous balance? $158.88
- 10. Last month's payment? $79.39
- 11. The purchase total? $62.05
- 12. The finance charge? $1.39
- 13. The new balance? $142.93
- 14. The minimum payment? $75.00
- 15. What is the Annual Percentage Rate (APR) finance charge for amounts up to $500? 21%
- 16. What is the Annual Percentage Rate (APR) finance charge for amounts over $500? 18%
- 17. How much was charged on 2/20? $9.00
- 18. How much was charged in January? $20.45

8.2 CREDIT FINANCE CHARGES

On your credit card statement, the **finance charge** is the amount you pay if the last balance has not been paid in full. The **unpaid balance** is any of the last balance that was not paid. The **new balance** is the amount you now owe.

NEW BALANCE = UNPAID BALANCE + FINANCE CHARGE + NEW CHARGES

EXAMPLE 1 Last month's balance was $285.00, of which $140.00 has been paid. Your new charges this month totaled $184.85. Your credit card company charges you 1.65% of the unpaid balance as a finance charge. Find the unpaid balance, the finance charge, and the new balance.

1. Subtract to find the unpaid balance. $285 − $140 = $145

The unpaid balance is $145.

2. Multiply to find the finance charge.
THINK: 1.65% = 0.0165 0.0165 × $145 = $2.3925

The finance charge is $2.39.

3. Add to find the new balance. $145 + $2.39 + $184.85 = $332.24

The new balance is $332.24.

Some credit cards have variable finance rates as shown below.

$1\frac{1}{2}$% on the first $500 of unpaid balance 1% on the unpaid balance above $500

EXAMPLE 2 The unpaid balance on Myra's account was $1,250. There were no new charges this month. Find the finance charge and the new balance.

1. Multiply to find the finance charge on the first $500. **THINK: $1\frac{1}{2}$% = 0.015** 0.015 × $500 = $7.50
2. Multiply to find the finance charge on the amount over $500.
THINK: $1,250 − $500 = $750 1% = 0.01 0.01 × $750 = $7.50
3. Add to find the total finance charge. $7.50 + $7.50 = $15.00

The total finance charge is $15.00.

4. Add to find the new balance. $1,250 + $15.00 = $1,265

The new balance is $1,265.

FOR DISCUSSION See TE side column.

What are some of the advantages and disadvantages of having a credit card?

OBJECTIVES

- Compute an unpaid credit balance and identify the new charges.
- Compute a credit finance charge using fixed or variable rates.
- Compute a new credit balance.

TEACHING THE LESSON

WARM-UP Write the following exercises on the chalkboard or an overhead transparency.

Compute.

1. $58.63 − $17.92

2. $207 − $118.98

3. $46.38 + $79.62 + $118.48

4. $208.49 + $7.69 + $131.08

Find the answer.

5. 1.5% of $200 **6.** $1\frac{1}{2}$% of $1,160

(**1.** $40.71 **2.** $88.02 **3.** $244.48 **4.** $347.26 **5.** $3 **6.** $17.40)

INTRODUCTION Ask students to name some reasons why you must pay a fee, or *finance charge,* to borrow money or extend credit payments over a period of time. Emphasize that, if something seems wrong on a statement, they should be able to compute the finance charges to find the error.

INSTRUCTION Use Example 1 to show students how to compute the unpaid balance, the finance charge, and the new balance. Discuss how the formula works. Refer students to the credit card statement on page 145 and point out where the results of these calculations appear on the statement. Remind students that they will probably want to use a calculator to check the computations on a monthly statement.

Use Example 2 to show students how to find the finance charge and the new balance when using variable finance rates. Reinforce the idea that finance charges apply only to the unpaid balance.

SKILLS BANK You may want to refer students to the following skills in the Skills Bank:

Add With Whole Numbers and Decimals, Skill 6

Subtract With Whole Numbers and Decimals, Skill 6

Equivalent Fractions, Decimals, and Percents, Skills 3, 5

Find the Percent of a Number, Skill 2

FOR DISCUSSION Students should explain that the advantages of having a credit card include the following: not having to carry cash to pay for purchases, being able to pay for purchases once monthly, and having a record of purchases. Students should explain that the disadvantages include the following: that it is easy to forget that charging is spending money, that it may be easy to exceed what you can afford to repay monthly, and paying finance charges.

EXTENSION Explain to students that credit-card companies often require a minimum payment. Point out the minimum payment on the credit card statement on page 145. Review the computation given.

ASSIGNMENTS

Level 1	Odd 1–9, 10–12, 15–16
Level 2	1–18, EX
Level 3	4–9, 15–18, EX

FOLLOW-UP ACTIVITY

APPLICATION Ask students to conduct a survey of their families' monthly credit card payments. Have them compute the total and average monthly payments.

SUPPLEMENTARY MATERIALS

TRP Practice, p. 51
TRP Reteaching, p. 43
TRP Transparency 26

PRACTICE EXERCISES

Remember to estimate whenever you use your calculator.

Find the unpaid balance, the finance charge, and the new balance to the nearest cent.

• 1. Last balance: $218.35
Payments: $0
Finance rate: 1.5%
New charges: $35.49
$218.35; $3.28; $257.12

• 2. Last balance: $80.45
Payments: $0
Finance rate: 1%
New charges: $99.85
$80.45; $0.80; $181.10

3. Last balance: $40.55
Payments: $0
Finance rate: 1.3%
New charges: $88.75
$40.55; $0.53; $129.83

4. Last balance: $143.50
Payments: $100.00
Finance rate: 1%
New charges: $0
$43.50; $0.44; $43.94

5. Last balance: $89.77
Payments: $19.00
Finance rate: 1.5%
New charges: $0
$70.77; $1.06; $71.83

6. Last balance: $530.85
Payments: $125.00
Finance rate: 1.2%
New charges: $0
$405.85; $4.87; $410.72

7. Last balance: $509.65
Payments: $350.00
Finance rate: $1\frac{1}{4}\%$
New charges: $199.89
$159.65; $2.00; $361.54

8. Last balance: $90.85
Payments: $55.00
Finance rate: $1\frac{1}{2}\%$
New charges: $345.90
$35.85; $0.54; $382.29

9. Last balance: $1,025.63
Payments: $750.00
Finance rate: $\frac{3}{4}\%$
New charges: $836.25
$275.63; $2.07; $1,113.95

Use the variable rates on page 147. Find the total finance charge on the given unpaid balance to the nearest cent.

• 10. $545.00 $7.95 11. $718.00 $9.68 12. $1,465 $17.15 13. $678.52 $9.29 14. $2,045.67 $22.96

Use the variable rates on page 147. Find the unpaid balance, the finance charge, and the new balance to the nearest cent.

	Last balance	Payments	New charges
• 15.	$234.00	$0	$345.00
16.	$1,089.00	$500.00	$0
17.	$2,534.58	$750.58	$234.00
18.	$345.68	$35.89	$38.34

15. $234.00; $3.51; $582.51 **16.** $589.00; $8.39; $597.39
17. $1,784.00; $20.34; $2,038.34 **18.** $309.79; $4.65; $352.78

EXTENSION Minimum Payments

Some credit card companies require a minimum payment on the new balance. Your new balance is $250. The minimum payment is 15% of the new balance. What is the minimum payment?

Multiply the new balance by 0.15. $0.15 \times \$250 = \$37.50$ minimum payment

Find the minimum payment.

1. The minimum payment is 12% of the new balance. The new balance is $456.
$54.72

2. The minimum payment is 15% of the new balance or $25, whichever is greater. The new balance is $158.
$25

148 CHAPTER 8

8.3 OVERDRAFT CHECKING

The bank agreed to lend you up to $500 to cover overdrafts. An **overdraft** is a check written for more than the balance in your account.

INTEREST = DAILY INTEREST RATE × SUM OF THE DAILY BALANCES

EXAMPLE 1 The sum of your daily balances for the month was $2,700. The **Annual Percentage Rate (APR)** is 17%. How much interest will you pay?

THINK: Look across from 17% to find the daily interest rate of 0.04657%.

Annual percentage rate	Daily interest rate
20%	0.05479%
19%	0.05205%
18%	0.04931%
17%	0.04657%

Multiply to find the interest.
THINK: 0.04657% = 0.0004657

0.0004657 × $2,700 = $1.25739

You will pay $1.26 in interest.

You can calculate the sum of the daily balances.

EXAMPLE 2 On February 1, the balance in your overdraft account was $245.80.
On February 16, you wrote a $212.50 check. New balance: $458.30
On February 23, you made a $50 payment. New balance: $408.30
The APR is 19%. Find the interest and the new balance.

1. Find the sum of the daily balances for February.

Dates	Balance	Number of days	Sum of the balances
Feb. 1–15	$245.80 ×	15	= $3,687.00
Feb. 16–22	$458.30 ×	7	= $3,208.10
Feb. 23–28	$408.30 ×	6	= $2,449.80
Total:		28	$9,344.90

2. Find the daily interest rate for an APR of 19%. 0.05205%

3. Multiply to find the interest. 0.0005205 × $9,344.90 = $4.8640204
THINK: 0.05205% = 0.0005205

The interest is $4.86.

4. Add to find the new balance. $408.30 + $4.86 = $413.16

The new balance is $413.16.

OBJECTIVE

- Compute an overdraft-checking finance charge (interest) and a new balance by using the daily interest rate and the sum of the daily balances.

TEACHING THE LESSON

WARM-UP Write the following exercises on the chalkboard or an overhead transparency.

Rename as a decimal.

1. 0.04931% **2.** 0.05479%

3. 0.04657%

Multiply.

4. 5 × $216.85 **5.** 18 × $308.19

(**1.** 0.0004931 **2.** 0.0005479 **3.** 0.0004657 **4.** $1,084.25 **5.** $5,547.42)

INTRODUCTION Explain to students that they are going to learn how banks and many credit-card companies compute interest. Remind students that they have already studied interest in Chapter 7. Review the formula and ask them to define interest and principal. As they work through this lesson, have them keep in mind that interest paid (not earned) is a finance charge.

INSTRUCTION Review the introductory situation. Discuss how the formula works. Explain that the daily interest rate is found by dividing the APR by 365 d.

Use Example 1 to show students how to find the interest when the sum of the daily balances is known. Invite students to use their calculators to verify that each APR in the table has been divided by 365 to get each daily interest rate.

Use Example 2 to show students how to find the sum of the daily balances, the interest, and the new balance.

Students often find this process mystifying, and you may want to try to explain further.

To find the interest, you multiply the balance each day by the daily interest rate. So, in Example 2, you would multiply $245.80 by 0.05205%. Then you would add this amount 15 times to find the interest for 15 d. That process is the same as multiplying $245.80 by 15 and then multiplying that product by 0.05205%.

SKILLS BANK You may want to refer students to the following skills in the Skills Bank:

Multiply With Decimals, Skills 4, 5

Find the Percent of a Number, Skill 2

FOR DISCUSSION Students should explain that the sum of the daily balances is $3,786. They should also explain that they found this sum by multiplying the average daily balance by 30, the number of days in September. Some banks and credit-card companies use this method to compute interest.

ASSIGNMENTS

Level	Exercises
Level 1	1–8
Level 2	1–11
Level 3	6–11

FOLLOW-UP ACTIVITY

COOPERATIVE LEARNING
Have students work in small heterogeneous groups to make up problems like the following:

The average daily balance is $480 for the month of July and the APR is 18%. How much interest would you pay?

Have the groups switch problems and try to solve them. Later, discuss any difficulties as a whole-class activity.

SUPPLEMENTARY MATERIALS

TRP Practice, p. 52
TRP Reteaching, p. 44
TRP Applications, p. 9
TRP Transparency 27

FOR DISCUSSION

See TE side column.

The **average daily balance** is the sum of the daily balances divided by the number of days in a month. Your average daily balance for September (30 d) was $126.20. What was the sum of the daily balances?

PRACTICE EXERCISES

Remember to estimate whenever you use your calculator.

Find the interest on the overdraft checking account.

	• 1. $1.53	2. $1.69	3. $1.90	4. $1.53	5. $1.14
Sum of daily balances	$2,945	$3,085.60	$4,074	$3,098.45	$2,453.89
Annual percentage rate	19%	20%	17%	18%	17%

Find the interest and the new balance for the account.

• **6.** The sum of the daily balances is $3,845 and the APR is 18%.
$1.90; $3,846.90

7. The sum of the daily balances is $7,086 and the APR is 20%.
$3.88; $7,089.88

• **8.** Sept. 1: Balance $858
Sept. 5: Made $175 payment
Sept. 19: Balance changed to $795
Sept. 20–30: No more activity
The APR is 20%.
$12.35; $807.35

9. March 1: Balance $908
March 8: Balance changed to $1,035
March 23: Made $295 payment
March 24–31: No more activity
The APR is 18%.
$14.22; $754.22

10. June 1: Balance $2,085
June 5: Made $475 payment
June 12: Balance changed to $1,985.75
June 23: Balance changed to $2,135.89
June 24–30: No more activity
The APR is 17%.
$27.26; $2,163.15

11. July 1: Balance $75
July 9: Balance changed to $525
July 10: Made $175 payment
July 21: Made $75.89 payment
July 22–31: No more activity
The APR is 19%.
$4.16; $278.27

8.4 TAKING OUT A LOAN

It is wise to compare different loan rates before you apply for a loan at a certain bank or credit union. The Annual Percentage Rate (APR) is the rate of interest you pay for loans and finance plans.

At a bank or a credit union, a loan officer gives you a loan application to fill out. You usually make monthly payments to pay back a loan.

EXAMPLE 1 You want to borrow $685 to pay off some debts. The loan officer at a bank tells you that the APR is 13.5%. She also tells you that you will pay off the loan in 12 monthly payments of $61.34. How much will you need to repay? What is the interest?

1. Multiply to find the amount you repay. $12 \times \$61.34 = \$736.08$

You will need to repay $736.08.

2. Subtract to find the interest. $\$736.08 - \$685.00 = \$51.08$

The interest is $51.08.

Banks and credit unions use rate tables to find out how much interest you will pay. The interest depends on the amount borrowed, the interest rate, and how long you will take to repay the loan.

EXAMPLE 2 Carlos takes out a $600 loan for a vacation trip. The APR is 14.5%. He will repay the loan in 9 mo. How much interest will he pay? How much will he repay each month?

THINK: Look across from 9 mo to find the interest rate per $100, or $6.138.

INTEREST PER $100

Months	14.5% APR
3	$2.426
6	$4.315
9	$6.138
12	$8.027
15	$9.937

1. Divide to find the number of $100 that Carlos is borrowing. $\$600 \div \$100 = 6$
2. Multiply to find the interest. $6 \times \$6.138 = \$36.828$

Carlos will pay $36.83 in interest.

3. Add to find the total amount he will repay. $\$600 + \$36.83 = \$636.83$
4. Divide to find the monthly payment. $\$636.83 \div 9 = \$70.758888$

Carlos will repay $70.76 per mo for 9 mo.

OBJECTIVES

- Compute the amount of interest for a loan.
- Compute the total amount to be repaid on a loan.
- Compute a monthly loan payment.

TEACHING THE LESSON

WARM-UP Write the following exercises on the chalkboard or an overhead transparency.

Multiply.

1. $9 \times \$68.72$ **2.** $12 \times \$80.52$

Divide.

3. $\$8{,}500 \div \$100$ **4.** $\$19{,}000 \div \$100$

Divide. Round to the nearest cent.

5. $\$540 \div 9$ **6.** $\$650 \div 6$

7. $\$968 \div 12$

(**1.** $618.48 **2.** $966.24 **3.** 85 **4.** 190 **5.** $60 **6.** $108.33 **7.** $80.67)

INTRODUCTION Ask students to name some reasons why people take out bank loans. You may wish to use the sample Loan Application Form offered in the Teacher's Resource Package and have students practice filling in the required information.

INSTRUCTION Review the introductory material. Mention to students that some banks or credit unions may advertise interest rates other than APR, but that they must be sure to compare only APRs. Explain the difference between interest rate and interest. Point out that the interest rate is a percent. The interest is the actual amount of money you pay back in addition to the amount borrowed.

Use Example 1 to show students how to find the total amount to be repaid and the interest. In this case, the monthly payments are known.

Use Example 2 to show students how to find the interest and the monthly payment from an interest rate table. Refer students to the table and ask why the interest per $100 goes up as the number of months to repay goes up. Encourage students to explain that the interest goes up because the money is being used for a longer time. Explain that they must find the number of $100 borrowed because the table shows interest per $100 borrowed.

Use Example 3 to show students how to find the amount to be repaid, the interest, and the monthly payment. In this case, a table that shows the monthly payment per $100 financed is used. Refer students to the table and ask why the payment per $100 goes down as the number of years increases. Encourage students to explain that if there are more payments (more years), each payment will usually be less. Emphasize the importance of locating the correct entry in the table. Ask students to give the monthly payment per $100 financed for each APR for each number of years given. Work through Example 3 very slowly, checking students' understanding at each step.

Each exercise is keyed to a particular Example.

Example 1: Exercises 1–6

Example 2: Exercises 7–16

Example 3: Exercises 17–28

You may want to assign some of the exercises immediately after an Example has been discussed and then assign the rest of the exercises at the end of the lesson.

SKILLS BANK You may want to refer students to the following skills in the Skills Bank:

Multiply With Decimals, Skill 5

Divide With Decimals, Skill 8

Banks also use tables to find monthly payments.

EXAMPLE 3 Victoria got a $7,000 loan. How much will she pay each month for a 10-y loan at $12\frac{1}{4}$%? How much interest will she pay?

THINK: Look across from 10 y under $12\frac{1}{4}$% to find the monthly payment for each $100 borrowed, or $1.499.

MONTHLY PAYMENT PER $100 FINANCED

Years	APR $10\frac{1}{2}$%	$12\frac{1}{4}$%	$13\frac{1}{2}$%
5	2.149	2.237	2.301
10	1.349	1.499	1.523
15	1.105	1.216	1.295

1. Divide to find the number of $100. $7,000 ÷ $100 = 70
2. Multiply to find the payment. 70 × $1.499 = $104.93

Victoria will pay $104.93 per mo.

3. Multiply to find the total amount to be repaid.
   THINK: 10 y = 10 × 12 mo = 120 mo 120 × $104.93 = $12,591.60
4. Subtract to find the interest. $12,591.60 − $7,000 = $5,591.60

Victoria will pay $5,591.60 in interest.

FOR DISCUSSION See TE side column.

A home equity loan is a **secured loan** since the value of your home guarantees payment. Would the interest rate on a secured loan usually be more or less than the interest rate on an unsecured loan? Why?

PRACTICE EXERCISES

Remember to estimate whenever you use your calculator.

Find the total amount to be repaid and the interest.

	Amount borrowed	Monthly payment	Number of payments	
1.	$ 385	$ 67.31	6	$403.86; $18.86
2.	$ 809	$ 96.08	9	$864.72; $55.72
3.	$ 580	$ 52.27	12	$627.24; $47.24
4.	$1,200	$ 75.41	18	$1,357.38; $157.38
5.	$1,385	$104.17	15	$1,562.55; $177.55
6.	$ 239	$ 82.13	3	$246.39; $7.39

Find the interest and the monthly payment.

	Amount Borrowed	APR	Months
• 7.	$ 500	13.5%	6
• 8.	$ 750	17.6%	9
9.	$ 186	13.5%	3
10.	$ 918	13.5%	12
11.	$1,020	17.6%	12
12.	$1,860	13.5%	18
13.	$1,238	17.6%	15
14.	$908.86	17.6%	18
15.	$1,087.95	13.5%	15
16.	$397.85	17.6%	9

INTEREST PER $100

Months	13.5% APR	17.6% APR
3	2.258	2.945
6	3.974	5.192
9	5.705	7.469
12	7.462	9.788
15	9.235	12.131
18	11.026	14.508

7. $19.87; $86.65 **8.** $56.02; $89.56
9. $4.20; $63.40 **10.** $68.50; $82.21
11. $99.84; $93.32 **12.** $205.08; $114.73
13. $150.18; $92.55 **14.** $131.86; $57.82
15. $100.47; $79.23 **16.** $29.72; $47.51

Find the monthly payment and the interest.

	Amount Borrowed	APR	Years
• 17.	$6,000	$13\frac{1}{4}\%$	5
• 18.	$8,200	$11\frac{1}{2}\%$	10
19.	$9,350	$15\frac{1}{2}\%$	15
20.	$8,725	$11\frac{1}{2}\%$	10
21.	$11,300	$13\frac{1}{4}\%$	5
22.	$9,890	$11\frac{1}{2}\%$	15
23.	$13,290	$15\frac{1}{2}\%$	15
24.	$6,780	$15\frac{1}{2}\%$	10
25.	$18,535	$13\frac{1}{4}\%$	5
26.	$23,265	$11\frac{1}{2}\%$	15

MONTHLY PAYMENT PER $100 FINANCED

Years	APR		
	$11\frac{1}{2}\%$	$13\frac{1}{4}\%$	$15\frac{1}{2}\%$
5	2.199	2.288	2.405
10	1.406	1.508	1.644
15	1.168	1.282	1.433

17. $137.28; $2,236.80 **18.** $115.29; $5,634.80
19. $133.99; $14,768.20 **20.** $122.67; $5,995.40
21. $258.54; $4,212.40 **22.** $115.52; $10,903.60
23. $190.45; $20,991.00 **24.** $111.46; $6,595.20
25. $424.08; $6,909.80 **26.** $271.74; $25,648.20

Solve.

27. Kim borrowed $18,565 for 15 y. How much more would her monthly payments have been if the APR were $15\frac{1}{2}\%$ instead of $13\frac{1}{4}\%$?
$28.04

28. David borrowed $13,675 for 5 y. How much more would he have repaid if the APR were $15\frac{1}{2}\%$ instead of $11\frac{1}{2}\%$?
$1,690.20

FOR DISCUSSION Students should explain that the interest rate for a secured loan is usually less than the interest rate for an unsecured loan. A bank takes less risk with a secured loan because the bank has something that it can repossess and sell to get at least some of its loan money back.

ASSIGNMENTS

Level 1	1–6, 7–11, 17–21
Level 2	1–27
Level 3	1–6, 12–16, 22–28

FOLLOW-UP ACTIVITY

ENRICHMENT Ask students to go to local banks to get copies of the rate tables for secured and unsecured loans. Use the rates from 2 different banks to find the monthly payments for each of these loans: (1) an unsecured loan of $1,800 to be paid back in 12 mo, and (2) a secured loan of $21,000 to be paid back in 15 y. Ask students to compare the monthly payments for the 2 banks and decide from which bank they would borrow.

SUPPLEMENTARY MATERIALS

TRP Practice, p. 53
TRP Reteaching, p. 45
TRP Lesson Aids, p. 9
TRP Transparency 27

OBJECTIVES

- Compute an installment price.
- Compute the finance charge for an installment plan.

TEACHING THE LESSON

WARM-UP Write the following exercises on the chalkboard or an overhead transparency.

Subtract.

1. \$75.50 − \$68.75 **2.** \$110 − \$89.60

Multiply.

3. 12 × \$8.38 **4.** 18 × \$9.08

(**1.** \$6.75 **2.** \$20.40 **3.** \$100.56 **4.** \$163.44)

INTRODUCTION Locate examples of current installment plans in newspapers or stores and write them on the chalkboard. Ask students if they have ever purchased anything on the installment plan. Explain that if something is being bought on the installment plan, we often say that it is being bought "on time."

INSTRUCTION Review the definitions of *installment price* and *finance charge.* Review the formula in terms of the units used. Ask students if the installment price will be more or less than the regular price. Remind students that the finance charge is often hidden in installment buying since only the monthly payments are given.

Use Example 1 to show students how to find the installment price and the finance charge when there is no down payment. Remind them how to use their calculators for multistep problems.

Use Example 2 to show students how to find the installment price and the finance charge when there is a down payment.

8.5 INSTALLMENT BUYING

You and your friend Jim saw just the watch you both want. You do not have enough money to buy it. But you think you can afford the **installment plan.** The **installment price** is the total of the installment payments. The **finance charge** is the amount you pay to use the plan.

FINANCE CHARGE = INSTALLMENT PRICE − REGULAR PRICE

EXAMPLE 1 You choose the installment plan without the down payment. Find the installment price and the finance charge.

1. Multiply to find the total monthly payments (the installment price). 12 × \$9.15 = \$109.80

The installment price is \$109.80.

2. Subtract to find the finance charge. \$109.80 − \$82.50 = \$27.30

The finance charge is \$27.30.

You may choose to pay a **down payment** when you get your purchase. In this case, the installment price is the total of the installment payments and the down payment.

EXAMPLE 2 Jim chooses the installment plan with the down payment. Find the installment price and the finance charge.

1. Multiply to find the total monthly payments. 12 × \$6.15 = \$73.80
2. Add to find the installment price. \$73.80 + \$20.00 = \$93.80

The installment price is \$93.80.

3. Subtract to find the finance charge. \$93.80 − \$82.50 = \$11.30

The finance charge is \$11.30.

FOR DISCUSSION See TE side column.

1. Why did the down payment in Example 2 decrease the finance charge?
2. If the watch were financed for 18 mo instead of 12 mo, would the monthly payments be more or less? Would the installment price be more or less? Why?

PRACTICE EXERCISES

Remember to estimate whenever you use your calculator.

Find the installment price and the finance charge. 1. $90; $5 2. $126; $24 3. $216; $17.35

	• 1.	• 2.	3.	4.	5.	6.	7.	8.
Regular price	$85	$102	$198.65	$305.76	$405	$598	$829.85	$932.96
Down payment	$0	$0	$0	$0	$80	$125	$175	$245
Monthly payment	$15	$14	$18	$27.50	$42	$44	$80.76	$64.27
Number of payments	6	9	12	12	9	12	9	12

4. $330; $24.24 5. $458; $53 6. $653; $55 7. $901.84; $71.99 8. $1,016.24; $83.28

• **9.** You buy a portable radio on the installment plan. It usually sells for $89.95. You pay $11 per mo for 9 mo.
$99; $9.05

10. You buy a television on the installment plan. It usually sells for $439.95. You pay $41 per mo for 12 mo.
$492; $52.05

11. The advertisement for a class ring reads "Nothing down and $17.50 a month for 6 months." The ring usually sells for $89.95.
$105; $15.05

12. A coat you want is advertised for "$25 down and $35.50 a month for 6 months." The coat usually sells for $220.95.
$238; $17.05

EXTENSION

Annual Percentage Rate (APR)

A lender should tell you what the APR is. But you should still figure it out for yourself. It might be more than you think. Use this formula to approximate the APR.

$$\text{Approximate APR} = \frac{24 \times \text{Finance charge}}{\text{Amount financed} \times (\text{Number of payments} + 1)}$$

You need to finance $400 for 12 mo toward the cost of a refrigerator. The finance charge is $80. What is the APR?

1. Substitute into the formula. Then solve.

$$\text{APR} = \frac{24 \times \$80}{\$400 \times (12 + 1)} = \frac{\$1{,}920}{\$400 \times 13} = \frac{\$1{,}920}{\$5{,}200} = 0.3692307$$

2. Round to the nearest thousandth and rename as a percent $0.369 = 36.9\%$

The APR is about 36.9%, much higher than you may have thought.

Find the approximate APR.

	1.	2.	3.	4.	5.	6.	7.	8.
Amount financed	$400	$34.86	$375	$750	$830	$1,010	$99.95	$65.60
Finance charge	$20	$2.64	$21	$78	$96.25	$119.50	$2.80	$8.75
Number of payments	12	6	12	18	15	18	3	9
	9.2%	26.0%	10.3%	13.1%	17.4%	14.9%	16.8%	32.0%

FOR DISCUSSION

1. Students should explain that the finance charge was less because less money was owed on installment. However, you cannot assume that the finance charge for one installment plan is less than in another plan just because there is a down payment.
2. Students should explain that the monthly payments would probably be less because there would be more payments. The installment price would probably be more because the money would be paid back over a longer time.

EXTENSION Explain that it is important to know the APR so that one installment plan can be compared to another. Remind students that operations inside parentheses are done first. The derivation of the formula is too complicated to be explained to students.

ASSIGNMENTS

Level 1	1–10
Level 2	1–12
Level 3	9–12, EX

FOLLOW-UP ACTIVITY

COOPERATIVE LEARNING

Have students work in small heterogeneous groups. Ask each group to devise 3 different installment plans for 1 product. The idea is for each group to devise plans that appear to be least expensive but are not. Have each group present its plans to the class and ask the class to choose the least expensive (lowest installment price).

SUPPLEMENTARY MATERIALS

TRP Practice, p. 54
TRP Reteaching, p. 46
TRP Transparency 27

OBJECTIVE

- Solve problems by finding patterns.

TEACHING THE LESSON

WARM-UP Write the following exercises on the chalkboard or an overhead transparency.

Find the pattern. Write the missing numbers.

1. 2, 4, 6, 8, ■, ■ (10, 12)

2. 27, 24, 21, ■, ■ (18, 15)

3. 6.5, 6.1, 5.7, 5.3, ■, ■ (4.9, 4.5)

4. 0.3, 0.6, 1.2, 2.4, ■, ■ (4.8, 9.6)

INTRODUCTION Pose this situation to the class: You are a substitute postal worker, delivering special mail in an unfamiliar neighborhood. As you approach a group of houses on 1 side of Oak Drive, you notice that the addresses are 4041, 4045, and 4049. The packages you want to deliver are for these addresses: 4065, 4076, and 4081. Which do you think are on that side of the street? Which are on the other side of the street? Tell how you decided on your answer.

Let students provide their responses and explanations. The idea of odd and even numbers and the word *pattern* will probably be mentioned. Use this problem to point out that often in daily life we use what we know about numbers to identify a pattern that may be useful in performing a task or solving a problem.

INSTRUCTION Present the problem situation to the class. Be sure the students understand that 2 steps are used to solve the problem: (1) First, use subtraction to find the pattern in the differences. (2) Next, apply what has been discovered to find the solution to the problem.

PROBLEM Solving STRATEGY

8.6 FINDING A PATTERN

Situation:

Ketti has a charge account at the Jaycee Department Store. She used her charge account to buy furniture and began paying for it monthly. With her type of charge account, she can pay a different amount each month. Her finance charges for the first 4 mo are shown below.

April	May	June	July
$3.84	$3.68	$3.36	$2.88

Suppose that Ketti makes no other purchases on her charge account and she continues to make payments according to the plan she has been using. What will be the finance charge in August?

Strategy:

Finding a pattern can help you solve some problems.

Applying the Strategy:

The finance charges have been decreasing from one month to the next. Find each difference.

April		May		June		July		August
$3.84		$3.68		$3.36		$2.88		■
	$0.16		$0.32		$0.48		?	

Look for a pattern. The differences are increasing by $0.16 each month. The next difference in the pattern will be $0.64, so the finance charge in August will be $2.88 − $0.64, or $2.24.

PRACTICE EXERCISES

Remember to estimate whenever you use your calculator.

Read the problem. Then answer the question.

1. Tony's charge account showed unpaid balances as follows: January—$85; February—$110; and March—$135. How did the increase from February to March compare with the increase from January to February?
   It was the same.
2. The finance charges on Beth's charge account were as follows: August—$1.81; September—$1.69; and October—$1.58. If the pattern continues, what will be the finance charge in December?
   $1.48

Solve the problem by finding and using a pattern.

• **3.** Ari used his charge account to buy audio equipment. With his type of charge account, he can pay a different amount each month. His finance charges for the first 4 mo were February—$4.05; March—$3.90; April—$3.60; and May—$3.15. Suppose that Ari makes no other purchases on his charge account and continues to make payments according to the plan he has been using. What will be the finance charge in June?

$2.55

4. Helga used her charge account to buy a computer. With her type of charge account, she can pay a different amount each month. Her finance charges for the first 4 mo were May—$13.60; June—$11.20; July—$8.80; and August—$6.40. Suppose that Helga makes no other purchases on her charge account and continues to make payments according to the plan she has been using. What will be the finance charge in September?

$4.00

5. During a 4-mo period, Stan kept a record of the unpaid balances on his charge account. The first month, the unpaid balance was $1,000. In each of the other months, the unpaid balance was 1.1 times as great as the unpaid balance the previous month. What were the unpaid balances for the third and fourth months?

$1,210; $1,331

6. Luveen kept a record of all of her charge-account payments for 3 y. The first year, her payments totaled $1,500. In each of the other years, the payments were 80% of those of the previous year. What was the total amount of her payments during the 3 y?

$3,660

In solving such problems independently, some students might simply find the pattern and give $0.64 as the answer. Emphasize that it is always necessary to look back at the original problem, note the question that is asked, and then make sure that they are answering that question.

SKILLS BANK You may want to refer students to the following skill in the Skills Bank:

Subtract With Whole Numbers and Decimals, Skill 6

ASSIGNMENTS

Level	Exercises
Level 1	1–4
Level 2	1–5
Level 3	3–6

FOLLOW-UP ACTIVITY

SITUATIONAL PROBLEM SOLVING Present students with the following problem: You and 2 friends have charge accounts at 3 different department stores. You want to compare the accounts and see whether there are significant differences between them.

Some questions to ask include: What types of information could you use to form some opinions? What questions might you want to have answered by each department store? How could you present the comparisons?

Some approaches to solving the problem include comparing your monthly charge account statements and account brochures, asking each store for additional information about their charge accounts, and showing your results in a table.

SUPPLEMENTARY MATERIALS

TRP Practice, p. 155
TRP Applications, p. 10
TRP Transparency 28

OBJECTIVE

- Decide, given relevant information, which credit plan to use.

TEACHING THE LESSON

WARM-UP Write the following exercises on the chalkboard or an overhead transparency.

Find the answer to the nearest cent.

1. $1\frac{1}{2}\%$ of \$350 **2.** 1.5% of \$675

Find the monthly payment.

3. You borrow \$10,000 for 5 y (60 mo). You pay \$2.169 per mo for each \$100 borrowed.

(**1.** \$5.25 **2.** \$10.13 **3.** \$216.90)

INTRODUCTION Present the problem situation to the class. Encourage students to name some of the factors they would use to make an informed decision about which credit plan to use and where this information is available. List the factors that are named on the chalkboard.

INSTRUCTION Read the features of the 4 credit plans that Ron has prepared. Ask students to discuss fully the features and to mention those features they find most appealing. Ask if there is any other information that Ron should have. Discuss the decision-making factors.

Have small groups of students complete Exercises 1–14 in the table in order to compare the 4 credit plans. If students have difficulty computing the monthly payments, it may be necessary to refer back to Lesson 8.4. Ask students to write statements that describe the data now in the table. Then ask them to write conclusions that can be drawn from the data. Discuss the appropriateness of these statements and conclusions and lead a discussion about interpretations and possible misinterpretations of the data.

DECISION MAKING

8.7 USING CREDIT WISELY

(Open-ended problem solving)

One of the most important decisions you will make is how and when to use credit. You should never borrow money unless you need it. When you do borrow money, choose the credit plan that is best for you.

PROBLEM

To borrow \$2,500 for his tuition, Ron can choose one of these credit plans.

Charge the tuition on his credit card.
Get an unsecured bank loan.
Use the school's tuition plan.
Ask his parents to get a home equity loan, which he will repay.

Ron listed the features of each plan to help him decide.

Credit Card:
Finance charge: $1\frac{1}{2}\%$ of the unpaid balance each month.
Minimum payment: \$100 monthly
The APR is 19.8%.
I plan to make the minimum monthly payment. I will not charge anything else on this card. It will take me 31 mo to pay off the loan. The interest will be \$597.52.

Unsecured Loan: The APR is 17.6%.
The bank will only give the loan for 18 mo.
The loan cannot be paid off any faster.

17.6% APR Interest Per \$100
18 mo. \$14.508

School's Tuition Plan:
I can pay the school 12 monthly payments of \$220.
The APR is 10.8%.
The loan cannot be paid off any faster.

Home Equity Loan:
The minimum term of the loan is 5 y (60 mo).
The APR is $11\frac{1}{2}\%$.
Interest paid is tax deductible.
My parents have the final responsibility for repaying the loan. The loan cannot be paid off any faster.

$11\frac{1}{2}\%$ APR Monthly Payment Per \$100
5 y \$2.199

DECISION-MAKING FACTORS

Annual Percentage Rate — Monthly payment — Number of payments
Interest paid — Other factors

DECISION-MAKING COMPARISONS **1.** 19.8% **2.** 10.8% **3.** $11\frac{1}{2}$% **4.** $100.00

Complete the table to compare the 4 credit plans. **5.** $159.04 **6.** $54.98 **7.** 18

Factor	Credit card	Unsecured loan	School's tuition plan	Home equity loan
APR	• 1. ■	17.6%	2. ■	3. ■
Monthly payment	• 4. ■	5. ■	$220.00	6. ■
Number of payments	31	7. ■	8. ■	60
Interest	$597.52	9. ■	10. ■	11. ■
Other factors	12. ■	Payments are fixed.	13. ■	14. ■

8. 12 **9.** $362.70 **10.** $140 **11.** $798.80 **12.** Payments not fixed.

MAKING THE DECISIONS **13.** Payments fixed. **14.** Payments fixed; interest tax deductible

Which credit plan would Ron choose if the only factor were:

•**15.** Lowest APR?
School's tuition plan

16. Lowest monthly payment?
Home equity loan

•**17.** Least amount of interest?
School's tuition plan

18. Getting a tax deduction?
Home equity loan

19. Which credit plan would allow Ron to adjust his monthly payments?
Credit card

20. If Ron wants to be totally responsible for repaying the loan, which plan should he eliminate?
Home equity loan

21. How much would Ron save in interest by choosing the unsecured loan instead of the home equity loan?
$436.10

22. How much would Ron save in interest by choosing the school's plan instead of the credit card?
$457.52

23. Which plans can Ron eliminate if he cannot afford to pay more than $150 per mo? Of the remaining plans, how much can Ron save by choosing the one with the lowest interest?
Unsecured loan, school's tuition plan; $198.80

24. Should you use APR alone to decide which credit plan to choose? Why or why not?
No; reasons may vary.

25. Which one of these credit plans would you choose? Why?
Answers may vary.

Have students complete Exercises 15–25 in small groups, asking students to explain and defend their answers and decisions. Remind students that in Exercise 25 there is no one right answer. Explain that choices will vary and that even those students with the same choice may have different reasons for their choices. Continually reinforce the fact that, in many consumer issues, best answers depend upon individual situations and preferences. Encourage all students to participate and express their opinions.

ASSIGNMENTS

Level 1	1–25
Level 2	1–25
Level 3	1–25

FOLLOW-UP ACTIVITY

APPLICATION Have students write their local rates for unsecured bank loans (from the Lesson 8.4 Follow-Up Activity) and some local store installment information (from store signs or advertisements) on the chalkboard. Write a credit card APR of 16.8% on the chalkboard. Have students discuss how each might be used to finance new living room furniture for 1 y. Besides a comparison of APRs, emphasize "Other Factors"—such as how much to finance, fixed or variable payments, ease of borrowing the money, consequences of not paying on time (penalties and/or effect on credit rating), and so forth.

SUPPLEMENTARY MATERIALS

TRP Practice, p. 56
TRP Transparency 28

OBJECTIVES

- Review vocabulary.
- Practice key chapter concepts and skills.

USING THE REVIEW

The Chapter Review is designed to help students prepare for taking the Chapter Test. The first section focuses on vocabulary. It requires students to select words to complete statements. The second section presents practice of key chapter skills.

Each item on the Chapter Review is referenced to the page on which the topic is taught in the Pupil's Edition. You may wish to have students refer to these pages to help review any concepts or skills they have not yet mastered.

It is suggested that students work in small-sized heterogeneous cooperative learning groups. Some cooperative learning methods that may be used are as follows:

1. After each student has independently completed the Chapter Review, a discussion should follow within each group about the solutions to the problems.
2. The group can complete the entire Chapter Review by working together to solve the problems.

End the lesson with an entire class discussion in which any questions brought up in group discussions are presented and answered.

CHAPTER REVIEW

Vocabulary Choose the letter of the word(s) that completes the sentence.

1. On a credit card statement, the ■ is any of the last balance that was not paid. [147] c
   a. Finance charge **b.** New balance **c.** Unpaid balance
2. A check written for more than the balance in your account is called ■. [149] c
   a. A payment **b.** The daily interest rate **c.** An overdraft
3. If you cannot afford the full price of an item, you may be able to use ■. [154] a
   a. An installment plan **b.** An APR **c.** A minimum payment

Skills Find the unpaid balance, the finance charge, and the new balance. [144, 147]

4. Last balance: \$106.84
   Payments: \$75.00
   Finance rate: $1\frac{1}{2}\%$
   New charges: \$56.28
   \$31.84; \$0.48; \$88.60
5. Last balance: \$318.60
   Payments: \$80.00
   Finance rate: 1.65%
   New charges: \$76.39
   \$238.60; \$3.94; \$318.93
6. Last balance: \$1,100
   Payments: \$100
   Finance rate: $1\frac{1}{2}\%$ on first \$500
   1% on over \$500
   New charges: \$65.40
   \$1,000; \$12.50; \$1,077.90
7. Last balance: \$850.68
   Payments: \$25.00
   Finance rate: $1\frac{1}{2}\%$ on first \$500
   1% on over \$500
   New charges: \$138.75
   \$825.68; \$10.76; \$975.19

Find the interest on the overdraft checking account to the nearest cent. [149]

8. Sum of daily balances: \$525
   Daily interest rate: 0.05205%
   \$0.27
9. Sum of daily balances: \$2,600
   Daily interest rate: 0.04931%
   \$1.28

Use the loan monthly payment table on page 152 for Exercises 10–13. [151]

10. Sue borrowed \$8,000 for 5 y at $10\frac{1}{2}\%$. What is the monthly payment?
    \$171.92
11. May borrowed \$14,000 for 10 y at $12\frac{1}{4}\%$. What is the monthly payment?
    \$209.86
12. Dan borrowed \$32,600 for 10 y at $13\frac{1}{2}\%$. How much interest will he pay?
    \$26,979.76
13. Sal borrowed \$50,000 for 15 y at $10\frac{1}{2}\%$. How much interest will he pay?
    \$49,450.00

Find the installment price and the finance charge. [154]

14. Regular price: \$130
    Down payment: \$0
    Monthly payment: \$17.30
    Number of payments: 9 \$155.70; \$25.70
15. Regular price: \$950.75
    Down payment: \$150
    Monthly payment: \$88.10 \$1,207.20;
    Number of payments: 12 \$256.45

Chapter Test

Find the finance charge and the new balance.

1. Last balance: $409.56
   Payments: $160.00
   Finance rate: 1.5%
   New charges: $350.49
   $3.74; $603.79

2. Last balance: $65.43
   Payments: $20.00
   Finance rate: $1\frac{1}{4}\%$
   New charges: $87.60
   $0.57; $133.60

3. Last balance: $375 Payments: $85
   Finance rate: $1\frac{1}{2}\%$ on first $500
   1% on over $500
   New charges: $94
   $4.35; $388.35

Use the daily interest rate table on page 149 to find the interest on the account to the nearest cent.

	4. $0.13	5. $0.25	6. $0.28	7. $0.38	8. $1.81	9. $4.94
Sum of daily balances	$270	$456.90	$509	$810	$3,895.67	$9,007.62
APR	17%	20%	20%	17%	17%	20%

Use the loan payment table on page 152 to find the monthly payment, the total amount to be repaid, and the interest to the nearest cent.

10. Borrow $500 for 5 y at $10\frac{1}{2}\%$.
    $10.75; $645; $145
11. Borrow $9,000 for 10 y at $10\frac{1}{2}\%$.
    $121.41; $14,569.20; $5,569.20
12. Borrow $4,675 for 10 y at $12\frac{1}{4}\%$.
    $70.08; $8,409.60; $3,734.60
13. Borrow $12,500 for 5 y at $13\frac{1}{2}\%$.
    $287.63; $17,257.80; $4,757.80

Find the installment price and the finance charge.

	Regular price	Down payment	Monthly payment	Number of payments
14.	$75	$0	$10.42	9
15.	$95.89	$0	$16.75	6
16.	$210	$25	$17.58	12
17.	$273.60	$85	$23.25	9

14. $93.78; $18.78 15. $100.50; $4.61 16. $235.96; $25.96 17. $294.25; $20.65

Your charge account finance charges for the first 4 mo on a single purchase were: Jan.–$8.50, Feb.–$7.55, Mar.–$6.84, and Apr.–$6.37.

18. By how much are the differences in finance charges increasing each month? $0.24
19. What will the difference be? $0.96
20. What should May's finance charge be? $6.10

Objective

- Evaluate the achievement of chapter objectives.

Using the Test

The Chapter Test may be used as a post-test to evaluate student achievement. However, you may wish to use the Chapter Post-Test offered in the Teacher's Resource Package or to design your own chapter test. If this page is not used as a test, you may want to assign it as additional review or practice.

The test items are correlated to the chapter objectives in the table below.

Chapter objectives	Test items
A. Compute a credit unpaid balance, finance charge, and new balance (fixed finance rate).	1–2
B. Compute a credit finance charge and new balance (variable finance rate).	3
C. Compute overdraft checking interest using the sum of the daily balances and a daily interest rate.	4–9
D. Compute loan interest, total to be repaid, and monthly payment using a monthly payment per $100.	10–13
E. Compute an installment price and finance charge.	14–17
F. Solve problems by looking for a pattern.	18–20

Supplementary Materials

TRP Ch. 8 Post-Test Form A, pp. 1–2
TRP Ch. 8 Post-Test Form B, pp. 1–2

OBJECTIVE

- Understand that buying on credit can be very risky if you do not put a limit on your credit spending.

USING THE PAGE

Begin discussion by asking students if they have ever asked friends to lend them money and then found it difficult to pay it back promptly. Also ask students if they know anyone who owes a lot of money on credit cards and discuss the pressures that situation creates.

Have students read the introductory material and carefully look over the budget formula. Have students work individually or in pairs to answer Exercises 1–3. Summarize the answers as follows: There is an easy way to figure out how much you can comfortably afford to use for buying on credit.

Point out to students that high finance charges can affect the amount of credit available for spending. Encourage them to call banks for information or ask people they know what annual finance charge rates they pay on their credit cards.

Have students work individually to answer Exercises 4–7. Then have them share their answers and reasons in a whole-class discussion.

ANSWERS/EXPLANATIONS

4. (a) Pay your total balance so no finance charges are incurred; or (b) charge no more than you can afford to pay back in the next month; or (c) open a credit account with a cash advance at a lower interest rate and pay off your credit card charges; or (d) take out a debt consolidation loan at a lower rate of interest.

6. (a) Send in a $300 payment instead of the minimum amount on your next monthly statement; or (b) send in an additional $75 each month for September through December.

MONEY TIPS

By calculating how much credit fits into your budget, you can avoid getting into debt.

LET'S LOOK AT THE FACTS

Charging purchases can be very tempting. It can, however, cause you to *spend* far beyond your means. The formula below is an easy way to figure out exactly how much you can buy on credit and pay back without getting deeper into debt. Imagine that the budget shown below is yours.

Monthly take-home pay	minus	Fixed expenses (rent, food, transportation, etc.)	minus	Other regular bills (phone, utilities, installment loans, etc.)	minus	High-priorty expenses (savings, insurance, medical care, recreation, etc.)	equals	Amount available for monthly repayment	multiply by 12 months, equals	Your credit limit for the year
For example: $1,400	–	$700	–	$150	–	$400	=	$150	× 12	= $1,800

LET'S DISCUSS WHY

1. If your monthly take-home pay is $1,400, what is your annual take-home pay?
   $16,800
2. What percent of your annual take-home pay does an $1,800 credit limit represent?
   10.7%
3. What percent of your monthly take-home pay is used for fixed expenses? Regular bills? High-priority expenses? Monthly credit repayment?
   50%; 10.7%; 28.6%; 10.7%

LET'S SEE WHAT YOU WOULD DO See TE side column for answers to Exercises 4 and 6.

4. Finance charges (which can range from 12% to 20% or more) are added to the amount you owe on a credit card and are compounded each month. Name some ways to lessen your credit debt.
5. Your rent goes up by $50, raising your fixed expenses to $750. Where could you cut back on spending to make up for this increase?
   food, transportation, telephone, recreation
6. In August, you see a terrific stereo system with a color chamber that changes colors as the music changes. It is on sale for $600 until the end of the year. However, you only have $300 left on your credit limit. Is there any way you could buy it without exceeding your credit limit or borrowing money from anyone?
7. Many experts say that, as a general rule, no more than about 10% to 11% of your take-home pay should be used for buying on credit. If your monthly take-home pay is $1,400, what is your monthly credit range? Your annual credit range?
   Monthly: $140–$154 Annual: $1,680–$1,848

OBJECTIVE

- Estimate differences, using front-end estimation.

USING THE PAGE

It should be clear to students that a simple front-end estimate of a difference may be either an underestimate or an overestimate. An underestimate occurs when the second greatest digit of the "top number" is greater than the corresponding digit of the "bottom number." In the first example on the page, 20,000 is an underestimate, since 8 is greater than 3.

	28,372
	− 3,480
First estimate:	20,000

An overestimate occurs when the second greatest digit of the top number is less than the corresponding digit of the bottom number. In the third example on the page, 0.2 is an overestimate, since 1 is less than 6.

	0.817
	−0.668
First estimate:	0.2

As shown in the lesson examples, an underestimate can be adjusted by addition. The positive difference between the second greatest digit in the top number and the corresponding digit in the bottom number is used to determine the amount to add. In the first example, 8 − 3 = 5, so 5,000 is added to 20,000

An overestimate is adjusted by subtraction. The *positive difference* between the digits is used to determine the amount to subtract. In the third example, 6 − 1 = 5, so 0.05 is subtracted from 0.2.

ESTIMATION SKILLS

FRONT-END ESTIMATION OF DIFFERENCES

One way to estimate differences is by using front-end estimation.

1. Find the greatest number.
2. Identify the place of its leading nonzero digit.
3. Subtract only the digits in that place.

Examples

	②8,372	$⑧.73	0.⑧17
	− 3,480	−⑤.39	−0.⑥68
	②	③	②
Estimate:	20,000	$3.00	0.2

To get a more precise estimate, front-end estimation can be adjusted upward or downward by *also* using the next greatest place. With subtraction, you will need to think of either a *plus* or a *minus* of your first estimate.

Examples

	②[8],372		$⑧.[7]3		0.⑧[1]7
	− [3],480		−⑤.[3]9		−0.⑥[6]8
First estimate:	②0,000	**First estimate:**	$③.00	**First estimate:**	0.②0
+ about	[5],000	+ about	0.[4]0	− about	0.0[5]
Final estimate:	25,000	**Final estimate:**	$3.40	**Final estimate:**	0.15

Use front-end estimation with adjusting to estimate the difference.

1. 573 − 128 = 450	**2.** 4,309 − 186 = 4,200	**3.** 86,245 − 5,017 = 81,000	**4.** 33,432 − 18,507 = 15,000	**5.** 95,880 − 8,141 = 87,000
6. 35,081 − 6,327 = 29,000	**7.** 24,973 − 8,789 = 16,000	**8.** $4.52 − 3.27 = $1.30	**9.** $27.50 − 2.89 = $25.00	**10.** $9.93 − 0.50 = $9.40
11. $62.85 − 37.44 = $25.00	**12.** $53.75 − 47.36 = $6.00	**13.** $47.23 − 9.72 = $38.00	**14.** $8.75 − 0.29 = $8.50	**15.** $2.09 − 0.45 = $1.60
16. 6.583 − 4.471 = 2.1	**17.** 0.489 − 0.273 = 0.21	**18.** 0.892 − 0.049 = 0.85	**19.** 0.607 − 0.152 = 0.45	**20.** 0.322 − 0.076 = 0.25
21. 0.873 − 0.425 = 0.45	**22.** 0.982 − 0.079 = 0.91	**23.** 0.902 − 0.353 = 0.55	**24.** 0.513 − 0.36 = 0.15	**25.** 0.73 − 0.472 = 0.26

OBJECTIVES

- Review and maintain previously taught concepts and skills.
- Practice taking tests in multiple-choice format.

USING THE REVIEW

Assign the Cumulative Review to all students. Provide students with an answer sheet to record their answers.

Each Cumulative Review gives students an opportunity to practice taking tests that are written in multiple-choice format.

Provide appropriate remedial help for students having difficulty with any of the skills and concepts on this page.

CUMULATIVE REVIEW

Choose the letter of the word(s) that completes the sentence.

1. The sum of a group of numbers divided by the number of addends is called the ■. a

   a. Mean **b.** Median **c.** Mode **d.** None of these

2. The sale price of an item is found by subtracting the ■ from the regular price. d

   a. Mark up **b.** Sales tax **c.** Sale price **d.** None of these

3. Money placed by you into a savings or a checking account is called ■. a

   a. A deposit **b.** A withdrawal
   c. Interest **d.** None of these

4. When interest is compounded quarterly, it is paid to you ■ times per year. c

   a. One **b.** Two **c.** Four **d.** None of these

5. On a credit card statement, the ■ is any of the last balance that was not paid. c

   a. Finance charge **b.** New balance
   c. Unpaid balance **d.** None of these

Select the best estimated answer.

6. 4.89 + 0.4 + 1.39 d **a.** 3 **b.** 4 **c.** 5.4 **d.** 6.4
7. 289.7 − 35.2 c **a.** 200 **b.** 240 **c.** 260 **d.** 340
8. 6 × 3.7 c **a.** 2.4 **b.** 10 **c.** 24 **d.** 30
9. $63.84 ÷ 19 c **a.** $2.00 **b.** $2.50 **c.** $3.00 **d.** $4.00

Compute.

10. 9.3 + 6.3 + 5.4 b

    a. 20 **b.** 21 **c.** 21.1 **d.** None of these

11. 5,700 − 923 a

    a. 4,777 **b.** 4,823 **c.** 5,200 **d.** None of these

12. 6 × 7.3 c

    a. 42 **b.** 42.8 **c.** 43.8 **d.** None of these

13. $3\overline{)9{,}270}$ b

a. 309 b. 3,090 c. 3,090 R 1 d. None of these

14. $4\overline{)6.8}$ d

a. 1.2 b. 1.4 c. 2 d. None of these

Solve.

15. This week, Rose worked 3 h on Monday, $5\frac{1}{2}$ h on Wednesday, and $4\frac{3}{4}$ h on Friday. Her hourly wage is $15. How much did she earn this week? b

a. $127.50 b. $198.75 c. $205 d. None of these

16. Jean bought a tee-shirt for $7.98, a tape for $6.95, a sweatshirt for $11.50, and a card for $0.85. Find the total cost of her purchases. d

a. $26.43 b. $27.25 c. $30 d. None of these

17. The regular price of a clock radio is $17.90. This week it is on sale for 20% off. Find the sale price. a

a. $14.32 b. $15.32 c. $17 d. None of these

18. Frank's savings account balance was $450. He withdrew $75 one week and deposited $100 the next week. He then withdrew $25. Then the bank posted interest of $8.36. What is his new balance? c

a. $400 b. $450 c. $458.36 d. None of these

19. The regular price of a sofa is $575. Leon bought it on an installment plan. With no down payment, he agreed to pay $55 per mo for 12 mo. What was his finance charge? b

a. $75 b. $85 c. $660 d. None of these

20. A package of pencils contains 12 pencils. There are 175 students taking a test. If each student gets 1 pencil, how many packages are needed? c

a. 10 b. 14 c. 15 d. None of these

THINKING ABOUT MATH See TE side column for answers.

1. It costs $5.50 to rent golf clubs and $3.75 to rent a cart at the golf course. The pro shop is selling a set of used clubs with bag for $50. If you plan to play golf at least 6 times this season, should you buy the used set of clubs?

2. Town A has a 6% sales tax on clothing. Town B in another state has no sales tax on clothing. If you plan to buy a jacket for $50, is it better to buy it in Town B? Explain why or why not.

THINKING ABOUT MATH The questions in this section can be used in discussions with individual students or with small groups. Guide students in conveying their ideas clearly and precisely.

Listed below are expected student answers. However, accept any reasonable answers.

1. The cost of renting clubs for 6 times would be $55.50. The used clubs only cost $50. If cost is the most important factor, it would be more sensible to buy the used clubs.
2. Since 6% of $50 is only $3.00, it may not be worth your time, gasoline, or other transporation costs to buy only a jacket in Town B.
3. If you finance a television for 24 mo, the monthly payments would be less. However, your finance charge will be greater than if you financed for only 18 mo.

SUPPLEMENTARY MATERIALS

TRP Cum. Test, Ch. 1–8, pp. 1–4

OBJECTIVE

- Test the prerequisite skills needed to learn the concepts developed in Chapter 9.

USING THE TEST

The Pre-Skills Test is designed to diagnose students' strengths and weaknesses on prerequisite skills necessary to study the mathematics in Chapter 9.

Assign the Pre-Skills Test. Allow students to work together in pairs or in small groups. Group members should help those who demonstrate a misunderstanding of a concept or a weakness in a skill.

The table below correlates the items on the Pre-Skills Test with the prerequisite skill and the lesson(s) in the chapter for which it is needed.

Item(s)	Prerequisite skill	Lesson(s)
1–6	Add and subtract money amounts.	All
7–12	Multiply money amounts.	9.2, 9.5
13–16	Find the percent of a number.	9.1
17–28	Divide whole numbers and decimals.	9.2, 9.3
29–34	Read and interpret a table.	9.1, 9.3, 9.4, 9.5

References to specific skills that are reviewed in the Skills Bank are given on the individual lessons themselves.

Pre Skills TEST

Add or subtract.

1. $145 + $567 $712
2. $116.85 + $203.97 $320.82
3. $409.87 + $90.16 $500.03
4. $746 − $219 $527
5. $378.26 − $45.13 $333.13
6. $906.47 − $514.53 $391.94

Multiply.

7. 6 × $225 $1,350
8. 15 × $204 $3,060
9. 9 × $213.05 $1,917.45
10. 16 × $305.56 $4,888.96
11. 27 × $209.83 $5,665.41
12. 51 × $519.76 $26,507.76

Find the answer to the nearest cent.

13. 7% of $95 $6.65
14. 15% of $140.60 $21.09
15. 8.5% of $219 $18.62
16. 16.4% of $319.65 $52.42

Divide. Round to the nearest tenth.

17. 214 ÷ 5 42.8
18. 317 ÷ 6 52.8
19. 403 ÷ 7 57.6
20. 347 ÷ 12 28.9
21. 518 ÷ 16 32.4
22. 607 ÷ 23 26.4
23. 287 ÷ 2.8 102.5
24. 417 ÷ 3.6 115.8
25. 235 ÷ 9.7 24.2
26. 2,087 ÷ 4.6 453.7
27. 1,897 ÷ 10.8 175.6
28. 4,387 ÷ 19.6 223.8

Use the price list. Find the cost.

29. 1 mirror $28.50
30. 1 set of bumper guards $89.95
31. 2 tail lights $59.50
32. 5 car deodorizers $18.75
33. 2 headlights and 1 set of floor mats $136.29
34. 4 seat cushions, 1 tail light, and 3 car deodorizers $116.80

AUTO SUPPLY SHOP - PRICE LIST	
Bumper guards	$89.95
Headlight	$47.85
Floor mats	$40.59
Mirror	$28.50
Tail light	$29.75
Seat cushion	$18.95
Car deodorizer	$ 3.75

Chapter 9 AUTOMOBILE EXPENSES

When buying a car, some people look at new as well as used cars. What information is needed before deciding which to buy?

OBJECTIVE

- Formulate problems, using text and data.

USING THE CHAPTER OPENER

Discuss students' answers to the question on the page. Possible answers for both cars include cost, size, and features. Also, a recent survey shows that consumers consider styling the second most important factor in choosing a new car. (Price is first.) For the used car, consumers need to know how much mileage the car has on it, the condition it's in, and if would be covered by any warranties.

Students should also be aware that people must consider their own means (how much they can afford to spend) and needs (how the car will be used).

COOPERATIVE LEARNING

Have students collect newspaper ads for new and used cars. Assign heterogeneous groups and relate the following situation to them.

A family of 5 is looking for a second car. They own a 2-year-old, mid-size model that Mrs. Smith uses to commute to work. (Mr. Smith takes the train.) The second car will be shared by Mr. Smith and his teenage son and daughter. They can afford a down payment of about $3,000 and monthly payments of about $125, but they would be happy to pay less. None of the Smiths is knowledgeable about cars, so all work on the car must be done at the garage.

Tell students that they are to choose a car for the family based on their understanding of the situation, figure their monthly payments, and tell why they chose the car they did. Have the groups compare their choices and discuss the different approaches.

SUPPLEMENTARY MATERIALS

TRP Group Projects, p. 9

OBJECTIVES

- Read and interpret a new car sticker.
- Compute the amount owed for a new car.

TEACHING THE LESSON

WARM-UP Write the following exercises on the chalkboard or an overhead transparency.

Add.

1. $12,340.85 + $850 + $38

2. $16,405.28 + $1,148 + $75

Find the answer.

3. 8% of $13,000 **4.** 7.5% of $11,900

(**1.** $13,228.85 **2.** $17,628.28
3. $1,040 **4.** $892.50)

INTRODUCTION Ask students to describe the type of new car they would like to buy. Stress that the most important feature of any car is the driver. Point out the importance of safe driving (wearing safety belts, never driving while impaired, etc.)

INSTRUCTION Use Example 1 to show students how to read and interpret a new car sticker. Direct students' attention to the Total Amount of $12,294.00. Explain that the Total Amount is often referred to as the *sticker price.* Point out how the Total Amount is computed. Review the options with students and mention that they can add other options for additional costs. Point out the fuel economy rating and emphasize that mileage estimates are most useful to compare the gas mileages of different cars.

Use Example 2 to show students how to find the amount they owe the dealer for a car. Emphasize that the price on the sticker is just an offering price by the dealer. A car dealer expects a potential buyer to negotiate a lower price. Explain that students should comparison shop at a number of car dealers before making a final selection.

9.1 BUYING A CAR

The time has come to buy a car. You go to a new-car dealer to look at cars. Each car has a large **sticker** on a side window. The sticker shows the suggested price of the car and the **options.**

EXAMPLE 1 Read the sticker on this new car.

VEHICLE IDENTIFICATION NUMBER: 1XY99XYZ987654321

MODEL: ARROW
DEALER: BILL'S AUTO SALES

MANUFACTURER'S SUGGESTED RETAIL PRICE OF THIS MODEL INCLUDING DEALER PREPARATION			$9,958.00
MANUFACTURER'S SUGGESTED RETAIL DELIVERED PRICE FOR OPTIONAL EQUIPMENT INSTALLED ON THIS VEHICLE BY MANUFACTURER			
TINTED GLASS	129.00	AIR CONDITIONER	875.00
WIDE SIDE MOLDING	26.00	VISOR VANITY MIRROR	10.00
FRONT CARPET SAVERS	21.00	REMOTE CONTROL MIRRORS	71.00
REAR CARPET SAVERS	16.00	TILT STEERING COLUMN	135.00
DOOR EDGE GUARDS	30.00	STEEL BELT W/W TIRES	70.00
3-SPEED WINDSHIELD WIPER	58.00	AM/FM STEREO RADIO	262.00
ELECTRIC REAR WINDOW DEFOGGER	170.00		
		OPTIONS SUBTOTAL	$1,873.00
		DESTINATION CHARGE	463.00
		TOTAL AMOUNT	$12,294.00

EPA FUEL ECONOMY RATING
CITY MPG 23 HIGHWAY MPG 32
ESTIMATED ANNUAL FUEL COST: $718
2.5 LITER L4 ENGINE
FUEL INJECTION, AUTO 3 SPD TRANS
ACTUAL MILEAGE WILL VARY WITH OPTIONS, DRIVING CONDITIONS, DRIVING HABITS, AND VEHICLE'S CONDITION.

The Suggested Retail Price Without Options

The Cost of the Options.

The Charge for Delivering the Car to the Dealer.

The Sticker Price (Suggested Retail Price + Cost of Options + Destination Charge).

Optional Equipment Installed in the Car with the Suggested Prices

The **sticker price** is just an offering price made by the dealer. You should be able to buy a car for a **negotiated price** of hundreds or thousands of dollars less. If you are getting a **trade-in allowance** on your old car, you subtract it from the negotiated price. You then pay the actual price plus the sales tax and the **vehicle-registration fee.**

EXAMPLE 2 You negotiate a price of $11,238 for the car with all of the options. You get a trade-in allowance of $1,250 for your old car. You pay a 6.5% sales tax and a $68 registration fee. What do you owe the dealer?

1. Subtract the trade-in allowance to find the actual price. $11,238 − $1,250 = $9,988
2. Multiply to find the tax. 0.065 × $9,988 = $649.22
3. Add to find the total cost. $9,988 + $649.22 + $68 = $10,705.22

You owe the dealer $10,705.22 for the car.

FOR DISCUSSION See TE side column.

1. Reports are available that show how much a dealer actually paid for a car. How could you use this information to negotiate a price?
2. How is buying a used car different from buying a new car?

PRACTICE EXERCISES

Remember to estimate whenever you use your calculator.

Use the sticker on page 168 for Exercises 1–4.

• 1. What is the suggested retail price before options?
$9,958.00

2. How much does the AM/FM stereo radio cost?
$262.00

3. What is the total cost for the carpet savers?
$37.00

4. What would the total sticker price be if the air conditioner were not included?
$11,419.00

Use the options price list at the right. Find the sticker price.

• 5. Suggested retail price: $10,889.00
Options: Air conditioner, AM/FM radio, front and rear mats
Destination charge: $475
$12,463

6. Suggested retail price: $14,903.00
Options: Rear defogger, AM/FM stereo radio, Tilt steering column, P185/75 tires
Destination charge: $464
$16,045

7. Suggested retail price: $12,008.00
Options: Cruise control, California emission package, Tinted glass, AM/FM stereo radio, P195/75 whitewall tires
Destination charge: $468
$13,343

Air Conditioner	$860
Cruise Control	$210
Defogger: Rear Elect	$174
Emission: CA Requirement	$118
Glass: Soft Ray Tinted	$142
Mats: Floor Front	$ 21
Mats: Floor Rear	$ 16
Radio: AM/FM	$202
Radio: AM/FM Stereo	$278
Radio: ETR AM/FM Stereo/Cass	$424
Steering: Tilt Column	$148
Tires: P185/75 R14 SBR W/W	$ 78
Tires: P195/75 R14 SBR B/W	$ 51
Tires: P195/75 R14 SBR W/W	$119

Find the amount owed to the dealer.

• 8. Negotiated price: $13,456.00
Sales tax: 8.5%
Registration fee: $88.00
$14,687.76

9. Negotiated price: $16,549.00
Sales tax: 9.6%
Registration fee: $36.50
$18,174.20

10. Negotiated price: $15,090.00
Trade-in allowance: $875.00
Sales tax: 7.3%
Registration fee: $103.75
$15,356.45

11. Negotiated price: $12,954.00
Trade-in allowance: $2,100.00
Sales tax: 8.25%
Registration fee: $87.65
$11,837.11

SKILLS BANK You may want to refer students to the following skills in the Skills Bank:

Add With Whole Numbers and Decimals, Skill 6

Subtract With Whole Numbers and Decimals, Skill 6

Find the Percent of a Number, Skill 2

FOR DISCUSSION

1. Students should explain that once they know how much the dealer paid, they can negotiate a fair price for a car. There are a number of books available that show the price the dealer paid for the car and options.
2. Students should explain that a used car is usually more likely than a new car to have mechanical problems, and that a used car does not have the same warranty as a new car. However, a used car is usually less expensive than the same type of new car.

ASSIGNMENTS

Level	Exercises
Level 1	1–4, Odd 5–11
Level 2	1–11
Level 3	5–11

FOLLOW-UP ACTIVITY

ENRICHMENT Ask students to visit a new car dealer. They should find a car that interests them, read the sticker, and write down the Suggested Retail Price, Cost of Options, Destination Charge, and the Total Amount. Have students compute the cost of the car, including sales tax and registration fees.

SUPPLEMENTARY MATERIALS

TRP Practice, p. 57
TRP Reteaching, p. 47
TRP Transparency 29

OBJECTIVES

- Compute the amount you need to finance to buy a new car.
- Compute the finance charge for an automobile loan.

TEACHING THE LESSON

WARM-UP Write the following exercises on the chalkboard or an overhead transparency.

Divide mentally.

1. \$7,482 ÷ \$100

2. \$12,872.56 ÷ \$100

Multiply. Round to the nearest cent.

3. 74.82 × \$2.36 **4.** 128.7256 × \$2.05

(**1.** \$74.82 **2.** \$128.7256 **3.** \$176.58 **4.** \$263.89)

INTRODUCTION Explain that almost everyone needs to borrow money when he or she buys a car. Use the introductory material to point out that the lender actually holds the title to a car until the loan is repaid. Explain to students that the word *finance,* in this situation, means *to borrow.* So the amount to be financed is the amount to be borrowed. Discuss the importance of affordable monthly payments and ask students what would happen if monthly payments could not be made.

INSTRUCTION Use Example 1 to show students how to find the amount they need to finance to buy a car. Point out that certain lenders will only finance a portion of the cost of a car. This may mean that a buyer will have to get additional money for a down payment before he or she can purchase a car.

Use Example 2 to show students how to find (1) the monthly payment, (2) the total amount to be repaid, (3) the finance charge for a car loan, and (4) the total cost of a car. Remind students that dividing a money amount by a money amount yields a decimal.

9.2 AUTOMOBILE LOANS

An **automobile loan** is a secured loan. A bank or other lender actually holds **title** to the car until the loan is repaid. When you buy a car, you need to determine how much you need to finance (borrow).

EXAMPLE 1 The cost of a car, with sales tax and registration fees, is \$12,036.47. You get a trade-in allowance of \$1,250 for your old car and you put \$2,000 down in cash. How much do you need to finance?

1. Add to find the down payment. \$1,250 + \$2,000 = \$3,250
2. Subtract to find the amount to be financed. \$12,036.47 − \$3,250 = \$8,786.47

You need to finance \$8,786.47.

You need to include the finance charge when determining the total cost of the car.

EXAMPLE 2 You borrow \$8,786.47 at $10\frac{1}{4}$% for 5 y. Find the finance charge (interest) and the total cost of the car.

THINK: Use the chart to find the monthly payments.

$10\frac{1}{4}$% APR	Monthly payments per \$100
5 y (60 mo) \$2.14	

1. Find the finance charge.
   - Find the number of \$100 to be financed. \$8,786.47 ÷ \$100 = 87.8647
   - Multiply to find the monthly payment. 87.8647 × \$2.14 = \$188.03045 ≈ \$188.03
   - Multiply to find the total amount to be repaid over 60 mo. 60 × \$188.03 = \$11,281.80
   - Subtract to find the finance charge. \$11,281.80 − \$8,786.47 = \$2,495.33

   The finance charge is \$2,495.33.
2. Add to find the total cost of the car.

   THINK: Cost from Example 1 was \$12,036.47. \$12,036.47 + \$2,495.33 = \$14,531.80

   The total cost of the car is \$14,531.80.

FOR DISCUSSION See TE side column.

Most new car dealers also arrange car loans. Suppose that a car dealer offers a very low interest rate. Does that necessarily mean that the total cost of the car will be less than if the interest rate were higher? Explain.

PRACTICE EXERCISES

Remember to estimate whenever you use your calculator.

Complete the table. See page 476 for answers to Exercises 1–32.

Cost of car	Down payment	Amount to be financed	Monthly payment	No. of months	Amount to be repaid	Finance charge	Total cost
$12,090	$3,400	$8,690	$187.86	60	$11,271.60	$2,581.60	$14,671.60
$13,789	$2,600	• 1. ■	$286.48	48	• 2. ■	3. ■	4. ■
$10,908	$5,650	5. ■	$171.64	36	6. ■	7. ■	8. ■
$11,678	$4,270	9. ■	$190.74	48	10. ■	11. ■	12. ■
$9,624	$700	13. ■	$193.84	60	14. ■	15. ■	16. ■
$12,476	$3,075	17. ■	$242.06	48	18. ■	19. ■	20. ■
$14,980	$5,690	21. ■	$303.26	36	22. ■	23. ■	24. ■
$13,678	$4,875	25. ■	$226.66	48	26. ■	27. ■	28. ■
$15,823	$6,950	29. ■	$289.65	60	30. ■	31. ■	32. ■

Use the monthly payment table. Find the finance charge and the total cost.

• **33.** Price: $12,876
Down payment: $2,855
Loan rate: $10\frac{1}{2}$% for 4 y
$2,292.92; $15,168.92

34. Price: $10,098
Down payment: $4,980
Loan rate: 7.9% for 3 y
$648.84; $10,746.84

35. Price: $15,708
Down payment: $1,800
Loan rate: 12.6% for 5 y
$4,867.80; $20,575.80

36. Price: $9,083
Down payment: $1,075
Loan rate: 10.5% for 4 y
$1,832.00; $10,915.00

MONTHLY PAYMENT PER $100

APR	3 y	4 y	5 y
7.9%	$3.13	$2.44	$2.02
$10\frac{1}{2}$%	$3.25	$2.56	$2.15
12.6%	$3.35	$2.66	$2.25

SKILLS BANK You may want to refer students to the following skills in the Skills Bank.

Multiply With Decimals, Skill 5

Divide With Decimals, Skills 1, 8

FOR DISCUSSION Students should explain that a lower loan-interest rate from a car dealer does not necessarily mean a lower total cost. For example, a car dealer could offer a low interest rate but charge more for the car. Students should explain that they should carefully check over any car-loan agreement. The agreement should show the amount financed, the interest rate, the number of payments, the monthly payment, and the total amount to be repaid.

ASSIGNMENTS

Level 1	1–16, 33–34
Level 2	1–36
Level 3	17–36

FOLLOW-UP ACTIVITY

COOPERATIVE LEARNING
Divide the class into pairs. Ask 1 member of each pair to draw up a financing agreement that shows the amount financed, the interest rate, the number of payments, the monthly payment, and the total amount to be repaid. Students should use the monthly-repayment table on page 171. Ask the person drawing up the agreement to make 1 or 2 deliberate errors. For example, the person might put down a finance rate less than the one actually being charged. The second person reviews the agreement and locates the error(s). Students then switch roles and continue the activity.

SUPPLEMENTARY MATERIALS

TRP Practice, p. 58
TRP Reteaching, p. 48
TRP Transparency 29

OBJECTIVES

- Estimate how many miles a car can go on a gallon of gas (mpg).
- Estimate the cost of gasoline for a trip.
- Estimate maintenance and repair costs.

TEACHING THE LESSON

WARM-UP Write the following exercises on the chalkboard or an overhead transparency.

Estimate the quotient.

1. 2,896 ÷ 98 **2.** 5,146 ÷ 210

Multiply. Round up to the nearest ten dollars.

3. 56.9 × $1.169 **4.** 70.6 × $1.089

(**1.** 30 **2.** 25 **3.** $70 **4.** $80)

INTRODUCTION Use the introductory material to discuss where students can find the estimated mpg for new cars. Explain that mpg means miles per gallon. Explain that a car with a lower mpg usually has lower gasoline costs.

INSTRUCTION Use Example 1 to show students how to estimate the mpg for a car. Explain how the formula works in terms of the units used. Emphasize that the measurements of mileage and gasoline used are often inexact, so an estimate of the mpg is sufficient.

Use Example 2 to show students how to estimate the cost of gasoline for a trip. Explain how the formula works and point out that the operation inside parentheses is always done first. Emphasize the correct substitution of values in the formula.

Use the introductory situation to Example 3 to explain the difference between maintenance and repairs. Discuss with students how maintenance can help save money on repairs.

9.3 AUTOMOBILE OPERATING EXPENSES

The **Environmental Protection Agency (EPA)** does tests to find out how many miles a car can go on a gallon of gasoline (mi per gal, or **mpg**). The estimated highway and city mileage ratings are shown on a new-car sticker. The actual gasoline mileage may be different from the EPA estimates. You can compute a car's mpg.

MPG = MILES DRIVEN ÷ GALLONS OF GASOLINE USED

EXAMPLE 1 Your odometer read 12,386 mi. You drove your car until the odometer read 16,282 mi and the car used 203 gal of gasoline. About how many miles per gallon has your car been getting?

1. Subtract to find how far you drove. 16,282 mi − 12,386 mi = 3,896 mi
2. Estimate the mpg.

   THINK: 3,896 mi is about 4,000 mi.
   203 gal is about 200 gal. 4,000 mi ÷ 200 gal = 20 mpg

Your car has been getting about 20 mpg.

You can compute the approximate cost of gasoline for a trip. Cost per gallon can be written as **cpg**.

COST = (MILES DRIVEN ÷ MPG) × CPG

EXAMPLE 2 You are planning to drive 672 mi on a trip. Your car gets about 20 mpg. Gasoline costs $1.129 per gal (cpg = $1.129). About how much will gasoline for the trip cost?

1. Divide to find number of gallons. 672 mi ÷ 20 mpg = 33.6 gal
2. Multiply to find the cost of the gasoline. Then round up to the nearest ten dollars. 33.6 gal × $1.129 cpg = $37.9344 ≈ $40.00

The gasoline for a 672-mi trip will cost about $40.00.

You will have to pay for preventive maintenance and repairs for your car. **Preventive maintenance** is periodic servicing to prevent more-costly repairs. **Repairs** are made when something goes wrong. The table on page 173 shows the approximate costs of preventive maintenance and some common repairs for 5 cars during the first 3 y of driving.

ESTIMATED MAINTENANCE AND REPAIR COSTS—FIRST 3 YEARS

	Arrow	Boltan	Decade	Elice	Van
Preventive maintenance	$620	$926	$745	$713	$978
Water pump	$196	$277	$103	$207	$236
Alternator	$248	$416	$309	$298	$497
Brake pads	$113	$89	$94	$87	$143
Starter	$318	$228	$239	$258	$209

EXAMPLE 3 Steve wants to compare the maintenance and repair costs of cars before he buys. He computes the total cost of maintenance and one of each repair for the Arrow and the Decade. Which car has the lower maintenance and repair cost?

1. Add the Arrow costs. $620 + $196 + $248 + $113 + $318 = $1,495

2. Add the Decade costs. $745 + $103 + $309 + $94 + $239 = $1,490

Because $1,490 < $1,495, the Decade has the lower cost.

FOR DISCUSSION

See TE side column.

1. Lighter cars and smaller engines usually mean better gas mileage. How else could the gas mileage of a car be improved?

2. Why is it important to get a written estimate before any repair work is done? What other steps could you take to ensure that maintenance and repairs are done correctly?

Use Example 3 to show students how to estimate maintenance and repair costs for a car. Point out that it may be worth paying more money for a car that has lower maintenance and repair costs. Explain that these estimates of repair costs can be found in several different publications. The Consumer's Union publishes yearly maintenance and repair records. This report can be a big help when deciding which car to buy.

SKILLS BANK You may want to refer students to the following skills in the Skills Bank:

Multiply With Decimals, Skill 5

Divide With Decimals, Skill 8

FOR DISCUSSION

1. Students should explain that features such as manual transmission, radial tires, and cruise controls usually improve gas mileage. Features such as air conditioning and power seats or windows reduce gas mileage. Fuel mileage will be lower on short trips of a few miles and in stop-and-go city traffic. A well-maintained car gets better gas mileage. Most additives and devices that are supposed to improve gas mileage do not work.

2. Students should explain that a written estimate prevents the mechanic from overcharging. Many states require that mechanics give written estimates and car owners often enjoy other protection under state laws. Students should explain that they should shop around for a good mechanic, rely on the mechanic's local reputation, and be sure that there are specific written instructions about the work to be done. They should look for a mechanic certified by the National Institute for Automotive Service Excellence and indicate an intention to return if they are satisfied with the work.

EXTENSION Use this example to show students how to estimate the value of a new car after several years of use. Point out that the answer can also be found using only 1 step. Explain that 100% minus 45% (depreciation) is equal to 55% (percent of original cost that car will be worth). Show students that 55% of $12,000, or 0.55 × $12,000 = $6,600. Remind students that they may be trading or selling a car after several years. So it may be worth spending more money for a car that depreciates slowly.

ASSIGNMENTS

Level 1	1–4, 7–10, 13–16, 19–22, 25–27
Level 2	1–28, EX
Level 3	4–6, 10–12, 16–18, 22–28, EX

FOLLOW-UP ACTIVITY

APPLICATION Write this chart on the chalkboard or an overhead transparency and ask students to copy it.

Car ________	Year ________
Second Reading ____	Gallons ____
First Reading ______	
Miles ____________	

Ask students to take the chart with them when they travel in the family car or another car they regularly use. On the first trip to the gas station, fill the gas tank and write the odometer reading next to First Reading. On the second trip to the gas station, fill the tank again. Write the odometer reading next to Second Reading and also record how many gallons and tenths it took to fill up the gas tank. Students should use this information to find the mpg for the car.

SUPPLEMENTARY MATERIALS

TRP Practice, p. 59
TRP Reteaching, p. 49
TRP Transparency 30

PRACTICE EXERCISES

Remember to estimate whenever you use your calculator.

Find the mileage. Estimate the mpg.

1. 199 2. 2,356 3. 3,056 4. 5,598 5. 8,507 6. 14,101 7. 20 8. 20 9. 15 10. 20 11. 18

	Car A	Car B	Car C	Car D	Car E	Car F
1st Odometer reading	2,367	10,567	18,972	34,502	68,087	106,587
2nd Odometer reading	2,566	12,923	22,028	40,100	76,594	120,688
Miles driven	•1. ■	•2. ■	3. ■	4. ■	5. ■	6. ■
Gallons of gasoline used	11	97	223	272	509	921
Mpg	7. ■	8. ■	9. ■	10. ■	11. ■	12. ■

12. 14

Find the approximate cost of gasoline for an 800-mi trip. Round up to the nearest ten dollars. Use the estimated mpg from the table above.

	Car A	Car B	Car C	Car D	Car E	Car F
Brand Y $0.939/gal	•13. ■	•14. ■	15. ■	16. ■	17. ■	18. ■
Brand Z $1.102/gal	19. ■	20. ■	21. ■	22. ■	23. ■	24. ■

13. $40 14. $40 15. $50 16. $40 17. $50 18. $60 19. $50 20. $50

Use the maintenance and repair table on page 173. Solve.

•25. Which car has the highest combined cost of maintenance and repairs?
Van

26. Which car has the lowest combined cost of maintenance and repairs?
Decade

27. Barbara takes her Van in to have the water pump and the brake pads fixed. There is a 7% sales tax on the cost of repairs. What is the total cost?
$405.53

28. Roy has the alternator and the starter fixed on his Decade. There is a 9% sales tax on the cost of repairs. What is the total cost?
$597.32

EXTENSION

Depreciation 21. $60 22. $50 23. $50 24. $70

Cars lose some of their value every month. This loss in value is called **depreciation.** The average car depreciates about 45% after 4 y.

A car originally cost $12,000. About what will it be worth after 4 y?

0.45 × $12,000 = $5,400 estimated amount of depreciation

$12,000 − $5,400 = $6,600 estimated value in 4 y

Find the estimated amount of depreciation and the depreciated value in 4 y.

1. Original cost: $14,500
$6,525; $7,975
2. Original cost: $12,380
$5,571; $6,809
3. Original cost: $18,087
$8,139.15; $9,947.85
4. Original cost: $10,666
$4,799.70; $5,866.30
5. Original cost: $15,075
$6,783.75; $8,291.25
6. Original cost: $24,840
$11,178; $13,662

9.4 AUTOMOBILE INSURANCE

For your own protection, you should not drive a car unless you have **automobile insurance.** Many states require that you be insured. An insurance company will pay, within limits, for the costs of an accident.

Liability insurance covers you for injury to others or for damage to their property. Liability insurance is identified by the amounts of coverage.

100/300/50

Bodily injury
$100,000 maximum to each person you injure

Property damage
$50,000 for damage to other's property

$300,000 **maximum** for all injuries in 1 accident

LIABILITY INSURANCE

Type	Amount	Yearly base premium
Bodily injury	25/50	$144.50
	50/100	$161.25
	100/300	$176.80
Property damage	25	$78.60
	50	$89.90
	100	$103.45

The **yearly base premium** depends on the coverage you choose.

EXAMPLE 1 You buy 100/300/50 liability insurance. What is the yearly base premium?

THINK: Locate the premiums in the table.

Add to find the total base premium. $176.80 + $89.90 = $266.70

The yearly base liability premium is $266.70.

Collision insurance pays for damage to your car caused by an accident. **Comprehensive insurance** pays for damage to or losses from your car caused by fire, theft, vandalism, and weather. Both types of insurance have **deductibles.** The premiums depend on the deductibles.

EXAMPLE 2 You buy $200-deductible collision insurance and $100-deductible comprehensive insurance. What is the total yearly premium?

THINK: Locate the premiums in the table.

Add to find the total base premium.

$141.71 + $56.90 = $198.61

The total yearly base premium is $198.61.

COLLISION/COMPREHENSIVE YEARLY BASE PREMIUMS

Deductible	Collision	Comprehensive
$50	$197.38	$72.65
$100	$183.42	$56.90
$200	$141.71	Not available

OBJECTIVES

- Identify the different types of automobile insurance.
- Compute yearly base insurance premiums.
- Compute rated yearly insurance premiums.

TEACHING THE LESSON

WARM-UP Write the following exercises on the chalkboard or an overhead transparency.

Find the number of thousands.

1. $50,000 **2.** $100,000 **3.** $300,000

Use the graph. Find the score.

4. Bob **5.** Liz **6.** Ron

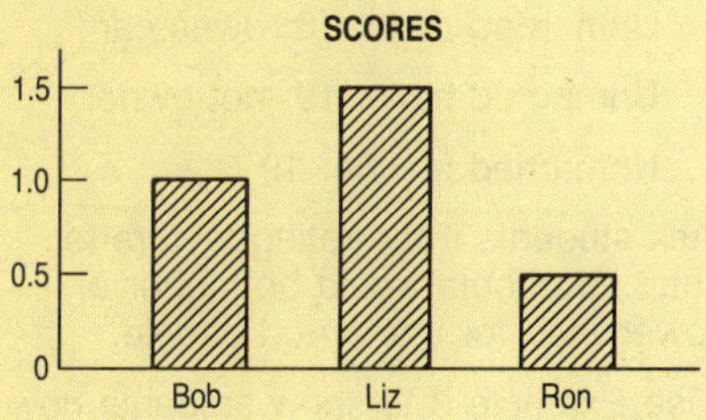

(**1.** 50 **2.** 100 **3.** 300 **4.** 1 **5.** 1.5 **6.** 0.5)

INTRODUCTION Discuss with students the costs of gas, maintenance, and repairs. Ask students to estimate how much they would need each month to cover these costs. Then ask students to estimate how much more they could afford each month to pay for car insurance.

INSTRUCTION Use the introductory situation to emphasize the importance of having insurance coverage. Review the definition of *liability insurance* and the meaning of the 3 numbers that describe the amount of coverage. Ask students to read aloud the yearly base premiums from the table.

Use Example 1 to show students how to find the yearly base premium for liability insurance. Explain to students that most people pay more than the base premium.

Use the introductory situation for Example 2 to explain *collision and comprehensive insurance.* Point out that the insurance company is not responsible for paying the deductible amount.

Use Example 2 to show students how to find the yearly base collision/comprehensive premium. Ask students why there is no coverage for $200 deductible comprehensive coverage.

Use the introductory material before the graphs to explain how insurance policies are rated. Refer students to the graph and have them identify the various rating factors along the side of the graph. Then ask students to identify the rating factors for the following drivers.

Unmarried male, 18, owns car

Unmarried male, 19, not owner

Unmarried female, 19

Ask students if the rating factors for married people would be higher or lower than for unmarried people.

Use Example 3 to show students how to find the total *rated* yearly liability/collision/comprehensive insurance premium. Show students that this sequence of keystrokes would yield the total yearly premium.

[1] [6] [1] [.] [2] [5]
[+] [7] [8] [.] [6]
[+] [1] [8] [3] [.] [4] [2]
[+] [7] [2] [.] [6] [5]
[×] [2] [.] [6] [5] [=]

Ask students how much George's total yearly premium would be if he were age 19 instead of 17.

SKILLS BANK You may want to refer students to the following skills in the Skills Bank:

Add With Whole Numbers and Decimals, Skill 6

Multiply With Decimals, Skill 5

Insurance policies are **rated** on a number of factors that can increase the cost of the base premium. These factors are your age, gender, driving record, and purpose for driving.

The graphs below show the ratings, by age and gender, for unmarried drivers under age 25.

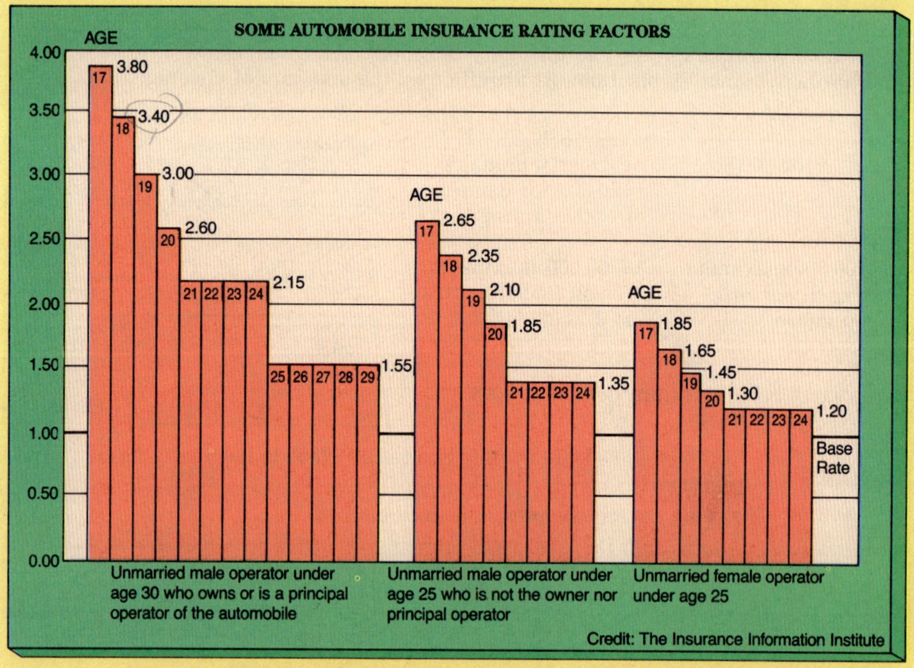

To find the total premium, multiply the base premium by the rating factor.

EXAMPLE 3 George purchased 50/100/25 liability insurance, $100-deductible collision insurance, and $50-deductible comprehensive insurance. George is a 17-year-old male who drives his parents' car. What is his total premium for those coverages?

1. Add to find the total base premium.
   THINK: Find the premiums in the tables on page 175.
   $161.25 + $78.60 + $183.42 + $72.65 = $495.92
2. Multiply the base premium by the rating factor.
   THINK: Find the rating factor on the graph. (2.65)
   2.65 × $495.92 = $1,314.188

George's total yearly premium is $1,314.19.

FOR DISCUSSION
See TE side column.

1. How would you decide whether to get a higher or a lower deductible for collision and comprehensive insurance?

2. Why do you think younger people and men have higher rating factors?

PRACTICE EXERCISES
Remember to estimate whenever you use your calculator.

Use the rate tables on page 175. Find the total yearly base premium.

• 1. Liability: 25/50/25
Collision: $100 deductible
Comprehensive: $50 deductible
$479.17

2. Liability: 50/100/50
Collision: $200 deductible
Comprehensive: $50 deductible
$465.51

3. Liability: 100/300/100
Collision: $50 deductible
Comprehensive: $100 deductible
$534.53

4. Liability: 50/100/25
Collision: $100 deductible
Comprehensive: $100 deductible
$480.17

Use the rating factors graph on page 176. Find the rating factor.

• 5. Unmarried male, 18
Drives his own car
3.40

• 6. Unmarried male, 19
Drives parents' car
2.10

7. Unmarried female, 17
1.85

8. Unmarried female, 19
1.45

9. Unmarried male, 24
Drives his own car
2.15

10. Unmarried male, 24
Drives parents' car
1.35

Complete the table. See page 476 for answers to Exercises 11–46.

Driver	Liability	Yearly base premium	Collision deductible	Yearly base premium	Compre-hensive deductible	Yearly base premium	Total base premium	Rating factor	Total premium
Male, 18, owns car	25/50/25	•11. ■	$100	•12. ■	$50	13. ■	14. ■	15. ■	16. ■
Female, 17	100/300/100	17. ■	$50	18. ■	$100	19. ■	20. ■	21. ■	22. ■
Male, 19, not owner	50/100/50	23. ■	$200	24. ■	$50	25. ■	26. ■	27. ■	28. ■
Female, 18	25/50/50	29. ■	$50	30. ■	$50	31. ■	32. ■	33. ■	34. ■
Male, 18, not owner	100/300/50	35. ■	$200	36. ■	$100	37. ■	38. ■	39. ■	40. ■
Female, 20	25/50/50	41. ■	$100	42. ■	$50	43. ■	44. ■	45. ■	46. ■

FOR DISCUSSION

1. Students should explain that a lower collision deductible is usually chosen for a new car. As cars get older and sustain more damage, higher collision deductibles are chosen. Usually, the comprehensive deductible is low. Comprehensive covers for theft and other losses that often have a value less than $100. A person might also choose a higher collision/comprehensive deductible because he or she wants to avoid the extra costs of lower-deductible policies.

2. Students should explain that the rating factors are determined by actual accident statistics. That is, if a particular group has more accidents, the rating factor for that group will be higher. Younger people, young men in particular, have the highest accident rates, so their rating factors are highest.

ASSIGNMENTS

Level 1	Odd 1–9, 11–22
Level 2	1–46
Level 3	11–46

FOLLOW-UP ACTIVITY

APPLICATION Obtain current yearly base rates and rating factors from a local insurance company or the Insurance Information Institute which has a toll-free (800) number. Write a sampling of base rates and rating factors on the chalkboard or an overhead transparency. Ask students to compute the yearly premium for these drivers.

1. Male, 19, owns car
2. Female, 20
3. Male, 18, not owner

SUPPLEMENTARY MATERIALS

TRP Practice, p. 60
TRP Reteaching, p. 50
TRP Transparency 30

OBJECTIVES

- Compute the cost for a rental car.
- Estimate rental car costs.

TEACHING THE LESSON

WARM-UP Write the following exercises on the chalkboard or an overhead transparency.

Subtract.

1. 22,472 − 21,396

2. 41,102 − 40,782

Multiply.

3. 89 × $0.35 **4.** 247 × $0.26

(**1.** 1,076 **2.** 320 **3.** $31.15 **4.** $64.22)

INTRODUCTION Ask students to describe when it might be necessary to rent a car. Discuss with students the difference between renting a car (for short periods of time) and leasing a car (usually for 3 or more y).

INSTRUCTION Point out that the rental cost depends on the type of rental car. Ask students to identify the different daily rates and mileage costs from the table. Point out that the renter gets 100 free miles for each day he or she rents the car.

Use Example 1 to show students how to compute the rental cost for a car. Remind students that they must subtract 100 from the number of miles to find the number of chargeable miles.

Use Example 2 to show students how to use estimated mileage to estimate the cost of renting a car. Explain that they will likely want to estimate rental costs before renting a car.

SKILLS BANK You may want to refer students to the following skills in the Skills Bank:

Add With Whole Numbers and Decimals, Skill 6

Subtract With Whole Numbers and Decimals, Skills 2, 3, 6

Multiply With Decimals, Skill 5

9.5 RENTING A CAR

One of your teachers rented a car for 1 d.

The cost of a car rental often includes a **daily rate** and a **mileage charge.**

DO DRIVE CAR RENTAL		
Car	Daily rate	Cost per mile
Compact	$29	$0.26
Midsize	$36	$0.31
Luxury	$41	$0.35
Van	$48	$0.41
100 FREE MILES EACH DAY		

EXAMPLE 1 The teacher rented a midsize car. The odometer read 38,546 mi when the car was taken and 38,724 mi when the car was brought back. What did the rental car cost?

THINK: Find the daily rate and the mileage charge for a midsize car.

1. Find the mileage charge.
   - Subtract to find the number of miles. 38,724 mi − 38,546 mi = 178 mi
   - Subtract to find the number of chargeable miles. 178 mi − 100 mi = 78 mi
   - Multiply to find the mileage charge. 78 × $0.31 = $24.18
2. Add to find the total cost. $36 + $24.18 = $60.18

The rental car cost $60.18.

To find approximate rental-car costs, estimate the number of miles you will travel.

EXAMPLE 2 The coach of the high-school debating team plans to rent a van for 4 d to take the team from Houston to Dallas and back. The estimated one-way mileage is 270 mi. About how much will it cost to rent the van?

1. Find the mileage charge.
   - Find the total mileage. 2 × 270 mi = 540 mi
   - Subtract to find the chargeable mileage.
     THINK: 4 d × 100 mi = 400 mi free 540 mi − 400 mi = 140 mi
   - Multiply to find the mileage charge. 140 × $0.41 = $57.40
2. Multiply to find the total daily-rate charge. 4 × $48 = $192
3. Add to find the total cost. Then round up. $192 + $57.40 = $249.40 ≈ $300.00

It will cost about $300.00 to rent the van.

FOR DISCUSSION See TE side column.

1. Why do rental-car companies usually refuse to rent cars to people under age 25?
2. Why do rental-car companies require a cash deposit or a blank, signed credit-card slip when a car is rented?

PRACTICE EXERCISES

Remember to estimate whenever you use your calculator.

Use the rate table on page 178. Find the rental cost.

• 1. Car: Compact
Days: 1
Mileage: 207
$56.82

2. Car: Midsize
Days: 2
Mileage: 623
$203.13

3. Car: Van
Days: 6
Mileage: 1,023
$461.43

4. Car: Luxury
Days: 3
Mileage: 678
$255.30

Complete the table. See page 476 for answers to Exercises 5–28.

Car	Odometer		Days	Daily cost	Chargeable mileage	Mileage charge	Total cost
	Begin	End					
Midsize	29,036	29,236	1	• 5. ■	• 6. ■	7. ■	8. ■
Compact	40,208	41,140	4	9. ■	10. ■	11. ■	12. ■
Van	51,208	53,040	2	13. ■	14. ■	15. ■	16. ■
Luxury	18,907	19,345	7	17. ■	18. ■	19. ■	20. ■
Van	9,308	10,506	5	21. ■	22. ■	23. ■	24. ■
Midsize	29,938	30,501	3	25. ■	26. ■	27. ■	28. ■

Use the rate table on page 178. Approximate the rental cost.

• 29. Liz plans to rent a compact car to travel from Los Angeles to San Francisco and back. The estimated one-way mileage is 500 mi. The trip will take 6 d.
about $300

30. Eric is going to rent a midsize car to travel from Cincinnati to St. Louis and back. The estimated one-way mileage is 450 mi. The trip will take 5 d. about $400

31. Lou wants to rent a luxury car to drive from Atlanta to Dallas and back. The estimated one-way mileage is 1,000 mi. The trip will take 7 d.
about $800

32. The athletic director needs to rent 2 vans to travel from Washington DC, to St. Louis and back. The estimated one-way mileage is 1,800 mi. The trip will take 5 d.
about $4000

FOR DISCUSSION

1. Students should explain that people under 25 have the highest accident rates. Rental car companies are unwilling to rent cars that are likely to be damaged. Refer students to the rating graphs on page 176.
2. Students should explain that rental car companies must have some way of guaranteeing that the car will be returned and the bill will be paid. In addition to positive identification, these companies require a deposit. Some rental car companies will only accept a signed credit card slip and will call the credit card company to set aside, in that account, the estimated cost of the rental.

ASSIGNMENTS

Level 1	1–16, 29–30
Level 2	1–32
Level 3	5–32

FOLLOW-UP ACTIVITY

ENRICHMENT Advertisements for rental cars can be found in the travel sections of most newspapers. Often, these advertisements will feature special weekly rates. Distribute advertisements showing weekly and daily rental costs to each student and ask him or her to compute the weekly rental costs, without mileage charges, from each ad. Ask students to discuss when and why rental companies offer less expensive rentals. Weekly car rentals can be less expensive because processing 1 weekly rental is less expensive than processing 7 daily rentals and because advertised rates are often for compact cars only,

SUPPLEMENTARY MATERIALS

TRP Practice, p. 61
TRP Reteaching, p. 51
TRP Applications, p. 11
TRP Transparency 31

OBJECTIVE

- Decide, given relevant information, which car to buy.

TEACHING THE LESSON

WARM-UP Write the following exercises on the chalkboard or an overhead transparency.

Find the answer to the nearest cent.

1. $\frac{1}{3}$ of $2,180 **2.** $\frac{1}{3}$ of $2,320

Use the loan payment table on page 171. Find the monthly payment.

3. You finance $10,804.60 for 4 y at $10\frac{1}{2}$%.

(**1.** $726.67 **2.** $773.33 **3.** $276.60)

INTRODUCTION Present the problem situation to the class. Encourage students to name some of the factors they would use to make an informed decision about which car to buy. List the factors that are named on the chalkboard.

INSTRUCTION Read the features of the 3 cars that Theresa has prepared. Ask students to discuss the features fully and to mention those features they find most appealing. Ask if there is any other information that Theresa should have. Discuss the decision-making factors.

Have small groups of students complete Exercises 1–14 in the table in order to compare the 3 cars. Refer students to Lesson 9.2 if they have difficulty computing monthly payments. Ask students to write statements that describe the data now in the table. Then ask them to write conclusions that can be drawn from the data. Discuss the appropriateness of these statements and conclusions and lead a discussion about interpretations and possible misinterpretations of the data.

DECISION MAKING

9.6 CHOOSING A NEW CAR

(Open-ended problem solving)

Comparison shopping can help you make sure that the car you buy fits within your budget and provides you with the level of safety you want and can afford.

PROBLEM

Theresa had been looking at new cars for several weeks. She found 3 cars that she liked. She listed the features of each car to help her decide which one to buy. Each car has comparable optional equipment.

CAR A
Price: $14,509.86
mpg: 19 Uses economy unleaded at $1.019 per gal
Estimated preventive maintenance and repair costs for 3 y: $2,180
Insurance costs: $1,653 per y
Safety: Good driver protection
Good passenger protection

CAR B
Price: $12,804.60
mpg: 22 Uses super unleaded at $1.149 per gal
Estimated preventive maintenance and repair costs for 3 y: $2,320
Insurance costs: $1,708 per y
Safety: Good driver protection
Moderate passenger protection

CAR C
Price: $10,987.00
mpg: 18.5 Uses regular unleaded at $1.039 per gal
Estimated preventive maintenance and repair costs for 3 y: $1,978
Insurance costs: $2,018 per y
Safety: Moderate driver protection
Moderate passenger protection

DECISION-MAKING FACTORS

Price	Monthly loan payment	Gasoline costs
Maintenance/repair costs	Cost per mile	Safety

DECISION-MAKING COMPARISONS

Compare the 3 cars by completing the table.

- Assume that Theresa will drive 15,000 mi per y.
- Theresa will put \$2,000 down and finance the rest with a 4-y loan at $10\frac{1}{2}$%. Use the monthly payment table on page 171.
- To find the total annual car expenses:
  Add one year's car payments, gasoline costs, insurance costs, and $\frac{1}{3}$ of the maintenance and repair costs.
- To find the cost per mile, divide the annual expenses by 15,000.

Factors	Car A	Car B	Car C
Price	•1. ■	\$12,804.60	2. ■ \$10,987.00
Monthly loan payment	\$320.25	•3. ■	4. ■ \$230.07
Gas costs for 1 y	5. ■	6. ■	\$842.43
Maintenance/repair costs for 1 y	7. ■	\$773.33	8. ■ \$659.33
Total annual expense	\$7,027.14	9. ■	10. ■ \$6,280.60
Cost per mile	\$0.47	11. ■	12. ■ \$0.42
Safety	13. ■	14. ■	Driver: Moderate Passenger: Moderate

1. \$14,509.86 **3.** \$276.60 **5.** \$804.47 **6.** \$783.41 **7.** \$726.67 **9.** \$6,583.94

MAKING THE DECISIONS **11.** \$0.44 **13.** Driver: Good; Pass.: Good

Which car should Theresa buy if the only factor were: **14.** Driver: Good; Pass.: Moderate

•**15.** Price? Car C
16. Monthly loan payment? Car C
17. Gasoline costs? Car B
18. Maintenance/repair costs? Car C
19. Cost per mile? Car C
20. Safety? Car A

•**21.** How much lower would Theresa's annual maintenance/repair costs be if she bought Car C instead of Car A? \$67.34

22. How much would she save each year by choosing the car with the lowest annual expense over the car with the highest annual expense? \$746.54

23. How much lower would her monthly payments be if she bought Car C instead of Car B? \$46.53

24. How much would she save on gasoline each year if she bought Car B instead of Car A? \$21.06

25. Should Theresa base her decision on the price of the car alone? Why or why not? Answers may vary.

26. Which car would you buy? Why? Answers may vary.

Have students complete Exercises 15–26 in small groups, asking students to explain and defend their answers and decisions. Remind students that there is no one right answer to either Exercise 25 or Exercise 26. Explain that choices will vary and even those with the same choice may have different reasons for making that choice. Continually reinforce that, in many consumer issues, best answers depend on individual situations and preferences. Encourage all students to participate and express their opinions.

ASSIGNMENTS

Level 1	1–26
Level 2	1–26
Level 3	1–26

FOLLOW-UP ACTIVITY

ENRICHMENT Distribute used-car advertisements to each student. Also distribute copies of suggested prices for used cars from a used-car price book. Show students how to find the suggested price for the basic car and the price for options. Finally distribute copies of repair-record reports from a recent edition of the Consumer's Union Annual Buying Guide.

Ask students to review 3 used-car advertisements and decide whether or not they would consider buying the car based on price and repair records. Have students discuss the ads they reviewed and the reasons for their choices. Explain to students that they should buy a used car based on a personal inspection and review by a mechanic, as well as on the price and repair record.

SUPPLEMENTARY MATERIALS

TRP Practice, p. 62
TRP Transparency 31

OBJECTIVES

- Review vocabulary.
- Practice key chapter concepts and skills.

USING THE REVIEW

The Chapter Review is designed to help students prepare for taking the Chapter Test. The first section focuses on vocabulary. It requires students to select words to complete statements. The second section presents practice of key chapter skills.

Each item on the Chapter Review is referenced to the page on which the topic is taught in the Pupil's Edition. You may wish to have students refer to these pages to help review any concepts or skills they have not yet mastered.

It is suggested that students work in small-sized heterogeneous cooperative learning groups. Some cooperative learning methods that may be used are as follows:

1. After each student has independently completed the Chapter Review, a discussion should follow within each group about the solutions to the problems.
2. The group can complete the entire Chapter Review by working together to solve the problems.

End the lesson with an entire class discussion in which any questions brought up in group discussions are presented and answered.

CHAPTER REVIEW

Vocabulary Choose the letter of the word(s) that completes the sentence.

1. The price you pay for a car after making a deal with the car dealer is the ■. [168] a

   a. Negotiated price b. Sticker price c. Trade-in allowance

2. The amount of money you can subtract from the price of a new car by turning in your old car is the ■. [168] c

   a. Negotiated price b. Sticker price c. Trade-in allowance

3. The type of automobile insurance that covers you for injury to others or for damage to their property is called ■. [175] a

   a. Liability b. Collision c. Comprehensive

Skills Find the amount owed for the car. [168]

4. Negotiated price: $12,750
   Sales tax: 8.5%
   Registration fee: $45
   $13,878.75

5. Negotiated price: $14,879
   Trade-in allowance: $985
   Sales tax: 7%
   Registration fee: $38.75
   $14,905.33

Find the answer.

6. Rafael drove 783 mi on 45 gal of gasoline. What is his estimated mpg? [172]
   17 mpg

7. Mary drove 1,853 mi on 80 gal of gasoline. What is her estimated mpg? [172]
   23 mpg

8. Louise spent $125 for a new water pump and $98 for new brake pads. There is a 6% sales tax on the cost of repairs. What was her total cost? [172]
   $236.38

9. Use the insurance rate tables on page 175. Abe purchased 25/50/100 liability insurance, $200 deductible collision insurance, and $100 deductible comprehensive insurance. What is his total yearly base premium? [175]
   $446.56

10. Use the insurance rate table on page 175 and the rating factors graph on page 176. How much would 100/300/100 liability insurance cost an 18-year-old who drives his parents' car? [175]
    $658.59

11. Use the car rental table on page 178. How much would it cost Dara to rent a compact car for 2 d and drive 800 mi? [178]
    $214

12. Use the car rental table on page 178. How much would it cost Victoria to rent a luxury car for 6 d and drive 1,256 mi? [178] $475.60

13. Use the car rental table on page 178. How much would it cost Paul to rent a van for 3 d and drive 709 mi? [178]
    $311.69

CHAPTER TEST

Find the amount owed for the car.

1. Negotiated price: $14,806.00
Sales tax: 7%
Registration fee: $55.00
$15,897.42

2. Negotiated price: $10,937.00
Trade-in allowance: $950.00
Sales tax: 6.7%
Registration fee: $87.50
$10,743.63

Estimate the mpg.

3. Miles driven: 883.2
Gallons of gasoline used: 28
30 mpg

4. Miles driven: 963.6
Gallons of gasoline used: 46.5
20 mpg

Use the insurance rate tables on page 175. Find the total yearly base premium.

5. Liability: 50/100/50
Collision: $100 deductible
Comprehensive: $50 deductible
$507.22

6. Liability: 100/300/100
Collision: $50 deductible
Comprehensive: $100 deductible
$534.53

Use the insurance rate tables on page 175 and the rating factors graph on page 176. Find the yearly total premium.

7. How much would 25/50/50 liability insurance cost a 20-year-old female driver?
$304.72

8. How much would $100-deductible collision insurance cost an 18-year-old male driver who drives his parents' car?
$431.04

Use the car-rental rate table at the right. Find the car-rental cost.

9. Car: Luxury
Days: 4
Mileage: 916
$446.40

10. Car: Midsize
Days: 7
Mileage: 1,308
$573.30

EASY DRIVE CAR RENTAL

Car	Daily rate	Cost per mile
Compact	$25	$0.28
Midsize	$34	$0.35
Luxury	$40	$0.40

50 FREE MILES EACH DAY

Solve.

11. Brenda drives a car that gets about 20 mpg. Gas costs $1.079 per gal. About how much will gas cost to drive 5,000 mi? Round the cost up.
about $300

12. Enrico bought a car for $16,489. He put down $2,800 and financed the rest. He will make a car payment of $307.20 each month for 48 mo. What is the total cost of the car?
$17,545.60

OBJECTIVE

- Evaluate the achievement of chapter objectives.

USING THE TEST

The Chapter Test may be used as a post-test to evaluate student achievement. However, you may wish to use the Chapter Post-Test offered in the Teacher's Resource Package or to design your own chapter test. If this page is not used as a test, you may want to assign it as additional review or practice.

The test items are correlated to the chapter objectives in the table below.

Chapter objectives	Test items
A. Compute the amount owed for a car.	1–2
B. Estimate mpg.	3–4
C. Compute a yearly base car insurance premium.	5–6
D. Compute a rated yearly car insurance premium.	7–8
E. Compute a car rental cost.	9–10
F. Estimate the cost of gasoline for a trip.	11
G. Compute the finance charge for an automobile loan.	12

SUPPLEMENTARY MATERIALS

TRP Ch. 9 Post-Test Form A, pp. 1–2
TRP Ch. 9 Post-Test Form B, pp. 1–2

OBJECTIVE

- Understand that people who do not own cars can rent them and often pay lower than regular rates.

USING THE PAGE

Read the information given in the table. Point out to students that the type of car and rental location also affect rates.

Have students work individually or in pairs to answer Exercises 1–6. Summarize the answers as follows: Mid-week rates and package rates can save you money and possibly enable you to rent a car for a longer period of time at a lower cost.

Have students work individually to answer Exercises 7–8. Then have them share their answers and reasons in a whole-class discussion.

ANSWERS/EXPLANATIONS

2. On weekdays: lower daily rates; for business trips

On weekends: for holiday travel or long weekend getaways; for moving

Other reasons for both: "emergency" travel; when regular car breaks down; when more than 1 family member needs a car

4. (a) $192 (4 × 48)
+ 225 (3 × 75)
$417

(b) $192 (4 × 48)
+ 150 (Weekender Special)
$342

(c) Week O' Wheels: $308
(a) costs the most
(c) costs the least

8. (a) 3 weekdays = $144
(b) 3 weekend days = $225
(c) Weekender Special = $150 for 3 d
(d) 4 weekdays = $192
(e) Weekender Special + Monday or Thursday = $198 for 4 d

Rental would cost the least with (a) for 3 d and (d) for 4 d.

MONEY TIPS

Renting a car can cost less if you take advantage of mid-week rates and package rates.

LET'S LOOK AT THE FACTS

You can save money by renting a car on certain days or for longer periods of time with special **packages.** Study the rates given below.

AMERICANA RENT-A-CAR COMPANY

Category	Days	Rate
Regular	Mon-Tues-Wed-Thurs	$48 a day
Regular	Fri-Sat-Sun	$75 a day
Weekender Special	Fri through Sun	$150
Week O' Wheels	Mon through Sun	$308

LET'S DISCUSS WHY See TE side column for answers to Exercises 2, 4, and 8.

1. Which days are considered mid-week days? Weekend days?
   Mid-week days: Mon., Tues., Wed., Thurs. Weekend days: Fri., Sat., Sun.
2. Why do people rent cars on weekdays? Why do they rent cars on weekends?
3. At regular rates, would it cost more to rent a car from Monday through Thursday or from Friday through Sunday? How much more?
   Fri. through Sun.; $33 more
4. Compare the cost of renting a car for 7 d by using all of the rates and packages shown above. Which package or combination of rates would cost the most? The least?
5. How much would you save by using the Weekender Special package instead of regular weekend day rates to rent a car for Friday, Saturday, and Sunday?
   $75
6. Rental rates cover basic rental only. What other considerations might increase or decrease your total rental cost?
   time of year, rental location, type of car, mileage, insurance, gasoline

LET'S SEE WHAT YOU WOULD DO

7. You have $150 and want to rent a car for the weekend. Which rate would give you the most rental time for your money?
   Weekender Special: $150
8. You need to rent a car to transport your belongings to the college dormitory. The move could take either 3 or 4 d. You need to reserve the car in advance. What is the least expensive way to rent the car for 3 d? For 4 d?

CALCULATOR THE CONSTANT FEATURE

Many calculators have a **constant arithmetic feature.** This feature allows you to repeatedly add, subtract, multiply, or divide a certain number each time you press the is-equal-to key, [=].

In addition, subtraction, and division, the number entered *after* the operation symbol becomes the constant.

Procedure	Calculator Entry	Calculator Display
Add: 7 + 4 + 4 + 4	[7] [+] [4] [=] [=] [=]	19.
Subtract: 72 − 9 − 9	[7] [2] [−] [9] [=] [=]	54.
Divide: 450 ÷ 5 ÷ 5 ÷ 5	[4] [5] [0] [÷] [5] [=] [=] [=]	3.6

Try each of the above examples on your calculator. In each case, do you get the same answer as shown?

In multiplication, the calculator works differently. The number entered *before* the operation symbol becomes the constant.

Procedure	Calculator Entry	Calculator Display
Multiply: 6 × 2 × 6 × 6	[6] [×] [2] [=] [=] [=]	432.
Multiply: 5 × 5 × 5 × 5 × 5	[5] [×] [=] [=] [=] [=]	3125.

Try each of the above multiplication examples on your calculator. In each case, do you get the same answer as shown?

Write the final display answer.

1. [3] [9] [+] [8] [=] [=] 55
2. [8] [.] [6] [−] [.] [5] [=] [=] [=] 7.1
3. [3] [×] [2] [.] [5] [=] [=] [=] [=] 202.5
4. [6] [4] [0] [÷] [4] [=] [=] [=] 10
5. [4] [.] [8] [3] [+] [.] [5] [=] [=] [=] [=] 6.83
6. [7] [−] [.] [3] [2] [=] [=] [=] [=] [=] 5.4
7. [8] [.] [6] [2] [÷] [.] [2] [=] [=] [=] 1077.5
8. [1] [.] [7] [×] [2] [.] [5] [=] [=] [=] [=] 20.88025

Use your calculator's constant feature to complete each pattern.

9. 15, 21, 27, ■, ■, ■ 33, 39, 45
10. 59, 48, 37, ■, ■, ■ 26, 15, 4
11. 3, 6, 18, ■, ■, ■ 54, 162, 486
12. 729, 243, 81, ■, ■, ■ 27, 9, 3

OBJECTIVE

- Use the constant arithmetic feature on a calculator.

USING THE PAGE

After students have read the introductory material on the page, point out that not all standard calculators have the constant feature and that the constant feature on some calculators may operate a little differently.

Have students work through the addition, subtraction, and division examples on their calculators to see if they get the same answers as indicated in the text. Remind them that, in these operations, it is the number entered just *after* the operation symbol (+, −, or ÷) that becomes the constant.

As students work through the multiplication examples, point out that the constant in each case is the factor that is entered just *before* the operation symbol. The second multiplication example has only one number entered, 5. It is both the initial factor and the constant. Point out that the answer, 3,125, can also be expressed as 5^5.

For Exercises 9–12, students will need to determine the pattern before they use their calculators.

Chapter 10 Transportation

OBJECTIVE

- Test the prerequisite skills needed to learn the concepts developed in Chapter 10.

USING THE TEST

The Pre-Skills Test is designed to diagnose students' strengths and weaknesses on prerequisite skills necessary to study the mathematics in Chapter 10.

Assign the Pre-Skills Test. Allow students to work together in pairs or in small groups. Group members should help those who demonstrate a misunderstanding of a concept or a weakness in a skill.

The table below correlates the items on the Pre-Skills Test with the prerequisite skill and the lesson(s) in the chapter for which it is needed.

Items	Prerequisite skill	Lesson(s)
1–6	Estimate sums and differences.	10.1, 10.7, 10.8
7–8	Add and subtract whole numbers.	10.1, 10.6, 10.7, 10.8
9–12	Estimate quotients.	10.1
13–20	Multiply and divide whole numbers and money amounts.	10.1, 10.4, 10.6, 10.7 10.8
21–24	Determine elapsed time.	10.3, 10.4, 10.8
25–28	Convert time between time zones.	10.4
29–36	Find the percent of a number.	10.6, 10.7

References to specific skills that are reviewed in the Skills Bank are given on the individual lessons themselves.

Pre-Skills Test

Estimate the sum or difference.

1. 593 + 435 1,000
2. 1,495 + 2,187 3,000
3. 3,288 + 1,788 5,000
4. 424 − 285 100
5. 3,832 − 675 3,300
6. 4,248 − 2,765 1,000

Find the answer.

7. How much greater is 1,754 + 854 than 2,487? 121
8. How much greater is 2,395 + 1,734 than 3,639? 490

Estimate the quotient.

9. 447 ÷ 50 9
10. 2,857 ÷ 50 60
11. 1,845 ÷ 60 30
12. 4,189 ÷ 52 80

Multiply or divide.

13. 4 × \$379 \$1,516
14. 3 × \$758 \$2,274
15. 15 × 80¢ \$12
16. 7 × 75¢ \$5.25
17. 800 ÷ 25 32
18. 750 ÷ 30 25
19. 1,500 ÷ 25 60
20. 480 ÷ 15 32

Find the elapsed time.

21. From 10:30 A.M. to 4:45 P.M. 6 h and 15 min
22. From 7:43 A.M. to 11:07 A.M. 3 h and 24 min
23. From 9:41 A.M. to 7:15 P.M. 9 h and 34 min
24. From 1:15 P.M. to 9:10 P.M. 7 h and 55 min

Complete. Use the Time Zones map and table in the Tables section at the back of this book.

25. At 10 A.M. Mountain time, it is ■ Central Time. 11 A.M.
26. At 6 P.M. Central time, it is ■ Pacific Time. 4 P.M.
27. At 12 noon Pacific Time, it is ■ Eastern Time. 3 P.M.
28. At 12 noon Eastern Time, it is ■ Pacific Time. 9 A.M.

Find the answer to the nearest cent.

29. 15% of \$11.75 \$1.76
30. 20% of \$15.35 \$3.07
31. 5% of \$82.60 \$4.13
32. 15% of \$23.89 \$3.58
33. 10% of \$39.50 \$3.95
34. 20% of \$52.50 \$10.50
35. 5% of \$39.00 \$1.95
36. 25% of \$36.40 \$9.10

Chapter 10

TRANSPORTATION

When planning a vacation, you need to think about transportation, meals, lodging, and entertainment. Suppose you have a limited budget. How could you keep your travel expenses down?

OBJECTIVE

- Formulate problems, using text and data.

USING THE CHAPTER OPENER

Discuss students' answers to the question on the page. Possible answers include traveling to places off-season when rates are lower; finding bargain motels, inns, or hostels; bringing and cooking your own food rather than eating out; sharing rooms or a vacation house with another family or group; camping in national parks; avoiding the usual tourist sites and finding free attractions; and traveling in a group at group rates.

COOPERATIVE LEARNING

Assign heterogeneous groups and tell students that they are to plan a 7-d vacation for their group. The only limitation they have is their budget of $500 per person. Help them gather travel brochures and newspaper ads for ideas and prices. Also encourage them to share any information they have with other groups so they all have as much data as possible.

They are to work together to come up with a budget and schedule, providing as much detail as possible. Be sure they consider all costs involved with travel, including film if they want to take pictures, souvenirs, appropriate clothing if they don't have it, and so on. Also, encourage them to be realistic in their approach to the problem. For example, if no one in the group has a driver's license, they should not plan to drive to their vacation spot. If driving is feasible, they should consider all costs involved with driving.

After the groups present their vacation plans, have the class determine which group got the most vacation for its money.

SUPPLEMENTARY MATERIALS

TRP Group Projects, p. 10

OBJECTIVES

- Use a mileage chart to estimate and compute travel distances.
- Estimate and compute travel times.

TEACHING THE LESSON

WARM-UP Write the following exercises on the chalkboard or an overhead transparency.

Estimate the sum or difference.

1. 879 + 623 **2.** 1,612 + 486

3. 1,287 − 477 **4.** 2,429 − 1,475

Estimate the quotient.

5. 894 ÷ 50 **6.** 1,185 ÷ 45

(**1.** 1,500 **2.** 2,500 **3.** 500 **4.** 1,000 **5.** 18 **6.** 20)

INTRODUCTION Ask students to think about recent automobile trips they have taken. Call on a student to tell when he or she left and to give the arrival time. Ask students to estimate the distance traveled on the trip. Discuss with students how they arrived at their estimates. Then ask students how useful estimates of travel time are in our daily lives to ensure that we arrive at places on time.

INSTRUCTION Use Example 1 to demonstrate how to use the distance mileage chart and how to estimate total mileage between more than 2 cities. Have students select a pair of cities and find the mileage between them. Use a map of the United States to plan trips entailing several cities. Again ask students to estimate the total driving distance and then compute the actual mileage.

Use Example 2 to show students how the mileage chart can also be used to compare distances. Again ask students to use a map of the United States and select alternate routes between 2 distant cities. Ask students to determine which route is shorter and by how much.

10.1 ESTIMATING DISTANCES AND TRAVEL TIMES

Mileage charts are very useful for planning trips. Suppose you are planning a trip from Detroit to Washington, DC, with a stop in Cincinnati.

EXAMPLE 1 How many miles will you need to drive?

	Estimate	Actual
1. Use the chart to find each distance.		
Detroit to Cincinnati: 259 mi	300 mi	259 mi
Cincinnati to Washington: 481 mi	+500 mi	+481 mi
2. Add to find the total mileage.	800 mi	740 mi

You will drive about 800 mi, or exactly 740 mi.

EXAMPLE 2 How much longer is the trip with the stop in Cincinnati than the direct route to Washington, DC?

	Estimate	Actual
1. Use the chart to find each distance.		
Detroit to Cincinnati to Washington: 740 mi	700 mi	740 mi
Detroit to Washington: 506 mi	−500 mi	−506 mi
2. Subtract to find the difference.	200 mi	234 mi

The trip is about 200 mi longer, or exactly 234 mi longer.

EXAMPLE 3 If you plan to drive at an average speed of 50 mi per h, about how long will it take to make the trip with the stop in Cincinnati?

THINK: DISTANCE = RATE × TIME, SO TIME = DISTANCE ÷ RATE

Round and divide to find the time. 700 ÷ 50 = 14

The trip should take about 14 h of driving time.

DRIVING MILEAGE CHART

	ATLANTA, GA	BOSTON, MA	CHICAGO, IL	CINCINNATI, OH	DALLAS, TX	DENVER, CO	DETROIT, MI	HOUSTON, TX	LOS ANGELES, CA	MIAMI, FL	NEW YORK, NY	PITTSBURGH, PA	ST. LOUIS, MO	SAN FRANCISCO, CA	SEATTLE, WA
ATLANTA, GA															
BOSTON, MA	1087														
CHICAGO, IL	674	963													
CINCINNATI, OH	440	840	287												
DALLAS, TX	795	1748	917	920											
DENVER, CO	1398	1949	996	1164	781										
DETROIT, MI	699	695	266	259	1143	1253									
HOUSTON, TX	788	1804	1067	1029	243	1019	1265								
LOS ANGELES, CA	2182	2978	2054	2179	1387	1059	2311	1538							
MIAMI, FL	655	1504	1329	1095	1300	2037	1352	1190	2687						
NEW YORK, NY	841	206	802	647	1552	1771	637	1608	2786	1308					
PITTSBURGH, PA	687	561	452	287	1204	1411	287	1313	2426	1200	368				
ST. LOUIS, MO	541	1141	289	340	630	857	513	779	1845	1196	948	588			
SAN FRANCISCO, CA	2496	3095	2142	2362	1753	1235	2399	1912	379	3053	2934	2578	2089		
SEATTLE, WA	2616	2976	2013	2300	2078	1307	2279	2274	1131	3273	2815	2465	2081	808	
WASHINGTON, DC	608	429	671	481	1319	1616	506	1375	2631	1075	233	221	793	2799	2684

FOR DISCUSSION See TE side column.

Why is the highway distance between 2 cities usually more than the air distance?

PRACTICE EXERCISES

Remember to estimate whenever you use your calculator.

Use the mileage chart to find the driving distance for the trip.

• 1. Los Angeles to Dallas 1,387 mi
2. Denver to St. Louis 857 mi
3. Detroit to Pittsburgh to Washington, DC 508 mi
4. Seattle to San Francisco to Los Angeles 1,187 mi

Use the mileage chart to estimate the driving distance for the trip.

• 5. New York to Pittsburgh to St. Louis 1,000 mi
6. Houston to Denver to San Francisco 2,000 mi
7. Boston to Cincinnati to Atlanta 1,200 mi
8. Washington, DC, to Dallas to Los Angeles 2,000 mi

Solve.

• 9. How much shorter is a direct trip from Boston to Chicago than one with a stop in New York? 45 mi

10. How much longer is a trip from St. Louis to San Francisco with a stop in Denver than a direct trip? 3 mi

11. How much shorter is a direct trip from Miami to Los Angeles than one with a stop in Houston? 41 mi

12. How much longer is a trip from Detroit to Seattle with a stop in Cincinnati than a direct trip? 280 mi

If you average 50 mi per h, estimate how long the trip will take.

• 13. Los Angeles to Dallas 20 h
• 14. Denver to St. Louis 18 h
15. New York to Pittsburgh to St. Louis 20 h
16. Boston to Cincinnati to Atlanta 24 h
17. Detroit to Pittsburgh to Washington, DC 10 h
18. Seattle to San Francisco to Los Angeles 24 h
19. Houston to Denver to San Francisco 40 h
20. Washington, DC, to Dallas to Los Angeles 40 h

Solve.

21. It took $5\frac{1}{4}$ h to drive to another city while averaging 52 mi per h. About how far did you drive? 250 mi

22. At 60 mi per h, which city is most likely to be reached driving for 11 h from Washington, DC? Atlanta

Use Example 3 to review transforming the distance formula to find time. Remind students to round the actual distance, 740 mi, to its greatest place to get 700 mi. If the rate were 55 mi per h, it would have been rounded to 60 mi per h.

SKILLS BANK You may want to refer students to the following skills in the Skills Bank:

Add With Whole Numbers and Decimals, Skills 1–2

Subtract With Whole Numbers and Decimals, Skills 1–2

Divide With Whole Numbers, Skills 1, 6–7

FOR DISCUSSION Students should explain that air distance is usually a straight line between the 2 cities. In contrast, highways often must curve and meander around rivers and mountains, and there are often no direct routes between 2 places, requiring indirect and longer highway distances.

ASSIGNMENTS

Level	Exercises
Level 1	1–10, 15–18
Level 2	1–22
Level 3	Odd 1–11, 17–22

FOLLOW-UP ACTIVITY

APPLICATION Have students plan a 10-d family vacation trip that includes at least 4 cities on the mileage chart. Ask them to compute the total driving time. Then ask them to compute the percent of vacation time spent driving.

SUPPLEMENTARY MATERIALS

TRP Practice, p. 63
TRP Reteaching, p. 52
TRP Transparency 32

OBJECTIVE

- Solve problems by using a map.

TEACHING THE LESSON

WARM-UP Write the following exercises on the chalkboard or an overhead transparency.

Add.

1. 4.5 + 3.0 **2.** 5.0 + 0.5

3. 3.5 + 0.5 + 1.5 **4.** 2.5 + 6.0 + 0.5

(**1.** 7.5 **2.** 5.5 **3.** 5.5 **4.** 9.0)

INTRODUCTION Ask students whether any of them have taken car trips with their families when road maps were used. Discuss the kinds of information that can be found on most road maps. Have a collection of road maps available for students to examine. Be sure that students are aware of the fact that all maps have a key in which the map symbols are explained, a scale (in miles and/or kilometers), and a direction finder.

INSTRUCTION Present the problem situation to the class. Draw a direction finder on the chalkboard. Review the directions north, south, east, and west. Help students understand that directions are always relative to another given point. For example, South Salt Lake is north of Taylorsville and driving from West Valley City to Millcreek, you would be driving east.

For Situation A, emphasize the use of the map scale. Distribute index cards and have students follow the directions for using the edge of the card as a ruler. Allow students to find each distance. Then add the 3 distances on the chalkboard.

For Situation B, have students refer to the map and trace the alternate route. Provide more practice in finding alternate routes by having students find 2 different routes to travel between Mt. Olympus and Millcreek. (215 to 171; 266 to 195 to 171)

PROBLEM *Solving* STRATEGY

10.2 USING A MAP

Situation:

Paula is driving from Kearns to Granite Park using Route 173, Interstate 15, and Route 171. How far does she need to travel? Is there an alternate route?

Strategy:

Reading a map can help you to solve a problem.

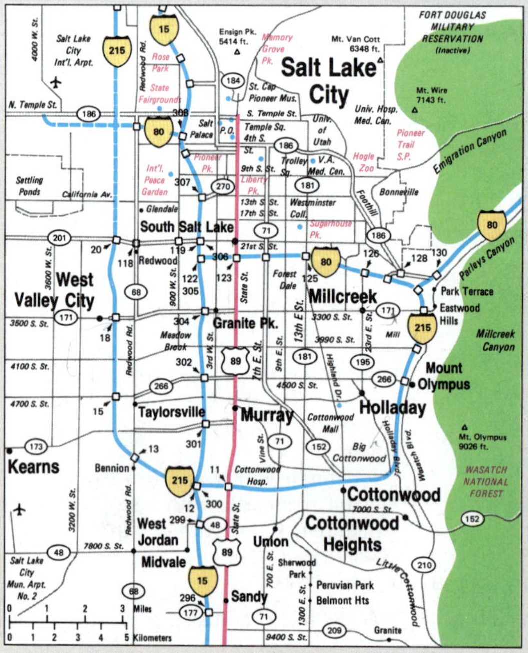

KEY

- Interstate Highway
- U.S. Highway
- State Highway
- Town
- Highway Exit Number
- Places to Visit

Applying the Strategy:

A. About how many miles is it from Kearns to Granite Park?

THINK: Copy the *Miles* **map scale** on the edge of an index card. Use the card like a ruler. Place the card along the highways to be measured.

East from Kearns along 173 to 15	→	about 5.0 mi
North on 15 to 171	→	about 3.0 mi
East on 171 to Granite Park	→	+about 0.5 mi
		about 8.5 mi

It is about 8.5 mi from Kearns to Granite Park.

190 CHAPTER 10

B. Paula hears on the radio that Exit 301 on (173) is closed for repairs. What other way could she travel to Granite Park?

THINK: Look for an alternate route.

1. Which other highway is near Granite Park? 89
2. In which direction would Paula travel on 89? North
3. When 89 meets (171), which way should she turn to get to Granite Park? Left, or West

PRACTICE EXERCISES

Read the map to solve the problem.

Tony traveled from Murray to Millcreek. He used 89, 80, and (195).

• **1.** About how far did he travel on 89? 4.0 mi

2. About how far did he travel on 80? 3.0 mi

3. About how far did he travel on (195)? 1.0 mi

4. About how far did he travel all together? 8.0 mi

Alice traveled from South Salt Lake directly south for about 3.5 mi. She then traveled directly east for about 5 mi.

• **5.** Which highway that runs north and south runs through South Salt Lake? 89

6. Which highway crosses 89 at about 3.5 mi south of South Salt Lake? (266)

7. Which town is located about 5 mi east of 89 on (266)? Mount Olympus

8. Which town was Alice in? Mount Olympus

About how far is it from:

• **9.** Millcreek to Holladay using (195)? 2.0 mi

10. Exit 15 to Exit 18 on 215? 2.0 mi

11. West Valley City to Granite Park using (171)? 3.0 mi

12. Cottonwood to (181) along Highland Drive? 3.5 mi

Solve.

13. Which highways would you use to travel from Granite Park to Taylorsville? (171), (68)

14. Bill wants to travel from West Jordan to West Valley City. He cannot use 15 because of an accident. Name 2 other ways he could get there. (68) to (171); (48) to 3200 W. St. to (171)

SKILLS BANK You may want to refer students to the following skills in the Skills Bank:

Add With Whole Numbers and Decimals, Skills 3, 5

Customary Units of Length, Skill 1

ASSIGNMENTS

Level	Exercises
Level 1	1–10
Level 2	1–13
Level 3	5–14

FOLLOW-UP ACTIVITY

SITUATIONAL PROBLEM SOLVING Present students with the following problem: Your family wants to take a car trip to a city approximately 500 mi from your home. You are in charge of planning the entire trip.

Some questions to ask include: Where can you get a road map of your state or region? Which places do you want to visit? Which roads will you take? What distance do you want to travel each day?

Some approaches to solving the problem include: Using a road map to guess and test various routes. Listing options. Drawing diagrams (sketched maps) of options.

SUPPLEMENTARY MATERIALS

TRP Practice, p. 64
TRP Transparency 32

OBJECTIVES

- Read and interpret a bus or train schedule.
- Determine elapsed time from a bus or train schedule.

TEACHING THE LESSON

WARM-UP Write the following exercises on the chalkboard or an overhead transparency.

Find the elapsed time.

1. From 9:15 A.M. to noon
2. From noon to 4:37 P.M.
3. From 7:40 P.M. to midnight
4. From midnight to 2:35 A.M.

(**1.** 2 h and 45 min **2.** 4 h and 37 min **3.** 4 h and 20 min **4.** 2 h and 35 min)

INTRODUCTION Remind students that the 2 most common means of public ground transportation between cities are buses and trains. Ask students to tell which they would prefer if they had a choice. Ask them how they would get additional information on local bus and train service. Discuss schedules, and list the types of information they could expect to find on such a schedule.

INSTRUCTION Ask students to look at the bus schedule, and help familiarize them with the format of the schedule. Use Example 1 to demonstrate how to read the bus schedule and how it is used to decide which bus to take. Explain the difference between locals and expresses and why the columns are read up for one direction and down for the other direction.

10.3 BUS AND TRAIN TRAVEL

Buses and trains are convenient means of transportation that are used by many people to commute to work as well as to travel comfortably from city to city. Bus and train **schedules** provide information on routes, arrival and departure times, and frequency of service.

EXAMPLE 1 If you are in Atlanta and want to be in Greensboro before 6:00 P.M., which bus should you take? When does it depart?

Check the schedule by reading up:

Bus #145 leaves Atlanta at 9:15 A.M. and arrives in Greensboro at 4:55 P.M.

Bus #204 leaves Atlanta at 10:20 A.M. but does not arrive in Greensboro until 8:35 P.M.

You should take Bus #145, departing from Atlanta at 9:15 A.M.

EXAMPLE 2 How long does it take Bus #106 to get from Washington, DC, to Atlanta?

1. Check the schedule by reading down:

   The bus leaves at 7:10 A.M. and arrives at 5:30 P.M.

2. THINK: 7:10 A.M. is 4 h and 50 min before noon, and
   5:30 P.M. is 5 h and 30 min after noon.

3. Add the times. 9 h and 80 min = 10 h and 20 min

It will take 10 h and 20 min to get to Atlanta.

BUS SCHEDULE

Bus #	# 106 Express	# 145 Express	# 167 Local	# 204 Local
	Read Down	Read Up	Read Down	Read Up
Washington, DC	7:10 a	7:05 p	11:30 a	11:55 p
Manassas, VA	↓	↑	12:10 p	11:10 p
Culpepper, VA	↓	6:10 p	12:45 p	10:50 p
Monroe, VA	8:20 a	↑	1:05 p	10:25 p
Danville, VA	↓	↑	1:40 p	9:45 p
Greensboro, NC	↓	4:55 p	2:35 p	8:35 p
Salisbury, NC	↓	↑	4:10 p	7:00 p
Charlotte, NC	11:55 a	3:05 p	5:50 p	5:15 p
Spartanburg, SC	↓	↑	8:15 p	3:00 p
Greenville, SC	2:35 p	↑	9:10 p	2:05 p
Clemson, GA	↓	11:10 a	10:25 p	12:45 p
Toccoa, GA	↓	↑	11:05 p	11:55 a
Atlanta, GA	5:30 p	9:15 a	12:45 a	10:20 a

Train and bus schedules are usually printed in similar ways.

EXAMPLE 3 You are in Philadelphia on a Friday. The earliest you can catch a train to Washington, DC, is noon. What is the earliest you can arrive in Washington?

Check the schedule:

Look across from Philadelphia to find the first train after noon. This is Train #222. Read down to find when this train arrives in Washington.

The #222 arrives in Washington at 1:55 P.M.

EXAMPLE 4 How long does it take Train #179 to get from New Haven to Philadelphia?

1. Check the schedule:

   Train #179 leaves New Haven at 9:25 A.M. and arrives in Philadelphia at 12:59 P.M.

2. **THINK:** 9:25 A.M. is 2 h and 35 min before noon and 12:59 P.M. is 1 min less than an hour after noon.

3. Add mentally. 2 h and 35 min plus 1 h less 1 min = 3 h and 34 min

The #179 will take 3 h and 34 min to get to Philadelphia.

TRAIN SCHEDULE

Train Number ▶		451	122	102	173	634	222	135	179	110	214	465
Days of Operation ▶		Ex Sa Su	Sa Su only	Ex Sa Su	Ex Sa Su	Sa Su only	Ex Sa Su	Sa only	Ex Su	Su only	Ex Sa	Ex Su
New Haven, CT	Dp	7 32A			8 28A	8 28A		9 25A	9 25A			10 30A
Bridgeport, CT					8 51A	8 51A						10 53A
Stamford, CT		8 20A			9 16A	9 16A		10 11A	10 11A			11 18A
Rye, NY					9 33A	9 27A						
New York, NY	Ar	9 15A			10 17A	10 13A		11 08A	11 08A			12 15P
New York, NY	Dp	9 30A	9 30A	10 00A	10 30A	10 30A	11 00A	11 30A	11 30A	11 30A	12 00N	12 30P
Newark, NJ		9 43A	9 45A	10 13A	10 43A	10 43A	11 13A	11 44A	11 44A	11 44A	12 13P	12 43P
Metropark, NJ		9 57A	9 59A	10 27A				11 58A	11 58A	11 58A		
New Brunswick, NJ												
Princeton Jct., NJ												
Trenton, NJ		10 23A	10 25A		11 18A	11 19A		12 23P	12 23P	12 23P		1 22P
N. Philadelphia, PA								12 48P	12 48P	12 48P		
Philadelphia, PA		10 54A	10 56A	11 16A	11 49A	11 50A	12 11P	12 59P	12 59P	12 59P	1 10P	1 55P
Wilmington, DE		11 18A	11 21A	11 37A	12 12P	12 14P	12 32P	1 23P	1 23P	1 23P	1 31P	2 19P
Newark, DE												
Aberdeen, MD												
Baltimore, MD		12 08P	12 09P	12 23P	1 02P	1 04P	1 20P	2 19P	2 19P	2 19P	2 17P	3 11P
BWI Airport Rail Sta., MD					1 14P	1 16P						
New Carrollton, MD		12 32P	12 42P		1 29P	1 31P	1 44P	2 41P	2 41P	2 41P		3 35P
Washington, DC EST	Ar	12 44P	12 54P	12 55P	1 41P	1 43P	1 55P	2 56P	2 56P	2 56P	2 49P	3 48P

Use Example 2 to show how to use the schedule to find the travel time for a scheduled bus trip. Show students how noon and midnight are convenient times to split a trip to simplify finding elapsed time. Spend as much time as is needed to ensure that students understand how to determine elapsed time. Ask students to select a bus and 2 cities along its route and to determine the travel time between the cities.

Ask students to look at the train schedule, and help familiarize them with its format. Ask students to describe how the train schedule is similar to the bus schedule and how it is different. Then use the train schedule and Examples 3 and 4 to review the process used with the train schedule to select an appropriate train and to determine travel time. Emphasize the mental computation in Example 4, Steps 2 and 3. Have students practice this method with other similar elapsed times.

FOR DISCUSSION

1. Students should explain that monthly passes are usually far less expensive than the equivalent number of individual tickets. In an average month, the monthly pass will be used for about 20 trips, and its cost is usually much less than 20 times the daily fare.
2. Students should explain that travel patterns are usually different on weekends since fewer people are commuting to work and, instead, people are visiting friends and relatives or traveling for recreational purposes.

ASSIGNMENTS

Level 1	1–6, 11–16
Level 2	1–19
Level 3	5–10, 15–20

FOLLOW-UP ACTIVITY

ENRICHMENT Ask students to plan a trip from their hometown or a nearby city to another city in another state served by both bus and train service. Ask students to gather information on fares, schedules, and distances from the bus company and the train line. Have students use the information they collect to compute the cost per minute and the cost per mile for both the bus trip and the train trip. Let students compare their results with those of other students and determine whether there is a pattern with respect to bus travel versus train travel. Have students discuss which trips are the most and the least expensive per mile and per minute.

SUPPLEMENTARY MATERIALS

TRP Practice, p. 65
TRP Reteaching, p. 53
TRP Transparency 33

FOR DISCUSSION See TE side column.

1. Why do many commuters purchase **monthly passes** for bus or train travel?
2. Why do many buses and trains run different schedules on weekends?

PRACTICE EXERCISES

Remember to estimate whenever you use your calculator.

Use the bus schedule to solve.

• 1. If you take Bus #106, what time will you arrive in Greenville?
2:35 P.M.

• 2. If you take Bus #204, what time will you arrive in Danville?
9:45 P.M.

3. If you leave Clemson at 11:10 A.M., what time does the bus get you into Culpepper?
6:10 P.M.

4. If you leave Salisbury at 4:10 P.M., what time does the bus get you into Toccoa?
11:05 P.M.

5. Which bus should you take from Washington, DC, if you need to arrive in Atlanta for an early meeting?
Bus #167

6. Which bus should you take from Charlotte if you want to arrive in Washington, DC, by 8:00 P.M.?
Bus #145

7. What time does the local to Washington, DC, leave Spartanburg?
3:00 P.M.

8. How long does it take Bus #204 to get from Spartanburg to Manassas?
8 h and 10 min

9. How long does it take Bus #167 to get from Washington, DC, to Atlanta?
13 h and 15 min

10. How much faster does Bus #106 get from Monroe to Charlotte than Bus #167?
1 h and 10 min

Use the train schedule to solve.

• 11. If you take Train #122, what time will you arrive in Trenton?
10:25 A.M.

• 12. If you take Train #110, what time will you arrive in Wilmington?
1:23 P.M.

13. If you leave Stamford at 9:16 A.M. on a Saturday, what time will you arrive in Baltimore?
1:04 P.M.

14. If you leave Newark at 10:13 A.M., what time will you arrive in Washington, DC?
12:55 P.M.

15. What is the latest train you can take from New York on a Tuesday to arrive in Washington, DC, by 3:00 P.M.?
Train #214 at 12:00 noon

16. Which train should you take from New Haven in order to arrive in New York by 10:00 A.M.?
Train #451

17. How long does it take Train #122 to get from New York to Washington, DC?
3 h and 24 min

18. How long does it take Train #179 to get from Metropark to Baltimore?
2 h and 21 min

19. How much faster does Train #222 get from New York to Washington, DC, than Train #179?
31 min

20. Which train and bus route gets you from New York to Washington, DC, to Atlanta in the shortest time? How long will the trip take?
Train #214, Bus #106; 12 h and 29 min

10.4 AIR TRAVEL

Pam Robbins lives in Dallas. She and her parents are planning a 1-wk vacation trip to Los Angeles.

EXAMPLE 1 If the Robbins want to arrive in Los Angeles by noon, which flight should they take? When does it depart?

Check the flight schedule:

The last flight between Dallas and Los Angeles to arrive before noon is Central Airlines Flight #651. It departs from Dallas at 10:50 A.M. and arrives at 11:54 A.M.

They should take Flight #651, departing from Dallas at 10:50 A.M.

FLIGHT SCHEDULE

From Dallas/Fort Worth (CST)

Leave	Arrive	Flight	Freq.
TO: Chicago, IL (CST)			
7:05 a	9:15 a	Central 580	Daily
8:30 a	10:33 a	Hemisphere 542	Daily
11:06 a	1:15 p	Central 180	x 7
1:00 p	2:51 p	Sierra 272	Daily
4:10 p	6:21 p	Hemisphere 716	x 6
TO: Los Angeles, CA (PST)			
8:40 a	9:35 a	Sierra 389	Daily
10:50 a	11:54 a	Central 651	Daily
12:01 p	12:45 p	Hemisphere 247	x 7
4:08 p	4:55 p	Central 533	Daily
6:42 p	7:29 p	Central 217	Daily
TO: New York, NY (EST)			
7:00 a	10:58 a	Central 244	Daily
10:19 a	2:25 p	Sierra 224	x 5
1:00 p	5:18 p	Central 164	Daily
4:20 p	8:40 p	Hemisphere 281	Daily
7:51 p	11:59 p	Sierra 466	Daily

EXAMPLE 2 How long will the flight take?

THINK: Departs: 10:50 A.M. Dallas Arrives: 11:54 A.M. Los Angeles
11:54 A.M. Los Angeles time is 1:54 P.M. Dallas time.

10:50 A.M. to 1:54 P.M. is 3 h and 4 min.

So the flight will take just over 3 h.

EXAMPLE 3 How much will it cost for airfare if they get the SuperSaver rates?

1. Check the Fare Schedule. The round-trip SuperSaver fare is $239.
2. Multiply to find the fare for 3 people. $3 \times \$239 = \$717$

The Robbins will pay $717 for SuperSaver airfare.

FROM: Dallas / Fort Worth, TX — ROUND-TRIP FARE SCHEDULE

TO:	Regular Coach	First Class	SuperSaver*
Chicago	$475	$625	$188
Los Angeles	$550	$675	$239
New York	$595	$725	$259

* SuperSaver fares — must stay over Saturday night

OBJECTIVES

- Read and interpret an airline flight schedule and a fare schedule.
- Determine elapsed time between time zones.

TEACHING THE LESSON

WARM-UP Write the following exercises on the chalkboard or an overhead transparency. Have students refer to the Time Zones map and table in the Tables section at the back of the Pupil's Edition.

Complete.

1. 5 P.M. EST = ■ MST (3 P.M.)
2. 11 A.M. PST = ■ CST (1 P.M.)
3. 1 P.M. CST = ■ MST (12 noon)
4. 10 P.M. PST = ■ EST (1 A.M.)

INTRODUCTION Ask students to relate their experiences with airline flights including confusion over time differences, fares, delays, making connections, close calls, and so on.

INSTRUCTION Ask students to examine the flight schedule. Point out that "x 5" means "Except Friday," "x 6" means "Except Saturday," and "x 7" means "Except Sunday."

Use Example 1 to show how to use the schedule to decide which flight is best. Explain that it is a common situation for more than 1 airline to fly between 2 cities.

Use Example 2 to demonstrate how to take differing time zones into account when figuring flight times from a schedule. Use the Time Zones map and table in the Pupil's Edition. Explain that it is easiest to convert the arrival time to its equivalent in the departure time zone to determine flight time.

Use Example 3 to show how to compute fares from a fare schedule. Explain the differences between First Class, Regular Coach, and SuperSaver. Distribute airline newspaper ads involving reduced fares to compute savings.

SKILLS BANK You may want to refer students to the following skills in the Skills Bank:

Multiply With Whole Numbers, Skills 2–3

FOR DISCUSSION

1. Students should explain that with time changes involving the gain or loss of 2 or more hours, it is often helpful to take time to become acclimated. It is also wise to anticipate travel delays.
2. Students should explain that fares are based on supply-and-demand market conditions. Since there are only a few First Class seats on each flight, they are far more expensive than other seats. Similarly, since many Coach fare passengers are traveling on business and have little flexibility on which flight to take or when to travel, they pay much more than SuperSaver passengers who tend to have far more flexibility in their travel plans.

ASSIGNMENTS

Level 1	1–9, 12–18
Level 2	1–21
Level 3	5–6, 9–14, 17–22

FOLLOW-UP ACTIVITY

APPLICATION Distribute copies of an airline flight schedule out of a nearby city. Provide students with a list of 5 meeting sites and times. (For example, "You must be in City X by 11:00 A.M.) Have students use the flight schedule to select the most appropriate flight to take. If possible, include cities that are only accessible via connecting flights.

SUPPLEMENTARY MATERIALS

TRP Practice, p. 66
TRP Reteaching, p. 54
TRP Transparency 33

FOR DISCUSSION See TE side column.

1. Why do people flying long distances try to arrive several hours before a scheduled meeting?
2. Why are Coach, First Class, and SuperSaver fares so different?

PRACTICE EXERCISES

Remember to estimate whenever you use your calculator.

Use the Flight Schedule to identify the best flight to take from Dallas in order to arrive in:

• 1. Chicago by 4:00 P.M. Sierra Airlines Flight #272
• 2. New York by noon. Central Airlines Flight #244
3. Los Angeles after 6:00 P.M. Central Airlines Flight #217
4. Chicago by 10:00 A.M. Central Airlines Flight #580
5. Los Angeles by 5:00 P.M. Central Airlines Flight #533
6. New York by 9:00 P.M. Hemisphere Airlines Flight #281

Find the actual flight time for the flight. Refer to the Time Zones map and table in the Tables section at the back of this book.

• 7. Leave 11:15 A.M. EST — Arrive 1:30 P.M. CST — 3 h and 15 min
8. Leave 2:15 P.M. CST — Arrive 4:45 P.M. PST — 4 h and 30 min
9. Leave 3:00 P.M. PST — Arrive 5:15 P.M. MST — 1 h and 15 min
10. Leave 10:00 A.M. MST — Arrive 4:40 P.M. EST — 4 h and 40 min
11. Leave 1:25 P.M. CST — Arrive 3:15 P.M. PST — 3 h and 50 min

Use the Flight Schedule to find the actual flight time for the flight.

12. Central #180 2 h and 9 min
13. Hemisphere #247 2 h and 44 min
14. Sierra #224 3 h and 6 min

Use the Fare Schedule to find the round-trip airfare for the trip.

• 15. Dallas to New York, First Class $725
16. Dallas to Los Angeles, Regular Coach $550
17. Dallas to Chicago, SuperSaver $188
18. Dallas to Los Angeles, First Class $675

Solve.

• 19. How much more expensive is a First Class round-trip ticket from Dallas to New York than a Regular Coach ticket? $130
20. How much can you save by flying round trip from Dallas to Chicago at SuperSaver rates rather than Regular Coach? $287
21. Jim and Phyllis flew from Dallas to Los Angeles and back at SuperSaver rates. How much over the Regular Coach rates did they save on the 2 tickets? $622
22. Richard flew round trip on business from Dallas to New York one week and from Dallas to Chicago the next week. If he flew First Class, how much did he spend on airfare? $1,350

10.5 USING SUBWAYS

One of the most convenient forms of transportation in many large cities is the subway. Subway systems usually have route maps in each station and in each car to help you find your way. Washington, DC, has one of the nation's newest subway systems.

Washington, DC Partial Subway Map

OBJECTIVES

- Use a subway map to determine routes.
- Read and interpret a subway fare schedule and a travel time chart.

TEACHING THE LESSON

INTRODUCTION Ask students to consider arriving in a large, strange city that has a complex, but convenient and safe, subway system. Ask students what information they would find helpful in order to use the subway system and how this information might best be presented. Discuss student answers and alternate display formats.

INSTRUCTION Remind students that maps, charts, and tables are convenient ways of displaying large amounts of information in an organized and accessible manner. Tell students to examine the maps and charts in this lesson, and help familiarize them with the formats. Compare the format for the subway fare schedule with the driving mileage chart in Lesson 10.1.

Use Step 1 in the Example to demonstrate how the map can be used to determine routes. Point out the different-colored subway lines and the transfer stations. Have students select other departure points and destinations and then use the map to determine the best routes.

Use Step 2 and the fare schedule to determine the fare for the trip. Discuss the difference between rush hour and non-rush hour. Ask students to determine the fares for other possible trips.

Use Step 3 and the Travel Time chart to determine the time for the trip. Discuss how this information can be helpful for subway riders. Ask students to determine the times for other possible trips.

Ask if anyone in the class has ever used the Washington, DC, subway system. If so, have the student(s) describe how the system works. Explain that the entire system is computerized and uses fare cards with magnetic strips that automatically deduct the correct fare when passengers leave the station.

EXAMPLE In Washington, DC, your hotel is located near the Dupont Circle station. Which subway route would you take to get to Arlington Cemetery? How much is the fare? How long should the trip take?

1. Locate Dupont Circle on the Red Line and Arlington Cemetery on the Blue Line. Find the transfer station between the lines at Metro Center.

To get to Arlington Cemetery, you would take the Red Line 2 stops to Metro Center, change to the Blue Line, and go 5 more stops.

2. Use the Fare Schedule to find the non–rush-hour fare for the trip.

   Locate Dupont Circle and Arlington Cemetery. Read down from Arlington and across from Dupont Circle to find .80.

The fare will be $0.80.

FARE SCHEDULE (Rush Hour: upper right; Non-Rush Hour: lower left)

	Addison Road	Archives	Arlington Cemetery	Ballston	Benning Road	Bethesda	Braddock Road	Brookland	Capitol Heights	Capitol South	Cheverly	Clarendon	Cleveland Park	Court House	Crystal City	Deanwood	Dupont Circle	Eastern Market	Eisenhower Avenue	Farragut North	Farragut West	Federal Center SW	Federal Triangle
Addison Road		1.45	1.80	2.15	.70	2.35	2.10	1.65	.80	1.30	1.25	2.05	1.90	2.00	1.80	1.05	1.70	1.25	2.30	1.65	1.65	1.40	1.55
Archives	1.00		.85	1.15	1.05	1.50	1.35	.90	1.35	.80	1.35	1.00	.90	.95	.90	1.10	.80	.80	1.55	.80	.80	.80	.80
Arlington Cemetery	1.10	.80		.80	1.40	1.65	1.20	1.30	1.70	.95	1.70	.80	1.15	.80	.80	1.45	.90	1.00	1.35	.85	.80	.85	.80
Ballston	1.10	.80	.80		1.75	1.80	1.50	1.55	2.05	1.35	2.05	.80	1.35	.80	1.10	1.80	1.15	1.40	1.60	1.15	1.00	1.25	1.10
Benning Road	.80	.80	1.00	1.10		1.95	1.75	1.25	.70	.85	.90	1.65	1.50	1.55	1.40	.75	1.25	.75	1.90	1.20	1.25	.95	1.10
Bethesda	1.10	1.00	1.10	1.10	1.10		2.30	1.65	2.20	1.65	2.15	1.70	.95	1.65	1.85	.95	1.25	1.70	2.40	1.35	1.45	1.60	1.45
Braddock Road	1.10	.80	.80	1.00	1.10	1.10		1.80	2.00	1.40	2.10	1.40	1.80	1.40	.85	1.85	1.55	1.45	.80	1.50	1.45	1.35	1.35
Brookland	1.10	.80	.80	1.00	.80	1.00	1.10		1.55	1.00	1.50	1.40	1.15	1.35	1.40	1.25	1.00	1.05	2.00	.95	.95	1.00	.90
Capitol Heights	.80	.80	1.10	1.10	.80	1.10	1.10	1.00		1.20	1.10	1.90	1.75	1.85	1.85	.90	1.55	1.10	2.15	1.50	1.50	1.25	1.40
Capitol South	.80	.80	.80	.80	.80	1.00	.80	.80	.80		1.20	1.20	1.10	1.10	.95	.95	.80	.80	1.55	.80	.80	.80	.80
Cheverly	.80	.80	1.10	1.10	.80	1.10	1.10	1.00	.80	.80		1.90	1.70	1.90	1.70	.70	1.55	1.15	2.25	1.50	1.50	1.30	1.40
Clarendon	1.10	.80	.80	.80	1.10	1.10	1.00	1.00	1.10	.80	1.10		1.20	.80	.95	1.65	1.00	1.25	1.55	.95	.85	1.10	.95
Cleveland Park	1.10	.80	.80	.80	1.00	.80	1.10	.80	1.10	.80	1.10	.80		1.15	1.35	1.50	.80	1.15	1.95	.80	.85	1.00	.85
Court House	1.10	.80	.80	.80	1.10	1.00	.80	.80	1.10	.80	1.10	.80	.80		.90	1.60	.95	1.20	1.50	.90	.80	1.05	.90
Crystal City	1.10	.80	.80	.80	1.00	1.10	.80	1.00	1.10	.80	1.10	.80	.80	.80		1.45	1.10	1.00	1.05	1.05	1.00	.85	.90
Deanwood	.80	.80	1.00	1.10	.80	1.10	1.10	.80	.80	.80	.80	1.10	1.00	1.10	1.00		1.30	.85	2.00	1.25	1.25	1.05	1.15
Dupont Circle	1.10	.80	.80	.80	.80	.80	1.00	.80	1.00	.80	1.00	.80	.80	.80	.80	1.00		.90	1.70	.80	.80	.80	.80
Eastern Market	.80	.80	.80	1.00	.80	1.10	1.00	.80	.80	.80	.80	.80	.80	.80	.80	.80	.80		1.60	.80	.80	.80	.80
Eisenhower Avenue	1.10	1.00	.80	1.00	1.10	1.10	.80	1.10	1.10	1.00	1.10	1.00	1.10	1.00	.80	1.10	1.10	1.00		1.65	1.65	1.50	1.55
Farragut North	1.00	.80	.80	.80	.80	.80	1.00	.80	1.00	.80	1.00	.80	.80	.80	.80	.80	.80	.80	1.00		.80	.80	.80
Farragut West	1.00	.80	.80	.80	.80	1.00	1.00	.80	1.00	.80	1.00	.80	.80	.80	.80	.80	.80	.80	1.00	.80		.80	.80
Federal Center SW	1.00	.80	.80	.80	.80	1.00	.80	.80	.80	.80	.80	.80	.80	.80	.80	.80	.80	.80	1.00	.80	.80		.80
Federal Triangle	1.00	.80	.80	.80	.80	1.00	.80	.80	.80	.80	1.00	.80	.80	.80	.80	.80	.80	.80	1.00	.80	.80	.80	

3. Use the Travel Time chart to find how long the trip will take.

   Read the chart in the same way you read the Fare Schedule to find 18.

The trip should take only 18 min.

TRAVEL TIME BETWEEN STATIONS (in minutes)

Addison Road	Archives	Arlington Cemetery	Ballston	Benning Road	Bethesda	Braddock Road	Brookland	Capitol Heights	Capitol South	Cheverly	Clarendon	Cleveland Park	Court House	Crystal City	Deanwood	Dupont Circle	Eastern Market	Eisenhower Avenue	Farragut North	Farragut West
26																				
32	16																			
44	28	16																		
6	20	26	38																	
45	25	31	37	39																
39	17	19	35	33	42															
37	15	31	29	31	26	32														
3	23	29	41	3	42	36	34													
14	12	18	24	8	31	22	23	11												
22	24	36	36	16	43	37	35	19	12											
40	24	12	4	34	33	31	25	37	20	32										
36	16	22	28	30	9	33	17	33	22	34	24									
38	22	10	6	32	31	29	23	35	18	30	2	22								
32	10	6	22	26	35	7	25	29	15	30	18	26	16							
20	22	34	34	14	41	35	33	17	10	2	30	32	28	28						
32	12	18	24	26	13	29	13	29	18	30	20	4	18	22	28					
12	14	20	26	6	33	24	25	9	2	10	22	24	20	17	8	20				
43	21	23	39	37	46	4	36	40	26	41	35	37	33	11	39	33	28			
30	10	16	22	24	15	27	11	27	16	28	18	6	16	20	26	2	18	31		
25	15	7	13	19	24	26	16	22	11	23	9	15	7	13	21	11	13	30	9	
16	10	16	22	10	29	20	21	13	2	14	18	20	16	13	12	16	4	24	14	9

FOR DISCUSSION See TE side column.

1. Why are subways often the fastest way to get from place to place in a city?
2. Why do many bus and subway systems charge a higher fare during rush hour?

PRACTICE EXERCISES

Remember to estimate whenever you use your calculator.
See page 477 for answers to Exercises 2, 3, and 6.

Use the Partial Subway Map. Describe the route you would take, including where you would change lines, if necessary, for the trip.

• 1. Union Station to Dupont Circle Red Line 5 stops
• 2. Union Station to Stadium Armory
3. Union Station to Arlington Cemetery
4. National Airport to Farragut West Blue line 7 stops
5. Gallery Place to the Pentagon Yellow line 3 stops
6. L'Enfant Plaza to Judiciary Square

Use the Fare Schedule to find the fare for the trip.

• 7. Ballston to Crystal City during non-rush hour $0.80
• 8. Bethesda to Dupont Circle during rush hour $1.25
9. Federal Center SW to Capital South during non-rush hour $0.80
10. Cheverly to Eisenhower Avenue during rush hour $2.25
11. Farragut North to Arlington Cemetery during rush hour $0.85
12. Cleveland Park to Brookland during non-rush hour $0.80

Use the Travel Time chart to find how long the trip will take.

• 13. Benning Road to Deanwood 14 min
• 14. Farragut North to Capital Heights 27 min
15. Archives to Eastern Market 14 min
16. Federal Center SW to Braddock Road 20 min
17. Court House to Clarendon 2 min
18. Addison Road to Eisenhower Avenue 43 min

Solve.

19. David lives near Crystal City and commutes 5 d per wk to his job near the Farragut North Station. How much money does he spend per week commuting to and from work? $10.50
20. Victoria lives near Eastern Market and commutes 5 d per wk to and from her job near Bethesda. How many minutes per week does she spend on the subway? How many hours is that? 330 min; $5\frac{1}{2}$ h

FOR DISCUSSION

1. Students should explain that since subways run on underground tracks, they avoid urban traffic jams and thus are often much faster than taxis or buses.
2. Students should explain that bus and subway systems attempt to reduce rush-hour passenger volume by offering lower fares during non-rush-hour times. This fare structure encourages people who do not have to travel during rush hour to use the transportation during other times.

ASSIGNMENTS

Level	Exercises
Level 1	1–18
Level 2	1–19
Level 3	1–20

FOLLOW-UP ACTIVITY

COOPERATIVE LEARNING
Have students work in small heterogeneous groups. Ask each group to imagine a 3-d trip to Washington, DC, where they stayed at a hotel near Dupont Plaza. During their 3-d trip, they went to the Capitol, the National Archives, and Arlington Cemetery. Have each group write a diary of the trip, describing (1) the subway routes, (2) the rush-hour fares, and (3) the traveling time of each day's trip.

SUPPLEMENTARY MATERIALS

TRP Practice, p. 67
TRP Reteaching, p. 55
TRP Transparency 34

OBJECTIVE

- Determine taxi fares including a tip.

TEACHING THE LESSON

WARM-UP Write the following exercises on the chalkboard or an overhead transparency.

1. How many $\frac{1}{4}$ mi are in $2\frac{1}{2}$ mi?
2. How many $\frac{1}{6}$ mi are in $2\frac{1}{2}$ mi?
3. How many $\frac{1}{8}$ mi are in $2\frac{1}{2}$ mi?
4. How many $\frac{1}{10}$ mi are in $2\frac{1}{2}$ mi?

Find the answer.

5. 10% of $7.50 6. 5% of $9.00

(**1.** 10 **2.** 15 **3.** 20 **4.** 25 **5.** $0.75 **6.** $0.45)

INTRODUCTION Begin discussion by asking students if they have ever ridden in a taxicab. Ask students to describe the meter and then explain that most taxis use a fare structure of a set amount for the first fraction of a mile plus a different set amount for each additional fraction of a mile.

INSTRUCTION Use Example 1 to demonstrate how taxi fares are computed. Remind students to reason that since each mile consists of 4 quarter miles, there are $(3 \times 4) + 2$, or 14 quarter miles, in $3\frac{1}{2}$ mi. Change the fare structure to $\frac{1}{6}$ mi and the number of passengers, and ask students to compute the fare for another taxi ride. Ask students which is the less expensive fare: 30¢ per $\frac{1}{6}$ mi or 30¢ per $\frac{1}{8}$ mi.

Use Example 2 to demonstrate how to mentally estimate a 15% tip. Orally present other meter fares and ask students to estimate a 15% tip and to explain their reasoning.

Explain that taxi fares in some cities are determined by zones, with one rate for any trip from one zone to another. Discuss the advantages of this alternate fare structure.

10.6 TAXI FARES

In many cities the most convenient but also the most expensive way of getting around is by taxicab.

EXAMPLE 1 You and a friend decide to take a taxi from your hotel to a movie theater $3\frac{1}{2}$ mi away. How much is the fare?

CITY CAB CO.	
1st $\frac{1}{4}$ mile	90¢
Each add'l $\frac{1}{4}$ mile	30¢
Each add'l passenger	$1.50

THINK: Each mile has 4 quarter miles.

1. Multiply to find how many $\frac{1}{4}$ miles. $3\frac{1}{2} \times 4 = 14$ mi.
2. Compute the fare.

1st $\frac{1}{4}$ mile = $0.90
13 additional $\frac{1}{4}$ miles at 30¢ each = $3.90
1 additional passenger = +$1.50
$6.30

The cab fare is $6.30.

It is common to tip your cab driver.

EXAMPLE 2 What is the cab fare to the movies including a 15% tip?

1. Estimate the tip.

THINK: 10% of $6.30 is about 60¢, so 5% is about 30¢. A 15% tip is about 60¢ + 30¢, or about 90¢.

2. Add the fare and the tip. $6.30 + $0.90 = $7.20

Including the tip, the cab fare is $7.20.

FOR DISCUSSION See TE side column.

1. Why do most taxis charge so much for the first fraction of a mile?
2. At many airports there are shuttle vans that charge low fares to take you to your hotel. Why might taxis be more convenient even though they usually cost much more?

PRACTICE EXERCISES

Remember to estimate whenever you use your calculator.

Compute the fare without a tip.

• **1.** A 5-mi ride at 90¢ for the first $\frac{1}{5}$ mi plus 30¢ for each additional $\frac{1}{5}$ mi
$8.10

• **2.** A 3.8-mi ride at 80¢ for the first $\frac{1}{5}$ mi plus 20¢ for each additional $\frac{1}{5}$ mi
$4.40

3. A $9\frac{2}{3}$-mi ride at 75¢ for the first $\frac{1}{3}$ mi plus 40¢ for each additional $\frac{1}{3}$ mi
$11.95

4. A 6.5-mi ride at 60¢ for the first $\frac{1}{4}$ mi plus 25¢ for each additional $\frac{1}{4}$ mi
$6.85

5. A $12\frac{1}{2}$-mi ride at $1.10 for the first $\frac{1}{6}$ mi plus 35¢ for each additional $\frac{1}{6}$ mi
$27.00

6. A 17-mi ride at 50¢ for the first $\frac{1}{8}$ mi plus 10¢ for each additional $\frac{1}{8}$ mi
$14.00

Estimate a 15% tip for the fare. Estimates may vary.

• **7.** $8.00 — $1.20
8. $4.80 — $0.75
9. $12.50 — $1.80
10. $17.40 — $2.55
11. $9.25 — $1.35

Use the Orange Cab Co. fares to solve. Include an estimated 15% tip.
Estimates for tips may vary.

ORANGE CAB CO.	
1st $\frac{1}{5}$ mile	$1.25
Each add'l $\frac{1}{5}$ mile	20¢
Each add'l passenger	$1.50

• **12.** What is the fare for a 7-mi trip from the hotel to the zoo for 1 passenger?
$8.05 + $1.20 tip = $9.25

13. What is the fare for a 9.8-mi trip to the museum for 4 passengers?
$15.35 + $2.25 tip = $17.60

14. If 20 city blocks is about 1 mi, about how much is the cab fare for 2 people to go 20 blocks?
$3.55 + $0.52 tip = $4.07

15. If 4 people share the fare for a 12-mi cab ride, what is each person's share?
$17.55 + $2.70 tip = $20.25 ÷ 4 = $5.06

EXTENSION Idle Time

Many taxis also charge for **idle time** when the taxi is stopped at traffic lights or caught in traffic. Gail took a taxi and was charged 90¢ for the first $\frac{1}{10}$ mi, 25¢ for each additional $\frac{1}{10}$ mi, and 40¢ for each minute of idle time. Her ride was 3.8 mi and included 3 min of idle time.

What was the fare for:

1. The first $\frac{1}{10}$ mi?
90¢

2. The remaining 3.7 mi?
$9.25

3. What was the cost of the idle time?
$1.20

4. What was the total fare?
$11.35

5. What did Gail pay for the ride including a 15% tip?
$11.35 + $1.65 tip = $13.00

SKILLS BANK You may want to refer students to the following skills in the Skills Bank:

Fractions and Mixed Numbers, Skills 3–4

Multiply With Fractions and Mixed Numbers, Skills 3–4

Find the Percent of a Number, Skill 2

FOR DISCUSSION

1. Students should explain that the large initial charge sets a minimum for every taxi ride.
2. Students should explain that, unlike buses and shuttle vans, taxis take passengers directly to their destination without side trips or stops for other passengers and that they leave passengers right at their destination.

EXTENSION Discuss how idle time protects the taxi driver when he or she gets stuck in heavy traffic and explain how idle time charges can result in variable fares for the same trip.

ASSIGNMENTS

Level	Exercises
Level 1	1–4, 7–13
Level 2	1–15, EX
Level 3	3–6, 9–15, EX

FOLLOW-UP ACTIVITY

ENRICHMENT Create a set of fare cards, each with the cost for the first fraction of a mile, each additional fraction of a mile, and the cost for each additional passenger. Create another set of trip cards, each with a distance traveled and the number of passengers. Have pairs of students select 1 card from each set and compute the fare, including the tip.

SUPPLEMENTARY MATERIALS

TRP Practice, p. 68
TRP Reteaching, p. 56
TRP Transparency 34

OBJECTIVE

- Determine the cost of a trip including gasoline or car rental, meals, and hotels.

TEACHING THE LESSON

WARM-UP Write the following exercises on the chalkboard or an overhead transparency.

Estimate. Then compute.

1. 2 × $38 × 7 **2.** 5 × $52 × 9

3. 400 ÷ 35 × $0.85

4. 1,200 ÷ 24 × $1.10

(**1.** $560; $532 **2.** $2,500; $2,340 **3.** $10; $9.71 **4.** $50; $55)

INTRODUCTION Begin discussion by asking students to imagine planning a week-long car trip to a distant city. Ask students to list what information they would need to know to estimate the total cost of the trip. Compile the list on the chalkboard, and then ask students to organize the items into categories such as food, lodging, gasoline, miscellaneous, and so on.

INSTRUCTION Use the introductory situation and the Example to demonstrate how to estimate the total cost of a trip.

For Step 1, explain how to use distance traveled, the gas mileage, and the cost of gasoline to estimate the automobile costs. Show how the various units cancel out to leave just dollars. Suggest an alternate method of determining the automobile cost by first computing the cost per mile of driving the car and then multiplying by the number of miles driven.

For Step 2, discuss how the lodging costs are determined by the number of rooms, the cost per night, and the number of nights.

For Step 3, discuss how the cost of meals is determined by the number of days, the cost per person per day, and the number of people.

10.7 DETERMINING THE COST OF A TRIP

Jim and Jonathan Owens and their parents are planning a trip to Denver, Colorado. They plan to stay 5 nights and drive the 450 mi each way. The family's car gets about 25 mi per gal, and they estimate that gasoline costs 90¢ per gal. The Owens will need 2 hotel rooms and estimate that each room rate will be $75 per night. They also estimate that meals will cost about $25 per person per d.

EXAMPLE

About how much will the Owens vacation cost?

1. Find the automobile cost.

THINK: 450 mi each way, 25 mi per gal, 90¢ per gal

- Divide to find the amount of gasoline needed.

900 mi ÷ 25 mi per gal = 36 gal

- Multiply to find the cost of the gasoline.

36 gal × 90¢ per gal = $32.40

2. Multiply to find the hotel cost.

2 rooms × $75 per night × 5 nights = $750

3. Find the cost of meals.

THINK: 5 nights mean 6 d including travel.

4 people × 6 d × $25 per person per d = $600

4. Add to find the total cost.

$32.40 + $750 + $600 = $1,382.40

The vacation will cost about $1,400.00.

202 CHAPTER 10

FOR DISCUSSION See TE side column.

1. Restaurant meals can be very expensive during a vacation. How can the cost of meals be reduced while traveling?
2. Many travel agents offer vacation **package deals** covering airfare, hotel, car rental, and breakfast. What are some advantages and disadvantages of package deals?

PRACTICE EXERCISES

Remember to estimate whenever you use your calculator.

Find the cost of gasoline for the trip. Assume that gasoline costs 90¢ per gal.

• **1.** 500 mi; car gets 40 mi per gal $11.25
• **2.** 270 mi; car gets 27 mi per gal $9.00
3. 680 mi; car gets 20 mi per gal $30.60
4. 846 mi; car gets 18 mi per gal $42.30
5. 1,485 mi; car gets 26 mi per gal $51.41
6. 1,374 mi; car gets 43 mi per gal $28.76

Use the Trip Planning Guidelines to estimate the cost.

• **7.** Car rental for 5 d $150
• **8.** Hotel room for 4 nights $300
9. 2 hotel rooms for 1 wk $1,050
10. Meals for 2 people for 2 d $100
11. Meals for 4 people for 5 d $500
12. Meals for 5 people for 1 wk $875
13. 2 hotel rooms for 3 d with meals for 2 people for 3 d $600
14. 3 hotel rooms for 1 wk with meals for 3 people for 1 wk $2,100

TRIP PLANNING GUIDELINES

Car rental—$30 per d

Hotel Room—$75 per night

Meals—$25 per person per d

Solve.

15. Tom and Becky Rogers are planning a trip that is about 700 mi each way in a car that gets 35 mi per gal. They will need a motel room for 6 nights and plan on 7 d of meals. They estimate 80¢ per gal for gas, $12 for tolls, $50 per night for motel rooms, and $20 per person per d for food. About how much will the trip cost?
$624, or about $600

16. Martha McNeal and her 2 teenagers are off to the beach for 2 wk. They plan to rent a car for $129 per wk and a cottage at the beach for $350 per wk. Since they will cook many of their own meals, they estimate food costs at $12 per person per d. About how much will the trip cost?
$1,462, or about $1,500

SKILLS BANK You may want to refer students to the following skills in the Skills Bank:

Add With Whole Numbers and Decimals, Skill 6

Multiply With Decimals, Skill 5

Divide With Whole Numbers, Skills 6–7

Divide With Decimals, Skill 8

FOR DISCUSSION

1. Students should explain that money can be saved by buying food in stores and preparing simple meals of sandwiches and fruit in the car or at picnic areas along the road. Also, less expensive restaurants can be used.
2. Students should explain that the advantage of vacation packages is that all the arrangements are made and that most of the cost is covered in 1 set price. However, packages are often less flexible in terms of side trips, accommodations, and meal choices.

ASSIGNMENTS

Level 1	1–4, 7–14
Level 2	1–16
Level 3	3–6, 9–16

FOLLOW-UP ACTIVITY

APPLICATION Provide students with trip-planning information. Tell them to use the information and compute the approximate cost of a trip they wish to take. Then tell students that they only have $1,000 to spend. Ask them to adjust their trip plans to live within the $1,000 budget and to describe how much of the trip was lost or gained.

SUPPLEMENTARY MATERIALS

TRP Practice, p. 69
TRP Reteaching, p. 57
TRP Applications, p. 12
TRP Transparency 35

OBJECTIVE

- Decide, given relevant information, which mode of transportation to select.

TEACHING THE LESSON

WARM-UP Write the following exercises on the chalkboard or an overhead transparency.

Complete.

1. 1 h and 15 min = ■ h $(1\frac{1}{4})$

2. 6 h and 30 min = ■ h $(6\frac{1}{2})$

Multiply.

3. 2 × 12 h **4.** 2 × 3 d

5. $2 \times 1\frac{1}{4}$ h **6.** $2 \times 6\frac{1}{2}$ h

(**3.** 24 h **4.** 6 d **5.** $2\frac{1}{2}$ h **6.** 13 h)

INTRODUCTION Present the problem situation to the class. Ask what key factors would influence the decision on which mode of transportation to select. List these factors on the chalkboard, and then discuss how to find out information on each of the factors. Ask students if there are other factors that might help to make the decision easier (for example, fear of flying, do not own a car, no airport near destination).

INSTRUCTION Read the notes that Carlos has prepared on the 3 options. Ask if there is any other information that he should have. Discuss the decision-making factors.

Read the notes that Carlos has prepared on the three options. Ask if there is any other information that he should have.

DECISION MAKING

10.8 THE MOST ECONOMICAL WAY TO TRAVEL

(Open-ended problem solving)

When planning a trip, one of the most important decisions you must make is determining which type of transportation to use. Should you fly? Should you take a bus or a train? Or should you drive?

PROBLEM

Carlos lives in Seattle and decided to go to San Diego for a week. He called the bus company, the airlines reservation number, and the automobile association for information before he decided whether to travel by bus, plane, or car. Carlos made the following notes from his calls.

Sierra Airlines: Leaves Seattle 9:30 A.M.
Arrives San Diego 1:15 P.M.
$339 round trip
Airport parking — $6 per d

Transcounty Bus: Leaves Seattle 12:30 P.M.
Arrives San Diego 9:30 A.M. the next morning
$159 round trip
Must pay for meals during travel

Automobile: 1,250 mi each way; about 2 d of driving
Car gets 28 mi per gal; gas — 90¢ per gal
Must pay for meals during travel and
1 night each way in motel

DECISION-MAKING FACTORS

Time Cost of Transportation Added Costs Flexibility

DECISION-MAKING COMPARISIONS

Compare the 3 options by completing the table.

Factor	Plane	Bus	Car
Round-trip travel time	• 1. ■	• 2. ■	3. ■
Round-trip cost for transportation	$339	4. ■	5. ■
Added costs	6. ■	Meals	7. ■
Flexibility—Good or Little	8. ■	Little	9. ■

$7\frac{1}{2}$ h; 42 h; 4 d

$159; $80.36

airport parking; meals, motel

little; good

MAKING THE DECISIONS

Which would Carlos choose if the only important factor were:

• **10.** Travel time? plane **11.** Transportation cost? car **12.** Flexibility? car

• **13.** If Carlos must pay for 7 d of airport parking and if his meals during the bus trip come to $30, how much would he really save by taking the bus?
$192

14. If instead of using his own car, Carlos borrows a car that gets 42 mi per gal, about how much would he save?
$26.80

15. If taking the bus saves Carlos the cost of 2 nights in a motel at $55 per night, how much less expensive would the bus trip be than using his car?
$31.36

16. Some experts claim that when all costs are considered, it costs 40¢ per mi to operate your car. At this rate, what would it cost Carlos to drive? Is driving still most economical?
$1,000; No

17. What would you decide to do? Why?
Answers may vary.

Have small groups of students complete Exercises 1–9 in the table in order to compare the 3 modes of transportation. Ask them to write statements that describe the data now in the table. Then ask them to write conclusions that can be drawn from the data. Discuss the appropriateness of these statements and conclusions and possible misinterpretations of the data.

Have students complete Exercises 10–17 in small groups, having students explain and defend their answers and their decisions. The answer to Exercise 17 involves a subjective decision and will therefore vary from student to student. Some students are reluctant to defend their point of view in front of their classmates. Continually reinforce the fact that, in many consumer issues, best answers depend on individual preferences and situations. Encourage all students to participate and express their opinions.

ASSIGNMENTS

Level	Exercises
Level 1	1–17
Level 2	1–17
Level 3	1–17

FOLLOW-UP ACTIVITY

COOPERATIVE LEARNING

Select a destination for a class trip in June. Decide what transportation options exist (plane, train, charter bus, public transportation bus). Gather information on each of the options including cost, availability, and time by assigning small groups of students specific tasks. Analyze each group's data, including added costs, and reach a class consensus on the best mode of transportation.

SUPPLEMENTARY MATERIALS

TRP Practice, p. 70
TRP Transparency 35

OBJECTIVES

- Review vocabulary.
- Practice key chapter concepts and skills.

USING THE REVIEW

The Chapter Review is designed to help students prepare for taking the Chapter Test. The first section focuses on vocabulary. It requires students to select words to complete statements. The second section presents practice of key chapter skills.

Each item on the Chapter Review is referenced to the page on which the topic is taught in the Pupil's Edition. You may wish to have students refer to these pages to help review any concepts or skills they have not yet mastered.

It is suggested that students work in small-sized heterogeneous cooperative learning groups. Some cooperative learning methods that may be used are as follows:

1. After each student has independently completed the Chapter Review, a discussion should follow within each group about the solutions to the problems.
2. The group can complete the entire Chapter Review by working together to solve the problems.

End the lesson with an entire class discussion in which any questions brought up in group discussions are presented and answered.

CHAPTER REVIEW

Vocabulary Choose the letter of the word(s) that completes the sentence.

1. You multiply rate (in miles per hour) by time (in hours) to find ■. [188] b
   a. Direct route **b.** Distance **c.** Average speed
2. Bus-, train-, and air-travel routes, arrival and departure times, and frequency of service can be found in ■. [192, 195] c
   a. Maps **b.** Fare tables **c.** Schedules
3. Daily commuters can save money by buying a ■ for bus or train travel. [192] b
   a. Route map **b.** Monthly pass **c.** Schedule

Skills Use the mileage chart on page 188 for Exercises 4–5. If you average 50 mi per h, estimate how long it will take to drive from: [188]

4. Denver to Boston.
   40 h
5. Dallas to Seattle.
   40 h

Find the answer.

6. Use the bus schedule on page 192. How long is the express-bus trip from Atlanta to Charlotte? [192]
   5 h and 50 min
7. Use the train schedule on page 193. Which train arrives in Washington, D.C., after 3:00 P.M.? [192]
   Train #465
8. Use the flight schedule on page 195. Which flight from Dallas to Chicago arrives closest to 3:00 P.M.? [195]
   Sierra Airlines Flight #272
9. Use the fares on page 195. What is the airfare for 2 First-Class tickets from Dallas to New York? [195]
   $1,450
10. Use the fares on page 195. How much can you save with SuperSaver over First Class on a round-trip flight from Dallas to Chicago? [195]
    $437
11. Use the fare schedule on page 198. What is the non–rush hour subway fare from Farragut North to Archives? [197]
    $0.80

Use the taxi fares below for Exercises 12–13. Compute the taxi fare. [200]

ACE TAXI	
1st $\frac{1}{4}$ mile	80¢
Each add'l $\frac{1}{4}$ mile	15¢
Each add'l passenger	$1.25

12. A 12-mi trip for 2 passengers
    $9.10
13. A 7.4-mi trip for 4 passengers
    $8.84

Use the trip-planning guidelines for Exercises 14–15. Find the cost. [202]

TRIP-PLANNING GUIDELINES
Hotel—$80 per night
Meals—$20 per person per day

14. 2 hotel rooms for 6 nights
    $960
15. Meals for 5 people for 9 d
    $900

CHAPTER TEST

Use the map on page 190. Estimate the traveling time.

1. Distance: Sandy to (186) on (89)
   Average rate: 50 mph 12 min (Estimates may vary: 12–15 min.)

Find the traveling time.

2. Leave: 10:15 A.M. (EST) Arrive: 7:02 P.M. (EST) 8 h and 47 min

Use the Bus and Train Schedules on pages 192 and 193. Find the answer.

3. How long does it take Bus #204 to travel from Clemson to Danville?
   9 h
4. How long does it take Train #451 to go from New Haven to Philadelphia?
   3 h and 22 min
5. How much faster does the Express Bus go from Washington to Charlotte than the Local Bus?
   1 h and 35 min
6. How much faster does Train #465 go from New York to Baltimore than Train #179?
   8 min

Use the Fare Schedules on pages 195 and 198. Find the answer.

7. What is the rush-hour subway fare for a trip from Braddock Road to Federal Triangle?
   $1.35
8. How much can you save with SuperSaver rather than Regular Coach on a round-trip flight from Dallas to Los Angeles?
   $311

Compute the taxi fare.

BLUEBIRD TAXI CO.

1st $\frac{1}{8}$ mile	75¢
Each add'l $\frac{1}{8}$ mile	15¢
Each add'l passenger	$1.50

9. A 5-mi trip for 3 passengers
   $9.60
10. A 12-mi trip for 2 passengers, including a 15% tip
    $16.50 + $2.40 tip = $18.90

Use the Trip Planning Guidelines. Find the answer.

TRIP PLANNING GUIDELINES

Hotel—$65 per night

Meals—$25 per person per d

11. About how much should be budgeted for 4 d of meals and 3 nights in a hotel for 1 person?
    $295

12. Sam and Jennifer Sawyer are planning a trip that is about 500 mi each way in a car that gets 30 mi per gal. They will need a hotel room for 5 nights and meals for 6 d. If gas is about 90¢ per gal, about how much will the trip cost?
    $655

OBJECTIVE

- Evaluate the achievement of chapter objectives.

USING THE TEST

The Chapter Test may be used as a post-test to evaluate student achievement. However, you may wish to use the Chapter Post-Test offered in the Teacher's Resource Package or to design your own chapter test. If this page is not used as a test, you may want to assign it as additional review or practice.

The test items are correlated to the chapter objectives in the table below.

Chapter objectives	Test items
A. Estimate and compute travel time.	1–2
B. Solve problems by using a map.	1
C. Use bus and train schedules.	3–6
D. Use subway and airline schedules.	7–8
E. Compute taxi fares.	9–10
F. Compute the cost of a trip.	11–12

SUPPLEMENTARY MATERIALS

TRP Ch. 10 Post-Test Form A, pp.1–2
TRP Ch. 10 Post-Test Form B, pp.1–2

OBJECTIVE

- Understand that you can take a fabulous vacation and save money on the trip by traveling during off-season months.

USING THE PAGE

Begin discussion by inviting students to talk about their favorite subject—vacations! Ask them to describe the most exciting trips they have taken. Also ask them to identify places they would like to visit and explain why. To stimulate interest in the lesson, suggest to students that they get brochures on St. John and Acapulco from a travel agency or read up on them in the library.

Read the information given in the table. Have students work individually or in pairs to answer Exercises 1–5. Summarize the answers as follows: You can enjoy all the features and luxuries of a warm-weather vacation spot but pay much less if you go during spring or summer months.

Have students work individually to answer Exercises 6–7. Then have them share their answers and reasons in a whole-class discussion.

ANSWERS/EXPLANATIONS

5. To take a break from work and have a mid-year vacation; to enjoy sun, surf, and warm-weather fun while cities back home are blasted by winter winds and snow; to come home with a great mid-winter tan
6. You: May 10
   $285 − $190 = $95
   Your friend: August 22
   $285 − $205 = $80
   You could save more.
7. Camera, film, leisurewear, sports clothes, beachwear, beach accessories (sunglasses, lotion, beach hat, raft, and so forth), sports equipment (scuba/snorkle gear, inflatable canoe or kayak, water skis, and so forth), luggage

MONEY TIPS

Trips to many warm-weather vacation spots are less expensive in the spring and summer.

LET'S LOOK AT THE FACTS

Vacations to many exciting places cost less during certain months of the year. Look at the rate-per-person table shown below. Read the information carefully.

	ST. JOHN, VIRGIN ISLANDS	ACAPULCO, MEXICO
Time Period	**Rate Per Week**	**Rate Per Week**
Off Season:		
SPRING 4/10–5/14	$190	$210
SPRING 5/15–6/9	$202	$225
SUMMER 6/10–7/15	$215	$215
SUMMER 7/16–8/31	$215	$205
Peak Season:		
FALL 9/1–10/31	$230	$230
FALL 11/1–12/20	$260	$260
WINTER 12/21–1/14	$285	$285
WINTER 1/15–2/28	$250	$260
WINTER 3/1–4/9	$235	$235

LET'S DISCUSS WHY See TE side column for answers to Exercises 5–7.

1. Which times of the year are considered off season? Peak season?
   Off season: Spring, Summer Peak season: Fall, Winter
2. When would it cost the most to stay in St. John? In Acapulco?
   St. John: 12/21–1/14
   Acapulco: 12/21–1/14
3. When would it cost the least to stay in St. John? In Acapulco?
   St. John: 4/10–5/14
   Acapulco: 7/16–8/31
4. Why do you think a warm-weather vacation might cost less in spring and summer? People would prefer to get away from the heat and go to cooler places.
5. Many people prefer to take a warm-weather vacation in December instead of in July, even though it costs more money. Why might they prefer this?

LET'S SEE WHAT YOU WOULD DO

6. You want to go to St. John on either May 10 or January 10. Your friend wants to go to Acapulco on either August 22 or December 22. To save money, when should each of you go? Which of you could save the most money?
7. With the money you could save by going to St. John on May 10 instead of January 10, what could you buy to take along on the trip? Think of things that might make the trip more comfortable, more enjoyable, or more memorable.

ESTIMATION SKILLS

OVERESTIMATES AND UNDERESTIMATES OF DIFFERENCES

The symbols $^+$ and $^-$ may be used to show that an actual difference is greater than or less than an estimate.

Examples

```
                   ③[6], 2 4 8
                 − ①[8], 8 9 7
First estimate:    ② 0 , 0 0 0
```

THINK: [6] < [8], so the actual difference is less than 20,000.

Final estimate: 20,000$^-$

```
                 $ ⑧[9] 3 . 0 5
                 −   [5] 9 . 7 6
First estimate:  $ ⑧ 0 0 . 0 0
```

THINK: [9] > [5], so the actual difference is greater than \$800.

Final estimate: \$800$^+$

```
                   0 . 0 0 ⑦[5] 8 6
                 − 0 . 0 0 ①[6] 8 7
First estimate:    0 . 0 0 ⑥ 0 0 0
```

THINK: [5] < [6], so the actual difference is less than 0.006.

Final estimate: 0.006$^-$

Estimate the difference. Use the symbol $^+$ or $^-$ to indicate whether the actual difference is greater than or less than the estimate.

1. 54,496 − 6,328 = 50,000$^-$	**2.** 85,605 − 12,736 = 70,000$^+$	**3.** 762,078 − 657,429 = 100,000$^+$	**4.** 554,712 − 73,960 = 500,000$^-$
5. 3,808,362 − 2,739,076 = 1,000,000$^+$	**6.** 9,242,720 − 3,054,945 = 6,000,000$^+$	**7.** 40,526,421 − 12,908,544 = 30,000,000$^-$	**8.** 61,062,715 − 8,266,046 = 60,000,000$^-$
9. \$375.25 − 153.16 = \$200$^+$	**10.** \$926.00 − 98.95 = \$900$^-$	**11.** \$806.42 − 618.75 = \$200$^-$	**12.** \$995.00 − 87.98 = \$900$^+$
13. \$1,376.50 − 285.87 = \$1,000$^+$	**14.** \$6,310.88 − 5,624.98 = \$1,000$^-$	**15.** \$28,495.00 − 14,916.85 = \$10,000$^+$	**16.** \$86,027.18 − 17,908.59 = \$70,000$^-$
17. 3.3654 − 0.0877 = 3$^+$	**18.** 54.7065 − 5.9071 = 50$^-$	**19.** 0.77260 − 0.63276 = 0.1$^+$	**20.** 0.97428 − 0.09339 = 0.9$^-$
21. 0.03405 − 0.00216 = 0.03$^+$	**22.** 0.04010 − 0.03354 = 0.01$^-$	**23.** 0.005166 − 0.001279 = 0.004$^-$	**24.** 0.009905 − 0.000768 = 0.009$^+$

OBJECTIVE

- Use the symbols $^+$ and $^-$ to indicate overestimated and underestimated differences.

USING THE PAGE

Students should use caution when they write symbols to show that an actual difference is greater or less than an estimated difference. When 2 digits are compared in the second step of the procedure, the actual difference *is less than* the first estimate if the digit in the "top number" *is less than* that in the "bottom number." On the other hand, the actual difference *is greater than* the first estimate if the digit in the top number *is greater than* that in the bottom number. You might wish to have students find the actual differences in the examples in order to get a better idea of how and why the method works.

The procedure described in the lesson can be adjusted for use in a case like this:

```
                  78,623
                 −18,906
First estimate:   60,000
```

THINK: The thousands' digits are the same, so compare the digits in the hundreds' place. 6 < 9, so the actual difference is less than 60,000.

Final estimate: 60,000$^-$

The symbol $^+$ may be used to show that an actual sum is greater than an estimated sum. For example, the sum of the numbers 315,460 and 432,075 is about 700,000$^+$. Note that the symbol $^-$ is not used with front-end estimated sums, since the actual sum is never less than the first estimate.

OBJECTIVE

- Test the prerequisite skills needed to learn the concepts developed in Chapter 11.

USING THE TEST

The Pre-Skills Test is designed to diagnose students' strengths and weaknesses on prerequisite skills necessary to study the mathematics in Chapter 11.

Assign the Pre-Skills Test. Allow students to work together in pairs or in small groups. Group members should help those who demonstrate a misunderstanding of a concept or a weakness in a skill.

The table below correlates the items on the Pre-Skills Test with the prerequisite skill and the lesson(s) in the chapter for which it is needed.

Item(s)	Prerequisite skill	Lesson(s)
1–6	Add money amounts.	All
7–12	Subtract money amounts.	11.2, 11.4–11.7
13–18	Rename percents as decimals.	11.4–11.5
19–24	Multiply decimals and money amounts.	11.4–11.5
25–27	Find the percent of a number.	11.4–11.5
28–35	Read a table.	11.2–11.3, 11.5, 11.7

References to specific skills that are reviewed in the Skills Bank are given on the individual lessons themselves.

Pre-Skills Test

Add.

1. $329 + $981 + $523.18 $1,833.18
2. $1,089 + $104 + $161 $1,354
3. $4,092.67 + $234.81 + $3,009.61 $7,337.09
4. $2,068.41 + $90.38 + $18.42 $2,177.21
5. $109.57 + $1,037.62 + $73.56 $1,220.75
6. $32,087 + $56.78 + $345.91 $32,489.69

Subtract.

7. $947 − $186 $761
8. $2,098 − $1,076 $1,022
9. $372.41 − $186.37 $186.04
10. $2,376.51 − $1,341.29 $1,035.22
11. $913.07 − $568.71 $344.36
12. $3,076.41 − $1,356.78 $1,719.63

Rename as a decimal.

13. 2% 0.02
14. 5% 0.05
15. 3.5% 0.035
16. 6.5% 0.065
17. 7.5% 0.075
18. 2.75% 0.0275

Multiply. Round to the nearest cent.

19. 0.02 × $5,000 $100
20. 0.05 × $3,000 $150
21. 0.075 × $7,000 $525
22. 0.05 × $486.34 $24.32
23. 0.065 × $4,520.38 $293.82
24. 0.035 × $28,962.41 $1,013.68

Find the answer.

25. 2% of $8,500 $170
26. 2.75% of $19,500 $536.25
27. 7.5% of $28,762.41 $2,157.18

Use the Taxable Income table at the right. In which interval does the taxable income lie?

28. $13,240
29. $13,012.50
30. $13,347.89
31. $13,428.99
32. $25,658
33. $25,799.45
34. $25,812.35
35. $25,942.75

28. At least $13,200, but less than $13,250
29. At least $13,000, but less than $13,050
30. At least $13,300, but less than $13,350
31. At least $13,400, but less than $13,450
32. At least $25,650, but less than $25,700
33. At least $25,750, but less than $25,800
34. At least $25,800, but less than $25,850
35. At least $25,900, but less than $25,950

TAXABLE INCOME

At least	But less than	At least	But less than
13,000	13,050	25,600	25,650
13,050	13,100	25,650	25,700
13,100	13,150	25,700	25,750
13,200	13,250	25,750	25,800
13,250	13,000	25,800	25,850
13,300	13,350	25,850	25,900
13,350	13,400	25,900	25,950
13,400	13,450	25,950	26,000

210 CHAPTER 11

A combination of local, state, and federal taxes pays for our public schools. How would education be affected if taxes were not used to provide public schools?

OBJECTIVE

- Formulate problems, using text and data.

USING THE CHAPTER OPENER

Discuss students' answers to the question on the page. It could be assumed that there would not be public education. Private schools could flourish, but many would have to find ways to get by with low tuition in order to attract students. It could also be assumed that a great many children would not be educated or would have a limited education. Religious organizations might take up some of the burden, or parents might become responsible for educating their children.

COOPERATIVE LEARNING

Assign heterogeneous groups and provide students with your school's annual budget. Then propose the following situation. (If yours is a private school, change the situation to the local public school.)

Suppose all funding for the school was cut off and there was no money available for this or any school in the country. How could you (they) raise enough money to keep the school operating?

Challenge the groups to come up with fund-raising ideas, keeping in mind that other schools in the community will be competing for funds also. Their plan for meeting the budget may include tuition and other fees as they see fit. Have them submit their plans and discuss each in terms of feasibility. They should come to realize that the budget would have to be reduced drastically, and programs would have to be cut. (Private-school students could also discuss the impact this situation would have on their school.)

SUPPLEMENTARY MATERIALS

TRP Group Projects, p. 11

OBJECTIVES

- Read W-2 and 1099-INT forms to identify compensation amounts withheld for taxes, Social Security, and interest earned.
- Compute gross income.

TEACHING THE LESSON

WARM-UP Write the following exercises on the chalkboard or an overhead transparency.

Add.

1. \$291 + \$78 **2.** \$1,096 + \$342

3. \$476.29 + \$5,273.19

(**1.** \$369 **2.** \$1,438 **3.** \$5,749.48)

INTRODUCTION Ask if any students have received W-2 forms from an employer. Remind students that they are liable for tax on any money they earn. Those who do not report income may be subject to penalties.

INSTRUCTION Use Example 1 to review the entries on the W-2 form. In addition to discussing the entries noted, point out the place for recording an employee's share of tips. Remind students that the employer contributes the same amount for Social Security that they do.

Use Example 2 to review the entries on the 1099-INT form. Point out that there are other types of 1099 forms for reporting other types of miscellaneous income (from freelance work, for example).

Use Example 3 to show students how to compute gross income. Emphasize that gross income is usually found by adding amounts from W-2 and 1099 forms. Point out that a person may have a number of W-2 and 1099 forms.

SKILLS BANK You may want to refer students to the following skill in the Skills Bank:

Add With Whole Numbers and Decimals, Skill 6

11.1 REPORTING INCOME

During the past year, you worked at Namca Industries. At the end of the year, the company sent you a **W-2 form.** The form shows how much you earned and how much was withheld for federal taxes, Social Security, and state and local taxes. The company sent copies of the W-2 form to federal and local tax agencies.

EXAMPLE 1 Read and interpret this W-2 form.

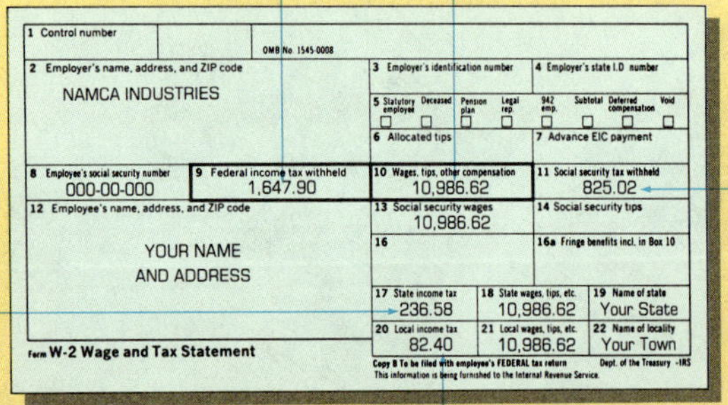

1 Control number | OMB No. 1545-0008
2 Employer's name, address, and ZIP code: NAMCA INDUSTRIES
3 Employer's identification number | 4 Employer's state I.D. number
5 Statutory employee | Deceased | Pension plan | Legal rep. | 942 emp. | Subtotal | Deferred compensation | Void
6 Allocated tips | 7 Advance EIC payment
8 Employee's social security number: 000-00-000
9 Federal income tax withheld: 1,647.90
10 Wages, tips, other compensation: 10,986.62
11 Social security tax withheld: 825.02
12 Employee's name, address, and ZIP code: YOUR NAME AND ADDRESS
13 Social security wages: 10,986.62 | 14 Social security tips
16 | 16a Fringe benefits incl. in Box 10
17 State income tax: 236.58 | 18 State wages, tips, etc.: 10,986.62 | 19 Name of state: Your State
20 Local income tax: 82.40 | 21 Local wages, tips, etc.: 10,986.62 | 22 Name of locality: Your Town
Form W-2 Wage and Tax Statement
Copy B To be filed with employee's FEDERAL tax return. Dept. of the Treasury - IRS
This information is being furnished to the Internal Revenue Service.

Last year, you also earned interest from your bank account. The bank sent a **1099 form** to show how much interest you earned.

EXAMPLE 2 Read and interpret this 1099 form.

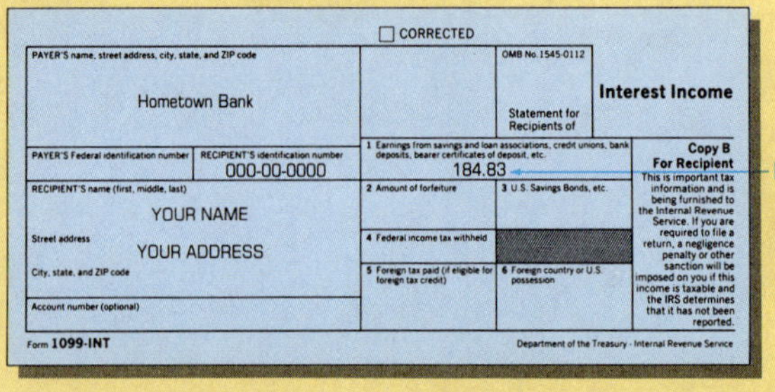

CORRECTED
PAYER'S name, street address, city, state, and ZIP code: Hometown Bank
OMB No. 1545-0112 | Statement for Recipients of | Interest Income
PAYER'S Federal identification number | RECIPIENT'S identification number: 000-00-0000
1 Earnings from savings and loan associations, credit unions, bank deposits, bearer certificates of deposit, etc.: 184.83
RECIPIENT'S name (first, middle, last): YOUR NAME
Street address: YOUR ADDRESS
City, state, and ZIP code
Account number (optional)
2 Amount of forfeiture | 3 U.S. Savings Bonds, etc.
4 Federal income tax withheld
5 Foreign tax paid (if eligible for foreign tax credit) | 6 Foreign country or U.S. possession
Copy B For Recipient. This is important tax information and is being furnished to the Internal Revenue Service. If you are required to file a return, a negligence penalty or other sanction will be imposed on you if this income is taxable and the IRS determines that it has not been reported.
Form 1099-INT Department of the Treasury - Internal Revenue Service

Your **gross income** is the total of earnings, interest, and dividends.

EXAMPLE 3 What was your gross income last year?

Add. \$10,986.62 + \$184.83 = \$11,171.45

Your gross income was \$11,171.45.

FOR DISCUSSION See TE side column.

1. Why are copies of the W-2 and 1099 forms sent to federal, state, and local tax agencies?
2. How could you check your W-2 form to be sure it is correct?

PRACTICE EXERCISES

Remember to estimate whenever you use your calculator.

Use the W-2 form on page 212 for Exercises 1–4. How much was withheld for

• **1.** Federal income tax? $1,647.90 **2.** Social Security tax? $825.02

3. State income tax? $236.58 **4.** Local income tax? $82.40

Find the gross income.

• **5.** Wages: $17,435
Interest: $296
$17,731

• **6.** Wages: $23,672
Interest: $408
$24,080

7. Wages: $12,009
Interest: $96
$12,105

8. Wages: $13,098
Wages: $18,456
Interest: $308
$31,862

9. Wages: $7,892
Tips: $8,623
Interest: $129
$16,644

10. Wages: $34,709
Interest: $728
Interest: $1,203
$36,640

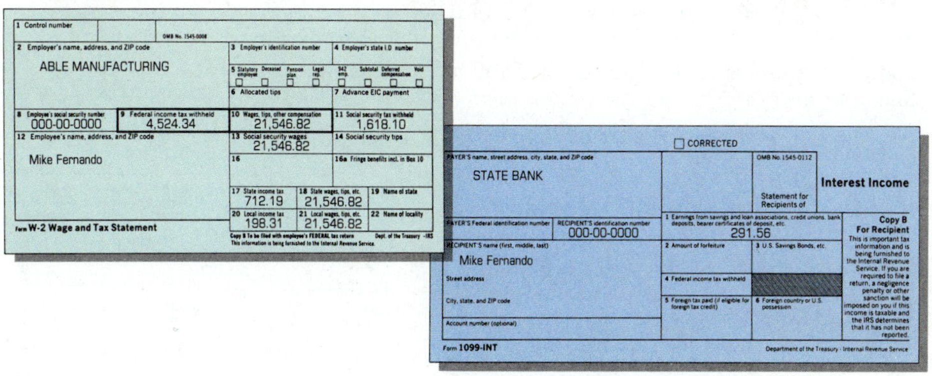
1 Control number
2 Employer's name, address, and ZIP code
ABLE MANUFACTURING
8 Employee's social security number 000-00-0000
9 Federal income tax withheld 4,524.34
10 Wages, tips, other compensation 21,546.82
11 Social security tax withheld 1,618.10
12 Employee's name, address, and ZIP code
Mike Fernando
13 Social security wages 21,546.82
17 State income tax 712.19
18 State wages, tips, etc. 21,546.82
20 Local income tax 198.31
21 Local wages, tips, etc. 21,546.82
Form W-2 Wage and Tax Statement

CORRECTED
PAYER'S name, street address, city, state, and ZIP code
STATE BANK
Interest Income
RECIPIENT'S identification number 000-00-0000
1 Earnings from savings and loan associations, credit unions, bank deposits, bearer certificates of deposit, etc. 291.56
RECIPIENT'S name (first, middle, last)
Mike Fernando
Copy B For Recipient
Form 1099-INT Department of the Treasury - Internal Revenue Service

Use Mike Fernando's W-2 and 1099 forms. Find these amounts.

• **11.** Wages $21,546.82 **12.** Federal tax withheld $4,524.34

13. Social Security withheld $1,618.10 **14.** State tax withheld $712.19

15. Local tax withheld $198.31 **16.** Interest income $291.56

17. Gross income $21,838.38

FOR DISCUSSION

1. Students should explain that W-2 and 1099 forms are sent to the federal, state, and local tax agencies so that these agencies can check the income reported by the individual. Tax agencies use sophisticated computers to check the forms sent by employers against those sent by employees. Often, the information from employers is sent directly to the computer, making checking even easier.
2. Students should explain that they can use their pay stubs to check their W-2 forms. The W-2 is worth checking to be sure that their contributions to Social Security or the tax withheld have not been understated.

ASSIGNMENTS

Level	Exercises
Level 1	1–17
Level 2	1–17
Level 3	5–17

FOLLOW-UP ACTIVITY

APPLICATION Write the following wage and tax information about a person on the chalkboard or an overhead transparency.
Earnings: $23,156
Federal tax withheld: $1,623.82
Social Security withheld: $1,739.02
State tax withheld: $529.61
Local tax withheld: $182.06

Ask students to number 8–22 on a piece of paper to stand for the numbered boxes on a W-2 form. Then ask students to enter the wage and tax amounts next to the correct number.

SUPPLEMENTARY MATERIALS

TRP Practice, p. 71
TRP Reteaching, p. 58
TRP Transparency 36

OBJECTIVES

- Compute Adjusted Gross Income.
- Find the federal tax in a tax table.
- Compute the income tax owed or the refund due.

TEACHING THE LESSON

WARM-UP Write the following exercises on the chalkboard or an overhead transparency.

Add or subtract.

1. \$3,500 + \$2,900

2. \$1,900 + \$2,540

3. \$962.34 − \$628.81

4. \$602.86 − \$483.17

(**1.** \$6,400 **2.** \$4,440 **3.** \$333.53 **4.** \$119.69)

INTRODUCTION Review the introductory information. Remind students that taxes are set by those whom we elect to office. Review the explanations of deductions, exemptions, AGI, and taxable income.

INSTRUCTION Use Example 1 to show students how to compute AGI and taxable income. Emphasize that adjustments are subtracted from gross income to find the AGI, while deductions and exemptions are subtracted from the AGI to find the taxable income. The AGI must be computed separately and is important in future tax computations.

Point out that the amount of an exemption used in this lesson was for the 1987–1988 tax year. In subsequent years, exemptions are as follows:
1988–1989 \$1,950
1989–1990 \$2,000

Use Example 2 to show students how to find the tax in a tax table. For the Practice Exercises, students will need to use the tax tables on pages 475–479 at the back of the book.

Use Example 3 to show students how to compute the amount of tax owed or the refund due them.

11.2 FEDERAL INCOME TAXES

The **Internal Revenue Service (IRS)** collects **federal income taxes.** Tax receipts are used primarily for social and defense programs, and to pay interest on the national debt.

Tax deductions and **tax exemptions** reduce the amounts of income to be taxed. **Adjusted Gross Income (AGI)** is gross income less any adjustments to income. **Taxable income** is AGI less any exemptions or deductions. Your federal taxes are based on your taxable income.

EXAMPLE 1 Last year, your gross income was \$11,171.45. You had no adjustments to income and you were entitled to 1 exemption of \$1,900. You had a standard deduction of \$2,540. What is your AGI and your taxable income?

THINK: There were no adjustments. Your Adjusted Gross Income (AGI) is the same as your gross income.

1. Add to find the total of exemptions and deductions. \$1,900 + \$2,540 = \$4,440
2. Subtract to find your taxable income. \$11,171.45 − \$4,440 = \$6,731.45

Your taxable income is \$6,731.45.

You use tax tables to find the tax due. You use your **filing status** to find your tax.

EXAMPLE 2 If you are single, how much tax do you owe on \$6,731.45?

THINK: Find your taxable income in the table. \$6,731.45 is between \$6,700 and \$6,750.

Look down the "Single" column to find your tax.

Your tax is \$937.

If taxable income is —		And you are —			
At least	But less than	Single	Married filing jointly	Married filing sepa-rately	Head of a house-hold
		Your tax is —			
6,000					
6,000	6,050	832	784	844	804
6,050	6,100	839	791	851	811
6,100	6,150	847	799	859	819
6,150	6,200	854	806	866	826
6,200	6,250	862	814	874	834
6,250	6,300	869	821	881	841
6,300	6,350	877	829	889	849
6,350	6,400	884	836	896	856
6,400	6,450	892	844	904	864
6,450	6,500	899	851	911	871
6,500	6,550	907	859	919	879
6,550	6,600	914	866	926	886
6,600	6,650	922	874	934	894
6,650	6,700	929	881	941	901
6,700	6,750	937	889	949	909
6,750	6,800	944	896	956	916
6,800	6,850	952	904	964	924
6,850	6,900	959	911	971	931
6,900	6,950	967	919	979	939
6,950	7,000	974	926	986	946

If too much tax was withheld, you will get a **refund.** If not enough tax was withheld, you will owe more taxes.

EXAMPLE 3 You had $752.15 withheld in federal taxes. Your federal tax is actually $937. How much do you owe or how much of a refund can you expect?

THINK: The tax withheld is less than the tax. You will owe money.

Subtract. $937 − $752.15 = $184.85

You owe $184.85.

FOR DISCUSSION See TE side column.

How can 2 people with the same taxable income pay different taxes?

PRACTICE EXERCISES

Remember to estimate whenever you use your calculator.

Find the Adjusted Gross Income (AGI) and taxable income.

1. Gross income: $20,876
   Adjustments: $1,500
   1 Exemption: $1,900
   Deduction: $2,540
   $19,376; $14,936
2. Gross income: $23,190
   Adjustments: $1,970
   2 Exemptions: $3,800
   Deduction: $2,870
   $21,220; $14,550
3. Gross income: $32,098
   Adjustments: $0
   2 Exemptions: $3,800
   Deduction: $3,760
   $32,098; $24,538

Use the tax tables on pages 475–479. Find the taxable income and the tax owed.

• 4. Single
   Gross income: $25,374
   Adjustments: $750
   1 Exemption: $1,900
   Deduction: $2,540
   $20,184; $3,393

5. Head of household
   Gross income: $43,092
   Adjustments: $3,762
   2 Exemptions: $3,800
   Deduction: $2,723
   $32,807; $6,101

6. Married, filing joint return
   Gross income: $30,926
   Adjustments: $1,500
   3 Exemptions: $5,700
   Deduction: $3,760
   $19,966; $2,876

7. Single
   Gross income: $19,723
   Adjustments: $0
   1 Exemption: $1,900
   Deduction: $2,540
   $15,283; $2,219

How much tax is owed or how much of a refund can be expected?

• 8. Tax withheld: $2,110.35
   Actual tax: $2,407
   owe $296.65

9. Tax withheld: $5,817.89
   Actual tax: $4,829
   refund $988.89

10. Tax withheld: $8,324.13
    Actual tax: $7,899
    refund $425.13

11. Tax withheld: $4,976.54
    Actual tax: $5,080
    owe $103.46

SKILLS BANK You may want to refer students to the following skills in the Skills Bank:

Add With Whole Numbers and Decimals, Skill 6

Subtract With Whole Numbers and Decimals, Skill 6

FOR DISCUSSION Students should explain that the same incomes can result in different taxes if the filing statuses are different. Students should refer to the tax table for the taxable income of $6,731.45 and show that each of the 4 different filing categories has a different tax. Students should note that the category "Married filing separately" has the highest tax while "Married filing jointly" has the lowest tax.

ASSIGNMENTS

Level 1	1–3, Even 4–10
Level 2	1–11
Level 3	1–11

FOLLOW-UP ACTIVITY

ENRICHMENT If taxable income exceeds $50,000, a taxpayer would use the tax rate schedule, not the tax tables. Write this portion of Schedule X for single filers on the chalkboard.

If taxable income is			
Over —	But not over —	Tax is	of the amount over —
27,000	54,000	5,304 + 35%	27,000
54,000	—	14,754 + 38.5%	54,000

Ask students to use the schedule to find the tax on:

1. $52,000 ($14,054)
2. $57,000 ($15,909)

SUPPLEMENTARY MATERIALS

TRP Practice, p. 72
TRP Reteaching, p. 59
TRP Transparency 36

OBJECTIVE

- Complete Form 1040EZ.

TEACHING THE LESSON

WARM-UP Write the following exercises on the chalkboard or an overhead transparency.

Add or subtract.

1. \$19,023.56 + \$319.62 (\$19,343.18)

2. \$25,023.58 − \$2,540 (\$22,483.58)

INTRODUCTION Ask if any students have filed an income tax form. Determine whether they filed Form 1040EZ or some other form. Ask these students to discuss their experiences filing tax forms. Be prepared to discuss your experiences if no students have filed tax forms. Remind students that tax forms must be filled out correctly and filed by a deadline, usually April 15.

The Internal Revenue Service has an Understanding Taxes Coordinator in each district office. Each district office can be reached by a toll-free number. Dial 1-800-555-1212 and ask the information operator for the number of the closest office. The Understanding Taxes Coordinator can send you blow-ups of Form 1040EZ for classroom use. The coordinator can also send you copies of Understanding Taxes and Understanding Taxes—Teacher's Guide. These publications contain up-to-date information about taxes and tax forms.

INSTRUCTION Example 1 consists of a completed Form 1040EZ and accompanying guidelines for preparing the form. Use Example 1 to show students how to file Form 1040EZ. Current tax forms may be different from those in the text. Alert students to any significant changes.

11.3 USING TAX FORM 1040EZ

Tax Form 1040EZ is often called the **short form.** This form can only be used if your filing status is single and you have no dependents. You use the information from the W-2 and 1099 forms to complete Form 1040EZ.

EXAMPLE Read and interpret Chad Blaire's Tax Form 1040EZ.

Form 1040EZ — Department of the Treasury • Internal Revenue Service — **Income Tax Return for Single filers with no dependents** — OMB No. 1545-0675

Name & address

Use the IRS mailing label. If you don't have one, please print.

Chad Blaire
Print your name above (first, middle, last)

33 Robin La.
Present home address (number and street). (If you have a P.O. box, see instructions.)

Hometown, USA 00000
City, town, or post office, state, and ZIP code

Please print your numbers like this: 1 2 3 4 5 6 7 8 9 0

Your social security number: 000 00 0000

If you filed a return in 1986 and this address is different, check here ▶ ☐

Presidential Election Campaign Fund — Do you want \$1 to go to this fund? (Note: Checking "Yes" will not change your tax or reduce your refund.) ▶ Yes ☐ No ☒

	Line		Dollars	Cents
Report your income	1	Total wages, salaries, and tips. This should be shown in Box 10 of your W-2 form(s). (Attach your W-2 form(s).)	21,050	37
	2	Taxable interest income of \$400 or less. If the total is more than \$400, you cannot use Form 1040EZ.	241	68
Attach Copy B of Form(s) W-2 here	3	Add line 1 and line 2. This is your **adjusted gross income.**	21,292	05
	4	Standard deduction. If you can be claimed as a dependent on another person's return, check here, ▶ and see line 4 instructions. Otherwise, enter 2,540.	2,540	00
	5	Subtract line 4 from line 3.	18,752	05
	6	Personal exemption. If you checked the box on line 4, enter 0. Otherwise, enter 1,900.	1,900	00
	7	Subtract line 6 from line 5. If line 6 is larger than line 5, enter 0 on line 7. This is your **taxable income.**	16,852	05
Figure your tax	8	Enter your Federal income tax withheld. This should be shown in Box 9 of your W-2 form(s).	2,809	37
	9	Use the **single** column in the tax table on pages 31-36 of the Form 1040A instruction booklet to find the **tax** on your taxable income on line 7. Enter the amount of tax.	2,469	00
Refund or amount you owe	10	If line 8 is larger than line 9, subtract line 9 from line 8. Enter the **amount of your refund.**	340	37
Attach tax payment here	11	If line 9 is larger than line 8, subtract line 8 from line 9. Enter the **amount you owe.** Attach check or money order for the full amount, payable to "Internal Revenue Service."		

Sign your return — **I have read this return. Under penalties of perjury, I declare that to the best of my knowledge and belief, the return is true, correct, and complete.**

Your signature: Chad Blaire Date: 4/12/–

For IRS Use Only — Please do not write in boxes below.

For Privacy Act and Paperwork Reduction Act Notice, see page 41. Form **1040EZ** (1987)

How to complete Form 1040EZ:

- Chad entered his name, address, and Social Security number. The form cannot be processed without the Social Security number.
- The IRS would prefer that you use the label they provide.
- Notice that dollar signs are not used on tax forms.

Line 1 Remember to attach a copy of your W-2 form(s). You must report earnings even if you don't get a W-2 Form.

Line 2 Do *not* attach a copy of Form 1099. Do not use Form 1040EZ if your interest income is over $400.

Line 4 Chad did not check the box since he is not taken as an exemption on his parents' tax form.

Line 6 Chad did not check the box on Line 4, so he is entitled to the personal exemption.

Line 8 Chad used information from the W-2 form(s) to enter the tax withheld.

Line 9 Chad used the amount from Line 7 to find his tax in the tax table. His taxable income falls between $16,850 and $16,900. He finds his tax in the "Single" column.

If taxable income is—		And you are—			
At least	But less than	Single	Married filing jointly	Married filing sepa-rately	Head of a house-hold
		Your tax is —			
16,000					
16,800	16,850	2,455	2,404	2,831	2,424
16,850	16,900	2,469	2,411	2,845	2,431
16,900	16,950	2,483	2,419	2,859	2,439
16,950	17,000	2,497	2,420	2,873	2,446

Line 10 If Chad is due a refund, the IRS will mail him a check a few months after his tax form is processed.

Line 11 If Chad still owed money, he would write his Social Security number on the check he sends to the IRS.

- Chad signed and dated his return.

Ask students to review the form and note the 4 major headings. In the *Name and Address* section, you give this information along with your Social Security number. In the *Report Your Income* section, you report income and subtract allowable deductions and exemptions. You actually *Figure Your Tax* in that section and compute the *Refund or Amount You Owe* in that section. Finally, you *Sign Your Return* in that section.

As you review the guidelines for completing Form 1040EZ, refer students to the appropriate sections of the form. Emphasize the importance of carefully entering and checking the information on tax forms. Here is more detailed information about some of the entries.

Line 1 You should report all income, except interest, on this line. Nontaxable income from some sources is not included here. Most often, nontaxable income includes child-support payments, gifts and inheritances, federal income tax refunds, most life insurance proceeds, most scholarships, some Social Security benefits, and some workmen's compensation benefits.

Students may not understand the requirements for reporting tips. Under current law, they must report tips to their *employers* if the tips total $20 or more in a month. In any event, all tips must be reported on the tax form.

Line 2 Certain interest payments, such as interest from state and municipal bonds, are nontaxable.

Line 6 Emphasize that if the student is claimed as a dependent on another return, such as a parent's, he or she does not qualify for the personal exemption on his or her own return. It is a common error for students to claim the exemption when they are not entitled to it.

Note that for Exercises 7–17, you may want to have students write the answers directly on a Form 1040EZ, available in the Lesson Aids.

FOR DISCUSSION

1. Students should explain that the exemption is denied because, otherwise, it would result in a person being deducted twice.
2. Students should explain that they should use calculators to complete and check their tax returns. If unsure, they should ask someone else to check their work.

SKILLS BANK You may want to refer students to the following skills in the Skills Bank:

Add With Whole Numbers and Decimals, Skill 6

Subtract With Whole Numbers and Decimals, Skill 6

ASSIGNMENTS

Level 1	1–17
Level 2	1–17
Level 3	7–17

FOLLOW-UP ACTIVITY

APPLICATION Distribute copies of Form 1040EZ, available in the Lesson Aids. Write the following wage and tax information on the chalkboard or an overhead transparency. Ask each student to complete a 1040EZ for this person.

Veronica Rac, single, not claimed on another return

Margaret's Restaurant
Wages $9,982.09
Tips $8,324.27
Federal tax withheld $1,232.51

Temporary Services
Wages $2,304.92
Federal tax withheld .... $ 302.71

Veteran's Bank
Interest .... $ 103.56

SUPPLEMENTARY MATERIALS

TRP Practice, p. 73
TRP Reteaching, p. 60
TRP Transparency 37
TRP Lesson Aids, p. 10

FOR DISCUSSION See TE side column.

1. Why are people claimed as dependents on another return denied a personal exemption on Form 1040EZ?
2. Arithmetic errors and unsigned forms are the most common mistakes. How could you avoid these mistakes on your tax forms?

PRACTICE EXERCISES Remember to estimate whenever you use your calculator.

Use the Form 1040EZ on page 216 for Exercises 1–6.

• 1. What is Chad's gross income? $21,292.05
• 2. What is Chad's AGI? $21,292.05
3. What is Chad's taxable income? $16,852.05
4. How much federal tax was withheld? $2,809.37
5. What is Chad's federal tax? $2,469.00
6. How much is Chad's refund? $340.37

Emira Santiago held 2 jobs during the past year. Below are copies of her W-2 and 1099 forms. She is single and not claimed on another person's tax return.

Use the tax tables on pages 475–479. To complete her 1040EZ form, what amount will she enter on:

• 7. Line 1? 11,253.82
• 8. Line 2? 319.56
9. Line 3? 11,573.38
10. Line 4? 2,540.00
11. Line 5? 9,033.38
12. Line 6? 1,900.00
13. Line 7? 7,133.38
14. Line 8? 580.52
15. Line 9? 997.00
16. Line 10? none
17. Line 11? 416.48

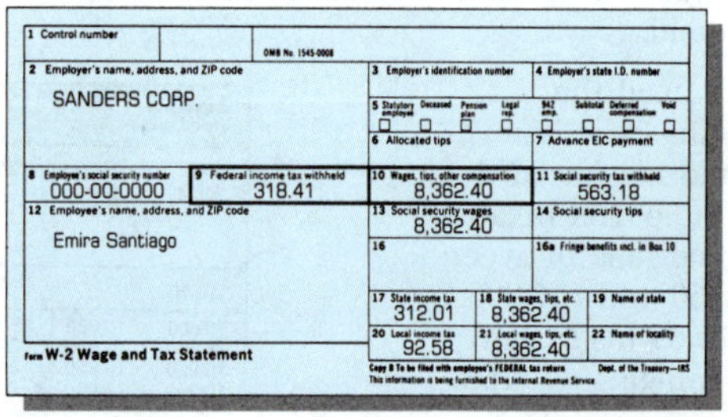
W-2 Wage and Tax Statement
2 Employer's name, address, and ZIP code: SANDERS CORP.
8 Employee's social security number: 000-00-0000
9 Federal income tax withheld: 318.41
10 Wages, tips, other compensation: 8,362.40
11 Social security tax withheld: 563.18
13 Social security wages: 8,362.40
12 Employee's name, address, and ZIP code: Emira Santiago
17 State income tax: 312.01
18 State wages, tips, etc.: 8,362.40
20 Local income tax: 92.58
21 Local wages, tips, etc.: 8,362.40

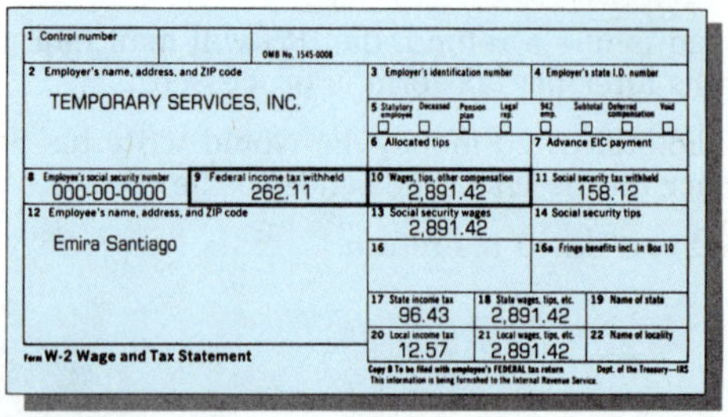
W-2 Wage and Tax Statement
2 Employer's name, address, and ZIP code: TEMPORARY SERVICES, INC.
8 Employee's social security number: 000-00-0000
9 Federal income tax withheld: 262.11
10 Wages, tips, other compensation: 2,891.42
11 Social security tax withheld: 158.12
13 Social security wages: 2,891.42
12 Employee's name, address, and ZIP code: Emira Santiago
17 State income tax: 96.43
18 State wages, tips, etc.: 2,891.42
20 Local income tax: 12.57
21 Local wages, tips, etc.: 2,891.42

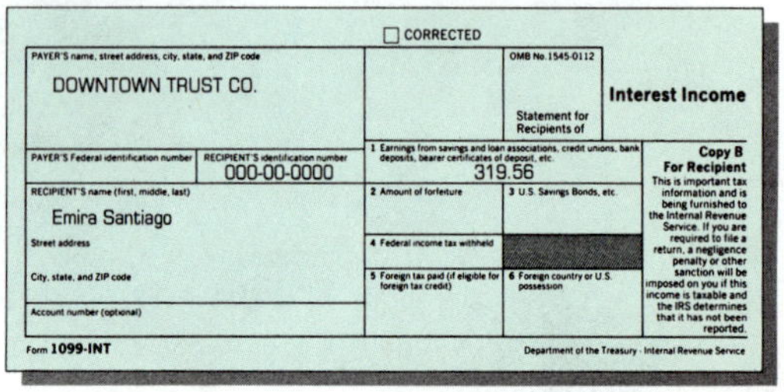
Form 1099-INT Interest Income
PAYER'S name, street address, city, state, and ZIP code: DOWNTOWN TRUST CO.
RECIPIENT'S identification number: 000-00-0000
1 Earnings from savings and loan associations, credit unions, bank deposits, bearer certificates of deposit, etc.: 319.56
RECIPIENT'S name: Emira Santiago

11.4 ITEMIZED DEDUCTIONS

When you file your income tax return, you may have **itemized deductions** that would reduce your tax. Use **Form 1040, Schedule A,** to itemize deductions.

EXAMPLE 1 Liz Ryan filed Form 1040 with Schedule A. Her Adjusted Gross Income (AGI) was $29,580 (on Form 1040, line 31). Read and interpret this Schedule A.

SCHEDULES A & B (Form 1040) Department of the Treasury, Internal Revenue Service — **Schedule A—Itemized Deductions** (Schedule B is on back) ▶ Attach to Form 1040. See Instructions for Schedules A and B (Form 1040). — OMB NO. 1545-0074, Attachment Sequence No. 07

Name(s) as shown on Form 1040 — Your social security number: 000 00 0000

Section	Line	Description	Line	Amount	Line	Total
Medical and Dental Expenses (Do not include expenses reimbursed or paid by others.)	1a	Prescription medicines and drugs, insulin, doctors, dentists, nurses, hospitals, insurance premiums you paid for medical and dental care, etc.	1a	961.00		
	b	Transportation and lodging	1b	23.00		
	c	Other (list-include hearing aids, dentures, eyeglasses, etc.) ▶	1c			
(See Instructions on page 19.)	2	Add lines 1a through 1c, and enter the total here	2	984.00		
	3	Multiply the amount on Form 1040, line 31, by 7.5% (.075)	3	2,219.00		
	4	Subtract line 3 from line 2. If zero or less, enter -0-. **Total** medical and dental ▶			4	
Taxes You Paid		Note: General sales taxes are no longer deductible.				
	5	State and local income taxes	5	865.00		
	6	Real estate taxes	6	913.00		
(See Instructions on page 20.)	7	Other taxes (list-include personal property taxes) ▶	7			
	8	Add the amounts on lines 5 through 7. Enter the total here. **Total** taxes ▶			8	1,778.00
Interest You Paid		Caution: If you paid interest on a loan incurred before August 16, 1986, that is secured by your home OR if you paid mortgage interest on more than two homes, see page 20 of the instructions.				
(See Instructions on page 20.)	9a	Deductible home mortgage interest you paid to financial institutions (report deductible points on line 10)	9a	1,256.00		
	b	Deductible home mortgage interest you paid to individuals (show that person's name and address) ▶	9b			
	10	Deductible points	10			
	11	Deductible investment interest	11			
	12a	Personal interest you paid (see page 20) 12a				
	b	Multiply the amount on line 12a by 65% (.65). Enter the result.	12b			
	13	Add the amounts on lines 9a through 11, and 12b. Enter the total here. **Total** interest ▶			13	1,256.00
Contributions You Made	14a	Cash contributions. (If you gave $3,000 or more to any one organization, report those contributions on line 14b	14a	407.00		
(See Instructions on page 21.)	b	Cash contributions totaling $3,000 or more to any one organization. (Show to whom you gave and how much you gave.) ▶	14b			
	15	Other than cash (You must attach Form 8283 if over $500.)	15	75.00		
	16	Carryover from prior year	16			
	17	Add the amounts on lines 14a through 16. Enter the total here. **Total** contributions ▶			17	482.00
Casualty and Theft Losses	18	Casualty or theft loss(es) (attach Form 4684). (See page 21 of the Instructions ▶			18	0
Moving Expenses	19	Moving expenses (attach Form 3903 or 3903F). (See page 22 of the Instructions.) ▶			19	0
Miscellaneous Deductions Subject to 2% AGI Limit	20	Unreimbursed employee business expenses (attach Form 2106)	20	683.00		
	21	Other expenses (list type and amount) ▶	21			
(See Instructions on Page 22.)	22	Add the amounts on lines 20 and 21. Enter the total	22	683.00		
	23	Multiply the amount on Form 1040, line 31, by 2% (.02). Enter the result here	23	592.00		
	24	Subtract line 23 from line 22. Enter the result (but not less than zero) ▶			24	91.00
Other Miscellaneous Deductions	25	Miscellaneous deductions not subject to 2% AGI limit (see page 22). (List type and amount.) ▶			25	0
Total Itemized Deductions	26	Add the amounts on lines 4, 8, 13, 17, 18, 19, 24, and 25. Enter your answer here and on Form 1040, line 32 ▶			26	3,607.00

For Paperwork Reduction Act Notice, see Form 1040 Instructions. — Schedule A (Form 1040) 1987

OBJECTIVES

- Identify allowable federal tax itemized deductions.
- Interpret Schedule A—Itemized Deductions.
- Given a situation, choose whether to use itemized deductions or the standard deduction.

TEACHING THE LESSON

WARM-UP Write the following exercises on the chalkboard or an overhead transparency.

Find the answer.

1. 2% of $14,800 **2.** 2% of $9,430

3. 7.5% of $14,800 **4.** 7.5% of $9,430

(**1.** $296 **2.** $188.60 **3.** $1,110 **4.** $707.25)

INTRODUCTION Review the 6 major categories of expenses that can be deducted on Schedule A. Ask students to name some of the expenses they think could be deducted in each category. Point out named deductions that are obviously not allowable and remind students that there are strict rules about which deductions are allowable.

The Internal Revenue Service's Understanding Taxes Coordinator can send you blow-ups of Schedule A for classroom use.

INSTRUCTION Example 1 consists of a complete Schedule A and accompanying guidelines for preparing the form. Use Example 1 to show students how to file Schedule A. Current tax forms may be different from those in the text. Alert students to any significant changes. Direct students' attention to the top of the form, where there is room to print your name and Social Security number. The right-hand column, which contains many shaded areas, shows the total deductions for each category.

As you review the guidelines for completing Schedule A, refer students to the appropriate sections of the form. Emphasize the importance of carefully entering and checking the information on tax forms. Here is more detailed information about some of the entries.

Line 3 Tax regulations require that only medical expenses in excess of 7.5% of the AGI can be deducted. Use this line to emphasize the difference between adjustments to income and deductions. A $2,000 adjustment would reduce the taxable income by $2,000, and reduce the AGI by this same amount. As a result, a person could deduct an additional $150 in medical expenses. So, a $2,000 adjustment can reduce taxable income by $2,150.

Line 4 There is no "negative tax." If the deduction is less than 7.5% of AGI, the deduction is simply not taken.

Line 23 Miscellaneous expenses must exceed 2% of AGI. So, a $2,000 adjustment could reduce taxable income by a maximum of $2,190.

Use Example 2 to show students how and when to take the standard deduction. Emphasize that students must compare itemized and standard deductions before using the itemized deductions.

Point out that the standard deductions given are for the 1987–1988 tax year. The Internal Revenue Service plans to use these standard deductions for the 1988–1989 tax year:

Single	$3,000
Married, joint return	$5,000
Married, separate return	$2,500

The standard deductions for the 1989–1990 tax year will be based on the inflation index.

FOR DISCUSSION

1. Students should explain that homeowners may have both mortgage interest and property taxes to deduct. In fact, 1 powerful argument for buying a home is the tax deductions it provides.

How to complete Schedule A—Itemized Deductions:

Line 1a Liz could not include health expenses paid for by insurance.

Line 3 Liz multiplied 0.075 × $29,580 (the AGI) to get $2,218.50. She rounded to the nearest dollar to make computation easier.

Line 4 Since Line 3 is greater than Line 2, Liz cannot take a deduction for medical and dental expenses.

Line 5 These amounts are from Liz's W-2 form, plus any additional taxes she paid. Federal income taxes are *not* deductible.

Line 6 Liz owns her house and pays real estate taxes.

Line 8 Shows the total deduction for taxes.

Line 9a The bank sent Liz a statement showing how much interest she had paid.

Line 13 Shows the total deduction for paid interest.

Line 15 This deduction is the value of books Liz donated to the library.

Line 17 Shows the total deduction for contributions.

Line 23 Liz multiplied 0.02 × $29,580 to get $592.

Line 24 Liz subtracted to find her total miscellaneous deductions.

Line 26 Shows the total of Liz's itemized deductions.

Everyone is entitled to a **standard deduction.** Only use itemized deductions if the total is more than the standard deduction. The standard deduction depends on filing status.

EXAMPLE 2 Liz is single. Should she itemize deductions or use the standard deduction?

Liz compared the total of her itemized deductions ($3,607) with her standard deduction ($2,540). The total itemized deductions are greater than the standard deduction.

Liz will itemize deductions.

STANDARD DEDUCTIONS

Filing status	Deduction
Single or head of household	$2,540
Married, filing joint return or widow(er) with dependent child	$3,760
Married, filing separate return	$1,880

FOR DISCUSSION See TE side column.

1. Why are homeowners likely to have more deductions than renters?
2. Why is it important to keep records of expenses you want to deduct on your tax return?

PRACTICE EXERCISES

Remember to estimate whenever you use your calculator.

Use the Schedule A on page 219 for Exercises 1–6.

• **1.** What was Liz's deduction for real-estate taxes?
$913

• **2.** What was Liz's deduction for contributions other than cash?
$75

3. What was the total of Liz's unreimbursed business expenses?
$683

4. How much of Liz's unreimbursed business expenses could be deducted?
$91

5. What is the total of Liz's itemized deductions?
$3,607

6. How much more than the standard deduction are Liz's itemized deductions?
$1,067

Find the difference between the itemized and the standard deduction. Which should the person use?

• **7.** Single
Total itemized deductions: $2,304
$236; standard

8. Married, joint return
Total itemized deductions: $3,256
$504; standard

9. Married, separate return
Itemized deductions: $2,409
$529; itemized

10. Head of household
Itemized deductions: $2,958
$418; itemized

Find the total itemized deductions. Should the person(s) use the itemized or standard deduction?

11. Lester Jefferson
Single
Adjusted Gross Income
(Form 1040A—Line 31): $27,500

Medical Expenses
(Line 2): $780

State and Local Taxes
(Line 5): $1,090

Other Taxes
(Line 7): $205

Cash Contributions
(Line 14a): $550

Casualty Loss
(Line 18): $600 $2,445; standard deduction

12. Jim and Marge Cochran
Married, Filing a Joint Return
Adjusted Gross Income
(Form 1040A—Line 31): $57,300

Medical Expenses
(Line 2): $2,456

State and Local Taxes
(Line 5): $2,035

Real Estate Taxes
(Line 6): $1,653

Mortgage Interest
(Line 9a): $3,025

Cash Contributions $7,913
(Line 14a): $1,200 itemized deduction

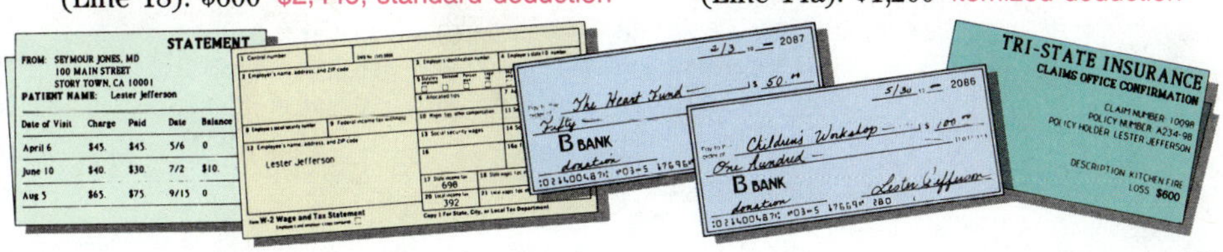

2. Students should explain that the Internal Revenue Service does not have to accept any deductions not supported by records.

SKILLS BANK You may want to refer students to the following skills in the Skills Bank:

Compare and Order Whole Numbers and Decimals, Skill 5

Multiply With Decimals, Skill 5

Find the Percent of a Number, Skill 2

ASSIGNMENTS

Level	Exercises
Level 1	1–10
Level 2	1–11
Level 3	7–12

FOLLOW-UP ACTIVITY

COOPERATIVE LEARNING
Distribute copies of Schedule A, available in the Lesson Aids. Review the steps for completing the schedule and then write the following on the chalkboard or an overhead transparency.

Rachel Almin, AGI	$26,908
Medical expenses	$1,802
State taxes	$ 408
Real estate taxes	$1,072
Federal taxes	$1,923
Sales taxes	$ 621
Mortgage interest	$1,438
Contributions	$ 823
Miscellaneous	$ 624

Assign students to work in heterogeneous groups of 2 or 3 to fill out Schedule A for Rachel. They should identify expenses in the listing that cannot be deducted and apply the appropriate limits to medical and miscellaneous expenses.

SUPPLEMENTARY MATERIALS

TRP Practice, p. 74
TRP Reteaching, p. 61
TRP Transparency 37
TRP Lesson Aids, p. 11

OBJECTIVE

- Compute state and city income taxes.

TEACHING THE LESSON

WARM-UP Write the following exercises on the chalkboard or an overhead transparency.

Multiply.

1. 0.02 × \$5,000 **2.** 0.04 × \$5,682

3. 0.065 × \$3,800

4. 0.0275 × \$5,600

(**1.** \$100 **2.** \$227.28 **3.** \$247 **4.** \$154)

INTRODUCTION Ask students to discuss whether their state or locality has an income tax. Briefly review with students the features of the state or local income tax, if one exists.

INSTRUCTION Use Example 1 to show students how to compute taxes for a *graduated income tax* based on Adjusted Gross Income. Explain that, for a graduated income tax, the tax rate goes up as Adjusted Gross Income goes up. Point out that the higher tax rate does not apply to a person's entire income, just that part above a given cut-off point.

Use Example 2 to show students how to find a graduated income tax based on taxable income after allowances for exemptions. Remind students that *dependent* refers to someone other than yourself or your spouse and yourself.

SKILLS BANK You may want to refer students to the following skills in the Skills Bank:

Multiply With Decimals, Skill 5

Find the Percent of a Number, Skill 2

FOR DISCUSSION

1. Students should explain that states or cities often have sales taxes and real estate taxes.

11.5 STATE AND CITY INCOME TAXES

Some **state and city income taxes** are a percent of a person's Adjusted Gross Income from their federal tax Form 1040.

STATE TAX

Adjusted Gross Income	Tax rate
First \$8,000	2%
Next \$7,000	3.5%
Next \$5,000	5%
Over \$20,000	6.5%

EXAMPLE 1 Bob's Adjusted Gross Income was \$22,250. How much state tax does he owe?

1. Multiply to find the tax.

First \$8,000 → 0.02 × \$8,000 = \$160
Next \$7,000 → 0.035 × \$7,000 = \$245
Next \$5,000 → 0.05 × \$5,000 = \$250
Over \$20,000 → 0.065 × \$2,250 = \$146.25

2. Add to find the total tax. \$801.25

Bob's state tax is \$801.25.

Some states and cities have tax rates that include personal exemptions.

EXEMPTIONS

Single	\$1,750
Married	\$3,400
Each Dependent	\$ 750

CITY TAX

Adjusted Gross Income	Tax rate
First \$9,000	1.5%
Next \$6,000	2.75%
Over \$15,000	3.5%

EXAMPLE 2 Kathy and Harry have a combined Adjusted Gross Income of \$29,540. They have 1 child. How much city income tax do they pay?

1. Add to find their total exemption. \$3,400 + \$750 = \$4,150
2. Subtract to find the taxable income. \$29,540 − \$4,150 = \$25,390
3. Multiply to find the tax.

First \$9,000 → 0.015 × \$9,000 = \$135
Next \$6,000 → 0.0275 × \$6,000 = \$165
Over \$15,000 → 0.035 × \$10,390 = \$363.65

4. Add to find the total tax. \$663.65

Kathy and Harry's city tax is \$663.65.

FOR DISCUSSION See TE side column.

1. What taxes other than income taxes might a state or city have?
2. Why would a city or state need to raise money through income taxes?

PRACTICE EXERCISES

Remember to estimate whenever you use your calculator.

Use the state tax table on page 222 to find the state tax on the Adjusted Gross Income.

• 1. $6,872 $137.44
• 2. $13,581 $355.34
3. $20,615 $694.98
4. $19,762 $643.10
5. $43,181 $2,161.77
6. $12,924 $332.34

Use the city tax table and exemptions on page 222 to find the city tax.

7. Single, 0 dependents
Adjusted Gross
Income: $9,324
$113.61

8. Married, 1 dependent
Adjusted Gross
Income: $12,618
$127.02

9. Married, 3 dependents
Adjusted Gross
Income: $37,861
$902.39

Complete the table to find the amount each person owes, or the refund they should get for state and city taxes. Use the tables on page 222.

	Mike, single	Teresa, married, 1 dependent	
Adjusted Gross Income	$19,468.12	$32,812.41	
State tax withheld	$698.13	$1,214.41	
State tax	• 10. ■	16. ■ $1,487.81	10. $628.41
State tax owed or refund	• 11. ■	17. ■ $273.40 (O)	11. $69.72 (R)
Exemptions (dollar value)	12. ■	18. ■ $4,150.00	12. $1,750.00
City taxable income	13. ■	19. ■ $28,662.41	13. $17,718.12
City tax withheld	$418.32	$709.21	
City tax	14. ■	20. ■ $778.18	14. $395.13
City tax owed or refund	15. ■	21. ■ $68.97 (O)	15. $23.19 (R)

2. Students should explain that a city or state will usually impose an income tax when sales taxes or real estate taxes are not raising enough money and when for whatever reason, the sales tax cannot be increased. A state or locality will often use income taxes to pay for social programs or for services such as police and fire protection. Usually, only the largest localities have income taxes.

ASSIGNMENTS

Level 1	1–15
Level 2	1–21
Level 3	10–21

FOLLOW-UP ACTIVITY

APPLICATION Show students how they can use their calculators to find state and local taxes. Refer students to page 222, Example 1, in the text. Explain that this series of keystrokes on a calculator with a memory would yield the state tax.

[.] [0] [2] [×] [8] [0] [0] [0] [=] [M+]
[.] [0] [3] [5] [×] [7] [0] [0] [0] [=] [M+]
[.] [0] [5] [×] [5] [0] [0] [0] [=] [M+]
[.] [0] [6] [5] [×] [2] [2] [5] [0] [=] [M+]

Then push [MRC], [MR], or [RM].

Ask students to use their calculators to find the city tax on page 222, Example 2. Then ask students to use their calculators and the tables on page 222 to find the state and city tax for the people below.

1. Lynne, single, no dependents
Adjusted Gross Income: $23,092
2. Larry, married, 2 dependents
Adjusted Gross Income: $34,982

SUPPLEMENTARY MATERIALS

TRP Practice, p. 75
TRP Reteaching, p. 62
TRP Transparency 38

OBJECTIVE

- Solve problems by projecting estimates.

TEACHING THE LESSON

WARM-UP Write the following exercises on the chalkboard or an overhead transparency.

1. \$600 ÷ 2	**2.** \$800 ÷ 4
3. \$1,000 ÷ 2	**4.** \$2,000 ÷ 4
5. \$3,500 ÷ 2	**6.** \$4,600 ÷ 4

(**1.** \$300 **2.** \$200 **3.** \$500 **4.** \$500 **5.** \$1,750 **6.** \$1,150)

INTRODUCTION Ask students if they know people who are self-employed. List the occupations of these people on the chalkboard. Some examples might be: freelance writers, artists, doctors, lawyers, dressmakers, and plumbers. Discuss the fact that when people are self-employed, federal, state, and local taxes are not deducted from their earnings. A self-employed worker must submit his or her own *estimated tax payments* quarterly to the federal, state, and local governments.

INSTRUCTION Present the problem situation and Section A to the class. Ask students how many payments Jake needs to make to the federal government this year (4 on the given dates). Discuss the fact that \$4,956 is rounded to \$5,000. Ask students what operation is used if you know the total amount and the number of payments and you need to find the amount of each payment (division). Have students complete the division using a calculator.

After discussing Section B, ask students to tell whether Jake's last 2 payments would be lower or higher than his first 2 payments. (higher because his income was greater) Have students complete each step of the solution using a calculator. Encourage them to check their work by adding up Jake's first 2 payments and his last 2 payments to see if they total \$6,500.

PROBLEM *Solving* STRATEGY

11.6 PROJECTING ESTIMATES

Situation:

Jake is a freelance writer. His paychecks have no taxes withheld from them. He must submit his own estimated tax payments quarterly to the federal, state, and local governments.

Quarterly Estimated Taxes Due

April 15	June 15
September 15	January 15

Strategy:

Projecting estimates can sometimes help you to solve a problem.

Applying the Strategy:

A. Last year, Jake paid \$4,956 in federal taxes. If Jake expects to earn about the same income this year as last year, how much estimated federal tax should Jake pay on April 15?

THINK: Last year, Jake paid about \$5,000 in federal taxes. This year, he expects to earn about the same amount of income. Therefore, he expects to pay about the same amount of taxes.

Divide by 4 to find the quarterly amount.

\$5,000 ÷ 4 = \$1,250

Jake should pay \$1,250 of estimated federal tax on April 15.

B. Suppose that in August, Jake realized that his income would be greater than he had expected for the year. He estimated that he should pay an additional \$1,500 to the federal government, for a total of \$6,500. He had already made the April 15 and June 15 payments of \$1,250 each. How much will Jake pay on September 15 and January 15?

1. Multiply to find the amount already paid.
   2 × \$1,250 = \$2,500
2. Subtract to find the amount still owed.
   \$6,500 − \$2,500 = \$4,000
3. Divide by 2 to find the amount for each of the last 2 payments.
   \$4,000 ÷ 2 = \$2,000

Jake will pay \$2,000 for each of the last 2 payments.

PRACTICE EXERCISES

Remember to estimate whenever you use your calculator.

Betty estimated that she will need to pay $3,200 in state taxes this year. She divided that amount by 4 and found that she should make 4 $800 payments during the year. In July, Betty realized that her income would be lower than she had expected. She estimated that she should pay $600 less to the state. She had already made the April 15 and June 15 payments.

• 1. In July, how much did Betty estimate her state taxes to be? $2,600

2. How much money had Betty already paid? $1,600

3. How much money did Betty have left to pay? $1,000

4. How many payments were left for Betty to make? 2

5. How much estimated state tax was due on September 15 and on January 15 if both payments were equal? $500

This table shows estimated taxes 3 freelance writers estimated they should pay this year.

Name	Estimated Taxes		
	Federal	State	City
Jed	$2,000	$ 850	$180
Irene	$8,000	$3,000	$600
Burt	$7,000	$2,400	$600

Use the facts in the table to solve the problem.

• 6. Jed will make 4 equal estimated tax payments to the state. How much should each payment be? $212.50

• 7. Jed will make 4 equal estimated tax payments to the city. How much should each payment be? $45

8. Irene had planned to make 4 $2,000 payments for her federal taxes. After 2 payments, she reduced her estimated federal tax due by $1,500. How much were each of Irene's last 2 payments? $1,250

9. Jed had planned to make 4 $500 payments for his federal taxes. After 1 payment, he increased his estimated federal tax due by $900. How much were each of Jed's last 3 payments? $800

10. Irene had planned to make 4 equal payments for her city taxes. After 2 payments, she lowered her estimated city taxes due by $100. How much were each of Irene's last 2 payments? $100

11. Burt planned to make 4 equal payments on his federal, state, and city taxes. He now wants to increase each estimated tax by 20%. How much money will Burt have to spend each quarter to pay all 3 estimated taxes? Federal: $2,100; State: $720; City: $180; Total: $3,000

You may wish to discuss Practice Exercises 1–5 aloud. In this situation, the amount of estimated income is decreasing so the last 2 payments will be lower than the first 2 payments.

SKILLS BANK You may want to refer students to the following skills in the Skills Bank:

Multiply With Decimals, Skill 5

Divide With Decimals, Skill 8

ASSIGNMENTS

Level 1	1–5, Even 6–10
Level 2	1–10
Level 3	6–11

FOLLOW-UP ACTIVITY

SITUATIONAL PROBLEM SOLVING Present students with the following problem. You want to start your own business in 10 y. What should you do now?

Some questions to ask include the following: What business would it be? What do you need to prepare for this? How much money will you need to start? Will you need to employ people? How will you get customers?

Some approaches to solving the problem include the following: estimating the amount of money needed and working backward to determine how to start accumulating it; developing a plan: product, employees, customers, and so forth.

SUPPLEMENTARY MATERIALS

TRP Practice, p. 76
TRP Transparency 38
TRP Application, p. 13

OBJECTIVE

- Decide, given relevant information, which tax form to use.

TEACHING THE LESSON

WARM-UP Write the following exercises on the chalkboard or an overhead transparency.

Find the AGI and taxable income.

1. Gross income: $34,982
   Adjustments: $2,560
   Deductions: $2,540
2. Gross income: $18,304
   Adjustments: $500
   Deductions: $3,760

(**1.** $32,422; $29,882 **2.** $17,804; $14,044)

INTRODUCTION Present the problem situation to the class. Encourage students to name some of the factors they would use to make an informed decision about which form to use. List on the chalkboard the factors that are named.

INSTRUCTION Read the features of each person's financial report. Ask students to discuss these features fully. Ask if there is any other information needed. Discuss the decision-making factors.

Have small groups of students complete Exercises 1–16 in the table in order to compare the 3 people. Ask students to write statements that describe the data now in the table. Then ask them to write conclusions that can be drawn from the data. Discuss the appropriateness of these statements and conclusions and lead a discussion about interpretations and possible misinterpretations of the data.

DECISION MAKING

11.7 CHOOSING THE CORRECT TAX FORM

(Open-ended problem solving)

You can use one of 3 tax forms to file your income taxes: FORM 1040EZ, FORM 1040A, or FORM 1040.

FORM 1040EZ can only be used by single people with no dependents, no adjustments, and no itemized deductions. Taxable interest cannot exceed $400.

FORM 1040A can be used by anyone with no limit on income. Adjustments are allowed only for Individual Retirement Accounts (IRAs). No itemized deductions are allowed.

FORM 1040 can be used by anyone. All adjustments are allowed and deductions can be itemized.

You want to use the simplest form and still pay the lowest allowable tax.

PROBLEM

You are a community tax consultant. Three different people come to you for advice on which tax form to use and how much tax they should pay.

	PERSON A	PERSON B	PERSON C
Filing status	Single	Married, joint return	Single
Exemptions	1 person, not claimed	2 adults	1 person, claimed on
	on another return	1 child	another return
Wages	$25,062	$19,390 and $27,430	$18,093 and
			$7,642 in tips
Interest	Bank A: $237	Bank: $319	Bank: $302
	Bank B: $95	Fund: $116	
Adjustments	IRA: $275	IRA: $2,500	None
Deductions			
Medical	$627	$1,079	$295
Taxes	$935	$618	$308
Interest	$239	$920	$0
Contributions	$285	$570	$185
Miscellaneous	$628	$315	$216

DECISION-MAKING FACTORS

Filing status	Wages	Interest income	Adjustments to income
Exemptions	AGI	Itemized deductions	Standard deduction

DECISION-MAKING COMPARISONS

Complete the table to compare the 3 people. Use the table on page 220 for Exercises 15–16.

Factors	Person A	Person B	Person C	
Filing status	Single	• 1. ■	2. ■	Married, joint return; Single
Wages	$25,062	• 3. ■	4. ■	$46,820; $25,735
Interest income	5. ■	6. ■	$302	$332; $435
Adjustments to income	7. ■	IRA: $2,500	8. ■	IRA: $275; 0
Adjusted Gross Income	9. ■	10. ■	$26,037	$25,119; $44,755
Exemptions ($1,900 per each allowable exemption)	11. ■	12. ■	$0—No exemptions	$1,900; $5,700
Itemized deductions	13. ■	14. ■	$1,004	$2,714; $3,502
Standard deduction	15. ■	16. ■	$2,540	$2,540; $3,760

• **17.** Which form should each person use if filing status were the only factor?
A: 1040EZ; B: 1040A; C: 1040EZ

• **18.** Which form should each person use if filing status and interest income were the only factors?
A: 1040EZ; B: 1040A; C: 1040EZ

19. Which form should each person use if filing status, interest income, and adjustments to income were the only factors?
A: 1040A; B: 1040A; C: 1040EZ

20. Should each person use itemized deductions or the standard deduction?
A: Itemized deductions; B: Standard deduction; C: Standard deduction

21. Which tax form should each person use? Explain.
See page 477 for answer.

22. What is the amount of the deduction for each person?
A: $2,714; B: $3,760; C: $2,540

23. What is the total of exemptions and deductions for each person?
A: $4,614; B: $9,460; C: $2,540

24. What is the taxable income for each person?
A: $20,505; B: $35,295; C: $23,497

25. What is the federal income tax for each person? Use the tax tables on pages 475–479.
A: $3,491; B: $6,117; C: $4,317

Ask students to complete Exercises 17–25 in small groups, having them explain and defend their answers and decisions. Remind students to give an answer for each person in each exercise. Remind students that, in each case, there is only 1 correct answer for each question.

ASSIGNMENTS

Level	Exercises
Level 1	1–25
Level 2	1–25
Level 3	1–25

FOLLOW-UP ACTIVITY

ENRICHMENT Distribute facsimile copies of Form 1040A to students. You may get copies of this form from the Understanding Taxes Coordinator at the nearest IRS district office. Refer students to the financial information about Person B, who should file Form 1040A. Guide them through the steps found on the left-hand side of the tax form.

SUPPLEMENTARY MATERIALS

TRP Practice, p. 77
TRP Transparency 39

OBJECTIVES

- Review vocabulary.
- Practice key chapter concepts and skills.

USING THE REVIEW

The Chapter Review is designed to help students prepare for taking the Chapter Test. The first section focuses on vocabulary. It requires students to select words to complete statements. The second section presents practice of key chapter skills.

Each item on the Chapter Review is referenced to the page on which the topic is taught in the Pupil's Edition. You may wish to have students refer to these pages to help review any concepts or skills they have not yet mastered.

It is suggested that students work in small-sized heterogeneous cooperative learning groups. Some cooperative learning methods that may be used are as follows:

1. After each student has independently completed the Chapter Review, a discussion should follow within each group about the solutions to the problems.
2. The group can complete the entire Chapter Review by working together to solve the problems.

End the lesson with an entire class discussion in which any questions brought up in group discussions are presented and answered.

CHAPTER REVIEW

Vocabulary Choose the letter of the word(s) that completes the sentence.

1. The W-2 form is used to report ■. [212] b
   a. Interest **b.** Wages **c.** Deductions
2. Gross income, less any adjustments, is called ■. [214] b
   a. Taxable Income **b.** Adjusted Gross Income **c.** Exemptions
3. Amounts that reduce the income to be taxed are called ■. [214] a
   a. Deductions **b.** Taxable income **c.** AGIs

Skills Find the answer.

Use the W-2 and the 1099 forms to find the amount. [212]

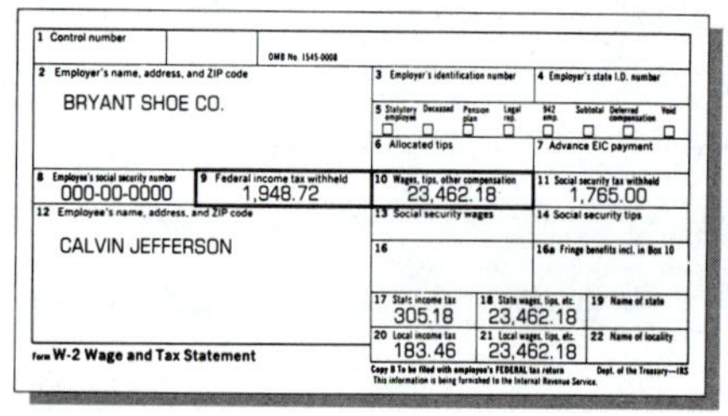

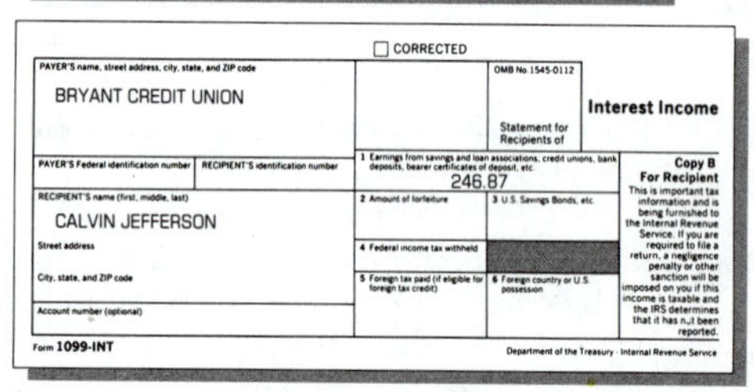

4. Wages $23,462.18
5. Federal tax withheld $1,948.72
6. State tax withheld $305.18
7. City tax withheld $183.46
8. Social Security tax withheld $1,765.00
9. Interest $246.87
10. Gross income $23,709.05

Solve.

11. Sarah's gross income is $27,345. She has $789 in adjustments to income. What is her Adjusted Gross Income? [214] $26,556
12. Martin's AGI is $19,456, with a $2,540 deduction and no exemptions. What is his taxable income? [214] $16,916
13. Lee Ann's taxable income is $18,932 and she is single. What is her federal tax? [214] $3,043
14. Ira files Form 1040EZ and is not claimed on another form. How much is his personal exemption? [216] $1,900
15. The Martinez's joint return has itemized deductions of $2,780. Should they take the itemized or the standard deduction? [219] Standard deduction
16. Diana's Adjusted Gross Income was $19,874 last year. How much state tax did she pay? (Use the tax table on page 222.) [222] $648.70

CHAPTER TEST

Ron and Beth are married and filing a joint return. They have 1 child and will take the standard deduction.

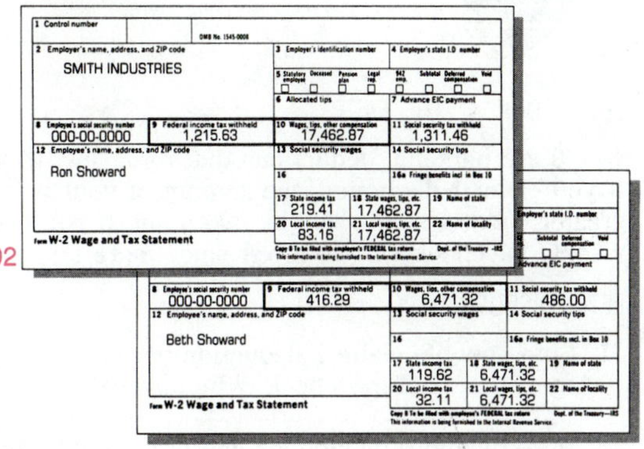
1 Control number
2 Employer's name, address, and ZIP code
SMITH INDUSTRIES
8 Employee's social security number 000-00-0000
9 Federal income tax withheld 1,215.63
10 Wages, tips, other compensation 17,462.87
11 Social security tax withheld 1,311.46
12 Employee's name, address, and ZIP code
Ron Showard
17 State income tax 219.41
18 State wages, tips, etc. 17,462.87
20 Local income tax 83.16
21 Local wages, tips, etc. 17,462.87
Form W-2 Wage and Tax Statement

8 Employee's social security number 000-00-0000
9 Federal income tax withheld 416.29
10 Wages, tips, other compensation 6,471.32
11 Social security tax withheld 486.00
12 Employee's name, address, and ZIP code
Beth Showard
17 State income tax 119.62
18 State wages, tips, etc. 6,471.32
20 Local income tax 32.11
21 Local wages, tips, etc. 6,471.32
Form W-2 Wage and Tax Statement

Use the W-2 forms to find the amount.

1. Total gross income $23,934.13
2. Total federal tax withheld $1,631.92
3. Total state tax withheld $339.03
4. Total city tax withheld $115.27
5. Total Social Security tax withheld $1,797.46

Find the Adjusted Gross Income, taxable income, and federal tax. Use the federal tax tables on pages 475–479.

6. Single, 1 exemption
   Gross income: $27,456
   Adjustments: $569
   Deductions: $2,540
   $26,887; $22,447; $4,023
7. Married, filing jointly, 3 exemptions
   Gross income: $32,709
   Adjustments: $1,750
   Deductions: $3,920
   $30,959; $21,339; $3,079

Use the W-2 forms at the top of the page. Use the federal tax tables on pages 475–479. Use the state and city tax tables on page 222.

8. What is the Showards' federal tax? $2,051
9. How much federal tax do the Showards owe, or what is their refund? owe $419.08
10. What is the Showards' state tax? $910.72
11. How much state tax do the Showards owe, or what is their refund? owe $571.69
12. What is the Showards' city tax? $467.44
13. How much city tax do the Showards owe, or what is their refund? owe $352.17

Howard's estimated taxes for the year were: Federal: $4,000; State: $1,670; and City: $290.

14. Howard will make 4 equal payments to each this year. How much should each payment to each agency be? Federal: $1,000; State: $417.50; Local: $72.50
15. After 2 payments, Howard lowered his estimated federal taxes due by $300. How much were each of his last 2 federal tax payments? $850

OBJECTIVE

- Evaluate the achievement of chapter objectives.

USING THE TEST

The Chapter Test may be used as a post-test to evaluate student achievement. However, you may wish to use the Chapter Post-Test offered in the Teacher's Resource Package or you may design your own chapter test. If this page is not used as a test, you may want to assign it as additional review or practice.

The test items are correlated to the chapter objectives in the table below.

Chapter objective	Test items
A. Read and interpret W-2 forms.	1–5
B. Find AGI, taxable income, and federal income tax.	6–7
C. Determine federal income tax and the amount owed or refund due.	8–9
D. Determine state income tax and the amount owed or refund due.	10–11
E. Determine city income tax and the amount owed or refund due.	12–13
F. Solve problems by projecting estimates.	14–15

SUPPLEMENTARY MATERIALS

TRP Ch. 11 Post-Test Form A, pp. 1–2
TRP Ch. 11 Post-Test Form B, pp. 1–2

OBJECTIVE

- Understand that, by claiming 1 tax deduction instead of 0, less tax will be taken our of your weekly paycheck, giving you more money throughout the year to spend, save, or invest.

USING THE PAGE

Begin discussion by telling students that everyone must pay taxes but that, as individuals, we have the option of deciding if we want more tax money taken out of our paychecks on a regular basis or not. Ask students if they have a relative or family friend who is an accountant or tax expert. If so, suggest that they talk to that person, or their parents, if they are still confused after completing this lesson.

Have students read the introductory material. Before proceeding with the Exercises, explain clearly the meaning of a *personal deduction*. Also mention that there are more legally allowable deductions that they may take in addition to a personal deduction.

Have students work individually or in pairs to answer Exercises 1–5. Summarize the answers as follows: It is important to consider your personal financial situation before deciding whether to claim 1 or 0 tax deduction and to consult a tax expert regarding how many other deductions to claim.

Have students work individually to answer Exercises 6–8. Then have them share their answers and reasons in a whole-class discussion.

ANSWERS/EXPLANATIONS

5. You are losing money because, by waiting until the following April to file your refund, you are letting the government earn interest on your money instead of you earning interest on it.

MONEY TIPS

You will have more money in each paycheck if you claim 1 tax deduction instead of 0.

LET'S LOOK AT THE FACTS

1 or 0 are personal deductions that refer only to you *as an individual.* If you claim 1 deduction, the amount of your weekly paycheck will be greater, since less tax will be taken out. If you claim 0 deductions, your weekly paycheck will be lower since more tax will be taken out.

LET'S DISCUSS WHY See TE side column for answer to Exercise 5.

1. Some people claim 1 deduction because they prefer to get more money in each paycheck. Why might they prefer this?
   To have more money available to them on a regular basis
2. If you claim 0 deductions, more tax will be taken out of your paycheck each week and you might be entitled to a refund. When would you receive this refund?
   Shortly after you file your taxes in April for the previous year
3. On January 1, you declared 1 deduction on your employment form. On July 1, you change jobs. You will earn the same salary but you change your deduction claim to 0. Next April, at tax time, are you likely to owe money? Get a refund? Neither?
   Neither. But other factors will affect the tax you owe.
4. Many people claim deductions for their spouse, their children, and their home mortgage payments. Tax experts warn that declaring too many deductions may require a large payment plus penalties in April. How can this be avoided?
   Work with a tax accountant to determine the correct number of deductions
5. Declaring too few deductions may result in a large refund. While this sounds appealing, you could be losing money by doing this. How is this possible?

LET'S SEE WHAT YOU WOULD DO

6. You must make car, student loan, and credit card payments every month. Would you prefer to receive more money in your weekly paycheck or receive a large refund later? Explain your decision.
   Answers may vary.
7. By claiming 1 deduction instead of 0, you receive $65 more in each weekly paycheck. What could you do with that extra money that might earn you even more money?
   Answers may vary. Invest it to earn interest, etc.
8. You are single and have no debts or financial obligations. Your accountant tells you that you will be getting a $5,200 refund this year. Make a list of all the things you might do with that money.
   Answers may vary.

CALCULATOR THE MEMORY KEYS

You can use calculators with **memory keys** to solve multi-step problems. These are the memory keys.

[M+]	[M−]	[MR] or [RM]	[MC] or [CM]
Adds a number to memory.	Subtracts a number from memory.	Recalls a number from memory.	Clears memory.

Using memory keys, compute: $(15 + 12) + (8 \times 5)$.

Procedure	Calculator Entry	Calculator Display
1. Perform the operation within the first set of parentheses.	[1] [5] [+] [1] [2] [=]	27.
2. Add the result to memory.	[M+]	M 27.
3. Perform the operation within the second set of parentheses.	[8] [×] [5] [=]	M 40.
4. Add the result to memory.	[M+]	M 40.
5. Recall the total in memory.	[MR]	M 67.

So $(15 + 12) + (8 \times 5) = 67$.

Always make sure you clear the memory before you use it.

Use a calculator to compute.

1. $(5 \times 6) + (3 \times 2)$ 36

2. $(8 \times 6) - (9 + 2)$ 37

3. $(11 + 5) + (7 - 3)$ 20

4. $(14.9 + 6) + (15.6 - 5.2)$ 31.3

5. $(9.2 \times 8.3) + (16.2 - 4.5)$ 88.06

6. $(248 \times 2.5) - (14.5 + 108)$ 497.5

7. $(4.6 \div 2) - (0.4 + 1.3)$ 0.6

8. $(0.7 \times 3) + (8 \div 0.2)$ 42.1

9. $(0.36 \div 0.3) + (0.8 \div 0.2)$ 5.2

Use a calculator to solve.

10. Zelda must pay a 16% tax on her taxable income of \$18,000. Her husband must pay a 19% tax on \$23,000. How much tax must Zelda and her husband pay? \$7,250

11. A state income tax is 1% on the first \$4,000 and 1.5% on the next \$4,000. What would be the total state tax for an income of \$7,800? \$97

OBJECTIVE

- Use the memory keys to solve multistep problems on a calculator.

USING THE PAGE

Four memory keys are described on the page. Not all calculators will have each of these 4 keys. Some will combine the functions of the [MR] and [MC] into 1 key. In this case, the first time the key is pressed it will perform the [MR] function—memory recall. The second time, it will perform the [MC] function—memory clear.

Point out that when the calculator is turned on, students should check that the memory is clear. Then have students enter the number 5 on their calculators and press the [M+] key to add the number to memory. Point out that the letter *M* on the display indicates that a number is now in memory. Then have students enter the number 6 and add it to memory. Point out that the sum of 11 has been stored in memory and is not displayed on the screen. In order to see it, have them press the [MR] key. Point out that the sum of 11 still remains in the memory even though it is also displayed. Now have students enter the number 3 and subtract it from the total in memory by pressing [M−]. To display the difference, they press [MR]. To clear the memory, they press [MC]. Tell students to clear both the memory and the display before beginning each problem. You may want to repeat the above process with other numbers before proceeding with the displayed example.

Have students work through the displayed example. After students have successfully completed it, you may wish to tell them that the use of the is-equal-to key may not be necessary.

OBJECTIVE

- Test the prerequisite skills needed to learn the concepts developed in Chapter 12.

USING THE TEST

The Pre-Skills Test is designed to diagnose students' strengths and weaknesses on prerequisite skills necessary to study the mathematics in Chapter 12.

Assign the Pre-Skills Test. Allow students to work together in pairs or in small groups. Group members should help those who demonstrate a misunderstanding of a concept or a weakness in a skill.

The table below correlates the items on the Pre-Skills Test with the prerequisite skill and the lesson(s) in the chapter for which it is needed.

Item(s)	Prerequisite skill	Lesson(s)
1–8	Add and subtract money amounts.	All
9–14	Multiply money amounts.	All
15–18	Find the percent of a number.	12.3–12.6
19–22	Divide money amounts.	12.1, 12.3, 12.5, 12.8
23–28	Solve proportions.	12.8
29–33	Find perimeter and area.	12.8

References to specific skills that are reviewed in the Skills Bank are given on the individual lessons themselves.

Pre-Skills Test

Add.

1. \$24,850 + \$18,920 \$43,770
2. \$57,350 + \$9,840 \$67,190
3. \$125,050 + \$78,360 \$203,410
4. \$225,100 + \$83,692 \$308,792

Subtract.

5. \$93,450 − \$71,130 \$22,320
6. \$70,545 − \$48,670 \$21,875
7. \$91,800 − \$56,072 \$35,728
8. \$130,985 − \$87,987 \$42,998

Multiply.

9. 9 × \$24,000 \$216,000
10. 4 × \$34,850 \$139,400
11. 6 × \$108,762 \$652,572
12. 0.6 × \$123,050 \$73,830
13. 0.4 × \$90,765 \$36,306
14. 0.8 × \$86,098 \$68,878.40

Find the answer.

15. 6% of \$20,000 \$1,200
16. 9% of \$34,500 \$3,105
17. $5\frac{1}{2}$% of \$58,050 \$3,192.75
18. $7\frac{1}{2}$% of \$102,085 \$7,656.38

Divide. Round the answer to the nearest cent.

19. \$45,000 ÷ 0.3 \$150,000
20. \$38,975 ÷ 0.28 \$139,196.43
21. \$58,975 ÷ 0.28 \$210,625
22. \$108,764 ÷ 0.28 \$388,442.86

Solve the proportion.

23. $\frac{1}{2} = \frac{n}{10}$ $n = 5$
24. $\frac{2}{5} = \frac{n}{10}$ $n = 4$
25. $\frac{1}{5} = \frac{n}{20}$ $n = 4$
26. $\frac{1}{4} = \frac{n}{12}$ $n = 3$
27. $\frac{1}{1.5} = \frac{n}{12}$ $n = 8$
28. $\frac{1}{1.5} = \frac{n}{18}$ $n = 12$

Use the diagram of the room.

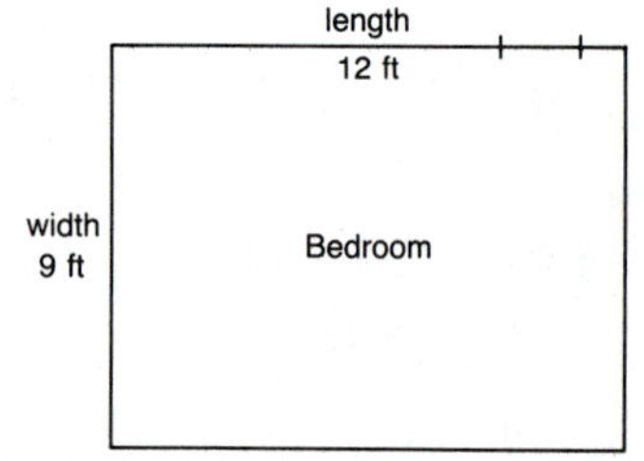

29. What is the width? 9 ft
30. What is the length? 12 ft
31. Find the perimeter in feet. 42 ft
32. Find the area in square feet. 108 square ft
33. Find the area in square yards. 12 square yd

Housing costs include a mortgage or rent, utilities, and the decoration or remodeling of the house or apartment. What costs were involved in furnishing this room?

OBJECTIVE

- Formulate problems, using text and data.

USING THE CHAPTER OPENER

Discuss students' answers to the question on the page. They should not have a problem enumerating the contents of the room, such as furniture, lighting, and accessories. They might also include the wall and floor covering, paint, and window treatment. If necessary, lead them to see that other costs might include labor for painting or other custom work, interior-decorating fees, finance charges for items bought on time, interest on a loan that may have been taken to do the redecorating, and so on.

COOPERATIVE LEARNING EXPERIENCE

Assign heterogeneous groups of students the task of furnishing a 1-room apartment. Tell them that the apartment is 15 ft by 25 ft and a complete kitchenette and bath are included. Their assignment is to develop a plan for furnishing the place. They should use newspapers and local stores to get costs. This would include painting or covering the walls, painting the windows and doors, and doing something with the grimy wood floors. Give half the groups a budget of $2,500 and the remaining groups a budget of $5,000. All groups are to do as much as they can within their budget to create a comfortable living space for a young person just starting out.

When the groups present their plans, have them compare their results in terms of the 2 budgets. Lead them to see that it is possible to do the job with limited funds if they spend wisely.

SUPPLEMENTARY MATERIALS

TRP Group Projects, p. 12

OBJECTIVES

- Compute the maximum amount that a person should be able to afford for rent.
- Compute the monthly costs for rent and utilities.
- Compute apartment move-in costs.

TEACHING THE LESSON

WARM-UP Write the following exercises on the chalkboard or an overhead transparency.

Find the answer.

1. 28% of $1,500 **2.** 28% of $1,750

Add.

3. $78 + $48 **4.** $407 + $56 + $79

(**1.** $420 **2.** $490 **3.** $126 **4.** $542)

INTRODUCTION Explain to students that it is important to set a limit on the amount of money to spend on rent. Discuss some of the possible consequences of spending too much on rent.

INSTRUCTION Explain that gross monthly pay can be found by dividing gross annual salary by 12.

Use Example 1 to show students how to find the maximum to be spent on rent each month. Point out that 28% of gross monthly pay is about 1 week's gross pay.

Use Example 2 to show students how to find the monthly costs for rent and utilities. Emphasize the importance of reading an advertisement carefully to find out whether or not utility costs are included in the rent.

Use Example 3 to show students how to compute the move-in costs for an apartment. Emphasize the importance of knowing the move-in costs before agreeing to rent an apartment. Remind students that they will have to pay any move-in costs *before* they occupy the apartment. Point out that the security deposit is usually returned, sometimes with interest, when a person moves out of an apartment.

12.1 RENTING AN APARTMENT

Before you look for an apartment, you should know the maximum amount you can afford to spend on **rent.** In general, you should be able to spend a maximum of 28% of your gross monthly pay on rent.

EXAMPLE 1 Your annual gross salary is $16,980. What is the maximum amount you should be able to spend on rent?

1. Divide to find your monthly gross pay. $16,980 ÷ 12 = $1,415
2. Multiply to find the maximum rent.

   THINK: 28% = 0.28 0.28 × $1,415 = $396.20

You should be able to spend a maximum of $396.20 on rent.

Utilities such as gas and electricity are usually not included in the rent.

EXAMPLE 2 You want to rent a 1-bedroom apartment in the Knickerbocker Square Apartments. *Plus Utilities* means that you must pay separately for electricity and gas. About how much will you pay each month for both rent and utilities?

KNICKERBOCKER SQUARE APARTMENTS
$375 1 bedroom
$460 2 bedrooms
$525 3 bedrooms
Plus Utilities

THINK: Rent for a 1-bedroom apartment: $375
Average monthly utility costs:
Electricity—$65 Gas—$43

1. Add to find cost of utilities. $65 + $43 = $108
2. Add to find the total cost. $375 + $108 = $483

You will pay about $483 per mo for both rent and utilities.

You may need to pay a **security deposit** and a **fee** to a rental agent when you move into your apartment. These payments are **move-in costs.**

EXAMPLE 3 Your rent is $375 per mo. You paid 60% of that as a rental fee and you paid twice that as a security deposit. What were your move-in costs?

1. Multiply to find the rental fee.

   THINK: 60% = 0.6 0.6 × $375 = $225
2. Multiply to find the security deposit. 2 × $375 = $750
3. Add to find the total move-in costs. $225 + $750 = $975

Your move-in costs were $975.

FOR DISCUSSION See TE side column.

1. What other costs might a person have when moving into his or her first apartment?

2. What other costs might a person have when he or she moves from one apartment to another one?

PRACTICE EXERCISES

Remember to estimate whenever you use your calculator.

Find the maximum monthly rent for the gross monthly pay.

1. $828 per mo $231.84
2. $2,110 per mo $590.80
3. $1,665 per mo $466.20

Find the gross monthly pay and the maximum monthly rent. For Exercises 7–9, the work week is 40 h, 50 wk per y.

• 4. $11,208 per y $934; $261.52
• 5. $14,940 per y $1,245; $348.60
6. $21,708 per y $1,809; $506.52
7. $7.30 per h $1,216.67; $340.67
8. $8.15 per h $1,358.33; $380.33
9. $11.80 per h $1,966.67; $550.67

Use the advertisements for Exercises 10–20.
Find the total monthly cost for rent, utilities, and extras.
Use these amounts for utilities.
Electricity—$49 per mo Gas—$37 per mo

• 10. THE OVERLOOK Studio $371
• 11. THE OVERLOOK 1 bedroom $446
12. HASTINGS HOUSE 1 bedroom $375
13. ECKERSON APTS. 2 bedrooms $709
14. ECKERSON APTS. 1 bedroom $544
15. THE OVERLOOK 2 bedrooms $501
16. HASTINGS HOUSE 3 bedrooms Parking $560
17. ECKERSON APTS. 2 bedrooms Health Club $744

THE OVERLOOK
Studio $285
1 bdrm $360
2 bdrm $415
Plus Utilities

ECKERSON APTS.
1 bdrm $495
2 bdrm $660
Plus elec.
Health Club —
$35 per month extra

HASTINGS HOUSE
1 bdrm $375
2 bdrm $485
3 bdrm $520
(utilities included)
Parking —
$40 per month

Find the move-in costs.

18. $990 19. $1,067 20. $1,650

Building	THE OVERLOOK	HASTINGS HOUSE	ECKERSON APTS.
Apartment type	1 bedroom	2 bedrooms	2 bedrooms
Rental agent fee	75% of 1 month's rent	10% of annual rent	1 month's rent
Security deposit	2 months' rent	1 month's rent	$1\frac{1}{2}$ month's rent

SKILLS BANK You may want to refer students to the following skills in the Skills Bank:

Add With Whole Numbers and Decimals, Skill 6

Multiply With Decimals, Skill 5

Divide With Whole Numbers, Skill 6

Find the Percent of a Number, Skill 2

FOR DISCUSSION

1. Students should explain that a person might need to buy furniture, appliances, curtains, and rugs when he or she first moves in. A person will also need sheets, pillowcases, a supply of food, and personal articles. These additional costs should be considered with the other move-in costs.
2. Students should explain that a person might need to pay for a mover and to buy new furniture or articles that fit in the new apartment.

ASSIGNMENTS

Level 1	1–6, 10–15, 18–19
Level 2	1–20
Level 3	4–9, 14–20

FOLLOW-UP ACTIVITY

APPLICATION Give each student a page of rental advertisements from a local newspaper. Ask students to choose the apartments they would like to rent. Students should determine whether or not utilities are included in the rent. Ask students to find the monthly costs of rent and utilities for their apartments. Use these utility costs—Electric: $63/mo; Gas: $42/mo.

SUPPLEMENTARY MATERIALS

TRP Practice, p. 78
TRP Reteaching, p. 63
TRP Transparency 39

OBJECTIVES

- Compute the maximum amount that a person should be able to afford to borrow when buying a house.
- Compute the maximum amount that a person should be able to afford to spend for a house.
- Compute the appreciated value of a house.

TEACHING THE LESSON

WARM-UP Write the following exercises on the chalkboard or an overhead transparency.

Compute.

1. \$45,995 + \$9,562 (\$55,557)

2. \$78,762 − \$54,693 (\$24,069)

3. 15% of \$35,000 (\$5,250)

4. 15% of \$65,750 (\$9,862.50)

INTRODUCTION Use the introductory material to point out that buying a home usually means borrowing money. Emphasize to students the importance of limiting the amount borrowed and the cost of the house to the amount they can afford.

INSTRUCTION Use Example 1 to show students how to find the amount they can afford to borrow and the amount they can afford to spend for a house. Remind students that this example shows the *maximum* they should be able to borrow or spend.

Use Example 2 to demonstrate how to find annual appreciation and the appreciated value of a house. It is a common mistake for students to think that the appreciated value after 2 y is 30% of the initial value. Emphasize in Step 3 that the second year's appreciation is based on the appreciated value of the house after 1 y. These calculator keystrokes will yield the appreciated value of the house after 2 y. Only on a scientific calculator must you use the [=].

[4] [3] [0] [0] [0] [+] [1] [5] [%] [+] [1] [5] [%]

12.2 BUYING A HOUSE

You know you will need to borrow money to buy a house. Before you look at houses, you should know the maximum you should be able to borrow and the maximum you should be able to spend for a house. In general, you should borrow no more than 2 times your annual gross income.

EXAMPLE 1 You have saved \$31,000 to buy a house. Your annual gross income is \$28,600. What is the maximum you should be able to borrow? What is the maximum you should be able to spend for a house?

1. Multiply to find the maximum you should borrow. $2 \times \$28{,}600 = \$57{,}200$

The maximum you should be able to borrow is \$57,200.

2. Add to find the maximum you can spend. $\$57{,}200 + \$31{,}000 = \$88{,}200$

The maximum you should be able to spend on a house is \$88,200.

You would like to buy a house that **appreciates** (increases in value) each year. The more the appreciation, the higher the future value of the house.

EXAMPLE 2 Fran bought a house for \$43,000. Her house appreciated 15% each year. What was the value of the house after 2 y? How much had it appreciated?

1. Multiply to find the first year's appreciation. THINK: 15% = 0.15 $0.15 \times \$43{,}000 = \$6{,}450$
2. Add to find the value after 1 y. $\$43{,}000 + \$6{,}450 = \$49{,}450$
3. Multiply to find the second year's appreciation. THINK: Use the value after 1 y. $0.15 \times \$49{,}450 = \$7{,}417.50$
4. Add to find the value after 2 y. $\$49{,}450 + \$7{,}417.50 = \$56{,}867.50$

Fran's house was valued at \$56,867.50 after 2 y.

5. Subtract to find the amount of appreciation. $\$56{,}867.50 - \$43{,}000 = \$13{,}867.50$

Fran's house was worth \$13,867.50 more than she paid for it.

FOR DISCUSSION See TE side column.

What factors do you think make a house appreciate in value?

PRACTICE EXERCISES

Remember to estimate whenever you use your calculator.

The person(s) want to buy a house. What is the maximum they can afford to borrow? What is the maximum they should be able to spend?

• 1. Fred Henderson's savings: $18,000
Gross salary: $41,000 per y
$82,000; $100,000

• 2. Denise Chin's savings: $41,600
Gross salary: $36,500 per y
$73,000; $114,600

3. Marge LaBeau's savings: $27,620
Gross salary: $785 per wk
$81,640; $109,260

4. Rafael Rodriguez's savings: $41,600
Gross salary: $2,850 per mo
$68,400; $110,000

5. Karen Goldblum's savings: $0
Gross salary: $3,760 per mo
$90,240; $90,240

6. The Militano's savings: $17,085
Gross salaries: Terri $37,200 per y
Frank $1,975 per mo
$121,800; $138,885

Complete the table to find the appreciation and the value of the house.

Value	Yearly rate of appreciation	Appreciation in 1 y	Value after 1 y
$50,000	10%	• 7. ■ $5,000	• 8. ■
$94,500	5%	9. ■ $4,725	10. ■
$76,845	8%	11. ■	12. ■
$80,395	9%	13. ■	14. ■
$88,340	7.5%	15. ■	16. ■

8. $55,000 **10.** $99,225 **11.** $6,147.60 **12.** $82,992.60 **13.** $7,235.55 **14.** $87,630.55 **15.** $6,625.50 **16.** $94,965.50

Solve.

17. Greg's house was valued at $45,500. The house appreciated 15% each year. What was the value after 2 y?
$60,173.75

18. Lillian's house was valued at $70,500. The house appreciated 6.5% each year. What was the value after 2 y?
$79,962.86

EXTENSION Appreciation

Just because a house costs more than another does not mean that it will be more valuable in the future. Ed bought a house for $53,900. It appreciated 13% each year. Eileen bought a house for $61,000. It appreciated 6% each year.

1. Which house was worth more after 2 y? Ed's house
2. If the houses keep appreciating in the same way, what will the difference in value be in 5 y? $17,675.49

SKILLS BANK You may want to refer students to the following skills in the Skills Bank:

Multiply With Decimals, Skill 5

Find the Percent of a Number, Skill 2

FOR DISCUSSION Students should explain that a house appreciates in value because of its location, its commuting distance to a city, its nearness to a recreation area or other attraction, its special features (a pool or a fireplace), and the availability of loans in the area.

EXTENSION Use this problem to show students that appreciation can be more important than purchase price when determining the future value of a house.

ASSIGNMENTS

Level 1	1–4, 7–10, 17
Level 2	1–18, EX
Level 3	3–6, 11–18, EX

FOLLOW-UP ACTIVITY

COOPERATIVE LEARNING
Divide students into pairs. One student plays the role of home buyer; the other plays the role of banker. The buyer tells the banker his or her gross annual income and the amount of savings. The banker advises the home buyer about the maximum he or she should be able to borrow and the maximum he or she should be able to spend. Students then switch roles.

SUPPLEMENTARY MATERIALS

TRP Practice, p. 79
TRP Reteaching, p. 64
TRP Transparency 40

OBJECTIVES

- Compute the monthly payment for a condominium.
- Compute the minimum gross monthly pay needed to afford a condominium.

TEACHING THE LESSON

WARM-UP Write the following exercises on the chalkboard or an overhead transparency.

Compute mentally.

1. $0.1 \times \$89,900$ **2.** $0.2 \times \$35,000$

3. $\$900 \div 0.3$ **4.** $\$600 \div 0.3$

(**1.** $8,990 **2.** $7,000 **3.** $3,000 **4.** $2,000)

INTRODUCTION Explain that a condominium is usually governed by an elected board. This board sets policy for individual condominium owners to follow, sets monthly maintenance fees, and decides how maintenance funds will be used. Ask students to discuss why it would be advantageous to be a board member.

INSTRUCTION Review the advertisement for the Luxury Condo. Tell students that the down payment is the amount of the purchase price they must pay *up front.* Usually, the down payment is a percent of the purchase price.

Use Example 1 to show students how to find the down payment and the monthly payment. Encourage students to compute the down payment mentally. Remind them that the loan payment is fixed but that the monthly maintenance fee may increase.

Use Example 2 to demonstrate the minimum gross monthly pay needed to afford a condominium. Use the THINK to explain to students how this formula was arrived at. Point out that an estimate gives a useful minimum pay. Remind students that they should earn more than the minimum before they purchase since the estimate does not include taxes and insurance.

12.3 BUYING A CONDOMINIUM

A **condominium** is a building in which apartments are owned. You purchase an apartment just as you would purchase a house. As the owner of an individual condo, or apartment, you pay an additional monthly **maintenance fee** for the care of the building and the grounds. The maintenance fee may include utilities.

> **LUXURY CONDO**
> **1 BR $67,000**
> **Maint. $345/mo**
> **Util. & Taxes**
> Magnificent View
>
> Financing Available
> 10% down
> $551/mo
>
> Parking
> Shopping on premises
>
> Sales office on site
> Call for appt.

EXAMPLE 1 You bought the luxury condo advertised in the newspaper. How much did you put down? How much will you pay each month?

THINK: Price: $67,000
Down payment: 10% = 0.1
Loan payment: $551 per mo
Maintenance fee: $345 per mo

1. Find the down payment. $0.1 \times \$67,000 = \$6,700$

You put down $6,700.

2. Add to find the total monthly payment. $\$551 + \$345 = \$896$

You will pay $896 per mo.

Remember that you should be able to spend 28% of your gross monthly pay for housing. When buying a condominium, that 28% includes maintenance, mortgage payment, taxes, and insurance.

You can "roughly" estimate the gross monthly pay you will need to afford the condominium by just using the loan payment and the maintenance fee.

EXAMPLE 2 The monthly payments for the condominium are $896. Roughly estimate the least you could earn each month to afford the condominium.

THINK: $896 is about $900. 0.28 is about 0.3.

If MONTHLY HOUSING COST = 0.28 × GROSS MONTHLY PAY,
then GROSS MONTHLY PAY = MONTHLY HOUSING COST ÷ 0.28
= $900 ÷ 0.3
= $3,000

To afford the condominium, your gross monthly pay would need to be at least about $3,000. It would definitely need to be more to cover taxes and insurance.

FOR DISCUSSION

See TE side column.

1. Why might someone choose to buy a condominium instead of a house?
2. Condominium maintenance fees are not fixed; they go up as expenses go up. What problems could this cause for the condominium owner?

PRACTICE EXERCISES

Remember to estimate whenever you use your calculator.

Find the down payment and the total monthly payment for the condominium

• 1. **PRESTIGE CONDOMINIUMS**
$178,000
Maint. $920/mo
Financing: 20% down
$1,260/mo

$35,600; $2,180

2. **CLIFFSIDE CONDOS**
$90,840
Maint. $385/mo
Financing: 10% down
$714/mo

$9,084; $1,099

3. **SCHOOLHOUSE CONDOMINIUMS**
$68,755
Maint. $205/mo
Financing: 5% down
$582/mo

$3,437.75; $787

Roughly estimate the minimum gross monthly pay needed to afford the condominium's total loan and maintenance payments.

• **4.** $550 $2,000
5. $948 $3,000
6. $685 $2,300
7. $535 $1,700
8. $1,050 $3,300

Solve.

• **9.** Pete bought a condominium for $87,600. He put 10% down and pays $705 per mo. The maintenance fee is $285 per mo. Find the down payment and the total monthly payment.
$8,760; $990

10. Lorraine's condo cost $63,400. She put 15% down and pays $573 per mo. The monthly maintenance fee is $340. Find the down payment and the total monthly payment.
$9,510; $913

11. Colleen wants to buy a condominium with monthly payments of $1,010. Roughly estimate the minimum gross monthly pay she needs to afford it.
about $3,300

12. Miguel wants to buy a condominium with monthly payments of $643. His current gross pay is $1,875 per mo. About how much more would he need to earn per month to afford the condominium?
about $125

EXTENSION Planning Ahead for Maintenance Fees

Brian wants to buy the condominium described in the advertisement. He thinks that the maintenance fee will go up 40% over the next 2 y. He wants to plan ahead. If Brian is right, what will his total monthly payments be in 2 y?
$1,228

CLOVERLEAF CONDOS
$87,450
Maint. $385/mo
Financing:
10% down $689/mo

SKILLS BANK You may want to refer students to the following skills in the Skills Bank:

Multiply With Decimals, Skill 5

Divide With Decimals, Skill 8

FOR DISCUSSION

1. Students should explain that someone might buy a condominium instead of a house because he or she does not want to be responsible for maintenance, cutting lawns, etc. Some find houses too large or too expensive to care for.
2. Students should explain that condominium owners can end up paying far more than they thought they would because maintenance fees can "skyrocket" and that the increased maintenance fees make the condominium unaffordable. People should find out the record of past increases and participate on the board.

EXTENSION Use this problem to show students that they must plan ahead for increases in maintenance costs.

ASSIGNMENTS

Level 1	Odd 1–11
Level 2	1–12, EX
Level 3	Even 2–8, 9–12, EX

FOLLOW-UP ACTIVITY

APPLICATION Pass out condominium advertisements from the local newspaper to each student in the class. Ask students to pick 5 condominiums in which they would like to live. Then ask students to find the minimum gross monthly pay needed to afford each condominium.

SUPPLEMENTARY MATERIALS

TRP Practice, p. 80
TRP Reteaching, p. 65
TRP Transparency 40

OBJECTIVES

- Use a mortgage payment table to find monthly payments.
- Compute mortgage closing costs, including points.

TEACHING THE LESSON

WARM-UP Write the following exercises on the chalkboard or an overhead transparency.

Divide.

1. $90,000 ÷ $1,000 (90)
2. $78,800 ÷ $1,000 (78.8)

Multiply.

3. 83.2 × $8.50 ($707.20)
4. 70.5 × $9.40 ($662.70)

INTRODUCTION Explain to students that a mortgage loan is a *secured* loan. The lender has the right to take possession of your home if the loan is not repaid. Discuss why it is particularly important to be sure that a person will be able to repay a mortgage.

INSTRUCTION Direct students' attention to the mortgage payment table. Point out that, as the interest rate increases, the monthly payment increases. Also point out that, as the number of payments increases, the monthly payment decreases.

Use Example 1 to show students how to find the monthly mortgage payment. Tell students that, in Step 3, they are answering the question, "How many thousand dollars are in $66,200?" The answer will thus be a decimal and not a money amount.

Use Example 2 to show students how to compute closing costs, including points. Point out that a *title fee* is paid to give the buyer the legal right of ownership, or deed, to the property. Explain that $2\frac{1}{2}$% can also be written as 2.5%. Then show students how to rename 2.5% as a decimal.

12.4 GETTING A MORTGAGE

The loan you get to buy a house or a condominium is called a **mortgage.** Your monthly mortgage payment depends on the amount you borrow, the interest rate, and the number of payments.

EXAMPLE 1 You bought a house for $82,750. You put 20% down. You borrowed the remainder at 10.5% for 30 y. What was the down payment and the mortgage amount? How much will you pay each month?

1. Multiply to find the down payment.
   THINK: 20% = 0.2
   0.2 × $82,750 = $16,550

The down payment was $16,550.

2. Subtract to find the mortgage amount.
   $82,750 − $16,550 = $66,200

The mortgage amount was $66,200.

3. Divide to find how many $1,000 you are borrowing.
   $66,200 ÷ $1,000 = 66.2

4. Multiply to find the monthly payment.
   THINK: Use the loan payment table. Find the monthly payment per $1,000, or $9.15.

   66.2 × $9.15 = $605.73

   You will pay $605.73 per mo.

MORTGAGE PAYMENTS PER $1,000

Interest rate	Monthly payment 20-y loan	25-y loan	30-y loan
9.0%	$9.00	$8.40	$8.05
9.5%	$9.33	$8.74	$8.41
10.0%	$9.66	$9.09	$8.78
10.5%	$9.99	$9.45	$9.15
11.0%	$10.33	$9.81	$9.53
11.5%	$10.66	$10.16	$9.90
12.0%	$11.01	$10.53	$10.29
12.5%	$11.36	$10.90	$10.67
13.0%	$11.72	$11.28	$11.06
13.5%	$12.07	$11.66	$11.45
14.0%	$12.44	$12.04	$11.85
14.5%	$12.80	$12.42	$12.25
15.0%	$13.17	$12.81	$12.64
15.5%	$13.54	$13.20	$13.05
16.0%	$13.92	$13.59	$13.45
16.5%	$14.29	$13.99	$13.86
17.0%	$14.67	$14.38	$14.26
17.5%	$15.05	$14.78	$14.67
18.0%	$15.44	$15.18	$15.08

The **closing** is the day on which you sign the mortgage papers and the contract. You will have to pay **closing costs** to the bank and others who helped process the mortgage. These closing costs may include **points.** Each point is 1% of the mortgage amount.

EXAMPLE 2 The mortgage amount was $66,200. Your closing costs were $2\frac{1}{2}$ points, $250 for the bank's attorney, and a $185 title fee. How much were your closing costs?

THINK: $2\frac{1}{2}$ **points** = $2\frac{1}{2}$**% = 0.025**

1. Multiply to find the points. 0.025 × $66,200 = $1,655
2. Add to find the closing costs. $1,655 + $250 + $185 = $2,090

Your closing costs were $2,090.

FOR DISCUSSION See TE side column.

1. How could you estimate the extra cost per month of borrowing $50,000 for 30 y at 11% instead of 10%?

2. Some people get an **adjustable rate mortgage (ARM).** As interest rates go up and down, the mortgage rates go up and down. What are the advantages and disadvantages of this type of mortgage?

PRACTICE EXERCISES

Remember to estimate whenever you use your calculator.

Find the monthly payment to the nearest cent.

Mortgage amount	Interest rate	Term	Monthly payment	
$45,000	11.5%	30 y	• 1. ■	$445.50
$89,000	10.5%	25 y	2. ■	$841.05
$76,500	9.5%	30 y	3. ■	$643.37
$67,300	10.0%	20 y	4. ■	$650.12
$98,700	10.5%	30 y	5. ■	$903.11

Complete the table to find the down payment, the mortgage amount, and the monthly payment. 6. $24,800 7. $99,200 8. $907.68 9. $8,970

Purchase price	Percent down	Down payment	Mortgage amount	Interest rate	Term	Monthly payment
$124,000	20%	• 6. ■	• 7. ■	10.5%	30 y	• 8. ■
$89,700	10%	9. ■	10. ■	10.0%	20 y	11. ■
$93,620	30%	12. ■	13. ■	11.0%	25 y	14. ■
$79,840	15%	15. ■	16. ■	9.5%	30 y	17. ■

10. $80,730 11. $779.85 12. $28,086 13. $65,534 14. $642.89 15. $11,976

Find the closing costs. 16. $67,864 17. $570.74

Mortgage amount	Points	Attorney's fees	Title fees	Closing costs	
$37,000	3	$750	$300	• 18. ■	$2,160
$57,000	1	$900	$250	19. ■	$1,720
$109,500	4	$840	$175	20. ■	$5,395
$96,450	$2\frac{1}{2}$	$575	$375	21. ■	$3,361.25
$88,750	$3\frac{1}{2}$	$465	$305	22. ■	$3,876.25

SKILLS BANK You may want to refer students to the following skills in the Skills Bank:

Divide With Decimals, Skill 1

Find the Percent of a Number, Skill 2

FOR DISCUSSION

1. Students should explain that the monthly payment per $1,000 for 11% ($9.53) is about $0.80 more than for 10% ($8.78). They can mentally multiply 50 × $0.80 to get $40 extra per month at 11%.
2. Students should explain that ARMs often have slightly lower interest rates than fixed-rate mortgages and protect the bank if interest rates increase, which is a real disadvantage. If interest rates go down, the monthly payment on this type of mortgage will decline, which is a real advantage.

ASSIGNMENTS

Level 1	1–11, 18–19
Level 2	1–22
Level 3	6–17, 21–22

FOLLOW-UP ACTIVITY

COOPERATIVE LEARNING
Divide the class into small heterogeneous groups. Give each group a page from a local newspaper that advertises houses for sale. Ask the group to choose 2 houses they would like to purchase. Explain that they will need to put 20% down and that they will get a 30-y, 10.5% mortgage for the rest. The closing costs will include 2 points and $675 in other fees. Have them compute the down payment, mortgage amount, monthly payment, and closing costs for each house.

SUPPLEMENTARY MATERIALS

TRP Practice, p. 81
TRP Reteaching, p. 66
TRP Transparency 41

OBJECTIVES

- Compute the assessed valuation of a property.
- Compute annual real estate tax.
- Compute the monthly mortgage and real estate payments to a bank.

TEACHING THE LESSON

WARM-UP Write the following exercises on the chalkboard or an overhead transparency.

Multiply.

1. 0.8 × $85,000 **2.** 0.6 × $93,500

Divide.

3. $85,000 ÷ $100 **4.** $78,560 ÷ $100

(**1.** $68,000 **2.** $56,100 **3.** 850 **4.** 785.6)

INTRODUCTION Explain to students that another sizable monthly payment for homeowners is real-estate taxes. Use the introductory material to discuss some of the services paid for by real estate taxes.

INSTRUCTION Review the definition of *assessed valuation.* Explain to students that different localities often decide on different assessment rates. So, 2 houses in different towns with the same market value might have different assessed valuations.

Use Example 1 to show students how to find the assessed valuation of a property and the annual real estate tax. Explain that the *market value* of a property is the amount for which it would probably sell. Point out that the real estate tax depends on the assessed valuation and the tax rate. Tell students that, in Step 2, they are answering the question, "How many hundred dollars are in $62,090?" The answer will thus be a decimal and not a money amount.

Use Example 2 to show students how to compute the combined monthly mortgage and real estate tax payments to a bank.

12.5 REAL ESTATE TAXES

Local governments collect **real estate taxes.** The taxes are used to pay for municipal services and schools. The real estate tax is based on the **assessed valuation** of a property. The assessed valuation is a percent of the property's **market value.** The tax rate is an amount per $100 of assessed valuation.

EXAMPLE 1 The market value of your property is $88,700. The assessment rate is 70%. What is the assessed valuation? If the real estate tax rate is $4.29 per $100, what is your annual real estate tax?

ASSESSED VALUATION = ASSESSMENT RATE × MARKET VALUE

1. Multiply to find the assessed valuation.
   THINK: 70% = 0.7 — 0.7 × $88,700 = $62,090

The assessed valuation of your property is $62,090.

2. Divide to find the number of $100 of assessed valuation. — $62,090 ÷ $100 = 620.9
3. Multiply to find the real estate tax. — 620.9 × $4.29 = $2,663.661

Your annual real estate tax is $2,663.66.

You usually pay $\frac{1}{12}$ of your annual real estate tax with your monthly mortgage payment. The bank then pays the taxes when they are due.

EXAMPLE 2 Your monthly mortgage payment is $487. Your annual real estate tax is $2,663.66. What is your combined monthy payment to the bank?

1. Divide to find the monthly tax payment. — $2,663.66 ÷ 12 = $221.97166 ≈ $221.97
2. Add to find the combined. payment. — $221.97 + $487 = $708.97

Your combined payment is $708.97.

FOR DISCUSSION See TE side column.

1. Why might the tax rate in a town increase?
2. How could a bank earn money for itself by collecting $\frac{1}{12}$ of the real estate taxes each month and holding them until the tax payment is due?

PRACTICE EXERCISES

Remember to estimate whenever you use your calculator.

Find the assessed valuation.

• 1. $76,000 2. $88,500 3. $72,900 4. $39,285 5. $76,791

Market value	$95,000	$118,000	$72,900	$87,300	$98,450
Assessment rate	80%	75%	100%	45%	78%

Find the annual real estate tax.

• 6. $2,391.96 7. $2,099.02 8. $1,975.68 9. $2,323.56 10. $5,242.56

Assessed valuation	$64,300	$51,700	$39,200	$80,400	$107,650
Tax rate per $100	$3.72	$4.06	$5.04	$2.89	$4.87

Find the combined monthly payment to the bank.

• 11. $721 12. $824 13. $878 14. $780 15. $945

Monthly mortgage payment	$413	$397	$516	$471	$507
Annual real estate tax	$3,696	$5,124	$4,344	$3,708	$5,256

Solve.

• 16. The market value of Serita's house is $125,000. The assessment rate is 70%. What is the assessed valuation? $87,500

• 17. The market value of Tony's house is $95,000. The assessment rate is 55%. What is the assessed valuation? $52,250

18. The market value of Ruben's house is $135,000. The assessment rate is 65%. The real estate tax rate is $3.29 per $100. What is the assessed valuation and his annual real estate tax? $87,750; $2,886.98

19. The market value of Kathy's house is $89,900. The assessment rate is 80%. The real estate tax rate is $4.23 per $100. What is the assessed valuation and her annual real estate tax? $71,920; $3,042.22

20. Ruben's monthly mortgage payment is $564. Use your answer to Exercise 18 to find his combined monthly payment to the bank. $804.58

21. Kathy's monthly mortgage payment is $459. Use your answer to Exercise 19 to find her combined monthly payment to the bank. $712.52

SKILLS BANK You may want to refer students to the following skills in the Skills Bank:

Multiply With Decimals, Skill 5

Divide With Whole Numbers, Skill 6

Divide With Decimals, Skill 1

FOR DISCUSSION

1. Students should explain that the tax rate in a town usually increases when costs increase or when a locality loses property from its tax rolls.
2. Students should explain that banks usually pay real estate taxes to the town twice a year. In the meantime, the bank earns interest on homeowners' money. Some people pay their taxes directly to the town or ask to receive the interest.

ASSIGNMENTS

Level 1	1–16
Level 2	1–19
Level 3	Odd 1–15, 18–21

FOLLOW-UP ACTIVITY

ENRICHMENT Present this problem. The average house in a town has a market value of $50,000. The assessment rate is 80% ($40,000 assessed valuation) and the tax rate is $4.00 per $100 ($1,600 annual taxes). The town is going to change to an assessment rate of 100%. But they still only want to collect $1,600 in taxes from the average house. What should the new tax rate be? The new assessed valuation for a $50,000 house is $50,000. Use calculators to compute $1,600 ÷ $50,000 = 0.032, or $3.20 per $100.)

SUPPLEMENTARY MATERIALS

TRP Practice, p. 82
TRP Reteaching, p. 67
TRP Transparency 41

OBJECTIVE

- Compute the coverages provided by homeowner's insurance policies.

TEACHING THE LESSON

WARM-UP Write the following exercises on the chalkboard or an overhead transparency.

Rename as a decimal.

1. 15% **2.** 23% **3.** 7%

Find the answer.

4. 20% of $45,000 **5.** 5% of $56,700

(**1.** 0.15 **2.** 0.23 **3.** 0.07 **4.** $9,000 **5.** $2,835)

INTRODUCTION Use the introductory material to review the coverages provided by homeowner's insurance. Point out that liability coverage is for when someone is injured on your property. Review this list of covered perils with students: (1) Fire or lightning; (2) Windstorm or hail; (3) Explosion; (4) Riot or civil commotion; (5) Aircraft; (6) Vehicles; (7) Smoke; (8) Vandalism; (9) Theft; (10) Falling objects; (11) Weight of ice, sleet, or snow; (12) Volcanic eruption; (13) Bursting of heat or water pipes; (14) Leakage of water from plumbing; and (15) Freezing of plumbing.

INSTRUCTION Review with students the other coverages provided when a home or condominium is covered for its full replacement value. Point out that each of the coverages is a percent of the replacement value. Use Example 1 to show students how to find the other coverage amounts.

Use Example 2 to show students how to find the amounts of the coverages for renter's insurance. Point out that the owner of the building or the condominium association takes out insurance on the rest of the property and pays for it through rent or maintenance fees. Discuss with students what might be included under additional living expenses.

12.6 HOMEOWNER'S INSURANCE

If you own a house or condominium or rent an apartment, you need **homeowner's insurance.** Homeowner's insurance pays for damage to your property and belongings and provides liability coverage in case someone is injured on your property. Most homeowner's policies cover loss or damage from all perils except flood, earthquake, and nuclear accident.

You should insure a home or condominium for its full **replacement value,** or the amount it would cost to reconstruct it if destroyed. If you insure for 100% replacement value, then your insurance company will provide these additional coverages.

Other structure (garage, etc.)	10% of house-replacement value
Personal property	50% of house-replacement value
Additional living expenses while house is being repaired or replaced	20% of house-replacement value
Trees, shrubs, plants	5% of house-replacement value

EXAMPLE 1 Your house is fully insured for its replacement value of $98,500. What other coverage would homeowner's insurance provide for losses or damages?

Multiply to find the amounts.

Other structures	(10%)	0.1 × $98,500 = $9,850
Personal property	(50%)	0.5 × $98,500 = $49,250
Living expenses	(20%)	0.2 × $98,500 = $19,700
Trees, shrubs, plants	(5%)	0.05 × $98,500 = $4,925

If you rent, you should get homeowner's insurance to cover the full replacement value of your personal property against losses due to fire or theft. Then the insurance company will provide these additional coverages.

Alterations to the apartment	10% of property-replacement value
Additional living expenses while apartment is being repaired	20% of property-replacement value

EXAMPLE 2 Gail rents and fully insures her personal property for $19,250. What other coverage would homeowner's insurance provide for losses or damage?

Multiply to find the amounts.

Alterations	(10%)	0.1 × $19,250 = $1,925
Living expenses	(20%)	0.2 × $19,250 = $3,850

FOR DISCUSSION See TE side column.

1. Statistics show that 96% of homeowners, but only 32% of renters, have insurance. Why is that?
2. What are the dangers of not having homeowner's insurance?

PRACTICE EXERCISES

Remember to estimate whenever you use your calculator.

Complete the table to show the coverages provided for the homeowner. The home is insured for its full replacement value.

Home-replacement value	Other structures	Personal property	Additional living expenses	Trees, shrubs, plants
$78,000	• 1. ■	• 2. ■	3. ■	4. ■
$109,500	5. ■	6. ■	7. ■	8. ■
$87,645	9. ■	10. ■	11. ■	12. ■

1. $7,800 2. $39,000 3. $15,600 4. $3,900 5. $10,950 6. $54,750

Complete the table to show the coverages provided for the renter. Personal property is insured for its full replacement value.

Personal property-replacement value	Alterations to apartment	Additional living expenses
$7,600	•13. ■	•14. ■
$18,250	15. ■	16. ■
$31,118	17. ■	18. ■

7. $21,900 8. $5,475 9. $8,764.50 10. $43,822.50 11. $17,529 12. $4,382.25

13. $760 14. $1,520 15. $1,825 16. $3,650 17. $3,111.80 18. $6,223.60

EXTENSION Fire Prevention

This circle graph shows the leading causes of fires in the United States.

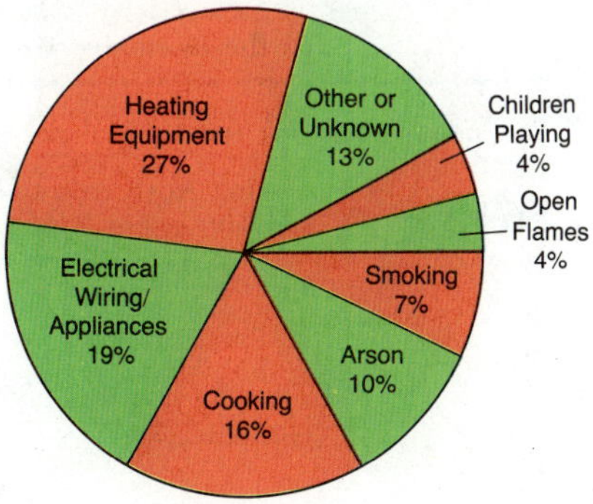

There were 868,000 fires in 1 y. About how many of them were caused by:

1. Heating equipment? 234,360
2. Cooking? 138,880
3. Children playing? 34,720
4. Smoking? 60,760
5. Based on these statistics, what steps could you take to reduce fire hazards in your home? Answers may vary.

SKILLS BANK You may want to refer students to the following skills in the Skills Bank:

Multiply With Decimals, Skill 5

Find the Percent of a Number, Skill 2

FOR DISCUSSION

1. Students should explain that most renters do not believe that they need insurance coverage because they think they are covered under the apartment's insurance policy or they do not realize how expensive it is to replace their property.
2. Students should explain that all they have worked for could be wiped out if loss or damage is uninsured. Discuss what a person would do to replace $20,000 worth of uninsured property.

EXTENSION Use this problem to make students aware of the causes of fires. Explain that the entire circle represents 100% and that the size of each section reflects the percent of fires attributed to that cause.

ASSIGNMENTS

Level 1	1–8, 13–16, EX
Level 2	1–18, EX
Level 3	5–12, 15–18, EX

FOLLOW-UP ACTIVITY

ENRICHMENT Explain that most insurance companies will replace a house if it is insured for at least 80% of its replacement value. A house has a replacement value of $80,000 and is insured for 80% of it ($64,000). The other percents remain the same, but they are based on $64,000. Ask students to find the other coverages.

SUPPLEMENTARY MATERIALS

TRP Practice, p. 83
TRP Reteaching, p. 68
TRP Transparency 42

OBJECTIVES

- Read an electric meter.
- Compute the cost of electricity.
- Compute the cost of gas.
- Compute the cost of water.

TEACHING THE LESSON

WARM-UP Write the following exercises on the chalkboard or an overhead transparency.

Estimate, then multiply. Round the answer to the nearest cent.

1. 2,967 × \$0.059 (\$180; \$175.05)

2. 3,958 × \$0.0702 (\$280; \$277.85)

INTRODUCTION Ask students to describe the utilities used in their homes or apartments. Find out how many students have gas, oil, or electric heat. Point out to students that an electric meter has a disc that spins around to show electrical usage. The faster the disc spins, the more electricity is being used. Ask students if they know where the meters are in their homes or apartments.

INSTRUCTION Explain to students that a Watt (named after James Watt, a Scottish inventor) is the basic unit of electrical energy. Ten 100-Watt light bulbs on for 1 h use 1 kWh of electricity.

Use Example 1 to show students how to read an electric meter. Explain that each dial on the meter represents a place value from 1 to 10,000. Point out that the 10,000's, 100's, and 1's dials move clockwise. The 1,000's and 10's dials move counterclockwise.

Use Example 2 to show students how to find the amount of electricity used and the cost of the electricity. Point out that in this and the following example, the utility cost is rounded to the nearest cent. Remind students that estimation to determine reasonableness is very important since the cost factor could easily be entered into the calculator incorrectly.

12.7 UTILITIES

The cost of **utilities** such as electricity, water, natural gas, or heating oil are part of your living costs. **Meters** keep track of how much electricity, water, and natural gas are used.

An electric meter shows the number of **kilowatt hours (kWh).**

1 kWh is 1,000 watts of electricity used for 1 h.

EXAMPLE 1 Read the electric meter. When the pointer is between 2 numbers, read the lower number.

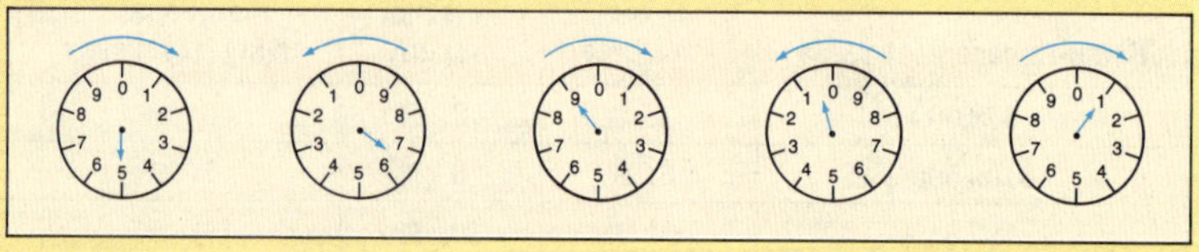

The meter reads 56,901 kWh.

You pay for electricity by the kWh.

EXAMPLE 2 On April 1, your electric meter read 54,095 kWh. On May 1, the meter read 56,901 kWh. A kWh costs \$0.0528. How much did it cost you for the electricity used from April 1 to May 1?

COST OF ELECTRICITY = COST PER KWH × KWH

1. Subtract to find the number of kWh used. 56,901 − 54,095 = 2,806
2. First estimate the cost. 3,000 × \$0.05 = \$150
3. Then multiply to find the exact cost. 2,806 × \$0.0528 = \$148.1568

The electricity cost \$148.16. The answer is reasonable since it is close to the estimate, \$150.

You usually pay for the natural gas you use in units of 100 cubic ft.

EXAMPLE 3 During April, you used 184 hundred cubic ft of gas. Gas costs \$0.6133 per 100 cubic ft. How much did it cost you for gas during April?

1. First estimate the cost. 200 × \$0.60 = \$120
2. Then multiply to find the exact cost. 184 × \$0.6133 = \$112.8472

The gas cost \$112.85. The answer is reasonable since it is close to the estimate, \$120.

You usually pay for the water you use in units of 1,000 cubic ft.

EXAMPLE 4 From July through August, you used 2,890 cubic ft of water. Water costs $14.38 per 1,000 cubic ft. How much did it cost you for water?

1. Divide to find the number of 1,000 cubic ft. $2,890 \div 1,000 = 2.89$
2. First estimate the cost. $3 \times \$14 = \$42$
3. Then multiply to find the exact cost. $2.89 \times \$14.38 = \$41.5582$

The water cost $41.56. The answer is reasonable since it is close to the estimate, $42.

FOR DISCUSSION

See TE side column.

1. How could you estimate the savings that would result from using less electricity?
2. Is more electricity used in summer or in winter? Why?

PRACTICE EXERCISES

Remember to estimate whenever you use your calculator.

Read the electric meter.

1. 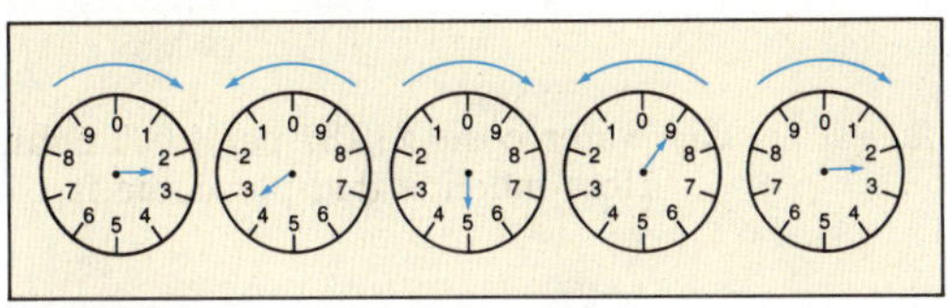

23,492 kWh

2. 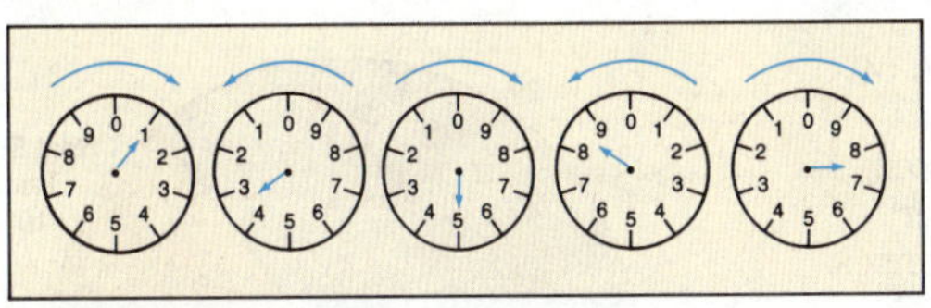

13,586 kWh

3. 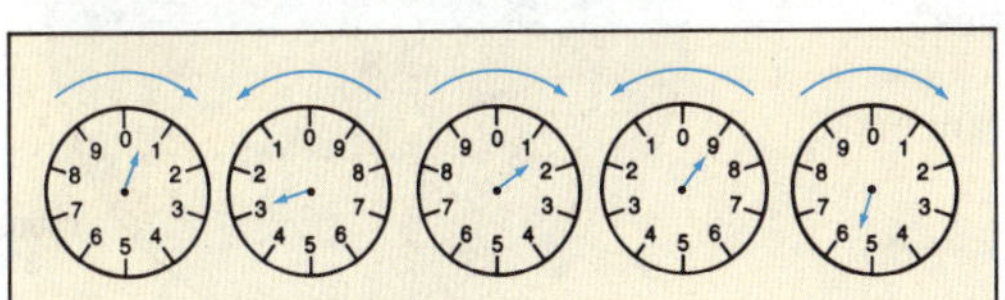

3,195 kWh

Use Example 3 to show students how to find the amount of gas used and the cost of the gas. Point out that gas is *sold* by 100 cubic ft. Emphasize the importance of estimating to be sure the answer is reasonable.

Use Example 4 to show students how to find the amount of water used and the cost of the water. Point out to students that water is *sold* by 1,000 cubic ft, and so they must divide by 1,000 before they find the cost.

SKILLS BANK You may want to refer students to the following skills in the Skills Bank:

The Decimal System, Skill 1

Round Whole Numbers and Decimals, Skill 4

Subtract With Whole Numbers and Decimals, Skill 2

Divide With Whole Numbers, Skill 1

FOR DISCUSSION

1. Students should explain that they could estimate electricity saved in kWh. For example, if a 100-Watt bulb that was left on all night (say, 10 h) is turned off during that period, then 1 kWh is saved. They should continue estimating the number of kWh they can save each day and multiply the daily savings by 30. Multiply the monthly kWh savings by the approximate cost of a kWh to estimate the monthly savings.

   The table below will help students estimate electrical energy saved.

Appliance	Watts
Air conditioner	1,350
Clothes dryer	4,850
Dishwasher	1,200
Fan	170
Hair dryer	1,000
Iron	1,100
Oven	12,000
Refrigerator	450
Television	225
Washing machine	550

2. Students should explain that more electricity is usually used in the summer because of the increased use of air conditioners and other refrigeration systems used to protect food, ice, and beverages from the heat.

EXTENSION Use this problem to help students review the relative monthly costs of owning a home. Ask students if they are surprised at any of the information in this circle graph. Point out the relatively low cost of insurance compared to the coverage it provides.

ASSIGNMENTS

Level 1	1–7, 12–19
Level 2	1–25, EX
Level 3	1–3, 8–15, 20–25, EX

FOLLOW-UP ACTIVITY

APPLICATION Obtain a representative rate for electricity from the local utility. Write this rate on the chalkboard and ask students to estimate the amount that electricity costs each day in their houses or apartments. Have each student write his or her estimate on a card and leave it with you. Then ask students to read their electric meters that night and again 24 h later. Students should compute the amount of electricity used and find the cost of the electricity. Ask students to compare the actual costs with their estimates. It may be impractical to ask students to read their own meters. In that case, base the activity on the electricity used in the school.

SUPPLEMENTARY MATERIALS

TRP Practice, p. 84
TRP Reteaching, p. 69
TRP Applications, p. 14
TRP Transparency 42

Complete the table to find the kWh used and the cost of the electricity.

Second reading	First reading	kWh used	Cost per kWh	Total cost
52,038 kWh	49,165 kWh	• 4. ■	$0.0493	• 5. ■
9,607 kWh	8,406 kWh	6. ■	$0.0603	7. ■
60,831 kWh	43,194 kWh	8. ■	$0.0238	9. ■
19,743 kWh	12,162 kWh	10. ■	$0.0537	11. ■

4. 2,873 5. $141.64 6. 1,201 7. $72.42 8. 17,637 9. $419.76 10. 7,581 11. $407.10

Find the cost of the gas.

	•12. $82.89	13. $150.76	14. $30.60	15. $153.93
100 Cubic ft used	141	159	933	222
Cost per 100 cubic ft	$0.5879	$0.9482	$0.0328	$0.6934

Complete the table to find the number of 1,000 cubic ft and the cost of the water.

Cubic ft used	1,000 Cubic ft used	Cost per 1,000 cubic ft	Total cost
3,051	•16. ■ 3.051	$14.38	•17. ■
2,134	18. ■ 2.134	$13.50	19. ■
4,892	20. ■ 4.892	$11.23	21. ■
4,809	22. ■ 4.809	$15.85	23. ■

17. $43.87 19. $28.81 21. $54.94 23. $76.22

Solve.

24. Gas costs $0.8523 per 100 cubic ft. How much is that per cubic ft? $0.008523

25. Water costs $13.87 per 1,000 cubic ft. How much is that per cubic ft? $0.01387

EXTENSION Monthly Costs of Owning a Home

This circle graph shows the percents of monthly payments the average homeowner spends on mortgage, taxes, utilities, and insurance. A homeowner spends about $1,450 per mo on expenses. About how much is spent on:

1. Mortgage? $1,073
2. Utilities? $188.50
3. Taxes? $145
4. Insurance? $43.50

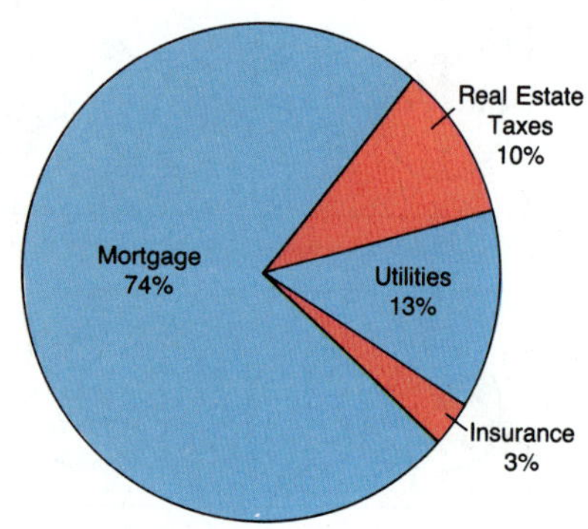

12.8 DECORATING AND REMODELING

You decide to paint the living room walls. You only want to make 1 trip to the paint store to get the paint, so you use the accepted guideline: 1 gal of paint covers about 400 square ft (including spaces for windows and doors).

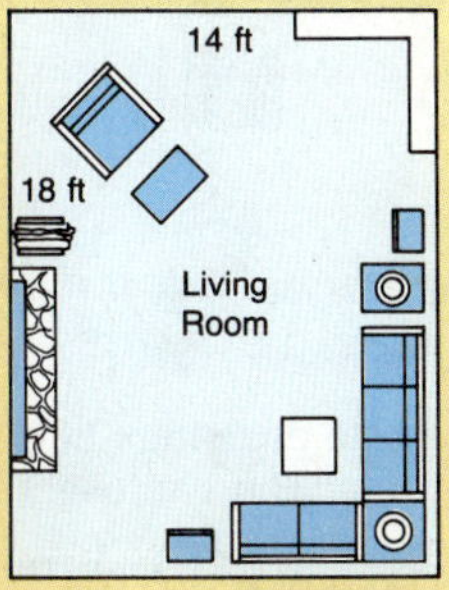

EXAMPLE 1 The ceiling in the room is 8 ft high. How many gallons of paint will you need to paint the walls? How much will the paint cost at $16.85 per gal (including tax)?

THINK: Identify the dimensions of the room.

Length = 18 ft Width = 14 ft Height = 8 ft

1. Add to find the perimeter. $14 + 18 + 14 + 18 = 64$ ft
2. Multiply to find the area of the 4 walls. $8 \times 64 = 512$ square ft
3. Divide to find the number of gallons. $512 \div 400 = 1.28$ gal

Since 1.28 gal > 1 gal, you will need 2 gal of paint.

4. Multiply to find the cost. $2 \times \$16.85 = \$33.70$

The paint will cost $33.70.

Denise hired a contractor to remodel her living room and porch. **Remodeling** means entirely rebuilding 1 or more rooms. **Contractors** often quote remodeling costs by the square foot.

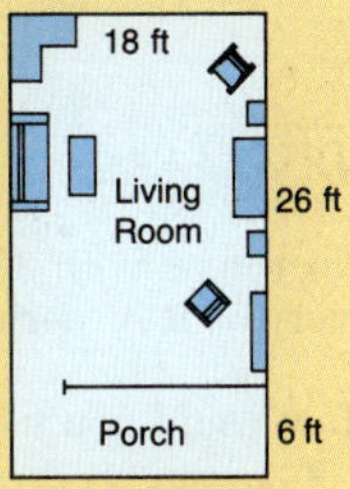

EXAMPLE 2 The contractor charged $18 per square ft to remodel the living room and $8 per square ft to remodel the porch. How much did the remodeling cost?

THINK: Identify the dimensions of the rooms.

Living room: 26 ft by 18 ft Porch: 18 ft by 6 ft

1. Multiply to find the area of each room. Living room: $26 \times 18 = 468$ square ft; Porch: $18 \times 6 = 108$ square ft
2. Multiply to find the cost of remodeling each room. Living room: $468 \times \$18 = \$8,424$; Porch: $108 \times \$8 = \$864$
3. Add to find the total cost. $\$8,424 + \$864 = \$9,288$

It cost $9,288 to remodel the living room and the porch.

OBJECTIVES

- Compute the amount of paint needed and the cost of the paint.
- Compute the cost of remodeling.
- Determine the dimensions of a scale drawing.

TEACHING THE LESSON

WARM-UP Write the following exercises on the chalkboard or an overhead transparency.

Divide.

1. $1,512 \div 400$ **2.** $660 \div 400$

Solve the proportion.

3. $\frac{1}{2} = \frac{n}{20}$ **4.** $\frac{1}{5} = \frac{n}{35}$

(**1.** 3.78 **2.** 1.65 **3.** $n = 10$ **4.** $n = 7$)

INTRODUCTION Explain to students that renters often decorate their apartments; homeowners often decorate and remodel (rebuild) their homes. Wise renters and homeowners alike need to know the projected costs of decorating and remodeling, and they will comparison shop to find the lowest price for quality materials and skilled designers and crafters.

INSTRUCTION Use Example 1 to show students how to find the number of gallons of paint they will need and the cost of the paint. Explain that they need to know the total area of all 4 walls. Point out that a wall is a rectangle, so Area = $l \times w$. Students already know that the "width" is 8 ft. Rather than finding the length of each wall, students can add the lengths of the walls to find the perimeter, as in Step 1. Then students can multiply the perimeter by 8 ft to find the combined area of the 4 walls.

Use Example 2 to show students how to determine the cost of remodeling. Entire walls or floors may be replaced and new fixtures or appliances may be installed. Remind students that they should get several price quotes. They should sign an agreement with the contractor stating exactly what will be done.

Use Example 3 to show students how to determine the dimensions of a scale drawing. Before they complete this example, students will need to review ratio and proportion. You may want to turn to the ratio and proportion sections listed below in the Skills Bank for a thorough review of ratios, rates, equal ratios, proportions, and solving proportions.

Remind students, in Step 1, that a proportion is a statement that 2 ratios are equal; that is, 1 is to 1.5 as *w* is to 9. Point out, in Step 2, that if ratios are equal, their cross products are equal. Finally, in Step 3, remind students that dividing both sides of an equation by the same quantity maintains the equality.

SKILLS BANK You may want to refer students to the following skills in the Skills Bank:

Ratios, Rates, and Equal Ratios, Skills 1–6

Proportions and Solving Proportions, Skills 1–4

FOR DISCUSSION

1. Students should explain that, in a room with 8-ft ceilings, you could buy 1 gal of paint for every 50 ft of perimeter. They should note that 8 ft by 50 ft is 400 square ft, the amount of coverage you can expect from 1 gal of interior paint.
2. Students should explain that they can make or obtain scale cutouts of furniture to match the scale drawing. The scale cutouts can be moved around the scale drawing to see if the furniture would fit. This approach can save a lot of time and unnecessary expense when buying or moving furniture. Many interior decorators (persons who help people plan decorating and remodeling) carry kits to help people decide where to put furniture.

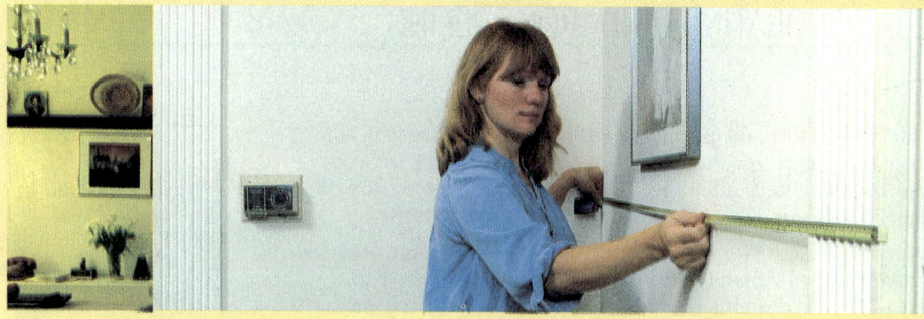

Scale drawings help people decide where to place furniture in a room. You can use a proportion to find scale dimensions.

EXAMPLE 3 A room is 12 ft long by 9 ft wide. You use a scale of 1 in. = 1.5 ft. What is the scale width of the room?

1. Write a proportion. $\frac{1}{1.5} = \frac{w}{9}$ ← **scale** ← **actual**
2. Cross multiply. $\frac{1}{1.5} = \frac{w}{9}$
   $1.5w = 9$
3. Solve for *n*. $w = 9 \div 1.5$
   $w = 6$

The scale width is 6 in.

FOR DISCUSSION See TE side column.

1. For rooms with 8-ft ceilings, why could you buy 1 gal of paint for every 50 ft of perimeter?
2. How could you use a scale drawing to decide where to put furniture?

PRACTICE EXERCISES Remember to estimate whenever you use your calculator.

Find the perimeter of the room.

1. 14 ft by 15 ft 58 ft
2. 9 ft by 12 ft 42 ft
3. 13 ft by 15 ft 56 ft
4. 12 ft by 18 ft 60 ft

How many gallons of paint are needed to paint the walls? How much will the paint cost?

5. Room: 14 ft by 15 ft
   Ceiling: 8 ft
   Paint: $13.95/gal
   2 gal; $27.90
6. Room: 16 ft by 20 ft
   Ceiling: 8 ft
   Paint: $17.05/gal
   2 gal; $34.10
7. Room: 24 ft by 38 ft
   Ceiling: 8 ft
   Paint: $10.65/gal
   3 gal; $31.95

How much will the remodeling cost?

• 8. Room: 9 ft by 12 ft
Cost: $14/square ft
$1,512

9. Garage: 14 ft by 16 ft
Cost: $9/square ft
$2,016

10. Basement: 12 ft by 24 ft
Cost: $10.50/square ft
$3,024

Find the scale dimensions of the room.

• 11. Room: 12 ft by 15 ft
Scale: 1 in. = 1.5 ft
l = 10 in.; *w* = 8 in.

12. Room: 15 ft by 20 ft
Scale: 2 in = 5 ft
l = 8 in.; *w* = 6 in.

13. Room: 12 ft by 15 ft
Scale: 5 cm = 2 ft
l = 37.5 cm; *w* = 30 cm

Use the scale drawing below for Exercises 14–19.

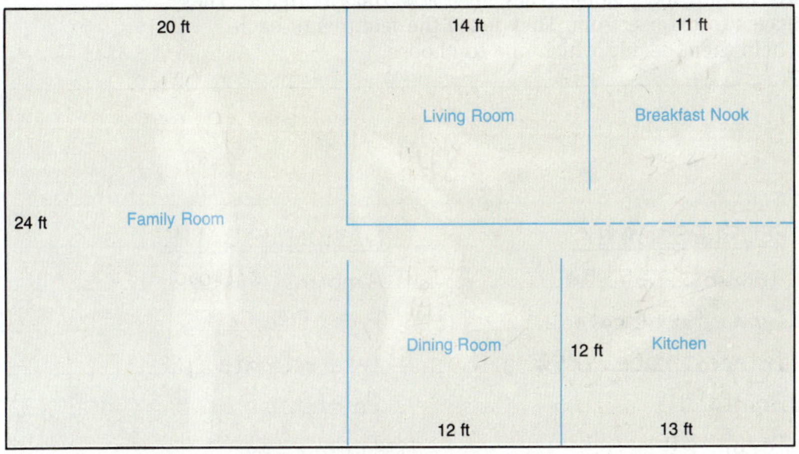

FIRST FLOOR All ceilings are 8 ft high.

14. How many gallons of paint should be bought to paint the living room walls? How much would the paint cost at $17.85 per gal?
2 gal; $35.70

15. How many gallons of paint should be bought to paint the walls of all of the rooms? How much would the paint cost at $16.35 per gal?
6 gal; $98.10

16. How much would it cost to remodel the dining room at $11.50 per square ft?
$1,656

17. How much would it cost to remodel the entire first floor at $9.20 per square ft?
$9,936

18. What would be the scale dimensions of the family room if the scale were 1 in. = 1.5 ft?
l = 16 in.; *w* = 13.3 in.

19. What would be the scale dimensions of the entire first floor if the scale were 5 cm = 2 ft?
l = 112.5 cm; *w* = 60 cm

Solve.

20. Alan wants to buy a rug for a 6 yd by 4 yd room. The rug costs $20.50 per square yd. How many square yards does he need? How much will the rug cost?
24 square yd; $492

21. Charlene wants to put a rug in her 12 ft by 15 ft bedroom. The rug costs $18.00 per square yd. How many square yards does she need? How much will the rug cost?
20 square yd; $360

ASSIGNMENTS

Level	Exercises
Level 1	Odd 1–17
Level 2	1–19
Level 3	Even 6–12, 14–21

FOLLOW-UP ACTIVITY

COOPERATIVE LEARNING
Divide the class into small heterogeneous groups. Have each group plan to decorate the classroom. Each group should find the dimensions of the room and determine how much paint would be needed to paint the walls one color and the ceiling another color. Next, they should determine the cost of the paint at $21.95 per gal. Finally, each group should determine how many square yards of carpet would be needed for the floor and the price of the carpet at $16.95 per square yd.

SUPPLEMENTARY MATERIALS

TRP Practice, p. 85
TRP Reteaching, p. 70
TRP Transparency 43

OBJECTIVE

- Decide, given relevant information, which mortgage to choose.

TEACHING THE LESSON

WARM-UP Write the following exercises on the chalkboard or an overhead transparency.

Multiply.

1. 82 × $6.37 **2.** 78.5 × $8.64

3. 300 × $608.32 **4.** 360 × $786.05

(**1.** $552.34 **2.** $678.24 **3.** $182,496 **4.** $282,978)

INTRODUCTION Present the problem situation to the class. Encourage students to name some of the factors they would use to make an informed decision about which mortgage to choose. List the factors that are named on the chalkboard.

INSTRUCTION Read the features of the 3 mortgages that Noel and Dave have prepared. Ask students to discuss the features fully and to mention those features they prefer. Ask if there is any other information that Noel and Dave should have. Discuss the decision-making factors.

Have small groups of students complete Exercises 1–13 in the table in order to compare the 3 mortgages. Ask students to write statements to describe the data now in the table. Then ask them to write conclusions that can be drawn from the data. Discuss the appropriateness of these statements and conclusions and lead a discussion about interpretations and possible misinterpretations of the data.

DECISION MAKING

12.9 CHOOSING A MORTGAGE

(Open-ended problem solving)

Banks and other lending institutions provide many different kinds of home mortgages. Listing the features of several mortgages can help you to decide which one to choose.

PROBLEM

Noel and Dave are buying a home. They need a $65,000 mortgage. They have 3 mortgages to choose from. They listed the features of each mortgage to help them decide which one to choose.

PEOPLES BANK
Amount: $65,000
Type: Fixed rate
Interest rate: 11.5%
Points: 1
Term: 30y
Other closing costs: $960

NATIONAL BANK
Amount: $65,000
Type: Fixed rate
Interest rate: 11%
Points: 2
Term: 25y
Other closing costs: $1,425

INTOWN BANK
Amount: $65,000
Type: Adjustable rate
(Interest can go up or down)
Interest rate: 10.5%
Points: 3
Term: 20y
Other closing costs: $640

DECISION-MAKING FACTORS

Annual Percentage Rate (APR)	Type of mortgage	Monthly payment
Closing costs (incl. points)	Term	Amount to be repaid

DECISION-MAKING COMPARISONS

Complete the table to compare the 3 mortgages. Use the table on page 240 to find the monthly payment.

Factors	Peoples Bank	National Bank	Intown Bank
APR	11.5%	• 1. ■ 11%	2. ■ 10.5%
Type	Fixed	• 3. ■	4. ■ Adjustable
Monthly payment	5. ■	6. ■	$649.35 (may vary)
Closing costs (incl. points)	7. ■	$2,725	8. ■ $2,590
Term	9. ■ 30 y	10. ■ 25 y	11. ■ 20 y
Amount to be repaid	12. ■	13. ■	$155,844 (may vary)

3. Fixed **5.** $643.50 **6.** $637.65 **7.** $1,610 **12.** $231,660 **13.** $191,295

MAKING THE DECISIONS

Which mortgage should Noel and David choose if they wanted

• **14.** The lowest APR? Intown Bank

15. The lowest monthly payment? National Bank

16. The shortest term? Intown Bank

17. To pay the least in closing costs? Peoples Bank

• **18.** Why might they choose an adjustable-rate mortgage?
The APR, and therefore the monthly payment, may decrease.

19. What are the dangers of choosing an adjustable rate mortgage?
The APR, and therefore the monthly payment, may increase.

20. How much less would they repay if they got their mortgage from the National Bank instead of the Peoples Bank?
$40,365

What would their monthly payments be if the adjustable rate mortgage:

21. Rose to $12\frac{1}{2}$%? $738.40

22. Fell to $9\frac{1}{2}$%? $606.45

23. Which mortgage would you choose? Why?
Answers may vary.

Have students complete Exercises 14–23 in small groups, having students explain and defend their answers and decisions. Remind students that in Exercise 23 there is no one right answer. Explain that choices will vary and even those with the same choice may have different reasons for their choices. Some students are very wary of defending their point of view among their classmates. Continually reinforce the fact that, in consumer issues, best answers depend on individual situations and preferences. Encourage all students to participate and to express their views.

ASSIGNMENTS

Level 1	1–23
Level 2	1–23
Level 3	1–23

FOLLOW-UP ACTIVITY

COOPERATIVE LEARNING
Assign students to small heterogeneous groups. Ask each group to contact a local bank or loan company for information about mortgages. Once all groups have gathered and summarized all the relevant information, each group reports its findings to the class. Then ask the class to choose the decision-making factors and organize the information about each bank or loan company under these factors. Finally, ask students to make individual decisions about which mortgage they would choose and to explain the basis for their decisions.

SUPPLEMENTARY MATERIALS

TRP Practice, p. 86
TRP Transparency 43

OBJECTIVES

- Review vocabulary
- Practice key chapter concepts and skills.

USING THE REVIEW

The Chapter Review is designed to help students prepare for taking the Chapter Test. The first section focuses on vocabulary. It requires students to select words to complete statements. The second section presents practice of key chapter skills.

Each item on the Chapter Review is referenced to the page on which the topic is taught in the Pupil's Edition. You may wish to have students refer to these pages to help review any concepts or skills they have not yet mastered.

It is suggested that students work in small-sized heterogeneous cooperative learning groups. Some cooperative learning methods that may be used are as follows:

1. After each student has independently completed the Chapter Review, a discussion should follow within each group about the solutions to the problems.
2. The group can complete the entire Chapter Review by working together to solve the problems.

End the lesson with an entire class discussion in which any questions brought up in group discussions are presented and answered.

CHAPTER REVIEW

Vocabulary Choose the letter of the word(s) that completes the sentence.

1. A building in which apartments are owned is called a ■. [238] c
   a. Rental **b.** House **c.** Condominium
2. The percent of the mortgage amount paid at the closing is called ■. [240] b
   a. Interest **b.** Points **c.** Appreciation
3. Utility companies sell electricity by the ■. [246] a
   a. Kilowatt hour **b.** Cubic foot **c.** Hour

Skills Find the answer.

4. Ed earns $1,500 per mo. What is the maximum monthly rent he should be able to afford? [234]
   $420
5. Maria's rent is $405 per mo. Utilities are $109. What is her total monthly cost? [234]
   $514
6. Lucy earns $26,500 per y. How much should she be able to borrow to buy a house? [236]
   $53,000
7. Martin's $55,000 house appreciated 15% over the last year. What is it now worth? [236]
   $63,250
8. Ron put 20% down on a $37,000 condo. How much was that? [238]
   $7,400
9. Liz put 20% down on a $75,000 house. What is her mortgage amount? [240]
   $60,000

Use the Mortgage Payments table on page 240 for Exercises 10–11.

10. Fran got a $56,000 mortgage for 25 y at 11%. How much will she pay each month? [240]
    $549.36
11. Carlos put 15% down on a $90,000 house. What is the mortgage amount and his monthly payment for a 30-y loan at 11.5%? [240]
    $76,500; $757.35
12. Sam's mortgage was $63,000. The closing costs were 2 points and $405 in fees. What were his closing costs? [240]
    $1,665
13. The market value of Monica's property is $88,000. The assessment rate is 80%. What is the assessed valuation? [242]
    $70,400
14. The assessed valuation of Jack's house is $45,000. The tax rate is $4.85 per $100. What is his annual real estate tax? [242]
    $2,182.50
15. Dawn's house used 2,093 kWh of electricity. How much did the electricity cost at $0.0427 per kWh? [246]
    $89.37
16. Diana used 172 hundred cubic ft of gas. How much did the gas cost at $0.7215 per 100 cubic ft? [246]
    $124.10
17. Carl's rectangular living room measures 18 ft by 21 ft by 8 ft high. How many gallons of paint are needed to paint the walls? [249]
    2 gal

CHAPTER TEST

Find the maximum monthly rent for the gross pay.

1. Gross monthly pay: $520
$145.60

2. Gross annual salary: $20,900
$487.67

Find the maximum to borrow for a house.

3. Gross annual income: $24,090
$48,180

4. Gross monthly income: $1,620
$38,880

Find the appreciation and the appreciated value of the house.

5. Initial value: $53,000
Appreciation: 15% per y for 1 y
$7,950; $60,950

6. Initial value: $60,900
Appreciation: 10% per y for 2 y
$12,789; $73,689

7. Find the total monthly cost.
Rent: $585 per mo
Electricity: $59 per mo
Gas: $42 per mo
Health club: $30 per mo
$716

8. Find the down payment and the total monthly payment.
Condo price: $100,980
Maintenance: $505 per mo
Financing: 15% down, $832 per mo
$15,147; $1,337

Use the Mortgage Payments table on page 240.

9. What is the monthly payment on a 25-y, $95,000 mortgage at 12%?
$1,000.35

10. The closing costs for an $83,000 mortgage were $1\frac{1}{2}$ points and $98 in fees. How much was that?
$1,343

Find the assessed valuation and the annual real estate tax.

11. Market value: $230,000
Assessment rate: 80%
Tax rate: $3.29 per $100
$184,000; $6,053.60

12. Market value: $198,900
Assessment rate: 75%
Tax rate: $4.09 per $100
$149,175; $6,101.26

Use the homeowner's insurance information on page 244.

13. A house is insured for its full replacement value of $137,000. What is the coverage for personal property?
$68,500

14. A renter's personal property is insured for its full replacement value of $23,500. What is the coverage for additional living expenses?
$4,700

Find the cost of the utility.

15. Electricity used: 3,905 kWh
Cost per kWh: $0.0398
$155.42

16. Water used: 4,287 cubic ft
Cost per 1,000 cubic ft: $12.97
$55.60

Find the answer.

17. A room measures 18 ft by 15 ft by 8 ft high. How many gallons of paint are needed to paint the walls?
2 gal

18. How much would it cost to remodel a 22 ft by 16 ft room at $8.50 per square ft?
$2,992

OBJECTIVE

- Evaluate the achievement of chapter objectives.

USING THE TEST

The Chapter Test may be used as a post-test to evaluate student achievement. However, you may wish to use the Chapter Post-Test offered in the Teacher's Resource Package or you may design your own chapter test. If this page is not used as a test, you may want to assign it as additional review or practice.

The test items are correlated to the chapter objectives in the table below.

Chapter objectives	Test items
A. Compute the maximum amount to be spent on rent.	1–2
B. Compute the maximum amount to borrow to buy a house.	3–4
C. Compute the appreciation and the appreciated value of a house.	5–6
D. Compute the total monthly cost for an apartment or a condominium.	7–8
E. Compute the monthly payments for a mortgage and the closing costs.	9–10
F. Compute the assessed valuation and the annual real estate tax of a property.	11–12
G. Compute homeowner's insurance coverages.	13–14
H. Compute the costs of utilities.	15–16
I. Compute decorating and remodeling costs.	17–18

SUPPLEMENTARY MATERIALS

TRP Ch. 12 Post-Test Form A, pp. 1–2
TRP Ch. 12 Post-Test Form B, pp. 1–2

OBJECTIVE

- Understand that you can pay less for telephone calls, depending on when you make them.

USING THE PAGE

Begin discussion by explaining to students that there are certain days and times of day, or *peak periods,* when most personal and business calls are made. Ask them if they know when these periods might be and why it might be easier and less expensive to make calls at other times.

Direct students to look over the map and 2 charts carefully. Have students work individually or in pairs to answer Exercises 1–3. Summarize the answers as follows: With reduced rates, telephone calls can cost less *and* you could make more calls or longer calls and still pay only the regular rate.

Have students work individually to answer Exercises 3–4. Then have them share their answers and reasons in a whole-class discussion.

ANSWERS/EXPLANATIONS

2. Poplar Grove to Rossi Park:
14.5¢ + (4.4¢ × 9 = 39.6¢) = 54.1¢ = 54¢ − 20% discount (10.8¢ or 11¢) = 43¢
Rossi Park to Poplar Grove:
17.5¢ + (5.9¢ × 9 = 53.1¢) = 70.6¢ = 71¢ − 40% discount (28.4¢ or 28¢) = 43¢
Both calls would cost the same.

4. Living in Mytown would be much cheaper than Edgeport in terms of telephone call costs. For example, a 4-min call would cost 9¢ in Mytown but would cost 46¢ in Edgeport (20¢ + 8.5¢ × 3).

MONEY TIPS

By calling at certain times on certain days, you can save money on local telephone calls.

LET'S LOOK AT THE FACTS

Many cities offer discount rates for local calls (when the caller and the person being called are in the same area). In some places, calls between nearby communities or counties are also considered local.

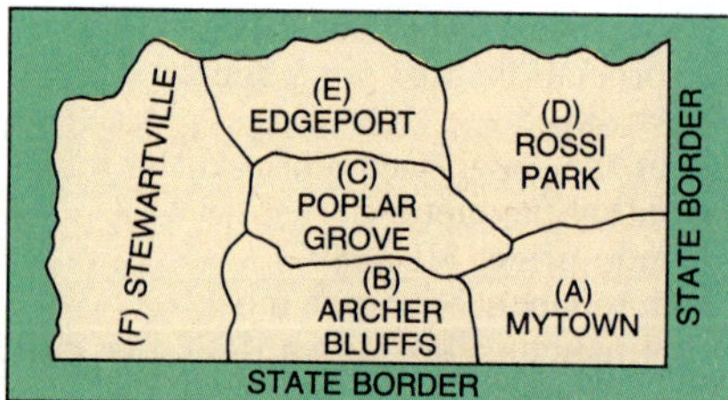

TIMED CHARGES FOR LOCAL CALLS
(Daytime Rates)

Calling Area	Initial Calling Period	Each Additional Minute
(A)	8.9¢ (4 min)	2.0¢
(B)	11.3¢ (1 min)	3.5¢
(C)	14.5¢ (1 min)	4.4¢
(D)	17.5¢ (1 min)	5.9¢
(E)	20.0¢ (1 min)	8.5¢
(F)	25.4¢ (1 min)	12.6¢

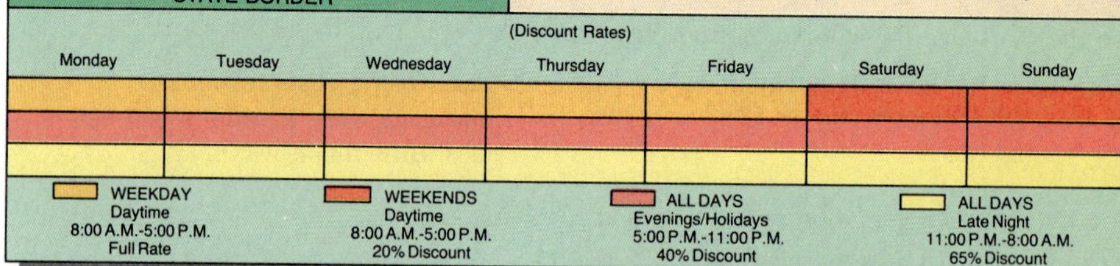

LET'S DISCUSS WHY See TE side column for answers to Exercises 2 and 4.

1. Why do you think telephone companies give discounts on local calls?
Local lines tied into central office; costs less to make the connections
2. Which would cost more: a 10-min call from Poplar Grove to Rossi Park on Saturday at 4:00 P.M. or a 10-min call from Rossi Park to Poplar Grove on Sunday at 8:00 P.M.?

LET'S SEE WHAT YOU WOULD DO

3. You live in Mytown and have your own consulting business at home. You have 1 telephone for business, as well as personal, calls. Most of your business calls are made on weekdays between 9:00 A.M. and 5:00 P.M. Should you use *timed service* or *untimed service,* which charges a flat monthly fee for a specific number of hours' calling time?
Untimed service
4. Your office is in south Edgeport. You live in Stewartville near the state border, which is 26 mi from work, so you decide to move. There is 1 apartment near your office in Edgeport and 1 just over the county line in Mytown. You make many personal phone calls. Considering this factor alone, which apartment would you take? Why?

ESTIMATION SKILLS

ESTIMATION OF QUOTIENTS USING COMPATIBLE NUMBERS

If one number can be divided by another with a remainder of 0, the numbers are called **compatible numbers.** You can estimate quotients using compatible numbers.

1. Find compatible numbers that are close in value to the given divisor and dividend.
2. Divide, using the compatible numbers.

Sometimes only the dividend needs to be changed.

Examples	4,695 ÷ 4		$93.68 ÷ 30
	↓		↓
THINK:	4,800 ÷ 4 = 1,200	**THINK:**	$90 ÷ 30 = $3
Estimate:	1,200	**Estimate:**	$3

At other times, it is necessary to change both the divisor and the dividend.

Examples	372,568 ÷ 232		762.84 ÷ 9.35
	↓ ↓		↓ ↓
THINK:	400,000 ÷ 200 = 2,000	**THINK:**	720 ÷ 9 = 80
Estimate:	2,000	**Estimate:**	80

Use compatible numbers to estimate the quotient.

1. 55 ÷ 3 20	**2.** 641 ÷ 2 300	**3.** 413 ÷ 4 100
4. 902 ÷ 32 30	**5.** 537 ÷ 61 10	**6.** 873 ÷ 22 40
7. 3,892 ÷ 16 200	**8.** 85,376 ÷ 416 200	**9.** 652,400 ÷ 822 800
10. $274.62 ÷ 5 $50	**11.** $31.45 ÷ 6 $5	**12.** $86.38 ÷ 3 $30
13. $912.98 ÷ 45 $20	**14.** $652.30 ÷ 72 $9	**15.** $4,362.00 ÷ 89 $50
16. $185.75 ÷ 12 $20	**17.** $172.30 ÷ 28 $6	**18.** $4,313.00 ÷ 22 $200
19. 45.2 ÷ 9 5	**20.** 203.4 ÷ 7 30	**21.** 131.35 ÷ 4 30
22. 23 ÷ 4.2 6	**23.** 492.6 ÷ 8.1 60	**24.** 30.2 ÷ 7.5 4
25. 0.461 ÷ 0.23 2	**26.** 28.474 ÷ 3.16 9	**27.** 29.06 ÷ 5.47 6

OBJECTIVE

- Estimate quotients, using compatible numbers.

USING THE PAGE

Compatible numbers are extremely useful when mental estimates of quotients are needed. However, there are no hard and fast rules for the procedure. For this reason, some students may have difficulty mastering the required techniques. Encourage them to practice estimating skills at every opportunity, in the classroom and elsewhere. They should be pleased to discover that as these procedures are used more and more often, they become easier and easier to apply.

Since different pairs of compatible numbers may be chosen in many division examples, any reasonable estimate should be accepted. For example, if 798.49 is divided by 9.35, either of the following estimates would be acceptable.

798.49 ÷ 9.35
THINK: 800 ÷ 10 = 80 Estimate: 80
THINK: 810 ÷ 9 = 90 Estimate: 90

It might be thought that the closest estimate is always obtained when both the divisor and the dividend are increased or both are decreased. However, this is not always the case. In the example above, the actual quotient is 85.4, so the closer of the 2 estimates is 90. However, note that the estimate 80 was obtained by increasing both the dividend and the divisor, while the estimate 90 was obtained by increasing the dividend but decreasing the divisor.

OBJECTIVES

- Review and maintain previously taught concepts and skills.
- Practice taking tests in multiple-choice format.

USING THE REVIEW

Assign the Cumulative Review to all students. Provide students with an answer sheet to record their answers.

Each Cumulative Review gives students an opportunity to practice taking tests that are written in multiple-choice format.

Provide appropriate remedial help for students having difficulty with any of the skills and concepts on this page.

CUMULATIVE REVIEW

Choose the letter of the word(s) that completes the sentence.

1. To determine which package of food is the better buy, compare the ■. a
   a. Unit prices　b. Package prices　c. Sizes　d. None of these
2. You multiply rate (in miles per hour) by time (in hours) to find ■. c
   a. Average speed　b. Mpg　c. Distance　d. None of these
3. Gross income, less any adjustments, is called ■. b
   a. Taxable income　b. Adjusted gross income (AGI)
   c. A deduction　d. None of these
4. The W-2 form is used to report ■. a
   a. Wages　b. Deductions　c. Interest　d. None of these
5. Utility companies sell electricity by the ■. b
   a. Hour　b. Kilowatt-hour　c. Cubic foot　d. None of these

Select the best estimated answer.

6. 5,672 + 6,154 + 705 d　a. 10,000　b. 11,000　c. 11,700　d. 12,700
7. 9,763 − 3,099 c　a. 5,000　b. 6,000　c. 7,000　d. 13,000
8. 61 × 398 c　a. 2,400　b. 18,000　c. 24,000　d. 240,000
9. 39,765 ÷ 21 b　a. 200　b. 2,000　c. 2,500　d. 3,000

Compute.

10. 6,437 + 9,804 + 351 c
    a. 15,592　b. 16,582　c. 16,592　d. None of these
11. 40.08 − 6.75 c
    a. 3.33　b. 33　c. 33.33　d. None of these
12. 8 × 679 b
    a. 4,879　b. 5,432　c. 5,600　d. None of these
13. $7\overline{)4.76}$ a
    a. 0.68　b. 0.71　c. 0.8　d. None of these

258 CHAPTER 12

14. 75% of 36 d

a. 24 b. 28 c. 32 d. None of these

Solve.

15. You buy milk for \$2.09 and 2.2 lb of meat at 85¢ per lb. You also have a "25¢ off" coupon for the milk. What is the total cost? b

a. \$1.62 b. \$3.71 c. \$3.96 d. None of these

16. On the first day of vacation, Monica drove 428.7 mi. She used 11 gal of gas. Estimate the mpg. b

a. 20 mpg b. 40 mpg c. 44 mpg d. None of these

17. A train leaves New York City at 9:17 A.M. (EST) and arrives in Fayetteville at 6:07 P.M. (EST). Find the traveling time. b

a. 8 h 47 min b. 8 h 50 min c. 9 h 50 min d. None of these

18. A room measures 23 ft × 16 ft × 8-ft high. If 1 gal of paint covers about 400 square ft, about how many gallons of paint are needed to paint the walls? b

a. 1 gal b. 2 gal c. 3 gal d. None of these

19. Jean has a part-time job. In 2 wk, she earned a total of \$120. During the first week, she earned \$20 more than the second week. How much did she earn in each of the 2 wk? a

a. \$70 and \$50 b. \$50 and \$70
c. \$80 and \$40 d. None of these

20. Enrico had planned to make 4 \$700 payments for his estimated federal taxes. After 1 payment, he increased his estimated federal tax due by \$1,200. How much were each of Enrico's last 3 payments? c

a. \$300 b. \$700 c. \$1,100 d. None of these

THINKING ABOUT MATH See TE side column for answers.

1. List some of the factors to consider before you buy a car. How can these factors help you decide which car to buy?

2. Pretend you are a salesperson who must travel from Orlando, Florida, to Austin, Texas. Which method of transportation should you use—plane, bus, or car? List some of the factors considered in your decision.

3. Why do people usually prefer to buy a home rather than to rent?

THINKING ABOUT MATH The questions in this section can be used in discussions with individual students or with small groups. Guide students in conveying their ideas clearly and precisely.

Listed below are expected student answers. However, accept any reasonable answers.

1. Some of the factors to consider are comfort, dealer cost, actual cost, appearance, gas mileage, safety, repair record, insurance costs, and resale value. If you know the dealer cost, you can negotiate a fair price for the car. If you know gas mileage, average repair costs, and approximate insurance costs, you can compare how much the car may cost to maintain on a yearly basis.
2. Answers may vary. Students should explain that the salesperson should consider time, total cost of transportation (ticket price, fuel, lodging, meals, and so on) and flexibility of schedule.

SUPPLEMENTARY MATERIALS

TRP Cum. Test, Ch. 1–12, pp. 1–4

Objective

- Test the prerequisite skills needed to learn the concepts developed in Chapter 13.

Using the Test

The Pre-Skills Test is designed to diagnose students' strengths and weaknesses on prerequisite skills necessary to study the mathematics in Chapter 13.

Assign the Pre-Skills Test. Allow students to work together in pairs or in small groups. Group members should help those who demonstrate a misunderstanding of a concept or a weakness in a skill.

The table below correlates the items on the Pre-Skills Test with the prerequisite skill and the lesson(s) in the chapter for which it is needed.

Items	Prerequisite skill	Lesson(s)
1–4	Multiply money amounts.	13.1–13.5, 13.7
5–11	Multiply fractions and mixed numbers.	13.1, 13.4
12–14	Estimate quotients	13.4
15–18	Divide whole numbers.	13.3–13.5
19–22	Add or subtract fractions and mixed numbers.	13.2, 13.3
23–30	Convert customary units of length.	13.1, 13.2
31–34	Find perimeter and area.	13.1, 13.4, 13.5

References to specific skills that are reviewed in the Skills Bank are given on the individual lessons themselves.

Chapter 13 Trade Industries

Pre-Skills Test

Multiply.

1. $145 \times \$50$ $7,250
2. $1{,}500 \times \$75$ $112,500
3. $180 \times 36¢$ $64.80
4. $225 \times 56¢$ $126
5. $20 \times 5\frac{1}{2}$ 110
6. $8 \times 6\frac{2}{3}$ $53\frac{1}{3}$
7. $150 \times 4\frac{2}{5}$ 660
8. $80 \times 5\frac{3}{4}$ 460
9. $2 \times \frac{2}{3} \times 16$ $21\frac{1}{3}$
10. $2 \times \frac{1}{3} \times 8$ $5\frac{1}{3}$
11. $4 \times \frac{1}{2} \times 14$ 28

Estimate the quotient.

12. $3{,}326 \div 400$ 8
13. $1{,}807 \div 290$ 6
14. $2{,}680 \div 300$ 9

Divide and round to the nearest tenth.

15. $750 \div 9$ 83.3
16. $3{,}873 \div 15$ 258.2
17. $2{,}580 \div 27$ 95.6
18. $1{,}900 \div 12$ 158.3

Add or subtract.

19. $4\frac{1}{2} + \frac{3}{8}$ $4\frac{7}{8}$
20. $\frac{5}{8} + 5\frac{3}{4}$ $6\frac{3}{8}$
21. $3\frac{1}{3} - \frac{2}{3}$ $2\frac{2}{3}$
22. $4\frac{5}{8} - \frac{3}{4}$ $3\frac{7}{8}$

Convert to feet.

23. 15 yd 45 ft
24. $5\frac{1}{3}$ yd 16 ft
25. 24 yd 2 ft 74 ft
26. 17 yd 8 ft 59 ft

Convert to yards.

27. 18 ft 6 yd
28. 288 ft 96 yd
29. 100 ft $33\frac{1}{3}$ yd
30. 85 ft $28\frac{1}{3}$ yd

Find the perimeter.

31.
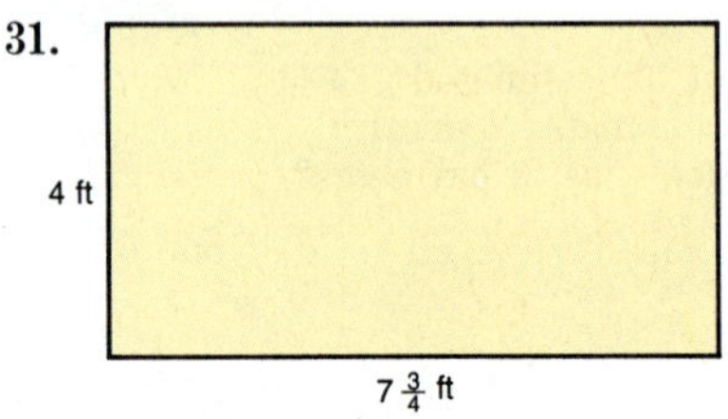

$23\frac{1}{2}$ ft

32.
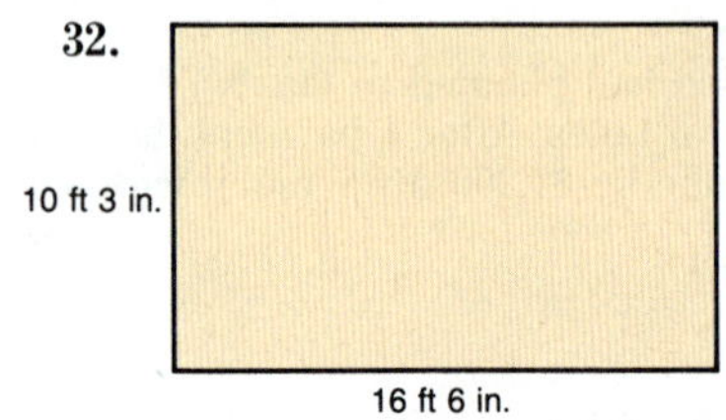

53 ft 6 in.

Find the area.

33.
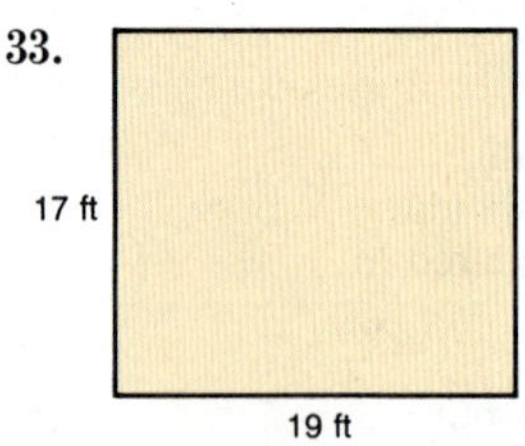

323 square ft

34.
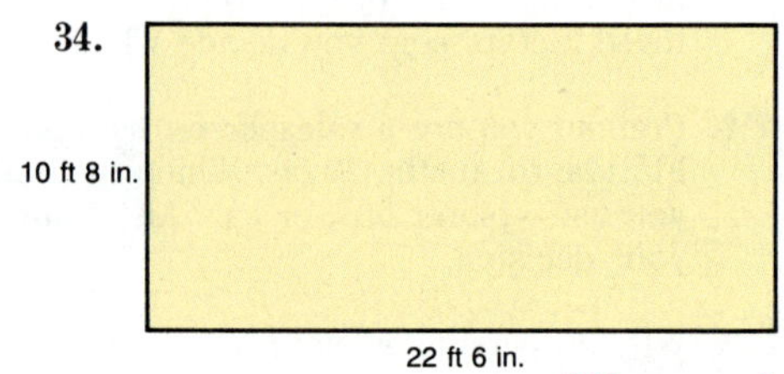

240 square ft

Chapter 13
TRADE INDUSTRIES

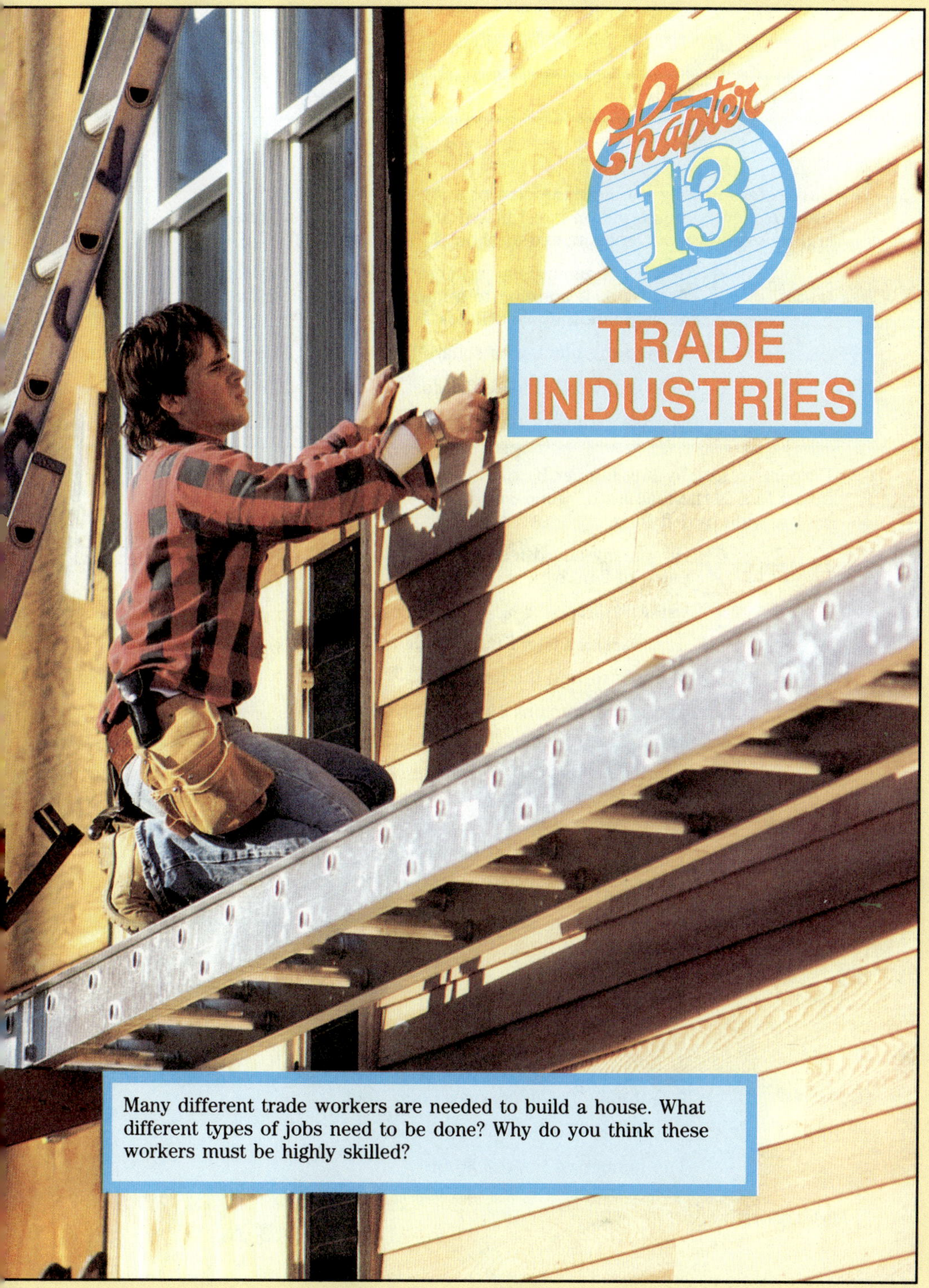

Many different trade workers are needed to build a house. What different types of jobs need to be done? Why do you think these workers must be highly skilled?

OBJECTIVE

- Formulate problems, using text and data.

USING THE CHAPTER OPENER

Discuss students' answers to the questions on the page. Some answers to the first question might include carpenters, electricians, plumbers, masons, and so on. The workers must be highly skilled because their work is part of a complex situation in which the parts are dependent on each other. Work must progress in a timely fashion so the next group can come in and do their work. In addition, the parts must all come together as planned, so workers must be able to follow plans precisely.

COOPERATIVE LEARNING EXPERIENCE

Assign heterogeneous groups and tell students they will be assuming the role of carpenters for this project. The job they have just received is to build a dog house for a Great Dane. The customers are very wealthy and want a large enough house that will look nice on their property near their large colonial-style home.

Each group is to design a dog house, draw some form of blueprint for it, estimate the amount of lumber needed, and prepare an estimate for the customer. Encourage them to be creative, but realistic, in their approach to the problem. They should use real prices for lumber and other supplies. You may want to remind them that they should be making a reasonable profit on the job.

Have the class decide which plan the customer would most likely choose and discuss why.

SUPPLEMENTARY MATERIALS

TRP Group Projects, p. 13

OBJECTIVES

- Estimate the costs of carpentry projects.
- Compute board feet and the costs of lumber orders.

TEACHING THE LESSON

WARM-UP Write the following exercises on the chalkboard or an overhead transparency.

Multiply.

1. $2 \times \frac{1}{2} \times 32$ **2.** $4 \times \frac{1}{3} \times 18$

3. $2 \times \frac{2}{3} \times 14$ **4.** $2 \times \frac{5}{6} \times 10$

5. $24 \times 5\frac{1}{2}$ **6.** $30 \times 18\frac{2}{3}$

(**1.** 32 **2.** 24 **3.** $18\frac{2}{3}$ **4.** $16\frac{2}{3}$ **5.** 132 **6.** 560)

INTRODUCTION Begin the discussion by asking students what a carpenter does. List the various tasks on the chalkboard. Then ask students how mathematics is needed to complete these tasks. Remind students that carpenters make frequent use of fractions and measurement.

INSTRUCTION Use Example 1 to show how carpenters use estimation to approximate the cost of a construction project. Explain that the \$50- and \$75-per-square-ft figures include labor and materials and must be adjusted depending on the quality and complexity of construction.

Ask students to check the accuracy of these approximations in their own neighborhood. Remind students that, at \$50 per square ft, a new 2,000-square-ft home should cost about \$100,000 plus the price of the land.

Use Example 2 to explain the concept of board feet. Explain that a *board foot* is a unit that allows lumber companies to price all similar type lumber uniformly, regardless of the size to which it is cut. Point out that a board foot is really a measure of volume. Review the conversion of inches to feet and the multiplication of fractions with students.

13.1 CARPENTRY

Carpenters usually estimate the cost of a job on a square-foot basis. New residential construction is estimated at about \$50 per square ft and renovation work is estimated at about \$75 per square ft.

EXAMPLE 1 Charles Atamian is a carpenter who submits an estimate to build a 24 ft by 30 ft addition to a house. At \$50 per square ft, how much does he estimate the job will cost?

1. Find the number of square ft (area). 24 ft × 30 ft = 720 square ft
2. Multiply to estimate the cost. 720 × \$50 = \$36,000

Charles estimates the addition will cost \$36,000.

Carpenters usually buy lumber by the **board foot.** A board foot is the amount of wood in a board 1-in. thick by 1-ft wide by 1-ft long.

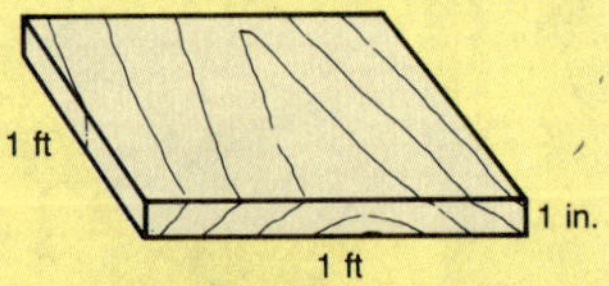

EXAMPLE 2 How many board feet are in 1 16-ft-long piece of 2 in. by 4 in. lumber?

THINK: Thickness: 2 in. Width: 4 in. Length: 16 ft

Since 4 in. = $\frac{1}{3}$ ft, the board is 2 in. by $\frac{1}{3}$ ft by 16 ft.

Multiply to find the board feet. $2 \text{ in.} \times \frac{1}{3} \text{ ft} \times 16 \text{ ft} = \frac{32}{3} = 10\frac{2}{3}$ board ft

The board contains $10\frac{2}{3}$ board ft.

EXAMPLE 3 Charles needs 36 16-ft-long 2 in. by 8 in. boards for the rafters and 60 8-ft-long 2 in. by 4 in. boards for the wall studs. At 36¢ per board ft, how much will the lumber cost?

1. Find the number of board feet in each size of lumber.

   $2 \text{ in.} \times 8 \text{ in.} \times 16 \text{ ft} = 2 \text{ in.} \times \frac{2}{3} \text{ ft} \times 16 \text{ ft} = \frac{64}{3} = 21\frac{1}{3}$ board ft

   $2 \text{ in.} \times 4 \text{ in.} \times 8 \text{ ft} = 2 \text{ in.} \times \frac{1}{3} \text{ ft} \times 8 \text{ ft} = \frac{16}{3} = 5\frac{1}{3}$ board ft

2. Multiply by the number of boards.

   36 boards × $21\frac{1}{3}$ board ft = 768 board ft

   60 boards × $5\frac{1}{3}$ board ft = 320 board ft

3. Add. Total = 1,088 board ft
4. Multiply by the cost per board foot.

   1,088 board ft × \$0.36 = \$391.68

The lumber will cost \$391.68.

FOR DISCUSSION See TE side column.

1. Why is the cost of renovation work often higher than the cost of new construction?

2. What is the difference between 1 board foot and 1 linear, or straight-measured, foot of lumber?

PRACTICE EXERCISES Remember to estimate whenever you use your calculator.

Find the estimated cost. Estimate new construction at $50 per square ft and renovation work at $75 per square ft.

• 1. New: 9 ft by 12 ft $5,400
2. New: 20 ft by 24 ft $24,000
3. New: 28 ft by 40 ft $56,000

• 4. Renovation: 10 ft by 15 ft $11,250
5. Renovation: 18 ft by 30 ft $40,500

6. Renovation: 20 ft by 45 ft $67,500
7. Renovation: 50 ft by 64 ft $240,000

Complete the table.

Board size	Length of board	Number of boards	Number of board feet	Cost per board foot	Total cost
1 in. by 6 in.	12 ft	2	• 8. ■	37.5¢	• 9. ■
2 in. by 4 in.	10 ft	24	10. ■	27¢	11. ■
2 in. by 6 in.	16 ft	15	12. ■	32.5¢	13. ■
1 in. by 10 in.	6 ft	8	14. ■	40¢	15. ■
2 in. by 8 in.	14 ft	30	16. ■	29.5¢	17. ■

8. 12 9. $4.50 10. 160 11. $43.20 12. 240 13. $78.00 14. 40 15. $16.00 16. 560 17. $165.20

Complete to find the total cost of the purchase.

RAY'S LUMBER COMPANY			
Qty.	Item	Unit Cost	Total
10	2 in. by 6 in. by 12 ft board	30¢/board foot	• 18. ■ $36
18	2 in. by 4 in. by 12 ft board	30¢/board foot	• 19. ■ $43.20
6	4 ft by 8 ft plywood	$14.95/sheet	20. ■ $89.70
4 pkg.	gray shingles	$19.75/pkg.	21. ■ $79
5 lb	common nails	62¢/lb	22. ■ $3.10
		Total	23. ■ $251

Use Example 3 to demonstrate how to find the cost of a large order of lumber. Remind students to convert each size of board to board feet.

SKILLS BANK You may want to refer students to the following skills in the Skills Bank:

Multiply With Fractions and Mixed Numbers, Skills 3–4

Customary Units of Length, Skill 3

FOR DISCUSSION

1. Students should explain that, since renovation work often includes removal of old work as well as replacement with new work, it is often more expensive and more time consuming.
2. Students should explain that a board foot is a unit of volume, while a linear foot is a unit of length. Since 1 board foot is the amount of wood in a 1-ft length of board 1 in. thick by 1 ft wide, a linear foot of either 1 in. by 12 in. or 2 in. by 6 in. is the same as a board foot.

ASSIGNMENTS

Level 1	1–13
Level 2	1–23
Level 3	Odd 1–7, 12–23

FOLLOW-UP ACTIVITY

COOPERATIVE LEARNING
Divide students into small heterogeneous groups. Provide each group with prices from a local lumber yard. Ask each group to plan the construction of a backyard deck or similar project. Ask them to make a complete list of the materials they will need and the total cost.

SUPPLEMENTARY MATERIALS

TRP Practice, p. 87
TRP Reteaching, p. 71
TRP Transparency 44

OBJECTIVES

- Compute inside and outside diameters of pipes.
- Compute the costs of plumbing supplies and labor charges.

TEACHING THE LESSON

WARM-UP Write the following exercises on the chalkboard or an overhead transparency.

Compute.

1. $2\frac{5}{8}+\frac{3}{4}+1\frac{1}{2}$ **2.** $3\frac{3}{4}-\frac{5}{16}$

3. $14\times 8\frac{3}{4}$ **4.** $27\times 12\frac{1}{2}$

(**1.** $4\frac{7}{8}$ **2.** $3\frac{7}{16}$ **3.** $122\frac{1}{2}$ **4.** $337\frac{1}{2}$)

INTRODUCTION Begin the discussion by asking students what a plumber does. List the various tasks on the chalkboard. Then ask students how mathematics is needed to complete these tasks. Remind students that plumbers make frequent use of fractions and measurement.

INSTRUCTION Use Example 1 and the pipe diagrams to show how to find inside and outside diameters of pipes. Remind students that the outside diameter is equal to the inside diameter plus twice the thickness.

Use Example 2 to show how plumbers might estimate the amounts of materials they need. Remind students always to round lengths up to assure enough material is ordered. Point out that when Rachel buys 3 coils, there will be some left over since 20 ft is an overestimate of $18\frac{1}{2}$ ft. Provide students with other examples, using different standard lengths to show how estimation makes ordering easier. (For example: How many 12-ft lengths should be ordered for a job requiring 20 5-ft lengths?)

Use Example 3 to demonstrate how to find the cost of larger orders of plumbing supplies plus labor charges. Point out to students how the Memory Add and Memory Recall keys on their calculators can be used to make this type of computation easier.

13.2 PLUMBING

Rachel O'Neal is a **plumber.** She installs and repairs sinks, showers, bath tubs, and toilets.

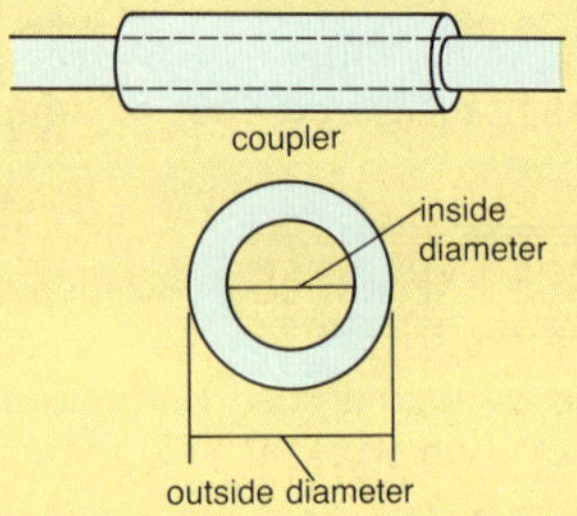

EXAMPLE 1 Rachel needs to connect 2 pieces of plastic drain pipe with a plastic coupler. The pipe has an inside diameter of $1\frac{5}{8}$ in. and a thickness of $\frac{1}{4}$ in. Find the required inside diameter of the coupler.

THINK: Inside diameter of the coupler = Outside diameter of the pipe = Inside diameter of the pipe plus twice the pipe's thickness

Add to find the diameter of the coupler.

$$1\frac{5}{8}\text{ in.}+\frac{1}{4}\text{ in.}+\frac{1}{4}\text{ in.}=1\frac{5}{8}\text{ in.}+\frac{2}{8}\text{ in.}+\frac{2}{8}\text{ in.}=1\frac{9}{8}\text{ in.}=2\frac{1}{8}\text{ in.}$$

The inside diameter of the coupler must be $2\frac{1}{8}$ in.

EXAMPLE 2 To install an underground sprinkler system, Rachel needs 15 $18\frac{1}{2}$-ft-long sections of plastic pipe. If the pipe comes in 100-ft coils, about how many coils will Rachel need?

THINK: $18\frac{1}{2}$ ft is about 20 ft.
$100 \div 20 = 5$, so Rachel can get about 5 sections per coil.

5 sections per coil means about 15 sections from 3 coils.
Rachel will need 3 100-ft coils of pipe.

EXAMPLE 3 Rachel needs 12 20-ft lengths of copper pipe, 20 copper couplers, and 8 elbow joints to install plumbing in a new house. Pipe costs 35¢ per ft, couplers cost 55¢ each, and elbows cost 85¢ each. She charges $30 per h for labor. How much will the 48-h job cost?

1. Multiply to find the cost of materials.	12×20 ft $\times$ \$0.35/ft = \$84.00
	$20 \times$ \$0.55 = \$11.00
	$8 \times$ \$0.85 = \$ 6.80
2. Add to find the total cost of materials.	\$101.80
3. Multiply to find the labor charges.	$48 \times$ \$30.00 = \$1,440
4. Add to find the total cost.	\$101.80 + \$1,440 = \$1,541.80

The total cost of the job will be $1,541.80.

FOR DISCUSSION See TE side column.

1. Why does a 10-ft piece of $\frac{3}{4}$-in. copper tubing cost more than a 10-ft piece of $\frac{1}{2}$-in. copper tubing?
2. Why does doubling the diameter of a pipe quadruple the pipe's capacity?

PRACTICE EXERCISES

Remember to estimate whenever you use your calculator.

Complete the table of pipe measurements.

Inside diameter	Thickness of pipe	Outside diameter	
$1\frac{1}{4}$ in.	$\frac{1}{16}$ in.	• 1. ■	$1\frac{3}{8}$ in.
$2\frac{1}{8}$ in.	$\frac{3}{16}$ in.	• 2. ■	$2\frac{1}{2}$ in.
$3\frac{3}{4}$ in.	$\frac{3}{8}$ in.	3. ■	$4\frac{1}{4}$ in.

Inside diameter	Thickness of pipe	Outside diameter	
4. ■	$\frac{1}{8}$ in.	$2\frac{1}{4}$ in.	2 in.
5. ■	$\frac{5}{16}$ in.	$3\frac{11}{16}$ in.	$3\frac{1}{16}$ in.
$\frac{3}{4}$ in.	6. ■	$1\frac{1}{8}$ in.	$\frac{3}{16}$ in.

Complete the table to estimate the coils of pipe needed.

Number of pieces needed	Length of each piece	Total length	Number of 20-ft sections needed	Number of 100-ft coils needed
12	$8\frac{1}{2}$ ft	• 7. ■	• 8. ■ 6	9. ■ 2
26	16 ft 9 in.	10. ■	11. ■ 22	12. ■ 5
15	26 ft 4 in.	13. ■	14. ■ 20	15. ■ 4

7. 102 ft 10. 435 ft 6 in. 13. 395 ft

Find the total cost of the job. Assume labor charges of $24.00 per h.

16. Materials: 50 ft copper pipe at 27¢/ft; 6 couplers at 59¢ each; 4 elbows at 79¢ each; 3 valves at $4.99 each
Labor: $3\frac{1}{2}$ h
$119.17

17. Materials: 250 ft plastic pipe at 12¢/ft; 80 ft copper pipe at 32¢/ft; 2 toilets at $120 each; 3 sinks at $85.50 each; 3 faucets at $27.50 each
Labor: 18 h
$1,066.60

EXTENSION Leaky Faucets

The kitchen faucet in the Powells' house drips once every 5 s. If there are 100 drops per oz, how many gallons of water can be wasted in 1 wk? In 1 y?
9.45 gal; 491.4 gal

SKILLS BANK You may want to refer students to the following skills in the Skills Bank:

Add With Fractions and Mixed Numbers, Skill 5

Customary Units of Length, Skill 3

FOR DISCUSSION

1. Students should explain that, since the $\frac{3}{4}$-in. tubing is thicker, there is more metal used per foot of pipe, thereby increasing the cost.
2. Students should explain that the pipe's capacity is determined by its cross-sectional area. Area is determined by the square of the radius. When the diameter of a circle doubles, so does the radius, and its area increases four times.

EXTENSION Use this problem to show how to create a series of equal rates that can be multiplied to solve a problem.

$$\frac{1 \text{ drop}}{5 \text{ s}} \times \frac{100 \text{ drops}}{1 \text{ oz}} \times \frac{60 \text{ s}}{1 \text{ min}} \times \text{etc.}$$

ASSIGNMENTS

Level 1	1–12, 16
Level 2	1–17
Level 3	4–17, EX

FOLLOW-UP ACTIVITY

ENRICHMENT Supply each student with a copy of a plumbing bill, including quantity of itemized materials, unit costs, and labor charges. Ask students to create 10 questions, each involving some computation that can be answered using the information on the bill. Have students trade their questions with another student and then check the accuracy of the answers.

SUPPLEMENTARY MATERIALS

TRP Practice, p. 88
TRP Reteaching, p. 72
TRP Transparency 44

OBJECTIVE

- Compute the quantities and costs of electrical supplies and labor charges.

TEACHING THE LESSON

WARM-UP Write the following exercises on the chalkboard or an overhead transparency.

Compute.

1. $(5 \times \$4.95) + (8 \times \$2.50)$
2. $(10 \times \$3.85) + (6 \times \$3.75)$
3. $6.5 \times \$24.50$
4. $4.75 \times \$19.40$

(**1.** \$44.75 **2.** \$61.00 **3.** \$159.25 **4.** \$92.15)

INTRODUCTION Begin the discussion by asking students what an electrical contractor does. List the various tasks on the chalkboard. Then ask students how mathematics is involved in any of these tasks. Look around the classroom and the corridor and list the devices that require electricity and must be wired by electrical contractors.

INSTRUCTION Use Example 1 to demonstrate how to estimate the amount of wire needed for a job. Remind students that they should round up to get an overestimate to be sure they have enough wire.

Use Example 2 to show how to compute the cost of electrical supplies. Remind students that the memory keys on their calculators are very useful for calculations like these, which involve the sum of products.

Use Example 3 to show how to compute the labor charges for electricians. Point out that fractional parts of hours should be renamed as decimals to compute labor charges on a calculator. You may also wish to point out that, while skilled labor is expensive, electricians and other skilled tradespeople must pay for their tools, vehicles, and insurance out of their labor charges.

13.3 ELECTRICAL CONTRACTING

Tom Williams is an **electrical contractor.** He wires houses and buildings to distribute the flow of electricity. Electricians estimate the materials they will need to complete a job.

EXAMPLE 1 If Tom averages 75 ft of wire per room when wiring a new home, how many 100-ft coils of wire should be purchased for a 9-room house?

1. Multiply to estimate the total amount of wire. $9 \times 75 \text{ ft} = 675 \text{ ft}$
2. Divide to find the number of coils. Round up to be sure there is enough. $675 \div 100 = 6.75$ coils; $6.75 \approx 7$ coils

Tom should purchase 7 coils of wire.

EXAMPLE 2 For an addition to a house, Tom ordered 200 ft of #14 electrical-cable wire, 9 outlet boxes, 3 switches, and 6 3-prong receptacles. Find the total cost of the materials if the wire costs \$39.50 per 100 ft, the outlet boxes are \$4.65 each, the switches are \$3.75 each, and the receptacles are \$3.25 each.

1. Multiply to find the total cost of each item.

wire	$2 \times \$39.50 =$	\$79.00
outlet boxes	$9 \times \$4.65 =$	\$41.85
switches	$3 \times \$3.75 =$	\$11.25
receptacles	$6 \times \$3.25 =$	\$19.50

2. Add to find the total cost. \$151.60

The materials will cost \$151.60.

EXAMPLE 3 For most jobs, Tom charges \$24.75 per h for labor for himself and \$15.45 per h for his assistant. What is the labor charge for a job on which they work together for $7\frac{1}{2}$ h?

1. Multiply to find the total charges.

electrician	$7.5 \text{ h} \times \$24.75/\text{h} = \$185.625 \approx \$185.63$
assistant	$7.5 \text{ h} \times \$15.45/\text{h} = \$115.875 \approx \$115.88$

2. Add to find the total. \$301.51

The total labor charge is \$301.51.

FOR DISCUSSION See TE side column.

1. Why do most building codes require that electrical work be done by licensed electricians?

2. What purpose do fuses and circuit breakers play in a building's electrical system?

PRACTICE EXERCISES

Remember to estimate whenever you use your calculator.

Complete the table.

Average number of feet of wire per room	Number of rooms	Number of 100-ft coils needed	Cost per 100 ft	Total cost	
75	3	• 1. ■ 3	$29.95	• 2. ■	$89.85
65	5	3. ■ 4	$32.50	4. ■	$130.00
90	6	5. ■ 6	$39.65	6. ■	$237.90
120	8	7. ■ 10	$45.25	8. ■	$452.50
80	11	9. ■ 9	$37.80	10. ■	$340.20

Complete the table. Assume that 1 d = 8 h.

Time worked	Number of electricians	Hourly labor rates	Total labor charge	
$6\frac{1}{2}$ h	1; 1 assistant	$21.00; $15.00	•11. ■	$234.00
15 h	1; 2 assistants	$18.73; $12.00	12. ■	$640.95
$12\frac{3}{4}$ h	1; 2 assistants	$37.25; $25.85	13. ■	$1,134.12
5 d	2; 2 assistants	$27.45; $18.50	14. ■	$3,676.00
2 d 3 h	2; 2 assistants	$38.35; $27.50	15. ■	$2,502.30

Find the total cost of the job. Assume labor charges of $28.50 per h for 1 electrician.

16. Parts: 800 ft #12 wire at $26.79 per 100 ft; 400 ft #8 wire at $19.85 per 100 ft; 12 outlet boxes at $7.45 each; 6 receptacles at $5.35 each; 6 outlet covers at $1.19 each

    Labor: $7\frac{3}{4}$ h

    $643.24

17. Parts: 300 ft #14 wire at $31.95 per 100 ft; 5 junction boxes at $9.50 each; 12 switches at $4.87 each; 2 dimmers at $3.95 each; 25 insulators at 16¢ each

    Labor: $5\frac{1}{2}$ h

    $370.44

SKILL BANKS You may want to refer students to the following skills in the Skills Bank:

Add With Whole Numbers and Decimals, Skill 6

Multiply With Decimals, Skill 5

FOR DISCUSSION

1. Students should explain that many house fires are caused by faulty and/or overloaded electrical circuits. It is therefore extremely important that all electrical work be done by licensed electricians.
2. Students should explain that fuses and circuit breakers are designed to cut off the flow of electricity if too much amperage is being drawn. Since drawing too much amperage can heat electrical lines to dangerous levels, fuses and circuit breakers are important safety features on all electrical systems.

ASSIGNMENTS

Level 1	1–6, 11–13, 16
Level 2	1–17
Level 3	5–10, 13–17

FOLLOW-UP ACTIVITY

APPLICATION Invite an electrician or an electrical contractor to speak to the class about how he or she uses mathematics on the job. Ask students to develop a list of questions before the visit.

SUPPLEMENTARY MATERIALS

TRP Practice, p. 89
TRP Reteaching, p. 73
TRP Transparency 45

OBJECTIVES

- Find the areas of walls.
- Determine the amount of paint needed for a job.
- Compute the cost of paint and labor charges.

TEACHING THE LESSON

WARM-UP Write the following exercises on the chalkboard or an overhead transparency.

Find the area.

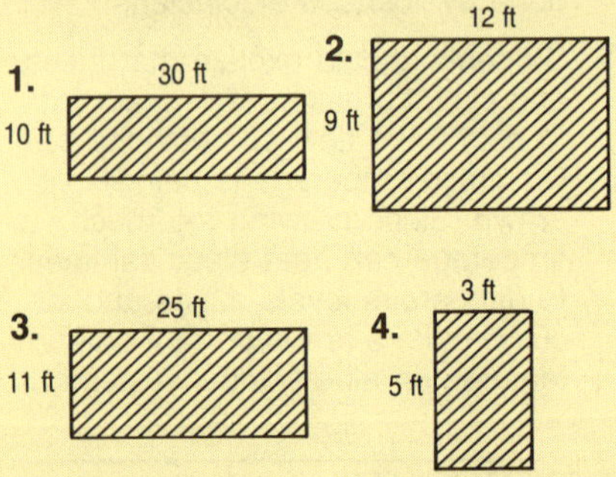

(**1.** 300 square ft **2.** 108 square ft
3. 275 square ft **4.** 15 square ft)

INTRODUCTION Begin the discussion by asking students what a painter does. List the various tasks on the chalkboard. Then ask students how mathematics is needed to complete these tasks.

INSTRUCTION Use Example 1 to demonstrate how to find the total area, the area to be painted, and the cost of paint. Point out that it would be helpful to draw a picture so that students can see that the areas of the windows must be subtracted from the total area to find the area that needs to be painted.

Remind students that paint-coverage data is used to estimate how much paint is needed. Remind students that 2 coats are usually needed when painting a lighter color over a darker color. Also point out that less expensive paints are more likely to require a second coat than more expensive paints.

13.4 PAINTING

Sam and Rick Owens are professional **painting contractors.** They need to order the correct amount of paint and to estimate the costs of jobs.

One gallon of most interior paint covers about 400 square ft with 1 coat.

EXAMPLE 1 Sam and Rick have been contracted to paint two 11 ft by 30 ft walls. Each wall contains two 3 ft by 5 ft windows and a 3 ft by 7 ft door. Sam and Rick will use 2 coats of paint. How many gallons of paint will they need? At $12.75 per gal, how much will the paint cost?

1. Find the area to be painted.
   - Find the total area. — $2 \times 11 \text{ ft} \times 30 \text{ ft} = 660$ square ft
   - Find the unpainted area. — $4 \times 3 \text{ ft} \times 5 \text{ ft} = 60$ square ft
     $2 \times 3 \text{ ft} \times 7 \text{ ft} = 42$ square ft
     102 square ft
   - Subtract to find the painted area. — $660 - 102 = 558$ square ft
   - Multiply to find the area for 2 coats of paint. — $2 \times 558 = 1{,}116$ square ft
2. Divide to find the number of gallons needed. — $1{,}116 \div 400 = 2.79$ gal

Sam and Rick will need to buy 3 gal of paint.

3. Multiply to find the cost. — $3 \times \$12.75 = \$38.25$

The paint will cost $38.25.

EXAMPLE 2 Sam and Rick usually work together and charge $20.00 per h each for their labor. It took them $3\frac{1}{2}$ h to paint the 2 walls. What was the labor cost?

THINK: 2 people $\times 3\frac{1}{2} \times \$20.00$ per h

Multiply. $2 \times 3\frac{1}{2} \times \$20 = \$140.00$

Their labor charge was $140.00.

FOR DISCUSSION See TE side column.

Why does a gallon of interior paint usually cover more area than a gallon of exterior paint?

PRACTICE EXERCISES

Remember to estimate whenever you use your calculator.
See page 477 for answers to Exercises 1–25.

Complete the table. Assume that each wall contains two 3 ft by 5 ft windows and a 3 ft by 7 ft door. Assume that paint costs $16.50 per gal.

Dimensions of wall(s)	Total area	Unpainted area	Painted area	Number of coats	Number of gallons needed	Cost of paint
one 9 ft by 12 ft	• 1. ■	• 2. ■	3. ■	1	4. ■	5. ■
one 15 ft by 25 ft	6. ■	7. ■	8. ■	1	9. ■	10. ■
one 8 ft by 16 ft	11. ■	12. ■	13. ■	2	14. ■	15. ■
two 24 ft by 40 ft	16. ■	17. ■	18. ■	2	19. ■	20. ■
two 40 ft by 70 ft	21. ■	22. ■	23. ■	2	24. ■	25. ■

Complete the table. Assume that 1 d = 8 h.

Number of painters	Time worked per painter	Cost per hour per painter	Total labor cost	
2	12 h	$14.00	• 26. ■	$336.00
5	$2\frac{1}{2}$ d	$16.50	27. ■	$1,650.00
3	4 d 3 h	$21.15	28. ■	$2,220.75
12	5 d	$18.75	29. ■	$9,000.00

Use the diagram of the barn for Exercises 30–34. What is the area of:

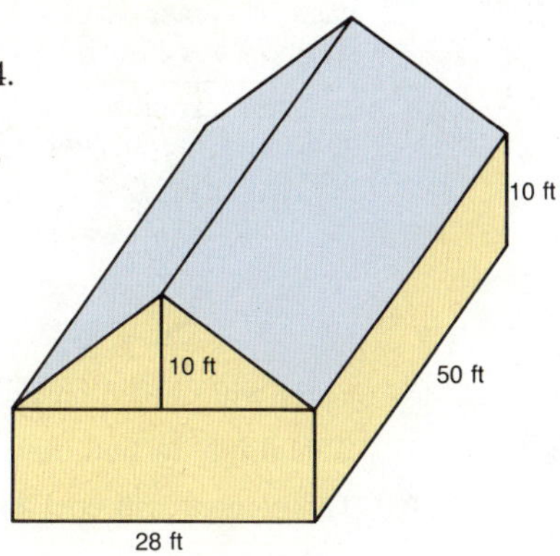

• **30.** the longer side? 500 square ft

31. the shorter side? 280 square ft

32. the triangular rise? 140 square ft

33. 2 longer sides, 1 shorter side, and the triangular rise? 1,420 square ft

34. One gallon of exterior paint covers about 250 square ft. How many gallons of paint will be needed to cover the area in Exercise 33? 6 gal

Use Example 2 to calculate the labor charges of a painting job. Remind students that fractional parts of hours could be renamed as decimals and easily entered into their calculators.

SKILLS BANKS You may want to refer students to the following skills in the Skills Bank:

Area, Skill 1

Multiply With Fractions and Mixed Numbers, Skill 3

FOR DISCUSSION Students should explain that exterior surfaces are usually more porous or rougher than interior surfaces, and therefore require more paint per square foot.

ASSIGNMENTS

Level 1	1–15, 26–28
Level 2	1–34
Level 3	16–25, 28–34

FOLLOW-UP ACTIVITY

COOPERATIVE LEARNING
Select a corridor in your school and ask students, working in heterogeneous groups of 2 or 3, to determine how much paint would be needed to repaint the corridor. Provide students with measuring tools and tell them to deduct doorways, locker spaces, and bulletin boards when finding the area. Assume that 1 gal of paint will cover about 400 square ft.

SUPPLEMENTARY MATERIALS

TRP Practice, p. 90
TRP Reteaching, p. 74
TRP Transparency 45

OBJECTIVES

- Compute the number of bricks and the amount of cement needed for a job.
- Compute the costs of bricks and cement.

TEACHING THE LESSON

WARM-UP Write the following exercises on the chalkboard or an overhead transparency.

Multiply.

1. 4 × \$6.75 **2.** 6 × \$2.75

3. 1,500 × \$0.15 **4.** 2,200 × \$0.15

(**1.** \$27.00 **2.** \$16.50 **3.** \$225.00 **4.** \$330.00)

INTRODUCTION Begin the discussion by asking students what a mason does. List the various tasks on the chalkboard. Then ask students how mathematics is needed to complete these tasks.

Tell students the dimensions of the standard brick pictured on the page. Ask students to draw the 3 different faces found on a brick with labeled dimensions. Discuss how masons can use these 3 different faces to create many interesting designs with brick.

INSTRUCTION Use Example 1 to demonstrate how to compute the number of bricks that are needed and how much they will cost.

Use Example 2 to show how a proportion can be used to determine the number of bags of cement needed. Make sure that students understand the equal ratios involved. Again remind students of the need to round up to assure enough materials.

Discuss the fact that concrete is a mixture of cement, sand, and gravel and the proportions of these ingredients vary, depending on the use of the concrete.

13.5 MASONRY

Carla Mendez is a **mason.** She uses bricks, cement blocks, and mortar to construct walls and fireplaces. Masons use mathematics to order materials, to compute the cost of jobs, and to complete projects accurately and according to plans.

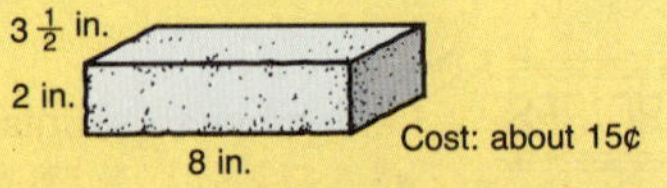

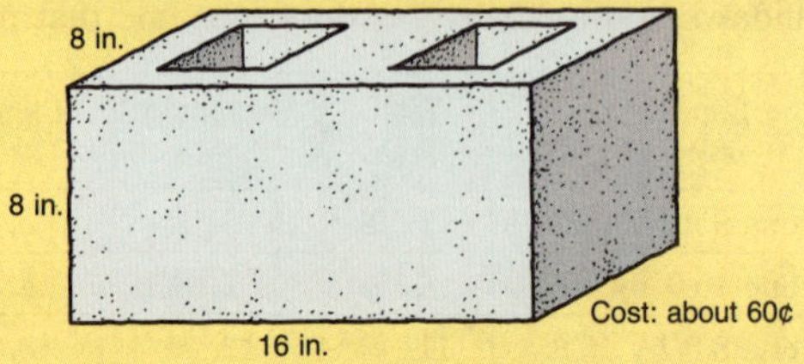

Masons use standard-sized bricks that measure 8 in. by 2 in. by $3\frac{1}{2}$ in. They use standard-sized concrete blocks that measure 16 in. by 8 in. by 8 in.

EXAMPLE 1 Carla is contracted to construct a 249-square-ft brick facade on the front of a new house. If Carla uses 9 bricks per square ft, how many bricks should she order? At 15¢ per brick, how much will the bricks cost?

1. Multiply to find the number of bricks.

249 square ft × 9 bricks/square ft = 2,241 bricks

Carla should order 2,241 bricks.

2. Multiply to find the cost. 2,241 × \$0.15 = \$336.15

The bricks will cost \$336.15.

EXAMPLE 2 Carla has found that mixing 2 bags of cement with 5 cubic ft of sand makes enough mortar for about 100 square ft of brickwork. How many bags of cement will she need for the 249 square ft-job in Example 1? At \$6.75 per bag, how much will the cement cost?

1. Use a proportion to find the number of bags. Round the answer up.

THINK: **2 bags is to 100 square ft as *n* bags is to 249 square ft.**

$$\frac{2 \text{ bags}}{100 \text{ square ft}} = \frac{n \text{ bags}}{249 \text{ square ft}}$$
$$100n = 2 \times 249$$
$$100n = 498$$
$$n = 498 \div 100 = 4.98 \approx 5$$

Carla will need 5 bags of cement.

2. Multiply to find the cost. 5 × \$6.75 = \$33.75

The cement will cost \$33.75.

270 CHAPTER 13

FOR DISCUSSION See TE side column.

1. Why do masons usually order more bricks than their computations indicate?

2. Why are many fireplaces constructed primarily of cement blocks with bricks placed only on the visible parts?

PRACTICE EXERCISES

Remember to estimate whenever you use your calculator.

Complete the table. Assume that concrete blocks cost 60¢ each and that you use 10 blocks per square yd.

Total area	Number of blocks needed	Cost of blocks	
12 square yd	• 1. ■ 120	• 2. ■	$72
38 square yd	3. ■ 380	4. ■	$228
327 square yd	5. ■ 3,270	6. ■	$1,962
1,624 square yd	7. ■ 16,240	8. ■	$9,744

Complete the table. Assume that bricks cost 15¢ each and that you use 9 bricks per square ft. Assume that you need 2 bags of cement for 100 square ft and that cement costs $7.25 per bag.

9. 1,080 **10.** $162

Total area	Number of bricks needed	Cost of bricks	Number of bags of cement needed	Cost of cement	
120 square ft	• 9. ■	• 10. ■	11. ■ 3	12. ■	$21.75
375 square ft	13. ■	14. ■	15. ■ 8	16. ■	$58
2,120 square ft	17. ■	18. ■	19. ■ 43	20. ■	$311.75
11,250 square ft	21. ■	22. ■	23. ■ 225	24. ■	$1,631.25

13. 3,375 **14.** $506.25 **17.** 19,080 **18.** $2,862 **21.** 101,250 **22.** $15,187.50

Use the diagram of the house for Exercises 25–27.

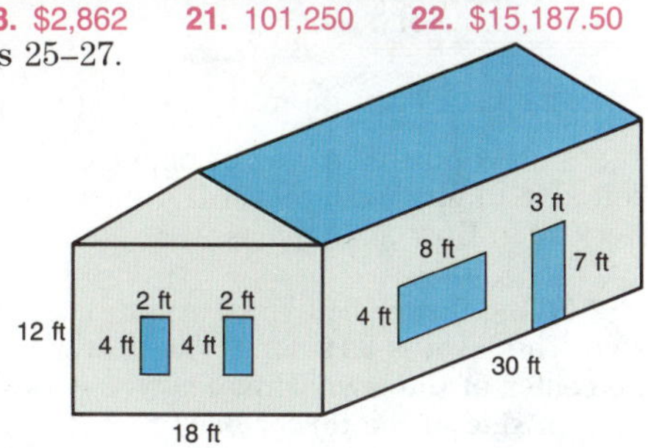

25. What is the area of the longer side of the house, excluding windows and doors? The shorter side?
307 square ft; 200 square ft

26. At 9 bricks per square ft, how many bricks will be needed to face the 2 visible sides?
4,563 bricks

27. At 2 bags of cement per 100 square ft, how many bags of cement will be needed for the job in Exercise 26?
11 bags

SKILLS BANK You may want to refer students to the following skills in the Skills Bank:

Multiply With Decimals, Skill 5

Ratios, Rates, and Equal Ratios Skills 1, 4, 5

Proportions and Solving Proportions, Skills 1, 3

FOR DISCUSSION

1. Students should explain that masons sometimes need to break bricks to fill in smaller places. By overestimating the number of bricks, a mason is also taking into account that many will become damaged.

2. Students should explain that concrete blocks are much less expensive, much stronger, and far easier to install. They are used to build a structurally stronger, less expensive fireplace. The bricks are then added for decoration.

ASSIGNMENTS

Level 1	1–4, 17–24
Level 2	1–26
Level 3	5–8, 17–27

FOLLOW-UP ACTIVITY

ENRICHMENT Provide students with a price list for bricks, concrete blocks, and cement. Tell students that they have been contracted to build a 4-ft high by 75-ft long by 1-ft wide retaining wall between the school and the athletic fields. Explain that the wall will have a concrete block interior and brick siding. Ask students to draw a diagram of the wall—including all dimensions—and to compute the number and cost of all materials.

SUPPLEMENTARY MATERIALS

TRP Practice, p. 91
TRP Reteaching, p. 75
TRP Transparency 46

OBJECTIVE

- Solve problems by drawing a diagram.

TEACHING THE LESSON

WARM-UP Write the following exercises on the chalkboard or an overhead transparency.

Compute.

1. $21 + 12$ (33) **2.** $45 - 12$ (33)

3. $8 + 14 + 8$ (30)

4. $18\frac{1}{2} + 18\frac{1}{2} + 18\frac{1}{2}$ $\left(55\frac{1}{2}\right)$

INTRODUCTION Provide practice in illustrating problem situations. Have students draw diagrams that show (1) a rectangle that measures 5 ft by 3 ft, and (2) a square that measures 6 yd on each side. Ask students to find the perimeter and area of each figure.

INSTRUCTION Present the problem situation to the class. Explain to students that it is often easier to "visualize" a problem by drawing a diagram of the situation. Repeat the 4 steps on the chalkboard. Have a student count the number of posts needed. (16) Emphasize the importance of labeling the diagram carefully.

SKILLS BANK You may want to refer students to the following skill in the Skills Bank:

Perimeter, Skill 1

FOR DISCUSSION

1. Students should explain that they would draw a series of rows of flowers (probably represented by dots). They would label or color each row according to the pattern: red, yellow, red, yellow, and so on.
2. Students should explain that they would continue the diagram to include 9 rows, the last row being red.

PROBLEM Solving STRATEGY

13.6 DRAWING A DIAGRAM

Situation:

Mike owns a landscape-contracting company. Mike's crew built a fence around a square piece of land that measured 4 yd on each side. The fence posts were placed 1 yd apart. How many posts did Mike need?

Strategy:

Drawing a diagram can help you to solve a problem.

Applying the Strategy:

1. Draw a square.
2. Mark the 4 corner posts.
3. Mark posts 1 yd apart.
4. Count the number of posts needed.

Mike needed 16 posts.

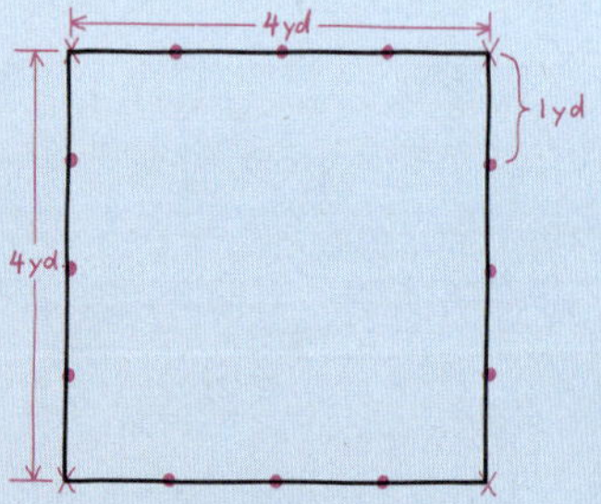

FOR DISCUSSION

See TE side column.

Mike designed a flower bed for a customer. The first row of flowers was red. The second row of flowers was yellow. This pattern was repeated.

1. What would you draw to help you see the flower bed?
2. What color were the flowers in the ninth row?

PRACTICE EXERCISES

Copy the diagram. Then fill in the missing numbers to solve the problem.

1. Mike's crew planted a row of pine trees on the 60-ft side of Mr. Wesley's yard. A pine was planted every 10 ft. How many pines were planted?
   7 pines

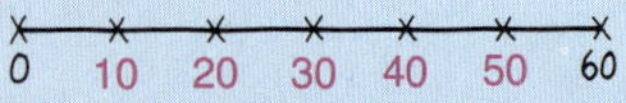

2. A 12-ft-long flower bed was cut into a rectangular lawn. The lawn is 32 ft long. The flower bed is in the center of the lawn. How many feet of lawn are on each side of the flower bed?
   10 ft

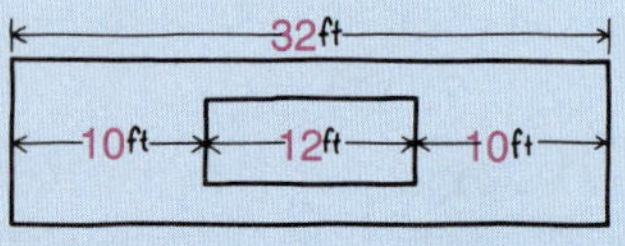

Solve the problem by drawing a diagram. Check students' diagrams.

3. Mike's crew built a fence around a square piece of land. The land was 9 yd on each side. The fence posts were placed 3 yd apart. How many posts were used?
12 posts

4. Mike planned a garden with 15 rows of flowers. The first row was blue, the second was white, the third was pink. How many times was this pattern repeated?
5 times

5. An azalea garden contains 49 azaleas, 7 in each row. The center and end azaleas in each row are pink. How many pink azaleas are there in all?
21 pink azaleas

6. A row of trees was planted along Maple Drive. The length of the row was 148 ft. A tree was planted every $18\frac{1}{2}$ ft. How many trees were planted?
9 trees

7. Mike built a fence around a rectangular garden that was 16 yd long and 12 yd wide. The posts were placed 2 yd apart. How many posts did he need?
28 posts

8. A rectangular lawn contains three 7-ft-square flower beds. There is a border of 3 ft between each flower bed and the perimeter of the plot. How long is the plot?
33 ft

ASSIGNMENTS

Level 1	1–5
Level 2	1–7
Level 3	3–8

FOLLOW-UP ACTIVITY

SITUATIONAL PROBLEM SOLVING Present students with the following problem: Your class is assigned the task of landscaping a new wing of classrooms for your school. You are chairperson of the project and responsible for reporting the class's ideas to the principal.

Some questions to ask include: What committees will be needed? How many trees and shrubs will you need? What will the project cost? How should you present your ideas to the principal?

Some approaches to solving the problem include measuring the distance around the new wing, drawing a diagram to show the placement of the trees and shrubs, assigning the committees the task of pricing the work you want done, and submitting the diagram and cost surveys to the principal.

SUPPLEMENTARY MATERIALS

TRP Practice, p. 92
TRP Applications, p. 15
TRP Transparency 46

OBJECTIVE

- Decide, given relevant information, what type of quality or quantity to purchase in light of price differences.

TEACHING THE LESSON

WARM-UP Write the following exercises on the chalkboard or an overhead transparency.

1. 15 × $11.99 **2.** 24 × $15.88

3. 750 × 8.25¢ **4.** 1,500 × 12.75¢

(**1.** $179.85 **2.** $381.12 **3.** $61.88 **4.** $191.25)

INTRODUCTION Present Problem A to the class. Encourage students to name some of the factors they would use to make an informed decision about which type of flooring to order. List the factors that are named on the chalkboard and discuss how to find out information on each of these factors.

INSTRUCTION Read the descriptions of the 3 types of flooring. Ask students to discuss the features of each type fully and to mention those features they find most appealing. Ask if there is any other information that Sandra should have. Discuss the decision-making factors.

Have small groups of students complete Exercises 1–8 in the table in order to compare the 3 options. Ask students to write statements that describe the data now in the table. Then ask them to write conclusions that can be drawn from the data. Discuss the appropriateness of these statements and conclusions and lead a discussion about interpretations and possible misinterpretations of the data.

DECISION MAKING

13.7 QUALITY AND QUANTITY VS. PRICE

(Open-ended problem solving)

Many goods are produced and sold in a variety of qualities. It is common to see items advertised as "Good" or "Everyday Grade" at lower prices and "Deluxe" or "Our Best" at higher prices.

PROBLEM A

Sandra's flooring contractor showed her samples of 3 grades of vinyl flooring for her 3 yd by 4 yd kitchen. She needs to decide which grade to order.

GOOD VINYL FLOORING	STYLISH VINYL FLOORING	DELUXE VINYL FLOORING
$8.99/square yd	$10.99/square yd No-Wax Finish	$12.99/square yd Never needs waxing 10-year guarantee

DECISION-MAKING FACTORS

Cost
Maintenance/upkeep
Quality
Guarantees

DECISION-MAKING COMPARISONS

Compare the 3 options by completing the table.

Factor	Good flooring	Stylish flooring	Deluxe flooring
Total cost	• 1. ■	$131.88	2. ■
Quality	Average	• 3. ■	4. ■
Maintenance	Needs waxing	5. ■	6. ■
Guarantees	7. ■	None	8. ■

$107.88; $155.88
Very good; Excellent
No-wax finish; never needs wax
None; 10 y

MAKING THE DECISIONS

Which flooring should Sandra order:

• **9.** If cost were the only factor? Good

• **10.** If quality were the only factor? Deluxe

11. If ease of maintenance *and* cost were important factors? Stylish

12. Which flooring could she expect to last the longest? Why? Deluxe; highest quality, 10-y guarantee

13. How much could Sandra save by using the Good flooring instead of the Deluxe flooring to do the kitchen? What could cause her not really to save this much over the next 10 y? $48; The Good may not last 10 y and she might need to replace it again at $8.99 or more per squ

14. Which type of flooring would you order? Why? Answers may vary.

PROBLEM B

Many contractors have the opportunity to save money by **buying in bulk.** A mason might find brick available at 3 different rates, depending on the quantity he orders. He must decide the best way to order the bricks based on the following pricing information.

LOOSE BRICKS—16¢ each	BRICKS ON A PALLET—$49.00 (400 bricks/pallet)	TRUCKLOAD OF BRICKS—$430.00 (10 pallets/truckload)

DECISION-MAKING FACTORS

Cost

Minimum purchase/number needed

Availability of storage space

Ease of transport

DECISION-MAKING COMPARISONS

Compare the 3 options by completing the table.

12.25¢; 10.75¢

400 bricks; 4,000 bricks

Factor	Loose	Pallet	Truckload
Cost per brick	16¢	• 15. ■	16. ■
Minimum purchase of bricks	None	• 17. ■	18. ■
Need for storage	19. ■	Moderate	20. ■
Ease of transport	21. ■	22. ■	Low

None; High

High; Moderate

MAKING THE DECISIONS

Which way should the mason order the bricks:

• 23. If cost were the only factor? Truckload

• 24. If he only needs 150 bricks? Loose

25. If transportation is not a factor and he needs 500 bricks? Pallet and Loose

26. How much can the mason save by ordering a pallet instead of 400 loose bricks? $15

27. How much can the mason save by ordering a truckload instead of 10 pallets? $60

28. What is the least expensive way to order 2,000 bricks? How much will they cost? Pallet; $245

29. Suppose the mason needs 3,600 bricks for a job. What is the best way to order the bricks? How much will they cost? Truckload; $430

30. What hidden costs might be involved before assuming that buying bricks by the truckload is least expensive? Storage, transportation, loading and unloading costs

31. Which way would you order the bricks if you needed 300? 3,000? 5,000? Why? Answers may vary. 300: loose; 3,000: 7 pallets + 200 loose; 5,000: 1 truckload + 2 pallets + 200 loose

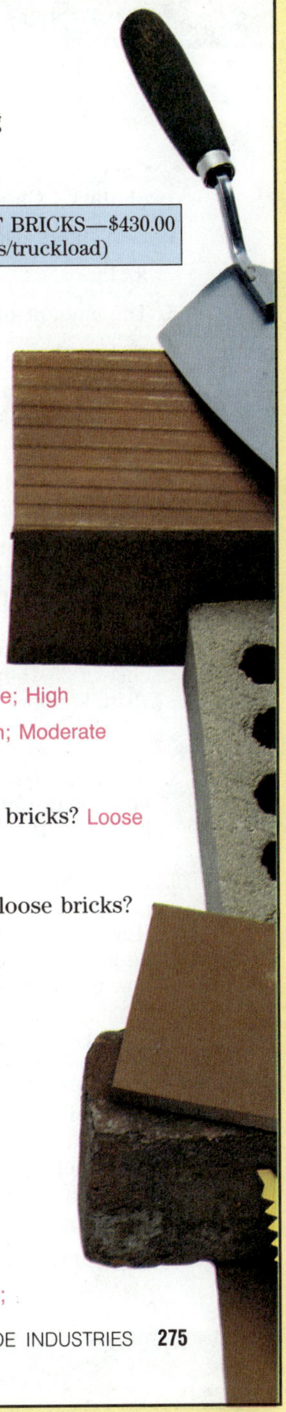

Ask students to complete Exercises 9–14 in small groups, having them explain and defend their answers and decisions. Remind students that in Exercise 14 there is no one right answer. Explain that choices will vary and that even those with the same choice may have different reasons for their choices. Continually reinforce the fact that, in many consumer issues, best answers depend on individual situations and preferences. Encourage all students to participate and to express their opinions.

Repeat this process for Problem B, which involves the purchase of variable quantities of brick.

ASSIGNMENTS

Level 1	1–31
Level 2	1–31
Level 3	1–31

FOLLOW-UP ACTIVITY

APPLICATION Ask students to collect advertisements that demonstrate price differences based on quality variation and price differences based on quantity variation. Ask students to mount the ads on paper and summarize their findings in a paragraph that describes how much can be saved under what conditions.

SUPPLEMENTARY MATERIALS

TRP Practice, p. 93
TRP Transparency 47

OBJECTIVES

- Review vocabulary.
- Practice key chapter concepts and skills.

USING THE REVIEW

The Chapter Review is designed to help students prepare for taking the Chapter Test. The first section focuses on vocabulary. It requires students to select words to complete statements. The second section presents practice of key chapter skills.

Each item on the Chapter Review is referenced to the page on which the topic is taught in the Pupil's Edition. You may wish to have students refer to these pages to help review any concepts or skills they have not yet mastered.

It is suggested that students work in small-sized heterogeneous cooperative learning groups. Some cooperative learning methods that may be used are as follows:

1. After each student has independently completed the Chapter Review, a discussion should follow within each group about the solutions to the problems.
2. The group can complete the entire Chapter Review by working together to solve the problems.

End the lesson with an entire class discussion in which any questions brought up in group discussions are presented and answered.

CHAPTER REVIEW

Vocabulary Choose the letter of the word(s) that completes the sentence.

1. Carpenters usually pay for lumber by the ■. [262] b

   a. Piece b. Board foot c. Ton

2. The amount of paint needed is most dependent on ■. [268] b

   a. Length b. Area c. Volume

3. A skilled person who works with bricks, cement blocks, and mortar is called a(n) ■. [270] c

   a. Plumber b. Electrical contractor c. Mason

Skills Find the answer.

4. At \$55 per square ft, how much will it cost to add a 16 ft by 24 ft playroom to a house? [262]
   \$21,120
5. At \$75 per square ft, how much will it cost to renovate 35,000 square ft of abandoned factory space? [262]
   \$2,625,000
6. At 32¢ per board ft, what will be the cost of 25 16-ft-long pieces of 2 in. by 10 in. lumber? [262]
   \$213.33
7. How thick is a pipe with an inside diameter of $2\frac{3}{8}$ in. and an outside diameter of $2\frac{3}{4}$ in.? [264] $\frac{3}{16}$ in.
8. If you need 25 12-ft sections of pipe, how many 100-ft coils do you need to buy? [264]
   3 100-ft coils
9. A plumber charges \$28 per h for labor. How much will a 24-h job cost? [264]
   \$672
10. An electrician averages 75 ft of wire per room when wiring a new home. How many 100-ft coils of wire should be purchased for a 6-room house? [266]
    5 coils
11. An electrician charges \$29.75 per h for his own labor and \$18.75 per h for his assistant's. What is the total labor charge for a $6\frac{1}{2}$-h job? [266]
    \$315.26
12. What is the area to be painted on two 16 ft by 30 ft walls that each contain one 3 ft by 5 ft window and a 3 ft by 7 ft door? [268]
    888 square ft
13. If a gallon of paint covers about 350 square ft, how much paint is needed to put 2 coats on the walls in Exercise 12? [268]
    6 gal
14. At 9 bricks per square ft, how many bricks are needed to construct a brick facade on a $9\frac{1}{2}$ ft by 27 ft wall? [270]
    2,309 bricks
15. At 15¢ per brick, how much will it cost to construct the wall in Exercise 14? [270]
    \$346.35

CHAPTER TEST

Find the cost.

1. 2,200 square ft of construction at $50 per square ft
   $110,000
2. 600 square ft of renovation at $75 per square ft
   $45,000
3. 10 18-ft-long pieces of 2 in. by 4 in. lumber at 47¢ per board ft
   $56.40
4. 15 20-ft sections of pipe at 15¢ per ft
   $45
5. 400 ft of electrical cable wire at $42.75 per 100 ft
   $171
6. 18 gal of paint at $13.99 per gal
   $251.82

Solve.

7. Find the inside diameter of a sewer pipe with a $6\frac{1}{4}$-in. outside diameter and a $\frac{5}{16}$-in. thickness? $5\frac{5}{8}$ in.
8. How many 100-ft coils of wire are needed for a 7-room house if an electrician uses an average of 75 ft per room?
   6 100-ft coils
9. An electrician charges $32.50 per h for her labor and $24.75 per h for her assistant's labor. What is the labor charge for a $9\frac{1}{2}$-job?
   $543.88
10. What is the area to be painted on two 9 ft by 20 ft walls that each contain two 3 ft by 5 ft windows and a 3 ft by 7 ft door?
    258 square ft
11. Paint costs $15.99 per gal, and 1 gal covers about 400 square ft. How much will it cost to put 2 coats on the walls in Exercise 10?
    $31.98
12. At 9 bricks per square ft and 18¢ per brick, what will it cost to add a brick facade to the 15 ft by 36 ft side of a house?
    $874.80

Find the cost of the materials.

13. 12 8-ft-long pieces of 2 in. by 4 in. lumber at 29.5¢ per board foot; 8 sheets walnut paneling at $8.99 per sheet; 2 lb nails at 85¢ per lb
    $92.50
14. 175 ft copper pipe at 18¢ per ft; 100 ft plastic pipe at 26¢ per ft; 8 plastic elbows at 69¢ each; 10 couplers at 49¢ each
    $67.92

Solve the problem by drawing a diagram.

15. Fencing was built around a 6 yd by 6 yd piece of land. The posts were 1 yd apart. How many posts were there?
    24 posts
16. There are 18 rows in a garden. The first row is white, the second is pink, and the third is red. This pattern is repeated. What color is the 15th row?
    red

OBJECTIVE

- Evaluate the achievement of chapter objectives.

USING THE TEST

The Chapter Test may be used as a post-test to evaluate student achievement. However, you may wish to use the Chapter Post-Test offered in the Teacher's Resource Package or to design your own chapter test. If this page is not used as a test, you may want to assign it as additional review or practice.

The test items are correlated to the chapter objectives in the table below.

Chapter objectives	Test items
A. Compute construction and renovation costs.	1–2, 11–14
B. Compute the costs of carpentry, plumbing, electrical, painting, or masonry supplies.	3–6
C. Compute the inside and outside diameters of pipes.	7
D. Compute the quantities of electrical supplies needed.	8
E. Compute labor charges.	9
F. Find the areas of walls to be painted.	10
G. Solve problems by drawing diagrams.	15–16

SUPPLEMENTARY MATERIALS

TRP Ch. 13 Post-Test Form A, pp. 1–2
TRP Ch. 13 Post-Test Form B, pp. 1–2

OBJECTIVE

- Understand that paying 1 annual fee for a service contract can cost much less than paying for service each time something goes wrong with appliances or other equipment.

USING THE PAGE

Begin discussion by asking students what items in their homes have needed repair, new parts, or reconditioning, and if they know how much those services cost. Point out to students the enormous amount of equipment used in offices, schools, hospitals, and so forth (copiers, typewriters, calculators, audiovisual equipment, and so forth) and how many people use (and misuse) these things every day. Ask them to imagine what it might cost if companies and institutions did not have service contracts.

Read the information given in the table. Stress the importance of buying brand-name items that include warranties and are made by companies that offer service contracts and use qualified service people who are familiar with the makes and models they sell.

Have students work individually or in pairs to answer Exercises 1–4. Summarize the answers as follows: The cost of an annual service contract is much less than what you might pay for parts, labor, service calls, and so forth over the course of a year, especially if appliances or equipment are older or used frequently.

Have students work individually to answer Exercises 5–6. Then have them share their answers and reasons in a whole-class discussion.

MONEY TIPS

Service contracts on many items can save you money on repairs and maintenance.

LET'S LOOK AT THE FACTS

Many new items include a **warranty** that covers defects and repairs during the first 90 d, the first year, and so forth. After the warranty expires, paying 1 annual **service contract** fee on a computer, for example, could save you money.

APT-4F WORD PROCESSOR/COMPUTER SYSTEM
(Service Contract—$350 per year)

Parts Replacement Charges:		Service & Labor Charges:	
Brightness control knob	$ 12	Service call	$ 75
Disc drive on/off switch	$ 15	Labor (repair time, per hour)	$ 35
Printer cable	$ 40	Maintenance check and system cleaning	$100
Keyboard adjustment	$ 55	Pickup & delivery for in-shop repairs	$ 50
Printer carriage	$ 60	Reinstallation of repaired system	$ 45
Interface unit	$ 90		
Disc drive	$200		

LET'S DISCUSS WHY

1. When would you be more likely to have ongoing or increasing problems with your computer?
   the longer you own it; the more frequently you use it
2. For every call to your home or office, you pay a service charge regardless of whether any work is done or not. How much would it cost just to call in a repair person 5 times in 1 y?
   $375
3. Your computer needs an overall cleaning twice a year. Each call requires $1\frac{1}{2}$ h of labor. What would this cost per year?
   $150 (2 service calls) + $200 (2 cleanings) + $105 (3 h labor) = $455
4. Your computer was repaired 10 times last year. You had bought a service contract for last year. Therefore, what was the most that each service call actually cost you?
   $350 ÷ 10 calls = $35 per call

LET'S SEE WHAT YOU WOULD DO

5. You bought a service contract on your computer for 2 y. In the first year, no repairs were needed. In the second year, all major parts were replaced, and the computer was cleaned twice and taken into the shop twice. Four service calls were made, requiring 12 h of labor. Compare the total cost with the cost of the service contract.
   Total cost: $1,492 Service contract: 2 y × $350 = $700
6. Look around your home, school, or office. On how many items could you take out a service contract?
   Answers may vary.

CALCULATOR DISCOUNTS AND ADD-ONS

You can use a calculator to find the total cost of an item selling at a discount or one that includes an add-on such as sales tax.

A drill that regularly sells for $64 is being sold at a 15% discount. What is the sale price?

Procedure	Calculator Entry	Calculator Display
1. Enter the regular price.	[6] [4]	64.
2. Enter the [−] key.	[−]	64.
3. Enter the discount rate.	[1] [5] [%]	54.4

The sale price is $54.40.

Try this on your calculator. On some calculators, you may need to enter the [=] key after the [%] key. If your calculator does not have a [%] key, multiply by 100 instead.

What is the total cost of a $325 lamp on which a 6% sales tax must be paid?

Procedure	Calculator Entry	Calculator Display
1. Enter the regular price.	[3] [2] [5]	325.
2. Enter the [+] key.	[+]	325.
3. Enter the sales tax rate.	[6] [%]	344.5

The total price is $344.50.

Try this on your calculator. On some calculators, you may need to enter the [=] key after the [%] key. If your calculator does not have a [%] key, multiply by 100 instead.

Use a calculator to compute.

1. Regular price: $300
   Discount rate: 12%
   Sale price: ■ $264
2. Regular price: $50
   Discount rate: 21.5%
   Sale price: ■ $39.25
3. Regular price: $175
   Sales tax rate: 6.5%
   Total cost: ■ $186.38

Use a calculator to solve.

4. A power saw that lists for $85 is on sale at a 25% discount. What is the sale price? $63.75
5. What is the total price of a $480 air conditioner in a state with a 7.5% sales tax? $516

OBJECTIVE

- Find costs with discount or add ons, using a calculator.

USING THE PAGE

Have a volunteer read the introductory material and the first example aloud to the class. Make sure students understand the difference between the discount, an amount, and the rate of discount, a percent. The sale price is the list price minus the discount.

Have students explain how they would solve the problem if they did not have a calculator. One way is to find 15% of $64 and then subtract this amount from $64. Another method would be to find 85% of $64 because a discount of 15% means the sale price will be 85% of the regular price. In either method, 2 steps are required.

On the calculator, only 1 step is required. Enter the regular price and subtract the rate of discount. Have students work through the first example on their calculator. Point out that the calculator will display an answer of 54.4, which must be converted to the final answer of $54.40. Point out that some calculators require pressing the is-equal-to key after pressing the percent key.

In the second example, students use the calculator to add on an amount of tax. Again, only 1 step is required when using the calculator.

Before assigning the exercises, you may want to repeat the 2 examples for students who do not have a percent key on their calculators.

OBJECTIVE

- Test the prerequisite skills needed to learn the concepts developed in Chapter 14.

USING THE TEST

The Pre-Skills Test is designed to diagnose students' strengths and weaknesses on prerequisite skills necessary to study the mathematics in Chapter 14.

Assign the Pre-Skills Test. Allow students to work together in pairs or in small groups. Group members should help those who demonstrate a misunderstanding of a concept or a weakness in a skill.

The table below correlates the items on the Pre-Skills Test with the prerequisite skill and the lesson(s) in the chapter for which it is needed.

Items	Prerequisite skill	Lesson(s)
1–3	Multiply mixed numbers and money amounts.	14.4
4–5	Multiply and add whole numbers and money amounts.	14.6
6–8	Divide money amounts.	14.1
9–11	Divide fractions.	14.1
12–14	Solve proportions.	14.3–14.5
15–20	Find the percent of a number.	14.4
21–26	Find metric equivalents.	14.5
27–30	Add and subtract units of time.	14.6
31–37	Interpret a line graph.	14.2

References to specific skills that are reviewed in the Skills Bank are given on the individual lessons themselves.

Pre-Skills TEST

Compute.

1. $8\frac{1}{2} \times \$12.20$ \$103.70
2. $17\frac{3}{4} \times \$18.50$ \$328.38
3. $56\frac{1}{2} \times \$25.74$ \$1,454.31
4. $(4 \times \$56.80) + (5 \times \$284.50)$ \$1,649.70
5. $(15 \times \$14.80) + (8 \times \$37.60)$ \$522.80

Divide. Round to the nearest whole number.

6. \$1,000 ÷ \$63.49 16
7. \$250 ÷ \$7.65 33
8. \$750 ÷ \$85.60 9
9. $8\frac{3}{4} \div 1\frac{3}{4}$ 5
10. $15 \div 1\frac{1}{2}$ 10
11. $13\frac{1}{2} \div 2\frac{1}{4}$ 6

Solve for n.

12. $\frac{15}{120} = \frac{n}{800}$ 100
13. $\frac{24}{90} = \frac{300}{n}$ 1,125
14. $\frac{2}{125} = \frac{n}{5{,}000}$ 80

Find the answer.

15. 1% of 2,400 24
16. 2% of 360 7.2
17. 5% of 6,700 335
18. 3% of 175 5.25
19. 15% of 4,000 600
20. $1\frac{1}{2}$% of 500 7.5

Copy and complete.

21. 1 liter = ■ milliliters 1,000
22. 500 milliliters = ■ liters 0.5
23. 50 cubic centimeters = ■ milliliters 50
24. 1,000 milligrams = ■ grams 1
25. 3 grams = ■ milligrams 3,000
26. 1,000 cubic centimeters = ■ liters 1

Add or subtract.

27. 3 min 37 s + 8 min 51 s 12 min and 28 s
28. 17 min 28 s + 4 min 43 s 22 min and 11 s
29. 30 min − 14 min 23 s 15 min and 37 s
30. 45 min − 8 min 12 s 36 min and 48 s

Use the line graph for Exercises 31–37. How many bathing suits were bought during the month?

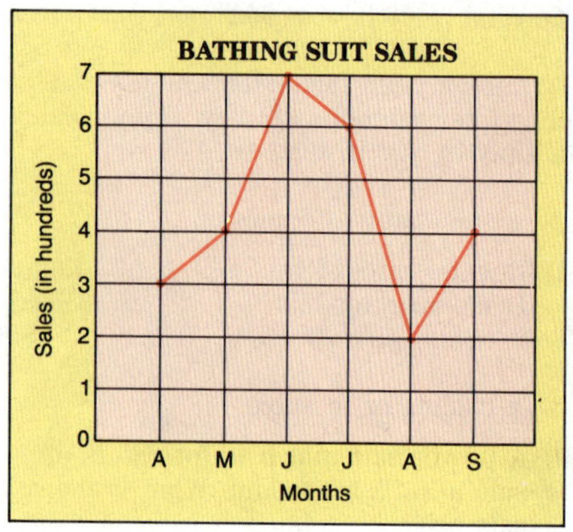

31. April 300
32. May 400
33. June 700
34. July 600
35. August 200
36. September 400
37. During which period of time was there the greatest increase in sales? May to June

Chapter 14 OTHER PROFESSIONS

Many different people are needed to run an office. What types of jobs are done in an office? Why do you think some jobs pay more than others?

OBJECTIVE

- Formulate problems, using text and data.

USING THE CHAPTER OPENER

Discuss students' answers to the questions on the page. The types of jobs found in an office might include managers, secretaries, word processors, accounting clerks, messengers, mail deliverers, bookkeepers, various types of assistants, receptionists, copy-machine operators, and others. The reasons some jobs pay more than others might include the level of responsibility, experience, importance of the work being done, amount of skill needed to do the work, and so on. The vast differences sometimes found in pay scales may be hard for some students to accept. Have them imagine themselves as managers with 20 y of experience hiring someone just out of high school for the same salary as theirs.

COOPERATIVE LEARNING

Assign heterogeneous groups and tell them that they are again to imagine that they (each group) are the manager with 20 y of experience. This manager has just been promoted and will head a new department. The first job will be to put together a pay budget for the following staff: 3 secretaries, 6 accounting clerks, 2 word processors, 1 gal/guy Friday, and 1 messenger. The manager's salary is $44,000 per y. The 2 assistant managers make $37,000 and $35,000.

Have students use the want ads in the local paper to find pay information about the positions listed. Be sure they consider how many experienced people they will need and how this affects the budget. Discuss the groups' budgets and have the class discuss any differences.

SUPPLEMENTARY MATERIALS

TRP Group Projects, p. 14

OBJECTIVES

- Compute the number of column-inches in a newspaper advertisement.
- Find the costs of placing newspaper ads.

TEACHING THE LESSON

WARM-UP Write the following exercises on the chalkboard or an overhead transparency.

Multiply or divide.

1. $3 \times 5\frac{1}{4}$ $(15\frac{3}{4})$ **2.** $4 \times 2\frac{1}{2}$ (10)

3. $5 \times 3\frac{3}{4}$ $(18\frac{3}{4})$ **4.** $6 \times 2\frac{3}{4}$ $(16\frac{1}{2})$

5. $10 \div 2\frac{1}{2}$ (4) **6.** $7 \div 1\frac{3}{4}$ (4)

INTRODUCTION Ask students how a newspaper makes money (sale of papers and advertising revenue). Hold open a page from the daily newspaper and ask what unit or units might be used to compute the cost of advertisements.

INSTRUCTION Use Example 1 to explain the unit of *column-inch* and how to use the rate per column-inch to compute the cost of ads. Point out that column-inch is used regardless of how many columns per page the newspaper runs and assures that all ads and copy will fit neatly within the columns on a page.

Use Example 2 to show how a particular-sized ad design is converted to column-inches. Explain that advertising designs are either enlarged or reduced in size to fit the required column widths. Given a particular size column width (for example, 2 in.), ask students to convert various-sized rectangles to column-inches and compute the cost of running the ads for 1 d.

Use Example 3 to show how an advertising budget might be spent. Discuss the commonplace tradeoff of size versus frequency, that is, running a smaller ad more times or a larger ad fewer times.

14.1 ADVERTISING

Mathematics is used extensively in the advertising industry in designing ads and computing the cost of placing ads in newspapers and magazines.

The cost of display ads in newspapers is usually computed based on **column-inches,** or the number of columns wide the ad is by the number of inches long the ad runs.

EXAMPLE 1 Fits-Well Shoe Store placed an ad for a sale in the local newspaper. The ad was 3 columns wide and ran $5\frac{1}{2}$ in. The newspaper charged $11.34 per column-inch. How much did it cost to run the ad for 1 d?

1. Multiply to find the column-inches. $3 \times 5\frac{1}{2}$ in. $= 16\frac{1}{2}$
2. Multiply to find the cost.

THINK: $16\frac{1}{2} = 16.5$ $16.5 \times \$11.34 = \$187.11$

The ad cost $187.11.

EXAMPLE 2 The Hoyt Agency designed an ad $8\frac{1}{2}$ in. wide by 11 in. long. How much did it cost to run the ad for 5 d if each column was $2\frac{1}{8}$ in. wide?

THINK: If each column was $2\frac{1}{8}$ in. wide, then 2 columns were $4\frac{1}{4}$ in. wide and 4 columns were $8\frac{1}{2}$ in. wide.
So an $8\frac{1}{2}$ in. $\times$ 11 in. ad used 4 columns.

1. Multiply to find the column-inches. 4×11 in. $= 44$
2. Multiply to find the cost. $44 \times \$11.34 \times 5$ d $= \$2,494.80$

It cost $2,494.80 to run the ad for 5 d.

Many newspapers have set rates for quarter-, half-, and full-page ads.

EXAMPLE 3 An ad agency has a $1,500 budget for a client. For how many days can they run a half-page ad? If the ad were reduced in size, for how many days could they run a quarter-page ad?

AD RATES	
Full page –	$1,326.12
Half page –	$667.53
Quarter page –	$341.86

1. Divide to find the half-page time. $\$1,500 \div \$667.53 = 2.24709$

They could run the half-page ad for 2 d.

2. Divide to find the quarter-page time. $\$1,500 \div \$341.86 = 4.387761$

They could run the quarter-page ad for 4 d.

FOR DISCUSSION See TE side column.

Why is the cost of a full-page ad often so much greater in one newspaper than in another?

PRACTICE EXERCISES Remember to estimate whenever you use your calculator.

Complete the table. Assume the cost per column-inch is \$16.87.

Number of columns	Number of inches	Number of column-inches	Cost to place ad	
1	$3\frac{1}{2}$	• 1. ■	• 2. ■	3.5; \$59.05
3	$7\frac{1}{4}$	3. ■	4. ■	21.75; \$366.92
2	$10\frac{3}{4}$	5. ■	6. ■	21.5; \$362.71
5	$14\frac{1}{4}$	7. ■	8. ■	71.25; \$1,201.99

Complete the table. Assume that each column is $1\frac{3}{4}$ in. wide and that the cost per column-inch is \$23.48.

Ad size (width by length)	Number of column-inches	Cost to place ad	
$3\frac{1}{2}$ in. by 8 in.	• 9. ■ 16	•10. ■	\$375.68
$1\frac{3}{4}$ in. by $4\frac{1}{2}$ in.	11. ■ 4.5	12. ■	\$105.66
$5\frac{1}{4}$ in. by $10\frac{3}{4}$ in.	13. ■ 32.25	14. ■	\$757.23
$8\frac{3}{4}$ in. by $8\frac{3}{4}$ in.	15. ■ 43.75	16. ■	\$1,027.25

Use the ad rates on page 282 to solve.

•17. With a \$2,500 budget, for how many days could you run a half-page ad?
3 d

18. With a \$4,000 budget, for how many days could you run a quarter-page ad?
11 d

19. Each page in a newspaper is 6 columns wide by 20 in. long. How many column-inches are there in a quarter-page ad? At \$14.87 per column-inch, how much will the ad cost?
30 column-inches; \$446.10

20. Each page in a local newspaper is 8 columns wide by 14 in. long. How many column-inches are in a half-page ad? At \$8.31 per column-inch, how much will the ad cost?
56 column-inches; \$465.36

SKILLS BANK You may want to refer students to the following skills in the Skills Bank:

Multiply With Decimals, Skills 4–5

Divide With Decimals, Skill 8

Multiply With Fractions and Mixed Numbers, Skill 3

FOR DISCUSSION Students should explain that advertising rates are based on the newspaper's circulation—that is, on the number of people expected to see the ad. The larger the circulation, the higher the cost of an ad.

ASSIGNMENTS

Level 1	1–12, 17–18
Level 2	1–20
Level 3	5–8, 13–20

FOLLOW-UP ACTIVITY

APPLICATION Provide each student with (1) a mock-up of a page from a newspaper, including columns, copy, and at least 5 numbered rectangular ads; (2) a ruler; and (3) the information that this paper charges \$24.85 per column-inch.

Ask students first to use the size of the ads to estimate which is least expensive, next least, and so on to the more expensive. Then ask students to use their rulers and compute both the size of the ads and the cost of running them.

SUPPLEMENTARY MATERIALS

TRP Practice, p. 94
TRP Reteaching, p. 76
TRP Transparency 47

OBJECTIVES

- Interpret line graphs and draw conclusions from data.
- Draw line graphs from tables of data.

TEACHING THE LESSON

WARM-UP Sketch the line graph below on the chalkboard or an overhead transparency.

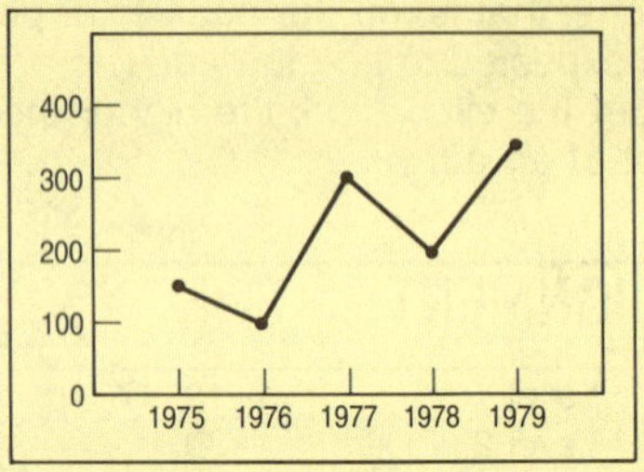

Ask students to identify in which year there was a maximum, in which year there was a minimum, and how much increase or decrease there was between any 2 y.

INTRODUCTION Ask students why graphs are so important. Discuss the value of the visual impact of a graph versus a table of numbers. Ask why salespersons would be particularly interested in using graphs to chart trends.

INSTRUCTION Sketch the TV Sales graph on the chalkboard or an overhead transparency. Ask students to write 5 statements that can be made about the data in the graph and help make the point that a picture is indeed worth a thousand words. Remind students that in addition to simply showing more or less, line graphs are valuable for showing trends over time.

Use Examples 1, 2, and 3 to show how the information in the graph can be interpreted, used to explain trends, and employed to develop strategies. Point out that the interpretation of graphs often involves the use of external information, such as when the ad campaign was conducted or when the holiday season occurred.

14.2 SALES

Salespeople in all industries frequently use graphs to display important sales data. These graphs provide information quickly and in a visually pleasing way. Graphs of sales data are often used to evaluate the success of past activities and to plan future activities.

Line graphs are very useful for showing trends over time.

A large discount appliance chain graphed the sales of 2 major brands of televisions during each month of the past year.

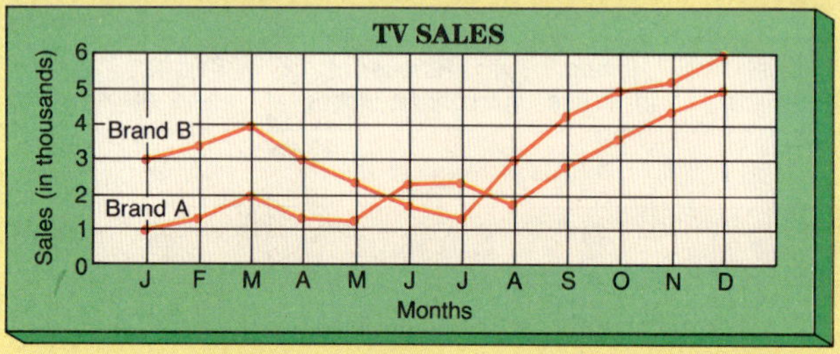

EXAMPLE 1 For which months were sales of Brand A greater than those of Brand B?

The Brand A line is higher than the Brand B line in only June and July.

EXAMPLE 2 The store ran a major advertising campaign for Brand A during May and June. Did the campaign appear to have been successful?

Since sales for Brand A surpassed those of Brand B following the campaign, it looks like the campaign was successful in promoting sales.

EXAMPLE 3 How would you explain the higher sales of both brands during November and December?

It is likely that many televisions were purchased as gifts during the holiday season, when many stores make 30% to 50% of their annual sales.

FOR DISCUSSION See TE side column.

1. What are some advantages of using a graph rather than a table with the same data?

2. Given the TV sales trends shown in the graph, when might you best schedule a major ad campaign to boost sales of Brand A? Brand B?

PRACTICE EXERCISES

Write 3 conclusions that can be drawn from the data in the graph. Answers may vary.

1.
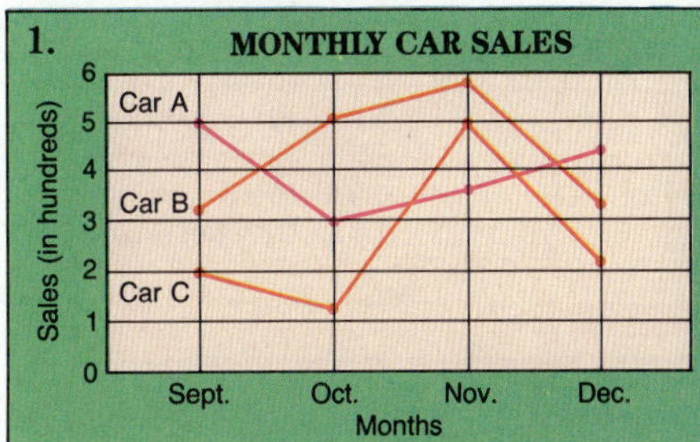

2.
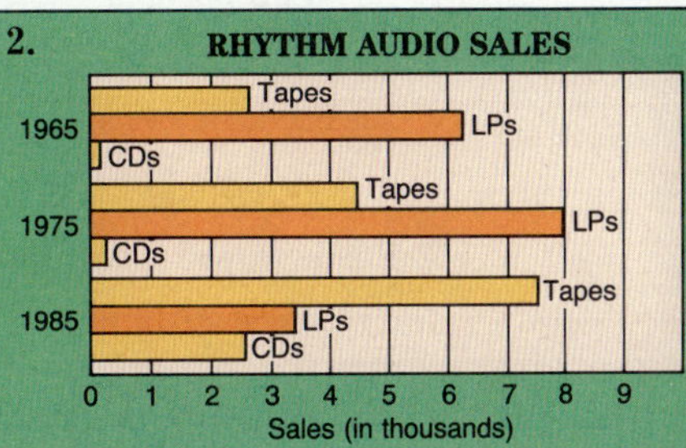

Suggest 1 action, including its timing, that you would use to boost sales in: Answers may vary.

3. Cars.

4. Tapes, LPs, or CDs.

Use graph paper. Make a line graph for the data in the table. Check students' graphs.

5. SALES

Week	Shoes	Sneakers
1	543	389
2	390	137
3	817	580
4	645	244
5	287	256

(Hint: Label the vertical scale in hundreds to 900.)

6. PURCHASES

Year	Hotdogs	Burgers
1985	4,385	3,189
1986	6,408	4,890
1987	5,724	5,836
1988	4,233	7,492
1989	4,879	8,943

(Hint: Label the vertical scale in thousands to 9,000.)

Write 3 conclusions that can be drawn from the data in:

7. Exercise 5.
8. Exercise 6.

Answers may vary.

Suggest 1 action, including its timing, that you would use to boost sales in:

9. Shoes.
10. Hot dogs.

Answers may vary.

SKILLS BANK You may want to refer students to the following skills in the Skills Bank:

Compare and Order Whole Numbers and Decimals, Skills 1–4

FOR DISCUSSION

1. Students should explain that upward and downward trends, as well as relative differences, are much easier to see in graph form than in table form.

2. Students should explain that sales and a major ad campaign might be tried in January and February to boost sales after the holiday rush. Students might also suggest heavy advertising for Brand B in June and July, and for Brand A in April and May, to boost sales when they are at their lowest.

ASSIGNMENTS

Level 1	1–5, 7, 9
Level 2	1–9
Level 3	1–9

FOLLOW-UP ACTIVITY

COOPERATIVE LEARNING
Assign pairs of students. Give 1 student in each pair a blank table with 5 consecutive years in the left column and 2 other columns labeled Item A and Item B. Give the other student in each pair a sheet of graph paper.

Ask the first student to make up data and fill in the table with sales figures. The second student should then convert the table to a graph. Then ask the first student to make 5 statements based on the graph. The second student should check the validity of each statement.

SUPPLEMENTARY MATERIALS

TRP Practice, p. 95
TRP Reteaching, p. 77
TRP Transparency 48

OBJECTIVES

- Convert dollar amounts to their equivalencies in other currencies.
- Convert money amounts in other currencies to dollars.

TEACHING THE LESSON

WARM-UP Write the following exercises on the chalkboard or an overhead transparency.

Solve for n. Round to the nearest tenth.

1. $\frac{2}{25} = \frac{n}{600}$ **2.** $\frac{1}{0.48} = \frac{n}{389}$

3. $\frac{5}{0.39} = \frac{n}{250}$ **4.** $\frac{1}{5.3} = \frac{n}{750}$

(**1.** 48 **2.** 810.4 **3.** 3,205.1 **4.** 141.5)

INTRODUCTION Ask if anyone in the class has traveled to a foreign country. Ask what the currency was in the foreign country and how he or she converted dollars.

INSTRUCTION Use Example 1 to demonstrate how to use proportions to convert dollar amounts to other currencies. Point out that the exchange rate chart is found in the financial section of most newspapers and that exchange rates fluctuate daily. Remind students that travel agents generally advise you on how much money to bring and whether to convert it before you travel abroad or after you arrive.

Use Example 2 to show how to evaluate the reasonableness of charges in other currencies by converting them to dollars. Remind students that, as is always the case with proportions, the units must agree; that is, if the first ratio is dollars to pounds, then the second ratio must also be dollars to pounds.

14.3 TRAVEL

International travelers and travel agents frequently turn to mathematics when they use **exchange rates** to convert dollars to other currencies, or other currencies to dollars. Travelers also use exchange rates to compute the equivalent value of an item in 2 different currencies.

EXAMPLE 1 Before leaving for France, you converted $600 in U.S. currency to francs. How many francs did you receive?

THINK: $1 U.S. = 6.076 francs

1. Use a proportion.

THINK: $1 is to 6.076 francs as $600 is to n francs.

$$\frac{\$1}{6.076 \text{ francs}} = \frac{\$600}{n \text{ francs}}$$

$$n = 600 \times 6.076 = 3{,}645.6$$

2. Round down. $3{,}645.6 \approx 3{,}645$

You received 3,645 francs.

Exchange rates	
$1 U.S. =	1.344 Canadian dollars
	6.076 French francs
	143.7 Japanese yen
	0.620 British pounds
	1.817 German marks

EXAMPLE 2 When you were in England, your hotel room was billed at a rate of 47 British pounds per night. How much was this in U.S. dollars?

THINK: $1 U.S. = 0.620 pounds

1. Use a proportion.

THINK: $1 is to 0.620 pounds as n dollars is to 47 pounds.

$$\frac{\$1}{0.620 \text{ pounds}} = \frac{n}{47 \text{ pounds}}$$

$$0.620n = 47$$

$$n = 47 \div 0.620 = 75.806451$$

2. Round up. $75.806451 \approx 75.81$

In U.S. dollars, the room was $75.81 per night.

FOR DISCUSSION See TE side column.

The dollar/yen exchange rate once fluctuated widely between 120 yen to the dollar and 160 yen to the dollar. When would have been the worst time to buy a Japanese import car? Why?

PRACTICE EXERCISES

Remember to estimate whenever you use your calculator.

What is the value of $750 U.S. in:

• 1. Francs? 4,557
2. Canadian dollars? 1,008
3. Yen? 107,775
4. Marks? 1,362
5. Pounds? 465

Complete the table.

Item	Value in U.S. dollars	Value in other currency
Watch	$75	• 6. ■ pounds 46
Dress	$260	7. ■ francs 1,579
Camera	$195	8. ■ yen 28,021
Jewelry	$39.95	9. ■ Canadian dollars 53
Clarinet	$550	10. ■ marks 999

Convert to dollars.

• 11. Perfume for 125 francs $20.57
• 12. Clothes for 79 Canadian dollars $58.78
13. An umbrella for 10 pounds $16.13
14. A television for 45,000 yen $313.15
15. A car for 23,000 marks $12,658.23
16. Furniture for 12,000 francs $1,974.98
17. Souvenirs for 101 Canadian dollars $75.15
18. A painting for 455 marks $250.41

EXTENSION The Effects of Changing Exchange Rates

An Army corporal stationed in West Germany is paid $320 per wk. What is the impact on the corporal's wages when the exchange rate decreases from 1.872 marks to 1.714 marks to the dollar?

Procedure: 1. Convert the wages to marks at the old rate. 599 marks

2. Convert the wages to marks at the new rate. 548 marks

3. Convert the decrease in marks to dollars at the old rate, that is, the loss in buying power expressed in dollars. $27.24

SKILLS BANK You may want to refer students to the following skills in the Skills Bank:

Proportions and Solving Proportions, Skills 1, 4

FOR DISCUSSION Students should explain that at 120 yen to the dollar, the imported car will be much more expensive than at 160 yen to the dollar. This is because at 120, the dollar buys fewer yen and thus has less value, meaning more dollars are needed to buy the car.

EXTENSION Use this problem to demonstrate how changes in exchange rates have direct impact on people. Solve the problem with the class to show how the same salary in dollars will convert into less actual buying power in marks.

ASSIGNMENTS

Level 1	1–12
Level 2	1–18, EX
Level 3	1–18, EX

FOLLOW-UP ACTIVITY

APPLICATION Provide students with a current foreign exchange rate chart and a list of items and their prices in another currency. Some of the prices should be reasonable and some should be absurd (for example, a shirt for 25 yen or a cassette player for 50 francs). Ask students to evaluate the likely accuracy of each price and note whether each price is reasonable or unreasonable. For those that are unreasonable, ask students to substitute another, more reasonable price in the same currency.

SUPPLEMENTARY MATERIALS

TRP Practice, p. 96
TRP Reteaching, p. 78
TRP Transparency 48

OBJECTIVES

- Determine the number of manufactured items to be tested for quality control.
- Predict the number of defective items on an assembly line.

TEACHING THE LESSON

WARM-UP Write the following exercises on the chalkboard or an overhead transparency.

Solve for *n*. Round to the nearest tenth.

1. $\frac{3}{250} = \frac{n}{12{,}800}$ **2.** $\frac{10}{875} = \frac{n}{30{,}000}$

Find the answer to the nearest tenth.

3. 2% of 1,300 **4.** 5% of 8,725

(**1.** 153.6 **2.** 342.9 **3.** 26 **4.** 436.3)

INTRODUCTION Ask students what is meant by the term *quality control.* Ask where it is used and why it is needed. Ask whether students have seen "Inspected by #X" tags on items they have purchased. Summarize by explaining that most manufacturing operations incorporate a quality-control component to maintain quality and consistency of production.

INSTRUCTION Use Example 1 to show how random testing of a set ratio of items would be implemented. Point out how to set up the proportion and remind students how to use a calculator to solve proportions efficiently.

Use Example 2 to demonstrate how quality-control testing can be used to project the total number of defective items in a production run. Point out that the underlying assumption to using this method is that it is equally likely for any particular item to be defective.

Use Example 3 to show how to combine these 2 ideas to make decisions about a production run. Explain that quality control is based on keeping below a preset, acceptable defective rate.

14.4 MANUFACTURING

A key component of all assembly lines and all manufacturing industries is **quality control**—that is, assuring the same high quality and consistency in each manufactured item. Quality control inspectors use **sampling techniques** and proportions in their ongoing efforts to assure quality goods for the buyer.

EXAMPLE 1 A quality-control inspector pulls 1 out of every 150 light bulbs from the production line for testing. If the line produces 27,000 bulbs per d, how many are pulled for testing?

Use a proportion.

THINK: 1 is to 150 as *n* is to 27,000. $\frac{1}{150} = \frac{n}{27{,}000}$

$$150n = 27{,}000$$
$$n = 27{,}000 \div 150 = 180$$

The inspector pulls 180 light bulbs for testing.

Inspectors use sampling to predict the number of defective items.

EXAMPLE 2 On a given day, a quality-control inspector found 2 defective bulbs out of 115 that were tested. If the day's total production were 22,500 bulbs, predict how many were defective.

Use a proportion.

THINK: 2 is to 115 as *n* is to 22,500. $\frac{2 \text{ defective}}{115 \text{ tested}} = \frac{n \text{ defective}}{22{,}500 \text{ produced}}$

$$115n = 2 \times 22{,}500 = 45{,}000$$
$$n = 45{,}000 \div 115 = 391.30434$$

You would predict that about 391 bulbs were defective.

EXAMPLE 3 A company that produces toaster ovens tries to inspect 5% of the 2,500 units produced each day. If the company can tolerate a defective rate of no more than 1%, how many defective toaster ovens can acceptably be found to be defective during a given day?

1. Multiply to find how many are tested.

   THINK: 5% = 0.05 $0.05 \times 2{,}500 = 125$

2. Multiply to find acceptable defects.

   THINK: 1% = 0.01 $0.01 \times 125 = 1.25$

So 1 defective toaster oven among every 125 tested would be acceptable. A second defective unit would suggest an unacceptably high defective rate.

FOR DISCUSSION See TE side column.

Company A tests 10% of the radios it produces. Company B tests 20% of the radios it produces. Can you conclude that Company B will have a higher likelihood of greater quality? Why?

PRACTICE EXERCISES

Remember to estimate whenever you use your calculator.

Complete the table.

Sampling criteria for inspection	Number of items produced	Number of items inspected
1 in 10	370	• 1. ▪ 37
Every 8th item	1,280	• 2. ▪ 160
3%	800	3. ▪ 24
1 in 50	3,750	4. ▪ 75
3 out of every 20	140	5. ▪ 21
Every 30th item	15,000	6. ▪ 500
$\frac{1}{2}$%	5,400	7. ▪ 27
2 out of every 15	780	8. ▪ 104

Complete the table.

Number of items found defective	Number sampled	Number produced	Predicted number of defective items
8	275	2,000	• 9. ▪ 58
1	60	150	•10. ▪ 3
13	850	7,400	11. ▪ 113
3	200	18,000	12. ▪ 270
2	185	1,500	13. ▪ 16
22	5,800	97,000	14. ▪ 368

Solve.

15. No more than $\frac{1}{2}$% of the rotor blades manufactured can be found to be defective. If 2% of the 13,500 blades are tested, how many defective blades can be acceptably found?
1 blade

16. A toy manufacturer inspects 5% of its monthly production of 4,350 toys. If the company can tolerate a 1% defective rate, how many defective toys can be acceptably found?
2 toys

SKILLS BANK You may want to refer students to the following skills in the Skills Bank:

Proportions and Solving Proportions, Skills 1, 4

Percent and Proportion, Skill 1

FOR DISCUSSION Students should explain that a higher proportion of radios tested does not mean a higher likelihood of quality. Company A might have much higher production standards and testing only 10% of the production run may identify far fewer defects then when Company B tests 20% of its run.

ASSIGNMENTS

Level 1	1–6, 9–12, 15
Level 2	1–16
Level 3	5–8, 12–16

FOLLOW-UP ACTIVITY

ENRICHMENT Ask the class to devise a simple product made out of paper that they can manufacture. The item must be cut, folded, hole punched, and striped in some fashion. Divide the class into 4-, 5-, or 6-member production teams. Give each team 5 min to plan the manufacturing process and then give each 10 min to produce the product. Then discuss and agree on a quality-control procedure, including testing some proportion of the product and an acceptable level of defect. Implement the quality-control procedures and determine which production team has the highest quality.

SUPPLEMENTARY MATERIALS

TRP Practice, p. 97
TRP Reteaching, p. 79
TRP Transparency 49

OBJECTIVES

- Determine appropriate dosages for administering medication.
- Compute the drop rate for intravenous fluids.

TEACHING THE LESSON

WARM-UP Write the following exercises on the chalkboard or an overhead transparency.

Copy and complete.

1. 3,000 mg = ■ g (3)
2. 2 L = ■ mL = ■ cc (2,000; 2000)
3. 200 mL = ■ L (0.2)
4. 30 g = ■ mg (30,000)

INTRODUCTION Begin the discussion by asking students how mathematics is used in the health profession. List the various mathematics skills used by doctors and nurses. Discuss the importance of accuracy in making these calculations and the consequences of even small errors.

INSTRUCTION Use Example 1 to demonstrate a common calculation in which a proportion is used to find how much of a solution should be injected. Use an actual syringe or a diagram of a syringe to help students see that 8 mg of medication are dissolved in each milliliter of liquid.

Use Example 2 to show how to compute the drop rates for intravenous fluids. Point out that *liters per hour* need to be converted to the more measurable *drops per minute.* Remind students that milliliters measure liquid capacity and that cubic centimeters measure volume.

14.5 HEALTH CARE

Nurses and other health-care professionals use mathematics when prescribing and administering medicine. **Standard dosages** must often be adjusted to fit a patient's age and condition.

EXAMPLE 1 A particular drug comes in an 8 mg per 1-mL solution. How many milliliters should be injected to provide a 20-mg dose?

Use a proportion.

THINK: **8 mg is to 1 mL as 20 mg is to *n* mL.**

$$\frac{8 \text{ mg}}{1 \text{ mL}} = \frac{20 \text{ mg}}{n \text{ mL}}$$

$$8n = 20$$

$$n = 20 \div 8 = 2.5$$

2.5 mL should be injected to provide a 20-mg dose.

Mathematics and measurement also play a critical role in an activity such as setting lifesaving **intravenous solution flow rates.**

EXAMPLE 2 A patient is prescribed a 1-L intravenous bottle of lactate solution every 8 h. How many drops per minute should be flowing if there are 12 drops per cc (cubic centimeter)?

1. Find the number of cc per minute.
   - Divide to find the number of milliliters per minute.
     THINK: **1 L = 1,000 mL**
     8 h = 8 × 60 = 480 min
     $1{,}000 \div 480 = 2.0833333 \approx 2.08$ mL
   - Find the number of cc per minute.
     THINK: **1 mL = 1 cc**
     2.08 mL per min = 2.08 cc per min
2. Multiply to find the number of drops per minute.
   $2.08 \times 12 = 24.96 \approx 25$

The drop rate should be 25 drops per min.

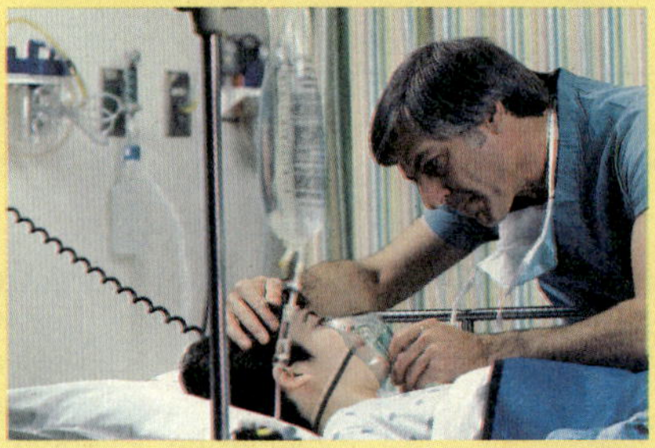

SKILLS BANK You may want to refer students to the following skills in the Skills Bank:

Proportions and Solving Proportions, Skills 1, 4

Metric Units of Capacity and Mass, Skills 2, 3

FOR DISCUSSION

1. Students should explain that, since 2 mL of the stronger solution contain as much medication as 5 mL of the weaker solution, the 2-mL dose is preferable because a smaller and probably less painful syringe could be used.
2. Students should explain that it is critical to monitor intravenous drop rates because it is the drop rate that determines the size of the dosage. If the rate decreases accidentally, the patient will not receive enough medication, and, conversely, it is possible to overdose if the rate increases accidentally.

ASSIGNMENTS

Level 1	1–7, 11–12
Level 2	1–14
Level 3	Odd 1–9, 11–14

FOLLOW-UP ACTIVITY

APPLICATION Invite a local pharmacist and a local nurse as guest speakers to discuss with the class how and when they use mathematics in their jobs. Before the visit, ask students to develop questions that can be shared with the guests.

SUPPLEMENTARY MATERIALS

TRP Practice, p. 98
TRP Reteaching, p. 80
TRP Transparency 49

FOR DISCUSSION See TE side column.

1. A drug comes in 10 mg per 1 mL strength and 25 mg per 1 mL strength. To provide a 50-mg dose, why might a nurse inject 2 mL of the stronger solution rather than 5 mL of the weaker solution?
2. Why is it critical that intravenous drop rates be closely monitored?

PRACTICE EXERCISES

Remember to estimate whenever you use your calculator.

Complete the table.

Available medication	Prescribed dose	Quantity to be administered
10 mg per 1 mL	30 mg	• 1. ■ mL 3
4 mg per 1 mL	10 mg	• 2. ■ mL 2.5
250 mg per 1 mL	150 mg	3. ■ mL 0.6
30 mg per 1 mL	200 mg	4. ■ mL 6.67
12 mg per 1 mL	20 mg	5. ■ mL 1.67
10 mg per 1 tablet	40 mg	6. ■ tablets 4
5 mg per 1 tablet	12.5 mg	7. ■ tablets 2.5
25 mg per 2 tablets	75 mg	8. ■ tablets 6
8 mg per 1 tablet	10 mg	9. ■ tablets 1.25
100 mg per 1 tablet	250 mg	10. ■ tablets 2.5

Solve.

• 11. A patient is prescribed 1 L of intravenous solution every 6 h. How many drops per minute should be administered if there are 10 drops per cc?
28 drops per min

12. A patient is prescribed 500 mL of intravenous solution every 2 h. How many drops per minute should be administered if there are 50 drops per cc? 209 drops per min

13. At 20 drops per cc, how many milliliters are being administered each hour if the drop rate is 30 drops per min?
90 mL

14. At 12 drops per cc, how many milliliters are being administered each hour if the drop rate is 20 drops per min?
100 mL

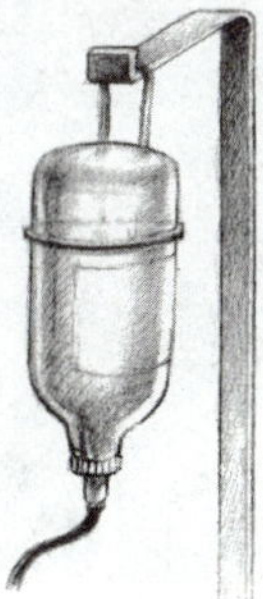

OBJECTIVES

- Schedule radio time.
- Compute radio and television advertising income.

TEACHING THE LESSON

WARM-UP Write the following exercises on the chalkboard or an overhead transparency.

Add or subtract.

1. 3 min and 26 s + 5 min and 49 s
2. 14 min and 19 s + 11 min and 53 s
3. 30 min − 4 min and 24 s
4. 15 min − 6 min and 36 s

(**1.** 9 min and 15 s **2.** 26 min 12 s **3.** 25 min and 36 s **4.** 8 min and 24 s)

INTRODUCTION Ask students how they think mathematics might be used in television and radio. List the various mathematics skills suggested. Explain that 1 important use of mathematics is in scheduling radio and television broadcasts.

INSTRUCTION Use Example 1 to show how 25 min of air time can be scheduled on the radio. Remind students that 1 min contains 60 s and not 100 s when adding or subtracting times. Point out and explain to students how these calculations could be done mentally.

Provide students with the times for a number of songs and ask them to compute the time remaining for introductions and ads in a 30-min block.

Use Example 2 to show how to compute advertising income. Discuss how ad rates are dependent on the station's audience, but that the more advertising time there is, the more income. Discuss how much time can reasonably be allocated to ads without losing an audience.

14.6 COMMUNICATIONS

Radio and television programmers use mathematics to do scheduling of programs, music, and advertising. Radio and television advertising executives use mathematics when they sell add time on their stations.

EXAMPLE 1 A radio program schedule calls for 4 songs between 10:05 and 10:30 A.M. The songs run (in minutes and seconds) 2:53, 3:21, 2:47, and 4:06, respectively. If the schedule allows 30 s for the disc jockey to introduce each number, how much time is left for advertising?

THINK: 10:05 A.M. to 10:30 A.M. = 25 min available.

1. Add to find the time for music.
   Minutes: 2 + 3 + 2 + 4 = 11 min
   Seconds: 53 + 21 + 47 + 6 = 127 s = 2 min 7 s
   11 min + 2 min 7 s = 13 min 7 s
2. Multiply to find the DJ time. $4 \times 30\text{ s} = 120\text{ s} = 2\text{ min}$
3. Add to find the total music and DJ time. 13 min 7 s + 2 min = 15 min 7 s
4. Subtract to find the time left for advertising.

$$\begin{array}{r} 25\text{ min} \\ -15\text{ min } 7\text{ s} \\ \hline \end{array} \rightarrow \begin{array}{r} 24\text{ min } 60\text{ s} \\ -15\text{ min } 7\text{ s} \\ \hline 9\text{ min } 53\text{ s} \end{array}$$

9 min and 53 s remain for advertising.

EXAMPLE 2 A radio station averages 1 30-s advertisement every 5 min. If they charge $239.50 for each 30-s spot, how much is the station's hourly advertising income?

THINK: 1 ad per 5 min = 60 ÷ 5, or 12 ads per h

Multiply to find the income. 12 × $239.50 = $2,874.00

The hourly advertising income is $2,874.

FOR DISCUSSION

See TE side column.

1. Why are audience ratings so important to radio and television stations?
2. About what percent of a prime-time TV hour is devoted to commercials? Why?

PRACTICE EXERCISES

Remember to estimate whenever you use your calculator.

Find the total time scheduled. (Remember: 4:30 means 4 min 30 s.)

• 1. News: 4:30
Weather: 0:45
Song 1: 3:13
Song 2: 4:27
Song 3: 2:51
3 60-s ads
6 30-s ads
3 15-s ads
22 min and 31 s

2. Song 1: 5:35
Song 2: 4:38
Song 3: 4:08
Song 4: 3:49
20-s intro for each song
7 30-s ads
23 min

3. News: 7:45
Sports: 4:15
Weather: 3:20
Traffic: 2:15
5 15-s ads
9 30-s ads
23 min and 20 s

How much time remains to fill a half-hour in:

4. Exercise 1? 7 min and 29 s

5. Exercise 2? 7 min

6. Exercise 3? 6 min and 40 s

Use the rate table to find the advertising income.

• 7. 5 15-s ads $178.75

8. 6 60-s ads $730.20

9. 4 15-s and 7 30-s ads $586.80

10. 6 60-s and 4 30-s ads $983.80

AD RATES	
15-s spots	$35.75
30-s spots	$63.40
60-s spots	$121.70

715.00
608.50

Use the rate table to solve.

•11. How many 30-s spots can run for $1,000?
15 spots

12. How many 60-s spots can be aired on a $2,000 ad budget?
16 spots

13. How much income will a radio station earn by running 10 15-s ads, 9 30-s ads, and 5 60-s ads?
$1,536.60

14. How much more will a radio station earn if it runs 20 15-s ads instead of 5 60-s ads?
$106.50

Ask students to compute hourly and daily advertising incomes for a station that, for example, runs 1 min of advertising out of every 5 min on the air and charges $14.50 per min.

SKILLS BANK You may want to refer students to the following skill in the Skills Bank:

Multiply With Decimals, Skill 5

FOR DISCUSSION

1. Students should explain that audience ratings for radio and television are like circulation figures for a newspaper—they are the figures on which advertising rates are set. The more audience, as for the Super Bowl, the higher the rates and the greater the income.
2. Students should explain that commercials consume about 7 min out of each half hour, or a little less than 25% of air time.

ASSIGNMENTS

Level 1	1–10
Level 2	1–14
Level 3	1–14

FOLLOW-UP ACTIVITY

ENRICHMENT Provide students with the times for 10 songs (between 3 and 5 min each) and directions for scheduling so much time for news, sports, weather, and advertising. Tell students to use this information to develop a schedule, including times, for 1 h of radio that includes all 10 songs and no less than 9 min of advertising time. Tell students to use "Announcer" to fill space between songs or between songs and ads.

SUPPLEMENTARY MATERIALS

TRP Practice, p. 99
TRP Reteaching, p. 81
TRP Applications, p. 16
TRP Transparency 50

OBJECTIVE

- Decide, given relevant information, which raise structure or promotion to accept.

TEACHING THE LESSON

WARM-UP Write the following exercises on the chalkboard or an overhead transparency.

Add.

1. \$150 + \$15 + \$15 + \$15

2. \$220 + \$10 + \$10 + \$10 + \$10

Multiply.

3. 26 × \$140 **4.** 26 × \$250

(**1.** \$195 **2.** \$260 **3.** \$3,640 **4.** \$6,500)

INTRODUCTION Present Problem A to the class. Encourage students to name some of the factors they would use to make an informed decision about which raise structure to accept. List on the chalkboard the factors that are named and discuss how to find out information on each of these factors. Point out to students that the starting salary is not always the best indicator of future salary and that the potential for raises should be considered an important factor when choosing a job.

INSTRUCTION Read the 2 job specifications. Ask students to discuss fully the features of each. Ask if there is any other information that Bess should have. Discuss the decision-making factors.

Have small groups of students complete Exercises 1–20 in the table in order to compare the 2 jobs. Ask students to write statements that describe the data now in the table. Then ask them to write conclusions that can be drawn from the data. Discuss the appropriateness of these statements and conclusions and lead a discussion about interpretations and possible misinterpretations of the data.

DECISION MAKING

14.7 EVALUATING RAISES AND PROMOTIONS

(Open-ended problem solving)

Many jobs are advertised at a particular starting pay but promise periodic **raises** for good performance. Sometimes a job with a lower starting pay eventually can provide a higher pay.

PROBLEM A

Bess has been offered 2 similar jobs with different pay structures.

Job 1: Offers \$200 per wk plus \$20 increases every 6 mo for the next 3 y.
Job 2: Offers \$215 per wk plus \$30 increases every year for the next 3 y.

Bess needs to decide which job to take.

DECISION-MAKING FACTORS

Starting pay Pay in 3 y Total income over 3 y

DECISION-MAKING COMPARISONS

Compare the 2 jobs by completing the table.

	JOB 1		JOB 2	
	Weekly pay	**Total income**	**Weekly pay**	**Total income**
1st 6 months	\$200	\$5,200	\$215	• 1. ■
2nd 6 months	• 2. ■	3. ■	\$215	4. ■
3rd 6 months	5. ■	6. ■	7. ■	8. ■
4th 6 months	\$260	9. ■	10. ■	11. ■
5th 6 months	12. ■	13. ■	\$275	14. ■
6th 6 months	15. ■	16. ■	17. ■	18. ■
Total income		19. ■		20. ■

1. \$5,590 **2.** \$220 **3.** \$5,720 **4.** \$5,590 **5.** \$240 **6.** \$6,240 **7.** \$245 **8.** \$6,370 **9.** \$6,760 **10.** \$245 **11.** \$6,370 **12.** \$280 **13.** \$7,280 **14.** \$7,150 **15.** \$300 **16.** \$7,800 **17.** \$275 **18.** \$7,150 **19.** \$39,000 **20.** \$38,220

MAKING THE DECISIONS

Which job should Bess take:

•21. If starting pay were the only factor? Job 2

22. If pay after 3 y were the only factor? Job 1

23. If total income over 3 y were the only factor? Job 1

24. Which job would you choose? Why? Answers may vary.

25. If Bess were offered a third job at \$240 per wk with no promised raises, should she take it over Job 1 and Job 2? Why?
No; After 3 years, the total income of the third job would be \$37,440.

A common form of **promotion** is from a job that pays an hourly wage to a more professional job that pays an annual salary.

PROBLEM B

You have been working a 35-h week in a job that pays $6.38 per h. You average 5 h per wk overtime at time and a half. You are offered a promotion to a 40-h per wk job that pays $12,900 per y with no overtime provision. Should you accept the promotion?

DECISION-MAKING FACTORS

Annual pay Hours per week Prestige Responsibility

DECISION-MAKING COMPARISONS

Compare the present job and the promotion by completing the table.

	Present job	Promotion
Annual pay, excluding overtime	•**26.** ■	•**27.** ■
Annual pay, including overtime	**28.** ■	$12,900
Hours per week, excluding overtime	35 h	**29.** ■
Hours per week, including overtime	**30.** ■	**31.** ■
Regular hourly wage	**32.** ■	**33.** ■
Overtime hourly wage	**34.** ■	**35.** ■

26. $11,611.60 **27.** $12,900 **28.** $14,099.80 **29.** 40 h **30.** 40 h **31.** 40 h **32.** $6.38 **33.** $6.20 **34.** $9.57 **35.** $0

MAKING THE DECISIONS

Would you take the promotion:

•**36.** If annual pay excluding overtime were the only factor?
Yes

•**37.** If annual pay including overtime were the only factor?
No

38. If regular hourly wage were the only factor?
No

39. If overtime hourly wage were the only factor?
No

40. Would you take the promotion if it were offered to you? Why?
Answers may vary.

41. Would you take the promotion if it paid $15,000 per y? Why?
Yes; yearly salary is greater.

Ask students to complete Exercises 21–25 in small groups, having students explain and defend their answers and decisions. Remind students that in Exercise 24, and later in Exercise 40, there is no one right answer. Explain that choices will vary and that even those with the same choice may have different reasons for their choice. Continually reinforce the fact that, in many consumer issues, best answers depend on individual situations and preferences. Encourage all students to participate and express their opinions.

Repeat this same procedure for Problem B, involving evaluating a promotion. Help students see that, when comparing annual salary with hourly wage, it is necessary to convert either hourly wage to its annual salary equivalent or annual salary to its hourly wage equivalent. Point out to students the impact of the overtime on the decision and how one's attitude toward overtime may be a critical factor in making the decision.

ASSIGNMENTS

Level 1	1–41
Level 2	1–41
Level 3	1–41

FOLLOW-UP ACTIVITY

ENRICHMENT Provide students with information on 2 or 3 jobs, including starting salary and raise structure, similar to those in Problem A. Ask students to evaluate the jobs based on the starting salary, potential salary at some future date, and total income over some period of time. Let students select the job and raise structure they would prefer and discuss their selection with others.

SUPPLEMENTARY MATERIALS

TRP Practice, p. 100
TRP Transparency 50

OBJECTIVES

- Review vocabulary.
- Practice key chapter concepts and skills.

USING THE REVIEW

The Chapter Review is designed to help students prepare for taking the Chapter Test. The first section focuses on vocabulary. It requires students to select words to complete statements. The second section presents practice of key chapter skills.

Each item on the Chapter Review is referenced to the page on which the topic is taught in the Pupil's Edition. You may wish to have students refer to these pages to help review any concepts or skills they have not yet mastered.

It is suggested that students work in small-sized heterogeneous cooperative learning groups. Some cooperative learning methods that may be used are as follows:

1. After each student has independently completed the Chapter Review, a discussion should follow within each group about the solutions to the problems.
2. The group can complete the entire Chapter Review by working together to solve the problems.

End the lesson with an entire class discussion in which any questions brought up in group discussions are presented and answered.

CHAPTER REVIEW

Vocabulary Choose the letter of the word(s) that completes the sentence.

1. Newspaper advertising is sold by the ■. [282] c
   a. Square inch **b.** Inch **c.** Column-inch
2. Trends over time can be easily seen on a ■ graph. [284] b
   a. Bar **b.** Line **c.** Circle
3. Currency is converted by means of the ■. [286] b
   a. Bank's formula **b.** Exchange rate **c.** Denomination factor
4. The manufacturing function that assures product consistency is called ■. [288] a
   a. Quality control **b.** Uniformity monitoring **c.** Variation inspection

Skills Find the answer.

5. What is the cost of a 3-column by 7-in. ad in a newspaper that charges $19.73 per column-inch? [282]
   $414.33
6. What is the cost of a $4\frac{1}{2}$-in. by 10-in. ad in a newspaper with $1\frac{1}{2}$-in. columns that charges $14.86 per column-inch? [282]
   $445.80
7. If $1 U.S. is equal to 6.147 francs, what is the value of $450 in francs? [286]
   2,766 francs
8. If $1 U.S. is equal to 142.9 yen, what is the value in dollars of an item priced at 3,895 yen? [286]
   $27.26
9. How many televisions will be tested if a quality-control inspector samples 1 out of every 25 televisions during a week when 1,575 were produced? [288]
   63
10. If a product inspector finds 3 defective items out of 80 that are tested, how many defective items are likely to be defective in a production run of 800? [288]
    30
11. A drug comes in a 12 mg per 1 mL solution. How much should be injected to provide a 30-mg dose? [290]
    2.5 mL
12. A medication comes in 5-mg tablets. How many tablets should be administered to provide a $7\frac{1}{2}$-mg dose? [290]
    $1\frac{1}{2}$ or 1.5 tablets
13. A radio schedule calls for songs that run 3:14, 5:24, and 4:48. How much time does this leave for ads during a 20-min segment? [292]
    6 min and 34 s
14. At $29.75 for a 15-s spot and $54.35 for a 30-s spot, how much income will be earned by 5 15-s spots and 8 30-s spots? [292]
    $583.55

CHAPTER TEST

Find the cost.

1. A 3-column by $10\frac{1}{2}$-in. ad in a newspaper that charges $19.47 per column-inch.
   $613.31
2. A 6-in. by 8-in. ad in a newspaper with 2-in. columns and a $15.38 per column-inch charge.
   $369.12
3. 15 30-s and 6 15-s radio spots on a station that charges $6.75 per 15-s ad and $10.35 per 30-s ad.
   $195.75
4. 10 60-s and 12 30-s radio spots on a station that has a flat charge of $29.38 per min.
   $470.08

Solve.

5. If $1 U.S. is equal to 0.637 pounds, what is the value of $1,500 in pounds?
   955 pounds
6. If $1 U.S. is equal to 1.793 marks, what is the value in dollars of an item priced at 735 marks?
   $409.93
7. If 2% of each production run is pulled for inspection, how many items will be inspected in a production run of 900 items?
   18 items
8. If an inspector finds 3 defective items out of 400 checked, how many defective items are likely to be found in a production run of 5,000 items?
   38 defective items
9. How many 8-mg tablets should be administered to provide a 20-mg dose?
   $2\frac{1}{2}$ or 2.5 tablets
10. A drug comes in a 3 mg per 1 mL solution. How much should be injected to provide a 12-mg dose?
    4 mL

Use the chart for Exercises 11–14. How much time is scheduled:

RADIO XYZ 9:00 A.M.–9:30 A.M.	
Music:	12:35
News:	5:15
Weather:	3:40

11. For music? 12 min and 35 s
12. All together? 21 min and 30 s
13. How much time is left for advertising?
    8 min and 30 s
14. All of the ad spots run for 30 s and cost $55.30 each. How much income will the radio station earn? $940.10
15. Use graph paper. Make a line graph for the data in the table. (Hint: Label the vertical scale in fifties—0, 50, 100, etc. to 550.)
    Check students' graphs.
16. Write 3 conclusions that can be drawn from the graph in Exercise 15.
    Answers may vary.

SALES

Week	Brand A	Brand B
1	375	505
2	253	425
3	520	371
4	408	287

OBJECTIVE

- Evaluate the achievement of chapter objectives.

USING THE TEST

The Chapter Test may be used as a post-test to evaluate student achievement. However, you may wish to use the Chapter Post-Test offered in the Teacher's Resource Package or to design your own chapter test. If this page is not used as a test, you may want to assign it as additional review or practice.

The test items are correlated to the chapter objectives in the table below.

Chapter objectives	Test items
A. Compute the costs of advertising.	1–4
B. Convert money amounts in other currencies to dollars.	5–6
C. Make quality-control predictions.	7–8
D. Compute medication dosages.	9–10
E. Schedule radio time.	11–14
F. Draw and interpret line graphs.	15–16

SUPPLEMENTARY MATERIALS

TRP Ch. 14 Post-Test Form A, pp. 1–2
TRP Ch. 14 Post-Test Form B, pp. 1–2

OBJECTIVE

- Understand that the more education you receive, the more likely you are to get a better job with better pay.

USING THE PAGE

Read the information given in the bar graph. Explain that median figures represent a midpoint between maximum and minimum. Have students work individually or in pairs to answer Exercises 1–5. Summarize the answers as follows: Education is worth time, effort, and expense because of future benefits.

Have students work individually to answer Exercises 6–7. Then have them share their answers and reasons in a whole-class discussion.

ANSWERS/EXPLANATIONS

1. $14,784; $26,432
More education makes people more desirable as employees, warranting higher salaries because of their added knowledge.

3. Noninterest, laziness, prefer to work and make money, problems at home, lack of money, need to work to help family. Answers may vary as to how to resolve these problems.

4. Education provides a broad knowledge base; higher degrees are required for law and medicine, and thus qualify students as professionals.

5. Loan alternatives: scholarships, part-time/summer jobs, on-campus work, selling your car, and so forth. Cutting expenses: eliminate unnecessary extras, trim amount spent on recreation, and so forth.

7. Cost of loan—$5,040 per y
Balance remaining—$26,460 ($2,205 per mo for living expenses)
In 3 y, you could earn $9,408 more ($7,056 net). The loan would be worth it since you could afford the monthly payment and could probably prepay the loan balance.

MONEY TIPS

Your level of education can affect the amount of money you earn.

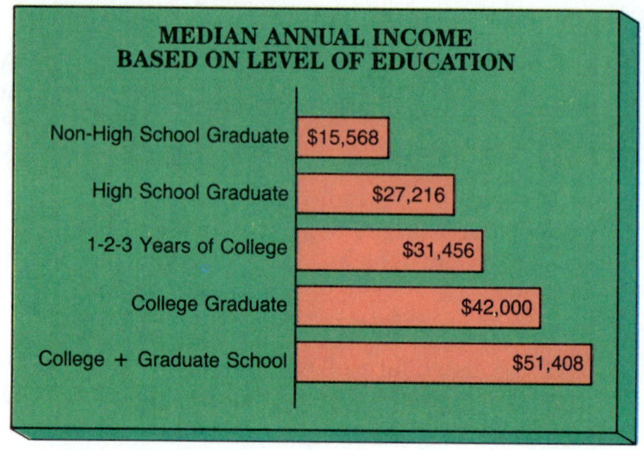

LET'S LOOK AT THE FACTS

Your academic degree can greatly influence your degree of financial success in today's sophisticated workplace. Look over the figures in the bar graph shown here. Remember that these are median figures.

LET'S DISCUSS WHY

See TE side column for answers to Exercises 1, 3–5, and 7.

1. How much more can a college graduate earn than a high school graduate? Than a non–high school graduate? Why?
2. How much more can a college graduate earn than someone with 3 y of college? What might be the reasons for this?
$10,544; distinction of receiving a diploma; completion of the curriculum, etc.
3. What might be some reasons for not completing high school? College? Could some of these problems be resolved? How?
4. Why do you think education affects how much you can earn?
5. Are there any alternatives to taking out a loan in order to further your education? Are there any ways to cut expenses so you would have more money for your education?

LET'S SEE WHAT YOU WOULD DO

6. You did not complete high school, are now 21 y old, and live with your aunt and uncle. You would like to have an apartment of your own, but the only place you can find costs $675 per mo. What steps could you take to improve your economic status?
Answers may vary, but should include getting high school equivalency diploma.
7. It will take 3 y to complete your Masters Degree. You work and earn a net pay of $31,500 per y (gross pay is $42,000). To attend graduate school, you would have to pay $420 per mo for a loan. Would the loan be worth it? Why? Keep in mind your current income and what your income might be in 3 y.

ESTIMATION SKILLS

ESTIMATION OF PRODUCTS OF DECIMALS

If one of two factors is close to 1, the other factor is a good estimate of the product.

Examples 0.976 × 4.32

THINK: 1 × 4.32 = 4.32
Estimate: 4.32

9,180 × 1.03

THINK: 9,180 × 1 = 9,180
Estimate: 9,180

Here is an estimation procedure to use when one of two factors is close to 0.5.

1. Use $\frac{1}{2}$ for the factor close to 0.5.
2. If the other factor is not an even number, round it up 1.
3. Multiply.

Examples 0.463 × 5.6

THINK: $\frac{1}{2}$ × 5.6
56 ÷ 2 = 2.8
Estimate: 2.8

375 × 0.53

THINK: 376 × $\frac{1}{2}$
376 ÷ 2 = 188
Estimate: 188

Estimate the product. Look for factors close to 1 or close to $\frac{1}{2}$.

1. 1.1 × 78 78
2. 84 × 0.52 42
3. 0.94 × 150 150
4. 6.7 × 1.072 6.7
5. 0.483 × 39 19 or 20
6. 6.2 × 0.509 3.1
7. 1.117 × 24.25 24.25
8. 0.048 × 0.899 0.048
9. 0.395 × 12.3 6.1 or 6.2
10. 0.996 × 0.06 0.06
11. 0.558 × 1.98 0.99 or 1.0
12. 0.471 × 18.9 9.4 or 9.5
13. 3.07 × 0.606 1.53 or 1.54
14. 0.002 × 0.9847 0.002
15. 0.498 × 0.008 0.004
16. 0.8743 × 0.0102 0.0102
17. 0.268 × 0.499 0.134
18. 12.317 × 0.9243 12.317
19. 0.50137 × 0.0106 0.0053
20. 1.299 × 0.8201 1
21. 0.469 × 0.00010 0.00005

OBJECTIVE

- Estimate products of decimals when one of two factors is close to 1 or to 0.5.

USING THE PAGE

The estimation procedures in this lesson are very useful when mental estimates of certain products are required. The closer one of the factors is to 1 or to 0.5, the closer the estimate will be to the actual product.

Most people find it easier to multiply by $\frac{1}{2}$ (that is, to divide by 2) than to multiply by 0.5. With this in mind, you might ask students to devise steps for estimating products when one factor is close to 0.2, 0.25, or 0.33. (Use $\frac{1}{5}$, $\frac{1}{4}$, or $\frac{1}{3}$, respectively, for the factors close to 0.2, 0.25, or 0.33. Then multiply.)

Chapter 15 Personal Finance

Objective

- Test the prerequisite skills needed to learn the concepts developed in Chapter 15.

Using the Test

The Pre-Skills Test is designed to diagnose students' strengths and weaknesses on prerequisite skills necessary to study the mathematics in Chapter 15.

Assign the Pre-Skills Test. Allow students to work together in pairs or in small groups. Group members should help those who demonstrate a misunderstanding of a concept or a weakness in a skill.

The table below correlates the items on the Pre-Skills Test with the prerequisite skill and the lesson(s) in the chapter for which it is needed.

Items	Prerequisite skill	Lesson(s)
1–10	Add and subtract money amounts.	All
11–16	Rename a decimal as a percent.	15.1
17–22	Multiply money amounts and decimals.	15.1
23–28	Estimate quotients.	15.4
29–40	Divide money amounts.	15.1–15.4

References to specific skills that are reviewed in the Skills Bank are given on the individual lessons themselves.

Pre-Skills Test

Add or subtract.

1. $475 + $819 $1,294
2. $631.18 + $47.52 $678.70
3. $789.56 + $123.48 $913.04
4. $827 − $351 $476
5. $303.50 − $126.40 $177.10
6. $838.51 − $629.48 $209.03
7. $134,097 + $238,692 $372,789
8. $256,240 + $109,378 $365,618
9. $309,472 − $168,341 $141,131
10. $260,480 − $48,396 $212,084

Rename as a percent.

11. 0.78 78%
12. 0.06 6%
13. 0.1924 19.24%
14. 0.5379321 53.79321%
15. 0.0824877 8.24877%
16. 0.9106025 91.06025%

Multiply.

17. 0.9 × $146 $131.40
18. 0.8 × $150 $120
19. 0.7 × $436 $305.20
20. 0.85 × $936 $795.60
21. 0.73 × $816 $595.68
22. 0.62 × $735 $455.70

Estimate.

23. $6,400 ÷ 3 $2,000
24. $19,201 ÷ 5 $4,000
25. $37,200 ÷ 4 $10,000
26. $197,600 ÷ 4 $50,000
27. $41,830 ÷ 5 $8,000
28. $76,921 ÷ 4 $20,000

Divide. Round to the nearest dollar.

29. $316 ÷ 7 $45
30. $407 ÷ 3 $136
31. $723 ÷ 8 $90
32. $573 ÷ 4 $143
33. $825 ÷ 7 $118
34. $999 ÷ 8 $125
35. $1,200 ÷ 6 $200
36. $2,600 ÷ 6 $433
37. $5,100 ÷ 6 $850
38. $7,800 ÷ 12 $650
39. $9,400 ÷ 12 $783
40. $16,000 ÷ 12 $1,333

Chapter 15 PERSONAL FINANCE

Prices for most items have changed drastically over the years! What would it cost you today to buy similar items? Why are the prices so different?

OBJECTIVE

- Formulate problems, using text and data.

USING THE CHAPTER OPENER

Discuss students' answers to the questions on the page. Jeans today have a very wide price range, and your students will certainly know the current rate in your area. Prices are different today (higher) because jeans, like everything else in today's economy, are affected by inflation, the shrinking dollar, and so on.

You may also want to ask students why they think they pay so much more for designer jeans and if they are worth the price relative to ordinary jeans.

COOPERATIVE LEARNING

Assign heterogeneous groups and tell them that they are going to do some comparative shopping. Ask them to visit several stores to gather information about at least 10 brands or styles of jeans. They are to get the price for each and rate each for fit and style. Then they are to prepare a consumer report indicating which is the best buy overall and rank the others in order after that. If possible, they should include brands they own and wear. For these, they should also evaluate durability.

When the groups present their reports, discuss why they think the reports differ. Elicit that personal preference or need, especially for fit, will play a major role in their evaluations. Also, ask if they would be influenced in their choice of jeans by a consumer report such as the one they just did.

SUPPLEMENTARY MATERIALS

TRP Group Projects, p. 15

OBJECTIVES

- Compute the inflation rate.
- Compute purchasing power after inflation.

TEACHING THE LESSON

WARM-UP Write the following exercises on the chalkboard or an overhead transparency.

Rename as a percent.

1. 0.52 **2.** 0.716 **3.** 0.608

Divide. Round to the nearest thousandth.

4. \$23 ÷ \$213 **5.** \$0.39 ÷ \$1.40

(**1.** 52% **2.** 71.6% **3.** 60.8%
4. 0.108 **5.** 0.279)

INTRODUCTION Use the introductory situation to review the meaning of *inflation.* Explain to students that the Consumer Price Index (CPI) is 1 way of measuring how expensive things are now compared to the past. Write a recent CPI on the chalkboard and compare it to the CPI in 1967 (100) and the CPI in 1985 (322.2).

INSTRUCTION Use Example 1 to show students how to find the inflation rate. In Step 2, remind students that they should divide by the original price, not the current price. Show students that this sequence of keystrokes would yield the inflation rate.

[1] [5] [9] [−] [1] [4] [3]
[÷] [1] [4] [3] [=]

Use Example 2 to show students that money has less purchasing power as the inflation rate goes up. Explain to students that a lower salary in the past may have had more purchasing power than a higher salary today.

SKILLS BANK You may want to refer students to the following skills in the Skills Bank:

Subtract With Whole Numbers and Decimals, Skill 6

Multiply With Decimals, Skill 5

15.1 PURCHASING POWER

It seems that you pay more for things each year. Indeed, most prices have risen steadily during the past decades. The **Consumer Price Index (CPI)** has gone up over 200% since it was first used in 1967. This steady increase in prices is called **inflation.** The **inflation rate** is the percent that prices increase.

EXAMPLE 1 Suppose that, in January, the average cost of food for your family was \$143 per wk. At the end of the year, the average cost of food was \$159 per wk. What was the inflation rate for your food costs?

INFLATION RATE = PRICE INCREASE ÷ ORIGINAL PRICE

1. Subtract to find the price increase. \$159 − \$143 = \$16
2. Divide to find the inflation rate. \$16 ÷ \$143 = 0.1118881
3. Rename as a percent. 11.2%

The inflation rate was about 11.2%.

As the inflation rate goes up, the **purchasing power** goes down.

EXAMPLE 2 Your net income was \$17,000 per y. During the year, the inflation rate was 8%. What was the purchasing power of your income after inflation?

1. Multiply to find the decrease in purchasing power.
   THINK: 8% = 0.08 — 0.08 × \$17,000 = \$1,360
2. Subtract to find the purchasing power after inflation. \$17,000 − \$1,360 = \$15,640

Your \$17,000 salary had the purchasing power of \$15,640 after inflation.

FOR DISCUSSION See TE side column.

1. How could you be sure that your income was keeping up with inflation?
2. Why is inflation particularly difficult for people with fixed incomes?

PRACTICE EXERCISES

Remember to estimate whenever you use your calculator.

Complete the table to find the inflation rate to the nearest tenth of a percent.

Item	Sports tickets	Weekly food	Monthly rent	Car rental	Gallon of gasoline	Slacks	Paper clips
Current price	$12.00	$183.00	$262.00	$104.00	$1.24	$33.89	$0.89
Original price	$10.00	$167.00	$257.00	$89.00	$0.98	$31.95	$0.63
Price increase	•1. ■	•2. ■	3. ■	4. ■	5. ■	6. ■	7. ■
Inflation rate	8. ■	9. ■	10. ■	11. ■	12. ■	13. ■	14. ■

1. $2.00 **2.** $16.00 **3.** $5.00 **4.** $15.00 **5.** $0.26 **6.** $1.94 **7.** $0.26 **8.** 20%
9. 9.6% **10.** 1.9% **11.** 16.9% **12.** 26.5% **13.** 6.1% **14.** 41.3%

Complete the table to find the purchasing power after inflation.

Net income	$19,000	$22,500	$24,600	$28,750	$30,325	$34,875
Inflation rate	10%	8%	4%	11%	6.5%	7.5%
Decrease in purchasing power	$1,900 •15. ■	$1,800 •16. ■	$984 17. ■	$3,162.50 18. ■	$1,971.13 19. ■	$2,615.63 20. ■
Purchasing power after inflation	$17,100 21. ■	$20,700 22. ■	$23,616 23. ■	24. ■	25. ■	26. ■

24. $25,587.50 **25.** $28,353.87 **26.** $32,259.37

Solve.

27. Originally, a box of your favorite cereal cost $1.29. Now the price is $1.47. What was the inflation rate to the nearest tenth of a percent?
14.0%

28. During the year when Andy's income was $26,480, the inflation rate was 7.5%. What was the purchasing power of his income after inflation?
$24,494

EXTENSION Deflation

A steady decrease in prices is called **deflation.** The prices of some goods and services have deflated during the past decades. Originally, a certain computer cost $1,900. A few years later, the same computer cost $1,100. What was the rate of deflation?

1. Subtract to find the price decrease. $1,900 − $1,100 = $800
2. Divide to find the deflation rate. $800 ÷ $1,900 = 0.4210526

The deflation rate was about 42.1%.

Find the deflation rate to the nearest tenth of a percent.

1. Current price: $350
Original price: $420
16.7%

2. Current price: $1,165
Original price: $1,830
36.3%

3. Current price: $0.99
Original price: $1.29
23.3%

FOR DISCUSSION

1. Students should explain that their income is keeping up with inflation if their percent of pay increase matches the inflation rate. They are moving behind inflation if their percent increase is less than the inflation rate.
2. Students should explain that people with fixed incomes often cannot keep up with inflation. Even retired people with inflation protection in their pensions usually fall behind the inflation rate.

EXTENSION Explain that the usual trend is for prices to move upward. However, there may be periods of *deflation* when prices move downward. Periods of deflation are not necessarily better because during these periods many people are out of work.

ASSIGNMENTS

Level	Exercises
Level 1	1–26
Level 2	1–28, EX
Level 3	5–7, 12–14, 18–20, 24–28, EX

FOLLOW-UP ACTIVITY

ENRICHMENT Write these CPIs on the chalkboard: Year 1—323.8; Year 2—341.7. Have students use the steps from Example 1 to find the inflation rate. Encourage students to use calculators. Students should subtract to find the difference (17.9) and then divide the difference by the Year 1 CPI to find the inflation rate (5.5%). Ask students to find the purchasing power of these salaries after 5.5% inflation:
$19,000; $23,500; $37,860. ($17,955; $22,207.50; $35,777.70)

SUPPLEMENTARY MATERIALS

TRP Practice, p. 101
TRP Reteaching, p. 82
TRP Applications, p. 17
TRP Transparency 51

OBJECTIVES

- Identify fixed, living, and annual expenses.
- Compute the amount to budget for fixed, living, and annual expenses.

TEACHING THE LESSON

WARM-UP Write the following exercises on the chalkboard or an overhead transparency.

Find the average to the nearest dollar.

1. \$49, \$65, \$72

2. \$318, \$296, \$342

Divide. Round to the nearest dollar.

3. \$468 ÷ 12 **4.** \$837 ÷ 12

(**1.** \$62 **2.** \$319 **3.** \$39 **4.** \$70)

INTRODUCTION Ask students if they ever have trouble making ends meet; that is, do they need or want to spend more money than they have. Discuss with students the importance of planning their expenses.

INSTRUCTION Use the introductory material to discuss what a budget is. Explain to students that the first step to effective budgeting is knowing how to budget the 3 different types of expenses. Review the meaning of fixed, living, and annual expenses.

Use Example 1 to show students how to budget variable living expenses. Explain that there is no perfect method for budgeting variable living expenses. However, the averaging method shown in this example is effective.

Use Example 2 to show students how to budget annual expenses. Explain that it is usually impossible to budget an entire yearly expense in a single month. So, a portion of the annual expense is set aside each month. Point out to students that dividing by 12 breaks the expense into equal monthly shares.

15.2 FIXED AND VARIABLE EXPENSES

You are working on your monthly budget. A **budget** is an organized plan for spending money. Your monthly budget has 3 types of expenses:

Fixed money expenses, which are the same each month.
Living expenses, which vary from month to month.
Annual expenses, which are amounts to be paid each year.

Fixed expenses include rent and car payments. When you budget fixed expenses, you use the actual amount of the expense.

Living expenses include food, utilities, and transportation. When you budget living expenses, you use an average, or mean, of past expenses.

EXAMPLE 1 In the last 3 mo, you spent \$29, \$19, and \$34 for transportation. How much should you budget next month for transportation?

1. Add to find the total of past expenses. \$29 + \$19 + \$34 = \$82
2. Divide to find the average expenses. \$82 ÷ 3 = \$27.333333

You should budget \$27 next month for transportation.

Annual expenses include insurance premiums, medical bills, and vacations. To budget the monthly share of annual expenses, divide the annual expense by 12.

EXAMPLE 2 You expect to spend \$500 per y on vacations. How much should you budget as the monthly share of the vacation expense?

Divide the annual expense by 12. \$500 ÷ 12 = \$41.666666

You should budget \$42 as the monthly share of vacation expenses.

FOR DISCUSSION See TE side column.

1. How could you predict monthly changes in some living expenses?
2. What problems might arise from budgeting part of an annual expense each month?

PRACTICE EXERCISES

Remember to estimate whenever you use your calculator.

Complete the table to find the amount to be budgeted for April.

Living expense	January	February	March	Amount budgeted for April
Food	$189	$236	$217	• 1. ▪ $214
Utilities	$178	$105	$116	• 2. ▪ $133
Clothing	$67	$83	$105	3. ▪ $85
Entertainment	$58	$46	$23	4. ▪ $42
Telephone	$72	$96	$48	5. ▪ $72
Personal expenses	$38	$46	$63	6. ▪ $49
Contributions	$13	$15	$18	7. ▪ $15

Find the monthly share of the annual expense.

• 8. Vacation: $900 $75
• 9. Medical: $450 $38
10. Car insurance: $1,050 $88
11. Life insurance: $315 $26
12. Car repairs: $285 $24
13. Home repairs: $165 $14
14. Dental: $1,005 $84
15. Health insurance: $1,260 $105
16. Health club: $750 $63

Solve.

• 17. Dave spent $45, $18, and $51 over the past 3 mo on entertainment. How much should he budget next month for entertainment? $38

• 18. The Monahans spent $467, $521, and $428 over the past 3 mo on food. How much should they budget next month for food? $472

19. Celia spent $300, $425, $295, and $375 over the past 4 mo on transportation. How much should she budget next month for transportation? $349

20. Carlos expects to spend $1,100 this year on home repairs. How much should he budget as the monthly share? $92

21. Martha expects to spend $89 this year on car registration and license fees. How much should she budget as the monthly share? $7

22. The Runningbears expect to spend $5,500 this year on medical expenses. How much should they budget as the monthly share? $458

SKILLS BANK You may want to refer students to the following skills in the Skills Bank:

Add With Whole Numbers and Decimals, Skill 6

Divide With Whole Numbers, Skills 3–4, 6

Round Whole Numbers and Decimals, Skill 4

FOR DISCUSSION

1. Students should explain that they can predict seasonal changes in living expenses. For example, when it gets colder, heat will be more expensive. When its gets warmer and air conditioners are used, electricity costs will go up. Students could also predict monthly changes based on past experience.
2. Students should explain that problems arise if the annual expense comes due before they have accumulated enough money to pay for it. In this case, the monthly share set aside for the annual expense would need to be increased.

ASSIGNMENTS

Level	Exercises
Level 1	1–18
Level 2	1–22
Level 3	Odd 1–15, 17–22

FOLLOW-UP ACTIVITY

APPLICATION Ask students to keep a record for several days of how they spend their money. The record should include the amounts they spend and the goods or services they purchase. Students should bring their records to class and estimate their monthly expenses. Ask students to analyze their spending habits.

SUPPLEMENTARY MATERIALS

TRP Practice, p. 102
TRP Reteaching, p. 83
TRP Transparency 51

OBJECTIVES

- Analyze a monthly budget sheet.
- Compute monthly living expenses, monthly fixed expenses, and the monthly share of annual expenses.
- Compute total monthly expenses and the monthly balance.

TEACHING THE LESSON

WARM-UP Write the following exercises on the chalkboard or an overhead transparency.

Add or subtract.

1. $38 + $74 + $62

2. $2,148 − $2,059

Divide. Round to the nearest dollar.

3. $1,246 ÷ 12 **4.** $2,036 ÷ 12

(**1.** $174 **2.** $89 **3.** $104 **4.** $170)

INTRODUCTION Use the introductory situation to explain to students that a budget with the same amount of income and expenses is called a *balanced budget.* A budget with more income than expenses has a *surplus.* A *budget deficit* means there are more expenses than income.

INSTRUCTION Use Example 1 to show students how to read and analyze a monthly budget sheet. Point out that the body of the budget sheet has 4 sections. The entire left part of the sheet shows the budget for living expenses. The fixed and annual expenses are on the right side of the sheet. The monthly summary on the bottom right of the sheet shows the total expenses and income.

Review the individual budgeted expense amounts, the monthly summary section, and the marginal notes. Explain to students that they will be asked to find the totals later. Remind students that they will divide the total annual expenses by 12 to find the monthly share. Emphasize that the balance should not be less than 0.

15.3 BUDGETING EXPENSES

You can use a **budget sheet** to plan how much to spend each month. You add the living expenses, fixed expenses, and monthly shares of annual expenses to find the total budget. You need to make sure that your total monthly budget does not exceed your monthly income.

EXAMPLE Read this budget sheet for Carl and Gloria Delcor.

BUDGET SHEET FOR THE MONTH OF *March*
FOR *Carl & Gloria Delcor*

LIVING EXPENSES	
Food	$386
HOUSEHOLD	
Electricity	$ 76
Natural Gas	$ 34
Telephone	$ 82
Water	
Other	
TRANSPORTATION	
Gas/Oil	$ 64
Parking/Tolls	$ 15
Repairs	$ 85
Bus/Train/Cab	$ 5
Other	
PERSONAL	
Clothing	$ 46
Expenses	$ 75
Credit Card	
ENTERTAINMENT	
Movies/Sports	$ 15
Eating Out	$ 20
TOTAL (LIVING EXPENSES)	

FIXED EXPENSES	
Rent	$275
Mortgage	
Car Payment	$108
Savings	$ 40
Contingency	$ 30
TOTAL	
ANNUAL EXPENSES	
INSURANCE	
Life	$105
Home	$156
Car	$780
Medical/Dental	$300
Vacation	$700
Property Taxes	
Home Repairs	
TOTAL	
MONTHLY SHARE	
MONTHLY SUMMARY	
NET INCOME	
TOTALS	
Living	
Fixed	
Annual	
TOTAL EXPENSES	
BALANCE	

These amounts may vary seasonally.

Use what you know about changing prices to budget this amount.

Add to find the total living expenses.

Savings are considered a fixed expense.

The contingency fund is for unexpected expenses.

List the annual amounts.

Divide the TOTAL by 12 to find the MONTHLY SHARE.

NET INCOME is take-home pay.

BALANCE = NET INCOME − TOTAL EXPENSES

The BALANCE should not be less than 0.

FOR DISCUSSION See TE side column.

1. What could you do if your monthly expenses were more than your monthly income?

2. How would the budget for a single person be different from the budget for a family?

PRACTICE EXERCISES

Remember to estimate whenever you use a calculator.

Use the budget sheet on page 306 to solve.

• 1. How much was budgeted for gas and oil?
$64

• 2. How much was budgeted for a car payment?
$108

3. How much was budgeted for car insurance?
$780

4. Find the total monthly budget for living expenses.
$903

5. Find the total monthly budget for fixed expenses.
$453

6. Find the total budget for annual expenses.
$2,041

7. Find the monthly share of annual expenses.
$170

8. Find the total monthly budget for all expenses.
$1,526

• 9. Carl and Gloria Delcor have a combined monthly net income of $1,703. What is their balance for this monthly budget?
$177

• 10. Suppose Carl and Gloria's net income were to drop to $1,512. How much more would they need to earn per month for the balance to be 0?
$14

11. Suppose Carl and Gloria's medical/dental costs increased to $2,860. Find the new annual expenses and the new monthly share of annual expenses. How much more than their $1,703 income would they need to earn to have a balance of 0?
$4,601; $383.42; $36.42

12. Suppose Carl and Gloria bought a house and their mortgage payment was $438 per mo. Natural gas cost $68 per mo and property taxes were $1,100 per y. Homeowner's insurance went up to $238. They stopped paying rent, and all other expenses were the same. Find the budget totals and the extra income they need.
For answer, see page 477.

13. Suppose Carl and Gloria had a credit card balance of $1,800. They budgeted to pay 10% of the remaining balance each month. How much will they pay the 1st month? About how much would they still owe after 6 mo? (Remember: After each month's payment, the remaining balance decreases.)
$180; $957

SKILLS BANK You may want to refer students to the following skills in the Skills Bank:

Add With Whole Numbers and Decimals, Skill 6

Divide With Decimals, Skill 8

Round Whole Numbers and Decimals, Skill 4

FOR DISCUSSION

1. Students should explain that there are just 2 approaches to this problem: reduce expenses or increase income. A person could choose a strategy that includes both approaches.
2. Students should explain that a budget for a single person is usually less than the budget for a family. The budget for a single person will certainly include lower food costs and probably have a higher proportion of personal and entertainment expenses.

ASSIGNMENTS

Level	Exercises
Level 1	1–10
Level 2	1–12
Level 3	4–13

FOLLOW-UP ACTIVITY

COOPERATIVE LEARNING
Another way to determine a budget is with a *circle graph.* Have students work in pairs to draw a circle graph with these expenditure percents: Housing, 31%; Food, 24%; Transportation, 12%; Clothing, 8%; Savings, 6%; Recreation, 6%; Medical, 5%; and Miscellaneous, 8%. Ask the pairs to use the graph to find the monthly budget for each category for these monthly net pays.

1. $1,860 **2.** $2,375 **3.** $3,090

SUPPLEMENTARY MATERIALS

TRP Practice, p. 103
TRP Reteaching, p. 84
TRP Transparency 52

Objectives

- Identify the long-term costs of raising a family.
- Compute the annual and total costs of raising a child.
- Compute the annual and total costs of paying for a college education.

Teaching the Lesson

WARM-UP Write the following exercises on the chalkboard or an overhead transparency.

Divide. Round to the greatest place.

1. \$32,000 ÷ 4 **2.** \$37,500 ÷ 5

3. \$279,500 ÷ 12 **4.** \$345,070 ÷ 12

(**1.** \$8,000 **2.** \$8,000 **3.** \$20,000 **4.** \$30,000)

INTRODUCTION Discuss with students all the costs of raising children, including food, clothes, shelter, recreation, allowance, and so forth. List the estimated costs on the chalkboard.

INSTRUCTION Use Example 1 to show students how to find the annual cost of raising a young child. Remind students that the total cost is an estimate and so the annual cost is an estimate as well.

Use Example 2 to show students how to find the annual cost of raising a child from age 5 to 18. Point out that it is usually more expensive to raise older children.

Use Example 3 to show students how to find the annual tuition for a child born today. Point out that government subsidies make public colleges less expensive.

Use Example 4 to show students how to find the total estimated cost of raising a child born today. Ask students to compute the costs if Nicole went to a public college (\$342,400) and the amount saved by attending a public college (\$80,100).

15.4 THE COSTS OF RAISING A FAMILY

When you raise a family, you need to plan now for expenses you will have later. One of your biggest expenses will be raising children.

According to some experts, it will cost about \$29,500 (not including child care) to raise a child born today until he or she reaches age 5.

EXAMPLE 1 Lisa and Gary just had a baby named Nicole. About how much will the average yearly cost be to raise Nicole until she is age 5?

Divide. \$29,500 ÷ 5 y = \$5,900

It will cost about \$6,000 per y to raise Nicole until she is age 5.

These same experts think that it will cost about \$265,000 to raise that child from age 5 to 18.

EXAMPLE 2 About how much will the average yearly cost be for Lisa and Gary to raise Nicole from age 5 until she is age 18?

THINK: There are 13 y from the 5th to the 18th birthday.

Divide. \$265,000 ÷ 13 y = \$20,384.62

It will cost about \$20,000 per y to raise Nicole from age 5 to age 18.

When Nicole is 18, it could cost \$128,000 to pay her tuition for 4 y at a private college or \$47,900 for 4 y at a public college.

EXAMPLE 3 About how much would Nicole's tuition be each year? Divide to find the annual tuition.

Private College	**Public College**
\$128,000 ÷ 4 = \$32,000	\$47,900 ÷ 4 = \$11,975

Tuition would be about \$32,000 for private college and \$12,000 for public college.

Raising a family can be expensive.

EXAMPLE 4 About how much would it cost Lisa and Gary to raise Nicole through age 18 and pay her tuition for 4 y at a private college?

Add. \$29,500 + \$265,000 + \$128,000 = \$422,500

It would cost Lisa and Gary about \$422,500.

FOR DISCUSSION See TE side column.

1. How does inflation affect the amount it costs to raise a child and pay for college tuition?
2. What steps could a person take to prepare for the costs of raising a family?

PRACTICE EXERCISES

Remember to estimate whenever you use your calculator.

Use the table below for Exercises 1–22.

ESTIMATED COSTS OF RAISING A FAMILY

Child born	Birth to 5 y	Ages 5 to 18	4 y Private college	4 y Public college
Last year	$27,730	$249,100	$120,300	$45,030
This year	$29,500	$265,000	$128,000	$47,900
Next year	$31,560	$283,500	$135,680	$51,520
2 y from now	$33,640	$297,750	$143,820	$54,830
3 y from now	$35,810	$312,600	$152,450	$58,470

Complete the table to find the average yearly costs. Round each to the nearest $1,000.
14. $24,000 15. $38,000 16. $15,000

Child born	Birth to 5 y	Ages 5 to 18	Private college	Public college
Last year	• 1. ■	• 2. ■	3. ■	4. ■
Next year	5. ■	6. ■	7. ■	8. ■
2 y from now	9. ■	10. ■	11. ■	12. ■
3 y from now	13. ■	14. ■	15. ■	16. ■

1. $6,000 2. $19,000 3. $30,000 4. $11,000 5. $6,000 6. $22,000 7. $34,000
8. $13,000 9. $7,000 10. $23,000 11. $36,000 12. $14,000 13. $7,000

Solve.

• 17. A child will be born next year. About how much will it cost to raise the child through age 18?
$315,060

• 18. About how much will it cost to raise a child born last year, from age 5 to 18, and pay 4 y public college tuition?
$294,130

19. About how much will it cost to raise a child who will be born 2 y from now and pay 4 y tuition for a private college?
$475,210

20. About how much will it cost to raise a child who will be born 3 y from now and pay 4 y tuition for a private college?
$500,860

21. About how much more expensive will it be to raise a child and pay 4 y private college tuition if the child is born 2 y from now and not this year?
$52,710

22. About how much less expensive will it be to pay 4 y public college instead of private college tuition for a child who will be born 3 y from now?
$93,980

SKILLS BANK You may want to refer students to the following skills in the Skills Bank:

Add With Whole Numbers and Decimals, Skill 6

Divide With Whole Numbers, Skills 2–4, 6

FOR DISCUSSION

1. Students should explain that much of the increase is caused by inflation. With an inflation rate of about 6%, something that costs $1,000 today would cost about $2,850 18 y from now.
2. Students should explain that most experts advise parents to establish a college fund as soon as a child is born.

ASSIGNMENTS

Level 1	1–18
Level 2	1–22
Level 3	9–22

FOLLOW-UP ACTIVITY

APPLICATION Distribute to students advertisements showing the costs of these items: crib (with mattress and sheets), chest, baby's toys, baby's bottles (6), premixed formula (20 cases), and jars of baby food (500). Ask students to determine the total of these costs. To that total they should add $1,000 for medical expenses not covered by insurance, $500 for other food, $1,100 for clothing, and $1,000 for other necessities to approximate the cost of raising a baby for the first year. Finally, students should add the estimated costs of 1 year's child care and a year's lost wages for 1 parent. The sum is the grand total of the first year costs of raising a child.

SUPPLEMENTARY MATERIALS

TRP Practice, p. 104
TRP Reteaching, p. 85
TRP Transparency 52

OBJECTIVES

- Identify assets and liabilities.
- Determine total assets and total liabilities.
- Compute net worth.

TEACHING THE LESSON

INTRODUCTION Ask students to make brief lists of all the things they own, including money, clothes, and so forth, along with the estimated value of each item. Then have students make brief lists of what they owe. They should include outstanding loans to family members and friends. Ask students to estimate the total value of what they own and the total amount they owe.

INSTRUCTION Use Example 1 to introduce students to the list of assets and show them how to find the total assets. Explain to students that assets are what a person owns. Emphasize that the value of a car and a house are estimates. Point out that the contents of the house include furniture, appliances, clothing, and so on.

Use Example 2 to introduce students to the list of liabilities and show them how to find the total liabilities. Explain to students that liabilities are what a person owes. Point out that liabilities tend to be more exact amounts than assets. Explain to students that a person's equity in a house is the value of the house less what they owe on the house.

Use Example 3 to show students how to find net worth. Remind students that net worth is an estimate.

SKILLS BANK You may want to refer students to the following skills in the Skills Bank:

Add With Whole Numbers and Decimals, Skill 6

Subtract With Whole Numbers and Decimals, Skill 6

15.5 DETERMINING NET WORTH

Maria and Roberto were trying to determine their **net worth.** They got out records that showed all they owned (**assets**) and all they owed (**liabilities**).

EXAMPLE 1 Maria and Roberto's assets are listed below. What is the total value of their assets?

ASSETS	Maria and Roberto Sanchez
Checking and Savings Account(s)	$ 6,432
Cash	$ 312
Stocks and Bonds	—
Car	$ 7,200
House	$ 93,000
Contents of House	$ 19,600
Collections, etc. Baseball Cards	$ 289
TOTAL ASSETS	$126,833

EXAMPLE 2 Maria and Roberto's liabilities are listed below. What is the total value of their liabilities?

LIABILITIES	Maria and Robert Sanchez
Balance on Mortgage	$46,830
Balance on Auto Loan	$14,800
Balance on Other Loans	$ 2,630
Balance on Credit Cards	$ 890
TOTAL LIABILITIES	$65,150

To determine net worth, you subtract liabilities from assets.

EXAMPLE 3 Find Maria and Roberto's net worth.

Subtract. $126,833 − $65,150 = $61,683

Maria and Roberto's net worth is $61,683.

FOR DISCUSSION See TE side column.

1. How would the net worth of a homeowner be different from the net worth of a renter?
2. Why is a person's net worth not a good indication of the amount of money he or she has available?

PRACTICE EXERCISES

Remember to estimate whenever you use your calculator.

Use the assets and liabilities listed below for Exercises 1–12.

ASSETS	Bill and Sally Jefferson
Checking/Savings/Cash	$ 3,892
Stocks/Bonds	$ 1,150
Car	$ 6,750
House	$63,900
House Contents	$12,850
Collections, etc.	$ 932
TOTAL ASSETS	

LIABILITIES	Bill and Sally Jefferson
Balance on Mortgage	$38,433
Balance on Auto Loan	$ 1,830
Balance on Other Loans	$ 2,850
Balance on Credit Cards	$ 1,263
TOTAL LIABILITIES	

• 1. How much do Bill and Sally have in checking, savings, and cash? $3,892

• 2. What is the balance on Bill and Sally's mortgage? $38,433

3. What is the total value of Bill and Sally's house and contents? $76,750

4. What is the total of Bill and Sally's assets? $89,474

5. What is the total of Bill and Sally's liabilities? $44,376

6. What is Bill and Sally's net worth? $45,098

Suppose Bill and Sally got a second mortgage of $45,000 to put an addition on their house. The value of the house increased by $47,600. The value of the house contents increased by $2,800. Their checking, savings, and cash dropped to $513. All their other assets and liabilities remained the same.

• 7. What is the value of their house with the addition? $111,500

• 8. What is the new value of the contents of the house? $15,650

9. What is the mortgage balance, including the second mortgage? $83,433

10. What is the new total of Bill and Sally's assets? $136,495

11. What is the new total of Bill and Sally's liabilities? $89,376

12. What is Bill and Sally's new net worth? $47,119

FOR DISCUSSION

1. Students should explain that the net worth of a homeowner will often be higher than the net worth of a renter. Houses often appreciate in value and, typically, the value of a house will continue to rise as years pass. Unless homeowners remortgage their property, the mortgage balances will slowly get smaller and equity in the house will grow. Of course, renters could have other real-estate holdings or other assets that would produce a higher net worth.
2. Students should explain that net worth is not a good indicator of available money since the assets often cannot be spent. For example, a couple may have a house worth $300,000 with a $100,000 mortgage, all the mortgage they can afford. The equity in the house is $200,000, but how do they spend it? They cannot afford another mortgage and if they sell the house, where will they live? At least this portion of their net worth is not available to spend.

ASSIGNMENTS

Level 1	1–6
Level 2	1–12
Level 3	4–12

FOLLOW-UP ACTIVITY

APPLICATION Ask students to find their own net worth. Have them list and total their assets and liabilities and then find the difference.

SUPPLEMENTARY MATERIALS

TRP Practice, p. 105
TRP Reteaching, p. 86
TRP Transparency 53

OBJECTIVE

- Decide, given relevant information, which plan to use to adjust a budget.

TEACHING THE LESSON

WARM-UP Write the following exercises on the chalkboard or an overhead transparency.

Find the answer.

Expense A: $561 Expense B: $908
You cut $183 from Expense A and added $75 to Expense B. What are the new amounts for Expense A and Expense B? (A: $378 B: $983)

INTRODUCTION Present the problem situation to the class. Encourage students to name some of the factors they would use to make an informed decision about how to adjust the budget.

INSTRUCTION Read the features of the 3 plans the Nampos have prepared. Ask students to discuss fully the features and to mention those features they find most appealing. Ask if there is any other information that the Nampos should have. Discuss the decision-making factors.

Have small groups of students complete Exercises 1–14 in the table in order to compare the 3 plans. Ask students to write statements that describe the data now in the table. Then ask them to write conclusions that can be drawn from the data. Discuss the appropriateness of these statements and conclusions. Lead a discussion about interpretations and possible misinterpretations of the data.

Ask students to complete Exercises 15–25 in small groups, having students explain and defend their answers and their decisions. For example, Solutions B and C could each be achieved by a wide range of budget cuts. In Solution C, the Nampos could have decided on a number of different ways to partition extra earnings and budget cuts.

DECISION MAKING

15.6 ADJUSTING A BUDGET

(Open-ended problem solving)

A **balanced budget** has enough income to meet expenses. It is important to have a balanced budget. You cannot continue to spend more than you earn.

PROBLEM

The Nampo family has a decision to make. Their income is $512 less than their expenses. They listed these 3 plans to help them decide which one to use.

PLAN A The Nampos could try to bring home $512 more per mo. They would need to earn about $750 to bring home $512.

PLAN B The Nampos could *cut* $512 from their budget. They would cut the $75 savings from fixed costs. The other cuts would have to come from living expenses.

PLAN C The Nampos could try to bring home $260 more per mo. They would need to earn about $380 to bring home $260. They would cut the other $252 from the budget. They would make all their cuts from living expenses.

BUDGET SHEET FOR THE MONTH OF April
FOR The Nampo Family

LIVING EXPENSES	
Food	$446
HOUSEHOLD	
Electricity	$93
Natural Gas	$61
Telephone	$72
Water	$11
Other	—
TRANSPORTATION	
Gas/Oil	$118
Parking/Tolls	$28
Repairs	$60
Bus/Train/Cab	$35
Other	—
PERSONAL	
Clothing	$90
Expenses	$150
Credit Card	—
ENTERTAINMENT	
Movies/Sports	$30
Eating Out	$154
TOTAL (LIVING EXPENSES)	$1,348

FIXED EXPENSES	
Rent	—
Mortgage	$516
Car Payment	$120
Savings	$75
Contingency	$40
TOTAL	$751

ANNUAL EXPENSES	
INSURANCE	
Life	$215
Home	$308
Car	$925
Medical/Dental	$670
Vacation	$900
Property Taxes	$1,855
Home Repairs	$110
TOTAL	$4,983
MONTHLY SHARE	$415

MONTHLY SUMMARY	
NET INCOME	$2,002
TOTALS	
Living	$1,348
Fixed	$751
Annual	$415
TOTAL EXPENSES	$2,514
BALANCE	−$512

DECISION-MAKING FACTORS

Living expenses | Fixed expenses | Total expenses
Net income | Balance | Extra hours worked | Other factors

DECISION-MAKING COMPARISONS

Compare the 3 plans by completing the table.

Factors	Plan A	Plan B	Plan C
Living expenses	$1,348	• 1. ■	• 2. ■
Fixed expenses	3. ■	4. ■	$751
Total expenses (including share of annual expenses)	5. ■	$2,002	6. ■
Net income	$2,514	7. ■	8. ■
Balance	9. ■	10. ■	0
Extra hours worked each month at $9.50 per h	79 h	11. ■	12. ■
Other factors	Away from home 20 h more per wk	13. ■	14. ■

$911; $1,096
$751; $676
$2,514; $2,262
$2,002; $2,262
$0; $0
0 h; 40 h

13. Eliminate savings; cut in living expenses

14. Away from home 10 h more per wk; cut in living expenses

MAKING THE DECISIONS

Which plan should the Nampos choose if the only factor were

•15. Having a higher net salary?
Plan A

16. Not reducing living costs?
Plan A

17. Having a balance of 0?
Plan A, B, or C

18. Not working additional hours?
Plan B

•19. How much lower would the Nampos' fixed expenses be if they choose Plan B instead of Plan C?
$75

20. How much lower would the Nampos' fixed expenses be if they choose Plan B instead of Plan A?
$75

21. How much higher would the Nampos' living expenses be if they choose Plan A instead of Plan C?
$252

22. About how many fewer hours would the Nampos work if they choose Plan C instead of Plan A?
10 h per wk

23. How would you advise the Nampos to cut living expenses if they choose Plan C?
Answers may vary.

24. How would you advise the Nampos to cut living expenses if they choose Plan B?
Answers may vary.

25. Which of these plans would you choose? Why?
Answers may vary.

Remind students that in Exercises 23–24 there is no one right answer. Explain that choices will vary and that even those with the same choice may have different reasons for their choice. Continually reinforce that, in many consumer issues, best answers depend on individual situations and preferences. Encourage students to participate and express their opinions.

ASSIGNMENTS

Level 1	1–25
Level 2	1–25
Level 3	1–25

FOLLOW-UP ACTIVITY

ENRICHMENT Draw this circle graph, which shows the budget for a family, on the chalkboard or an overhead transparency.

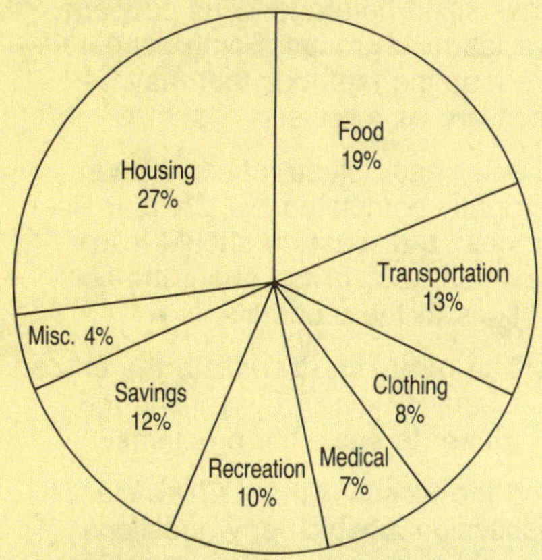

Tell students that the family's net monthly income is $1,900 and that they use all of their money for expenses. Explain that the food budget must be $399 per mo, and that the current budget is less than that. Ask students to devise a plan to adjust the budget allocations on the circle graph so that the monthly food budget is $399.

SUPPLEMENTARY MATERIALS

TRP Practice, p. 106
TRP Transparency 53

OBJECTIVES

- Review vocabulary.
- Practice key chapter concepts and skills.

USING THE REVIEW

The Chapter Review is designed to help students prepare for taking the Chapter Test. The first section focuses on vocabulary. It requires students to select words to complete statements. The second section presents practice of key chapter skills.

Each item on the Chapter Review is referenced to the page on which the topic is taught in the Pupil's Edition. You may wish to have students refer to these pages to help review any concepts or skills they have not yet mastered.

It is suggested that students work in small-sized heterogeneous cooperative learning groups. Some cooperative learning methods that may be used are as follows:

1. After each student has independently completed the Chapter Review, a discussion should follow within each group about the solutions to the problems.
2. The group can complete the entire Chapter Review by working together to solve the problems.

End the lesson with an entire class discussion in which any questions brought up in group discussions are presented and answered.

CHAPTER REVIEW

Vocabulary Choose the letter of the word(s) that completes the sentence.

1. The percent that prices increase is called the ■. [302] a
   a. Inflation rate **b.** CPI **c.** Price difference
2. The expenses that vary from month to month are called ■. [304] a
   a. Living expenses **b.** Annual expenses **c.** Net pay
3. An organized plan for spending money is called a ■. [304, 306] c
   a. Charge account **b.** Annual expense **c.** Budget

Skills Find the answer.

4. A price went from $87 to $93. What was the inflation rate? [302] 6.9%
5. A price went from $7.39 to $7.99. What was the inflation rate? [302] 8.1%
6. What is the purchasing power of $17,800 after 8% inflation? [302] $16,376
7. What is the purchasing power of $30,296 after 9% inflation? [302] $27,569.36
8. What should be budgeted for food if past expenses have been $396, $408, and $472? [304] $425
9. What should be budgeted for gas if past expenses have been $93, $72, and $86? [304] $84
10. What is the monthly share of an annual $2,088 expense? [304] $174
11. What is the monthly share of an annual $5,832 expense? [304] $486

Use the table of estimated costs for raising a family on page 309 for Exercises 12–13.

12. What will it cost to raise a child that will be born 3 y from now, from birth to age 18? [308] $348,410
13. What will it cost to raise a child that will be born 2 y from now, from age 5 to age 18 and pay for 4 y of private college tuition? [308] $441,570

Don's house and its contents are worth $156,300. He has $8,362 in checking/savings/cash, and his car is worth $8,364. Don has balances of $64,382 on his mortgage; $5,380 on his auto loan; and $3,172 on his credit card.

Find Don's: [310]

14. Total assets. $173,026
15. Total liabilities. $72,934
16. Net worth. $100,092

Chapter Test

Find the inflation rate.

1. Original price: $153
   Price now: $167
   9.2%

2. Original price: $2.39
   Price now: $2.89
   20.9%

3. Original price: $1,500
   Price now: $1,785
   19%

Find the purchasing power after inflation.

4. Amount: $19,300
   Inflation rate: 9%
   $17,563

5. Amount: $27,078
   Inflation rate: 7%
   $25,182.54

6. Amount: $52,800
   Inflation rate: 5%
   $50,160

Find the amount to budget for May.

7. Past transportation expenses:
   Feb., $48; March, $52; April, $71
   $57

8. Past food expenses:
   Feb., $432; March, $416; April, $427
   $425

Use the monthly budget for Exercises 9–11.

MONTHLY BUDGET	
Living expenses	$ 894
Fixed expenses	$ 739
Annual expenses	$7,092
Net pay	$2,309

9. Find the monthly share of the annual expenses.
   $591

10. Find the total monthly expenses.
    $2,224

11. Find the balance. $85

Use the table of estimated costs for raising a family on page 309 for Exercises 12–13.

12. A child will be born next year. About how much will it cost to raise the child from birth to age 18?
    $315,060

13. A child will be born next year. About how much will it cost to raise the child from age 5 to 18 and pay 4 y tuition at a public college?
    $335,020

Use the assets and liabilities listed below for Exercises 14–16.

ASSETS	Hal and Carol Roberts
Checking/Savings/Cash	$ 7,309
Stocks/Bonds	$ 2,360
Car	$ 12,980
House/House Contents	$ 138,900
TOTAL ASSETS	

LIABILITIES	Hal and Carol Roberts
Balance on Mortgage	$78,976
Balance on Auto Loan	$ 4,670
Balance on Other Loans	$ 9,456
Balance on Credit Cards	$ 2,098
TOTAL LIABILITIES	

Find the Roberts':

14. Total assets. $161.549

15. Total liabilities. $95,200

16. Net worth. $66,349

OBJECTIVE

- Evaluate the achievement of chapter objectives.

USING THE TEST

The Chapter Test may be used as a post-test to evaluate student achievement. However, you may wish to use the Chapter Post-Test offered in the Teacher's Resource Package or to design your own chapter test. If this page is not used as a test, you may want to assign it as additional review or practice.

The test items are correlated to the chapter objectives in the table below.

Chapter objective	Test items
A. Compute the inflation rate.	1–3
B. Compute purchasing power after inflation.	4–6
C. Compute the budget for living expenses.	7–8
D. Find the monthly share of an annual expense.	9
E. Compute the total monthly expenses and the monthly balance.	10–11
F. Compute the costs of raising children.	12–13
G. Compute total assets, total liabilities, and net worth.	14–16

SUPPLEMENTARY MATERIALS

TRP Ch. 15 Post-Test Form A, pp. 1–2
TRP Ch. 15 Post-Test Form B, pp. 1–2

OBJECTIVE

- Understand that some collectibles can be worth a great deal of money.

USING THE PAGE

Read over the list of categories. Have students work individually or in pairs to answer Exercises 1–4. Summarize the answers as follows: Many of today's cast-offs could be tomorrow's treasures!

Have students work individually to answer Exercises 5–6. Then have them share their answers and reasons in a whole-class discussion.

ANSWERS/EXPLANATIONS

3. Demand exceeds supply; item has special significance (rare or notable because of its owner/creator or historical value or it represents a certain culture, style, era)
4. Suggestions to present:
   - 8: laser sword from *Star Wars;* Sylvester Stallone's boxing gloves;
   - 9: McGuffy Readers; first editions
   - 16: photos of JFK in Dallas in 1963; photos of 1986 Challenger/crew
   - 17: Hand-written manuscripts of famous works, original music scores; letters written by famous people
   - 18: Michael Jackson's glove; Elvis's scarf, guitar strap; home run baseball from a World Series game
5. Have items appraised by Disneyana experts, professional collectors, dealers or magazine editors. If not sold, put valuable items in a vault or safety deposit box.
6. Value projected by experts in the year 2000: original Coca-Cola cans—\$2 each; Ferraro buttons—\$10 each; 1983 Cabbage Patch Kids (in box with birth certificate)—\$300; toy robots—\$25 to \$500 depending on size.

MONEY TIPS

Investing in collectibles may help you fight inflation and realize a profit.

LET'S LOOK AT THE FACTS

Many people collect things, not simply as a hobby, but because certain things tend to increase in value over time. Following is a list of categories that could include **collectible items.**

1. Coins
2. Stamps
3. Records
4. Fine Art
5. Comic Books
6. Autographs
7. Old Flags
8. Film Props
9. Rare Books
10. Classic Cars
11. Tiffany Glass
12. Foreign Currency
13. Disney Memorabilia
14. Ancient Porcelain
15. Diamonds/Gemstones
16. Historical Photos
17. Original Manuscripts
18. Items Belonging to Celebrities

LET'S DISCUSS WHY

See TE side column for answers to Exercises 3–6.

1. There is only 1 1856 British Guinea one penny stamp remaining in the entire world. In 1962, it was valued at \$55,000. In 1987, it was valued at \$1,250,000. What is the percent of increase?
   2,172.7272%
2. A drawing done by a famous artist was certified authentic by an art expert. What else might make it even more valuable?
   signed by artist; in good condition
3. Why would some items increase in value while others would not?
4. Think of examples for categories 8, 9, 16, 17, and 18. What do you think determines the value of each?

LET'S SEE WHAT YOU WOULD DO

5. You inherit a large wooden chest from your aunt. Inside is a Mickey Mouse watch; movie posters; buttons; porcelain Disney figurines marked *limited edition;* a Fantasia record album; 2 ticket stubs from Disneyland dated July 17, 1955; and a complete Mouseketeer outfit. How would you find out their value? If they are valuable but you do not want to sell them yet, what would you do with them?
6. In the early 1980s, some people thought that original Coca-Cola cans, Geraldine Ferraro campaign buttons, Cabbage Patch Kids, and toy robots would be big collectibles in the next 10–15 y. Have you heard of these items? If not, ask friends and family what they know about them.

CALCULATOR FIND WHAT PERCENT ONE NUMBER IS OF ANOTHER

You can use a calculator to find what percent one number is of another. Remember, on some calculators you may need to enter the [=] key after the [%] key.

60 is what percent of 75?

Procedure	Calculator Entry	Calculator Display
1. Enter 60.	[6] [0]	60.
2. Enter the [÷] key.	[÷]	60.
3. Enter 75 and the [%] key.	[7] [5] [%]	80.

So 60 is 80% of 75.

If your calculator does not have a [%] key, multiply by 100 instead.

90 is what percent of 40?

Procedure	Calculator Entry	Calculator Display
1. Divide 90 by 40.	[9] [0] [÷] [4] [0] [=]	2.25
2. Multiply by 100.	[×] [1] [0] [0] [=]	225.

So 90 is 225% of 40.

Use a calculator to find the answers.

1. 45 is what percent of 90?
   50%
2. 5 is what percent of 20?
   25%
3. 250 is what percent of 50?
   500%
4. 110 is what percent of 550?
   20%
5. 27 is what percent of 0.45?
   6,000%
6. $68.40 is what percent of $22.80?
   300%
7. $3.25 is what percent of $125?
   2.6%
8. $18.60 is what percent of $2.50?
   744%

Use a calculator to solve.

9. Julio earns $1,200 per mo. He pays $300 rent. What percent of his monthly pay goes for rent?
   25%
10. Sally has budgeted $480 for travel expenses. She has already spent $60. What percent has she spent?
    12.5%
11. Frank saved $2.25 of each $25 he earned. What percent of his income did he save?
    9%
12. Of 25 students enrolled in a course, 4 transferred out. What percent of the students remained in the course?
    84%

OBJECTIVE

- Find the percent one number is of another, using a calculator.

USING THE PAGE

Ask students to explain how they would solve the first example without a calculator. (Form a ratio with the 2 numbers and rename the ratio as a percent.)

Tell students that they will be following the same process on the calculator. First, they will form a ratio. Then, the percent key will rename the ratio as a percent. Remind students that a ratio indicates division. The key step in setting up the ratio is to determine which number goes in the numerator and which goes in the denominator. The ratio here is $\frac{60}{75}$. Although 60 is the first number entered in the calculator, students should not think that the first number in a problem is always entered first. The example could also have been written as "What percent of 75 is 60?"

The second example demonstrates a similar exercise, but on a calculator that does not have a percent key. In this case, the decimal value of the ratio must be multiplied by 100.

Have students use their calculators to work through the examples in the text. Discuss the use of the is-equal-to key on their calculators.

When solving Exercises 9–12, have students estimate each answer first. Then have them check if their final calculator answer makes sense.

OBJECTIVE

- Test the prerequisite skills needed to learn the concepts developed in Chapter 16.

USING THE TEST

The Pre-Skills Test is designed to diagnose students' strengths and weaknesses on prerequisite skills necessary to study the mathematics in Chapter 16.

Assign the Pre-Skills Test. Allow students to work together in pairs or in small groups. Group members should help those who demonstrate a misunderstanding of a concept or a weakness in a skill.

The table below correlates the items on the Pre-Skills Test with the prerequisite skill and the lesson(s) in the chapter for which it is needed.

Item(s)	Prerequisite skill	Lesson(s)
1–4	Subtract money amounts.	16.1
5–8	Subtract mixed numbers.	16.3
9–12	Rename mixed numbers as decimals.	16.3
13–14	Multiply mixed numbers.	16.3
15–16	Multiply decimals.	16.1, 16.2, 16.4, 16.5
17–22	Divide by mixed numbers.	16.3
23–28	Find the percent of a number.	16.2, 16.4, 16.6
29–32	Find the percent one number is of another.	16.4

References to specific skills that are reviewed in the Skills Bank are given on the individual lessons themselves.

Chapter 16 Investments

Pre-Skills Test

Subtract.

1. \$1,287.50 − \$1,000 \$287.50
2. \$87.23 − \$50 \$37.23
3. \$564.32 − \$500 \$64.32
4. \$7,384.24 − \$5,000 \$2,384.24
5. $39\frac{1}{8} - 28\frac{5}{8}$ $10\frac{1}{2}$
6. $84\frac{1}{2} - 68\frac{3}{4}$ $15\frac{3}{4}$
7. $7\frac{7}{8} - 3\frac{3}{4}$ $4\frac{1}{8}$
8. $74\frac{1}{4} - \frac{5}{8}$ $73\frac{5}{8}$

Rename as a decimal.

9. $43\frac{1}{4}$ 43.25
10. $27\frac{5}{8}$ 27.625
11. $14\frac{3}{4}$ 14.75
12. $8\frac{7}{8}$ 8.875

Multiply.

13. $80 \times 17\frac{3}{8}$ 1,390
14. $150 \times 50\frac{1}{2}$ 7,575
15. 250×24.62 6,155
16. 450×3.875 1,743.75

Divide. Round down to the nearest whole number.

17. \$1,000 ÷ $7\frac{3}{4}$ \$129
18. \$5,000 ÷ $24\frac{7}{8}$ \$201
19. \$300 ÷ $5\frac{5}{8}$ \$53
20. \$500 ÷ $28\frac{1}{2}$ \$17
21. \$2,000 ÷ $63\frac{1}{8}$ \$31
22. \$7,500 ÷ $13\frac{1}{4}$ \$566

Find the answer to the nearest cent.

23. 8.75% of \$2,500 \$218.75
24. 7.739% of \$459.50 \$35.56
25. 5.438% of \$9,873.50 \$536.92
26. 6.75% of \$5,000 \$337.50
27. $5\frac{3}{4}$% of \$4,000 \$230
28. $6\frac{1}{2}$% of \$7,420.80 \$482.35

Find the answer to the nearest tenth of a percent.

29. \$43.85 is what percent of \$250? 17.5%
30. \$983.50 is what percent of \$1,000? 98.4%
31. \$135.38 is what percent of \$500? 27.1%
32. \$1,218.49 is what percent of \$1,000? 121.8%

Millions of shares of stock are traded each day on the New York Stock Exchange. Why do people invest their money in stocks? How do stockbrokers help them?

Objective

- Formulate problems, using text and data.

Using the Chapter Opener

Discuss students' answers to the questions on the page. Some possible answers might include the following: stocks offer an opportunity to earn a lot of money on your investment, stocks offer a diversified way to invest money, and major investors may buy up enough stock to gain control of a company. Stockbrokers offer advice about stocks that is useful to people who are not experts, they buy and sell the stocks for people (for a fee), they keep stockholders informed of changes that are occurring with the stocks, and so on.

Cooperative Learning

Assign heterogeneous groups and explain that they are going to "invest" in the stock market. Each group will have $5,000 to spend.

They are to choose their stocks from those listed on the New York Stock Exchange. Explain that the project will last for 1 mo (or shorter if this is not practical), and their goal is to make the greatest profit they can. During the month, they may sell stocks and buy others, but they must keep a record of all transactions. For this project, they need not take into account brokerage fees, but they should be aware that each time they buy and sell, they would normally be charged such a fee.

Have the groups with the greatest profit and the least profit explain their strategies and discuss why one worked and the other did not.

Supplementary Materials

TRP Group Projects, p. 16

OBJECTIVES

- Compute the cost of U.S. Savings Bonds.
- Compute the redemption value of U.S. Savings Bonds.
- Compute the interest earned on a U.S. Savings Bond.

TEACHING THE LESSON

WARM-UP Write the following exercises on the chalkboard or an overhead transparency.

Multiply.

1. $\frac{1}{2} \times \$20(\$10)$ **2.** $\frac{1}{2} \times \$50(\$25)$

3. $\frac{1}{2} \times \$75(\$37.50)$ **4.** $\frac{1}{2} \times \$100(\$50)$

5. $\frac{1}{2} \times \$200(\$100)$ **6.** $\frac{1}{2} \times \$500(\$250)$

INTRODUCTION Begin discussion by asking if anyone has purchased or owns a U.S. Savings Bond. Discuss who gets the money when you buy a bond, how the money is used, and why people buy savings bonds. Lead the discussion to a comparison and contrast of savings bonds and savings accounts.

INSTRUCTION Use Example 1 to reinforce the idea that you can buy a savings bond for $\frac{1}{2}$ of its face value. When students ask why a $50 bond is not called a $25 bond since that is what it costs, you might explain that it is called a $50 bond because that is its full value at maturity.

Use Example 2 to demonstrate how to compute the value of any denomination bond held for up to 10 y. Examine and discuss the redemption-value table to show how interest accumulates and increases the value of the bond over time. Point out that, although the table stops at 10 y, the bonds continue to earn interest for as long as you hold them. Explain how to find the redemption value of a $200 bond after 6 y. Provide some additional practice exercises for larger-sized bonds held for varying lengths of time.

16.1 U.S. SAVINGS BONDS

A popular form of investing money is through the purchase of **Series EE United States Savings Bonds.** Many employers offer payroll deduction plans to purchase savings bonds, and many people give bonds as gifts. Series EE Savings Bonds are available with **face values** of $50; $75; $100; $200; $500; $1,000; $5,000; and $10,000. They can be purchased for $\frac{1}{2}$ of their face value.

EXAMPLE 1 What is the cost of a $500 Series EE Savings Bond?

THINK: The purchase price is $\frac{1}{2}$ of the face value.

Multiply. $\frac{1}{2} \times \$500 = \$250$

A $500 savings bond will cost $250.

Savings bonds earn variable rates of interest but guarantee 5.5% compounded semiannually for the first year, increasing by $\frac{1}{4}\%$ each half year to a maximum of 7.5% after 5 y. Savings bonds can be redeemed or cashed in at any time after 6 mo at values derived from the table at the right.

GUARANTEED REDEMPTION VALUES ON $50 BONDS

Years held	Value	Years held	Value
0.5	$25.69	5.5	37.48
1.0	26.40	6.0	38.89
1.5	27.22	6.5	40.34
2.0	28.14	7.0	41.86
2.5	29.16	7.5	43.43
3.0	30.29	8.0	45.06
3.5	31.54	8.5	46.75
4.0	32.92	9.0	48.50
4.5	34.44	9.5	50.32
5.0	36.14	10.0	52.20

EXAMPLE 2 How much will you receive if you redeem a $200 savings bond after 6 y?

1. Use the table. After 6 y, a $50 bond can be redeemed for $38.89.
2. Divide to find how many $50 are in $200.

   $\$200 \div \$50 = 4$
3. Multiply.

   $4 \times \$38.89 = \$155.56$

The $200 bond can be redeemed for $155.56.

EXAMPLE 3 How much interest was earned on the $200 savings bond redeemed after 6 y?

THINK: INTEREST EARNED = REDEMPTION VALUE − PURCHASE PRICE

Subtract. $155.56 − $100 = $55.56

You earned $55.56 interest.

FOR DISCUSSION

See TE side column.

1. Who pays the interest on U.S. Savings Bonds?
2. Why are U.S. Savings Bonds considered to be "patriotic" investments?

PRACTICE EXERCISES

Remember to estimate whenever you use your calculator.

Find the cost of buying the bonds.

	•1.	•2.	3.	4.	5.	6.
Face value of bonds	$100	$75	$50	$500	$1,000	$5,000
Number of bonds	1	1	4	2	3	5

1. $50 2. $37.50 3. $100 4. $500 5. $1,500 6. $12,500

Find the redemption value of the bond.

	•7.	•8.	9.	10.	11.	12.
Face value of bond	$500	$500	$5,000	$1,000	$100	$75
Years held	3.0	4.5	5.0	8.0	2.5	9.0

7. $302.90 8. $344.40 9. $3,614 10. $901.20 11. $58.32 12. $72.75

Complete the table.

Bonds held	Years held	Redemption value	Interest earned
3 $500 bonds	$3\frac{1}{2}$	•13. ■	•14. ■
2 $100 bonds	8	15. ■	16. ■
2 $1,000 bonds	$5\frac{1}{2}$	17. ■	18. ■
5 $50 bonds	$7\frac{1}{2}$	19. ■	20. ■
8 $5,000 bonds	6	21. ■	22. ■
6 $75 bonds	1	23. ■	24. ■
7 $10,000 bonds	$4\frac{1}{2}$	25. ■	26. ■

13. $946.20
14. $196.20
15. $180.24
16. $80.24
17. $1,499.20
18. $499.20
19. $217.15
20. $92.15
21. $31,112.00
22. $11,112.00
23. $237.60
24. $12.60
25. $48,216.00
26. $13,216.00

Use Example 3 to show how to compute the actual interest earned on a bond. Emphasize that, as is the case with all investments, your earnings are how much you make beyond the initial investment and not simply the sale price or redemption value.

SKILLS BANK You may want to refer students to the following skill in the Skills Bank:

Multiply With Fractions and Mixed Numbers, Skill 3

FOR DISCUSSION

1. Students should explain that interest on savings bonds is paid by the federal government out of the money collected in federal taxes.
2. Students should explain that U.S. Savings Bonds are considered "patriotic" investments because when you buy a savings bond you are really lending money to the federal government.

ASSIGNMENTS

Level	Exercises
Level 1	Odd 1–11, 13–18
Level 2	1–26
Level 3	1–6, 13–26

FOLLOW-UP ACTIVITY

ENRICHMENT Ask students to imagine that they are given a $100 bond on their first birthday and each birthday thereafter until they are 10 y old. Ask students to use their calculators to compute (1) the total cost of this investment and (2) the total value of these investments on their 11th birthday. Ask students to write a paragraph describing why the difference between their answers is so great.

SUPPLEMENTARY MATERIALS

TRP Practice, p. 107
TRP Reteaching, p. 87
TRP Transparency 54

OBJECTIVE

- Compute the interest earned and the yield on a certificate of deposit.

TEACHING THE LESSON

WARM-UP Write the following exercises on the chalkboard or an overhead transparency.

Find the answer.

1. 6.73% of \$2,000 (\$134.60)
2. 9.02% of \$5,000 (\$451)
3. 5.38% of \$1,500 (\$80.70)
4. 7.87% of \$3,000 (\$236.10)

INTRODUCTION Begin discussion with a bank's newspaper ad advertising certificates of deposit. Point out that CDs are special forms of savings accounts for which you get a certificate instead of a passbook and incur a penalty if you withdraw your money too soon. Discuss the relationship between term and interest rate. Also discuss the difference between the annual interest rate and the annual effective yield. Use a calculator to show that this difference is the result of compounding the interest and then expressing it as simple interest.

INSTRUCTION Use Example 1 to show how to use the table to find the total interest earned on a 1-y CD.

Use Example 2 to explain how to figure the interest on a CD whose term is less than 1 y. Point out that terms of 30 d, 60 d, 90 d, and 180 d correspond to $\frac{1}{12}$, $\frac{1}{6}$, $\frac{1}{3}$, and $\frac{1}{2}$ of a year.

Point out that many people who buy CDs roll them over at the end of the term and continue to compound the interest.

SKILLS BANK You may want to refer students to the following skills in the Skills Bank:

Divide With Whole Numbers, Skill 3

Find the Percent of a Number, Skill 2

16.2 CERTIFICATES OF DEPOSIT

You can purchase another investment called a **certificate of deposit,** or **CD,** at your bank. Since money invested in a CD must be left on deposit for a set period of time to avoid a penalty charge, CDs earn higher interest than regular savings accounts.

CDs can usually be purchased for \$500 or more at terms of 30 d, 90 d, 6 mo, 1 y, and 5 y. The longer the term, the higher the interest rate, but the longer you do not have ready access to your money.

The table shows how 1 bank advertises its CDs. Since interest is compounded daily, you can use the **annual effective yield** to find the actual interest earned in a year. The annual effective yield can also be used to compare CDs with the returns on other investments.

CERTIFICATES OF DEPOSIT

Term	Annual rate	Annual effective yield
30 d	6%	6%
90 d	7.15%	7.38%
6 mo	7.30%	7.54%
1 y	7.65%	7.92%
5 y	7.75%	8.03%

EXAMPLE 1 How much interest will you earn on a 1-y \$5,000 CD?

1. Use the table.

   The annual effective yield for a 1-y CD is 7.92%.

2. Multiply to find the interest.

   THINK: 7.92% = 0.0792 $\quad$ 0.0792 × \$5,000 = \$396.00

You will earn \$396 interest on the CD.

The annual effective yield must be divided by 2 for 6-mo terms, by 4 for 90-d terms, and by 12 for 30-d terms.

EXAMPLE 2 How much interest will you earn on a 90-d \$2,000 CD?

1. Use the table.

   The annual effective yield for a 90-d CD is 7.38%.

2. Multiply to find the annual interest.

   THINK: 7.38% = 0.0738 $\quad$ 0.0738 × \$2,000 = \$147.60

3. Divide to find the 90-d interest.

   THINK: 90 d is $\frac{1}{4}$ of a year. $\quad$ \$147.60 ÷ 4 = \$36.90

The CD will earn \$36.90 interest.

FOR DISCUSSION See TE side column.

1. Why do CDs earn higher interest than regular savings accounts?
2. Why are some people hesitant to purchase longer-term CDs?

PRACTICE EXERCISES

Remember to estimate whenever you use your calculator.

Use the table on page 322. What is the annual effective yield for the CD?

• 1. 1-y CD 7.92% 2. 90-d CD 7.38% 3. 30-d CD 6% 4. 5-y CD 8.03% 5. 6-mo CD 7.54%

Find the interest earned over the term.

	• 6.	• 7.	8.	9.	10.	11.
Amount invested	\$2,500	\$1,000	\$7,500	\$5,000	\$1,500	\$10,000
Term of CD	1 y	1 y	6 mo	30 d	90 d	6 mo

6. \$198 7. \$79.20 8. \$282.75 9. \$25 10. \$27.68 11. \$377

Solve.

• 12. Michael invested \$15,000 in a 6-mo CD. After 6 mo, he took his interest and reinvested the \$15,000 in another 6-mo CD. How much interest did he earn in all?
\$1,131

13. Nancy invested \$15,000 in a 1-y CD. How much interest did she earn? Did the 1-y CD earn more or less than the 2 6-mo CDs in Exercise 12? How much more or less?
\$1,188; more; \$57

14. Sonya invested \$5,000 in a 90-d CD. After 90 d, she took her interest and reinvested the \$5,000 in another 90-d CD. How much interest did she earn in all?
\$184.50

15. Manuel invested \$5,000 in a 6-mo CD. How much interest did he earn? Did the 6-mo CD earn more or less than the 2 90-d CDs in Exercise 14? How much more or less?
\$188.50; more; \$4

EXTENSION Investment Yield

The annual yield of any investment is calculated by dividing the interest earned in 1 y by the amount invested. For example, if you earned \$163.85 interest on an investment of \$2,000, your annual yield was \$163.85 ÷ \$2,000, or 0.081925, or 8.2%.

You earned \$397.37 interest on a \$5,000 CD and \$203.75 interest on a \$3,000 savings account. Which investment provided the greater yield?
\$397.37 interest on a \$5,000 CD

FOR DISCUSSION

1. Students should explain that CDs earn higher interest than regular savings accounts because CDs are for set periods of time and the bank knows how long it will have access to these funds and how long before it will have to redeem the CD. Money in savings accounts can be withdrawn at any time.
2. Students should explain that some people are hesitant because, while their money is tied up in a CD, interest rates may go up and they would incur a penalty if they withdraw their money early.

EXTENSION Use the problem to explain that all investments can be compared by the percent that the interest earned is of the initial investment.

ASSIGNMENTS

Level 1	1–11
Level 2	1–14, EX
Level 3	Odd 1–11, 12–15, EX

FOLLOW-UP ACTIVITY

COOPERATIVE LEARNING

Present the class with this situation: Beginning with your 20th birthday, you buy a \$2,500 1-y CD that pays 8%. You continue to roll over all of your CDs and buy the same type of CD on each of your next 40 birthdays, to the age of 60.

Divide the class into small heterogeneous groups to work on smaller parts of the problem and review how to use a calculator to compound interest. Ask the class to answer this question: Will you be a millionaire at age 60?

SUPPLEMENTARY MATERIALS

TRP Practice, p. 108
TRP Reteaching, p. 88
TRP Transparency 54

OBJECTIVES

- Read a stock-exchange listing.
- Compute the value of shares of stock.
- Compute gains and losses on stock transactions.
- Compute stock dividends and yield.

TEACHING THE LESSON

WARM-UP Write the following exercises on the chalkboard or an overhead transparency.

Multiply. Round to the nearest cent.

1. $35 \times \$3\frac{7}{8}$ **2.** $250 \times \$14\frac{1}{4}$

3. $150 \times \$18\frac{1}{2}$ **4.** $3{,}000 \times \$100\frac{3}{4}$

(**1.** \$135.63 **2.** \$3,562.50 **3.** \$2,775 **4.** \$302,250)

INTRODUCTION Begin discussion by asking the class what a *share of stock* is. Explain that a share of stock is a piece of ownership in a corporation and that shares are usually sold to raise money or capital to expand the corporation. Also explain that the money the corporation makes is shared with its stockholders in the form of *dividends,* which are paid on each share of stock.

INSTRUCTION Use Example 1 to examine and discuss the stock listing in the text or one from a local newspaper. Point out the meaning of each column of the listing and how the information in each column is used by an investor. Ask the class a series of questions that can be answered by reading the stock listing and that clarify the meaning of each column.

16.3 CORPORATE STOCKS

Another form of investment is buying **corporate stocks** that are traded on a stock exchange. A **share** of stock is a piece of ownership in a corporation. Owners of stocks are paid **dividends** on their shares when the corporation makes a profit.

Investors also make or lose money when they sell their stocks, depending on whether the price has gone up or down. Stocks can be purchased or sold through stockbrokers and you can keep track of how your stocks are doing by reading the stock tables in the financial sections of most newspapers.

EXAMPLE 1 Part of a stock exchange listing is shown below. Identify the column headings.

Highest and lowest prices over past year

Annual dividend per share in dollars and cents

Yield = Dividend ÷ Price

Number of shares sold in 100s

Increase (+) or decrease (–) in closing price from the previous day

Closing price on previous day

Highest and lowest prices of the day

52-week High	52-week Low	Stock	Div	Yld	Sales 100s	High	Low	Last	Chg.
37⅜	27⅝	EGH	.56	1.6	683	34⅞	34⅛	34¾	+ ½
11¾	9¾	EQT N	1.05	10.2	97	10⅜	10¼	10¼	– ⅛
17½	13⅝	EQT P	1.66	12.1	115	14	13¾	13¾	– ¼
16½	9½	ERN		–	108	14	13½	13½	– ⅜
39⅝	29½	ER Syst	.50	1.6	1142	32	31¼	31¾	– ½
53	29¾	EagleT	1.12	2.4	109	45¾	45	45¾	+ ½
33¼	24½	EastProd	1.30	4.9	318	26½	26	26½	+ ⅜
40½	30¼	East Util	2.30	7.2	213	32¼	31¾	31⅞	– ⅛
85⅛	52⅛	EGadw	2.52	2.9	1286	87	84¼	87	+ 2⅛
94	63	Eatone	2.00	2.1	742	94	92¾	93⅛	– ⅜
25⅜	14⅞	Echmer	.56	3.3	1532	17	16¾	17	—
31½	20¾	Ecotab	.58	2.0	212	29⅜	28⅞	28⅞	– ⅝
44	33	EdisWr	1.80	4.8	60	37¼	36¾	37¼	+ ¾
19⅞	14⅜	EDA	.28	1.8	141	15½	15¼	15⅜	+ ⅛
14¼	9⅛	EdMat	.16	1.3	71	12	11¾	12	+ ¼
38⅞	21⅝	Edmond	.68	2.1	130	32⅜	31¾	32	– ⅜
13¾	6⅛	Eldar	.22	1.6	124	13½	13¼	13½	—
13⅛	4	ElecDr		–	30	4⅜	4¼	4¼	– ⅛
31¾	13⅝	Elctspt	.08	.3	479	30⅝	30¼	30⅝	– ⅛
18	12½	Elgon		–	70	14⅜	14	14¼	– ¼
3	1⅜	Elsant		–	166	2¼	2⅛	2¼	—
10⅞	8⅞	Emrit	1.20	12.0	83	10	9⅞	10	+ ⅛
110¼	78½	EmrsAl	2.88	2.7	1450	107¾	105¼	107¾	+ 1¼
12¾	6¼	ERan		–	955	7⅛	6⅞	7	– ⅛
18	10½	EmprA		–	971	16	15⅝	15⅞	– ⅜
43¼	30¾	Enhart	1.40	3.2	594	43¾	42⅝	43⅜	+ ¾
29¼	17¼	EnglF	.52	2.1	1085	25	24½	24⅝	– ¾
22⅝	14⅜	EntB		–	287	22	21	22	+ 1⅝
50⅝	37⅞	Enrin	2.48	5.5	630	45	44¾	45	+ ¼
26	13⅛	Encrsh	.80	3.3	922	24½	24	24¼	– ¼
18½	10⅛	EnsExt	1.20	8.1	194	15	14½	14¾	– ⅛
8⅞	4½	Entora		–	271	8½	8⅛	8½	+ ¼
6⅛	3⅛	EntexB	.60	14.5	66	4¼	4⅛	4⅛	– ⅛
19⅝	12⅜	EntexTr	.35	2.4	1020	15	14⅛	14⅞	+ ¼
24½	14¾	EnwSys		–	368	20⅛	19¾	20	+ ¼
26	18	EnwSy	1.75	7.3	29	24⅛	24	24⅛	+ ⅛
28½	19½	Equipax	.68	3.0	105	23	22¾	22¾	– ¼
6⅝	4⅛	EquiMk		–	435	4⅝	4½	4⅝	—
10⅞	8¾	EqtBl	.50	5.3	134	9½	9⅜	9⅜	– ¼
46⅞	28	EqtBs	1.20	2.8	101	44	43½	43½	– ⅜
10	6	Equitax	.16	2.0	33	8	7⅞	7⅞	– ¼
35	19	ErbSys	.52	1.8	71	29¼	29	29	– ⅜
42⅜	29	EssWt	.72	1.7	41	42½	42	42½	+ ⅝
35	20½	EsxTh	.60	2.4	205	28	25	25	– 3¼
20¼	10¼	Estren		–	41	16½	16¼	16¼	– ½
32¼	16	EthyPr	.40	1.3	1081	29⅞	29⅜	29¾	—
92	57½	ExetRf	3.60	4.0	6132	89⅝	89⅛	89½	+ ⅛

EXAMPLE 2 You decide to buy 50 shares of Empire Airfreight (EmprA). How much will the shares cost?

1. Use the stock listing to find that the last price for EmprA was $15\frac{7}{8}$.
2. Rename the price as a decimal. $\$15\frac{7}{8} = \$15.875$
3. Multiply to find the cost. $50 \times \$15.875 = \$793.75$

The stock will cost $793.75.

EXAMPLE 3 You purchased 40 shares of Ethyl Products (EthyPr) at $16\frac{7}{8}$. How much profit will you make if you sell the shares today?

1. Use the stock listing to find that the last price for Ethyl was $29\frac{3}{4}$.
2. Subtract to find the increase for each share. $\$29\frac{3}{4} - \$16\frac{7}{8}$
   $\$29.75 - \$16.875 = \$12.875$
3. Multiply to find the profit. $40 \times \$12.875 = \$515$

You will make a profit of $515.00.

EXAMPLE 4 How much did you receive in dividends over the past year from your 250 shares of Exeter Refining (ExetRf)? What was your rate of return (yield) on your investment over the past year?

1. Use the stock listing to find that Exeter paid a dividend of $3.60 per share and that the last price was $89\frac{1}{2}$.
2. Multiply to find your earnings. $250 \times \$3.60 = \$900$
3. Multiply to find the value of your investment. $250 \times \$89\frac{1}{2}$
   $250 \times \$89.50 = \$22{,}375$
4. Divide to find your return. $\$900 \div \$22{,}375 = 0.0402234 = 4\%$

You received $900 in dividends, representing a 4% return on your investment.

FOR DISCUSSION See TE side column.

1. What causes stocks to go up or down in price?
2. What is meant by a "bull" market and a "bear" market?

Use Example 2 to find the cost of buying some stock. Review with students the conversion of quarters and eighths to decimals. Point out that the "last price" used in the example would apply first thing in the morning when the market opens, but that your actual cost per share would depend on the trading price at the time your order was processed.

Use Example 3 to calculate the profit on a stock-market investment. Explain that, while the gains can be great, playing the stock market can be risky because there is always the possibility of a significant loss. After demonstrating how to calculate the gain, ask students to calculate the loss on a $10,000 investment in a stock purchased at $\$7\frac{1}{2}$ that is now selling at $\$2\frac{1}{4}$—a common situation for smaller, newer, and riskier companies.

Use Example 4 to calculate dividends and return (yield) on investments. Emphasize that the return on an investment is based on the original cost of the investment—in this case, the cost of buying the stock in the first place. Point out that a 4% return is a little misleading because dividend income is over and beyond any potential capital gains that might accrue.

SKILLS BANK You may want to refer students to the following skills in the Skills Bank:

Multiply With Decimals, Skill 2

Subtract With Fractions and Mixed Numbers, Skills 1–5

FOR DISCUSSION

1. Students should explain that the prices of stocks fluctuate in response to supply and demand. When more people want to buy than to sell, the price goes up. Similarly, when more people want to sell than to buy, the price goes down.
2. Students should explain that a "bull" market is a strong market with prices moving up, while a "bear" market is a weak market with prices moving down.

ASSIGNMENTS

Level 1	1–7, 8–10, 14–15, 18–23
Level 2	1–29
Level 3	1–7, 11–13, 16–17, 24–29

FOLLOW-UP ACTIVITY

APPLICATION Invite a stockbroker to speak to the class about the stock market and about how he or she uses mathematics on the job. Ask students to develop a list of questions before the visit.

SUPPLEMENTARY MATERIALS

TRP Practice, p. 109
TRP Reteaching, p. 89
TRP Transparency 55

PRACTICE EXERCISES

Remember to estimate whenever you use your calculator.

Use the stock listing to answer.

• 1. What was the highest price Elgon was traded at over the past year? $18

• 2. What was the lowest price Enhart was traded at over the past year? $$30\frac{3}{4}$

3. How many shares of ERN were traded on this particular day? 10,800

4. By how much did a share of Equitax decline from its 52-wk high? $2\frac{1}{8}$

5. How much did Eastern Gadwick (EGadw) pay per share in dividends? $2.52

6. What was the day's high price for Enrin? $45

7. Which 3 stocks closed at their highest prices in a year? EGadw, Enhart, EssWt

Find the current cost of the shares.

	•8.	•9.	10.	11.	12.	13.
Stock	Eldar	EntB	East Util	EDA	Estren	EagleT
Number of shares held	75	30	125	65	250	2,300

8. $1,012.50 9. $660 10. $3,984.38 11. $999.38 12. $4,062.50 13. $105,225

Complete the table.

Purchase price	Last price	Number of shares	Amount of Profit (P) or loss (L)	
$14\frac{1}{2}$	$21\frac{3}{4}$	30	•14. ■	(P) $217.50
$25\frac{7}{8}$	$19\frac{1}{4}$	80	15. ■	(L) $530.00
$84\frac{3}{8}$	$125\frac{1}{2}$	50	16. ■	(P) $2,056.25
64	$42\frac{7}{8}$	250	17. ■	(L) $5,281.25

Complete the table.

Stock owned	Dividend per share	Total annual dividend	Rate of return
10 shares of EGH	•18. ■ $0.56	•19. ■ $5.60	20. ■ 1.6%
40 shares of EdisWr	21. ■ $1.80	22. ■ $72.00	23. ■ 4.8%
75 shares of EssWt	24. ■ $0.72	25. ■ $54.00	26. ■ 1.7%
500 shares of EDA	27. ■ $0.28	28. ■ $140.00	29. ■ 1.8%

16.4 CORPORATE AND MUNICIPAL BONDS

Another form of investment is buying **corporate bonds** or **municipal bonds.** When corporations need to raise large sums of money for such things as plant expansions or when states and cities need money for such things as roads or schools, they issue bonds. When you buy a bond, you are really loaning the corporation or the government your money based on an agreement that you will be paid interest over the life of the bond and repaid the full amount when the bond matures.

The face value of a bond, often $1,000, is called the **par value.** The **maturity date** of the bond, often 5, 10, or 20 y, is when the bond can be redeemed at par value. However, many investors buy and sell bonds on the bond market before they reach maturity at the bond's **market price** that may be higher or lower than the par value.

OBJECTIVES

- Compute the percent of par for which bonds are selling.
- Compute the annual interest on bonds.
- Compute the current yields on bonds.

TEACHING THE LESSON

WARM-UP Write the following exercises on the chalkboard or an overhead transparency.

Find the answer to the nearest tenth of a percent.

1. $797 is what percent of $1,000?

2. $1,214 is what percent of $1,000?

3. $64.82 is what percent of $950?

4. $15.75 is what percent of $200?

(**1.** 79.7% **2.** 121.4% **3.** 6.8% **4.** 7.9%)

INTRODUCTION Use the introductory material to explain that corporations have 2 ways to raise money—they can issue stock or they can issue bonds. Explain that when people invest in bonds, they are lending the corporation money for which they are paid interest and then repaid the original amount of the bond when it matures at some point in the future. Also explain that few people keep their bonds until maturity; rather, bonds are bought and sold on the bond market, which works like the stock market. Explain the difference between *par value* and *market value*.

Examine and discuss the bond listing in the text, including what the figures mean. Point out that state and local governments issue bonds that are very popular because, although they usually pay lower interest, they are tax free.

INSTRUCTION Use Example 1 to reinforce the notion of par value and that bonds sell at prices both above and below their par values. Remind students that these kinds of calculations involving division by 1,000 should be done mentally.

Use Example 2 to demonstrate interest calculations on bonds. Remind students that interest is always paid on the bond's par value and not on its market value.

Use Example 3 to show how to find the current yield on a bond. Remind students that, when finding yield, they are looking at how much they will make as it relates to how much they spend. Therefore, while interest is based on par value, yield is based on market value.

Select some other bond issues from the business pages of a newspaper. Calculate the current yields on several published issues and decide which are the best investments.

SKILLS BANK You may want to refer students to the following skills in the Skills Bank:

Find the Percent of a Number, Skills 1–2

Find the Percent One Number Is of Another, Skills 1–2

BONDS

Bond	Maturity date	Interest rate	Market price
Excelsior Corporation	1994	$5\frac{1}{2}\%$	\$825
Roadway Company	1997	$7\frac{1}{4}\%$	\$1,075
State of Ohio	1992	$6\frac{3}{4}\%$	\$937

EXAMPLE 1 At what percent of par are the Excelsior Corporation bonds currently selling?

THINK: \$1,000 par value bonds are selling for \$825.

Divide to find the percent. \$825 ÷ \$1,000 = 0.825 = 82.5%

The bonds are currently selling at 82.5% of par.

EXAMPLE 2 How much interest would you receive per year if you owned 8 State of Ohio \$1,000 bonds?

THINK: Interest is paid on the par value, not on the market price.

1. Multiply to find the interest on 1 bond.

   $6\frac{3}{4}\%$ of \$1,000 = 0.0675 × \$1,000 = \$67.50

2. Multiply to find the interest on 8 bonds.

   8 × \$67.50 = \$540

You would earn \$540 annually in interest.

EXAMPLE 3 What is the current yield on a \$1,000 Roadway Company bond?

THINK: CURRENT YIELD = ANNUAL INTEREST ÷ MARKET PRICE

1. Multiply to find the annual interest.

   $7\frac{1}{4}\%$ × \$1,000 = 0.0725 × \$1,000 = \$72.50

2. Divide to find the current yield.

   \$72.50 ÷ \$1,075 = 0.0674418 = 6.7%

The current yield is 6.7%.

FOR DISCUSSION See TE side column.

1. How are bonds different from stocks?
2. How are municipal bonds different from corporate bonds?

PRACTICE EXERCISES

Remember to estimate whenever you use your calculator.

BONDS			
Chapman International	1990	$8\frac{3}{4}$%	\$975
City of Yuma	1995	$5\frac{3}{8}$%	\$950
Dover Products	1991	$7\frac{3}{4}$%	\$1,050
Eastern Metals	1997	$9\frac{1}{2}$%	\$837.50
Finch Township	1996	$6\frac{5}{8}$%	\$1,087.50

Use the bond listing above.
Which bond(s):

• 1. Has the earliest maturity? Chapman International

• 2. Is currently the most expensive? Finch Township

3. Has the lowest interest rate? City of Yuma

4. Has the highest interest rate? Eastern Metals

5. Are selling below par? Chapman International, City of Yuma, Eastern Metals

6. Are selling above par? Dover Products, Finch Township

Find the current cost of the bonds.

• 7. 4 Dover bonds $4,200

8. 8 Finch bonds $8,700

9. 20 Eastern Metals bonds $16,750

10. 15 City of Yuma bonds $14,250

At what percent of par is the $1,000 bond currently selling?

• 11. Market price: $950 95%

• 12. Market price: $875 87.5%

13. Market price: $837.50 83.75%

14. Market price: $1,500 150%

15. Market price: $1,050 105%

16. Market price: $1,125 112.5%

Find the annual interest you would earn each year from the bonds.

• 17. 6 Chapman $1,000 bonds $525

18. 10 City of Yuma $1,000 bonds $537.50

19. 30 Eastern $1,000 bonds $2,850

20. 50 Finch $1,000 bonds $3,312.50

Find the current yield on the bond.

• 21. Chapman 9.0%

22. City of Yuma 5.7%

23. Dover 7.4%

24. Finch 6.1%

FOR DISCUSSION

1. Students should explain that stocks represent a piece of ownership in a corporation, while bonds represent only a loan to a corporation; stocks pay dividends, while bonds pay interest.
2. Students should explain that the primary difference between most municipal bonds and most corporate bonds is that the municipal bonds are tax free—that is, you pay no income tax on the interest earned.

ASSIGNMENTS

Level	Exercises
Level 1	1–8, 11–12, 17–18, 21–22
Level 2	1–24
Level 3	1–6, Odd 7–23

FOLLOW-UP ACTIVITY

APPLICATION Provide each student with a copy of the bond listing from a newspaper. Ask each student to select a bond and pretend to invest $5,000 in the bond. At periodic intervals over the next month, provide each student with an updated bond listing so that students can monitor the market value of their bonds. At the end of a month, ask all students to sell their bonds and calculate their net earning or loss.

SUPPLEMENTARY MATERIALS

TRP Practice, p. 110
TRP Reteaching, p. 90
TRP Transparency 55

OBJECTIVE

- Find the values of shares of mutual funds.

TEACHING THE LESSON

WARM-UP Write the following exercises on the chalkboard or an overhead transparency.

Multiply.

1. 50 × $7.37 **2.** 220 × $15.84

3. 150 × $25.53 **4.** 500 × $46.83

(**1.** $368.50 **2.** $3,484.80 **3.** $3,829.50 **4.** $23,415)

INTRODUCTION Begin discussion by explaining that mutual funds are pools of money from many investors that are professionally invested in a range of investments. Explain that, rather than you having to pick and choose stocks and bonds for your own personal portfolio, a mutual fund allows you to benefit from the wisdom of a few professional investors and the buying power of many investors.

INSTRUCTION Examine and discuss the mutual fund listing in the textbook or from a local newspaper. Discuss how this listing is similar to and different from the stock listing studied earlier. Emphasize the difference between the *buy price* and the *sell price*.

Use Example 1 to show how to find the price of 80 shares of a mutual fund. Explain that 1 reason there is a difference between the buy price and the sell price is to cover a service charge when you first buy many mutual funds. Explain that those funds marked NL have "no load," or service charge, and thus no difference between their buy price and their sell price.

Use Example 2 to compute the value of shares of a mutual fund. Point out that these calculations are exactly the same as those used for buying stocks and bonds.

16.5 MUTUAL FUNDS

Rather than selecting your own mixture of stocks and other investments, you can pool your money with other investors and buy shares in a **mutual fund.** When the fund earns money on its investments, the profits are paid to investors in the fund in the form of **dividends.**

Most mutual funds have a **sell price** that represents the value of a share and a **buy price** that is usually higher than the sell price since it includes a sales charge.

According to the mutual fund listing, a share of the Corvette Fund can be sold for $23.17, it can be purchased for $24.76, and its value increased by 6¢ a share from the previous day.

MUTUAL FUNDS

Fund	Sell	Buy	Chg.
Allegiance	10.31	10.74	+.09
Americana	15.07	NL	+.07
Cornell	13.98	15.28	–.04
Corvette	23.17	24.76	–.06
Criteria	17.18	NL	+.14
Dayton	6.95	7.24	–.03
Forecast	7.38	7.85	+.07
Future	24.12	NL	–.11
Generation	10.25	11.01	+.02
Guard	36.54	38.12	+.16
Hanover	16.45	NL	–.03
International	12.83	14.12	+.08
Keymark	9.56	NL	–.14

NL means No Load
(Buy price is equal to sell price.)

EXAMPLE 1 What will it cost to buy 80 shares of the Hanover Fund?

THINK: Since the fund is NL, the buy price is the same as the sell price.

Multiply to find the price.

80 × $16.45 = $1,316

80 shares will cost $1,316.00.

EXAMPLE 2 An investor sells 250 shares of the Forecast Fund. What is the current value of these shares?

THINK: The sell price for a share of the fund is $7.38.

Multiply to find the value. 250 × $7.38 = $1,845

The 250 shares are currently worth $1,845.00.

FOR DISCUSSION See TE side column.

What are the advantages of investing in a mutual fund rather than buying individual stocks or bonds?

PRACTICE EXERCISES

Remember to estimate whenever you use your calculator.

Use the mutual funds listing on page 330. Complete the table.

Fund	Cost to buy 1 share	Cost to buy 50 shares	Cost to buy 250 shares
Guard	• 1. ■ $38.12	• 2. ■ $1,906	3. ■ $9,530
Criteria	4. ■ $17.18	5. ■ $859	6. ■ $4,295
Cornell	7. ■ $15.28	8. ■ $764	9. ■ $3,820
Future	10. ■ $24.12	11. ■ $1,206	12. ■ $6,030

Find the current value of the mutual fund shares.

•**13.** 400 shares of Allegiance
$4,124

•**14.** 1,500 shares of Keymark
$14,340

15. 550 shares of Generation
$5,637.50

16. 40 shares of Dayton
$278

17. 2,200 shares of Americana
$33,154

18. 750 shares of International
$9,622.50

Solve.

•**19.** An investor has $4,000 to invest in a mutual fund. How many shares of the Forecast Fund can be bought?
509 shares

20. An investor has $12,000 to invest in a mutual fund. How many shares of the Hanover Fund can be bought?
729 shares

21. An investor bought 400 shares of the Generation Fund at $8.17 per share. If the shares are sold at today's price, how much profit will be made?
$832

22. An investor bought 650 shares of the Criteria Fund at $21.74 per share. If the shares are sold at today's price, how much of a loss will there be?
$2,964

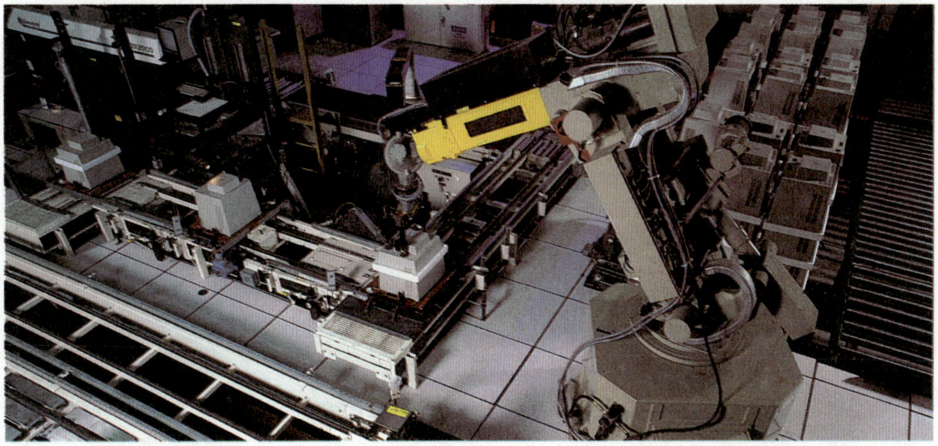

You may want to point out that mutual funds are very popular forms of investment because they often offer competitive rates of return and built-in diversification.

SKILLS BANK You may want to refer students to the following skill in the Skills Bank:

Multiply With Decimals, Skill 5

FOR DISCUSSION Students should explain that, since mutual funds represent a broad array of investments, investing in mutual funds is usually less risky than buying and selling individual stocks. Moreover, since they are professionally managed, their earnings, or yields are often higher than most individuals would be likely to get by themselves.

ASSIGNMENTS

Level	Exercises
Level 1	1–18
Level 2	1–20
Level 3	7–20

FOLLOW-UP ACTIVITY

COOPERATIVE LEARNING
Divide the class into pairs and give each pair $2,000 to spend on mutual funds. Provide each pair with a listing of mutual funds and tell each pair to decide into which 3 funds they would like to buy. Tell each pair to make a buy order for shares of exactly 3 mutual funds without going over the $2,000 limit. Let students, still working in pairs, monitor the progress of the mutual fund market and, at the end of the week, see who earns the most and the least.

SUPPLEMENTARY MATERIALS

TRP Practice, p. 111
TRP Reteaching, p. 91
TRP Transparency 56

OBJECTIVE

- Compute annual pension benefits.

TEACHING THE LESSON

WARM-UP Write the following exercises on the chalkboard or an overhead transparency.

Multiply.

1. 1.20% × 18 × $9,450 ($2,041.20)

2. 1.32% × 15 × $12,400 ($2,455.20)

3. 1.85% × 24 × $20,755 ($9,215.22)

INTRODUCTION Explain that pensions and pension funds are used to provide income to workers after they retire. Point out that many employees have pension plans in addition to Social Security. Explain that most pension funds combine employee contributions from payroll deductions with employer contributions to create a large pool of money that is invested to earn interest and dividends.

INSTRUCTION Use the Example to show how to use the formula and benefit table to compute annual pension benefits when you retire. Explain that many people do these kinds of calculations to decide when they can retire with enough income to maintain their lifestyle.

Review the components of the formula and point out that the longer you work, the greater your benefit, and, the more you make, the greater your benefit.

Provide students with other situations—including the worker's age, years of service, and annual income—and ask students to compute the annual and monthly pension benefits.

SKILLS BANK You may want to refer students to the following skills in the Skills Bank:

Multiply With Decimals, Skill 5

Divide With Decimals, Skill 8

16.6 PENSION PLANS

Many employees are automatically enrolled in a **pension plan** that provides them with a **pension** after they retire. Most pension plans combine employee payroll deductions with employer contributions that are invested in stocks, bonds, or mutual funds.

Many pension benefits can be calculated using the formula and the benefit percent table below.

$$\text{ANNUAL PENSION BENEFIT} = \text{BENEFIT PERCENT} \times \text{YEARS OF SERVICE} \times \text{AVERAGE ANNUAL EARNINGS}$$

BENEFIT PERCENT FOR EACH YEAR OF SERVICE

Years of Service	Age 55	56	57	58	59	60	61	62	63	64	65+
5 to 9	0	0	0								
10	1.00	1.06	1.12	1.20	1.30	1.40	1.52	1.64	1.76	1.88	2.00
11	1.07	1.12	1.18	1.25	1.35	1.44	1.55	1.66	1.78	1.89	2.00
12	1.13	1.18	1.24	1.31	1.39	1.48	1.58	1.69	1.79	1.90	2.00
13	1.20	1.25	1.29	1.36	1.44	1.52	1.62	1.71	1.81	1.90	2.00
14	1.26	1.31	1.35	1.41	1.48	1.56	1.65	1.74	1.82	1.91	2.00
15	1.33	1.37	1.41	1.47	1.53	1.60	1.68	1.76	1.84	1.92	2.00
16	1.40	1.43	1.47	1.52	1.58	1.64	1.71	1.78	1.86	1.93	2.00
17	1.46	1.49	1.53	1.57	1.62	1.68	1.74	1.81	1.87	1.94	2.00
18	1.53	1.56	1.58	1.62	1.67	1.72	1.78	1.83	1.89	1.94	2.00
19	1.59	1.62	1.64	1.68	1.71	1.76	1.81	1.86	1.90	1.95	2.00
20	1.66	1.68	1.70	1.73	1.76	1.80	1.84	1.88	1.92	1.96	2.00
21	1.73	1.74	1.76	1.78	1.81	1.84	1.87	1.90	1.94	1.97	2.00
22	1.79	1.80	1.82	1.84	1.85	1.88	1.90	1.93	1.95	1.98	2.00
23	1.86	1.87	1.87	1.89	1.90	1.92	1.94	1.95	1.97	1.98	2.00
24	1.93	1.93	1.93	1.94	1.94	1.96	1.97	1.98	1.98	1.99	2.00
25 or over	2.00	2.00	2.00	2.00	2.00	2.00	2.00	2.00	2.00	2.00	2.00

EXAMPLE A worker retires at age 59 with 18 y of service in the company. If her average annual earnings were $35,000, what is her annual pension benefit? How much will she receive per month?

1. Use the table. The benefit percent for a 59-year-old with 18 y of service is 1.67.
2. Use the formula to find the annual benefit. 1.67% × 18 × $35,000 = $10,521
3. Divide to find the monthly benefit. $10,521 ÷ 12 = $876.75

This worker will receive a pension of $10,521 per y, or $876.75 per mo.

FOR DISCUSSION See TE side column.

Why does an employee's contributions to his or her pension fund represent only a small part of the eventual pension benefits?

PRACTICE EXERCISES

Remember to estimate whenever you use your calculator.

Use the benefit percent table on page 332. Complete the table.

Age at retirement	Years of service	Average annual earnings	Benefit percent	Annual pension	Monthly pension
65	20	$24,000	• 1. ■	• 2. ■	3. ■
57	15	$18,900	4. ■	5. ■	6. ■
58	23	$12,400	7. ■	8. ■	9. ■
63	12	$39,850	10. ■	11. ■	12. ■
69	34	$24,375	13. ■	14. ■	15. ■

1. 2.0 **2.** $9,600 **3.** $800 **4.** 1.41 **5.** $3,997.35 **6.** $333.11 **7.** 1.89 **8.** $5,390.28 **9.** $449.19 **10.** 1.79 **11.** $8,559.78 **12.** $713.32 **13.** 2.0 **14.** $16,575 **15.** $1,381.25

EXTENSION IRAs

IRAs, or **Individual Retirement Accounts,** allow many people who are not covered by company pension plans to save for their retirement on their own.

Person(s)	Adjusted Gross Income	Full annual tax deduction	Tax-deferred investment income
Single	below $25,000	Yes; $2,000	All
Single	above $25,000	No	All
Married couple; 1 earner	below $40,000	Yes; $2,250	All
Married couple; 2 earners	below $40,000	Yes; $4,000	All
Married couple	above $40,000	No	All

1. Why are people not covered by company pension plans allowed to deduct their annual IRA contributions from their taxes?
To help them afford to have their own retirement pension plans.

At what income level do the person(s) lose their ability to deduct their entire IRA contributions from their taxes?

2. Single person $25,000
3. Married couple $40,000

4. How are IRAs beneficial to all of the persons listed in the table?
Investment income is also tax deferred.

FOR DISCUSSION Students should explain that, since pension fund contributions are invested and earn interest, the fund grows as much from ongoing employee contributions as from investment income.

EXTENSION Explain how IRAs work and use the table to illustrate how different people may or may not benefit from maintaining one each year. Review and explain the terms *tax deduction* and *tax-deferred income*.

ASSIGNMENTS

Level 1	1–15, EX
Level 2	1–15, EX
Level 3	1–15, EX

FOLLOW-UP ACTIVITY

ENRICHMENT Describe a simulation of a worker whose earnings over 40 y increase $500 per y and grow from $5,000 to $25,000. Assume that the worker contributes 3% of his or her annual earnings to a pension fund. Also assume that the worker's pension benefit provides $1\frac{1}{2}$% of the last year's income for each year of service. Ask students to compute the worker's annual pension benefit and compare pension benefits with pension contributions.

SUPPLEMENTARY MATERIALS

TRP Practice, p. 112
TRP Reteaching, p. 92
TRP Transparency 56

OBJECTIVE

- Solve problems by working backward.

TEACHING THE LESSON

WARM-UP Write the following exercises on the chalkboard or an overhead transparency.

Find the answer.

1. $\frac{1}{2}$ of 10,000 (5,000)
2. $\frac{1}{2}$ of 24,000 (12,000)
3. $\frac{1}{3}$ of 21,000 (7,000)
4. $\frac{1}{3}$ of 75,000 (25,000)

INTRODUCTION Discuss situations in which the solutions to problems can be found by working backward. One example might be deciding which bus to take in order to get to the movie theater on time for the 7:30 P.M. show. Elicit from students that they must work backward from the time the movie starts and subtract the amount of time for the bus ride. They must also consider which bus stops in their town and the town the movie theater is in near to the correct time.

INSTRUCTION Present the problem situation to the class. Review the facts given by asking students:

- How much money does Edie want? ($24,000)
- When does she need it? (in 18 y)
- How does she expect her money to grow in this mutual fund? (double every 9 y)
- What do you need to find? (the amount of money she should initially invest)

Read each step of the solution. If students are using calculators for this lesson, emphasize the idea that multiplying a number by $\frac{1}{2}$ is the same as dividing that number by 2.

PROBLEM Solving STRATEGY

16.7 WORKING BACKWARD

Situation:

Edie wants to make 1 investment now so that she will have $24,000 to take a luxury cruise when she retires in 18 y. She invests in a mutual fund in which her money is expected to double every 9 y. How much money should Edie initially invest in order to have $24,000 in 18 y?

Strategy:

Working backward can help you find a solution.

Applying the Strategy:

1. What amount should be in Edie's fund at the end of the 9th y so it would double to be $24,000 by the end of the 18th y?

   THINK: $\frac{1}{2}$ of \$24,000 = $\frac{1}{2}\times$ \$24,000 = \$12,000

2. What amount should Edie invest initially so that at the end of the 9th y it would double to be $12,000?

   THINK: $\frac{1}{2}$ of \$12,000 = $\frac{1}{2}\times$ \$12,000 = \$6,000

   Edie should initially invest $6,000.

FOR DISCUSSION

See TE side column.

Roy invests money once in a real-estate partnership. It is expected that his money will triple every 7 y. He wants to have $108,000 in 21 years to set up his own business.

1. Explain how you can work backward to determine how much money Roy should initially invest.
2. What is the solution?

PRACTICE EXERCISES

Remember to estimate whenever you use your calculator.

Solve by working backward.

Alice wants to have \$40,000 saved to help pay for her daughter's college education 18 y from now. She invests in a corporate bond fund in which her money is expected to double every 9 y.

1. What amount should be in Alice's fund at the end of the 9th y so it would double to be \$40,000 by the end of the 18th y?
\$20,000

2. What amount should Alice initially invest so that at the end of the 9th y it would double to be \$20,000?
\$10,000

Ben invests money once in a stock income fund. It is expected that the money in his account will triple every 8 y. Ben wants to have \$81,000 in 24 y to supplement his retirement income.

• 3. What amount should be in Ben's account at the end of the 16th y so it would triple to be \$81,000 by the end of the 24th y?
\$27,000

• 4. What amount should be in Ben's account at the end of the 8th y so it would triple to be \$27,000 by the end of the 16th y?
\$9,000

5. What amount should Ben initially invest so that at the end of the 8th y it would triple to be \$9,000?
\$3,000

6. Susan wants to have \$100,000 saved to begin investing in real estate 20 y from now. She invests in a fund she expects to double her money every 10 y. How much money should Susan invest in the fund?
\$25,000

7. Beth wants to have \$36,000 saved to start her own business 12 y from now. She invests money once in a bond fund. She expects her money to double every 6 y. How much money should Beth initially invest?
\$9,000

8. Roberto buys a zero coupon bond for his newborn grandson. It is expected that the money will double every 12 y. How much should Roberto spend on the bond so that his grandson will have \$75,000 in 36 y?
\$9,375

9. Alicia wants to have \$35,000 saved to help pay for some property in 15 y. Her mutual funds are expected to double her money every $7\frac{1}{2}$ y. How much should Alicia initially invest in the funds?
\$8,750

10. Helga invests money in a fund in which her money is expected to double every 7.5 y. How much money should Helga invest now if she wants to have \$200,000 for retirement in 30 y?
\$12,500

11. Martin wants to have \$270,000 saved for his retirement $37\frac{1}{2}$ y from now. He invests in a stock and bond portfolio in which his money is expected to triple every $12\frac{1}{2}$ y. What should Martin initially invest in the portfolio?
\$10,000

FOR DISCUSSION

1. Students should explain that they need to find the amount Roy has:
   (a) At the end of the 14th y, so it would triple to be \$108,000 at the end of the 21st y.
   (b) At the end of the 7th y, so it would triple to be \$36,000 at the end of the 14th y.
   (c) At the beginning of the 1st y, so it would triple to be \$12,000 at the end of the 7th y.
2. Students should explain that Roy must initially invest \$4,000.

SKILLS BANK You may want to refer students to the following skill in the Skills Bank:

Multiply With Fractions and Mixed Numbers, Skill 3

ASSIGNMENTS

Level 1	1–7
Level 2	1–9
Level 3	3–11

FOLLOW-UP ACTIVITY

SITUATIONAL PROBLEM SOLVING Present students with the following problem: You are offered 2 money-making options.
Option 1: \$1,000 per d for 1 mo (30 d)
Option 2: 1 penny on the first day of the month. Each day of the month after that the money would double.
Which option would you choose?

Some questions to ask include the following: How much would you have at the end of the month under each option? Should you work forward or backward to solve this problem?

Some approaches to solving the problem include the following: working forward to find each daily amount, comparing the monthly totals.

SUPPLEMENTARY MATERIALS

TRP Practice, p. 113
TRP Applications, p. 18
TRP Transparency 57

OBJECTIVE

- Decide, given relevant information, in which stock(s) to invest.

TEACHING THE LESSON

WARM-UP Write the following exercises on the chalkboard or an overhead transparency.

Divide. Round to the nearest hundredth.

1. \$1.20 ÷ \$42.50 **2.** \$0.75 ÷ \$24.25

Find the answer to the nearest tenth of a percent.

3. \$5.75 is what percent of \$35.62?

4. \$1.25 is what percent of \$18.75?

(**1.** 0.03 **2.** 0.03 **3.** 16.1% **4.** 6.7%)

INTRODUCTION Present the situation to the class. Encourage students to name some of the factors they would use to make an informed decision about in which stock(s) to invest. List on the chalkboard the factors that are named and discuss how to find out information on each of these factors.

INSTRUCTION Read the information about the 6 stocks under consideration. Ask which ones look most appealing and least appealing on the surface. Ask students to discuss fully their reasons for liking or not liking particular stocks. Ask if there is any other information that you should have. Discuss the decision-making factors.

Have small groups of students complete Exercises 1–25 in the table in order to compare the 6 stocks. Ask students to write statements that describe the data now in the table. Then ask them to write conclusions that can be drawn from the data. Discuss the appropriateness of these statements and conclusions and lead a discussion about interpretations and possible misinterpretations of the data.

DECISION MAKING

16.8 BUYING STOCKS

(Open-ended problem solving)

Louis and Rebecca O'Rourke decide it would be fun, and perhaps profitable, to invest some of their money in the stock market.

PROBLEM

Louis and Rebecca decide to start with a \$3,000 investment. They visit a stockbroker and ask for the names of a variety of "high-tech" companies. They take the list and use the stock listings from their newspaper and information from their broker to prepare the following table. In which stock or stocks should they invest?

New Era	\$37.50/share	20¢ dividend	Up \$5/share in past year	High risk
Geoparts	\$87.37/share	\$2.75 dividend	Down \$20/share in past year	Moderate risk
Technics	\$3.87/share	No dividend	Up 50¢/share in past year	Moderate risk
Laserpert	\$15.25/share	\$1.75 dividend	Up \$4.50/share in past year	Low risk
Jetstorm	\$117.75/share	\$4.50 dividend	Down \$2/share in past year	Low risk
Phantom	\$21.50/share	\$1.00 dividend	Up \$3.00/share in past year	High risk

DECISION-MAKING FACTORS

Yield (Dividends ÷ Price) Growth prospects Risk

DECISION-MAKING COMPARISONS

1. \$37.50 **2.** 0.5% **3.** 15.4% (I) **4.** High **5.** \$87.37 **6.** \$2.75 **7.** 3.1% **8.** 18.6% (D) **9.** None **10.** 0% **11.** 14.8% (I) **12.** Moderate

Compare the 6 stocks by completing the table.

Stock	Price per share	Dividend per share	Yield	Percent Increase (I) or decrease (D) in price over past year	Risk
New Era	• 1. ■	\$0.20	• 2. ■	3. ■	4. ■
Geoparts	5. ■	6. ■	7. ■	8. ■	Moderate
Technics	\$3.87	9. ■	10. ■	11. ■	12. ■
Laserpert	13. ■	14. ■	11.5%	15. ■	16. ■
Jetstorm	17. ■	18. ■	19. ■	1.7% (D)	20. ■
Phantom	21. ■	22. ■	23. ■	24. ■	25. ■

13. \$15.25 **14.** \$1.75 **15.** 41.9% (I) **16.** Low **17.** \$117.75 **18.** \$4.50 **19.** 3.8% **20.** Low **21.** \$21.50 **22.** \$1.00 **23.** 4.7% **24.** 16.2% (I) **25.** High

MAKING THE DECISIONS

Which 1 stock should Louis and Rebecca buy if the only factor were:

• **26.** Lowest price? Technics

• **27.** Highest dividend? Jetstorm

28. Highest yield? Laserpert

29. Greatest recent percent growth? Laserpert

30. Which stock should be purchased if Louis and Rebecca wish to minimize risk and maximize yield?
Laserpert

31. Which stock should be purchased if Louis and Rebecca are willing to take high risks and wish to maximize the yield?
Phantom

32. Which stock(s) are probably the safest investments? Why?
Laserpert, Jetstorm; low risk

33. Louis and Rebecca decide to place $\frac{1}{2}$ of their investment in a low-risk, low-yield investment and half of their investment in a high-risk, high-yield investment. How many shares of which stocks should they buy?
12 shares of Jetstorm and 69 shares of Phantom

34. How would you invest the $3,000? Why? Answers may vary.

Ask students to complete Exercises 26–34 in small groups, having students explain and defend their answers and decisions. Remind students that in Exercise 34 there is no one right answer. Some students are very wary of defending their point of view among their classmates. Continually reinforce the fact that in many consumer issues, best answers depend on individual situations and preferences. Encourage all students to participate and express their opinions.

ASSIGNMENTS

Level 1	1–34
Level 2	1–34
Level 3	1–34

FOLLOW-UP ACTIVITY

APPLICATION Select 5 or 6 large corporations with plants or offices located in or near your community and with stock traded on the New York Stock Exchange. Provide students with the stock listing from the local newspaper and tell them that they have $1,000 to invest in 1 or more of the listed corporations. Ask students to use the information in the stock listing to decide how to invest the $1,000. Ask students to submit their decisions in writing, including (1) how many shares of which stock or stocks will be purchased, and (2) a paragraph justifying for the decision.

SUPPLEMENTARY MATERIALS

TRP Practice, p. 114
TRP Transparency 57

OBJECTIVES

- Review vocabulary.
- Practice key chapter concepts and skills.

USING THE REVIEW

The Chapter Review is designed to help students prepare for taking the Chapter Test. The first section focuses on vocabulary. It requires students to select words to complete statements. The second section presents practice of key chapter skills.

Each item on the Chapter Review is referenced to the page on which the topic is taught in the Pupil's Edition. You may wish to have students refer to these pages to help review any concepts or skills they have not yet mastered.

It is suggested that students work in small-sized heterogeneous cooperative learning groups. Some cooperative learning methods that may be used are as follows:

1. After each student has independently completed the Chapter Review, a discussion should follow within each group about the solutions to the problems.
2. The group can complete the entire Chapter Review by working together to solve the problems.

End the lesson with an entire class discussion in which any questions brought up in group discussions are presented and answered.

CHAPTER REVIEW

Vocabulary Choose the letter of the word(s) that completes the sentence.

1. Investments that allow you to pool your money and invest in a variety of investments are called ■. [330] b

   a. Stocks **b.** Mutual funds **c.** Corporate bonds

2. Investments that are really loans to the federal government are called ■. [320] a

   a. Savings bonds **b.** Government stocks **c.** Certificates of deposit

3. Investments are best compared based on their ■. [322, 336] c

   a. Dividends **b.** Interest rate **c.** Yield

Skills Find the answer.

4. Warren redeemed a \$500 U.S. savings Bond for \$389.67. How much interest did he earn on the bond? [320]
   \$139.67

5. How much interest will Jennifer earn on a 6-mo \$2,000 CD that has an annual effective yield of 7.784%? [322]
   \$77.84

6. What is the cost of 60 shares of a stock selling at $35\frac{5}{8}$? [324]
   \$2,137.50

7. What is the value of 500 shares of a stock trading at $56\frac{1}{8}$? [324]
   \$28,062.50

8. How much profit did Myra make when she bought 30 shares of a stock at $17\frac{3}{4}$ and sold than at $31\frac{3}{8}$? [324]
   \$408.75

9. What was the loss when Kevin was forced to sell 150 shares of stock at $27\frac{1}{2}$ that he purchased at $41\frac{3}{4}$? [324]
   \$2,137.50

10. What is the yield on a stock that sells for $46\frac{1}{2}$ and pays \$2.75 in dividends? [324]
    5.9%

11. What is the yield on a bond that costs \$953 and pays \$68.93 in interest annually? [327]
    7.2%

12. How much interest is earned on a \$1,000 bond that pays 6.492%? [327]
    \$64.92

13. How much will Chita earn on 240 shares of a mutual fund that pays \$5.90 per share in dividends? [330]
    \$1,416.00

14. A pension plan pays 2% per y for each year of service. How much will the annual pension benefit be for a woman with 24 y of service and average annual earnings of \$37,500? [332]
    \$18,000

15. A pension plan pays 1.5% per y for each year of service. How much will the annual pension benefit be for a man with 17 y of service and average annual earnings of \$26,000? [332]
    \$6,630

CHAPTER TEST

Complete the table.

Type of investment	Purchase price	Interest or dividend	Yield
A share of stock	$27.50	$1.15 dividend	1. ■
A corporate bond	$935.00	$89.47 interest	2. ■
A share of a mutual fund	$17.50	42¢ dividend	3. ■
A municipal bond	$1,000	$57.90 interest	4. ■

1. 4.2% **2.** 9.6% **3.** 2.4% **4.** 5.8%

Solve.

5. If a $50 U.S. Savings Bond can be redeemed for $38.89 after 6 y, how much will a $1,000 bond be worth after 6 y?
   $777.80

6. How much interest is earned on a $3,000 6-mo CD that offers an effective annual yield of 6.86%?
   $102.90

7. What is the cost of 125 shares of a stock selling at $59\frac{3}{8}$?
   $7,421.88

8. How much profit did you make when you bought 750 shares of a stock at $12\frac{1}{2}$ and sold them at $19\frac{3}{4}$?
   $5,437.50

9. How much was lost when you sold 25 shares of stock at $25\frac{1}{8}$ that you purchased at $34\frac{3}{4}$?
   $240.63

10. How much interest is earned on a $1,000 bond that pays 5.831%?
    $58.31

11. A pension plan pays 2% per y for each year of service. How much will the annual pension benefit be for a worker with 28 y of service and average annual earnings of $18,000?
    $10,080

12. You want to have $48,000 saved in 18 y. The fund you invest in expects to double your money after 9 y. How much should you initially invest in the fund?
    $12,000

You decide to invest $1,000 in either:

a. A stock selling for $14.50 that pays $1.00 in dividends.
b. A corporate bond that pays 8.52%.
c. A mutual fund selling for $28.75 that pays 25¢ in dividends.
d. A 1-y CD with an effective annual yield of 7.41%.

13. What are the annual dividends paid on $1,000 of the stock?
    $68.97

14. What are the annual dividends paid on $1,000 of the mutual fund?
    $8.70

15. Which offers the higher return, the stock or the mutual fund?
    stock

16. How much more interest will the corporate bond pay per year than the CD?
    $11.10

OBJECTIVE

- Evaluate the achievement of chapter objectives.

USING THE TEST

The Chapter Test may be used as a post-test to evaluate student achievement. However, you may wish to use the Chapter Post-Test offered in the Teacher's Resource Package or to design your own chapter test. If this page is not used as a test, you may wish to assign it as additional review or practice.

The test items are correlated to the chapter objectives in the table below.

Chapter objectives	Test items
A. Compute the yield on an investment.	1–4
B. Compute the value of a U.S. Savings Bond.	5
C. Compute the interest earned on a CD.	6
D. Compute the value of shares of stock.	7
E. Compute the profit or loss on a stock transaction.	8–9
F. Compute the interest earned on a bond.	10
G. Compute pension fund benefits.	11
H. Solve problems by working backward.	12
I. Compare dividends and interest on various investments.	13–16

SUPPLEMENTARY MATERIALS

TRP Ch. 16 Post-Test Form A, pp. 1–2
TRP Ch. 16 Post-Test Form B, pp. 1–2

OBJECTIVE

- Understand that a retirement plan for self-employed persons enables them to save money.

TEACHING THE LESSON

Begin discussion by telling students that anyone earning any self-employed income can open his or her own Keogh profit-sharing plan.

Instruct students to study the formulas. Have students work individually or in pairs to answer Exercises 1–5. Summarize the answers as follows: Self-employed persons can invest in a retirement plan that can earn them money *and* save them tax money now and in the future.

Have students work individually to answer Exercises 6–7. Then have them share their answers and reasons in a whole-class discussion.

ANSWERS/EXPLANATIONS

1. To provide for a more comfortable retirement; to shelter hard-earned money from taxes; to invest in a plan that may accumulate and compound at a rate higher than other investments would offer
4. Less earned income (self-employed persons' income usually fluctuates from year to year, often dramatically); higher expenses; money used for more immediate needs (car, home repair or remodeling, a new baby)
5. Self-employed persons deduct business expenses from gross income, as federal tax is based on a lower amount and tax due is also lower.
6. (a) Invest the 10%, pay your expenses, and live on a more limited budget, or
   (b) Invest the 10% and cut down expenses where possible, or
   (c) Invest less, say 5%, to ease the financial burden, or
   (d) Invest nothing for that year.
   Answers will vary as to which option to choose.

MONEY TIPS

Self-employed persons can invest in and manage their own retirement plans.

LET'S LOOK AT THE FACTS

Most companies have retirement plans for their employees. Self-employed persons can establish similar plans for themselves. One type is called a **Keogh profit-sharing plan,** which allows you to invest varying amounts up to 15% of your Net Earned Income (NEI). The formulas shown below can be used to calculate your NEI and your contribution amounts.

$$\text{NET EARNED INCOME} = \frac{\text{NET PROFITS (GROSS INCOME MINUS EXPENSES)}^*}{1 + \%\ \text{OF CONTRIBUTION}}$$

$$\text{PLAN CONTRIBUTION} = \text{NET EARNED INCOME (NEI)} \times \%\ \text{OF CONTRIBUTION}$$

***Gross income minus expenses, shown on Schedule C of your 1040 tax return.**

LET'S DISCUSS WHY See TE side column for answers to Exercises 1 and 4–6.

1. Why do you think it would be a good idea for self-employed persons to invest in a retirement plan?
2. You can invest as much as 15% of your NEI but no more than $30,000 in any one year. If $30,000 was 15% of your NEI, what would your total annual NEI be? $200,000
3. If your annual NEI is $40,000, what would your plan contribution be if you invest 5% of it? 8% of it? 12% of it? $2,000; $3,200; $4,800
4. You are not required to contribute to a retirement plan every year. What might be some reasons for not contributing in some years?
5. Contributions to a Keogh profit-sharing plan are considered business expense deductions. How would this affect the taxes you pay now?

LET'S SEE WHAT YOU WOULD DO

6. You are a freelance artist. Last year, you earned $35,000 and you invested 5% of it in a Keogh profit-sharing plan. This year you will earn about the same amount of money but would like to invest 10% of it. Your expenses this year, however, are higher. What are your options, and which do you think would be the most financially sound option to choose?
7. You have been investing in a tax-deferred Keogh profit-sharing plan since age 38. You are now age 64, healthy, and able to continue working productively. Discuss the financial pros and cons of retiring. Answers may vary.

ESTIMATION SKILLS

OVERESTIMATES AND UNDERESTIMATES OF PRODUCTS AND QUOTIENTS

The symbols $^{+}$ and $^{-}$ may be used to show that an actual product is greater than or less than an estimate.

Examples

38×295 ↓ ↓ $40 \times 300 = 12{,}000$ **First estimate:** 12,000	**THINK:** Each factor was rounded up, so the actual product is less than 12,000. **Final estimate:** $12{,}000^{-}$
9.2×5.327 ↓ ↓ $9 \times 5 = 45$ **First estimate:** 45	**THINK:** Each factor was rounded down, so the actual product is greater than 45. **Final estimate:** 45^{+}
$18 \times \$74.45$ ↓ ↓ $20 \times \$70 = \$1{,}400$ **First estimate:** $1,400	**THINK:** One factor was rounded up and one was rounded down. It is not immediately clear whether $1,400 is high or low. Keep the first estimate. **Final estimate:** $1,400

Estimated quotients can be adjusted in a similar way.

Examples

$822 \div 21$ ↓ ↓ $800 \div 20 = 40$ **First estimate:** 40	**THINK:** $21 \times 40 = 840$ Since $840 > 822$, the actual quotient is less than 40. **Final estimate:** 40^{-}
$\$142.98 \div 28$ ↓ ↓ $\$150 \div 30 = \$5$ **First estimate:** $5	**THINK:** $28 \times \$5 = \$140$ Since $\$140 < \$142.98$, the actual quotient is greater than $5. **Final estimate:** $\$5^{+}$

Estimate the product or quotient. Where appropriate, use the symbol $^{+}$ or $^{-}$ to indicate whether the actual answer is greater than or less than the estimate.

1. 49×66 $3{,}500^{-}$	**2.** 32×325 $9{,}000^{+}$	**3.** 58×315 18,000	**4.** $27 \times \$2.98$ $\$90^{-}$
5. $18 \times \$2.29$ $40	**6.** $31 \times \$21.50$ $\$600^{+}$	**7.** 4.2×9.3 36^{+}	**8.** 6.7×19.38 140^{-}
9. 5.25×7.854 40	**10.** $616 \div 32$ 20^{-}	**11.** $855 \div 42$ 20^{+}	**12.** $4{,}316 \div 81$ 50^{+}
13. $\$38.60 \div 20$ $\$2^{-}$	**14.** $\$74.16 \div 25$ $\$3^{-}$	**15.** $\$560.98 \div 72$ $\$8^{-}$	**16.** $9.15 \div 2.8$ 3^{+}

OBJECTIVE

- Use the symbols $^{+}$ and $^{-}$ to indicate overestimated and underestimated products and quotients.

USING THE PAGE

Students should not find it difficult to understand that rounding each of 2 factors *up* leads to an overestimate, while rounding both factors *down* leads to an underestimate. The suggested procedure for determining overestimated and underestimated quotients, however, is somewhat less easy to apply and to understand. This method is based on the fact that division and multiplication are inverse operations.

To illustrate, the reasoning used in the first division example on the page, $822 \div 21$, is as follows:

If $822 \div 21 = q$, then $21 \times q = 822$. The estimate for q is 40, but $21 \times 40 = 840$. Since 840 is greater than 822, 40 is an overestimate.

Note that in the last example on the page, compatible numbers were used to determine the first estimate, $5. (See page 257 for a discussion of compatible numbers in division.) Students may use various pairs of compatible numbers to find their estimates in Exercises 10–16. It cannot be emphasized too much that estimates given in this Teacher's Edition are not the only reasonable estimates. If time permits, you might ask students with different (but reasonable) estimates to explain their procedures to the class.

OBJECTIVES

- Review and maintain previously taught concepts and skills.
- Practice taking tests in multiple-choice format.

USING THE REVIEW

Assign the Cumulative Review to all students. Provide students with an answer sheet to record their answers.

Each Cumulative Review gives students an opportunity to practice taking tests that are written in multiple-choice format.

Provide appropriate remedial help for students having difficulty with any of the skills and concepts on this page.

CUMULATIVE REVIEW

Choose the letter of the word(s) that completes the sentence.

1. The ■ calculator key can be used when you have entered the wrong number into the calculator. b

   a. Clear **b.** Clear entry **c.** Memory **d.** None of these

2. A check written for more than the balance in your account is called ■. c

   a. New balance **b.** A payment **c.** An overdraft **d.** None of these

3. The percent of the mortgage amount paid at the closing is called ■. a

   a. Points **b.** Mortgage rate **c.** Appreciation **d.** None of these

4. An organized plan for spending money is called ■. c

   a. An annual expense **b.** A charge account **c.** A budget **d.** None of these

5. Investments that are really loans to the federal government are called ■. b

   a. Stocks **b.** Savings bonds **c.** Mutual funds **d.** None of these

Select the best estimated answer.

6. 85¢ + \$3.18 + \$4.96 d **a.** \$7.00 **b.** \$7.50 **c.** \$8.00 **d.** \$8.90
7. 0.42 − 0.089 b **a.** 0.2 **b.** 0.3 **c.** 0.4 **d.** 0.5
8. 82 × \$0.36 c **a.** \$2.40 **b.** \$3.20 **c.** \$32 **d.** \$320
9. 416.3 ÷ 7.89 c **a.** 5 **b.** 6 **c.** 50 **d.** \$500

Compute.

10. 38.7 + 42.9 + 16.34 c

    a. 96.94 **b.** 97.9 **c.** 97.94 **d.** None of these

11. 6,504 − 4,397 a

    a. 2,107 **b.** 2,297 **c.** 10,901 **d.** None of these

12. 9 × 5.73 b

    a. 45.37 **b.** 51.57 **c.** 54 **d.** None of these

13. $8\overline{)7.76}$ b

    a. 0.9 **b.** 0.97 **c.** 97 **d.** None of these

14. $33\frac{1}{3}\%$ of 54 b

a. 5.4 b. 18 c. 27 d. None of these

Solve.

15. Sandy works a 40-h week at an hourly rate of $7.45. Her deductions total $93.50. What is her net pay? b

a. $167.25 b. $204.50 c. $391.50 d. None of these

16. During the billing period, you used 4,236 kWh of electricity. The utility company charges $0.0375 per kWh. What was the total cost of electricity during this period? b

a. $127.08 b. $158.85 c. $168.85 d. None of these

17. If 3% of each production run is pulled for inspection, how many items will be inspected in a production run of 800 items? a

a. 24 b. 30 c. 80 d. None of these

18. The average price of a house was $70,000. The average price is now $77,000. Find the inflation rate. c

a. 1% b. 7% c. 10% d. None of these

19. Fencing was installed to enclose a yard that measures 6 yd by 3 yd. The fence posts are 1 yd apart. How many posts are there? a

a. 18 b. 20 c. 162 d. None of these

20. You want to have $63,000 saved in 18 y. The fund you invest in expects to triple your money every 9 y. How much should you initially invest in the fund to reach your goal? d

a. $6,000 b. $8,000 c. $10,000 d. None of these

THINKING ABOUT MATH See TE side column for answers.

1. Why do masons usually order more bricks than their computations indicate?

2. Why is inflation particularly difficult for people with fixed incomes?

3. How would the budget for a single person differ from the budget for a family?

4. Why are some people hesitant to purchase long-term certificates of deposit?

THINKING ABOUT MATH The questions in this section can be used in discussions with individual students or with small groups. Guide students in conveying their ideas clearly and precisely.

Listed below are expected student answers. However, accept any reasonable answers.

1. Students should explain that more construction material is ordered than needed because of possible breakage, change in plans, or need to break bricks to fit spaces.
2. Inflation is difficult for people on fixed incomes because purchasing power goes down as the inflation rate goes up.
3. Though fixed expenses are approximately the same for a single person and a family, living expenses (food, household, personal, and so on) are greatly increased for a family. Some annual expenses (medical, dental, and vacation) may be much more for a family.
4. Some people are hesitant to purchase long-term CDs because, even though the interest rate is higher than on a short-term one, they do not have access to the money for a longer period. Also, some people want flexibility so that, as rates go up or down, they can change the investment without a penalty.

SUPPLEMENTARY MATERIALS

TRP Cum. Test, Ch. 1–16, pp. 1–4

OBJECTIVE

- Test the prerequisite skills needed to learn the concepts developed in Chapter 17.

USING THE TEST

The Pre-Skills Test is designed to diagnose students' strengths and weaknesses on prerequisite skills necessary to study the mathematics in Chapter 17.

Assign the Pre-Skills Test. Allow students to work together in pairs or in small groups. Group members should help those who demonstrate a misunderstanding of a concept or a weakness in a skill.

The table below correlates the items on the Pre-Skills Test with the prerequisite skill and the lesson(s) in the chapter for which it is needed.

Item(s)	Prerequisite skill	Lesson(s)
1–4	Write fractions in lowest terms.	All
5–8	Multiply fractions.	17.3–17.5
9–12	Rename fractions as percents.	17.5
13–16	Round decimals.	17.5
17–20	Find the percent of a number.	17.5
21–29	Identify possible outcomes.	17.1–17.4, 17.6

References to specific skills that are reviewed in the Skills Bank are given on the individual lessons themselves.

Pre-Skills Test

Write in lowest terms.

1. $\frac{5}{10}$ $\frac{1}{2}$
2. $\frac{8}{12}$ $\frac{2}{3}$
3. $\frac{36}{100}$ $\frac{9}{25}$
4. $\frac{24}{72}$ $\frac{1}{3}$

Multiply. Write the answer in lowest terms.

5. $\frac{1}{4} \times \frac{3}{4}$ $\frac{3}{16}$
6. $\frac{2}{3} \times \frac{5}{8}$ $\frac{5}{12}$
7. $\frac{3}{10} \times \frac{4}{9}$ $\frac{2}{15}$
8. $\frac{2}{5} \times \frac{3}{7}$ $\frac{6}{35}$

Rename as a percent.

9. $\frac{1}{5}$ 20%
10. $\frac{34}{50}$ 68%
11. $\frac{6}{20}$ 30%
12. $\frac{12}{30}$ 40%

Round to the nearest whole number.

13. 14.86 15
14. 3.1242 3
15. 837.9284 838
16. 25.395 25

Find the answer.

17. 20% of 350 70
18. 10% of 283 28.3
19. $33\frac{1}{3}\%$ of 660 220
20. 60% of 750 450

Look at a standard number cube.

21. How many possible numbers can come up for 1 roll? 6
22. How many numbers are greater than 2? 4
23. How many numbers are less than 3? 2
24. How many numbers are even? 3
25. How many numbers are odd? 3

Look at a standard deck of playing cards.

26. How many cards are there in the deck? 52
27. How many spades are there in the deck? 13
28. How many 6s are there in the deck? 4
29. How many queens are there in the deck? 4

Meteorologists study climate and weather conditions. Weather predictions are made on the basis of past weather patterns. How reliable are the daily predictions of a 5-day weather forecast?

OBJECTIVE

- Formulate problems, using text and data.

USING THE CHAPTER OPENER

Discuss students' answers to the question on the page. Answers will vary, but most people believe that forecasts are usually wrong—especially long-range forecasts. Be sure students understand that forecasts are only predictions; and that, while meteorology is a science, it is a science with a great many variables that cannot be controlled.

You may also want to discuss why accurate forecasts are important to some people. These might include farmers, construction-company supervisors, people who work outdoors, and so on.

COOPERATIVE LEARNING

Assign heterogeneous groups the task of tracking a 5-d weather forecast. It would be useful if each group got the forecast from a different source, such as radio, television, or newspapers. Have students be sure to note if the forecast is provided by the station, the paper itself, or a weather service. In checking the accuracy of the prediction, they should not rely on the same source, but on their own observations. Also, they should continue to monitor their source to keep track of how the forecasts change during the period.

At the end of the period, compare findings to see which forecast came closest to the actual weather.

SUPPLEMENTARY MATERIALS

TRP Group Projects, p. 17

OBJECTIVES

- Find the probabilities of simple events.
- Define probabilities of 0 and 1.

TEACHING THE LESSON

WARM-UP Write the following exercises on the chalkboard or an overhead transparency.

Write the fraction in lowest terms.

1. $\frac{4}{6}$ **2.** $\frac{14}{28}$ **3.** $\frac{8}{8}$

4. $\frac{12}{20}$ **5.** $\frac{25}{75}$ **6.** $\frac{0}{4}$

(**1.** $\frac{2}{3}$ **2.** $\frac{1}{2}$ **3.** 1 **4.** $\frac{3}{5}$ **5.** $\frac{1}{3}$ **6.** 0)

INTRODUCTION Begin discussion by asking students to name situations that involve uncertainty. The list might include such things as the weather, future events, the gender of your next child, and so forth. Explain that probability is used to help make sense of this uncertainty and to make predictions about what might happen.

INSTRUCTION Divide students into pairs and give 1 number cube to each pair. Tell students to roll the cube 10 times and record the outcome of each roll. Ask students how many different outcomes were possible when they rolled the cube. Use Example 1 to explain what is meant by the *probability* of an event. Point out that since rolling a 4 was 1 out of 6 possible outcomes, the probability is $\frac{1}{6}$. Tally the class's data. Compare the experimental results with the theoretical result.

Use Example 2 to reinforce the definition of probability and show how to deal with multiple favorable outcomes.

Use Example 3 to discuss probabilities of events *not* occurring. Point out that not rolling a 6 is the same as actually rolling anything but a 6. Point out that $P(6)$ plus $P(\text{not } 6)$ equals 1.

Use Examples 4 and 5 to explain that *impossible events* have a probability of 0 and *certain events* have a probability of 1.

17.1 SIMPLE PROBABILITY

Probability is used to make educated guesses about the likelihood of events. The probability, or chance, of an event, or **outcome,** occurring is a ratio:

$$P(\text{Event}) = \frac{\text{Number of favorable outcomes}}{\text{Number of possible outcomes}}$$

Suppose you roll a number cube. You can roll either a 1, 2, 3, 4, 5, or 6. These are the 6 **possible outcomes** for a roll. Each is **equally likely** to occur.

EXAMPLE 1 Find the probability of rolling a 4.

THINK: There is only 1 favorable outcome—a 4—out of the 6 possible outcomes.

The probability of rolling a 4 or $P(4) = \frac{1}{6}$.

EXAMPLE 2 Find the probability of rolling an even number.

THINK: There are 3 favorable outcomes—a 2, 4, or 6—out of the 6 possible outcomes.

The probability of rolling an even number or $P(\text{even number}) = \frac{3}{6}$, or $\frac{1}{2}$ in simplest form.

EXAMPLE 3 Find the probability of *not* rolling a 6.

THINK: There are 5 favorable outcomes—a 1, 2, 3, 4, or 5—out of the 6 possible outcomes.

The probability of not rolling a 6 or $P(\text{not } 6) = \frac{5}{6}$.

EXAMPLE 4 Find the probability of rolling an 8.

THINK: There are no favorable outcomes possible.

The probability of rolling an 8 or $P(8) = \frac{0}{6}$, or 0.

An **impossible event** has a probability of 0.

EXAMPLE 5 Find the probability of rolling a number less than 10.

THINK: All possible outcomes are favorable.

The probability of rolling a number less than 10 or $P(\text{less than } 10) = \frac{6}{6}$, or 1.

A **certain event** has a probability of 1.

FOR DISCUSSION

See TE side column.

Why can P(event) never be greater than 1?

PRACTICE EXERCISES

Find the probability of the event in simplest form.

You pick 1 marble at random from the bag. What is:

1. P(red)? $\frac{1}{2}$
2. P(blue)? $\frac{1}{3}$
3. P(black)? 0
4. P(red or yellow)? $\frac{2}{3}$
5. P(not blue)? $\frac{2}{3}$
6. P(not yellow)? $\frac{5}{6}$

You pick a number at random from a deck of cards numbered 1–20. What is:

7. $P(7)$? $\frac{1}{20}$
8. P(4, 5, or 6)? $\frac{3}{20}$
9. P(15 or greater)? $\frac{3}{10}$
10. P(greater than 15)? $\frac{1}{4}$
11. P(even)? $\frac{1}{2}$
12. P(not even)? $\frac{1}{2}$
13. P(odd or less than 10)? $\frac{7}{10}$
14. P(odd and less than 10)? $\frac{1}{4}$
15. $P(25)$? 0
16. P(positive whole number)? 1

You pick a single card at random from a deck of playing cards. What is:

17. P(jack of spades)? $\frac{1}{52}$
18. P(red ten)? $\frac{1}{26}$
19. P(club)? $\frac{1}{4}$
20. P(king)? $\frac{1}{13}$
21. P(not a heart)? $\frac{3}{4}$
22. P(not an ace)? $\frac{12}{13}$
23. P(picture card)? $\frac{3}{13}$
24. P(red spade)? 0

SKILLS BANK You may want to refer students to the following skill in the Skills Bank:

Lowest Terms, Skill 2

FOR DISCUSSION Students should explain that P(event) can never be greater than 1 because P represents the ratio of some or all of the favorable outcomes over all possible outcomes. The greatest ratio possible is $\frac{n}{n}$, or 1, or 100%.

ASSIGNMENTS

Level	Exercises
Level 1	1–16
Level 2	1–24
Level 3	7–24

FOLLOW-UP ACTIVITY

ENRICHMENT First pull a jack, queen, king, and ace from a deck of playing cards. Turn them face down and ask students to select 1 card. Record the card picked. Repeat this activity in small groups or pairs until approximately 400 games have been played. Tally the class results and compare the experimental outcomes with the predicted outcomes; that is, was each card selected $\frac{1}{4}$ of the time?

Now ask each student to pick 1 of the cards mentally—either a jack, queen, king, or ace. Ask students to try this with 10 other students. Tally the class results for this experiment and compare them with the predicted outcome.

Use the notion of *randomness* to discuss why the first experiment is more likely to approach predicted outcomes than the second experiment.

SUPPLEMENTARY MATERIALS

TRP Practice, p. 115
TRP Reteaching, p. 93
TRP Transparency 58

OBJECTIVES

- Draw tree diagrams and list sample spaces.
- Use the fundamental counting principle.

TEACHING THE LESSON

WARM-UP Write the following exercises on the chalkboard or an overhead transparency.

List the possible outcomes of:

1. A flipped coin.
2. A rolled number cube.
3. Drawing from a set of cards marked A, 5, B, 7, and F.
4. Spinning a spinner with sections labeled red, green, pink, and orange.

(**1.** heads, tails **2.** 1, 2, 3, 4, 5, 6 **3.** A, 5, B, 7, F **4.** red, green, pink, orange)

INTRODUCTION Ask students what might happen if you were to flip a coin 3 times in a row. Students should respond that you could get no heads, 1 head, 2 heads, or 3 heads. Ask them whether flipping a head, then a tail, and then a head (2 heads) is different from flipping a tail, then a head, and then another head (also 2 heads). Students should respond that although both occurrences entail 2 heads, they are different.

Use a *tree diagram* to show how to display all of the possible ways that you could flip a coin 3 times. Use the tree to show the difference between head/tail/head and tail/head/head. Also use the tree to show that 1 and 2 heads are more likely than no or 3 heads.

INSTRUCTION Use Example 1 to reinforce the idea of using a tree diagram to illustrate possible outcomes of multiple events. Point out that tree diagrams are based on thinking "This could happen first, and then this could happen, and so forth"—until all possible outcomes are listed.

17.2 SAMPLE SPACE

One way to determine the total number of possible outcomes of an event is to make a list or a table. This list of all possible outcomes in a given situation is called a **sample space.**

A common method for generating the sample space for an event is using a **tree diagram.** Each **branch** represents a choice or possible outcome.

EXAMPLE 1 You flip a coin with 1 hand and roll a number cube with the other. What are the possible outcomes of this activity?

Since the coin can come up either heads or tails and the number cube can be rolled as either 1, 2, 3, 4, 5, or 6, you can use a tree diagram to make a list of all possible outcomes.

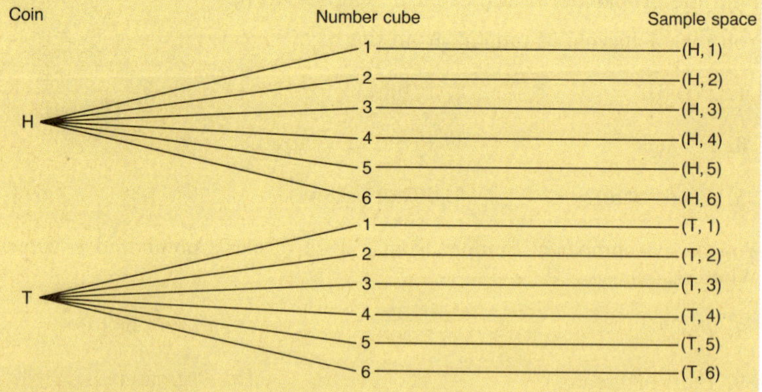

Sometimes when you have a series of events, you need to find only the number of possible outcomes and not what each of them is. You can multiply the number of ways each event can occur. This is called the **counting principle.**

EXAMPLE 2 A menu offers 3 types of salads, 4 different entrees, and 3 desserts. How many different meals could be ordered?

THINK: You could make a tree diagram, but it is much easier to use the counting principle.

3 types of salad × 4 entrees × 3 desserts = 36 different meals

You could order 36 different meals.

FOR DISCUSSION See TE side column.

In Example 1, why are each of the sample space outcomes equally likely to occur?

PRACTICE EXERCISES

Remember to estimate whenever you use your calculator.
See pages 477–478 for answers to Exercises 1–8. Check students' tree diagrams.

Make a tree diagram and then list the sample space for the event.

• 1. Flipping 2 quarters

• 2. Flipping 3 dimes

3. Rolling 2 number cubes

4. Spinning 2 spinners numbered 1–8

5. Selecting 2 marbles at random from a bag containing 2 red, 2 blue, and 1 green marble

6. Selecting 1 card from each group

2 4 6 and 10 20 30

7. Spinning both of these spinners

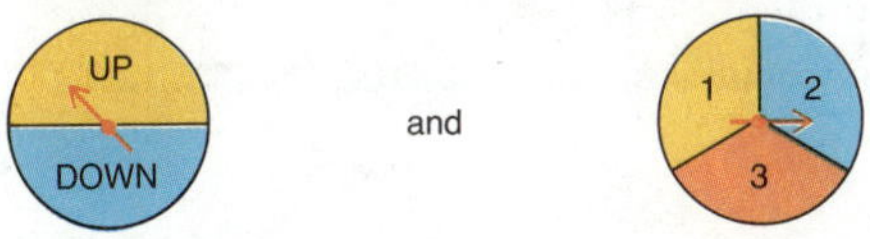

8. Picking a card and rolling a number cube

A B C D and

Solve.

• 9. Josh has 5 shirts and 3 pairs of slacks. How many different outfits can he create with 1 shirt and 1 pair of slacks?
15 outfits

10. Cindy has 7 sweaters and 5 skirts. How many different outfits can she create with 1 sweater and 1 skirt?
35 outfits

11. Frozen yogurt sundaes can be ordered with 3 types of fruit toppings, 2 types of nuts, and 1 of 6 flavors of yogurt. How many different types of sundaes can be ordered? 36 sundaes

12. Lunch includes 5 different types of sandwiches, 3 types of bread, and 5 types of beverages. How many different lunches can be ordered?
75 lunches

Show how the tree diagram is then converted into a listing of the *sample space.*

Use Example 2 to illustrate the power of the fundamental counting principle. First make a tree diagram to show all the possible orders and then show how the same result arises out of simply multiplying.

SKILLS BANK You may want to refer students to the following skill in the Skills Bank:

Multiply With Whole Numbers, Skill 2

FOR DISCUSSION Students should explain that for each outcome the same coin and number cube are used. Since the conditions are the same for each flip and roll, the 12 sample space outcomes are equally likely to occur each time.

ASSIGNMENTS

Level 1	Odd 1–11
Level 2	1–12
Level 3	1–12

FOLLOW-UP ACTIVITY

ENRICHMENT Explain to students that different states use different combinations of numbers and letters for license plates. Tell students to determine the total number of different license plates possible in a state that uses:

a. 2 letters and 3 numbers

b. 2 letters and 4 numbers

c. 3 letters and 2 numbers

d. 3 letters and 3 numbers

Ask how many different plates are possible in your state.

SUPPLEMENTARY MATERIALS

TRP Practice, p. 116
TRP Reteaching, p. 94
TRP Transparency 58

OBJECTIVE

- Find the probability of 2 independent events both occurring.

TEACHING THE LESSON

WARM-UP Write the following exercises on the chalkboard or an overhead transparency.

Multiply.

1. $\frac{1}{3} \times \frac{1}{5}$ **2.** $\frac{1}{4} \times \frac{2}{3}$

3. $\frac{1}{2} \times \frac{2}{5} \times \frac{5}{8}$ **4.** $\frac{4}{9} \times \frac{3}{8}$

(**1.** $\frac{1}{15}$ **2.** $\frac{1}{6}$ **3.** $\frac{1}{8}$ **4.** $\frac{1}{6}$)

INTRODUCTION Ask students what the probability is that, if you have 3 children, they will all be girls. Suggest that they draw a tree diagram to find the total number of possible outcomes and establish that the probability is $\frac{1}{8}$. Then ask them to consider the probability of having a girl for the first child $\left(\frac{1}{2}\right)$, a girl for the second child $\left(\frac{1}{2}\right)$, and a girl for the third child $\left(\frac{1}{2}\right)$. Explain that since the gender of each child is independent of the gender of any other children, you also get $\frac{1}{8}$ by multiplying the individual probabilities. $\left(\frac{1}{2} \times \frac{1}{2} \times \frac{1}{2} = \frac{1}{8}\right)$

INSTRUCTION Use Example 1 to reinforce the method for finding the probability of *independent events* occurring together. Show students how to construct a sample space for 2 rolls of the number cube and how much easier it is to multiply the individual probabilities to get the probability of $\frac{1}{36}$. Discuss other situations in which events are independent.

Use Example 2 to show how the process works for 3 independent events. Again ask students to use a tree diagram and to list the sample space so that they see that the method of multiplying individual probabilities works.

17.3 INDEPENDENT EVENTS

You roll a number cube. Then you roll it again. Since the outcome of the second roll is not affected in any way by the first roll, we call these **independent events.**

To find the probability of a series of independent events occurring together, you can multiply the probabilities of the individual events.

EXAMPLE 1 Find the probability of rolling a 2 on the first roll and a 4 on the second roll.

THINK: $P(2) = \frac{1}{6}$ $\quad P(4) = \frac{1}{6}$

Multiply. $P(2, \text{then } 4) = P(2) \times P(4)$

$= \frac{1}{6} \times \frac{1}{6} = \frac{1}{36}$

The probability of rolling a 2 and then a 4 is $\frac{1}{36}$.

You spin each of the spinners below.

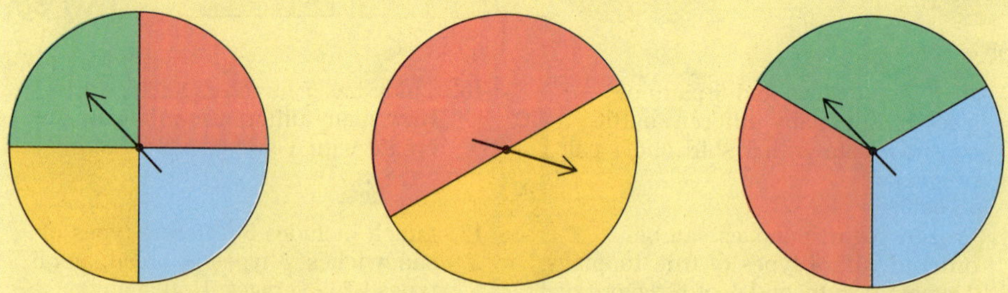

EXAMPLE 2 What is the probability that each of the spinners will point to red?

THINK: Each spinner is independent of the other spinners, so each spin is an independent event.

Multiply. $P(\text{red, red, and red}) = P(\text{red}) \times P(\text{red}) \times P(\text{red})$

$= \frac{1}{4} \times \frac{1}{2} \times \frac{1}{3} = \frac{1}{24}$

The probability that each spinner will point to red is $\frac{1}{24}$.

FOR DISCUSSION See TE side column.

When rolling a number cube, why does the first roll *not* affect further rolls?

PRACTICE EXERCISES

Remember to estimate whenever you use your calculator.

Find the probability.

Event: Rolling a number cube and then rolling it again.

• 1. $P(3,3)$ $\frac{1}{36}$

• 2. $P(8,2)$ 0

3. P(odd number, even number) $\frac{1}{4}$

4. P(not 5, not 3) $\frac{25}{36}$

5. P(1, 2, or 3, and then 4, 5, or 6) $\frac{1}{4}$

6. P(odd, greater than 2) $\frac{1}{3}$

Event: Selecting 1 marble at random, replacing it, and selecting another from a bag containing 3 blue, 5 red, and 2 yellow marbles.

• 7. P(red, red) $\frac{1}{4}$

• 8. P(red, blue) $\frac{3}{20}$

9. P(yellow, yellow) $\frac{1}{25}$

10. P(blue or yellow, blue or red) $\frac{2}{5}$

11. P(not blue, not blue) $\frac{49}{100}$

12. P(black) 0

Event: Randomly selecting a card from a deck of playing cards, replacing it, and selecting another card.

• 13. P(black, black) $\frac{1}{4}$

• 14. P(red, black) $\frac{1}{4}$

15. P(black, heart) $\frac{1}{8}$

16. P(diamond, spade) $\frac{1}{16}$

17. P(picture card, picture card) $\frac{9}{169}$

18. P(ace of hearts, ace of hearts) $\frac{1}{2,704}$

Event: Spinning the spinner and then flipping the coin.

• 19. P(odd number and a head) $\frac{3}{10}$

• 20. P(2 or 3 and a tail) $\frac{1}{5}$

21. P(5 and a head) $\frac{1}{10}$

22. P(not a 4 and a tail) $\frac{2}{5}$

23. P(not a 1 or a 2 and not a head) $\frac{3}{10}$

24. P(3, 4, or 5 and either a head or a tail) $\frac{3}{5}$

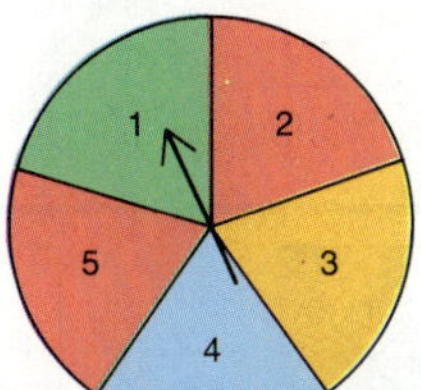

Have students construct their own spinners and try to replicate this experiment. Tally the class's results and compare experimental outcomes with predicted probabilities.

SKILLS BANK You may want to refer students to the following skills in the Skills Bank:

Multiply With Fractions and Mixed Numbers, Skills 1–2

FOR DISCUSSION Students should explain that after the first roll, there are still the same 6 numbers (possible outcomes) available for further rolls. Therefore, the first roll has no effect on further rolls.

ASSIGNMENTS

Level 1	1–10, 13–16, 19–21
Level 2	1–22
Level 3	1–24

FOLLOW-UP ACTIVITY

APPLICATION Give each student a 5-item multiple-choice test in an imaginary, unintelligible language that requires each student to guess on each item. Before marking the test, ask students to predict how many students in the class would get all the items correct, 4 correct, 3 correct, and so forth. Discuss these predictions. Then announce your preconceived answers to the test items and have students score their tests. Tally the results and record them on the chalkboard. Compare the predictions with the actual results.

Now use a tree diagram (first item right or wrong, second item right or wrong, and so forth) to find the actual probabilities of 5, 4, 3, 2, 1, and none correct. Compare the actual results with the theoretical probabilities.

SUPPLEMENTARY MATERIALS

TRP Practice, p. 117
TRP Reteaching, p. 95
TRP Transparency 59

OBJECTIVE

- Find the probability of 2 dependent events occurring.

TEACHING THE LESSON

WARM-UP Write the following exercises on the chalkboard or an overhead transparency. (Check students' tree diagrams.)

1. Draw a tree diagram showing all the possible outcomes for picking 1 card, replacing it, and then picking another card from a set of cards labeled A, B, C, and D.
2. Draw a tree diagram showing all the possible outcomes for picking 1 card, not replacing it, and then picking another card from a set labeled A, B, C, and D.

INTRODUCTION Use the 2 Warm-Up tree diagrams to show the difference between independent events (replacing the card in #1) and dependent events (not replacing the card in #2). Ask students to look at the probability of picking card A first and B second in each of the situations.

INSTRUCTION Use Example 1 to reinforce the method for finding the probability of *dependent events* occurring together. Show students how to construct a sample space for picking the cards and how much easier it is to multiply the probabilities of the events. Discuss other situations in which events are dependent.

Use Example 2 to show how the process works when each outcome is not unique—2 triangles, for example. Again ask students to use a tree diagram and to list the sample space so they see that the method of multiplying individual probabilities works.

Have students construct their own cards and try to replicate this experiment. Tally the class's results and compare experimental outcomes with predicted probabilities.

17.4 DEPENDENT EVENTS

Suppose you are selecting from among 5 cards numbered 1, 2, 3, 4, and 5. The probability of selecting a 3, for example, is $\frac{1}{5}$. But if the 3 is *not replaced*, then the probability of next selecting a 5, for example, is $\frac{1}{4}$ because only 4 cards remain. Since the outcome of the second event is affected by the outcome of the first event, the events are called **dependent events.**

To find the probability of dependent events, you must take into account the affect of the first event on the second event.

EXAMPLE 1 Find the probability of selecting a 1, not replacing it, and then selecting a 2.

THINK: $P(1) = \frac{1}{5}$ Since 1 of 4 remaining cards is a 2, $P(2) = \frac{1}{4}$.

Multiply. $P(1, \text{ then } 2) = P(1) \times P(2 \text{ after } 1)$

$= \frac{1}{5} \times \frac{1}{4} = \frac{1}{20}$

The probability is $\frac{1}{20}$.

EXAMPLE 2 Find the probability of selecting the triangle card, not replacing it, and then selecting the other triangle card.

THINK: $P(\text{first } \triangle) = \frac{2}{5}$ Since 1 of 4 remaining cards is a $\triangle$, $P(\text{second } \triangle) = \frac{1}{4}$.

Multiply. $P(\triangle, \triangle) = P(\triangle) \times P(\triangle \text{ after } \triangle)$

$= \frac{2}{5} \times \frac{1}{4} = \frac{2}{20} = \frac{1}{10}$

The probability is $\frac{1}{10}$.

FOR DISCUSSION See TE side column.

When randomly selecting objects and not replacing them, why does the first selection affect further selections? How does the second selection affect further selections?

PRACTICE EXERCISES

Remember to estimate whenever you use your calculator.

Identify whether the events are dependent or independent.

• 1. Rolling a number cube and spinning a spinner independent

2. Picking a card, replacing it, and then picking another card independent

3. Picking a card, not replacing it, and then picking another card dependent

4. Flipping a coin and then flipping it again independent

5. First child is a boy, guessing the gender of the second child independent

Find the probability.

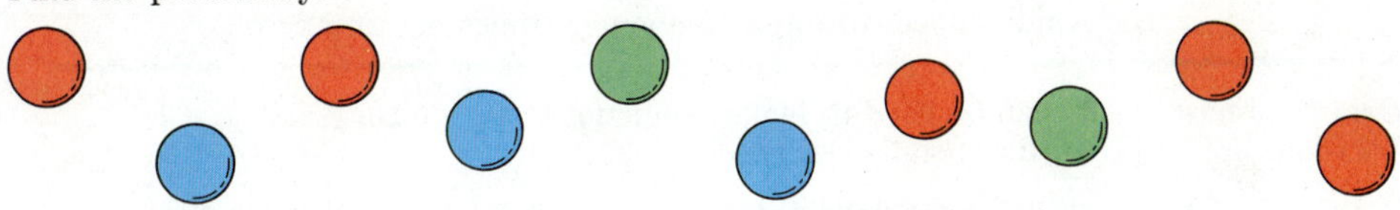

Event: Selecting 1 marble, not replacing it, and then selecting another.

• 6. P(red, red) $\frac{2}{9}$

• 7. P(red, blue) $\frac{1}{6}$

8. P(green, green) $\frac{1}{45}$

9. P(blue or green, and then red) $\frac{5}{18}$

10. P(not blue, not blue) $\frac{7}{15}$

11. P(not red, not red) $\frac{2}{9}$

Event: Picking a card, not replacing it, and then picking another card.

• 12. P(even number, even number) $\frac{3}{14}$

• 13. P(even, 5) $\frac{1}{14}$

14. P(odd number, even number) $\frac{2}{7}$

15. P(1, 2, or 3, and then 6, 7, or 8) $\frac{9}{56}$

16. P(4, 4) 0

17. P(8, and then a 1 or a 2) $\frac{1}{28}$

Event: Randomly selecting a card from a deck of playing cards, not replacing it, and then selecting another card.

• 18. P(black, black) $\frac{25}{102}$

• 19. P(red, black) $\frac{13}{51}$

20. P(black, heart) $\frac{13}{102}$

21. P(diamond, spade) $\frac{13}{204}$

22. P(picture card, picture card) $\frac{11}{221}$

23. P(ace of hearts, ace of clubs) $\frac{1}{2,652}$

SKILLS BANK You may want to refer students to the following skills in the Skills Bank:

Multiply With Fractions and Mixed Numbers, Skills 1–2

FOR DISCUSSION Students should explain that after the first selection, there is 1 fewer object (possible outcome) for the second selection. After the second selection, there are 2 fewer objects (possible outcomes) for further selection. Therefore, each selection affects the next selection.

ASSIGNMENTS

Level	Exercises
Level 1	1–17
Level 2	1–23
Level 3	6–23

FOLLOW-UP ACTIVITY

COOPERATIVE LEARNING Pair students and provide each pair with the following 6 cards:
jack, queen, and king of hearts
jack, queen, and king of spades

Tell students that they are to pick a card, not replace it, and then pick a second card each time they play. Tell them that they are going to play the game 20 times.

Before playing, ask students to:

1. Estimate the number of games in which they predict a heart will be chosen first and a spade second.
2. Compute the theoretical probability of selecting a heart first and then a spade.

After playing 20 games, compare and contrast the estimate, the theoretical probability, and the experimental outcomes. Ask students to describe what happened in 2 or 3 paragraphs.

SUPPLEMENTARY MATERIALS

TRP Practice, p. 118
TRP Reteaching, p. 96
TRP Transparency 59

OBJECTIVE

- Use probability to make predictions about outcomes.

TEACHING THE LESSON

WARM-UP Write the following exercises on the chalkboard or an overhead transparency.

Find the percent.

1. 10% of 480 **2.** 25% of 800

3. 3.7% of 1,000 **4.** 5% of 10,000

(**1.** 48 **2.** 200 **3.** 37 **4.** 500)

INTRODUCTION Begin discussion by asking what it means when a company states that there is a 2% chance of defect in its products. Discuss what this means if the company produces 100 items; 1,000 items; or 1,000,000 items. Ask students how they think the 2% chance figure was arrived at.

INSTRUCTION Use Example 1 to show how probabilities are applied to situations involving many repetitions of the same event. Point out that while probabilities are usually calculated on the basis of 1 event, they are most often used for making predictions on what will happen over many events. Also remind students that for many of the experiments done in this chapter, the more events, the closer the experimental outcomes should have been to the theoretical probability.

Use Example 2 to describe a real example of how probabilities are used by the insurance industry to set insurance rates. Reinforce the idea that while probability can tell you nothing about the likelihood of 1 particular person dying, it can be used to predict very accurately what will happen among a large population.

Discuss the mortality rates as a form of probability where the rates are really experimentally generated from past experience or events.

17.5 MAKING PREDICTIONS

Probability is most often used to **predict** the number of times a particular event will occur.

EXAMPLE 1 Suppose you roll 2 number cubes. What is the probability of rolling a sum of 12? If you rolled the 2 number cubes 500 times, how many times would you expect to roll a 12?

1. 12 can only be rolled when each number cube is a 6.

$$P(12) = P(6, 6) = P(6) \times P(6) = \frac{1}{6} \times \frac{1}{6} = \frac{1}{36}$$

The probability of rolling a 12 is $\frac{1}{36}$.

2. Multiply the number of trials by the probability to make a prediction.

$$\frac{1}{36} \times 500 \text{ trials} = \frac{500}{36} = 13.888888 \approx 14$$

You would expect to roll a 12 about 14 times out of 500 rolls.

Data from past events can be used to make predictions. Many business decisions are based on these statistical predictions.

EXAMPLE 2 An insurance company's experience is that 2.37 out of every 100 62-year-old males will die before reaching age 63. What is the probability that a 62-year-old male will die before reaching 63? If the insurance company insures 38,300 62-year-old males, how many would you expect to die before age 63?

1. Convert the data to a fraction or percent.

$$2.37 \text{ out of } 100 = \frac{2.37}{100}, \text{ or } 2.37\%$$

There is a 2.37% probability of dying.

2. Multipy to predict.

$$2.37\% \times 38{,}300 = 0.0237 \times 38{,}300 = 907.71 \approx 900$$

You would expect about 900 of the 38,300 62-year-old males to die.

FOR DISCUSSION See TE side column.

1. How does probability help insurance companies decide what rates to charge people for life-insurance policies?
2. How can probability help amusement-park owners decide what rides to construct?

PRACTICE EXERCISES

Remember to estimate whenever you use your calculator.

Situation: Historical records show the probabilities at the right for a July day's weather.

Sunny and beautiful	26.7%	Cloudy	19.2%
Hazy, hot, and humid	39.5%	Rainy	14.6%

How many days during this July would you predict:

• 1. It will be sunny and beautiful? 8 d

• 2. It will be cloudy? 6 d

3. It will not rain? 26 d

4. It will be cloudy or rainy? 10 d

5. It will not be beautiful? 23 d

6. You can schedule the perfect picnic? 8 d

Situation: George's hitting statistics for his last 100 at bats are shown at the right.

Home runs	4	Triples	2
Doubles	9	Singles	18
Walks	7	Outs	60

At this pace, if George records 538 at bats this season, predict:

• 7. How many doubles he will hit. 48 doubles

• 8. How many home runs he will hit. 22 home runs

9. How many times he will walk. 38 times

10. How many outs he will get. 323 outs

11. Compute George's on-base percent. 40%

12. Without counting walks, what is George's batting average for his last 100 at bats? 0.330

Situation: A game awards 1 point 70% of the time, 2 points 15% of the time, 3 points 10% of the time, and 4 points 5% of the time.

If the game is played 60 times, how many games would you predict:

• 13. Will get 1 point? 42 games

14. Will get 4 points? 3 games

15. Will get less than 3 points? 51 games

16. Will get more than 1 point? 18 games

17. If it takes 3 or 4 points to win, how many winners would you predict if the game were played 1,000 times? 150 winners

SKILLS BANK You may want to refer students to the following skills in the Skills Bank:

Find the Percent of a Number, Skills 1–2

FOR DISCUSSION

1. Students should explain that by using mortality rates based on past experience, insurance companies charge older people, men, and smokers higher rates for policies since the chances are higher of the company paying out the money to beneficiaries sooner.
2. Students should explain that by using data based on the past attendance of each type of ride, amusement park owners can decide which are the most profitable to construct.

ASSIGNMENTS

Level 1	1–10
Level 2	1–16
Level 3	1–17

FOLLOW-UP ACTIVITY

APPLICATION Provide students with a mortality table from a reference book or an insurance company. Discuss with students the meaning of the data in the table and how to convert data such as 2.4 per 1,000 to percents.

Develop a worksheet that requires students to predict the number of deaths for large numbers of people (100,000 or half a million) at different ages.

SUPPLEMENTARY MATERIALS

TRP Practice, p. 119
TRP Reteaching, p. 97
TRP Transparency 60

OBJECTIVE

- Solve problems by making an organized list.

TEACHING THE LESSON

WARM-UP Write the following exercises on the chalkboard or an overhead transparency.

1. Write all of the 2-digit numbers you can using the digits 2, 4, and 6 each only once.
   (24, 26, 42, 46, 62, 64)
2. Write all of the 3-digit numbers you can using the digits 2, 4, and 6 each only once.
   (246, 264, 426, 462, 624, 642)

INTRODUCTION Discuss situations in which ordered *arrangements,* or *permutations,* are important. (For example, naming the first-, second-, and third-place winners of a contest.) Discuss situations in which order is not important, or *combinations.* (For example, selecting 4 students from your school to represent your school in a contest.)

INSTRUCTION Present the problem situation to the class. Ask students whether or not order is important in naming different ways these players could be organized to bowl. (Yes, order is important.) Emphasize that they are finding arrangements.

As you read each question, use the chalkboard to show students how to make a list for organizing the answers.

Note that in Exercises 1 and 2, the number of arrangements is given, but students must list them. In Exercises 3 and 4, students must first list the possible arrangements and then tell how many there are. Note that in Exercises 5 and 6, the number of combinations to be made is given, but students must list them. In Exercises 7 and 8, students must first list the possible combinations and then tell how many there are.

PROBLEM *Solving* STRATEGY

17.6 MAKING AN ORGANIZED LIST

Situation:

John, Sheila, and Yuri are members of the Strikers bowling team. They will bowl in the first 3 positions of the team's bowling order. In how many different ways could the manager arrange these 3 players?

Strategy:

Making an organized list can sometimes help you to solve a problem.

Applying the Strategy:

	Bowling Order		
	First	**Second**	**Third**
1. How many different bowling orders are possible if John is first?	John John	Sheila Yuri	Yuri Sheila
2. How many different bowling orders are possible if Sheila is first?	Sheila Sheila	John Yuri	Yuri John
3. How many different bowling orders are possible if Yuri is first?	Yuri Yuri	John Sheila	Sheila John
4. How many different bowling orders are possible in all?	6		

The manager can make 6 arrangements of the first-, second-, and third-place bowlers.

FOR DISCUSSION

See TE side column.

Pia, Dale, and Ed all want to be on the Starlites bowling team. Only 2 new members are needed. Which list shows all the possible combinations of 2 new members that can be formed? Explain.

a. Pia and Dale
Pia and Ed
Dale and Ed

b. Pia and Dale — Dale and Pia
Pia and Ed — Ed and Pia
Dale and Ed — Ed and Dale

PRACTICE EXERCISES

Remember to estimate whenever you use your calculator.

Make a list to solve the problem. See pages 478–479 for answers to Exercises 1 and 3–8.

• **1.** Martin, Jane, and Eli are members of the Blazers bowling team. The manager plans to have them bowl in the first 3 positions of the bowling order. List all 6 possible arrangements.

• **2.** Joe is picking a 2-digit number for the back of his bowling shirt. He must choose from the digits 4, 8, and 9, without using a digit more than once. List all 6 possible arrangements. 48, 49, 84, 89, 94, 98

3. Jason, Meg, Ben, and Doris are members of the Ridgemont bowling team. They will bowl in the first 4 positions of the team's bowling order. List all possible arrangements. How many are there?

4. Ben is picking a 3-digit number for his locker in the bowling alley. He must choose from the digits 3, 5, 7, and 9, without using a digit more than once. List all possible arrangements. How many are there?

5. Sally, Kim, Dave, and Shaun all want to be on the Avery bowling team. Only 2 new members are needed. List all 6 possible combinations of 2 new members that can be formed.

6. Before the tournament begins, each of the 5 members of the Dalewood bowling team shakes hands with each of the other members once. List all 10 possible combinations of the handshakes made.

7. There are 6 bowlers at the bowling alley who do not belong to a team. The manager wants to make a team having 4 bowlers to bowl against the Bradford team. List all the possible combinations. How many are there?

8. The Starlites won 4 ribbons at the state tournament. At least 2 of their ribbons were 1st-place ribbons. List all the possible combinations of 1st-, 2nd-, and 3rd-place ribbons they could have won.

FOR DISCUSSION After reading the situation, ask students whether or not order is important in naming the new members of the team. (No, order is not important.) Emphasize that they are being asked to find combinations. Students should explain that list A is correct because list B shows each combination twice. For example, the team of Pia and Dale is the same as the team of Dale and Pia.

ASSIGNMENTS

Level	Exercises
Level 1	1, 2, 5, 6
Level 2	1–7
Level 3	3, 4, 7, 8

FOLLOW-UP ACTIVITY

SITUATIONAL PROBLEM SOLVING Present students with the following problem: You have won first prize in a state bowling tournament. You have a choice of spending a week in Hawaii or a week in New York City with all expenses paid. Which vacation will you choose?

Some questions to ask include the following: What would you want to see in each place? What activities could you do in each place? What are the pros and cons of each place?

Some approaches to solving the problem include the following: making a list of sightseeing and activity opportunities for each place and making a list of the pros and cons of each place.

SUPPLEMENTARY MATERIALS

TRP Practice, p. 120
TRP Applications, p. 19
TRP Transparency 60

OBJECTIVES

- Review vocabulary.
- Practice key chapter concepts and skills.

USING THE REVIEW

The Chapter Review is designed to help students prepare for taking the Chapter Test. The first section focuses on vocabulary. It requires students to select words to complete statements. The second section presents practice of key chapter skills.

Each item on the Chapter Review is referenced to the page on which the topic is taught in the Pupil's Edition. You may wish to have students refer to these pages to help review any concepts or skills they have not yet mastered.

It is suggested that students work in small-sized heterogeneous cooperative learning groups. Some cooperative learning methods that may be used are as follows:

1. After each student has independently completed the Chapter Review, a discussion should follow within each group about the solutions to the problems.
2. The group can complete the entire Chapter Review by working together to solve the problems.

End the lesson with an entire class discussion in which any questions brought up in group discussions are presented and answered.

CHAPTER REVIEW

Vocabulary Choose the letter of the word(s) that complete the sentence.

1. The likelihood of an event occurring is called the ■. [346] a
   a. Probability **b.** Occurrence rate **c.** Sample space
2. The list of all possible outcomes of an event is called a ■. [348] c
   a. Super list **b.** Tree **c.** Sample space
3. Two unrelated events are called ■. [350] b
   a. Dependent **b.** Independent **c.** Interdependent

Skills Find the answer. See page 479 for answers to Exercises 8–9.

4. What is the probability of rolling a 4 on a single roll of a number cube? [346] $\frac{1}{6}$
5. What is the probability of selecting an even number from a deck of cards numbered 1–25? [346] $\frac{12}{25}$
6. What is the probability of rolling a number less than 10 on a number cube? [346] 1
7. What is the probability of selecting a red counter from a box containing 2 white, 3 blue, 3 green, and 4 red? [346] $\frac{1}{3}$

Event: Picking 1 card and flipping the coin.

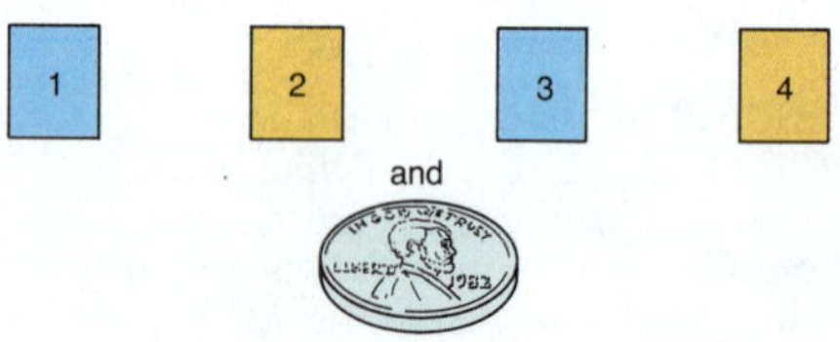

8. Make a tree diagram for the event. [348]
9. List the sample space for the event. [348]
10. What is the probability of picking a 2 and flipping a head? [350] $\frac{1}{8}$
11. What is the probability of picking an even number and flipping a tail? [350] $\frac{1}{4}$

Event: Picking 1 card, not replacing it, and picking another card.

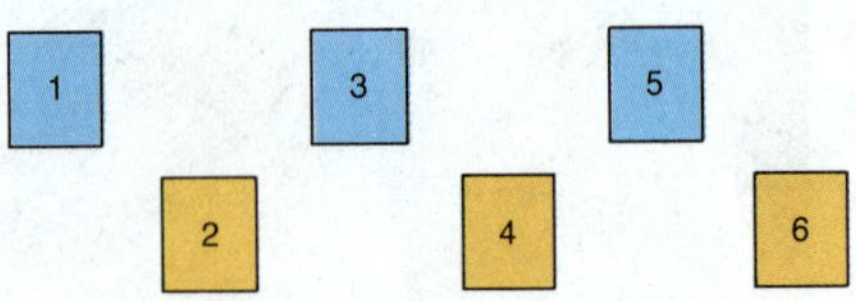

12. Find $P(1, 6)$. [352] $\frac{1}{30}$
13. Find P(even number, odd number). [352] $\frac{3}{10}$
14. Find P(2 or 4, 1 or 3). [352] $\frac{2}{15}$
15. How many heads would you expect if you flip a coin 400 times? [354] 200 heads
16. How many 3s would you expect if you roll a number cube 1,000 times? [354] 167 3s

Chapter Test

A bag contains 3 red, 4 green, and 8 yellow marbles.

See page 479 for answers to Exercises 13–14.

If you randomly pick 1 marble:

1. What is P(red)? $\frac{1}{5}$
2. What is P(green or yellow)? $\frac{4}{5}$
3. What is P(blue)? 0
4. What is P(not yellow)? $\frac{7}{15}$

If you randomly pick 1 marble, replace it, and then pick another:

5. What is P(yellow, yellow)? $\frac{64}{225}$
6. What is P(red, yellow)? $\frac{8}{75}$
7. What is P(green, green)? $\frac{16}{225}$
8. What is P(not red, not yellow)? $\frac{28}{75}$

If you randomly pick 1 marble, do not replace it, and then pick another:

9. What is P(yellow, yellow)? $\frac{4}{15}$
10. What is P(red, green)? $\frac{2}{35}$
11. What is P(yellow, then red or green)? $\frac{4}{15}$
12. What is P(green, not yellow)? $\frac{4}{35}$

13. Make a tree diagram for picking a card and flipping the coin.

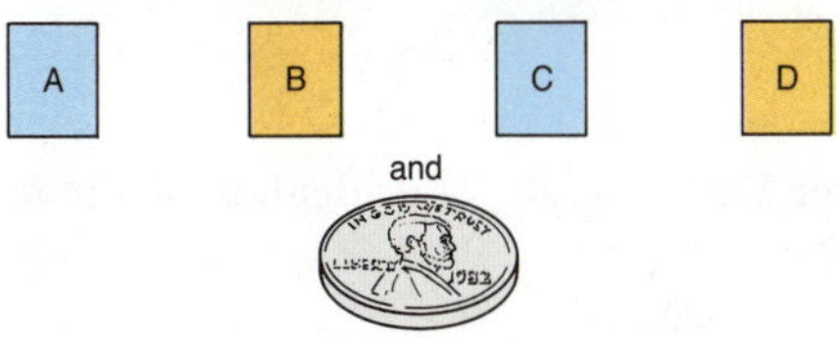

14. List the sample space for rolling the number card and picking a card.

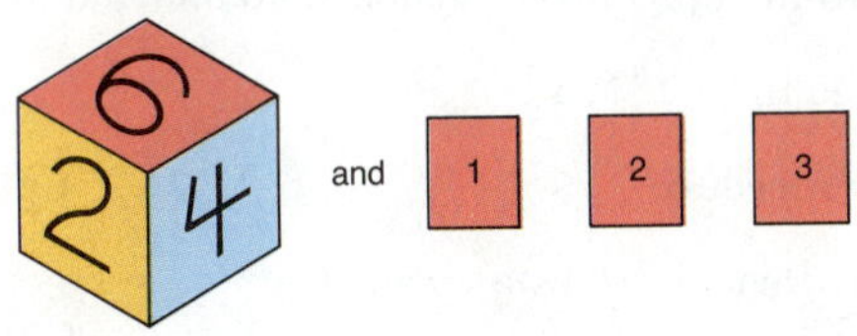

Solve.

15. How many different salads can you make if there are 3 types of salad and 4 types of dressing?
12 salads
16. How many different outfits can you wear if you have 5 shirts, 4 ties, and 3 pairs of pants?
60 outfits
17. The probability of winning a game is 13%. How many wins would you expect if you played the game 450 times?
59 wins
18. The probability of Rob's getting a hit is $\frac{3}{5}$. How many hits would you expect in 159 at bats?
95 hits
19. List all of the possible arrangements of the digits 4, 6, and 8 to make a 3-digit number, without using a digit more than once. How many are there?
468, 486, 648, 684, 846, 864; 6 arrangements
20. Ray, Carla, and Jennifer all want to be on the chess team. Only 2 new members are needed. List all 3 combinations of 2 new members that can be formed.
Ray and Carla, Ray and Jennifer, Carla and Jennifer

Objective

- Evaluate the achievement of chapter objectives.

Using the Test

The Chapter Test may be used as a post-test to evaluate student achievement. However, you may wish to use the Chapter Post-Test offered in the Teacher's Resource Package or to design your own chapter test. If this page is not used as a test, you may wish to assign it as additional review or practice.

The test items are correlated to the chapter objectives in the table below.

Chapter objectives	Test items
A. Find the probabilities of simple events.	1–4
B. Find the probability of 2 independent events.	5–8
C. Find the probability of 2 dependent events.	9–12
D. Draw tree diagrams or list sample spaces.	13–14
E. Use the counting principle to find possible outcomes.	15–16
F. Use probability to predict outcomes.	17–18
G. Solve problems by making an organized list.	19–20

Supplementary Materials

TRP Ch. 17 Post-Test Form A, pp. 1–2
TRP Ch. 17 Post-Test Form B, pp. 1–2

OBJECTIVE

- Multiply fractions, using a calculator.

USING THE PAGE

To multiply fractions or mixed numbers on a standard calculator, you must rename them as decimals. The product displayed on the calculator will also be a decimal. Therefore, you may wish to review with students how to rename decimals as fractions. First, write the decimal as a fraction with a denominator of 10, 100, 1,000, and so on. Then, divide the numerator and denominator by the greatest common factor to give a fraction in lowest terms.

Have students carefully work through the first example. In the first step, $\frac{3}{5}$ is renamed as the decimal, 0.6. If the second factor is a mixed number, as in the second example, the memory key needs to be used. However, since it is a fraction here, multiplying by the numerator, 5, and then dividing by the denominator, 8, will give the correct answer.

The second example contains 2 mixed numbers as factors. In this case, the first factor must be stored in memory after it has been renamed as a decimal. Remind students that when renaming a mixed number as a decimal on a calculator, they should first rename the fraction as a decimal and then add the whole number.

Before assigning the exercises, have students use their calculators to work through the examples in the text. You may also want to discuss with students how to multiply 3 fractions on a calculator.

CALCULATOR MULTIPLY FRACTIONS

You have learned that a fraction can be renamed as a decimal by dividing the numerator by the denominator. To multiply fractions or mixed numbers using a calculator, first rename them as decimals.

Multiply: $\frac{3}{5} \div \frac{5}{8}$.

Procedure	Calculator Entry	Calculator Display
1. Rename $\frac{3}{5}$ as a decimal and press the [×] key.	[3] [÷] [5] [×]	0.6
2. Rename $\frac{5}{8}$ as a decimal and press the [=] key.	[5] [÷] [8] [=]	0.375

So $\frac{3}{5} \times \frac{5}{8} = 0.375$.

To multiply mixed numbers, you should use the memory keys.

Multiply: $2\frac{3}{4} \times 1\frac{1}{2}$.

Procedure	Calculator Entry	Calculator Display
1. Rename $2\frac{3}{4}$ as a decimal and enter it in memory.	[3] [÷] [4] [+] [2] [M+]	M 2.75
2. Rename $1\frac{1}{2}$ as a decimal and press the [×] key.	[1] [÷] [2] [+] [1] [×]	M 1.5
3. Recall the number in memory and press the [=] key.	[MR] [=]	M 4.125

So $2\frac{3}{4} \times 1\frac{1}{2} = 4.125$.

Use a calculator to compute. Round the answer to the nearest thousandth.

1. $\frac{3}{4} \times \frac{2}{5}$ 0.3
2. $\frac{1}{2} \times \frac{3}{4}$ 0.375
3. $\frac{7}{10} \times \frac{2}{3}$ 0.467
4. $\frac{4}{5} \times \frac{1}{7}$ 0.114
5. $2\frac{3}{5} \times \frac{1}{5}$ 0.52
6. $2\frac{1}{4} \times 1\frac{1}{2}$ 3.375
7. $\frac{4}{5} \times 1\frac{3}{4}$ 1.4
8. $3\frac{1}{2} \times 2\frac{1}{14}$ 7.25
9. $4\frac{1}{4} \times 2\frac{3}{8}$ 10.094

ESTIMATION SKILLS

FRONT-END ESTIMATION OF SUMS OF MIXED NUMBERS

Front-end estimation can be used to estimate sums of fractions and mixed numbers.

1. Add the whole numbers for a first estimate.
2. Look for fractions with a sum of about 1.
3. Adjust the first estimate.

Examples

$2\frac{1}{6} + 6\frac{2}{3} + 4$

First estimate: $2 + 6 + 4 = 12$

THINK: $\frac{1}{6} + \frac{2}{3}$ is about 1.

$12 + 1 = 13$

Final estimate: 13

$1\frac{7}{8} + 3\frac{1}{4} + 5\frac{1}{4} + \frac{1}{2}$

First estimate: $1 + 3 + 5 = 9$

THINK: $\frac{7}{8}$ is about 1.

$\frac{1}{4} + \frac{1}{4} + \frac{1}{2}$ is equal to 1.

$9 + 1 + 1 = 11$

Final estimate: 11

Use front-end estimation with adjusting to estimate the sum.

1. $6\frac{1}{2} + 2\frac{5}{8}$ 9
2. $4\frac{1}{3} + 7\frac{2}{3}$ 12
3. $1\frac{1}{4} + \frac{2}{3}$ 2

4. $4\frac{3}{8} + 5\frac{1}{2}$ 10
5. $8\frac{4}{5} + 3\frac{1}{6}$ 12
6. $\frac{7}{10} + 6\frac{1}{5}$ 7

7. $8\frac{3}{4} + 9\frac{3}{10}$ 18
8. $\frac{1}{4} + 9\frac{7}{12}$ 10
9. $\frac{5}{12} + 7\frac{1}{2}$ 8

10. $7 + 3\frac{1}{2} + 2\frac{3}{5}$ 13
11. $1\frac{2}{3} + \frac{3}{8} + 8$ 10
12. $1\frac{1}{4} + 1\frac{1}{2} + 2\frac{1}{8}$ 5

13. $1\frac{9}{10} + \frac{2}{5} + 3\frac{3}{5}$ 6
14. $3\frac{1}{2} + 10\frac{4}{5} + 5\frac{1}{2}$ 20
15. $\frac{2}{3} + 4\frac{3}{4} + 1\frac{1}{2}$ 7

16. $2 + 1\frac{1}{2} + 2\frac{1}{4} + 3\frac{3}{8}$ 9
17. $\frac{1}{2} + 1\frac{1}{5} + 1\frac{7}{10} + 2\frac{2}{5}$ 6
18. $1\frac{1}{2} + 1\frac{1}{4} + \frac{1}{8} + \frac{1}{16}$ 3

19. $1\frac{5}{6} + \frac{1}{3} + \frac{1}{3} + 2\frac{2}{3}$ 5
20. $3\frac{3}{4} + 3\frac{3}{10} + 1\frac{1}{10} + \frac{3}{4}$ 9
21. $1\frac{1}{8} + 1\frac{3}{8} + 3\frac{5}{8} + \frac{3}{4}$ 7

22. $4\frac{1}{2} + 3\frac{1}{4} + 4\frac{2}{3} + 1\frac{2}{3}$ 14
23. $\frac{1}{6} + \frac{1}{2} + \frac{5}{12} + \frac{11}{12}$ 2
24. $\frac{1}{2} + 2\frac{3}{4} + 3\frac{2}{5} + 6\frac{1}{4}$ 13

OBJECTIVE

- Estimate the sums of mixed numbers, using front-end estimation.

USING THE PAGE

In order to do the exercises on this page, students should make mental estimates; that is, they should "picture" the fractions. They should not find common denominators, rename, and add.

The standard circular region is often useful in the picturing process. To illustrate, in the first example on the page, students can be encouraged to think of the fractional parts as shown below.

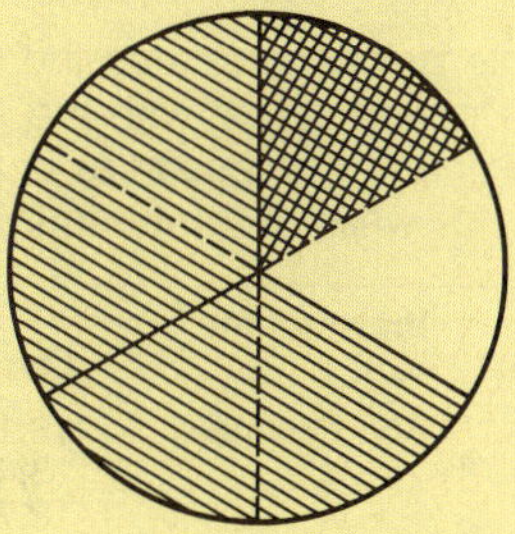

$\frac{1}{6} + \frac{2}{3} \approx 1$

Note that if a closer estimate is required, the procedure shown in the lesson can be refined by considering fractions with a sum of about 1 *and* fractions with a sum of about $\frac{1}{2}$. For example, estimate the following sum:

$3\frac{1}{4} + 4\frac{2}{3} + 6\frac{1}{3} + 7\frac{1}{6}$

First estimate: $3 + 4 + 6 + 7 = 20$

THINK: $\frac{2}{3} + \frac{1}{3}$ is equal to 1.

$\frac{1}{4} + \frac{1}{6}$ is about $\frac{1}{2}$.

$20 + 1 + \frac{1}{2} = 21\frac{1}{2}$

Final estimate: $21\frac{1}{2}$

OBJECTIVE

- Test the prerequisite skills needed to learn the concepts developed in Chapter 18.

USING THE TEST

The Pre-Skills Test is designed to diagnose students' strengths and weaknesses on prerequisite skills necessary to study the mathematics in Chapter 18.

Assign the Pre-Skills Test. Allow students to work together in pairs or in small groups. Group members should help those who demonstrate a misunderstanding of a concept or a weakness in a skill.

The table below correlates the items on the Pre-Skills Test with the prerequisite skill and the lesson(s) in the chapter for which it is needed.

Item(s)	Prerequisite skill	Lesson(s)
1–6	Add decimals.	18.1, 18.3, 18.4, 18.6, 18.7
7–12	Subtract decimals.	18.1, 18.3, 18.4, 18.6, 18.7
13–20	Multiply fractions and whole numbers (reciprocals).	18.1, 18.5, 18.6
21–28	Divide whole numbers and decimals (product = 1).	18.5, 18.6
29–32	Rename percents as decimals.	18.2, 18.3, 18.5, 18.6
33–36	Find the percent of a number.	18.3, 18.5, 18.6
37–40	Find the square of a number.	18.1, 18.3
41–48	Solve simple equations.	18.4, 18.5

References to specific skills that are reviewed in the Skills Bank are given on the individual lessons themselves.

Pre-Skills Test

Add.

1. $0.2 + 0.9$ 1.1 **2.** $4.3 + 0.7$ 5.0 **3.** $9.7 + 5.6$ 15.3

4. $0.8 + 19.3$ 20.1 **5.** $16.2 + 9.03$ 25.23 **6.** $7.08 + 9.93$ 17.01

Subtract.

7. $0.6 - 0.3$ 0.3 **8.** $8.2 - 0.4$ 7.8 **9.** $4.3 - 2.7$ 1.6

10. $17.3 - 12.1$ 5.2 **11.** $20.8 - 7.9$ 12.9 **12.** $9.07 - 3.8$ 5.27

Multiply.

13. $\frac{1}{5} \times 5$ 1 **14.** $\frac{1}{7} \times 7$ 1 **15.** $\frac{1}{2} \times 2$ 1 **16.** $\frac{1}{4} \times 4$ 1

17. $\frac{1}{6} \times 6$ 1 **18.** $\frac{1}{3} \times 3$ 1 **19.** $\frac{1}{10} \times 10$ 1 **20.** $\frac{1}{12} \times 12$ 1

Divide.

21. $9 \div 9$ 1 **22.** $22 \div 22$ 1 **23.** $16 \div 16$ 1 **24.** $53 \div 53$ 1

25. $0.2 \div 0.2$ 1 **26.** $0.7 \div 0.7$ 1 **27.** $1.25 \div 1.25$ 1 **28.** $0.65 \div 0.65$ 1

Rename as a decimal.

29. 80% 0.8 **30.** 16% 0.16 **31.** 9% 0.09 **32.** 12.5% 0.125

Find the answer.

33. 80% of 25 20 **34.** 16% of 30 4.8 **35.** 9% of 70 6.3 **36.** 12.5% of 90 11.25

37. 2^2 4 **38.** 3^2 9 **39.** 6^2 36 **40.** 8^2 64

Find the missing number.

41. $n + 3 = 9$ 6 **42.** $n - 4 = 11$ 15 **43.** $n + 6 = 22$ 16 **44.** $n - 10 = 33$ 43

45. $3 \times n = 21$ 7 **46.** $n \div 2 = 8$ 16 **47.** $8 \times n = 72$ 9 **48.** $n \div 10 = 7$ 70

Some computers are able to simulate how a new product will react under certain conditions. For what kinds of products do you think computer simulation would be valuable? Why?

OBJECTIVE

- Formulate problems, using text and data.

USING THE CHAPTER OPENER

Discuss students' answers to the questions on the page. Some products that they might mention include all forms of aircraft, automobiles, spacecraft, missiles, reactors, and so on. Generally, simulators are used to test products without actually building them and possibly destroying them. This, of course, is especially valuable when the product could cost millions of dollars to build.

There are many other types of computer programs that run simulations. If your students have never seen one, you may want to check if your school's science department has one that could be demonstrated for the class. Another type of simulator would be any video game that involves driving or maneuvering along a course.

COOPERATIVE LEARNING

Assign heterogeneous groups and tell them they are to work together to write a proposal for a computer simulation program. To do this, they will first choose something they think could be simulated and would be of some value. If this does not spark ideas for them, tell them the simulation could be a video game.

Their proposal should be as detailed as possible. For example, if the simulation was for a stunt-pilot course, they would indicate every type of maneuver that would be required, sequences, speeds, emergencies, interferences from outside sources, and so on.

Have the class decide which proposal is the most useful and which is the most original.

SUPPLEMENTARY MATERIALS

TRP Group Projects, p. 18

OBJECTIVE

- Use the correct order of operations to simplify numerical expressions.

TEACHING THE LESSON

WARM-UP Write the following exercises on the chalkboard or an overhead transparency.

Find the answer.

1. 6.5 + 1.9 (8.4) **2.** 7.1 − 2.8 (4.3)

3. 4.2 × 3 (12.6) **4.** $\frac{92}{4}$ (23)

INTRODUCTION Write the expression $3^2 + 6 \cdot 4$ on the chalkboard. Show students how "adding first" and "multiplying first" yield 2 different answers. Discuss the need for an order of operations on which everyone agrees.

INSTRUCTION Review the rules for order of operations and the use of the raised dot to stand for the operation of multiplication. Explain that students should first raise any number to the power or exponent given. Then they should work through an expression from "left to right" and multiply or divide as they come across these operations. When all the multiplications and divisions are complete, they should work through the expression again and add or subtract as they come across these operations.

Use Example 1 to show students how to simplify the introductory expression with exponents, addition, and multiplication. Note that, although addition comes before multiplication in the expression, the multiplication is done first.

Use Example 2 to show students that work inside parentheses is always done first. Then the rest of the order of operations is followed. Review the occasional use of parentheses to indicate the operation of multiplication.

Use Example 3 to show students that the division bar is a grouping symbol that indicates the operation of division.

18.1 ORDER OF OPERATIONS

To **simplify the expression** to the right, you need to use the correct **order of operations.**

$3^2 + 6 \cdot 4$
↓
The raised dot means multiply.

Order of Operations
- Raise to a power.
- Multiply or divide in order from left to right.
- Add or subtract in order from left to right.

EXAMPLE 1 Simplify. $3^2 + 6 \cdot 4$

1. Raise to a power. $9 + 6 \cdot 4$
2. Multiply. $9 + 24$
3. Add. 33

So $3^2 + 6 \cdot 4 = 33$.

To simplify expressions with **parentheses,** do the work within the parentheses first.

EXAMPLE 2 Simplify. $5 + 4(1.8 - 0.6 \div 3)$

1. Work within parentheses.

 THINK: Divide first, then subtract. $5 + 4(1.8 - 0.2)$
 $5 + 4(1.6)$

2. Multiply.

 THINK: 4(1.6) means $4 \cdot 1.6$. $5 + 6.4$

3. Add. 11.4

So $5 + 4(1.8 - 0.6 \div 3) = 11.4$.

To simplify expressions with a **division bar,** do the work first above the bar, then below the bar.

EXAMPLE 3 Simplify. $\frac{3 + 2}{30 \div 5 - 1}$

1. Do work above bar.

 THINK: Add. $\frac{5}{30 \div 5 - 1}$

2. Do work below bar.

 THINK: Divide, then subtract. $\frac{5}{6 - 1} = \frac{5}{5}$

3. Divide. 1

So $\frac{3 + 2}{30 \div 5 - 1} = 1$.

FOR DISCUSSION

See TE side column.

How do different calculators use the rules for order of operations?

PRACTICE EXERCISES

Remember to estimate whenever you use your calculator.

Simplify the expression mentally.

• 1. $3 + 8 - 6$ 5
• 2. $4 \cdot 7 - 7$ 21
3. $8 - 3 + 5$ 10
4. $2^2 + 12 \div 3$ 8
5. $7(3 + 2)$ 35
6. $3(3^2 + 2)$ 33
7. $6 \cdot 2^3 + 3$ 51
8. $(3 \cdot 4 - 6) + 7$ 13
9. $(17 - 5 \cdot 3) + 2^2$ 6

Simplify the expression.

• 10. $40 - 10 - 19$ 11
11. $47 - 19 \cdot 2$ 9
12. $28 \div 7 \cdot 8$ 32
13. $21 - 24 \div 8$ 18
14. $4^2 + 3 \cdot 2$ 22
15. $5^2 - 16 \div 4$ 21
• 16. $3(14 - 2) + 6$ 42
17. $19 + (6 \cdot 3 + 2)$ 39
18. $3 + 2(1.8 - 0.4)$ 5.8
19. $7 - 5(0.8 - 0.4)$ 5
20. $\frac{8}{2} + 12 - 7$ 9
21. $\frac{6 - 3 + 5}{8 - 6}$ 4
22. $\frac{8 - 3 \cdot 2 + 3}{10}$ 0.5
23. $\frac{6}{15 - 4 \cdot 3}$ 2
24. $\frac{12 - 3(4) + 6}{3(18 - 16)}$ 1
25. $\frac{2^2 \div 2 - 3(0.5)}{0.6 - 0.5}$ 5
26. $\frac{6 + 6 - 3^2}{4(9.5 - 6 \div 3)}$ 0.1
27. $\frac{2^5 \div (7 - 5 + 6)}{8 + 3(2^2)}$ 0.2

EXTENSION

Computers and Order of Operations

Computers use different symbols to stand for some operations. **Computer-programming languages** use the rules for order of operations.

Arithmetic symbol	Computer symbol
$6 + 2$	$6 + 2$
$6 - 2$	$6 - 2$
6×2	$6 * 2$
$6 \div 2$	6/2
6^2	$6 \wedge 2$

Find the computer's answer.

1. $4 * 3 + 6$ 18
2. $(5 - 2) * 7$ 21
3. $6/3 + 4$ 6
4. $5 + 8/2$ 9
5. $4 \wedge 2 + 1$ 17
6. $7 + 3 \wedge 2$ 16
7. $4 * 5 \wedge 2$ 100
8. $27/3 \wedge 2$ 3
9. $(6 - 2) \wedge 2$ 16
10. $(7 + 5)/(4 * 3)$ 1
11. $4 + 3 \wedge 2 * 5$ 49
12. $(3 * 12)/3 \wedge 2$ 4

SKILLS BANK You may want to refer students to the following skills in the Skills Bank:

Multiply With Decimals, Skill 4

Divide With Decimals, Skills 1–5

Write Fractions as Decimals, Skill 1

FOR DISCUSSION Students should explain that some calculators follow the rules for order of operations and some do not. Students should enter this simple expression.

[3] [+] [4] [×] [5] [=]

If the answer is 23, then their calculators follow the rules for order of operations.

If the answer is 35, the calculators do not follow the rules and they could not just enter an expression from left to right, but would have to enter the operations according to the rules.

EXTENSION Use the Extension to show students how computers use special symbols and the order of operations.

ASSIGNMENTS

Level 1	Odd 1–25
Level 2	1–24, EX
Level 3	Odd 1–9, 10–27, EX

FOLLOW-UP ACTIVITY

APPLICATION Write the following expressions on the chalkboard.

1. $35 - 6 \times 2$ (23)
2. $9 + 8 \div 2 - 5$ (8)
3. $15 + (7 + 18 \div 3) \times 1$ (28)
4. $\frac{63 \div 7 - 2}{18 - (2 \times 8)}$ (3.5)

Ask students to use their calculators to simplify each expression.

SUPPLEMENTARY MATERIALS

TRP Practice, p. 121
TRP Reteaching, p. 98
TRP Transparency 61

OBJECTIVES

- Identify algebraic expressions, variables, and equations.
- Translate words into algebraic expressions and equations.

TEACHING THE LESSON

INTRODUCTION Ask students to come to the chalkboard and write words for numbers and operations. After you have accumulated a balanced list of about 15 words, have students write the operation symbol or the numeral next to each word. Point out how it was possible to translate words into symbols and numerals.

Refer students to the box showing the other clue words and phrases that can help them decide which operations to use when translating words into expressions and equations.

Also refer students to the introductory material, which defines the terms *algebraic expression* and *variable*.

INSTRUCTION Use Example 1 to show students how to translate word phrases directly into algebraic expressions.

Use Example 2 to present the definition of *equation* and to show students how to translate word statements into equations. Point out that *is* usually means *is equal to*. In **b,** point out that we use the variable *a* to represent age because we do not know how old the person is.

Use Example 3 to show students how to write an equation for a word problem. Point out that students could rewrite a problem and underline the key words, or list the key words separately, before writing the equation. Point out that the key words often do not appear in the problem in the same order that they appear in the equation.

18.2 EXPRESSIONS AND EQUATIONS

An **algebraic expression** consists of 1 or more variables with or without operational signs. A **variable** is a letter used to stand for a number.

You can translate word phrases into algebraic expressions. The table below lists some words commonly used to describe operations.

Add (+):	sum of, total, and, increased by, more than
Subtract (−):	difference of, decreased by, fewer, less
Multiply (·):	of, product of, at, times
Divide (÷):	quotient of, for each, shared equally, ratio

EXAMPLE 1 Write algebraic expressions for these word phrases.

a. 9 less than w

$w - 9$

b. 7 times y plus 3

$7 \cdot y + 3$, or

$7y + 3$

c. $\frac{2}{5}$ of y

$\frac{2}{5} \cdot y$

$\frac{2}{5}y$

An **equation** is a mathematical sentence that states that 2 expressions are equal.

EXAMPLE 2 Write equations for these statements.

a. 2 times the sum of 5 and n is 34. $2 \cdot (5 + n) = 34$, or $2(5 + n) = 34$

b. Your age is 6 more than $\frac{1}{2}$ of 22. $a = \frac{1}{2} \cdot 22 + 6$, or $a = \frac{1}{2}(22) + 6$

You can write an equation for a word problem.

EXAMPLE 3 Write an equation for this problem.

Chad bought a computer for $10 more than $\frac{1}{2}$ of its regular price. Chad paid $304. What was the regular price of the computer?

THINK: $10 more than $\frac{1}{2}$ the regular price is $304.

Write. $10 + $\frac{1}{2}r$ = $304

366 CHAPTER 18

FOR DISCUSSION See TE side column.

Describe the equation in words.

1. $4 + n = 13$
2. $16(5 - 3) = x$
3. $\frac{8 - 2y}{2} = 3$

PRACTICE EXERCISES

Write an algebraic expression for the word phrase.

• 1. The sum of x and 9 $x + 9$
• 2. 15 divided by x $15 \div x$, or $\frac{15}{x}$
3. 19 less than y $y - 19$
4. The product of 3 and x $3x$
5. 8 more than y $y + 8$
6. 12% of y $0.12y$
7. 8 fewer than y $y - 8$
8. 6 more than the difference of x and 5 $(x - 5) + 6$
9. 6 more than the product of y and 9 $9y + 6$
10. 10 increased by 3 less than y $10 + (y - 3)$

Write an equation for the statement.

• 11. 10 increased by 6 is equal to y. $10 + 6 = y$
• 12. The product of 6 and 3 is x. $6(3) = x$
13. 10% of y is 42. $0.1y = 42$
14. 16 is 11 more than y. $16 = y + 11$
15. 4 times the sum of 6 and y is 19. $4(6 + y) = 19$
16. 40 is the sum of x and 16. $40 = x + 16$
17. 6 less than the product of x and 2 is 42. $2x - 6 = 42$
18. Your test score is 23 divided by 25. $t = 23 \div 25$, or $t = \frac{23}{25}$

Write the letter of the equation that matches the word problem.

19. Claire bought 9 pads for \$0.64 each. How much did she spend? c
20. Eight students shared the cost of a \$20 gift. How much did each pay? b
21. The \$20 model less the discount costs \$8. What is the discount? d

a. $a = 9 + \$0.64$

b. $\frac{\$20}{8} = p$

c. $9(\$0.64) = s$

d. $\$20 - d = \$8$

e. $\frac{\$20}{d} = \$8$

Write an equation for the problem.

• 22. The \$40 cost of a taxi ride was shared equally by 5 people. How much did each person pay? $p = \$40 \div 5$, or $p = \frac{\$40}{5}$
23. Ryan increased the 9 models in his collection by 13. How many models are there in his collection now? $m = 9 + 13$
24. Sandy bought a dress for 60% of its regular price. She paid \$42. What was the regular price? $0.6p = \$42$
25. Six friends each paid \$12 as their share of a meal. They added a \$15 tip to the cost of the meal. How much did the meal cost the group all together? $c = (6 \times \$12) + \$15$

FOR DISCUSSION Students should explain that it is possible to use words to describe equations. Students should respond with (1) 4 plus n is 13, (2) 16 times the quantity 5 minus 3 is x, and (3) the quantity 8 minus 2 times y divided by 2 is 3.

Students should also explain that using words to describe an equation can often help you understand the equation.

ASSIGNMENTS

Level 1	1–18, Odd 19–23
Level 2	1–24
Level 3	11–25

FOLLOW-UP ACTIVITY

COOPERATIVE LEARNING Divide the class into small heterogeneous groups of 2 or 3 students. In turn, each student in the group writes an equation on a piece of paper. The other student(s) writes the equation in words and then reads the words aloud. The student who wrote the equation comments on the response.

SUPPLEMENTARY MATERIALS

TRP Practice, p. 122
TRP Reteaching, p. 99
TRP Transparency 61

OBJECTIVE

- Evaluate algebraic expressions and formulas.

TEACHING THE LESSON

WARM-UP Write the following exercises on the chalkboard or an overhead transparency.

Simplify. Use the correct order of operations.

1. 6(9) − 6 **2.** 4(8.9) − 6(4)

3. $\frac{4+7-2}{4}$ **4.** $\frac{7(4+5)}{3}$

(**1.** 48 **2.** 11.6 **3.** 2.25 **4.** 21)

INTRODUCTION Write the formula for the area of a square on the chalkboard ($A = s \times s$). Ask students to make a table that shows the areas of squares when the lengths of the sides are 1, 2, and up to 10 in. Discuss with students that the formula for the area of a square is always the same, but that the actual area depends on the length.

INSTRUCTION Use Example 1 to show students how to evaluate an expression with 1 variable. Throughout the Examples on this page, use the term "second color" to show students how each replacement is made. Encourage students to replace y with other values and note the results.

Use Example 2 to show students how to evaluate an expression with more than 1 variable. Remind students that bh means "b times h."

Use Example 3 to remind students that they must use the correct order of operations when evaluating expressions. Ask students to explain that $(x - 4)$ must be done first because it is in parentheses and that $a(x - 4)$ means "a times the quantity x minus 4."

Use Example 4 to show students how to evaluate the formula that converts degrees Celsius to degrees Fahrenheit. Remind students that a formula is a particular type of equation that expresses a relationship or a rule.

18.3 EVALUATING EXPRESSIONS AND FORMULAS

To **evaluate an expression,** first replace the variable with a given value. Then simplify the resulting expression.

EXAMPLE 1 Evaluate $7y + 3$ for $y = 6$

1. Replace y with 6. $7y + 3 = 7(6) + 3$
2. Simplify. $42 + 3$

45

There may be more than 1 variable in the expression.

EXAMPLE 2 Evaluate: $\frac{bh}{2}$ for $b = 3$ and $h = 12$

1. Replace b with 3 and h with 12. $\frac{bh}{2} = \frac{3(12)}{2}$
2. Simplify. $\frac{36}{2}$

18

Remember to follow the rules for order of operations.

EXAMPLE 3 Evaluate: $\frac{a(x-4)}{y}$ for $a = 3$, $x = 12$, and $y = 6$

1. Replace the variables with the given values. $\frac{a(x-4)}{y} = \frac{3(12-4)}{6}$
2. Simplify. $\frac{3(8)}{6} = \frac{24}{6}$

4

You can also evaluate formulas. A **formula** is an equation that expresses a relationship between 2 or more quantities involving more than 1 variable. This formula shows how to convert from Celsius to Fahrenheit temperature readings. $F = \frac{9}{5}C + 32$

EXAMPLE 4 Evaluate the formula to solve this problem. The temperature is 25° Celsius. What is the reading in degrees Fahrenheit?

1. Replace C with 25. $F = \frac{9}{5}C + 32 = \frac{9}{5}(25) + 32$
2. Simplify. $F = \frac{9}{\cancel{5}_1}(\cancel{25}^5) + 32 = 45 + 32 = 77$

A temperature reading of 25°C is 77°F.

FOR DISCUSSION
See TE side column.

What formula would you use to:

1. Find the area of a wall in your classroom?
2. Find the distance you can drive in a given time?

PRACTICE EXERCISES
Remember to estimate whenever you use your calculator.

Evaluate for $x = 8$.

• 1. $x + 3$ 11
• 2. $10 - x$ 2
3. $11x$ 88
4. $\frac{x}{2}$ 4
5. $3x - 2$ 22
6. $\frac{x}{4} + 2$ 4
7. $19 - 2x$ 3
8. $23 + \frac{x}{8}$ 24

Evaluate for $a = 2$ and $b = 4$.

• 9. $a + b$ 6
• 10. $b - a$ 2
11. ab 8
12. $\frac{b}{a}$ 2
13. $2b + a$ 10
14. $(12 - b) + a$ 10
15. $6 - \frac{b}{a}$ 4
16. $4a - 2b$ 0

Evaluate.

• 17. $3(x - 4)$ for $x = 7$ 9
• 18. $\frac{12}{x} - 2$ for $x = 3$ 2
19. $(x + 2)6$ for $x = 4$ 36
20. $\frac{24}{x} + 5$ for $x = 12$ 7
21. $x(y - 5)$ for $x = 2, y = 7$ 4
22. $x\left(\frac{y}{8}\right)$ for $x = 6, y = 56$ 42
23. $x - y + z$ for $x = 7, y = 2, z = 4$ 9
24. $x(y - z)$ for $x = 3, y = 9, z = 2$ 21
25. $x + \frac{y}{z} - 2$ for $x = 12, y = 21, z = 3$ 17
26. $\frac{x(y + 3)}{(3 + y)z}$ for $x = 12, y = 1, z = 6$ 2

Evaluate the formula to solve the problem.

• 27. The temperature is 30° Celsius. What is the Fahrenheit reading?
(Use $F = \frac{9}{5}C + 32$.)
86°F

28. The temperature is 86° Fahrenheit. What is the Celsius reading?
[Use $C = \frac{5}{9}(F - 32)$.]
30°C

29. A car traveled at a rate of 45 mph for $2\frac{1}{2}$ h. What distance was that?
(Use $d = rt$.)
112.5 mi

30. You painted a square wall 13 ft on a side. How many square feet was that? (Use $A = s^2$.)
169 square ft

SKILLS BANK You may want to refer students to the following skills in the Skills Bank:

Multiply With Whole Numbers, Skill 2

Multiply With Decimals, Skill 2

Divide With Whole Numbers, Skills 1–2

Divide With Decimals, Skills 2–3

FOR DISCUSSION

1. Students should explain that they should first determine the shape of the wall. Most will be rectangles. Next, they should write the formula for the area of a rectangle, $A = l \times w$ or $A = lw$.
2. Students should explain that distance, rate, and time are related to each other in this formula: $d = r \times t$ or $d = rt$.

ASSIGNMENTS

Level	Exercises
Level 1	1–16, Odd 17–29
Level 2	1–30
Level 3	17–30

FOLLOW-UP ACTIVITY

ENRICHMENT The formula below shows about how many feet it would take a car to stop on a dry road. The variable m stands for miles per hour.

$$f = \frac{m + m^2}{20}$$

Write the formula on the chalkboard. Ask students to complete the table.

Mph	25	40	55	60	65
Feet to stop	(32.5)	(82)	(154)	(183)	(214.5)

Discuss the implications of the table results in terms of safe driving.

SUPPLEMENTARY MATERIALS

TRP Practice, p. 123
TRP Reteaching, p. 100
TRP Transparency 62

OBJECTIVE

- Solve addition and subtraction equations.

TEACHING THE LESSON

WARM-UP Write the following exercises on the chalkboard or an overhead transparency.

What number would you add or subtract to get the result 0?

1. 15 **2.** 3.9 **3.** −7 **4.** −11.2

(**1.** −15 **2.** −3.9 **3.** 7 **4.** 11.2)

INTRODUCTION Use the introductory information to discuss with students why addition and subtraction are inverse operations. Point out that the result of adding and then subtracting the same number (or the reverse) is 0.

INSTRUCTION Explain to students that they will be using the inverse of a number attached to the variable. The idea is to unattach the number from the variable so that the variable stands alone. Take time here to explain that adding or subtracting the same number from both sides of an equation results in another true equation. Demonstrate by starting with the equation 16 + 9 = 25 and then successively adding and subtracting the same number from both sides. At each step, students should confirm that the equation is still true.

Use Example 1 to show students how to use addition to unattach a number and solve an equation. Remind students that addition undoes subtraction. Emphasize the importance of checking the solution to any equation.

Use Example 2 to show students how to use subtraction to unattach a number and solve an equation. Remind students that subtraction is used to undo addition. Emphasize again the importance of checking the solution.

18.4 SOLVING ADDITION AND SUBTRACTION EQUATIONS

Addition and subtraction are **inverse operations.** One undoes the other. To solve an equation, use an inverse operation to get the variable by itself.

When a number is attached to a variable by subtraction, you can unattach it by adding that number to both sides of the equation.

EXAMPLE 1 The price of your weekly train ticket has been decreased by \$9 to \$15. What was the original cost of the ticket?

1. Write an equation.

THINK: **Original cost less \$9 is \$15.** $c - \$9 = \$15$

2. Solve the equation.

THINK: **\$9 is attached by subtraction. Add \$9 to both sides.**

$c - \$9 + \$9 = \$15 + \$9$

$c = \$24$

3. Check the solution.

THINK: **Substitute the value of *c* into the original equation.**

$\$24 - \$9 \stackrel{?}{=} \$15$

$\$15 = \$15$ It checks.

The original cost of the ticket was \$24.

When a number is attached to a variable by addition, you can unattach it by subtracting that number from both sides of the equation.

EXAMPLE 2 You increased your weekly pay by \$3.80 to \$73.30. How much had you been earning?

1. Write an equation.

THINK: **Weekly pay plus \$3.80 is \$73.30.** $p + \$3.80 = \$73.30$

2. Solve the equation.

THINK: **\$3.80 is attached by addition. Subtract \$3.80 from both sides.**

$p + \$3.80 - \$3.80 = \$73.30 - \$3.80$

$p = \$69.50$

3. Check the solution.

THINK: **Substitute the value of *p* into the original equation.**

$\$69.50 + \$3.80 \stackrel{?}{=} \$73.30$

$\$73.30 = \$73.30$ It checks.

You had been earning \$69.50 per wk.

SKILLS BANK You may want to refer students to the following skills in the Skills Bank:

Add With Whole Numbers and Decimals, Skills 1–4

Subtract With Whole Numbers and Decimals, Skills 1–4

FOR DISCUSSION Students should explain that −9.5 is the additive inverse of +9.5 because the result of adding them is 0. So, when a number is attached to a variable by addition, we could subtract to unattach it; or, we could add the additive inverse of the number. The result will be the same.

ASSIGNMENTS

Level 1	Odd 1–19, 25–27
Level 2	1–30
Level 3	13–30

FOLLOW-UP ACTIVITY

APPLICATION Give students this problem: The temperature increased from −6° to 7°. How much did the temperature rise?

Ask students to write and solve an equation for this problem. The problem involves writing and solving equations with negative numbers. Students must see that they need to use the additive inverse of −6 to solve the equation.
(Equation: $-(6) + t = 7$;
Solution: $t = 13$)

Ask students to devise and solve similar problems.

SUPPLEMENTARY MATERIALS

TRP Practice, p. 124
TRP Reteaching, p. 101
TRP Transparency 62

FOR DISCUSSION See TE side column.

Why is "− 9.5" called the **additive inverse** of "+ 9.5"?

PRACTICE EXERCISES

Remember to estimate whenever you use your calculator.

Solve. Check your solution.

• **1.** $x - 15 = 45$ 60 • **2.** $x - 2.8 = 1.6$ 4.4 **3.** $x - 7 = 0$ 7 **4.** $2.8 = x - 3.7$ 6.5

5. $405 = x - 35$ 440 **6.** $5 + x = 92$ 87 **7.** $x + 23 = 50$ 27 **8.** $x + 4.7 = 7.2$ 2.5

9. $x - 6 = 0$ 6 **10.** $31 = x + 18$ 13 **11.** $12.9 = x + 7.1$ 5.8 **12.** $29 = x + 12$ 17

13. $23 = a + 18$ 5 **14.** $b - 4.3 = 7.8$ 12.1 **15.** $24 + c = 90$ 66 **16.** $3.2 = d - 1.5$ 4.7

17. $23.6 = m - 28$ 51.6 **18.** $3.6 + n = 8.9$ 5.3 **19.** $p + 6 = 504$ 498 **20.** $7.8 + r = 9.3$ 1.5

21. $-(7) + n = 21$ 28 **22.** $n - 7 = -3$ 4 **23.** $-(9) + a = 6$ 15 **24.** $-(3.5) - d = 2$ −5.5

Write an equation and solve.

• **25.** The store manager reduced the price by \$6 to \$38. What was the original price?
$p - \$6 = \$38$; \$44

• **26.** Ron improved his test score by 8 points to 89 points. What was his original test score?
$t + 8 = 89$; 81 points

27. The price of a share of Zeno stock increased from $34\frac{1}{4}$ to $37\frac{1}{2}$. By how much did the stock increase?
$34\frac{1}{4} + x = 37\frac{1}{2}$; $3\frac{1}{4}$

28. The price of a share of Naman stock decreased from 19 to $17\frac{1}{4}$. By how much did the stock decrease?
$19 - x = 17\frac{1}{4}$; $1\frac{3}{4}$

29. Your \$128 regular pay, plus commission, was \$237. How much was your commission?
$\$128 + c = \$237$; \$109

30. Your \$983 monthly pay, less expenses, was \$47. What were your expenses?
$\$983 - e = \$47$; \$936

OBJECTIVE

- Solve multiplication and division equations.

TEACHING THE LESSON

WARM-UP Write the following exercises on the chalkboard or an overhead transparency.

What number would you multiply by to get the result 1?

1. $\frac{1}{2}$ **2.** $\frac{1}{7}$ **3.** $\frac{1}{5}$ **4.** $\frac{1}{10}$

What number would you divide by to get the result 1?

5. 2 **6.** 8 **7.** 3.5 **8.** 7.4

(**1.** 2 **2.** 7 **3.** 5 **4.** 10 **5.** 2 **6.** 8 **7.** 3.5 **8.** 7.4)

INTRODUCTION Use the introductory information to discuss with students why multiplication and division are inverse operations. Point out that the result of dividing and then multiplying the same number (or the reverse) is 1.

INSTRUCTION Explain to students that they will be using the inverse of a number attached to the variable. The idea is to unattach the number from the variable so that the variable stands alone. Take time here to explain that multiplying or dividing the same number from both sides of an equation results in another true equation. Demonstrate by starting with the equation $5 \times 20 = 100$ and then successively multiplying and dividing the same number from both sides. At each step, students should confirm that the equation is still true.

Use Example 1 to show students how to use multiplication to unattach a number and solve an equation. Remind students that multiplication undoes division. Emphasize the importance of checking the solution to any equation.

18.5 SOLVING MULTIPLICATION AND DIVISION EQUATIONS

Multiplication and division are inverse operations.

When a number is attached to a variable by division, you can unattach it by multiplying both sides of the equation by that number.

EXAMPLE 1 You prepaid \$92, which was $\frac{1}{2}$ of the cost of your trip. How much will the trip cost?

1. Write an equation.

THINK: **$\frac{1}{2}$ of the cost is \$92.** $\frac{c}{2} = \$92$

2. Solve the equation.

THINK: **2 is attached by division. Multiply both sides by 2.** $2\left(\frac{c}{2}\right) = 2(\$92)$

$c = \$184$

3. Check the solution.

THINK: **Substitute the value of *c* into the original equation.** $\frac{\$184}{2} \stackrel{?}{=} \$92$

$\$92 = \$92$ It checks.

The trip will cost \$184.

When a number is attached to a variable by multiplication, you can unattach it by dividing both sides of the equation by that number.

EXAMPLE 2 You are trying to guess a number your friend is thinking of. You know that 0.8 times the number is 24. What is the number?

1. Write an equation.

THINK: **0.8 times the number is 24.** $0.8n = 24$

2. Solve the equation.

THINK: **0.8 is attached by multiplication. Divide both sides by 0.8.** $\frac{0.8n}{0.8} = \frac{24}{0.8}$

$n = 30$

3. Check the solution.

THINK: **Substitute the value of *n* into the original equation.** $0.8(30) \stackrel{?}{=} 24$

$24 = 24$ It checks.

Your friend is thinking of the number 30.

FOR DISCUSSION See TE side column.

Why does $\frac{x}{7}$ mean the same as $\frac{1}{7}(x)$?

PRACTICE EXERCISES

Remember to estimate whenever you use your calculator.

Solve. Check your solution.

• **1.** $\frac{x}{4} = 8$ 32

• **2.** $\frac{x}{4} = 12$ 48

3. $\frac{x}{7} = 6$ 42

4. $7 = \frac{x}{0.2}$ 1.4

5. $34 = \frac{x}{1.5}$ 51

6. $\frac{x}{3} = 42$ 126

7. $4x = 16$ 4

8. $0.5x = 12$ 24

9. $42 = 1.2x$ 35

10. $3x = 7.5$ 2.5

11. $1.25x = 15$ 12

12. $32 = 0.02x$ 1,600

13. $3a = 12$ 4

14. $8b = 56$ 7

15. $\frac{c}{5} = 12$ 60

16. $21 = \frac{d}{3}$ 63

17. $14 = 0.1p$ 140

18. $\frac{n}{7} = 12$ 84

19. $2 = \frac{y}{9}$ 18

20. $0.7r = 21$ 30

21. $2.5s = 55$ 22

22. $32 = \frac{t}{0.04}$ 1.28

23. $3.2 = 1.25n$ 2.56

24. $160 = \frac{n}{0.7}$ 112

Write an equation and solve.

• **25.** One-fifth of a number is 9. What is the number?
$\frac{n}{5} = 9$; 45

• **26.** Seven times a number is 56. What is the number?
$7n = 56$; 8

27. Ralph paid \$55, $\frac{1}{5}$ of the cost of the car rental. How much did it cost to rent the car?
$\frac{c}{5} = \$55$; \$275

28. The profits of a sale were shared equally by 3 partners. Each one received \$305. What were the profits?
$\frac{p}{3} = \$305$; \$915

29. Profits in December were 4 times the profits in November. December's profits were \$2,492. What were November's profits?
$n = \frac{\$2,492}{4}$; \$623

30. Lynne bought her plane ticket for \$432, which was 90% of the original price. What was the original price?
$0.9p = \$432$; \$480

Use Example 2 to show students how to use division to unattach a number and solve an equation. Remind students that division is used to undo multiplication. Emphasize again the importance of checking the solution.

SKILLS BANK You may want to refer students to the following skills in the Skills Bank:

Multiply With Whole Numbers, Skill 2

Multiply With Decimals, Skill 2

Divide With Whole Numbers, Skills 1–2

Divide With Decimals, Skills 2–3

FOR DISCUSSION Students should explain that these are just 2 different ways of naming the same number. This series of steps helps establish that both these notations mean the same thing.

$$\frac{1}{7}(x) = \frac{1}{7}\left(\frac{x}{1}\right) = \frac{1(x)}{7(1)} = \frac{x}{7}$$

ASSIGNMENTS

Level	Exercises
Level 1	Odd 1–23, 25–27
Level 2	1–30
Level 3	13–30

FOLLOW-UP ACTIVITY

ENRICHMENT Involve students in a game of backward equations. Assign each student one of the practice exercises on page 373. Do not reveal each student's assignment to the class. Each student is to write a word problem that will go along with his or her equation. After the problems are written, ask students to read them aloud while other students in the class respond with the matching equation.

SUPPLEMENTARY MATERIALS

TRP Practice, p. 125
TRP Reteaching, p. 102
TRP Transparency 63

OBJECTIVE

- Solve 2-step equations.

TEACHING THE LESSON

WARM-UP Write the following exercises on the chalkboard or an overhead transparency.

Tell how you would unattach the number from the variable.

1. $6a$ **2.** $b - 12$ **3.** $\frac{c}{4}$ **4.** $7(d)$

(**1.** divide by 6 **2.** add 12 **3.** multiply by 4 **4.** divide by 7)

INTRODUCTION Lead students in a review of the ways they solved equations in the previous 2 lessons. Emphasize the notion of inverse operations and ask students to detail how they would unattach numbers attached to a variable by (1) addition, (2) subtraction, (3) multiplication, and (4) division.

INSTRUCTION Explain to students that they will be combining the methods they learned in the previous 2 lessons for unattaching numbers from variables. Emphasize that, almost without exception, they will be adding or subtracting first when they solve these equations.

Use Example 1 to show students how to solve a 2-step equation. Emphasize that they should subtract first to solve the equation. Emphasize the check to make sure that a solution is correct.

Use Example 2 to show students how to simplify before they solve a 2-step equation. Point out that students should add and subtract to simplify each side of the equation before solving it.

SKILLS BANK You may want to refer students to the following skills in the Skills Bank:

Add With Whole Numbers and Decimals, Skills 1–4

Subtract With Whole Numbers and Decimals, Skills 1–4

18.6 SOLVING 2-STEP EQUATIONS

Sometimes you need to use 2 steps to solve an equation.

EXAMPLE 1 You bought 7 boxes of cassette tapes and 5 individual tapes for a total of 89 tapes. How many tapes were in each box?

1. Write an equation.
   THINK: 7 boxes plus 5 tapes is 89 tapes. $7b + 5 = 89$
2. Solve the equation.
   THINK: Unattach first by subtraction, then by division.
   $7b + 5 - 5 = 89 - 5$
   $7b = 84$
   $\frac{7b}{7} = \frac{84}{7}$
   $b = 12$
3. Check the solution.
   THINK: Substitute the value of b into the original equation.
   $7(12) + 5 \stackrel{?}{=} 89$
   $89 = 89$ It checks.

There were 12 tapes in each box.

Sometimes you may need to simplify before you solve.

EXAMPLE 2 You earned \$40, plus $\frac{1}{5}$ of your sales, less \$15 for withholding, to equal a net pay of \$130. What were your sales?

1. Write an equation.
   THINK: \$40 plus $\frac{1}{5}$ of sales minus \$15
   $\$40 + \frac{s}{5} - \$15 = \$130$
2. Simplify the equation.
   THINK: \$40 − \$15 = \$25
   $\$25 + \frac{s}{5} = \$130$
3. Solve the equation.
   THINK: Unattach first by subtraction, then by multiplication.
   $\$25 - \$25 + \frac{s}{5} = \$130 - \$25$
   $\frac{s}{5} = \$105$
   $5\left(\frac{s}{5}\right) = 5(\$105)$
   $s = \$525$
4. Check the solution.
   THINK: Substitute the value of s into the original equation.
   $\$40 + \frac{\$525}{5} - \$15 \stackrel{?}{=} \$130$
   $\$40 + \$105 - \$15 \stackrel{?}{=} \$130$
   $\$130 = \$130$ It checks.

Your sales were \$525.

FOR DISCUSSION See TE side column.

Why do we add or subtract first when solving a 2-step equation?

PRACTICE EXERCISES

Remember to estimate whenever you use your calculator.

Simplify the equation.

1. $13 + \frac{n}{2} - 9 = 54$ $4 + \frac{n}{2} = 54$, or $\frac{n}{2} + 4 = 54$
2. $3 + 5x + 7 = 70$ $10 + 5x = 70$, or $5x + 10 = 70$
3. $18 + 3t - 7 = 51$ $11 + 3t = 51$, or $3t + 11 = 51$

Solve. Check your solution.

• 4. $9s + 8 = 71$ 7
• 5. $4y - 6 = 42$ 12
6. $54 = 7x + 12$ 6
7. $38 = \frac{t}{4} + 8$ 120
8. $4z - 0.8 = 4$ 1.2
9. $19 + 8s = 91$ 9
10. $70 = \frac{a}{6} - 10$ 480
11. $\frac{b}{8} - 2.6 = 15.2$ 142.4
12. $87 = 15 + \frac{z}{10}$ 720
13. $19 + \frac{x}{3} - 3 = 17$ 3
14. $15 + 6y - 2 = 31$ 3
15. $31 = 0.08c - 21$ 650
16. $34 + \frac{d}{5} - 19 = 42$ 135
17. $1.2a - 79 = 17$ 80
18. $14 + 2z - 23 = 43$ 26
19. $-(80) + 0.4b = 92$ 430
20. $-(27) + 3z + 51 = 42$ 6
21. $-(4) + \frac{y}{8} + 23 = 36$ 136

Write an equation and solve.

• 22. Five more than three times a number is 32. What is the number?
$3n + 5 = 32$; 9

• 23. Six less than $\frac{2}{5}$ of a number is 49. What is the number?
$\frac{2}{5}n - 6 = 49$; 137.5

24. A number divided by 7, plus 6, is 18. What is the number?
$\frac{n}{7} + 6 = 18$; 84

25. Thirty percent of a number, minus 8, is 4. What is the number?
$0.3n - 8 = 4$; 40

26. Angelo bought 3 boxes of erasers, plus 3 more erasers, for a total of 57. How many erasers were in a box?
$3b + 3 = 57$; 18 erasers

27. Julie bought $\frac{1}{3}$ case of granola bars. She gave away 3 of the bars and had 15 left. How many granola bars were in a case?
$\frac{c}{3} - 3 = 15$; 54 bars

28. Manuel drove the same distance on each of 5 d. Then he drove 35 mi more for a total of 675 mi. How many miles did he drive on each of the 5 d?
$5d + 35 = 675$; 128 mi

29. Claire delivered 28 papers, 4 less than 80% of the papers she needs to deliver. How many papers does Claire usually deliver?
$0.8p - 4 = 28$; 40 papers

30. Sally earned \$80, plus $\frac{1}{8}$ of her sales, less \$20 withholding, for a total net of \$192. What were her sales?
$\$80 + \frac{s}{8} - \$20 = \$192$; \$1,056

31. Enrico earned \$70, plus 15% of his sales, less \$42 withholding, for a total net of \$166.45. What were his sales?
$\$70 + 0.15s - \$42 = \$166.45$; \$923

Multiply With Whole Numbers, Skill 2

Multiply With Decimals, Skill 2

Divide With Whole Numbers, Skills 1–2

Divide With Decimals, Skills 2–3

FOR DISCUSSION Students should explain that adding or subtracting first results in a simpler equation. Addition and subtraction can be applied to a single number in the equation. Multiplication and division are applied to the entire side of the equation and multiplying or dividing first usually results in a more complicated equation. Students should give examples to show how complicated multiplying or dividing first could be.

ASSIGNMENTS

Level 1	1–3, Even 4–18, 22–25
Level 2	1–30
Level 3	13–31

FOLLOW-UP ACTIVITY

APPLICATION Show students how they could use calculators to solve equations. Write these examples on the chalkboard.

1. $x + 9 = 21$ 2. $3x = 15$
3. $\frac{y}{3} + 2 = 12$ 4. $3x - 7 = 41$

Show these calculator keystrokes for solving the equations.

1. [2] [1] [−] [9] [=] (12)
2. [1] [5] [÷] [3] [=] (5)
3. [1] [2] [−] [2] [×] [3] [=] (30)
4. [4] [1] [+] [7] [÷] [3] [=] (16)

SUPPLEMENTARY MATERIALS

TRP Practice, p. 126
TRP Reteaching, p. 103
TRP Transparency 63

OBJECTIVE

- Solve multistep problems by using rounded numbers in simpler problems (subproblems).

TEACHING THE LESSON

WARM-UP Write the following exercises on the chalkboard or an overhead transparency.

Round to the greatest place.

1. \$5,628.95 (\$6,000)
2. \$752.14 (\$800)
3. \$2,734.58 (\$3,000)
4. \$349.99 (\$300)

INTRODUCTION Pose this situation to the class: Susie bought one ring for \$287.55 and another ring for \$442.76. How much money did she receive as change from a \$1,000 bill?

Ask, before you can determine the money she will receive, what must you know? (How much the 2 rings cost in all) If the clerk handed Susie \$180 and some coins, should she accept that as correct? Why or why not? (No; That is not enough, since the total purchase is about \$700, not \$800.)

Have students give their answers *without* using pencil and paper. Point out that rounded numbers can often help many people to analyze or visualize various problems.

INSTRUCTION Present the problem situation to the class. Discuss the facts given in the problem. It is important that all students recognize that the problem does *not* tell what Bernard's income from investments is *this* year; rather, it tells how much less than last year's it is.

As you work through Subproblems 1 and 2, involving rounded numbers, emphasize that in these steps students are not solving the given problem. They are solving an easier, but very similar, problem.

PROBLEM *Solving* STRATEGY

18.7 SOLVING A SIMPLER PROBLEM

Situation:

Bernard invests in stocks. Last year, his income from investments was \$4,176.85. This year, his income from investments was \$643.25 less than it was last year. What was the total income from his investments for the 2 y?

Strategy:

Sometimes it helps to solve a simpler problem. Using **rounded numbers** can make the situation seem easier. Often, you can also "break" a problem into subproblems and write an equation for each subproblem.

Applying the Strategy:

You do not know this year's income from investments. So, first you need to find that income. Then you can find the total income for the 2 y.

Subproblem 1: Let x represent the investment income this year.

THINK: \$643.25 is about \$600. \$4,176.85 is about \$4,000.

INCOME LAST YEAR − \$600 = INCOME THIS YEAR

$\$4{,}000 - \$600 = x$

$\$3{,}400 = x$

Bernard's income this year was about \$3,400.

Subproblem 2: Let t represent the total income from investments.

INCOME LAST YEAR + INCOME THIS YEAR = TOTAL INCOME

$\$4{,}000 + \$3{,}400 = t$

$\$7{,}400 = t$

Bernard's total investment income for the 2 y was about \$7,400.

Now solve the problem using the actual numbers.

Subproblem 1: $\$4,176.85 - \$643.25 = x$
$\$3,533.60 = x$

Subproblem 2: $\$4,176.85 + \$3,533.60 = t$
$\$7,710.45 = t$

Bernard's total investment income for the 2 y was $7,710.45.

PRACTICE EXERCISES

Remember to estimate whenever you use your calculator.

Write the letter of the better choice for simplifying the problem.

1. Don invests in stocks. Last year, his income from investments was $2,630.50. This year, his income from investments was $792.25 more than it was last year. What was the total income from his investments in the 2 y? b

   a. $\$2,600 - \$800 = x$
   $\$1,800 = x$
   $\$2,600 + \$1,800 = t$
   $\$4,400 = t$

   b. $\$2,600 + \$800 = x$
   $\$3,400 = x$
   $\$2,600 + \$3,400 = t$
   $\$6,000 = t$

2. Stella earns $21,216 per y. This is 2.6 times as much as her younger sister, Tina, earns. What is the difference between their incomes? a

   a. $\$21,000 = 3t$
   $\$7,000 = t$
   $\$21,000 - \$7,000 = d$
   $\$14,000 = d$

   b. $t = 3 \times \$21,000$
   $t = \$63,000$
   $\$63,000 - \$21,000 = d$
   $\$42,000 = d$

Solve by writing a simpler problem.

• 3. Al and Ben are brothers. They are saving to buy a stereo. Al saves $13.75 per wk. Ben saves $2.75 less per wk than Al saves. In all, how much money do the 2 brothers save each week?
$24.75

4. Winona invests in stocks. Last year, her income from investments was $2,750. This year, her income from investments was 2.1 times as great as it was last year. What was the total income from her investments in the 2 y?
$8,525

5. Karen Ching earns $50,232 per y. This is 4.6 times as much as her son earns. What is the difference between their incomes?
$39,312

6. For 3 y, Marianne has been depositing money in a special bank account. During the third year, she deposited $3,960. This amount was $935 more than she deposited during the second year. The third-year amount was 2.4 times as much as she deposited during the first year. What was the total amount that she deposited during the 3 y?
$8,635

After finding the answer to the easier problem, they will need to go back and solve the actual problem. You may also wish to mention that their answer for the original problem should not differ greatly from the one they obtained using rounded numbers.

SKILLS BANK You may want to refer students to the following skills in the Skills Bank:

Round Whole Numbers and Decimals, Skills 2–4

Add With Whole Numbers and Decimals, Skill 6

Subtract With Whole Numbers and Decimals, Skill 6

ASSIGNMENTS

Level 1	1–4
Level 2	1–5
Level 3	3–6

FOLLOW-UP ACTIVITY

SITUATIONAL PROBLEM SOLVING Present students with the following problem: The student body and the Parents and Teachers Association of your school want to sponsor a concert to raise money for the school yearbook. The PTA will match and then double any amount that is donated by the students themselves. What are some combinations of donated funds and the total amount for each combination?

Some questions to ask include the following: How much do you think the students can raise? How much is double that amount?

Some approaches to solving the problem include the following: making a list of combinations and totals and using rounded numbers.

SUPPLEMENTARY MATERIALS

TRP Practice, p. 127
TRP Applications, p. 20
TRP Transparency 64

OBJECTIVES

- Review vocabulary.
- Practice key chapter concepts and skills.

USING THE REVIEW

The Chapter Review is designed to help students prepare for taking the Chapter Test. The first section focuses on vocabulary. It requires students to select words to complete statements. The second section presents practice of key chapter skills.

Each item on the Chapter Review is referenced to the page on which the topic is taught in the Pupil's Edition. You may wish to have students refer to these pages to help review any concepts or skills they have not yet mastered.

It is suggested that students work in small-sized heterogeneous cooperative learning groups. Some cooperative learning methods that may be used are as follows:

1. After each student has independently completed the Chapter Review, a discussion should follow within each group about the solutions to the problems.
2. The group can complete the entire Chapter Review by working together to solve the problems.

End the lesson with an entire class discussion in which any questions brought up in group discussions are presented and answered.

CHAPTER REVIEW

Vocabulary Choose the letter of the word(s) that completes the sentence.

1. A variable is a letter used to stand for ■. [366] b
   a. Addition **b.** A number **c.** An expression
2. A statement with an = sign is called ■. [366] c
   a. A translation **b.** An expression **c.** An equation
3. To solve the equation $3x - 7 = 20$, you would ■ first. [374] a
   a. Add **b.** Subtract **c.** Multiply

Skills Simplify the expression. [364]

4. $37 - 6 \cdot 4$ 13
5. $24 \div 2 - 3^2 - 1$ 2
6. $\frac{4(6-3)}{2(20-2)}$ $\frac{1}{3}$

Write an equation for the statement. [366]

7. 15 fewer than y is 6. $y - 15 = 6$
8. 12% of x is 20. $0.12x = 20$
9. 32 divided by y is 8. $32 \div y = 8$, or $\frac{32}{y} = 8$
10. 4 times the sum of 2 and x is 44. $4(2 + x) = 44$

Evaluate. [368]

11. $3x + 9$ for $x = 7$ 30
12. $A = lw$ for $l = 6$, $w = 9$ 54

Solve.

13. $x - 8 = 32$ [370] 40
14. $x + 4.7 = 7.9$ [370] 3.2
15. $3x = 18$ [372] 6
16. $\frac{y}{5} = 6$ [372] 30
17. $7y + 9 = 44$ [374] 5
18. $0.8x - 10 = 30$ [374] 50

Write an equation and solve.

19. The temperature decreased 9° from 76°. What is the new temperature? [370]
    $t = 76 - 9$; 67°
20. Your hourly wage increased \$1.50 to \$6.35. What was your original wage? [370]
    $w + \$1.50 = \$6.35$; \$4.85
21. A number multiplied by 9 is 36. What is the number? [372]
    $9n = 36$; 4
22. Ed answered 60% of the questions correctly for a score of 15. How many questions were on the test? [372]
    $0.6q = 15$; 25 questions
23. Rex bought 6 boxes of erasers and 7 additional erasers for a total of 151. How many erasers were in a box? [374] $6b + 7 = 151$; 24 erasers
24. Sandra earned \$80, plus 20% of her sales, less \$28 withholding, for a total net of \$128. What were Sandra's sales? [374] $\$80 + 0.2s - \$28 = \$128$; \$380

CHAPTER TEST

Simplify the expression.

1. $29 - 6 \cdot 3$ 11
2. $2(6 - 3)$ 6
3. $20 \div (3 + 2^3 - 1)$ 2
4. $\frac{2(3 + 7) - 4}{4(7 - 5)}$ 2

Write an expression or an equation.

5. 10 increased by x $10 + x$
6. 3 more than the product of 7 and x $7x + 3$
7. 8 less 5 is equal to x. $8 - 5 = x$
8. 5 times the difference of x and 3 is 42. $5(x - 3) = 42$

Evaluate.

9. $4(y + 3)$ for $y = 9$ 48
10. $2x + (3 - y) + z$ for $x = 9$, $y = 2$, $z = 7$ 26

Solve.

11. $x - 15 = 7$ 22
12. $17 + y = 32$ 15
13. $5y = 30$ 6
14. $\frac{x}{2} = 38$ 76
15. $0.6z = 4.8$ 8
16. $9x - 3 = 24$ 3
17. $5 + 4x = 29$ 6
18. $23 + 6y - 15 = 62$ 9

Write an equation and solve.

19. The temperature increased from 3° to 12°. How much did the temperature rise?
$3 + t = 12$; 9°
20. Lori decreased her number of work hours by 7 to 29. How many hours had Lori been working?
$h - 7 = 29$; 36 h
21. One-third of a number is 16. What is the number?
$\frac{n}{3} = 16$; 48
22. Les read 42 pages, which is 60% of the pages he has to read. How many pages does he have to read in all?
$0.6p = 42$; 70 pages
23. Six buses left the terminal. Five buses had all the seats taken, and 1 bus had 18 seats taken, for a total of 253. How many seats were on each bus?
$5s + 18 = 253$; 47 seats
24. Joanne earned \$95, plus 15% of her sales, less \$72 withholding, for a total of \$398. What were her sales?
\$95 + 0.15s − \$72 = \$398; \$2,500
25. Ira invests in stocks. Last year, his income from investments was \$2,385.75. This year, his income from investments was \$757.95 more than it was last year. What was the total income from his investments for the 2 y?
\$5,529.45

OBJECTIVE

- Evaluate the achievement of chapter objectives.

USING THE TEST

The Chapter Test may be used as a post-test to evaluate student achievement. However, you may wish to use the Chapter Post-Test offered in the Teacher's Resource Package or you may design your own chapter test. If this page is not used as a test, you may want to assign it as additional review or practice.

The test items are correlated to the chapter objectives in the table below.

Chapter objectives	Test items
A. Simplify numerical expressions.	1–4
B. Translate words into algebraic expressions or equations.	5–8
C. Evaluate algebraic expressions.	9–10
D. Write and solve 1-step equations.	11–15, 19–22
E. Write and solve 2-step equations.	16–18 23–24
F. Solve problems by using rounded numbers in simpler problems.	25

SUPPLEMENTARY MATERIALS

TRP Ch. 18 Post-Test Form A, pp. 1–2
TRP Ch. 18 Post-Test Form B, pp. 1–2
TRP Final Test, pp. 1–8

OBJECTIVE

- Solve equations, using a calculator.

USING THE PAGE

Remind students that an equation is solved by performing 1 or more inverse operations that *undo* operations within the equation to isolate the variable.

In the first example, the solution requires 1 step, subtracting 1.8 from both sides. Have students use their calculators to perform the subtraction on the right side of the equation. Have them check their result with the display shown in the text.

The second example is a 2-step equation. Students must subtract and then divide. Have them use their calculators to solve the equation. Point out that solving multistep equations requires *undoing* the operations in the *reverse* of the algebraic order of operations, that is, adding or subtracting before multiplying or dividing.

Have students complete the exercises and word problems. As a challenge, have them use their calculators to write as many equations as they can in 1 minute that contain 2.5 on the left side and 30 on the right side of the is-equal-to sign; for example, $2.5n + 10 = 30$. Then have students solve each other's equations.

CALCULATOR SOLVE EQUATIONS

You can use a calculator to solve equations.

Solve for n: $n + 1.8 = 3.4$.

THINK: Subtract 1.8 from both sides of the equation.

Procedure	Calculator Entry	Calculator Display
$n + 1.8 = 3.4$		
$n + 1.8 - 1.8 = 3.4 - 1.8$		
$n = 3.4 - 1.8$	[3] [.] [4] [−] [1] [.] [8] [=]	1.6
$n = 1.6$		

You can also solve 2-step equations on a calculator.

Solve for n: $6n + 8.4 = 36$.

THINK: First subtract, and then divide.

Procedure	Calculator Entry	Calculator Display
$6n + 8.4 = 36$		
$6n + 8.4 - 8.4 = 36 - 8.4$		
$6n = 36 - 8.4$	[3] [6] [−] [8] [.] [4] [=]	27.6
$6n = 27.6$		
$6n \div 6 = 27.6 \div 6$	[÷] [6] [=]	4.6
$n = 4.6$		

Use a calculator to solve for n.

1. $n + 14.7 = 33.9$ 19.2
2. $n - 18 = 2.6$ 20.6
3. $n + 0.06 = 1.29$ 1.23
4. $6n = 33$ 5.5
5. $9.3n = 0.2418$ 0.026
6. $4.5n = 37.35$ 8.3
7. $\frac{n}{12} = 2.6$ 31.2
8. $\frac{n}{3.2} = 4.7$ 15.04
9. $\frac{n}{0.04} = 800$ 32
10. $4n + 8 = 22$ 3.5
11. $6n + 3.2 = 26$ 3.8
12. $2.5n - 12.5 = 17.5$ 12

Write an equation and then use your calculator to solve.

13. A bookkeeper worked 35 h and earned $367.50. What did he earn per hour?
    $35n = \$367.50$; $n = \$10.50$
14. The Jacksons ordered a bed priced at $692.95 and 2 matching chairs. The bill came to $1,213.95. What did each chair cost?
    $\$692.95 + 2n = \$1,213.95$; $n = \$260.50$

ESTIMATION SKILLS

ESTIMATING PRODUCTS AND QUOTIENTS OF FRACTIONS AND MIXED NUMBERS

Here is one way to estimate the product of a mixed number and a fraction close to $\frac{1}{2}$.

1. Round the fraction to $\frac{1}{2}$.
2. Round the mixed number to the nearest even number.
3. Multiply.

Examples $8\frac{3}{4} \times \frac{2}{5}$ ↓ ↓

THINK: $8 \times \frac{1}{2}$

$8 \div 2 = 4$

Estimate: 4

$\frac{5}{8} \times 17\frac{1}{3}$ ↓ ↓

THINK: $\frac{1}{2} \times 18$

$18 \div 2 = 9$

Estimate: 9

You can use compatible numbers when you estimate the quotient of 2 mixed numbers.

1. Find compatible whole numbers that are close in value to the given divisor and dividend.
2. Divide, using the compatible numbers.

Examples $12\frac{3}{4} \div 3\frac{1}{3}$ ↓ ↓

THINK: $12 \div 3 = 4$

Estimate: 4

$7\frac{5}{8} \div 4\frac{3}{5}$ ↓ ↓

THINK: $8 \div 4 = 2$

Estimate: 2

Estimate the product. Look for factors close to $\frac{1}{2}$.

1. $\frac{2}{5} \times 6\frac{1}{4}$ 3 **2.** $10\frac{1}{2} \times \frac{3}{8}$ 5 **3.** $\frac{4}{9} \times 12\frac{1}{3}$ 6

4. $\frac{5}{8} \times 9\frac{3}{4}$ 5 **5.** $15\frac{3}{8} \times \frac{3}{5}$ 8 **6.** $\frac{7}{15} \times 2\frac{1}{2}$ 1

7. $23\frac{1}{10} \times \frac{5}{9}$ 12 **8.** $\frac{7}{12} \times 13\frac{3}{4}$ 7 **9.** $31\frac{2}{5} \times \frac{9}{16}$ 16

Estimate the quotient. Use compatible numbers.

10. $9\frac{1}{4} \div 1\frac{3}{4}$ 5 **11.** $10\frac{1}{3} \div 4\frac{2}{5}$ 2 **12.** $37\frac{1}{2} \div 9\frac{1}{8}$ 4

13. $17\frac{3}{8} \div 2\frac{4}{5}$ 6 **14.** $14\frac{1}{4} \div 2\frac{3}{5}$ 5 **15.** $28\frac{3}{4} \div 5\frac{1}{2}$ 5

16. $23\frac{7}{8} \div 5\frac{1}{2}$ 4 **17.** $26\frac{3}{4} \div 3\frac{5}{8}$ 7 **18.** $38\frac{1}{2} \div 7\frac{5}{8}$ 5

OBJECTIVE

- Estimate the products and quotients of fractions and mixed numbers.

USING THE PAGE

In the teaching notes on page 299, it is mentioned that it is generally easier to multiply by $\frac{1}{2}$ (that is, divide by 2) than to multiply by 0.5, especially when an answer is to be obtained mentally. Similarly, it is easier to multiply by $\frac{1}{2}$ than by a fraction with a numerator greater than 1. This lesson's procedure for estimating products makes use of this fact.

Extending the ideas of the lesson, you might emphasize the convenience of multiplying by any unit fraction.

$\frac{1}{3} \times 8\frac{3}{4} \approx \frac{1}{3} \times 9$, or 3
$\frac{1}{4} \times 8\frac{3}{4} \approx \frac{1}{4} \times 8$, or 2
$\frac{1}{5} \times 8\frac{3}{4} \approx \frac{1}{5} \times 10$, or 2
$\frac{1}{8} \times 8\frac{3}{4} \approx \frac{1}{8} \times 8$, or 1

The technique for estimating the quotient of 2 mixed numbers is similar to that used with decimals on page 257. Keep in mind the fact that there may be more than 1 reasonable answer. In Exercise 16, for example, the estimate 4 is based on this thought process:
$23\frac{7}{8} \div 5\frac{1}{2} \approx 24 \div 6$, or 4.
However, it is also reasonable to give the estimated quotient as 5:
$23\frac{7}{8} \div 5\frac{1}{2} \approx 25 \div 5$, or 5.

SKILLS BANK

TABLE OF CONTENTS

PART A **NUMERATION**

PART B **WHOLE-NUMBER, DECIMAL, AND INTEGER COMPUTATION**

PART C **FRACTIONS AND MIXED NUMBERS**

PART D **RATIO, PROPORTION, AND PERCENT**

PART E **MEASUREMENT**

383

THE DECIMAL SYSTEM

Skill 1: Identify whole-number place value

The table shows the names of the places through the thousands **period.** You can use the table to identify **place values** and the values of the digits in a number.

Name the place and the value of 7 in the number 792,834.

THOUSANDS			ONES		
hundred-thousands	ten-thousands	thousands	hundreds	tens	ones
7	9	2	8	3	4

In the number 792,834, the 7 is in the hundred-thousands place. The value of 7 is 7 × 100,000, or 700,000.

Name the place of the underlined digit.

1. $\underline{7}86$ hundreds
2. $\underline{4}2{,}051$ ten-thousands
3. $6\underline{5}{,}319$ thousands
4. $1\underline{8}9{,}231$ ten-thousands

Name the value of the underlined digit.

5. $\underline{6}{,}513$ 6,000
6. $\underline{8}2{,}045$ 80,000
7. $68{,}\underline{3}61$ 300
8. $\underline{5}01{,}632$ 500,000

Skill 2: Read and write whole numbers

You can read numbers by using the **period names** and **commas.**

Read the number 792,834.
The number 792,834 is read as 792 thousand, 834.

To write a number from a word name, start with the digit in the greatest place. Then write the digit for each place to its right. Write zeros as placeholders when necessary.

Write the number for fifty-six thousand, two hundred nine.

THOUSANDS		ONES		
ten-thousands	thousands	hundreds	tens	ones
5	6	2	0	9

The number is 56,209.

Write the number.

9. 6 thousand, 242 6,242
10. 16 thousand, 17 16,017
11. 23 thousand, 504 23,504
12. 418 thousand, 984 418,984
13. nine hundred eighty-one 981
14. sixty-three thousand, forty 63,040
15. four thousand, three hundred nineteen 4,319
16. three hundred forty thousand, two hundred ten 340,210

THE DECIMAL SYSTEM

Skill 3: Identify decimal place value

This table shows the names of the decimal places through hundred-thousandths. You can use the table to identify decimal place values and the values of the digits in a decimal.

Find the value of the 6 in the number 2.456.

ones	.	tenths	hundredths	thousandths	ten-thousandths	hundred-thousandths
2	.	4	5	6		

In the number 2.456, 6 is in the thousandths place. The value of 6 is 6 thousandths, or 0.006.

Name the place of the underlined digit.

17. 0.4<u>3</u> hundredths **18.** 0.20<u>7</u> thousandths **19.** 0.413<u>5</u> ten-thousandths **20.** 14.9850<u>7</u> hundred-thousandths

Name the value of the underlined digit.

21. 0.<u>9</u> 0.9 **22.** 0.3<u>8</u> 0.08 **23.** 1.76<u>2</u>5 0.002 **24.** 29.1876<u>2</u> 0.00002

Skill 4: Read and write decimals

When you read decimals, first read the whole-number part. Then read the decimal part. Use the name of the place of the last decimal digit. Remember to use the word *and* for the decimal point.

Read the decimal 2.456.

The number 2.456 is read as 2 and 456 thousandths.

You can use a table to write numbers from short word names and word names. Remember to use zeros as placeholders when necessary.

Write the number for two hundred seventeen ten-thousandths.

THINK: Ten-thousandths means 4 decimal places.

ones	.	tenths	hundredths	thousandths	ten-thousandths
0	.	0	2	1	7

Write the number.

25. 23 hundredths 0.23

26. 417 thousandths 0.417

27. 685 ten-thousandths 0.0685

28. 69 hundred-thousandths 0.00069

29. nine and seven tenths 9.7

30. twenty-four and eight hundredths 24.08

31. sixty and seven hundred fifty-three thousandths 60.753

32. thirteen and four hundred eighteen ten-thousandths 13.0418

COMPARE AND ORDER WHOLE NUMBERS AND DECIMALS

Skill 1: Compare whole numbers

To compare whole numbers, write the numbers in a column, aligning the places. The whole number with the greater number of digits is the greater number. If both numbers have the same number of digits, compare the digits in order from left to right.

Remember: = means *is equal to*
$>$ means *is greater than*
$<$ means *is less than*

Compare 4,523 and 4,517.
Use $>$, $<$, or $=$.

4,	5	2	3
↕	↕	↕	
4,	5	1	7

2 tens $>$ 1 ten, so
4,523 $>$ 4,517

Compare. Write $>$, $<$, or $=$.

1. 317 ● 324 $<$ **2.** 4,812 ● 946 $>$ **3.** 6,509 ● 6,549 $<$ **4.** 547,213 ● 547,138 $>$

Skill 2: Order whole numbers

To order whole numbers, write the numbers in a column, aligning the places. In the example, 80,016 has the greatest number of digits. It is the greatest number. Now compare the digits from left to right in the numbers 8,916 and 8,016. Since 0 hundreds $<$ 9 hundreds, 8,016 $<$ 8,916.

Write in order from least to greatest: 8,916; 8,016; and 80,016.

8,916
8,016
80,016 ← **greatest**

8,	9	1	6	
↕	↕			
8,	0	1	6	← **least** **THINK: 0 $<$ 9**

The ordered numbers are 8,016; 8,916; and 80,016.

Write in order from least to greatest.

5. 85, 80, 850 80, 85, 850

6. 1,243; 1,236; 236 236; 1,236; 1,243

7. 1,436; 1,514; 1,504; 15,006
1,436; 1,504; 1,514; 15,006

8. 6,217; 6,207; 6,200; 6,102
6,102; 6,200; 6,207; 6,217

9. 27,026; 127,020; 29,002; 27,206
27,026; 27,206; 29,002; 127,020

10. 158,216; 58,214; 59,006; 58,137
58,137; 58,214; 59,006; 158,216

Write in order from greatest to least.

11. 4,016; 4,006; 4,106 4,106; 4,016, 4,006

12. 8,612; 86,012; 86,112 86,112; 86,012; 8,612

13. 87,395; 87,935; 87,305; 87,359
87,935; 87,395; 87,359; 87,305

14. 93,406; 93,604; 93,064; 93,046
93,604; 93,406; 93,064; 93,046

15. 314,654; 314,564; 314,645; 314,456
314,654; 314,645; 314,564; 314,456

16. 612,253; 612,532; 612,325; 612,352
612,532; 612,352; 612,325; 612,253

COMPARE AND ORDER WHOLE NUMBERS AND DECIMALS

Skill 3: Compare decimals

To compare decimals, write the numbers in a column. Line up the **decimal points** and **places.** Compare the digits from left to right. In the example, note that the decimal with the greater number of digits is not the greater number.

Compare 0.4317 and 0.452.

0.	4	3	1	7
↕	↕	↕		
0.	4	5	2	

3 hundredths $<$ 5 hundredths, so
$0.4317 < 0.452$

Compare. Write $>$, $<$, or $=$.

17. 0.41 ● 0.81 $<$ **18.** 2.784 ● 2.84 $<$ **19.** 0.406 ● 0.409 $<$ **20.** 0.86956 ● 0.8692 $>$

Skill 4: Order decimals

To order decimals, write the numbers in a column, aligning the decimal points and places. Write additional zeros as needed so that each number will have an equal number of decimal places. Compare the digits from left to right.

Write in order from least to greatest: 0.54, 0.546, and 0.12.

0.	5	4	0	
↕	↕	↕	↕	
0.	5	4	6	← **greatest**
↕	↕			
0.	1	2	0	← **least**

The ordered decimals are 0.12, 0.54, and 0.546.

Write in order from least to greatest.

21. 0.35, 0.31, 0.3 0.3, 0.31, 0.35

22. 0.07, 0.007, 0.7 0.007, 0.07, 0.7

23. 0.216, 0.206, 0.26, 2.1606
0.206, 0.216, 0.26, 2.1606

24. 0.761, 7.6101, 7.0101, 7.61015
0.761, 7.0101, 7.6101, 7.61015

Skill 5: Order money amounts

To order money amounts, write them in a column, aligning the **cents points.** Then compare the digits from left to right as you did with decimals. In the example, $2.31 has the least number of digits. It is the least. Now compare $23.14 and $23.59.

Write in order from least to greatest: $23.14, $23.59, and $2.31.

$2	3.	1	4	
↕	↕	↕		
$2	3.	5	9	← **greatest**
$	2.	3	1	← **least**

The ordered money amounts are $2.31, $23.14, and $23.59.

Write in order from least to greatest.

25. $0.21, $2.12, $0.12 $0.12, $0.21, $2.12

26. $3.45, $3.54, $3.05 $3.05, $3.45, $3.54

27. $16.41, $1.64, $16.40 $1.64, $16.40, $16.41

28. $45.86, $45.80, $45.26 $45.26, $45.80, $45.86

ROUND WHOLE NUMBERS AND DECIMALS

Skill 1: Round whole numbers using the number line

You can use the number line to round numbers. For example, 110 is nearer to 100 than to 200. So 110 rounds to 100. Notice that 180 is nearer to 200 than to 100. So 180 rounds to 200. When a number is located halfway between 2 numbers, it is rounded to the greater number. So 150 rounds to 200.

Round 110, 150, and 180 to the nearest hundred.

110 rounds to 100.
180 rounds to 200.
150 rounds to 200.

Round to the nearest hundred. Use the number line.

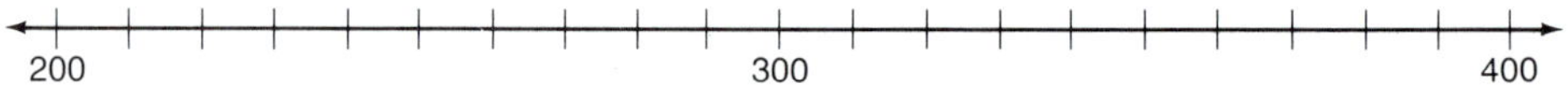

1. 220 200 **2.** 250 300 **3.** 290 300 **4.** 330 300 **5.** 370 400

Skill 2: Round any whole number

When you round greater numbers to a certain place, ring the digit to the right of that place. If the ringed digit is less than 5, round down. If the ringed digit is 5 or greater, round up.

Round 1,237 to the nearest hundred.

1, 2 ③ 7 ← **Ring the digit to the right of the hundreds place.**
↓ ↓ ↓ ↓
1, 2 0 0 ← **Since $3 < 5$, the hundreds place stays the same. Replace the tens and ones with zeros.**

1,237 rounds to 1,200.

Round 3,546 to the nearest thousand.

3, ⑤ 4 6 ← **Ring the digit to the right of the thousands place.**
↓ ↓ ↓ ↓
4, 0 0 0 ← **Since $5 = 5$, add 1 to the thousands place. Replace the hundreds, tens, and ones with zeros.**

3,546 rounds to 4,000.

Round to the nearest hundred.

6. 236 200 **7.** 695 700 **8.** 853 900 **9.** 907 900 **10.** 3,852 3,900

Round to the nearest thousand.

11. 4,281 4,000 **12.** 6,929 7,000 **13.** 7,723 8,000 **14.** 9,583 10,000 **15.** 65,100 65,000

Round to the nearest ten thousand.

16. 10,599 10,000 **17.** 25,099 30,000 **18.** 42,999 40,000 **19.** 58,924 60,000 **20.** 151,000 150,000

ROUND WHOLE NUMBERS AND DECIMALS

Skill 3: Round any decimal

When you round decimals, follow the same rounding rules used for whole numbers. When you round to a certain place, look at the digit to its right.

Remember:

If this digit is less than 5, round down. If this digit is 5 or greater, round up.

Round 0.423 to the nearest hundredth.

0. 4 2 ③ ← **3 is to the right.**
↓ ↓ ↓
0. 4 2 ← **Drop all digits to the right.**

0.423 rounds to 0.42.

Round 0.658 to the nearest tenth.

0. 6 ⑤ 8 ← **5 is to the right.**
↓ ↓
0. 7 ← **Add 1 to the tenths place. Drop all digits to the right.**

0.658 rounds to 0.7.

Round to the nearest whole number.

21. 1.6 2 **22.** 2.3 2 **23.** 17.5 18

Round to the nearest tenth.

24. 0.42 0.4 **25.** 0.65 0.7 **26.** 0.89 0.9

Round to the nearest hundredth.

27. 0.865 0.87 **28.** 0.679 0.68 **29.** 0.005 0.01

Round to the nearest thousandth.

30. 0.4456 0.446 **31.** 0.6981 0.698 **32.** 0.3105 0.311

Skill 4: Round money amounts

When you round money amounts to the nearest ten dollars, dollar, or cent, follow the same rounding rules used for whole numbers and decimals.

Round $46.915 to the nearest dollar.

$4 6. ⑨ 1 5 ← **9 is to the right.**
↓ ↓
$4 7. 0 0 ← **Add 1 to the dollars place.**

$46.915 rounds to $47.00.

Round $46.915 to the nearest cent.

$4 6. 9 1 ⑤ ← **5 is to the right.**
↓ ↓ ↓ ↓
$4 6. 9 2 ← **Add 1 to the cents place.**

$46.915 rounds to $46.92.

Round to the nearest ten dollars.

33. $26.49 $30.00 **34.** $35.79 $40.00 **35.** $81.99 $80.00

Round to the nearest dollar.

36. $6.42 $6.00 **37.** $8.96 $9.00 **38.** $10.82 $11.00

Round to the nearest cent.

39. $0.865 $0.87 **40.** $1.426 $1.43 **41.** $19.981 $19.98 **42.** $0.4231 $0.42 **43.** $1.8926 $1.89

USE INTEGERS

Skill 1: Graph integers

Integers to the right of 0 are called **positive integers.** The set of positive integers is written as +1, +2, +3, +4,... (positive 1, positive 2, positive 3, positive 4, and so on).

Integers to the left of 0 are called **negative integers.** The set of negative integers is written as −1, −2, −3, −4,... (negative 1, negative 2, negative 3, negative 4, and so on).

Zero is an integer that is neither positive nor negative.

Graph *A* at −2 and *B* at +4.

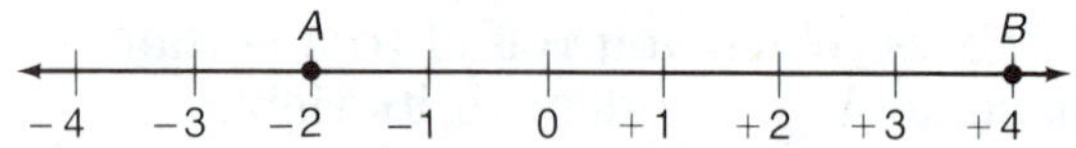

Point *A* is the graph of −2.
−2 is the **coordinate** of Point *A*.

Point *B* is the graph of +4.
+4 is the coordinate of Point *B*.

Draw a number line from −5 to +5. Then graph the point. Check students' number lines.

1. *A* at −5 **2.** *B* at −3 **3.** *C* at 0 **4.** *D* at +2 **5.** *E* at +3 **6.** *F* at +5

Skill 2: Find opposites

On the number line, the coordinate of point *C* is −2. The coordinate of point *D* is +2. −2 and +2 are called **opposites.** They are the same distance from 0, but on opposite sides.

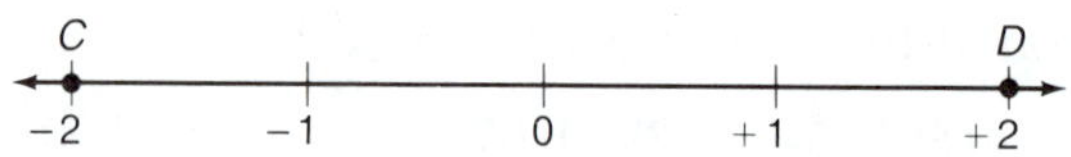

Write the opposite.

7. +6 −6 **8.** −7 +7 **9.** 0 0 **10.** +15 −15 **11.** −25 +25 **12.** +31 −31

13. −9 +9 **14.** +66 −66 **15.** +72 −72 **16.** −115 +115 **17.** +200 −200 **18.** −650 +650

Skill 3: Compare integers

You can use the number line to compare integers. Notice that, on the number line, numbers become greater as you go from left to right.

Compare −3 and −7.

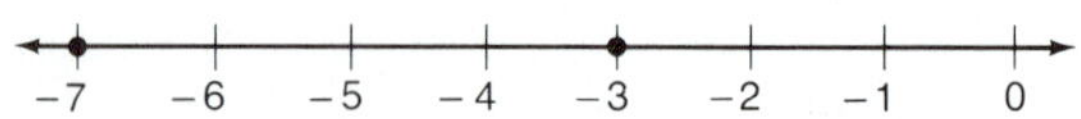

The integer to the right is always the greater number. So −3 is greater than −7. Write $-3 > -7$.

Compare. Write $>$, $<$, or $=$.

19. +7 ● +5 $>$ **20.** −2 ● −3 $>$ **21.** 0 ● +4 $<$ **22.** +8 ● +5 $>$

23. −11 ● −15 $>$ **24.** −14 ● +13 $<$ **25.** +22 ● +32 $<$ **26.** −38 ● −30 $<$

USE INTEGERS

Skill 4: Order integers

To order integers, graph the integers on a number line. Remember that numbers become greater as you go from left to right.

Order +1, −5, −4, and +3.

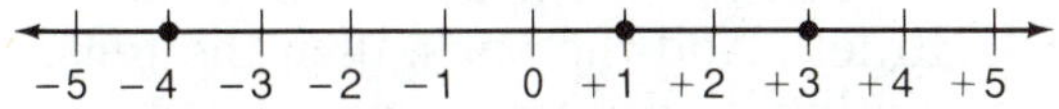

The order from least to greatest is −5, −4, +1, and +3.

The order from greatest to least is +3, +1, −4, and −5.

Order the integers from least to greatest.

27. +6, −3, 0, −4 −4, −3, 0, +6
28. +5, −3, −4, −1 −4, −3, −1, +5
29. +6, +3, +1, −4 −4, +1, +3, +6
30. 0, −5, −6, −7 −7, −6, −5, 0
31. +1, −1, −3, −7 −7, −3, −1, +1
32. 0, −6, +2, −3 −6, −3, 0, +2
33. +1, +3, −8, +2 −8, +1, +2, +3
34. +5, 0, −8, −9 −9, −8, 0, +5
35. −1, 0, −11, +6 −11, −1, 0, +6
36. +2, −5, −9, −6 −9, −6, −5, +2
37. +3, +5, 0, −8 −8, 0, +3, +5
38. −9, −7, +1, −5 −9, −7, −5, +1
39. 0, −6, +3, −5 −6, −5, 0, +3
40. +3, −6, +2, +4 −6, +2, +3, +4
41. −9, −7, −6, −8 −9, −8, −7, −6
42. +2, +8, +11, −6 −6, +2, +8, +11
43. +6, +1, 0, −12 −12, 0, +1, +6
44. −13, −9, +2, 0 −13, −9, 0, +2
45. −6, −2, −9, −4 −9, −6, −4, −2
46. +4, +2, −9, −7 −9, −7, +2, +4

Order the integers from greatest to least.

47. +6, 0, −3, +4 +6, +4, 0, −3
48. −7, −8, +1, +6 +6, +1, −7, −8
49. 0, +5, −6, −2 +5, 0, −2, −6
50. +3, +5, −4, −7 +5, +3, −4, −7
51. +4, +13, +12, 0 +13, +12, +4, 0
52. +1, +6, +3, +4 +6, +4, +3, +1
53. +8, +11, 0, −9 +11, +8, 0, −9
54. +2, 0, +4, −1 +4, +2, 0, −1
55. −7, −9, +2, +1 +2, +1, −7, −9
56. −8, +8, −2, +1 +8, +1, −2, −8
57. −6, +2, −3, −1 +2, −1, −3, −6
58. 0, +5, +2, −4 +5, +2, 0, −4
59. −6, −11, −4, −1 −1, −4, −6, −11
60. +9, −5, −2, +3 +9, +3, −2, −5
61. +8, 0, +1, −2 +8, +1, 0, −2
62. −4, +5, −3, −1 +5, −1, −3, −4
63. −8, 0, +2, +5 +5, +2, 0, −8
64. +3, −6, −8, +7 +7, +3, −6, −8
65. −9, +5, +4, +6 +6, +5, +4, −9
66. −7, −11, 0, +3 +3, 0, −7, −11

Skill 5: Find absolute value

The **absolute value** of a number is the number of units that number is from 0. The absolute value of −3 is 3 because −3 is 3 units from 0 on the number line. This is written as $|-3| = 3$.

Find the absolute values of $|-2|$ and $|+3|$. First graph the numbers.

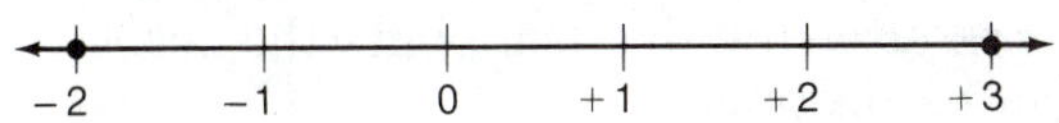

$|-2| = 2$ because −2 is 2 units from 0.

$|+3| = 3$ because +3 is 3 units from 0.

Find the absolute value.

67. $|-9|$ 9 **68.** $|+11|$ 11 **69.** $|-16|$ 16 **70.** $|-23|$ 23 **71.** $|+31|$ 31 **72.** $|+46|$ 46

ADD WITH WHOLE NUMBERS AND DECIMALS

Skill 1: Add whole numbers with no regrouping

When you add numbers, write the addends in columns. Then work from right to left. Add the ones, then the tens, then the hundreds, and so on.

Add: 4,123 + 615.

	th	h	t	o
	4,	1	2	3
+		6	1	5
	4,	7	3	8

Add.

1. 256 + 123 379
2. 6,534 + 163 6,697
3. 7,053 + 1,426 8,479
4. 12,435 + 7,520 19,955
5. 35,081 + 40,718 75,799
6. 642,537 + 56,412 698,949
7. 235,014 + 162,873 397,887

Skill 2: Add whole numbers with regrouping

Sometimes you must **regroup** when you add numbers. For example, 3 + 9 = 12 ones. Regroup 12 ones as 1 ten 2 ones. Write 2 in the ones column. Regroup 1 ten. Then continue to add, regrouping as necessary.

Add: 3,453 + 1,679.

	th	h	t	o
	1	1	1	
	3,	4	5	3
+	1,	6	7	9
	5,	1	3	2

Add.

8. 352 + 419 771
9. 674 + 387 1,061
10. 296 + 417 713
11. 585 + 805 1,390
12. 3,684 + 198 3,882
13. 7,046 + 3,497 10,543
14. 2,308 + 4,746 7,054
15. 85,721 + 69,852 155,573
16. 98,453 + 2,597 101,050
17. 21,532 + 17,009 38,541
18. 345,892 + 89,407 435,299
19. 736,495 + 573,889 1,310,384
20. 430,098 + 56,917 487,015

Skill 3: Add decimals with the same number of places

When you add decimals, align the decimal points in the addends. Place the decimal point in the sum. Then add as with whole numbers. If the sum ends in 1 or more zeros after the decimal point, the zeros can be dropped.

Add: 62.54 + 37.26.

```
       1
  6 2 .5 4
 +3 7 .2 6
 ---------
  9 9 .8 0 = 99.8
```

Add.

21. 4.3 + 1.6 5.9
22. 5.2 + 4.1 9.3
23. 8.9 + 7.3 16.2
24. 23.2 + 16.8 40
25. 7.35 + 1.83 9.18
26. 92.1 + 18.59 110.69
27. 43.36 + 29.84 73.2
28. 6.428 + 2.837 9.265
29. 26.017 + 51.984 78.001
30. 35.606 + 11.794 47.4
31. 1.3265 + 8.7316 10.0581
32. 51.0604 + 27.8594 78.9198
33. 44.0845 + 27.8186 71.9031

ADD WITH WHOLE NUMBERS AND DECIMALS

Skill 4: Add decimals with different numbers of places

When you add decimals with different numbers of places, align the decimal points in the numbers. Place the decimal point in the sum. Write additional zeros as needed and then add as with whole numbers.

Add: 8.236 + 27.9.

$$\begin{array}{r} {}^{1\,1} \\ 8.236 \\ +27.900 \\ \hline 36.136 \end{array}$$

Add.

34. 8.94 + 0.2 9.14 **35.** 29.8 + 4.98 34.78 **36.** 43.76 + 16.5 60.26 **37.** 6.987 + 4.5 11.487

38. 86.907 + 4.26 91.167 **39.** 234.895 + 83.63 318.525 **40.** 7.8643 + 1.907 9.7713

41. 9.31 + 27.0604 36.3704 **42.** 763.5 + 1.8168 765.3168 **43.** 37.245 + 416.9855 454.2305

Skill 5: Add more than 2 numbers

When you add more than 2 numbers, remember to align the numbers properly.

Add: 4.3 + 23.7 + 36.

$$\begin{array}{r} {}^{1\,1} \\ 4.3 \\ 23.7 \\ +36.0 \\ \hline 64.0 \end{array} = 64$$

Add.

44. 38 + 42 + 18 98 **45.** 5,937 + 89 + 351 6,377 **46.** 6,436 + 2,831 + 439 9,706

47. 8.6 + 0.9 + 1.5 11 **48.** 7.8 + 23.41 + 35.16 66.37 **49.** 9.356 + 12.607 + 2.51 24.473

50. 24.617 + 0.5 + 6.54 31.657 **51.** 16.64 + 8.9 + 0.7256 26.2656 **52.** 5.642 + 13.096 + 2.7314 21.4694

Skill 6: Add money amounts

When you add money amounts, align the cents points in the addends. Place the dollar sign and cents point in the sum. Then add.

Add: \$657.89 + \$56.13.

$$\begin{array}{r} {}^{1\,1\,1\;\;1} \\ \$657.89 \\ +\quad 56.13 \\ \hline \$714.02 \end{array}$$

Add.

53. \$5.86 + \$7.83 \$13.69 **54.** \$46.24 + \$32.53 \$78.77 **55.** \$77.10 + \$3.09 \$80.19 **56.** \$132.76 + \$67.99 \$200.75

57. \$456.67 + \$39.03 + \$2.55 \$498.25 **58.** \$1,436.43 + \$156 + \$32.01 \$1,624.44

59. \$67.45 + \$0.23 + \$35.86 \$103.54 **60.** \$25.25 + \$6.87 + 1,468.31 \$1,500.43

61. \$57 + \$23,460.26 + \$6,400.19 \$29,917.45 **62.** \$2.07 + \$32,461 + \$146,800.73 \$179,263.80

SUBTRACT WITH WHOLE NUMBERS AND DECIMALS

Skill 1: Subtract whole numbers with no regrouping

When you subtract numbers, write the numbers in columns. Then work from right to left. Subtract the ones, then the tens, then the hundreds, and so on.

Subtract: 6,345 − 4,102.

	th	h	t	o
	6,	3	4	5
−	4,	1	0	2
	2,	2	4	3

Subtract.

1. 516 − 403 113 **2.** 8,394 − 273 8,121 **3.** 7,896 − 6,245 1,651 **4.** 28,403 − 6,100 22,303

5. 67,214 − 56,113 11,101 **6.** 439,453 − 28,113 411,340 **7.** 672,435 − 501,332 171,103

Skill 2: Subtract whole numbers with regrouping

Sometimes you must regroup when you subtract numbers. For example, you cannot subtract 8 ones from 5 ones. Regroup 3 tens 5 ones as 2 tens 15 ones. Subtract: 15 ones − 8 ones = 7 ones. Write 7 in the ones column. Then continue to subtract, regrouping as necessary.

Subtract: 8,435 − 6,548.

	th	h	t	o
		13	12	
	7	~~3~~	2	15
	~~8~~,	~~4~~	~~3~~	~~5~~
−	6,	5	4	8
	1,	8	8	7

Subtract.

8. 865 − 629 236 **9.** 4,563 − 475 4,088 **10.** 9,523 − 7,684 1,839 **11.** 36,413 − 9,534 26,879

12. 46,438 − 39,159 7,279 **13.** 845,126 − 39,429 805,697 **14.** 892,453 − 128,698 763,755

Skill 3: Subtract across zeros

When subtracting with zeros, regroup carefully. For example, when subtracting 2,987 from 4,000, you cannot subtract 7 ones from 0 ones. You must regroup thousands, hundreds, tens, and ones.

Subtract: 4,000 − 2,987.

	th	h	t	o
		9	9	
	3	~~10~~	~~10~~	10
	~~4~~,	~~0~~	~~0~~	~~0~~
−	2,	9	8	7
	1,	0	1	3

Subtract.

15. 500 − 198 302 **16.** 8,000 − 289 7,711 **17.** 9,006 − 8,857 149 **18.** 36,000 − 5,762 30,238

19. 50,006 − 48,195 1,811 **20.** 760,004 − 89,475 670,529 **21.** 810,000 − 768,541 41,459

SUBTRACT WITH WHOLE NUMBERS AND DECIMALS

Skill 4: Subtract decimals with no regrouping

When you subtract decimals, align the decimal points. Place the decimal point in the answer. Then subtract as with whole numbers. If the answer ends in 1 or more zeros after the decimal point, the zeros can be dropped.

Subtract: 79.43 − 58.13.

$$\begin{array}{r} 79.43 \\ -58.13 \\ \hline 21.30 \end{array} = 21.3$$

Subtract.

22. 5.32 − 4.01 1.31 **23.** 61.8 − 31.6 30.2 **24.** 79.54 − 9.13 70.41 **25.** 8.436 − 5.106 3.33

26. 56.636 − 42.513 14.123 **27.** 235.48 − 120.18 115.3 **28.** 415.123 − 303.021 112.102

29. 362.421 − 201.310 161.111 **30.** 752.846 − 31.021 721.825 **31.** 504.816 − 3.602 501.214

Skill 5: Subtract decimals with regrouping

Sometimes you must regroup when you subtract decimals. Place the decimal point in the answer. Write additional zeros as needed and then subtract as with whole numbers.

Subtract: 52 − 41.5.

$$\begin{array}{r} \overset{1}{\cancel{2}}\ \overset{10}{\cancel{0}} \\ 52.0 \\ -41.5 \\ \hline 10.5 \end{array}$$

Subtract.

32. 6.1 − 0.8 5.3 **33.** 81.43 − 16.59 64.84 **34.** 80 − 10.84 69.16 **35.** 8.431 − 4.952 3.479

36. 14.003 − 11.98 2.023 **37.** 129.043 − 118.95 10.093 **38.** 23.4982 − 10.0095 13.4887

39. 50.105 − 6.97 43.135 **40.** 231.6425 − 75.316 156.3265 **41.** 25.73 − 7.8265 17.9035

Skill 6: Subtract money amounts

When you subtract money amounts, align the cents points. Place the dollar sign and the cents point in the answer. Then subtract.

Subtract: \$829.43 − \$76.51.

$$\begin{array}{r} 7\ 12\ 8\ \ 14 \\ \$829.43 \\ -\ \ 76.51 \\ \hline \$752.92 \end{array}$$

Subtract.

42. \$19.85 − \$7.13 \$12.72 **43.** \$54.76 − \$23.14 \$31.62 **44.** \$60 − \$42.98 \$17.02 **45.** \$138.41 − \$56.83 \$81.58

46. \$146.83 − \$37.98 \$108.85 **47.** \$359.86 − \$186.49 \$173.37 **48.** \$1,800 − \$186.43 \$1,613.57

49. \$2,000 − \$1,643.25 \$356.75 **50.** \$700.01 − \$283.99 \$416.02 **51.** \$7,002 − \$3,420.01 \$3,581.99

52. \$23,468 − \$5,401.76 \$18,066.24 **53.** \$984.77 − \$16.89 \$967.88 **54.** \$22,000.06 − \$721.97 \$21,278.09

MULTIPLY WITH WHOLE NUMBERS

Skill 1: Multiply with multiples of 10, 100, or 1,000

When multiplying powers of 10, count the number of end zeros in each factor. For example, when multiplying 2 × 4,000, write 3 zeros in the answer, then multiply 2 × 4. Write 8,000 as the answer.

Multiply: 2 × 4,000.

```
  4,000 ← 3 zeros
×     2
-------
  8,000 ← 3 zeros
```

When both factors are a power of 10, remember to count the number of zeros in each factor. For example, the first factor, 6,000, has 3 zeros and the second factor, 100, has 2 zeros. Write 5 zeros in the answer, then multiply 1 × 6. Write 600,000 as the answer.

Multiply: 100 × 6,000.

```
    6,000 ← 3 zeros
×     100 ← 2 zeros
---------
  600,000 ← 5 zeros
```

Multiply.

1. 2 × 40 80
2. 3 × 200 600
3. 3 × 800 2,400
4. 8 × 6,000 48,000
5. 10 × 20 200
6. 10 × 300 3,000
7. 200 × 800 160,000
8. 50 × 4,000 200,000
9. 30 × 2,000 60,000
10. 60 × 7,000 420,000
11. 700 × 2,000 1,400,000
12. 400 × 8,000 3,200,000
13. 5,000 × 3,000 15,000,000
14. 8,000 × 4,000 32,000,000
15. 9,000 × 3,000 27,000,000
16. 6,000 × 5,000 30,000,000
17. 1,000 × 8,000 8,000,000
18. 300 × 8,000 2,400,000
19. 20 × 5,000 100,000
20. 400 × 4,000 1,600,000
21. 7,000 × 6,000 42,000,000
22. 70 × 7,000 490,000
23. 40 × 9,000 360,000
24. 10 × 6,000 60,000

Skill 2: Multiply by 1-digit numbers with no regrouping

When multiplying by a 1-digit number, multiply the ones, then the tens, then the hundreds, and so on.

Multiply: 2 × 4,321.

	th	h	t	o
	4,	3	2	1
×				2
	8,	6	4	2

Multiply.

25. 2 × 314 628
26. 2 × 401 802
27. 3 × 201 603
28. 3 × 122 366
29. 3 × 2,110 6,330
30. 4 × 2,002 8,008
31. 2 × 3,102 6,204
32. 2 × 4,233 8,466
33. 3 × 31,210 93,630
34. 4 × 12,020 48,080
35. 2 × 44,421 88,842
36. 5 × 10,001 50,005
37. 2 × 433,202 866,404
38. 3 × 132,231 396,693
39. 4 × 122,021 488,084
40. 2 × 341,002 682,004

MULTIPLY WITH WHOLE NUMBERS

Skill 3: Multiply by 1-digit numbers with regrouping

When multiplying with regrouping, remember you must first multiply and then add the regrouped number.

Multiply: 7 × 1,354.

	th	h	t	o
	2	3	2	
	1,	3	5	4
×				7
	9,	4	7	8

In the example:

- Multiply the ones: 7 × 4 ones = 28 ones. Regroup 28 ones as 2 tens 8 ones. Write 8 in the ones column. Regroup the 2 tens.
- Multiply the tens and add the regrouped number: (7 × 5) + 2 = 37. Write 7 in the tens column. Regroup the 3.
- Then continue to multiply, regrouping as necessary.

Multiply.

41. 2 × 353 706
42. 4 × 613 2,452
43. 8 × 916 7,328
44. 4 × 108 432
45. 6 × 872 5,232
46. 9 × 427 3,843
47. 9 × 162 1,458
48. 6 × 843 5,058
49. 3 × 1,423 4,269
50. 4 × 2,621 10,484
51. 2 × 4,704 9,408
52. 5 × 1,017 5,085
53. 6 × 5,872 35,232
54. 9 × 8,427 75,843
55. 9 × 7,162 64,458
56. 6 × 7,843 47,058
57. 3 × 23,574 70,722
58. 6 × 14,817 88,902
59. 8 × 32,061 256,488
60. 7 × 42,136 294,952
61. 4 × 123,065 492,260
62. 2 × 236,456 472,912
63. 7 × 141,219 988,533
64. 5 × 215,076 1,075,380

Skill 4: Multiply by 2-digit numbers

When multiplying greater numbers, work in columns. Multiply the ones and then the tens in the multiplier. Be sure to line up all numbers correctly. You may want to write zeros to help. Then add the partial products.

Multiply: 43 × 986.

```
   3 2
   2 1
   9 8 6
 ×   4 3
 2  9 5 8  ← 3 × 986
39  4 4 0  ← 40 × 986
42, 3 9 8
```

Multiply.

65. 22 × 42 924
66. 31 × 32 992
67. 12 × 13 156
68. 33 × 33 1,089
69. 36 × 84 3,024
70. 36 × 29 1,044
71. 28 × 43 1,204
72. 65 × 56 3,640
73. 23 × 133 3,059
74. 42 × 203 8,526
75. 13 × 233 3,029
76. 24 × 113 2,712
77. 47 × 486 22,842
78. 53 × 309 16,377
79. 63 × 425 26,775
80. 75 × 468 35,100
81. 31 × 2,702 83,762
82. 42 × 4,116 172,872
83. 22 × 2,543 55,946
84. 32 × 1,833 58,656

MULTIPLY WITH WHOLE NUMBERS

Skill 5: Multiply by 3-digit numbers

When you multiply by a 3-digit number, multiply by the ones, then the tens, and then the hundreds in the multiplier. When you write the partial products, remember to align the numbers properly. Then add.

Multiply: 461 × 839.

```
      1 3
      2 3
      8 3 9
     ×4 6 1
     ------
      8 3 9  ← 1 × 839
   5 0 3 4 0  ← 60 × 839
 3 3 5 6 0 0  ← 400 × 839
 -----------
 3 8 6,7 7 9
```

Multiply.

85. 112 × 213 23,856 **86.** 221 × 224 49,504 **87.** 136 × 458 62,288 **88.** 342 × 563 192,546

89. 132 × 121 15,972 **90.** 212 × 123 26,076 **91.** 416 × 218 90,688 **92.** 428 × 295 126,260

93. 426 × 1,325 564,450 **94.** 734 × 2,068 1,517,912 **95.** 574 × 6,616 3,797,584 **96.** 212 × 3,844 814,928

97. 555 × 6,734 3,737,370 **98.** 735 × 8,206 6,031,410 **99.** 452 × 3,119 1,409,788 **100.** 817 × 6,044 4,937,948

101. 189 × 7,016 1,326,024 **102.** 313 × 2,910 910,830 **103.** 745 × 8,013 5,969,685 **104.** 952 × 9,604 9,143,008

Skill 6: Omit zeros in partial products

You can save time when you multiply with zeros by omitting them in the partial products. For example, when 0 is in the multiplier, omit the partial product of zeros. However, watch the alignment of the next partial product.

Multiply: 206 × 751.

```
      1                       1
      3                       3
    7 5 1                   7 5 1
   ×2 0 6                  ×2 0 6
   ------                  ------
  4 5 0 6                 4 5 0 6
  0 0 0                 1 5 0 2
1 5 0 2                 -----------
-----------             1 5 4,7 0 6
1 5 4,7 0 6
```

Multiply.

105. 105 × 238 24,990 **106.** 208 × 359 74,672 **107.** 409 × 123 50,307 **108.** 606 × 987 598,122

109. 707 × 346 244,622 **110.** 202 × 435 87,870 **111.** 609 × 231 140,679 **112.** 807 × 983 793,281

113. 501 × 3,604 1,805,604 **114.** 708 × 5,039 3,567,612 **115.** 201 × 8,670 1,742,670 **116.** 709 × 3,005 2,130,545

117. 302 × 6,502 1,963,604 **118.** 803 × 5,070 4,071,210 **119.** 604 × 1,623 980,292 **120.** 801 × 7,110 5,695,110

121. 107 × 4,301 460,207 **122.** 407 × 7,040 2,865,280 **123.** 601 × 8,302 4,989,502 **124.** 909 × 6,030 5,481,270

MULTIPLY WITH DECIMALS

Skill 1: Multiply decimals by 10, 100, or 1,000

When you multiply a decimal by 10, move the decimal point 1 place to the right.

Multiply: 10 × 4.2.

10 × 4.2 → 4.2 → 42

When you multiply a decimal by 100, move the decimal point 2 places to the right.

Multiply: 100 × 8.931.

100 × 8.931 → 8.931 → 893.1

When you multiply a decimal by 1,000, move the decimal point 3 places to the right. Remember to write additional zeros when necessary. For example, when multiplying 0.16 by 1,000, write 1 additional zero before you move the decimal point.

Multiply: 1,000 × 0.16.

1,000 × 0.16 → 0.160 → 160

Multiply.

1. 10 × 0.8 8
2. 10 × 1.56 15.6
3. 10 × 23.43 234.3
4. 10 × 136.425 1,364.25
5. 100 × 0.43 43
6. 100 × 1.896 189.6
7. 100 × 14.8631 1,486.31
8. 100 × 185.2 18,520
9. 1,000 × 0.871 871
10. 1,000 × 1.0433 1,043.3
11. 1,000 × 21.1 21,100
12. 1,000 × 361.5 361,500

Skill 2: Multiply decimals by whole numbers

To multiply a decimal by a whole number, multiply as with whole numbers. The product should have the same number of decimal places as the decimal factor.

Multiply: 43 × 3.172.

```
    3.172 ← 3 decimal places
  ×    43
   ------
    9 516
  126 88
  -------
  136 396 ← 3 decimal places
```

If the product ends in 1 or more zeros after the decimal point, the zeros can be dropped.

Multiply: 86 × 4.25.

```
    4.25 ← 2 decimal places
  ×   86
   -----
   25 50
  340 0
  ------
  365 50 ← 2 decimal places
```

365.50 = 365.5

Multiply.

13. 9 × 8.9 80.1
14. 18 × 24.6 442.8
15. 5 × 23.52 117.6
16. 67 × 76.32 5,113.44
17. 8 × 59.041 472.328
18. 65 × 95.434 6,203.21
19. 7 × 0.8236 5.7652
20. 6 × 7.5954 45.5724
21. 6 × 2.9308 17.5848
22. 24 × 0.876 21.024
23. 51 × 72.394 3,692.094
24. 26 × 3.505 91.13

MULTIPLY WITH DECIMALS

Skill 3: Multiply decimals by 0.1, 0.01, or 0.001

When you multiply a decimal by 0.1, move the decimal point 1 place to the left. Remember to write a zero in the ones place.

Multiply: 0.1×8.6.

$0.1 \times 8.6 \rightarrow 8.6 \rightarrow 0.86$

When you multiply a decimal by 0.01, move the decimal point 2 places to the left.

Multiply: 0.01×14.92.

$0.01 \times 14.92 \rightarrow 14.92 \rightarrow 0.1492$

When you multiply a decimal by 0.001, move the decimal point 3 places to the left. Remember to write additional zeros when necessary.

Multiply: 0.001×3.57.

$0.001 \times 3.57 \rightarrow 003.57 \rightarrow 0.00357$

Multiply.

25. 0.1×0.5 0.05
26. 0.1×41.83 4.183
27. 0.1×251.436 25.1436
28. 0.01×4.954 0.04954
29. 0.01×36.4235 0.364235
30. 0.001×0.421 0.000421
31. 0.001×61.7 0.0617
32. 0.001×254.6 0.2546

Skill 4: Multiply 2 decimals

To multiply decimals, multiply as with whole numbers. Add the numbers of decimal places in the factors to find the number of decimal places in the product. Write zeros in the product when necessary.

Multiply: 0.2×0.306.

```
   0.306 ← 3 decimal places
 ×   0.2 ← 1 decimal place
 0 0612 ← 4 decimal places
```

Multiply.

33. 0.8×0.9 0.72
34. 0.8×0.43 0.344
35. 0.12×0.85 0.102
36. 0.49×0.78 0.3822
37. 0.86×1.47 1.2642
38. 0.03×1.004 0.03012
39. 1.59×3.809 6.05631
40. 0.2×0.4305 0.0861

Skill 5: Multiply money amounts

When you multiply money amounts, follow the rules for multiplying with decimals. When necessary, round the product to the nearest cent.

Multiply: $4.6 \times \$23.84$.

```
   $23.84 ← 2 decimal places
 ×    4.6 ← 1 decimal place
   14 304
   95 36
 $109.664 ← 3 decimal places
```

$\$109.664 \approx \$109.66$

Multiply. Round the product to the nearest cent when necessary.

41. $16 \times \$17.86$ $285.76
42. $0.8 \times \$23.45$ $18.76
43. $1.23 \times \$0.49$ $0.60
44. $13.82 \times \$41.89$ $578.92

DIVIDE WITH WHOLE NUMBERS

Skill 1: Divide with multiples of 10, 100, or 1,000

When dividing powers of 10 by a number less than 10, think of the related multiplication sentence.

Divide: 1,200 ÷ 6.

THINK: 200 × 6 = 1,200, so 1,200 ÷ 6 = 200

When you divide with 2 powers of 10, cross out an equal number of zeros in the divisor and the dividend. Then divide.

Divide: 860 ÷ 10.

86~~0~~ ÷ 1~~0~~ = 86

Divide: 900 ÷ 30.

90~~0~~ ÷ 3~~0~~ = 30

Divide.

1. 400 ÷ 4 100
2. 8,000 ÷ 4 2,000
3. 5,100 ÷ 10 510
4. 83,000 ÷ 10 8,300
5. 900 ÷ 100 9
6. 6,400 ÷ 100 64
7. 71,000 ÷ 100 710
8. 80,000 ÷ 400 200
9. 7,000 ÷ 1,000 7
10. 23,000 ÷ 1,000 23
11. 817,000 ÷ 1,000 817
12. 600,000 ÷ 2,000 300

Skill 2: Divide by 1-digit numbers

Divide in order from left to right. **THINK:** Divide, multiply, subtract, bring down. If the difference is greater than the divisor, the partial quotient is an **underestimate.** Align the digits in the quotient over the digits in the dividend. Remember to include the remainder as part of the answer.

Divide: 6,123 ÷ 5.

```
   1,224 R3
 5)6,123
   5
   1 1
   1 0
     12
     10
      23
      20
       3
```

Divide.

13. 354 ÷ 3 118
14. 721 ÷ 3 240 R1
15. 8,616 ÷ 4 2,154
16. 9,523 ÷ 6 1,587 R1
17. 25,926 ÷ 2 12,963
18. 93,827 ÷ 4 23,456 R3
19. 976,624 ÷ 4 244,156
20. 564,267 ÷ 5 112,853 R2

Skill 3: Locate the first digit in a quotient

Sometimes the divisor will not divide into the first digit in the dividend. For example, 9 does not go into 4. It goes into 46. Place the first digit of the quotient above the 6 in the hundreds place.

Divide: 4,613 ÷ 9.

```
     512 R5
 9)4,613
   4 5
     11
      9
     23
     18
      5
```

DIVIDE WITH WHOLE NUMBERS

Divide.

21. 288 ÷ 8 36 **22.** 256 ÷ 5 51 R1 **23.** 7,512 ÷ 8 939 **24.** 4,139 ÷ 7 591 R2

25. 19,614 ÷ 6 3,269 **26.** 66,419 ÷ 8 8,302 R3 **27.** 325,116 ÷ 4 81,279 **28.** 133,916 ÷ 9 14,879 R5

Skill 4: Divide with zeros in a quotient

Sometimes there is a zero in the quotient. Be careful to align the digits in the quotient over the digits in the dividend.

Divide: 634 ÷ 7.

```
   9 0 R4
7)6 3 4
  6 3↓
      4
      0
      4
```

Divide.

29. 315 ÷ 3 105 **30.** 143 ÷ 7 20 R3 **31.** 8,020 ÷ 4 2,005 **32.** 4,041 ÷ 5 808 R1

33. 42,091 ÷ 7 6,013 **34.** 48,302 ÷ 6 8,050 R2 **35.** 234,009 ÷ 9 26,001 **36.** 728,034 ÷ 8 91,004 R2

Skill 5: Divide using a shortcut

You can use a shortcut when you divide by 1-digit numbers. For example, when dividing 456 by 3, divide 4 by 3. Write 1 over the 4. Multiply and subtract mentally. THINK: 1 × 3 = 3 and 4 − 3 = 1. Write 1 next to the 5 in 56. Continue this process until you complete the division.

Divide: 456 ÷ 3.

```
  1 5 2
3)4 ¹5 6
```

Divide: 7,654 ÷ 4.

```
  1 , 9 1 3 R2
4)7 , ³6 5 ¹4
```

Divide, using the shortcut method.

37. 576 ÷ 3 192 **38.** 670 ÷ 5 134 **39.** 1,218 ÷ 2 609 **40.** 3,314 ÷ 4 828 R2

41. 76,206 ÷ 6 12,701 **42.** 81,512 ÷ 7 11,644 R4 **43.** 911,628 ÷ 9 101,292 **44.** 456,127 ÷ 8 57,015 R7

Skill 6: Divide by 2-digit numbers

Divide in order from left to right. Estimate first. In the example, use the 4 in tens place in the divisor to estimate. Then divide, multiply, subtract, and bring down. Align the digits in the quotient over the digits in the dividend. Remember to include the remainder as part of the answer.

Divide: 8,899 ÷ 41.

```
       2 1 7 R2
  4 1)8 8 9 9  ← Estimate: 4)8. Try 2 × 41.
      8 2↓ |
        6 9    ← Estimate: 4)6. Try 1 × 41.
        4 1 ↓
        2 8 9  ← Estimate: 4)28. Try 7 × 41.
        2 8 7
            2
```

DIVIDE WITH WHOLE NUMBERS

Divide.

45. 860 ÷ 40 21 R20 **46.** 8,750 ÷ 50 175 **47.** 891 ÷ 42 21 R9 **48.** 8,965 ÷ 55 163

49. 4,189 ÷ 63 66 R31 **50.** 49,975 ÷ 31 1,612 R3 **51.** 58,474 ÷ 61 958 R36 **52.** 183,663 ÷ 54 3,401 R9

Skill 7: Divide when estimates are too great

Sometimes your first estimate may be too great. In this case, adjust your estimate and try again.

Divide: 3,476 ÷ 44.

```
       8
 44)3,476 ← Estimate: 4)34. Try 8 × 44.
    3 52  ← Cannot subtract.

      79
 44)3,476
    3 08↓ ← Try 7 × 44.
      396 ← Estimate: 4)39. Try 9 × 44.
      396
        0
```

Divide.

53. 367 ÷ 17 21 R10 **54.** 290 ÷ 42 6 R38 **55.** 2,642 ÷ 57 46 R20 **56.** 2,870 ÷ 78 36 R62

57. 21,143 ÷ 29 729 R2 **58.** 62,115 ÷ 35 1,774 R25 **59.** 721,054 ÷ 48 15,021 R46 **60.** 642,219 ÷ 73 8,797 R38

Skill 8: Divide by 3-digit numbers

When you divide by a 3-digit number, follow the same rules as dividing by a 2-digit number. In the example, use the 2 in the hundreds place in the divisor to estimate. Remember to adjust your estimate when necessary.

Divide: 82,198 ÷ 214.

```
          4
 214)82,198 ← Estimate: 2)8. Try 4 × 214.
     85 6   ← Cannot subtract.

        384 R22
 214)82,198
     64 2↓ |  ← Try 3 × 214.
     17 99 |  ← Estimate: 2)17. Try 8 × 214.
     17 12 ↓
        878   ← Estimate: 2)8. Try 4 × 214.
        856
         22
```

Divide.

61. 73,521 ÷ 389 189 **62.** 42,729 ÷ 436 98 R1 **63.** 35,264 ÷ 341 103 R141 **64.** 32,509 ÷ 563 57 R418

65. 576,081 ÷ 194 2,969 R95 **66.** 243,167 ÷ 231 1,052 R155 **67.** 412,018 ÷ 176 2,341 R2 **68.** 108,432 ÷ 251 432

DIVIDE WITH DECIMALS

Skill 1: Divide decimals by multiples of 10, 100, or 1,000

When you divide a decimal by 10, move the decimal point 1 place to the left.

Divide: 6.3 ÷ 10.

6.3 ÷ 10 → 6.3 → 0.63

When you divide a decimal by 100, move the decimal point 2 places to the left.

Divide: 34.96 ÷ 100.

34.96 ÷ 100 → 34.96 → 0.3496

When you divide a decimal by 1,000, move the decimal point 3 places to the left. Remember to write additional zeros when necessary.

Divide: 1.23 ÷ 1,000.

1.23 ÷ 1,000 → 001.23 → 0.00123

Divide.

1. 0.9 ÷ 10 0.09
2. 2.13 ÷ 10 0.213
3. 14.58 ÷ 10 1.458
4. 234.13 ÷ 10 23.413
5. 0.89 ÷ 100 0.0089
6. 1.214 ÷ 100 0.01214
7. 36.34 ÷ 100 0.3634
8. 138.4 ÷ 100 1.384
9. 0.23 ÷ 1,000 0.00023
10. 2.43 ÷ 1,000 0.00243
11. 43.8 ÷ 1,000 0.0438
12. 489.6 ÷ 1,000 0.4896

Skill 2: Divide decimals by whole numbers

When dividing by a whole number, place the decimal point in the quotient directly above the decimal point in the dividend. Then divide as with whole numbers. Sometimes you must write additional zeros in the quotient as placeholders. Remember, if the quotient is less than 1, write 0 in the ones place.

Divide: 36.42 ÷ 6.

```
     6 .0 7
  6)3 6 .4 2
    3 6
        4
        0
        4 2
        4 2
          0
```

Divide: 0.208 ÷ 4.

```
     0 .0 5 2
  4)0 .2 0 8
       2 0
           8
           8
           0
```

Divide.

13. 25.5 ÷ 3 8.5
14. 102.2 ÷ 7 14.6
15. 187.72 ÷ 19 9.88
16. 858.636 ÷ 36 23.851
17. 20.36 ÷ 4 5.09
18. 0.544 ÷ 8 0.068
19. 310.52 ÷ 28 11.09
20. 0.8366 ÷ 47 0.0178

Skill 3: Divide by tenths

When you divide by tenths, multiply both the divisor and the dividend by 10 to get a whole-number divisor. Move each decimal point 1 place to the right. Then divide as with whole numbers.

Divide: 3.224 ÷ 5.2.

```
          0 .6 2
  5.2)3 .2 2 4
      3  1 2
         1 0 4
         1 0 4
             0
```

or

```
        0 .6 2
  5 2)3 2 .2 4
      3 1  2
         1 0 4
         1 0 4
             0
```

DIVIDE WITH DECIMALS

Write an equivalent example that has a whole-number divisor.

21. 3.24 ÷ 0.4
32.4 ÷ 4

22. 0.4132 ÷ 0.8
4.132 ÷ 8

23. 84.58 ÷ 1.2
845.8 ÷ 12

24. 0.1865 ÷ 2.5
1.865 ÷ 25

Divide.

25. 1.888 ÷ 0.8
2.36

26. 0.0486 ÷ 0.6
0.081

27. 28.914 ÷ 6.1
4.74

28. 0.2852 ÷ 9.2
0.031

Skill 4: Divide by hundredths

When you divide by hundredths, move the decimal point in the divisor and the dividend 2 places to the right. Then divide as with whole numbers.

Divide: 0.0144 ÷ 0.03.

```
          0.48
0.03)0.0144
        12↓
         24
         24
          0
```

or

```
   0.48
3)1.44
  1 2↓
    24
    24
     0
```

Write an equivalent example that has a whole-number divisor.

29. 0.1235 ÷ 0.03
12.35 ÷ 3

30. 0.2356 ÷ 0.04
23.56 ÷ 4

31. 4.2314 ÷ 0.36
423.14 ÷ 36

32. 39.128 ÷ 3.59
3,912.8 ÷ 359

Divide.

33. 0.0085 ÷ 0.05
0.17

34. 0.0927 ÷ 0.09
1.03

35. 1.8936 ÷ 0.24
7.89

36. 12.985 ÷ 2.45
5.3

Skill 5: Divide by thousandths

When you divide by thousandths, move the decimal point in the divisor and the dividend 3 places to the right. Then divide as with whole numbers.

Divide: 0.0432 ÷ 0.072.

```
              0.6
0.072)0.0432
          43 2
             0
```

or

```
     0.6
72)43.2
   43 2
      0
```

Divide.

37. 0.0054 ÷ 0.009
0.6

38. 0.0279 ÷ 0.031
0.9

39. 0.0212 ÷ 0.106
0.2

40. 1.02193 ÷ 1.123
0.91

Skill 6: Write additional zeros in dividends

Sometimes you need to write additional zeros in the dividend. For example, when dividing 2.7 by 0.15, the decimal points must be moved 2 places to the right.

Sometimes additional zeros may also be needed in a dividend in order to continue dividing. For example, when dividing 7.3 by 0.2, 1 additional zero is needed to get a zero remainder.

Divide: 2.7 ÷ 0.15.

```
         18.
0.15)2.70
     1 5↓
     1 20
     1 20
        0
```

Divide: 7.3 ÷ 0.2.

```
     3 6.5
0.2)7.3 0
    6↓
    1 3
    1 2
      1 0
      1 0
        0
```

DIVIDE WITH DECIMALS

Divide.

41. 301 ÷ 4.3 70 **42.** 0.8 ÷ 0.16 5 **43.** 46 ÷ 1.15 40 **44.** 52.54 ÷ 0.071 740

45. 8.26 ÷ 4 2.065 **46.** 4.21 ÷ 25 0.1684 **47.** 0.0144 ÷ 0.45 0.032 **48.** 108.84 ÷ 0.24 453.5

Skill 7: Round quotients

To find a quotient to the nearest tenth, divide through the hundredths place. Then round the quotient to tenths.

To find a quotient to the nearest hundredth, divide through the thousandths place. Then round the quotient to hundredths.

Divide to the nearest tenth: 1.04 ÷ 5.6.

```
        0 18 ≈ 0.2
5.6)1.0 40
      5 6
      4 80
      4 48
        32
```

Divide to the nearest tenth.

49. 14.03 ÷ 16 0.9 **50.** 1.09 ÷ 0.8 1.4 **51.** 3.4626 ÷ 0.76 4.6 **52.** 14.9805 ÷ 1.83 8.2

53. 25.374 ÷ 11 2.3 **54.** 9.321 ÷ 1.77 5.3 **55.** 62.596 ÷ 0.88 71.1 **56.** 5.1362 ÷ 4.54 1.1

Divide to the nearest hundredth.

57. 18.43 ÷ 23 0.80 **58.** 5.0376 ÷ 0.6 8.40 **59.** 25.3 ÷ 0.75 33.73 **60.** 59.8312 ÷ 1.69 35.40

61. 22.6 ÷ 0.77 29.35 **62.** 3.521 ÷ 6.9 0.51 **63.** 7.3425 ÷ 2.24 3.28 **64.** 0.873 ÷ 41.9 0.02

Skill 8: Divide money amounts

When you divide money amounts, follow the rules for dividing decimals. When necessary, round the quotient to the nearest cent by dividing through 1 place to the right of the cents place. Then round to the nearest cent.

Divide: $10.89 ÷ 0.4.

```
       $2 7  2 2 5 ≈ $27.23
0.4)$1 0.8  9 0 0
      8
      2 8
      2 8
        0 9
          8
          1 0
            8
            2 0
            2 0
              0
```

Divide. Round to the nearest cent when necessary.

65. $103.80 ÷ 5 $20.76 **66.** $124.65 ÷ 6 $20.78 **67.** $521.16 ÷ 8 $65.15 **68.** $999.99 ÷ 2 $500.00

69. $2.48 ÷ 0.8 $3.10 **70.** $386.45 ÷ 1.13 $341.99 **71.** $654.21 ÷ 5.3 $123.44 **72.** $936.07 ÷ 0.234 $4,000.30

EXPONENTS

Skill 1: Write powers of 10 in exponent form

Powers of 10 can be written in **exponent form** to show repeated multiplication. The **exponent** tells the numbers of zeros in the product.

$1 = 10^0$	$10 = 10^1$	$100 = 10^2$

Write 1,000 in exponent form.

THINK: 3 zeros $1{,}000 \rightarrow 10^3$ ← **exponent** (10 is the **base**)

Write the product of 10^5.

THINK: $10 \times 10 \times 10 \times 10 \times 10$, or 5 zeros
$10^5 \rightarrow 100{,}000$

Write in exponent form.

1. 10,000,000 10^7
2. 100,000,000 10^8
3. 1,000,000,000 10^9
4. $10 \times 10 \times 10 \times 10$ 10^4
5. $10 \times 10 \times 10$ 10^3
6. $10 \times 10 \times 10 \times 10 \times 10 \times 10$ 10^6

Write the product.

7. 10^3 1,000
8. 10^4 10,000
9. 10^2 100
10. 10^6 1,000,000
11. 10^7 10,000,000
12. 10^9 1,000,000,000

Skill 2: 2 as an exponent

When a number is used as a factor 2 times, the exponent is 2. For example, 4^2 is read as 4 **to the second power** or 4 **squared.**

Write 4×4 in exponent form.

THINK: 2 fours 4 is the base.
2 is the exponent.

$4 \times 4 \rightarrow 4^2$

Write in exponent form.

13. 3×3 3^2
14. 7×7 7^2
15. 17×17 17^2
16. 20×20 20^2

Write the product.

17. 4^2 16
18. 6^2 36
19. 9^2 81
20. 12^2 144
21. 14^2 196
22. 21^2 441

Skill 3: 3 as an exponent

When a number is used as a factor 3 times, the exponent is 3. For example, 2^3 is read as 2 **to the third power** or 2 **cubed.**

Write $2 \times 2 \times 2$ in exponent form.

THINK: 3 twos 2 is the base.
3 is the exponent.

$2 \times 2 \times 2 \rightarrow 2^3$

Write in exponent form.

23. $1 \times 1 \times 1$ 1^3
24. $4 \times 4 \times 4$ 4^3
25. $13 \times 13 \times 13$ 13^3
26. $25 \times 25 \times 25$ 25^3

Write the product.

27. 2^3 8
28. 3^3 27
29. 5^3 125
30. 6^3 216
31. 7^3 343
32. 11^3 1,331

ADD WITH INTEGERS

Skill 1: Add integers using the number line

You can add integers using the number line. Always start at 0. Moving to the right means adding a positive integer. Moving to the left means adding a negative integer.

Add: $-2 + (-1)$.

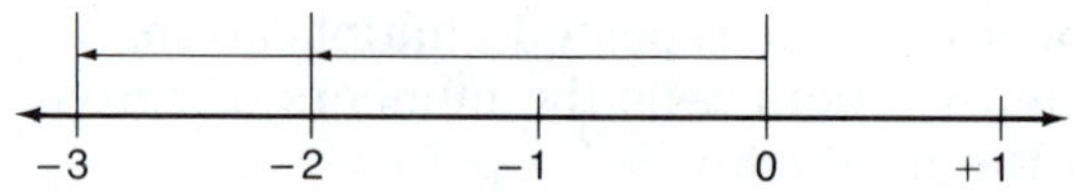

$-2 + (-1) = -3$

Since positive 3 can be written +3 or 3, write 3 (without the plus sign) from now on. Also, when a negative number follows an operation sign, use parentheses. For example, $-2 + (-1)$.

Add: $3 + (-1)$.

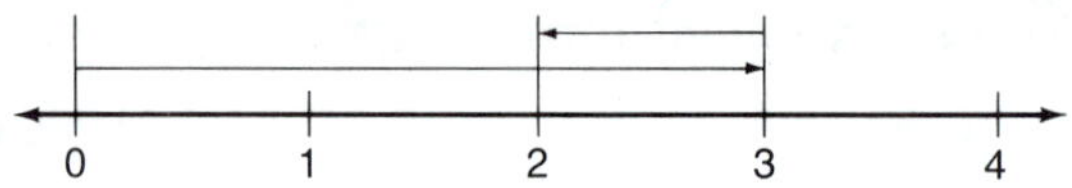

$3 + (-1) = 2$

Add. Use the number line.

1. $5 + 1$ 6 **2.** $-3 + (-2)$ −5 **3.** $5 + (-2)$ 3 **4.** $-1 + (-3)$ −4
5. $6 + (-3)$ 3 **6.** $5 + 5$ 10 **7.** $8 + (-4)$ 4 **8.** $-6 + (-2)$ −8

Skill 2: Add integers with like signs

The sum of 2 positive integers is a positive integer. The sum of 2 negative integers is a negative integer.

Add: $2 + 1$.

$2 + 1 = 3$

Add: $-4 + (-3)$.

$-4 + (-3) = -7$

Add.

9. $3 + 4$ 7 **10.** $4 + 2$ 6 **11.** $21 + 5$ 26 **12.** $18 + 17$ 35
13. $-1 + (-1)$ −2 **14.** $-3 + (-5)$ −8 **15.** $-11 + (-3)$ −14 **16.** $-23 + (-25)$ −48
17. $-18 + (-18)$ −36 **18.** $-12 + (-32)$ −44 **19.** $-28 + (-13)$ −41 **20.** $-35 + (-23)$ −58

Skill 3: Add integers with unlike signs

When you add integers with unlike signs, first subtract the absolute values of the integers (ignoring the signs). The answer has the same sign as the addend with the greater absolute value.

Add: $-5 + 2$.

THINK: $|-5| = 5$ $|2| = 2$ $5 - 2 = 3$
Since $|-5| > |2|$, the answer is negative.

$-5 + 2 = -3$

Add.

21. $-3 + 1$ −2 **22.** $-8 + 4$ −4 **23.** $-16 + 14$ −2 **24.** $-28 + 34$ 6
25. $5 + (-4)$ 1 **26.** $6 + (-7)$ −1 **27.** $23 + (-21)$ 2 **28.** $34 + (-16)$ 18
29. $16 + (-20)$ −4 **30.** $30 + (-22)$ 8 **31.** $33 + (-17)$ 16 **32.** $27 + (-38)$ −11

SUBTRACT WITH INTEGERS

Skill 1: Subtract integers with like signs

When you subtract an integer, add its opposite. For example, when you subtract $-4 - (-3)$, note that 3 is the opposite of -3. So add 3 to -4.

When you add $-4 + 3$, remember the rules for adding integers with unlike signs. First subtract the absolute values of the integers. The answer has the same sign as the addend with the greater absolute value.

Remember that positive 3 can be written as 3 and to use parentheses when a negative number follows an operation sign.

Subtract: $-4 - (-3)$.

$-4 - (-3) = -4 + 3$ (−3 and 3 are **opposites**)

THINK: $|-4| = 4$ $|3| = 3$ $4 - 3 = 1$
Since $|-4| > |3|$, the answer is negative.

$-4 - (-3) = -4 + 3 = -1$

Subtract.

1. $5 - 3$ 2	**2.** $5 - 9$ −4	**3.** $9 - 6$ 3	**4.** $3 - 7$ −4
5. $11 - 12$ −1	**6.** $21 - 36$ −15	**7.** $14 - 32$ −18	**8.** $16 - 11$ 5
9. $-8 - (-4)$ −4	**10.** $-3 - (-6)$ 3	**11.** $-17 - (-18)$ 1	**12.** $-35 - (-43)$ 8
13. $-9 - (-1)$ −8	**14.** $-2 - (-8)$ 6	**15.** $-4 - (-2)$ −2	**16.** $-6 - (-7)$ 1

Skill 2: Subtract integers with unlike signs

When you subtract an integer, add its opposite. For example, when you subtract $-5 - 6$, note that -6 is the opposite of 6. So add $-5 + (-6)$.

When you add $-5 + (-6)$, remember the rules for adding integers with like signs. The sum of 2 positive integers is positive. The sum of 2 negative integers is negative.

Subtract: $-5 - 6$.

$-5 - 6 = -5 + (-6)$ (6 and −6 are **opposites**)

THINK: $5 + 6 = 11$

$-5 - 6 = -5 + (-6) = -11$

Subtract: $3 - (-7)$.

$3 - (-7) = 3 + 7$
$3 - (-7) = 3 + 7 = 10$

Subtract.

17. $-5 - 7$ −12	**18.** $-1 - 7$ −8	**19.** $-5 - 9$ −14	**20.** $-9 - 4$ −13
21. $-14 - 15$ −29	**22.** $-54 - 39$ −93	**23.** $8 - (-3)$ 11	**24.** $4 - (-9)$ 13
25. $6 - (-3)$ 9	**26.** $2 - (-9)$ 11	**27.** $13 - (-24)$ 37	**28.** $41 - (-34)$ 75
29. $-16 - 9$ −25	**30.** $-21 - 26$ −47	**31.** $9 - (-4)$ 13	**32.** $13 - (-15)$ 28

MULTIPLY WITH INTEGERS

Skill 1: Multiply integers with like signs

When you multiply integers with like signs, the product is a positive number.

Remember that positive 4 can be written as 4 and to use parentheses when a negative number follows an operation sign.

Multiply: 4×4.

$4 \times 4 = 16$

Multiply: $-3 \times (-5)$.

$-3 \times (-5) = 15$

Multiply.

1. 5×3 15	**2.** 9×2 18	**3.** 6×7 42	**4.** 8×8 64
5. 11×2 22	**6.** 4×12 48	**7.** 25×3 75	**8.** 16×7 112
9. 14×8 112	**10.** 19×5 95	**11.** 7×26 182	**12.** 8×28 224
13. 39×8 312	**14.** 45×7 315	**15.** 38×8 304	**16.** 6×24 144
17. $-9 \times (-9)$ 81	**18.** $-3 \times (-26)$ 78	**19.** $-24 \times (-3)$ 72	**20.** $-1 \times (-72)$ 72
21. $-3 \times (-11)$ 33	**22.** $-6 \times (-12)$ 72	**23.** $-18 \times (-4)$ 72	**24.** $-8 \times (-50)$ 400
25. $-5 \times (-21)$ 105	**26.** $-4 \times (-16)$ 64	**27.** $-21 \times (-9)$ 189	**28.** $-11 \times (-12)$ 132
29. $-6 \times (-4)$ 24	**30.** $-4 \times (-1)$ 4	**31.** $-5 \times (-5)$ 25	**32.** $-7 \times (-8)$ 56
33. $-17 \times (-5)$ 85	**34.** $-6 \times (-11)$ 66	**35.** $-21 \times (-4)$ 84	**36.** $-24 \times (-9)$ 216

Skill 2: Multiply integers with unlike signs

When you multiply integers with unlike signs, the product is negative.

Multiply: -6×3.

$-6 \times 3 = -18$

Multiply: $7 \times (-4)$.

$7 \times (-4) = -28$

Multiply.

37. -4×1 -4	**38.** -3×9 -27	**39.** -7×6 -42	**40.** -9×9 -81
41. -13×3 -39	**42.** -4×17 -68	**43.** -6×12 -72	**44.** -24×7 -168
45. -5×9 -45	**46.** -11×5 -55	**47.** -2×45 -90	**48.** -5×61 -305
49. -17×4 -68	**50.** -24×8 -192	**51.** -26×9 -234	**52.** -3×22 -66
53. $3 \times (-12)$ -36	**54.** $16 \times (-1)$ -16	**55.** $34 \times (-4)$ -136	**56.** $15 \times (-5)$ -75
57. $7 \times (-21)$ -147	**58.** $23 \times (-5)$ -115	**59.** $47 \times (-8)$ -376	**60.** $5 \times (-61)$ -305
61. $6 \times (-30)$ -180	**62.** $28 \times (-7)$ -196	**63.** $36 \times (-4)$ -144	**64.** $15 \times (-11)$ -165
65. $5 \times (-1)$ -5	**66.** $7 \times (-3)$ -21	**67.** $6 \times (-4)$ -24	**68.** $2 \times (-9)$ -18
69. $16 \times (-2)$ -32	**70.** $7 \times (-14)$ -98	**71.** $13 \times (-5)$ -65	**72.** $32 \times (-4)$ -128

DIVIDE WITH INTEGERS

Skill 1: Divide integers with like signs

When you divide integers with like signs, the quotient is a positive number.

Remember that positive 8 can be written as 8 and to use parentheses when a negative number follows an operation sign.

Divide: $8 \div 4$.

$8 \div 4 = 2$

Divide $-9 \div (-3)$.

$-9 \div (-3) = 3$

Divide.

1. $4 \div 2$ 2 **2.** $5 \div 1$ 5 **3.** $12 \div 6$ 2 **4.** $18 \div 3$ 6

5. $45 \div 15$ 3 **6.** $60 \div 10$ 6 **7.** $75 \div 5$ 15 **8.** $84 \div 12$ 7

9. $9 \div 3$ 3 **10.** $60 \div 12$ 5 **11.** $90 \div 30$ 3 **12.** $45 \div 9$ 5

13. $44 \div 11$ 4 **14.** $100 \div 10$ 10 **15.** $65 \div 13$ 5 **16.** $100 \div 25$ 4

17. $-12 \div (-2)$ 6 **18.** $-72 \div (-9)$ 8 **19.** $-70 \div (-10)$ 7 **20.** $-64 \div (-8)$ 8

21. $-64 \div (-16)$ 4 **22.** $-84 \div (-21)$ 4 **23.** $-40 \div (-5)$ 8 **24.** $-600 \div (-6)$ 100

25. $-65 \div (-13)$ 5 **26.** $-36 \div (-6)$ 6 **27.** $-81 \div (-9)$ 9 **28.** $-378 \div (-7)$ 54

29. $-6 \div (-3)$ 2 **30.** $-8 \div (-2)$ 4 **31.** $-15 \div (-5)$ 3 **32.** $-14 \div (-7)$ 2

33. $-54 \div (-2)$ 27 **34.** $-75 \div (-15)$ 5 **35.** $-80 \div (-4)$ 20 **36.** $-90 \div (-5)$ 18

Skill 2: Divide integers with unlike signs

When you divide integers with unlike signs, the quotient is negative.

Divide: $-6 \div 2$.

$-6 \div 2 = 3$

Divide: $4 \div (-1)$.

$4 \div (-1) = -4$

Divide.

37. $-6 \div 1$ −6 **38.** $-6 \div 3$ −2 **39.** $-8 \div 4$ −2 **40.** $-15 \div 5$ −3

41. $-60 \div 5$ −12 **42.** $-68 \div 4$ −17 **43.** $-78 \div 2$ −39 **44.** $-96 \div 32$ −3

45. $-9 \div 9$ −1 **46.** $-8 \div 2$ −4 **47.** $-20 \div 4$ −5 **48.** $-86 \div 2$ −43

49. $-88 \div 4$ −22 **50.** $-35 \div 1$ −35 **51.** $-72 \div 18$ −4 **52.** $-45 \div 15$ −3

53. $16 \div (-8)$ −2 **54.** $28 \div (-7)$ −4 **55.** $80 \div (-16)$ −5 **56.** $88 \div (-44)$ −2

57. $48 \div (-6)$ −8 **58.** $81 \div (-9)$ −9 **59.** $77 \div (-11)$ −7 **60.** $40 \div (-4)$ −10

61. $64 \div (-8)$ −8 **62.** $121 \div (-11)$ −11 **63.** $60 \div (-12)$ −5 **64.** $84 \div (-14)$ −6

65. $9 \div (-3)$ −3 **66.** $10 \div (-1)$ −10 **67.** $14 \div (-7)$ −2 **68.** $20 \div (-2)$ −10

69. $50 \div (-25)$ −2 **70.** $55 \div (-11)$ −5 **71.** $90 \div (-15)$ −6 **72.** $93 \div (-3)$ −31

ORDER OF OPERATIONS

Skill 1: Simplify expressions with no parentheses

To simplify expressions with no parentheses:

- First multiply and divide in order from left to right.
- Then add and subtract in order from left to right.

Compute: $6 + 4 \times 6$.

THINK: Multiply first.
$4 \times 6 = 24$

$6 + 4 \times 6 = 6 + 24 = 30$

Compute: $4 \times 2 - 15 \div 5$.

THINK: Multiply and divide in order first.
$4 \times 2 = 8 \quad 15 \div 5 = 3$

$4 \times 2 - 15 \div 5 = 8 - 3 = 5$

Compute.

1. $4 + 2 + 3$ 9
2. $5 \times 2 + 3$ 13
3. $3 + 6 \times 4$ 27
4. $5 - 4 + 2$ 3
5. $25 - 12 \times 2$ 1
6. $20 \div 5 \times 2$ 8
7. $4 \times 3 - 6 \times 1$ 6
8. $25 \div 5 + 4 \times 1$ 9
9. $6 + 3 - 2 + 4$ 11
10. $8 \div 4 - 4 \div 2$ 0
11. $8 + 6 \div 2 + 1$ 12
12. $16 \div 8 \times 4 + 1$ 9
13. $6 \times 4 \div 12 - 8$ −6
14. $8 + 3 \div 3 - 2$ 7
15. $20 - 9 \times 2 + 4$ 6

Skill 2: Simplify expressions with parentheses

The following are the rules for the **Order of Operations:**

1. Do operations inside parentheses.
2. Multiply and divide in order from left to right.
3. Add and subtract in order from left to right.

Compute: $(5 + 4) \times 2$.

THINK: Compute inside parentheses first.
$5 + 4 = 9$

$(5 + 4) \times 2 = 9 \times 2 = 18$

Compute: $(7 + 1) \div (7 - 5)$.

THINK: Compute inside parentheses first.
$7 + 1 = 8 \quad 7 - 5 = 2$

$(7 + 1) \div (7 - 5) = 8 \div 2 = 4$

Compute.

16. $(3 + 2) \times 4$ 20
17. $(6 - 4) \times 7$ 14
18. $11 - (9 \div 3)$ 8
19. $12 \div (4 + 2)$ 2
20. $8 \times (9 - 5)$ 32
21. $25 \div (4 + 1)$ 5
22. $(6 + 2) \div (4 - 2)$ 4
23. $(7 - 1) \times (9 \div 3)$ 18
24. $(16 \div 4) \times (8 \times 1)$ 32
25. $(5 \times 3) - (10 \div 5)$ 13
26. $(7 + 6) + (20 \times 2)$ 53
27. $(13 + 5) \div (21 \div 7)$ 6
28. $(14 - 8) \div (12 \div 4)$ 2
29. $(27 \div 9) \times (15 - 6)$ 27
30. $(7 \times 7) + (24 \div 4)$ 55

FRACTIONS AND MIXED NUMBERS

Skill 1: Write fractions and mixed numbers for shaded regions

A **fraction** can be used to describe part of a whole. The **numerator** (top) of a fraction shows how many parts are shaded. The **denominator** (bottom) shows the total number of parts.

A **mixed number** consists of a whole number part and a fraction part.

For example, $1\frac{1}{3}$ is a mixed number.

Write a fraction for the shaded region.

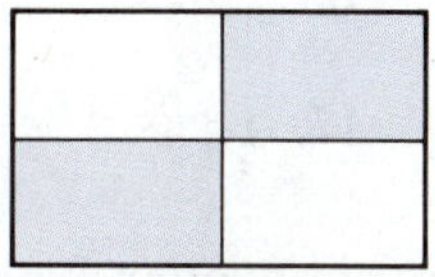

$\frac{\textbf{parts shaded}}{\textbf{total parts}} \begin{matrix} \rightarrow 2 \\ \rightarrow 4 \end{matrix}$

Write a mixed number for the shaded regions.

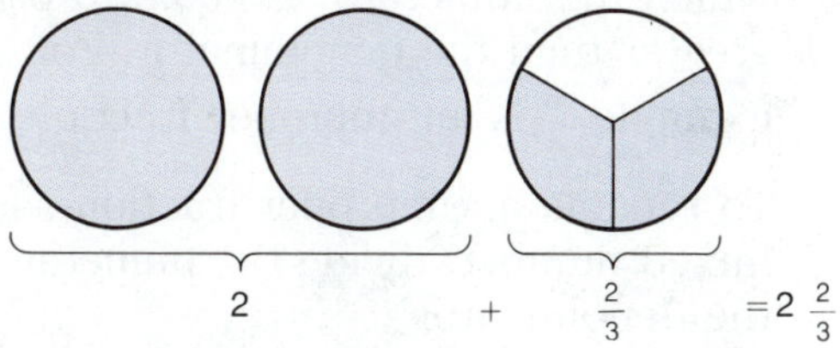

Write a fraction for the shaded region.

1. 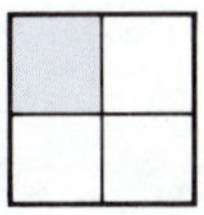 $\frac{1}{4}$

2. 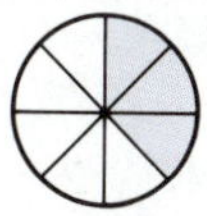 $\frac{3}{8}$

3. 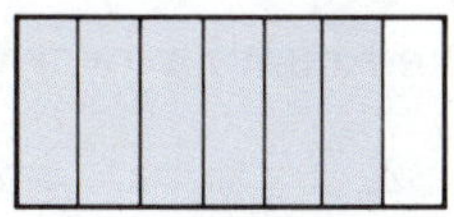 $\frac{6}{7}$

4. 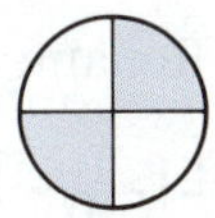 $\frac{2}{4}$

Write a mixed number for the shaded regions.

5. 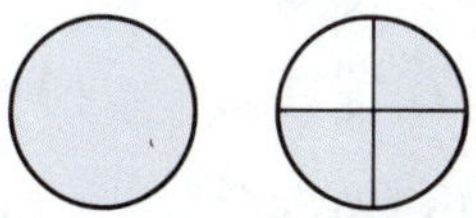 $1\frac{3}{4}$

6. 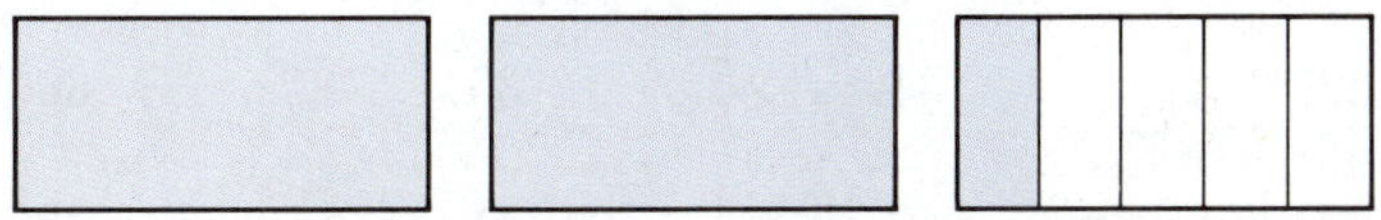 $2\frac{1}{5}$

7.  $2\frac{3}{8}$

8.  $2\frac{3}{4}$

9. 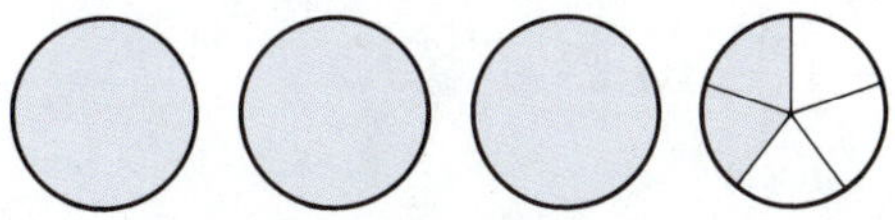 $3\frac{2}{5}$

10. 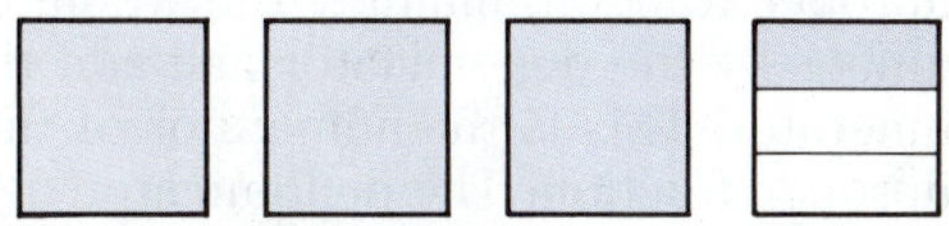 $3\frac{1}{3}$

Skill 2: Round fractions and mixed numbers using the number line

To round fractions and mixed numbers, you can use the number line. First graph the number on the number line. Then determine to which whole number it is closer.

Round $\frac{3}{4}$ and $1\frac{1}{2}$ using a number line.

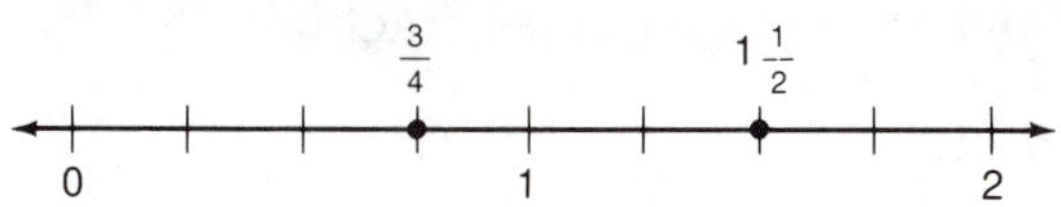

$\frac{3}{4}$ is closer to 1 than 0. Round $\frac{3}{4}$ to 1.

$1\frac{1}{2}$ is halfway between 1 and 2. Round $1\frac{1}{2}$ to 2.

FRACTIONS AND MIXED NUMBERS

Round using the number line.

11. $\frac{1}{2}$ 1　**12.** $\frac{1}{3}$ 0　**13.** $\frac{2}{3}$ 1　**14.** $\frac{1}{4}$ 0　**15.** $\frac{1}{8}$ 0

16. $1\frac{1}{3}$ 1　**17.** $1\frac{3}{4}$ 2　**18.** $2\frac{1}{2}$ 3　**19.** $2\frac{7}{8}$ 3　**20.** $3\frac{1}{4}$ 3

Skill 3: Rename an improper fraction as a mixed number

An **improper fraction** is a fraction in which the numerator is equal to or greater than the denominator. For example, $\frac{6}{4}$ is an improper fraction.

To rename an improper fraction as a mixed number, divide the numerator by the denominator.

Rename $\frac{6}{4}$ as a mixed number.

$$\frac{6}{4} \rightarrow \begin{array}{r} 1 \\ 4\overline{)6} \\ \underline{4} \\ 2 \end{array} \rightarrow 1\frac{2}{4} \begin{array}{l} \leftarrow \textbf{remainder} \\ \leftarrow \textbf{divisor} \end{array}$$

$\frac{6}{4}$ is renamed as $1\frac{2}{4}$.

Rename as a mixed number or a whole number.

21. $\frac{5}{4}$ $1\frac{1}{4}$　**22.** $\frac{7}{3}$ $2\frac{1}{3}$　**23.** $\frac{7}{5}$ $1\frac{2}{5}$　**24.** $\frac{6}{2}$ 3　**25.** $\frac{9}{3}$ 3　**26.** $\frac{60}{5}$ 12

27. $\frac{10}{4}$ $2\frac{2}{4}$　**28.** $\frac{34}{8}$ $4\frac{2}{8}$　**29.** $\frac{75}{9}$ $8\frac{3}{9}$　**30.** $\frac{25}{10}$ $2\frac{5}{10}$　**31.** $\frac{40}{12}$ $3\frac{4}{12}$　**32.** $\frac{60}{14}$ $4\frac{4}{14}$

33. $\frac{5}{2}$ $2\frac{1}{2}$　**34.** $\frac{12}{5}$ $2\frac{2}{5}$　**35.** $\frac{24}{7}$ $3\frac{3}{7}$　**36.** $\frac{7}{6}$ $1\frac{1}{6}$　**37.** $\frac{10}{8}$ $1\frac{2}{8}$　**38.** $\frac{9}{2}$ $4\frac{1}{2}$

39. $\frac{9}{6}$ $1\frac{3}{6}$　**40.** $\frac{10}{3}$ $3\frac{1}{3}$　**41.** $\frac{32}{10}$ $3\frac{2}{10}$　**42.** $\frac{22}{5}$ $4\frac{2}{5}$　**43.** $\frac{26}{6}$ $4\frac{2}{6}$　**44.** $\frac{42}{12}$ $3\frac{6}{12}$

Skill 4: Rename a mixed number as an improper fraction

To rename a mixed number as an improper fraction, multiply the whole number by the denominator and add the numerator. This is the numerator of the improper fraction. The denominator of the improper fraction is the same as the original denominator.

Rename $1\frac{3}{4}$ as an improper fraction.

$$1\frac{3}{4} \rightarrow 1\frac{3}{4}\; 4 + 3 \rightarrow \frac{7}{4}$$
$$\mathbf{1 \times 4}$$

$1\frac{3}{4}$ is renamed as $\frac{7}{4}$.

Rename as an improper fraction.

45. $1\frac{1}{3}$ $\frac{4}{3}$　**46.** $2\frac{2}{5}$ $\frac{12}{5}$　**47.** $3\frac{4}{5}$ $\frac{19}{5}$　**48.** $4\frac{3}{5}$ $\frac{23}{5}$　**49.** $9\frac{1}{3}$ $\frac{28}{3}$　**50.** $11\frac{3}{5}$ $\frac{58}{5}$

51. $3\frac{2}{10}$ $\frac{32}{10}$　**52.** $4\frac{4}{12}$ $\frac{52}{12}$　**53.** $6\frac{5}{15}$ $\frac{95}{15}$　**54.** $7\frac{1}{4}$ $\frac{29}{4}$　**55.** $9\frac{1}{2}$ $\frac{19}{2}$　**56.** $14\frac{1}{2}$ $\frac{29}{2}$

57. $7\frac{2}{3}$ $\frac{23}{3}$　**58.** $3\frac{6}{11}$ $\frac{39}{11}$　**59.** $5\frac{4}{5}$ $\frac{29}{5}$　**60.** $9\frac{7}{10}$ $\frac{97}{10}$　**61.** $13\frac{1}{6}$ $\frac{79}{6}$　**62.** $8\frac{7}{12}$ $\frac{103}{12}$

63. $16\frac{2}{5}$ $\frac{82}{5}$　**64.** $11\frac{5}{6}$ $\frac{71}{6}$　**65.** $10\frac{3}{4}$ $\frac{43}{4}$　**66.** $8\frac{5}{6}$ $\frac{53}{6}$　**67.** $12\frac{3}{10}$ $\frac{123}{10}$　**68.** $10\frac{1}{5}$ $\frac{51}{5}$

EQUIVALENT FRACTIONS

Skill 1: Rename fractions to higher terms

To rename fractions to higher terms, multiply the numerator and denominator of the fraction by the same nonzero number.

Rename: $\frac{2}{3} = \frac{\blacksquare}{15}$.

THINK: $3 \times \blacksquare = 15$ $3 \times 5 = 15$

$\frac{2}{3} = \frac{2 \times 5}{3 \times 5} = \frac{10}{15}$ So $\frac{2}{3} = \frac{10}{15}$.

Rename.

1. $\frac{1}{2} = \frac{\blacksquare}{4}$ 2
2. $\frac{3}{5} = \frac{\blacksquare}{10}$ 6
3. $\frac{5}{6} = \frac{\blacksquare}{18}$ 15
4. $\frac{2}{9} = \frac{\blacksquare}{36}$ 8

5. $1\frac{1}{2} = 1\frac{\blacksquare}{10}$ 5
6. $2\frac{1}{3} = 2\frac{\blacksquare}{9}$ 3
7. $3\frac{4}{5} = 3\frac{\blacksquare}{15}$ 12
8. $5\frac{1}{6} = 5\frac{\blacksquare}{12}$ 2

Skill 2: Find the LCD when there are common factors

To rename fractions using the **least common denominator (LCD),** first see if the denominators have a common factor. If they do, list the multiples of both denominators until you find a common multiple.

Rename $\frac{1}{4}$ and $\frac{5}{6}$ using the LCD.

THINK: Do 4 and 6 have a common factor? Yes, 2.

List the multiples.
4, 8, 12, 16, 20, ⋯ 6, 12, ⋯
12 is the least common multiple of 4 and 6. 12 is the LCD.

$\frac{1}{4} = \frac{1 \times 3}{4 \times 3} = \frac{3}{12}$ $\frac{5}{6} = \frac{5 \times 2}{6 \times 2} = \frac{10}{12}$

Rename both fractions using the LCD.

9. $\frac{1}{2}, \frac{3}{4}$ $\frac{2}{4}, \frac{3}{4}$
10. $\frac{2}{3}, \frac{4}{9}$ $\frac{6}{9}, \frac{4}{9}$
11. $\frac{1}{4}, \frac{7}{8}$ $\frac{2}{8}, \frac{7}{8}$
12. $\frac{2}{6}, \frac{5}{12}$ $\frac{4}{12}, \frac{5}{12}$

13. $\frac{1}{6}, \frac{3}{4}$ $\frac{2}{12}, \frac{9}{12}$
14. $\frac{1}{6}, \frac{7}{8}$ $\frac{4}{24}, \frac{21}{24}$
15. $\frac{1}{9}, \frac{5}{12}$ $\frac{4}{36}, \frac{15}{36}$
16. $\frac{1}{10}, \frac{2}{15}$ $\frac{3}{30}, \frac{4}{30}$

Skill 3: Find the LCD when there are no common factors

To rename fractions using the LCD, first find if the denominators have a common factor. If they do not, multiply the 2 denominators to find the LCD.

Rename $\frac{1}{5}$ and $\frac{5}{6}$ using the LCD.

THINK: Do 5 and 6 have a common factor? No.

$5 \times 6 = 30$ Use 30 as the LCD.

$\frac{1}{5} = \frac{1 \times 6}{5 \times 6} = \frac{6}{30}$ $\frac{5}{6} = \frac{5 \times 5}{6 \times 5} = \frac{25}{30}$

Rename the fractions using the LCD.

17. $\frac{1}{2}, \frac{2}{3}$ $\frac{3}{6}, \frac{4}{6}$
18. $\frac{1}{4}, \frac{2}{5}$ $\frac{5}{20}, \frac{8}{20}$
19. $\frac{1}{2}, \frac{3}{7}$ $\frac{7}{14}, \frac{6}{14}$
20. $\frac{5}{7}, \frac{3}{4}$ $\frac{20}{28}, \frac{21}{28}$

21. $\frac{1}{2}, \frac{2}{9}$ $\frac{9}{18}, \frac{4}{18}$
22. $\frac{2}{5}, \frac{7}{9}$ $\frac{18}{45}, \frac{35}{45}$
23. $\frac{2}{7}, \frac{3}{8}$ $\frac{16}{56}, \frac{21}{56}$
24. $\frac{5}{9}, \frac{7}{10}$ $\frac{50}{90}, \frac{63}{90}$

LOWEST TERMS

Skill 1: Find the GCF

The **greatest common factor (GCF)** of 2 or more numbers is the greatest number that is a factor of each number. To find the GCF, first list the factors of each number. Then list all the common factors.

Find the GCF of 12 and 18.

Factors of 12: 1, 2, 3, 4, 6, 12
Factors of 18: 1, 2, 3, 6, 9, 18
Common factors: 1, 2, 3, 6

The GCF of 12 and 18 is 6.

Find the GCF.

1. 2 and 6 2
2. 4 and 8 4
3. 10 and 12 2
4. 16 and 24 8
5. 2, 4, and 8 2
6. 3, 12, and 24 3
7. 4, 8, and 12 4
8. 12, 20, and 40 4

Skill 2: Rename a fraction in lowest terms

A fraction is in **lowest terms** when the only common factor of the numerator and denominator is 1. To write a fraction in lowest terms, divide the numerator and denominator by the GCF.

Write $\frac{4}{12}$ in lowest terms.

Factors of 4: 1, 2, 4
Factors of 12: 1, 2, 3, 4, 6, 12
The GCF of 4 and 12 is 4.

$\frac{4}{12} = \frac{4 \div 4}{12 \div 4} = \frac{1}{3}$

$\frac{4}{12}$, in lowest terms, is $\frac{1}{3}$.

Write the fraction in lowest terms.

9. $\frac{2}{4}$ $\frac{1}{2}$
10. $\frac{3}{9}$ $\frac{1}{3}$
11. $\frac{4}{24}$ $\frac{1}{6}$
12. $\frac{7}{21}$ $\frac{1}{3}$
13. $\frac{9}{36}$ $\frac{1}{4}$
14. $\frac{8}{12}$ $\frac{2}{3}$
15. $\frac{9}{21}$ $\frac{3}{7}$
16. $\frac{12}{20}$ $\frac{3}{5}$
17. $\frac{25}{35}$ $\frac{5}{7}$
18. $\frac{28}{36}$ $\frac{7}{9}$

Skill 3: Simplify mixed numbers

To **simplify** mixed numbers, write the fraction part in lowest terms.

Simplify: $1\frac{15}{20}$.

The GCF of 15 and 20 is 5.

$1\frac{15}{20} = 1\frac{15 \div 5}{20 \div 5} = 1\frac{3}{4}$

Simplify: $3\frac{3}{2}$.

$3\frac{3}{2} = 3 + \frac{3}{2}$
$= 3 + 1\frac{1}{2} = 4\frac{1}{2}$

Simplify.

19. $1\frac{4}{16}$ $1\frac{1}{4}$
20. $2\frac{7}{21}$ $2\frac{1}{3}$
21. $2\frac{8}{20}$ $2\frac{2}{5}$
22. $3\frac{10}{25}$ $3\frac{2}{5}$
23. $14\frac{6}{27}$ $14\frac{2}{9}$
24. $1\frac{4}{3}$ $2\frac{1}{3}$
25. $2\frac{5}{4}$ $3\frac{1}{4}$
26. $3\frac{9}{7}$ $4\frac{2}{7}$
27. $12\frac{12}{9}$ $13\frac{1}{3}$
28. $16\frac{20}{15}$ $17\frac{1}{3}$

COMPARE AND ORDER FRACTIONS AND MIXED NUMBERS

Skill 1: Compare fractions and mixed numbers

When you compare fractions and mixed numbers, the fractions must have the same denominator. If they do not, rename the fractions using the LCD. Then compare the numerators.

Compare $1\frac{1}{2}$ and $1\frac{1}{3}$ using >, <, or =.

$1\frac{1}{2} \bullet 1\frac{1}{3}$ ← **unlike denominators**

Use 6 as the LCD.

$\frac{1}{2} = \frac{3}{6}$ $\qquad \frac{1}{3} = \frac{2}{6}$

$\frac{3}{6} > \frac{2}{6}$ ← **The numerator 3 is greater than 2.**

So $1\frac{1}{2} > 1\frac{1}{3}$.

Compare. Write >, <, or =.

1. $\frac{1}{4} \bullet \frac{1}{8}$ >
2. $\frac{3}{4} \bullet \frac{9}{16}$ >
3. $\frac{1}{8} \bullet \frac{1}{3}$ <
4. $\frac{5}{12} \bullet \frac{7}{8}$ <
5. $1\frac{1}{2} \bullet 1\frac{1}{4}$ >
6. $2\frac{1}{5} \bullet 2\frac{3}{10}$ <
7. $8\frac{1}{4} \bullet 8\frac{3}{12}$ =
8. $11\frac{2}{3} \bullet 11\frac{3}{4}$ <

Skill 2: Order fractions and mixed numbers

To order fractions and mixed numbers, the fractions must have the same denominator. If they do not, rename the fractions using the LCD. Then compare the numerators.

Order $1\frac{1}{4}$, $1\frac{1}{2}$, and $1\frac{3}{8}$ from least to greatest.

$1\frac{1}{4}, 1\frac{1}{2}, 1\frac{3}{8}$ ← **unlike denominators**

Use 8 as the LCD.

$1\frac{1}{4} = 1\frac{2}{8}$ $\qquad 1\frac{1}{2} = 1\frac{4}{8}$ $\qquad 1\frac{3}{8}$

$1\frac{2}{8}, 1\frac{3}{8}, 1\frac{4}{8}$ ← **Since the whole-number parts are the same, order the numerators.**

$1\frac{1}{4}$, $1\frac{3}{8}$, and $1\frac{1}{2}$ are the ordered numbers.

Order from least to greatest.

9. $\frac{1}{2}, \frac{3}{4}, \frac{7}{8}$ — $\frac{1}{2}, \frac{3}{4}, \frac{7}{8}$
10. $\frac{2}{3}, \frac{5}{6}, \frac{4}{9}$ — $\frac{4}{9}, \frac{2}{3}, \frac{5}{6}$
11. $\frac{1}{4}, \frac{1}{3}, \frac{1}{5}$ — $\frac{1}{5}, \frac{1}{4}, \frac{1}{3}$
12. $\frac{5}{6}, \frac{5}{12}, \frac{5}{18}$ — $\frac{5}{18}, \frac{5}{12}, \frac{5}{6}$
13. $1\frac{1}{3}, 1\frac{1}{6}, 1\frac{5}{12}$ — $1\frac{1}{6}, 1\frac{1}{3}, 1\frac{5}{12}$
14. $2\frac{3}{4}, 2\frac{1}{2}, 2\frac{3}{8}$ — $2\frac{3}{8}, 2\frac{1}{2}, 2\frac{3}{4}$
15. $9\frac{1}{6}, 9\frac{3}{8}, 9\frac{5}{12}$ — $9\frac{1}{6}, 9\frac{3}{8}, 9\frac{5}{12}$
16. $11\frac{4}{7}, 11\frac{5}{21}, 11\frac{7}{14}$ — $11\frac{5}{21}, 11\frac{7}{14}, 11\frac{4}{7}$

Order from greatest to least.

17. $\frac{1}{2}, \frac{3}{8}, \frac{1}{4}$ — $\frac{1}{2}, \frac{3}{8}, \frac{1}{4}$
18. $\frac{1}{5}, \frac{2}{3}, \frac{7}{15}$ — $\frac{2}{3}, \frac{7}{15}, \frac{1}{5}$
19. $\frac{1}{2}, \frac{1}{3}, \frac{1}{4}$ — $\frac{1}{2}, \frac{1}{3}, \frac{1}{4}$
20. $\frac{3}{5}, \frac{5}{6}, \frac{7}{20}$ — $\frac{5}{6}, \frac{3}{5}, \frac{7}{20}$
21. $1\frac{1}{4}, 1\frac{1}{3}, 1\frac{5}{12}$ — $1\frac{5}{12}, 1\frac{1}{3}, 1\frac{1}{4}$
22. $2\frac{1}{3}, 2\frac{5}{6}, 2\frac{7}{18}$ — $2\frac{5}{6}, 2\frac{7}{18}, 2\frac{1}{3}$
23. $8\frac{2}{5}, 8\frac{4}{7}, 8\frac{1}{2}$ — $8\frac{4}{7}, 8\frac{1}{2}, 8\frac{2}{5}$
24. $13\frac{2}{3}, 13\frac{1}{4}, 13\frac{5}{6}$ — $13\frac{5}{6}, 13\frac{2}{3}, 13\frac{1}{4}$

MULTIPLY WITH FRACTIONS AND MIXED NUMBERS

Skill 1: Multiply 2 fractions

When you multiply fractions, multiply the numerators and then multiply the denominators. Write the answer in lowest terms.

Multiply: $\frac{1}{3} \times \frac{3}{5}$.

$$\frac{1}{3} \times \frac{3}{5} = \frac{1 \times 3}{3 \times 5} = \frac{3}{15}$$

$$\frac{3}{15} = \frac{3 \div 3}{15 \div 3} = \frac{1}{5} \leftarrow \textbf{lowest terms}$$

So $\frac{1}{3} \times \frac{3}{5} = \frac{1}{5}$.

Multiply. Write the answer in lowest terms.

1. $\frac{1}{2} \times \frac{1}{4}$ $\frac{1}{8}$	**2.** $\frac{2}{5} \times \frac{2}{3}$ $\frac{4}{15}$	**3.** $\frac{4}{9} \times \frac{4}{5}$ $\frac{16}{45}$	**4.** $\frac{5}{12} \times \frac{1}{3}$ $\frac{5}{36}$
5. $\frac{3}{4} \times \frac{1}{3}$ $\frac{1}{4}$	**6.** $\frac{1}{5} \times \frac{5}{6}$ $\frac{1}{6}$	**7.** $\frac{5}{6} \times \frac{2}{7}$ $\frac{5}{21}$	**8.** $\frac{7}{8} \times \frac{8}{9}$ $\frac{7}{9}$
9. $\frac{2}{3} \times \frac{1}{2}$ $\frac{1}{3}$	**10.** $\frac{3}{5} \times \frac{2}{3}$ $\frac{2}{5}$	**11.** $\frac{4}{7} \times \frac{1}{2}$ $\frac{2}{7}$	**12.** $\frac{1}{3} \times \frac{6}{7}$ $\frac{2}{7}$
13. $\frac{7}{10} \times \frac{4}{5}$ $\frac{14}{25}$	**14.** $\frac{2}{3} \times \frac{7}{8}$ $\frac{7}{12}$	**15.** $\frac{5}{6} \times \frac{1}{2}$ $\frac{5}{12}$	**16.** $\frac{3}{4} \times \frac{2}{3}$ $\frac{1}{2}$
17. $\frac{1}{9} \times \frac{3}{4}$ $\frac{1}{12}$	**18.** $\frac{4}{5} \times \frac{3}{4}$ $\frac{3}{5}$	**19.** $\frac{8}{9} \times \frac{1}{2}$ $\frac{4}{9}$	**20.** $\frac{7}{8} \times \frac{4}{5}$ $\frac{7}{10}$

Skill 2: Multiply 2 fractions using a shortcut

You can use a multiplication shortcut when there is a common factor in any numerator and denominator. First find the common factor(s). Simplify by dividing by the common factor(s). Then multiply.

For example, the common factor of 3 and 27 is 3 and the common factor of 7 and 14 is 7.

Multiply: $\frac{3}{7} \times \frac{14}{27}$.

$$\frac{\overset{1}{\cancel{3}}}{\underset{1}{\cancel{7}}} \times \frac{\overset{2}{\cancel{14}}}{\underset{9}{\cancel{27}}} = \frac{1 \times 2}{1 \times 9} = \frac{2}{9}$$

So $\frac{3}{7} \times \frac{14}{27} = \frac{2}{9}$.

Multiply using the shortcut.

21. $\frac{1}{2} \times \frac{6}{11}$ $\frac{3}{11}$	**22.** $\frac{2}{3} \times \frac{1}{4}$ $\frac{1}{6}$	**23.** $\frac{2}{9} \times \frac{6}{7}$ $\frac{4}{21}$	**24.** $\frac{3}{4} \times \frac{14}{17}$ $\frac{21}{34}$
25. $\frac{1}{3} \times \frac{6}{8}$ $\frac{1}{4}$	**26.** $\frac{1}{2} \times \frac{10}{15}$ $\frac{1}{3}$	**27.** $\frac{7}{9} \times \frac{3}{5}$ $\frac{7}{15}$	**28.** $\frac{4}{5} \times \frac{5}{7}$ $\frac{4}{7}$
29. $\frac{3}{4} \times \frac{8}{9}$ $\frac{2}{3}$	**30.** $\frac{5}{6} \times \frac{3}{5}$ $\frac{1}{2}$	**31.** $\frac{7}{8} \times \frac{12}{21}$ $\frac{1}{2}$	**32.** $\frac{4}{9} \times \frac{15}{28}$ $\frac{5}{21}$
33. $\frac{2}{8} \times \frac{4}{6}$ $\frac{1}{6}$	**34.** $\frac{4}{5} \times \frac{5}{6}$ $\frac{2}{3}$	**35.** $\frac{3}{4} \times \frac{12}{15}$ $\frac{3}{5}$	**36.** $\frac{4}{9} \times \frac{3}{8}$ $\frac{1}{6}$
37. $\frac{6}{9} \times \frac{3}{4}$ $\frac{1}{2}$	**38.** $\frac{2}{7} \times \frac{14}{20}$ $\frac{1}{5}$	**39.** $\frac{7}{8} \times \frac{2}{14}$ $\frac{1}{8}$	**40.** $\frac{3}{4} \times \frac{8}{15}$ $\frac{2}{5}$

MULTIPLY WITH FRACTIONS AND MIXED NUMBERS

Skill 3: Multiply a fraction and a whole number

To multiply a fraction and a whole number, rename the whole number as a fraction. For example, rename 5 as $\frac{5}{1}$. Then multiply and simplify the answer when possible.

Multiply: $\frac{1}{2} \times 5$.

$\frac{1}{2} \times 5 = \frac{1}{2} \times \frac{5}{1} = \frac{5}{2} = 2\frac{1}{2}$

Multiply: $\frac{5}{6} \times 8$.

$\frac{5}{6} \times 8 = \frac{5}{6} \times \frac{8}{1} = \frac{5}{\cancel{6}_3} \times \frac{\cancel{8}^4}{1} = \frac{20}{3} = 6\frac{2}{3}$

Multiply. Write the answer in simplest form.

41. $\frac{1}{3} \times 7$ $2\frac{1}{3}$ **42.** $\frac{1}{4} \times 13$ $3\frac{1}{4}$ **43.** $\frac{1}{2} \times 4$ 2 **44.** $\frac{1}{5} \times 20$ 4

45. $\frac{2}{5} \times 25$ 10 **46.** $\frac{3}{7} \times 21$ 9 **47.** $\frac{3}{8} \times 18$ $6\frac{3}{4}$ **48.** $\frac{9}{10} \times 35$ $31\frac{1}{2}$

Skill 4: Multiply a fraction and a mixed number

When you multiply a fraction and a mixed number, rename the mixed number as a fraction. For example, rename $1\frac{1}{3}$ as $\frac{4}{3}$. Then multiply and simplify the answer when possible.

Multiply: $\frac{7}{8} \times 1\frac{1}{3}$.

$\frac{7}{8} \times 1\frac{1}{3} = \frac{7}{8} \times \frac{4}{3} = \frac{7}{\cancel{8}_2} \times \frac{\cancel{4}^1}{3} = \frac{7}{6} = 1\frac{1}{6}$

Multiply. Write the answer in simplest form.

49. $\frac{1}{3} \times 1\frac{1}{4}$ $\frac{5}{12}$ **50.** $2\frac{1}{3} \times \frac{1}{4}$ $\frac{7}{12}$ **51.** $\frac{1}{3} \times 1\frac{1}{2}$ $\frac{1}{2}$ **52.** $1\frac{1}{5} \times \frac{1}{2}$ $\frac{3}{5}$

53. $\frac{4}{13} \times 1\frac{3}{10}$ $\frac{2}{5}$ **54.** $\frac{2}{7} \times 1\frac{3}{4}$ $\frac{1}{2}$ **55.** $6\frac{3}{4} \times \frac{8}{15}$ $3\frac{3}{5}$ **56.** $\frac{2}{3} \times 3\frac{3}{8}$ $2\frac{1}{4}$

57. $\frac{3}{4} \times 1\frac{1}{3}$ 1 **58.** $2\frac{1}{10} \times \frac{5}{7}$ $1\frac{1}{2}$ **59.** $2\frac{1}{4} \times \frac{2}{3}$ $1\frac{1}{2}$ **60.** $\frac{5}{6} \times 1\frac{4}{5}$ $1\frac{1}{2}$

Skill 5: Multiply 2 mixed numbers

When you multiply 2 mixed numbers, rename the mixed numbers as fractions. Then multiply and simplify the answer when possible.

Multiply: $1\frac{1}{3} \times 2\frac{5}{6}$.

$1\frac{1}{3} \times 2\frac{5}{6} = \frac{4}{3} \times \frac{17}{6} = \frac{\cancel{4}^2}{3} \times \frac{17}{\cancel{6}_3} = \frac{34}{9} = 3\frac{7}{9}$

Multiply. Write the answer in simplest form.

61. $2\frac{1}{4} \times 1\frac{1}{4}$ $2\frac{13}{16}$ **62.** $3\frac{1}{4} \times 2\frac{1}{2}$ $8\frac{1}{8}$ **63.** $3\frac{3}{5} \times 4\frac{2}{3}$ $16\frac{4}{5}$ **64.** $4\frac{2}{5} \times 5\frac{1}{2}$ $24\frac{1}{5}$

65. $1\frac{1}{2} \times 1\frac{1}{3}$ 2 **66.** $6\frac{2}{3} \times 4\frac{1}{2}$ 30 **67.** $4\frac{2}{7} \times 9\frac{1}{3}$ 40 **68.** $5\frac{2}{5} \times 6\frac{2}{3}$ 36

DIVIDE WITH FRACTIONS AND MIXED NUMBERS

Skill 1: Find the reciprocal of a fraction

Two numbers whose product is 1 are called **reciprocals.** For example, 2 and $\frac{1}{2}$ are reciprocals. To find the reciprocal of a whole or mixed number, first rename it as a fraction.

Find the reciprocal of $1\frac{1}{3}$.

THINK: $1\frac{1}{3} = \frac{4}{3}$

The reciprocal of $\frac{4}{3}$ is $\frac{3}{4}$ because $\frac{4}{3} \times \frac{3}{4} = 1$.

Find the reciprocal.

1. $\frac{3}{4}$ $\frac{4}{3}$ **2.** $\frac{7}{8}$ $\frac{8}{7}$ **3.** $\frac{1}{3}$ $\frac{3}{1}$ **4.** 4 $\frac{1}{4}$ **5.** $1\frac{1}{2}$ $\frac{2}{3}$ **6.** $2\frac{1}{4}$ $\frac{4}{9}$

Skill 2: Divide 2 fractions

When you divide by a fraction, write the reciprocal of the divisor. For example, $\frac{3}{2}$ is the reciprocal of the divisor, $\frac{2}{3}$. Then multiply and write the answer in lowest terms.

Divide: $\frac{4}{7} \div \frac{2}{3}$.

$\frac{4}{7} \div \frac{2}{3} = \frac{4}{7} \times \frac{3}{2} = \frac{\overset{2}{\cancel{4}}}{7} \times \frac{3}{\underset{1}{\cancel{2}}} = \frac{6}{7}$

Divide. Write the answer in lowest terms.

7. $\frac{1}{2} \div \frac{4}{7}$ $\frac{7}{8}$ **8.** $\frac{1}{4} \div \frac{2}{3}$ $\frac{3}{8}$ **9.** $\frac{3}{4} \div \frac{1}{3}$ $2\frac{1}{4}$ **10.** $\frac{7}{8} \div \frac{4}{5}$ $1\frac{3}{32}$

11. $\frac{1}{12} \div \frac{3}{4}$ $\frac{1}{9}$ **12.** $\frac{3}{10} \div \frac{1}{5}$ $1\frac{1}{2}$ **13.** $\frac{4}{9} \div \frac{2}{3}$ $\frac{2}{3}$ **14.** $\frac{9}{14} \div \frac{3}{7}$ $1\frac{1}{2}$

Skill 3: Divide a fraction and a whole number

When you divide a fraction and a whole number, rename the whole number as a fraction. For example, rename 2 as $\frac{2}{1}$. Then write the reciprocal of the divisor and multiply. Simplify the answer when possible.

Divide: $\frac{1}{4} \div 2$.

$\frac{1}{4} \div 2 = \frac{1}{4} \div \frac{2}{1} = \frac{1}{4} \times \frac{1}{2} = \frac{1}{8}$

Divide: $3 \div \frac{6}{7}$.

$3 \div \frac{6}{7} = \frac{3}{1} \div \frac{6}{7} = \frac{\overset{1}{\cancel{3}}}{1} \times \frac{7}{\underset{2}{\cancel{6}}} = \frac{7}{2} = 3\frac{1}{2}$

Divide. Write the answer in simplest form.

15. $\frac{1}{3} \div 4$ $\frac{1}{12}$ **16.** $\frac{5}{6} \div 10$ $\frac{1}{12}$ **17.** $\frac{8}{9} \div 12$ $\frac{2}{27}$ **18.** $\frac{10}{11} \div 20$ $\frac{1}{22}$

19. $8 \div \frac{1}{2}$ 16 **20.** $9 \div \frac{3}{5}$ 15 **21.** $10 \div \frac{2}{5}$ 25 **22.** $14 \div \frac{3}{4}$ $18\frac{2}{3}$

23. $6 \div \frac{6}{7}$ 7 **24.** $5 \div \frac{10}{11}$ $5\frac{1}{2}$ **25.** $\frac{3}{4} \div 9$ $\frac{1}{12}$ **26.** $\frac{5}{6} \div 15$ $\frac{1}{18}$

DIVIDE WITH FRACTIONS AND MIXED NUMBERS

Skill 4: Divide a fraction and a mixed number

When you divide a fraction and a mixed number, rename the mixed number as a fraction. Then write the reciprocal of the divisor and multiply. Simplify the answer when possible.

Divide: $3\frac{1}{2} \div \frac{4}{5}$.

$$3\frac{1}{2} \div \frac{4}{5} = \frac{7}{2} \div \frac{4}{5} = \frac{7}{2} \times \frac{5}{4} = \frac{35}{8} = 4\frac{3}{8}$$

Divide: $\frac{7}{8} \div 1\frac{1}{4}$.

$$\frac{7}{8} \div 1\frac{1}{4} = \frac{7}{8} \div \frac{5}{4} = \frac{7}{\cancel{8}_{2}} \times \frac{\cancel{4}^{1}}{5} = \frac{7}{10}$$

Divide. Write the answer in simplest form.

27. $\frac{5}{8} \div 1\frac{1}{3}$ $\frac{15}{32}$	**28.** $\frac{4}{9} \div 2\frac{1}{5}$ $\frac{20}{99}$	**29.** $\frac{3}{4} \div 1\frac{1}{2}$ $\frac{1}{2}$	**30.** $\frac{7}{8} \div 2\frac{1}{4}$ $\frac{7}{18}$
31. $1\frac{1}{3} \div \frac{1}{2}$ $2\frac{2}{3}$	**32.** $2\frac{5}{6} \div \frac{1}{5}$ $14\frac{1}{6}$	**33.** $3\frac{1}{6} \div \frac{7}{8}$ $3\frac{13}{21}$	**34.** $4\frac{7}{8} \div \frac{3}{4}$ $6\frac{1}{2}$
35. $1\frac{3}{7} \div \frac{5}{7}$ 2	**36.** $\frac{6}{9} \div 1\frac{1}{3}$ $\frac{1}{2}$	**37.** $\frac{3}{4} \div 3\frac{1}{2}$ $\frac{3}{14}$	**38.** $4\frac{1}{3} \div \frac{5}{9}$ $7\frac{4}{5}$
39. $3\frac{1}{3} \div \frac{2}{5}$ $8\frac{1}{3}$	**40.** $\frac{7}{10} \div 1\frac{1}{5}$ $\frac{7}{12}$	**41.** $2\frac{1}{3} \div \frac{7}{9}$ 3	**42.** $\frac{2}{3} \div 1\frac{1}{5}$ $\frac{5}{9}$
43. $\frac{2}{5} \div 3\frac{1}{3}$ $\frac{3}{25}$	**44.** $1\frac{1}{5} \div \frac{7}{10}$ $1\frac{5}{7}$	**45.** $\frac{5}{9} \div 4\frac{1}{3}$ $\frac{5}{39}$	**46.** $1\frac{1}{5} \div \frac{2}{3}$ $1\frac{4}{5}$

Skill 5: Divide 2 mixed numbers

To divide with mixed numbers, rename the mixed numbers as fractions. Then write the reciprocal of the divisor and multiply. Simplify the answer when possible.

Divide: $2\frac{1}{2} \div 1\frac{1}{4}$.

$$2\frac{1}{2} \div 1\frac{1}{4} = \frac{5}{2} \div \frac{5}{4} = \frac{\cancel{5}^{1}}{\cancel{2}_{1}} \times \frac{\cancel{4}^{2}}{\cancel{5}_{1}} = \frac{2}{1} = 2$$

Divide. Write the answer in simplest form.

47. $2\frac{1}{3} \div 1\frac{1}{4}$ $1\frac{13}{15}$	**48.** $6\frac{1}{2} \div 2\frac{1}{3}$ $2\frac{11}{14}$	**49.** $9\frac{1}{2} \div 1\frac{1}{4}$ $7\frac{3}{5}$	**50.** $8\frac{7}{9} \div 2\frac{2}{3}$ $3\frac{7}{24}$
51. $3\frac{2}{3} \div 2\frac{4}{9}$ $1\frac{1}{2}$	**52.** $5\frac{3}{5} \div 2\frac{1}{10}$ $2\frac{2}{3}$	**53.** $3\frac{1}{5} \div 1\frac{11}{25}$ $2\frac{2}{9}$	**54.** $6\frac{5}{6} \div 3\frac{5}{12}$ 2
55. $1\frac{1}{3} \div 1\frac{1}{9}$ $1\frac{1}{5}$	**56.** $1\frac{1}{7} \div 1\frac{1}{3}$ $\frac{6}{7}$	**57.** $2\frac{3}{4} \div 3\frac{1}{2}$ $\frac{11}{14}$	**58.** $3\frac{1}{2} \div 1\frac{1}{6}$ 3
59. $1\frac{5}{7} \div 1\frac{1}{3}$ $1\frac{2}{7}$	**60.** $4\frac{2}{3} \div 1\frac{1}{6}$ 4	**61.** $3\frac{1}{2} \div 1\frac{3}{4}$ 2	**62.** $2\frac{1}{4} \div 1\frac{1}{8}$ 2
63. $1\frac{1}{3} \div 1\frac{5}{7}$ $\frac{7}{9}$	**64.** $1\frac{1}{6} \div 4\frac{2}{3}$ $\frac{1}{4}$	**65.** $1\frac{3}{4} \div 3\frac{1}{2}$ $\frac{1}{2}$	**66.** $1\frac{1}{8} \div 2\frac{1}{4}$ $\frac{1}{2}$

ADD WITH FRACTIONS AND MIXED NUMBERS

Skill 1: Add like fractions with sums less than 1

Fractions with the same denominator are called **like fractions.** To add like fractions, add the numerators.

For example, 2 + 1 = 3. Use the same denominator, 5. Write the answer in lowest terms.

Add: $\frac{2}{5} + \frac{1}{5}$.

$$\frac{2}{5} + \frac{1}{5} = \frac{3}{5}$$

Add: 2 + 1

Add: $\frac{1}{8} + \frac{3}{8}$.

$$\frac{1}{8} + \frac{3}{8} = \frac{4}{8} = \frac{1}{2}$$

Add: 1 + 3

Add. Write the answer in lowest terms.

1. $\frac{1}{3} + \frac{1}{3}$ $\frac{2}{3}$	**2.** $\frac{1}{7} + \frac{2}{7}$ $\frac{3}{7}$	**3.** $\frac{1}{9} + \frac{4}{9}$ $\frac{5}{9}$	**4.** $\frac{4}{11} + \frac{3}{11}$ $\frac{7}{11}$
5. $\frac{1}{8} + \frac{3}{8}$ $\frac{1}{2}$	**6.** $\frac{3}{10} + \frac{1}{10}$ $\frac{2}{5}$	**7.** $\frac{9}{12} + \frac{1}{12}$ $\frac{5}{6}$	**8.** $\frac{1}{16} + \frac{1}{16}$ $\frac{1}{8}$
9. $\frac{2}{6} + \frac{2}{6}$ $\frac{2}{3}$	**10.** $\frac{2}{8} + \frac{4}{8}$ $\frac{3}{4}$	**11.** $\frac{1}{4} + \frac{1}{4}$ $\frac{1}{2}$	**12.** $\frac{3}{9} + \frac{3}{9}$ $\frac{2}{3}$
13. $\frac{6}{10} + \frac{2}{10}$ $\frac{4}{5}$	**14.** $\frac{3}{5} + \frac{1}{5}$ $\frac{4}{5}$	**15.** $\frac{2}{12} + \frac{4}{12}$ $\frac{1}{2}$	**16.** $\frac{3}{16} + \frac{5}{16}$ $\frac{1}{2}$
17. $\frac{6}{14} + \frac{4}{14}$ $\frac{5}{7}$	**18.** $\frac{2}{7} + \frac{3}{7}$ $\frac{5}{7}$	**19.** $\frac{1}{8} + \frac{1}{8}$ $\frac{1}{4}$	**20.** $\frac{5}{12} + \frac{3}{12}$ $\frac{2}{3}$

Skill 2: Add like fractions with sums equal to or greater than 1

Sometimes when you add like fractions, the sum will be equal to or greater than 1. Follow the rules for adding like fractions. Remember to simplify the answer.

Since $\frac{10}{8}$ is an improper fraction, rename it as a mixed number in simplest form.

Add: $\frac{5}{8} + \frac{5}{8}$.

$$\frac{5}{8} + \frac{5}{8} = \frac{10}{8} = 1\frac{2}{8} = 1\frac{1}{4}$$

Add: 5 + 5

Add. Write the answer in simplest form.

21. $\frac{1}{4} + \frac{3}{4}$ 1	**22.** $\frac{3}{8} + \frac{5}{8}$ 1	**23.** $\frac{8}{11} + \frac{7}{11}$ $1\frac{4}{11}$	**24.** $\frac{4}{15} + \frac{13}{15}$ $1\frac{2}{15}$
25. $\frac{5}{6} + \frac{5}{6}$ $1\frac{2}{3}$	**26.** $\frac{7}{8} + \frac{3}{8}$ $1\frac{1}{4}$	**27.** $\frac{5}{12} + \frac{11}{12}$ $1\frac{1}{3}$	**28.** $\frac{11}{16} + \frac{9}{16}$ $1\frac{1}{4}$
29. $\frac{2}{3} + \frac{1}{3}$ 1	**30.** $\frac{3}{5} + \frac{4}{5}$ $1\frac{2}{5}$	**31.** $\frac{6}{8} + \frac{6}{8}$ $1\frac{1}{2}$	**32.** $\frac{1}{7} + \frac{6}{7}$ 1
33. $\frac{7}{9} + \frac{5}{9}$ $1\frac{1}{3}$	**34.** $\frac{2}{4} + \frac{3}{4}$ $1\frac{1}{4}$	**35.** $\frac{4}{6} + \frac{4}{6}$ $1\frac{1}{3}$	**36.** $\frac{3}{12} + \frac{11}{12}$ $1\frac{1}{6}$
37. $\frac{11}{15} + \frac{4}{15}$ 1	**38.** $\frac{9}{11} + \frac{4}{11}$ $1\frac{2}{11}$	**39.** $\frac{6}{10} + \frac{8}{10}$ $1\frac{2}{5}$	**40.** $\frac{8}{9} + \frac{7}{9}$ $1\frac{2}{3}$

ADD WITH FRACTIONS AND MIXED NUMBERS

Skill 3: Add mixed numbers with like fractions

When you add mixed numbers with like fractions, add the fractions first. Then add the whole numbers. Simplify the answer when possible.

Add: $1\frac{1}{8} + 1\frac{3}{8}$.

$$\begin{array}{r} 1\frac{1}{8} \\ +1\frac{3}{8} \\ \hline 2\frac{4}{8} = 2\frac{1}{2} \end{array}$$

Add: $2\frac{3}{4} + 1\frac{3}{4}$.

$$\begin{array}{r} 2\frac{3}{4} \\ +1\frac{3}{4} \\ \hline 3\frac{6}{4} = 3 + 1\frac{2}{4} = 4\frac{2}{4} = 4\frac{1}{2} \end{array}$$

Add. Write the answer in simplest form.

41. $2\frac{2}{5} + 1\frac{1}{5}$ $3\frac{3}{5}$ **42.** $1\frac{2}{7} + 1\frac{3}{7}$ $2\frac{5}{7}$ **43.** $3\frac{1}{8} + 1\frac{1}{8}$ $4\frac{1}{4}$ **44.** $4\frac{1}{6} + 2\frac{1}{6}$ $6\frac{1}{3}$

45. $2\frac{1}{3} + 1\frac{2}{3}$ 4 **46.** $4\frac{7}{8} + 2\frac{3}{8}$ $7\frac{1}{4}$ **47.** $3\frac{7}{10} + 2\frac{9}{10}$ $6\frac{3}{5}$ **48.** $5\frac{7}{12} + 3\frac{11}{12}$ $9\frac{1}{2}$

Skill 4: Add unlike fractions

Fractions with different denominators are called **unlike fractions.** To add unlike fractions, first rename the fractions using the LCD. Then add the numerators and use the same denominator. Simplify the answer when possible.

Add: $\frac{1}{2} + \frac{1}{3}$.

$$\begin{array}{r} \frac{1}{2} = \frac{3}{6} \\ +\frac{1}{3} = \frac{2}{6} \\ \hline \frac{5}{6} \end{array}$$

Add: $\frac{2}{3} + \frac{5}{6}$.

$$\begin{array}{r} \frac{2}{3} = \frac{4}{6} \\ +\frac{5}{6} = \frac{5}{6} \\ \hline \frac{9}{6} = 1\frac{3}{6} = 1\frac{1}{2} \end{array}$$

Add. Write the answer in simplest form.

49. $\frac{1}{2} + \frac{1}{6}$ $\frac{2}{3}$ **50.** $\frac{3}{5} + \frac{1}{15}$ $\frac{2}{3}$ **51.** $\frac{1}{2} + \frac{3}{4}$ $1\frac{1}{4}$ **52.** $\frac{7}{8} + \frac{3}{16}$ $1\frac{1}{16}$

53. $\frac{1}{4} + \frac{2}{5}$ $\frac{13}{20}$ **54.** $\frac{3}{8} + \frac{1}{5}$ $\frac{23}{40}$ **55.** $\frac{1}{2} + \frac{3}{5}$ $1\frac{1}{10}$ **56.** $\frac{5}{12} + \frac{7}{20}$ $\frac{23}{30}$

Skill 5: Add mixed numbers with unlike fractions

When you add mixed numbers with unlike fractions, first rename the fractions using the LCD. Then add the numerators and use the same denominator. Simplify the answer when possible.

Add: $2\frac{5}{6} + 1\frac{1}{3}$.

$$\begin{array}{r} 2\frac{5}{6} = 2\frac{5}{6} \\ +1\frac{1}{3} = 1\frac{2}{6} \\ \hline 3\frac{7}{6} + 3 + 1\frac{1}{6} = 4\frac{1}{6} \end{array}$$

Add. Write the answer in simplest form.

57. $1\frac{3}{8} + 1\frac{1}{4}$ $2\frac{5}{8}$ **58.** $3\frac{3}{10} + 1\frac{1}{5}$ $4\frac{1}{2}$ **59.** $1\frac{1}{2} + 1\frac{1}{3}$ $2\frac{5}{6}$ **60.** $13\frac{3}{8} + 1\frac{1}{10}$ $14\frac{19}{40}$

61. $3\frac{3}{4} + 1\frac{1}{8}$ $4\frac{7}{8}$ **62.** $5\frac{5}{6} + 4\frac{7}{12}$ $10\frac{5}{12}$ **63.** $8\frac{7}{12} + 5\frac{3}{4}$ $14\frac{1}{3}$ **64.** $14\frac{9}{10} + 3\frac{1}{2}$ $18\frac{2}{5}$

SUBTRACT WITH FRACTIONS AND MIXED NUMBERS

Skill 1: Subtract like fractions

To subtract **like fractions,** subtract the numerators. For example, $7 - 3 = 4$. Use the same denominator, 8. Write the answer in lowest terms.

Subtract: $\frac{7}{8} - \frac{3}{8}$.

$$\begin{array}{r} \frac{7}{8} \\ -\frac{3}{8} \\ \hline \frac{4}{8} = \frac{1}{2} \end{array}$$

Subtract: 7 − 3

Subtract. Write the answer in lowest terms.

1. $\frac{2}{3} - \frac{1}{3}$ $\frac{1}{3}$ **2.** $\frac{4}{5} - \frac{2}{5}$ $\frac{2}{5}$ **3.** $\frac{8}{9} - \frac{3}{9}$ $\frac{5}{9}$ **4.** $\frac{7}{11} - \frac{4}{11}$ $\frac{3}{11}$

5. $\frac{5}{6} - \frac{1}{6}$ $\frac{2}{3}$ **6.** $\frac{9}{10} - \frac{3}{10}$ $\frac{3}{5}$ **7.** $\frac{11}{12} - \frac{7}{12}$ $\frac{1}{3}$ **8.** $\frac{7}{20} - \frac{3}{20}$ $\frac{1}{5}$

9. $\frac{3}{4} - \frac{1}{4}$ $\frac{1}{2}$ **10.** $\frac{5}{7} - \frac{3}{7}$ $\frac{2}{7}$ **11.** $\frac{5}{10} - \frac{3}{10}$ $\frac{1}{5}$ **12.** $\frac{6}{8} - \frac{4}{8}$ $\frac{1}{4}$

Skill 2: Subtract unlike fractions

To subtract **unlike fractions,** first rename the fractions using the LCD. Then subtract the numerators and use the same denominator. Write the answer in lowest terms.

Subtract: $\frac{11}{12} - \frac{1}{4}$.

$$\begin{array}{r} \frac{11}{12} = \frac{11}{12} \\ -\frac{1}{4} = \frac{3}{12} \\ \hline \frac{8}{12} = \frac{2}{3} \end{array}$$

Subtract. Write the answer in lowest terms.

13. $\frac{4}{9} - \frac{1}{3}$ $\frac{1}{9}$ **14.** $\frac{7}{8} - \frac{1}{4}$ $\frac{5}{8}$ **15.** $\frac{1}{3} - \frac{1}{4}$ $\frac{1}{12}$ **16.** $\frac{5}{6} - \frac{1}{5}$ $\frac{19}{30}$

17. $\frac{11}{14} - \frac{2}{7}$ $\frac{1}{2}$ **18.** $\frac{11}{15} - \frac{2}{5}$ $\frac{1}{3}$ **19.** $\frac{3}{4} - \frac{1}{6}$ $\frac{7}{12}$ **20.** $\frac{5}{6} - \frac{3}{10}$ $\frac{8}{15}$

21. $\frac{3}{5} - \frac{2}{10}$ $\frac{2}{5}$ **22.** $\frac{3}{4} - \frac{1}{2}$ $\frac{1}{4}$ **23.** $\frac{2}{3} - \frac{2}{5}$ $\frac{4}{15}$ **24.** $\frac{4}{5} - \frac{1}{2}$ $\frac{3}{10}$

Skill 3: Subtract mixed numbers with no renaming

To subtract mixed numbers, first subtract the fractions. For example, when subtracting $\frac{3}{5}$ and $\frac{1}{3}$, find the LCD first. Then subtract the whole numbers. Simplify the answer when possible.

Subtract: $3\frac{3}{5} - 1\frac{1}{3}$.

$$\begin{array}{r} 3\frac{3}{5} = 3\frac{9}{15} \\ -1\frac{1}{3} = 1\frac{5}{15} \\ \hline 2\frac{4}{15} \end{array}$$

SUBTRACT WITH FRACTIONS AND MIXED NUMBERS

Subtract. Write the answer in simplest form.

25. $5\frac{2}{3} - 2\frac{1}{3}$ $3\frac{1}{3}$ **26.** $6\frac{4}{5} - 1\frac{1}{5}$ $5\frac{3}{5}$ **27.** $6\frac{5}{12} - 4\frac{1}{12}$ $2\frac{1}{3}$ **28.** $9\frac{11}{16} - 3\frac{9}{16}$ $6\frac{1}{8}$

29. $2\frac{8}{9} - 1\frac{1}{3}$ $1\frac{5}{9}$ **30.** $7\frac{7}{16} - 1\frac{1}{4}$ $6\frac{3}{16}$ **31.** $9\frac{4}{5} - 2\frac{1}{2}$ $7\frac{3}{10}$ **32.** $8\frac{5}{6} - 2\frac{1}{8}$ $6\frac{17}{24}$

33. $4\frac{5}{6} - 2\frac{1}{3}$ $2\frac{1}{2}$ **34.** $5\frac{11}{12} - 1\frac{2}{3}$ $4\frac{1}{4}$ **35.** $5\frac{3}{4} - 1\frac{1}{2}$ $4\frac{1}{4}$ **36.** $3\frac{9}{10} - 2\frac{3}{5}$ $1\frac{3}{10}$

37. $11\frac{2}{3} - 3\frac{1}{2}$ $8\frac{1}{6}$ **38.** $9\frac{3}{4} - 3\frac{1}{3}$ $6\frac{5}{12}$ **39.** $7\frac{7}{12} - 6\frac{1}{4}$ $1\frac{1}{3}$ **40.** $11\frac{3}{5} - 2\frac{1}{15}$ $9\frac{8}{15}$

Skill 4: Rename whole numbers and mixed numbers

Sometimes you need to rename a whole number or a mixed number before you can subtract. For example, when subtracting 5 and $3\frac{2}{3}$, you must rename 5 as $4\frac{3}{3}$.

Rename: $5 = 4\frac{\blacksquare}{3}$.

$5 = 4 + 1 = 4 + \frac{3}{3} = 4\frac{3}{3}$

Rename: $2\frac{1}{3} = 1\frac{\blacksquare}{3}$.

$2\frac{1}{3} = 1 + 1 + \frac{1}{3} = 1 + \frac{3}{3} + \frac{1}{3} = 1\frac{4}{3}$

Rename.

41. $3 = 2\frac{\blacksquare}{3}$ 3 **42.** $4 = 3\frac{\blacksquare}{5}$ 5 **43.** $6 = 5\frac{\blacksquare}{7}$ 7 **44.** $8 = 7\frac{\blacksquare}{11}$ 11 **45.** $12 = 11\frac{\blacksquare}{15}$ 15

46. $8\frac{1}{2} = 7\frac{\blacksquare}{2}$ 3 **47.** $9\frac{3}{4} = 8\frac{\blacksquare}{4}$ 7 **48.** $10\frac{7}{8} = 9\frac{\blacksquare}{8}$ 15 **49.** $11\frac{1}{12} = 10\frac{\blacksquare}{12}$ 13 **50.** $15\frac{2}{5} = 14\frac{\blacksquare}{5}$ 7

Skill 5: Subtract mixed numbers with renaming

When you subtract mixed numbers, first find the LCD, if necessary. Then subtract the fractions. Rename when necessary. Finally, subtract the whole numbers. Simplify the answer when possible.

Subtract: $4\frac{1}{3} - 2\frac{1}{2}$.

$$\begin{array}{r} 4\frac{1}{3} = 4\frac{2}{6} = 3\frac{8}{6} \\ -2\frac{1}{2} = 2\frac{3}{6} = 2\frac{3}{6} \\ \hline 1\frac{5}{6} \end{array}$$

Subtract. Write the answer in simplest form.

51. $4 - 2\frac{1}{2}$ $1\frac{1}{2}$ **52.** $5 - 3\frac{4}{5}$ $1\frac{1}{5}$ **53.** $7 - 5\frac{2}{9}$ $1\frac{7}{9}$ **54.** $18 - 13\frac{5}{11}$ $4\frac{6}{11}$

55. $3\frac{1}{3} - 1\frac{2}{3}$ $1\frac{2}{3}$ **56.** $5\frac{1}{6} - 3\frac{5}{6}$ $1\frac{1}{3}$ **57.** $8\frac{3}{8} - 6\frac{7}{8}$ $1\frac{1}{2}$ **58.** $11\frac{1}{16} - 8\frac{3}{16}$ $2\frac{7}{8}$

59. $2\frac{1}{2} - 1\frac{2}{3}$ $\frac{5}{6}$ **60.** $6\frac{1}{3} - 3\frac{4}{5}$ $2\frac{8}{15}$ **61.** $8\frac{1}{4} - 5\frac{1}{3}$ $2\frac{11}{12}$ **62.** $11\frac{1}{2} - 3\frac{9}{10}$ $7\frac{3}{5}$

63. $11\frac{1}{2} - 6\frac{4}{5}$ $4\frac{7}{10}$ **64.** $10\frac{1}{4} - 6\frac{2}{3}$ $3\frac{7}{12}$ **65.** $12\frac{1}{5} - 7\frac{5}{6}$ $4\frac{11}{30}$ **66.** $14\frac{1}{6} - 9\frac{2}{3}$ $4\frac{1}{2}$

RENAME FRACTIONS AS DECIMALS

Skill 1: Rename fractions as decimals

Some fractions can be renamed as decimals by first writing an equivalent fraction or mixed number with a denominator of 10, 100, 1,000, or so on.

Rename $\frac{3}{4}$ and $\frac{3}{2}$ as decimals.

$\frac{3}{4} = \frac{75}{100} = 0.75 \qquad \frac{3}{2} = 1\frac{1}{2} = 1\frac{5}{10} = 1.5$

Rename as a decimal.

1. $\frac{1}{10}$ 0.1	2. $\frac{7}{10}$ 0.7	3. $\frac{31}{100}$ 0.31	4. $\frac{47}{100}$ 0.47	5. $\frac{89}{100}$ 0.89	6. $\frac{123}{1,000}$ 0.123	7. $\frac{463}{1,000}$ 0.463	8. $\frac{620}{1,000}$ 0.620
9. $\frac{1}{2}$ 0.5	10. $\frac{1}{4}$ 0.25	11. $\frac{2}{5}$ 0.4	12. $\frac{4}{5}$ 0.8	13. $\frac{7}{20}$ 0.35	14. $\frac{8}{25}$ 0.32	15. $\frac{16}{250}$ 0.064	16. $\frac{37}{500}$ 0.074
17. $\frac{5}{2}$ 2.5	18. $\frac{5}{4}$ 1.25	19. $\frac{7}{5}$ 1.4	20. $\frac{11}{10}$ 1.1	21. $\frac{24}{20}$ 1.2	22. $\frac{55}{50}$ 1.1	23. $\frac{105}{100}$ 1.05	24. $\frac{125}{100}$ 1.25

Skill 2: Rename fractions as terminating decimals

To rename any fraction as a decimal, divide the numerator by the denominator. If you can divide until the remainder is 0, the decimal is called a **terminating decimal.** For example, 0.375 is a terminating decimal.

Rename $\frac{3}{8}$ as a decimal.

```
        0.375
3/8 → 8)3.000 ← Write additional
        2 4↓     zeros if necessary.
          60
          56↓
           40
           40
            0
```

Rename as a decimal.

25. $\frac{1}{8}$ 0.125	26. $\frac{5}{8}$ 0.625	27. $\frac{1}{16}$ 0.0625	28. $\frac{14}{16}$ 0.875	29. $\frac{3}{40}$ 0.075	30. $\frac{9}{40}$ 0.225	31. $\frac{1}{80}$ 0.0125	32. $\frac{7}{80}$ 0.0875
33. $\frac{11}{8}$ 1.375	34. $\frac{33}{8}$ 4.125	35. $\frac{22}{16}$ 1.375	36. $\frac{24}{16}$ 1.5	37. $\frac{45}{40}$ 1.125	38. $\frac{87}{40}$ 2.175	39. $\frac{90}{80}$ 1.125	40. $\frac{105}{80}$ 1.3125

Skill 3: Rename fractions as decimals rounded to given places

Sometimes it is more convenient to use a decimal rounded to a given place. In this case, divide the numerator by the denominator to 1 more place than the given place. Then round. For example, $0.833 \approx 0.83$ when rounded to the nearest hundredth.

Rename $\frac{5}{6}$ as a decimal to the nearest hundredth.

```
        0.833 ≈ 0.83
5/6 → 6)5.000 ← Write 3 additional
        4 8↓     zeros.
          20
          18↓
           20
           18
            2
```

RENAME FRACTIONS AS DECIMALS

Rename as a decimal to the nearest tenth.

41. $\frac{1}{9}$ 0.1 42. $\frac{2}{11}$ 0.2 43. $\frac{5}{11}$ 0.5 44. $\frac{7}{15}$ 0.5 45. $\frac{4}{21}$ 0.2 46. $\frac{2}{30}$ 0.1 47. $\frac{9}{40}$ 0.2 48. $\frac{25}{60}$ 0.4

49. $\frac{11}{9}$ 1.2 50. $\frac{14}{13}$ 1.1 51. $\frac{19}{15}$ 1.3 52. $\frac{21}{20}$ 1.1 53. $\frac{31}{25}$ 1.2 54. $\frac{35}{30}$ 1.2 55. $\frac{45}{40}$ 1.1 56. $\frac{70}{60}$ 1.2

Rename as a decimal to the nearest hundredth.

57. $\frac{4}{9}$ 0.44 58. $\frac{6}{11}$ 0.55 59. $\frac{8}{15}$ 0.53 60. $\frac{1}{19}$ 0.05 61. $\frac{2}{21}$ 0.10 62. $\frac{6}{27}$ 0.22 63. $\frac{7}{40}$ 0.18 64. $\frac{8}{75}$ 0.11

65. $\frac{8}{7}$ 1.14 66. $\frac{13}{11}$ 1.18 67. $\frac{15}{14}$ 1.07 68. $\frac{37}{30}$ 1.23 69. $\frac{60}{35}$ 1.71 70. $\frac{67}{40}$ 1.68 71. $\frac{65}{60}$ 1.08 72. $\frac{80}{75}$ 1.07

Skill 4: Rename fractions as repeating decimals

When you rename a fraction as a decimal by dividing, a digit or group of digits may repeat. For example, $\frac{1}{3}$ is a **repeating decimal** and can be written as 0.3333 . . . or $0.\overline{3}$. Both notations mean that the 3 repeats without end.

Rename $\frac{1}{3}$ and $\frac{13}{11}$ as a decimal.

$0.3333 \approx 0.\overline{3}$

$\frac{1}{3} \rightarrow 3\overline{)1.0000}$ ← **Write 4 additional zeros.**

$$\begin{array}{r} 9 \\ \hline 10 \\ 9 \\ \hline 10 \\ 9 \\ \hline 10 \\ 9 \\ \hline 1 \end{array}$$

$1.1818 \approx 1.\overline{18}$

$\frac{13}{11} \rightarrow 11\overline{)13.0000}$ ← **Write 4 additional zeros.**

$$\begin{array}{r} 11 \\ \hline 2\,0 \\ 1\,1 \\ \hline 90 \\ 88 \\ \hline 20 \\ 11 \\ \hline 90 \\ 88 \\ \hline 2 \end{array}$$

Rename as a repeating decimal.

73. $\frac{2}{3}$ $0.\overline{6}$ 74. $\frac{1}{6}$ $0.1\overline{6}$ 75. $\frac{5}{6}$ $0.8\overline{3}$ 76. $\frac{5}{9}$ $0.\overline{5}$ 77. $\frac{1}{11}$ $0.\overline{09}$ 78. $\frac{5}{11}$ $0.\overline{45}$ 79. $\frac{1}{12}$ $0.08\overline{3}$ 80. $\frac{5}{12}$ $0.41\overline{6}$

81. $\frac{4}{3}$ $1.\overline{3}$ 82. $\frac{7}{6}$ $1.1\overline{6}$ 83. $\frac{10}{9}$ $1.\overline{1}$ 84. $\frac{13}{9}$ $1.\overline{4}$ 85. $\frac{13}{11}$ $1.\overline{18}$ 86. $\frac{17}{12}$ $1.41\overline{6}$ 87. $\frac{20}{12}$ $1.\overline{6}$ 88. $\frac{40}{30}$ $1.\overline{3}$

427

RATIOS, RATES, AND EQUAL RATIOS

Skill 1: Write ratios

A **ratio** is used to compare 2 quantities. A ratio is usually written using a fraction or a colon. The ratio $\frac{3}{4}$ or 3:4 can be read as 3 to 4, 3 for 4, or 3 out of 4.

There are 15 girls and 14 boys. Write each ratio in 2 ways.

Girls to boys: $\frac{15}{14}$ or 15:14

Girls to students: $\frac{15}{29}$ or 15 to 29

Write the ratio in 2 ways.

1. 7 girls and 11 boys; ratio of girls to boys $\frac{7}{11}$ or 7:11
2. 25 records and 16 tapes; ratio of tapes to records $\frac{16}{25}$ or 16:25
3. 8 African elephants and 4 Asian elephants; ratio of African elephants to all elephants $\frac{8}{12}$ or 8:12
4. 16 club members and 23 nonmembers; ratio of club members to total people $\frac{16}{39}$ or 16:39

Skill 2: Simplify ratios

To **simplify a ratio,** write it as a fraction with similar units. Then simplify the ratio by writing the fraction in lowest terms.

Simplify the ratio: 400 cm to 8 m.

THINK: **You need the same unit, cm.**
8 m = 800 cm

$\frac{400}{800} = \frac{1}{2}$

Simplify the ratio.

5. 16 m to 2 m $\frac{8}{1}$
6. 18 oz to 6 oz $\frac{3}{1}$
7. 18 in. to 12 in. $\frac{3}{2}$
8. 23 min to 46 min $\frac{1}{2}$
9. 15 mL to 25 mL $\frac{3}{5}$
10. 10 qt to 8 qt $\frac{5}{4}$
11. 24 min to 1 h $\frac{2}{5}$
12. 12 oz to 1 lb $\frac{3}{4}$
13. 54 pennies to 5 dimes $\frac{27}{25}$
14. 18 in. to 1 ft $\frac{3}{2}$
15. 15 in. to 1 yd $\frac{5}{12}$
16. 1 dime to 1 dollar $\frac{1}{10}$

Skill 3: Compare ratios

To compare ratios, write both as fractions. Then use the **cross-product rule** to test if they are equal, that is, if they have the same value.

Write = or ≠.

3:4 ● 4:5 or $\frac{3}{4}$ ● $\frac{4}{5}$

$3 \times 5 \overset{?}{=} 4 \times 4$

$15 \neq 16$ So $\frac{3}{4} \neq \frac{4}{5}$.

Write = or ≠.

17. 1:2 ● 3:4 ≠
18. 2:3 ● 4:6 =
19. 6:8 ● 3:4 =
20. 5:10 ● 1:3 ≠
21. 7:10 ● 8:11 ≠
22. 1:3 ● 2:6 =

RATIOS, RATES, AND EQUAL RATIOS

Skill 4: Find equal ratios

Ratios that have the same value are called **equal ratios.** To find equal ratios, write the ratios as fractions. Then find the equal ratios by writing the fractions with a common denominator.

Which are equal ratios?
2:3, 1:5, 4 to 5, 10 to 15

$\frac{2}{3} = \frac{10}{15}$ $\frac{1}{5} = \frac{3}{15}$ $\frac{4}{5} = \frac{12}{15}$ $\frac{10}{15} = \frac{10}{15}$

So 2:3 = 10:15.

Which are equal ratios?

23. 1:5, 1:3, 3:15 1:5 = 3:15

24. 1:2, 2:4, 3:5 1:2 = 2:4

25. 2:3, 5 to 6, 4 to 6
2:3 = 4 to 6

26. 3 to 4, 7 to 8, 12 to 16
3 to 4 = 12 to 16

27. 1 to 2, 3 to 5, 6 to 10, 1 to 5
3 to 5 = 6 to 10

28. 3 to 12, 1 to 4, 2 to 6, 1 to 2
3 to 12 = 1 to 4

Skill 5: Write rates

A **rate** compares 2 quantities with unlike units. Rates, like ratios, can be written using a fraction or a colon.

Write the rate in 2 ways.

30 mi per h: $\frac{30}{1}$ or 30:1

Write the rate in 2 ways.

29. \$25 for 2 tapes $\frac{\$25}{2}$ or \$25:2

30. 35 mi per h $\frac{35}{1}$ or 35:1

31. 25 mi per gal $\frac{25}{1}$ or 25:1

32. 3 pens for 79¢
$\frac{3}{\$0.79}$ or 3:\$0.79

33. 2 jars for \$1.53
$\frac{2}{\$1.53}$ or 2:\$1.53

34. 30 revolutions per min
$\frac{30}{1}$ or 30:1

Skill 6: Find unit rates

The rate for 1 unit is called the **unit rate.** To find a unit rate, write the rate as a fraction. Then divide the numerator by the denominator.

Find the unit rate.
Five pieces of gum cost 55¢.
How much does 1 piece cost?

cost → 55¢
number → 5

$\frac{55¢}{5}$ 55¢ ÷ 5 = 11¢

The unit rate is 11¢ per piece.

Find the unit rate.

35. You traveled 165 mi in 3 h. How many miles did you average per hour?
55 mi per h

36. If 5 cucumbers cost \$1.20, how much does 1 cost?
\$0.24 per cucumber

37. Roy serviced 150 cars in 3 h. About how many did he service each hour?
50 cars per h

38. Gail used 28 gal of gas this week. About how many gallons of gas did she use each day?
4 gal per d

PROPORTIONS AND SOLVING PROPORTIONS

Skill 1: Write proportions

A **proportion** is a statement that 2 ratios are equal. For example, $\frac{1}{2}$ and $\frac{2}{4}$ are equal ratios, so $\frac{1}{2} = \frac{2}{4}$ is a proportion.
Proportions can be written in 2 ways, using fractions or colons.

Write the proportion in 2 ways:
1 is to 3 as 2 is to 6.

Using fractions: $\frac{1}{3} = \frac{2}{6}$

Using colons: $1:3 = 2:6$

Write the proportion in 2 ways.

1. 4 is to 8 as 8 is to 16 $\frac{4}{8} = \frac{8}{16}$ or $4:8 = 8:16$
2. 1 is to 4 as 3 is to 12 $\frac{1}{4} = \frac{3}{12}$ or $1:4 = 3:12$
3. 4 is to 10 as 40 is to 100 **3.** $\frac{4}{10} = \frac{40}{100}$ or $4:10 = 40:100$
4. 0.2 is to 0.4 as 0.3 is to 0.6 **4.** $\frac{0.2}{0.4} = \frac{0.3}{0.6}$ or $0.2:0.4 = 0.3:0.6$
5. 1 is to 16 as 2 is to 32 $\frac{1}{16} = \frac{2}{32}$ or $1:16 = 2:32$
6. 1 is to 7 as 2 is to 14 $\frac{1}{7} = \frac{2}{14}$ or $1:7 = 2:14$
7. 3 is to 5 as 9 is to 15 $\frac{3}{5} = \frac{9}{15}$ or $3:5 = 9:15$
8. 1 is to 2 as 25 is to 50 $\frac{1}{2} = \frac{25}{50}$ or $1:2 = 25:50$
9. 2 is to 3 as 6 is to 9 $\frac{2}{3} = \frac{6}{9}$ or $2:3 = 6:9$
10. 2 is to 5 as 4 is to 10 $\frac{2}{5} = \frac{4}{10}$ or $2:5 = 4:10$
11. 3 is to 10 as 30 is to 100 **11.** $\frac{3}{10} = \frac{30}{100}$ or $3:10 = 30:100$
12. 3 is to 4 as 12 is to 16 $\frac{3}{4} = \frac{12}{16}$ or $3:4 = 12:16$
13. 0.6 is to 1.2 as 1.8 is to 3.6 **13.** $\frac{0.6}{1.2} = \frac{1.8}{3.6}$ or $0.6:1.2 = 1.8:3.6$
14. 0.5 is to 1.5 as 2 is to 6 $\frac{0.5}{1.5} = \frac{2}{6}$ or $0.5:1.5 = 2:6$

Skill 2: Determine if 2 ratios form a proportion

Two ratios form a proportion if the **cross products** are equal. To determine whether you have a proportion, write the proportion using fractions. Then check if the cross products are equal.

Is this a proportion?
$3:6 \stackrel{?}{=} 4:8$

$\frac{3}{6} \times \frac{4}{8}$

$3 \times 8 \stackrel{?}{=} 4 \times 6$
$24 = 24$

Yes, it is a proportion.

Is it a proportion? Write *yes* or *no*.

15. $3:15 \stackrel{?}{=} 4:20$ yes
16. $6:8 \stackrel{?}{=} 7:10$ no
17. $\frac{2}{5} \stackrel{?}{=} \frac{3}{10}$ no
18. $\frac{3}{7} \stackrel{?}{=} \frac{6}{14}$ yes
19. $\frac{1}{0.2} \stackrel{?}{=} \frac{5}{1.0}$ yes
20. $\frac{3}{0.2} \stackrel{?}{=} \frac{6}{0.4}$ yes
21. $3:4 \stackrel{?}{=} 6:8$ yes
22. $9:27 \stackrel{?}{=} 14:52$ no
23. $2:7 \stackrel{?}{=} 6:21$ yes
24. $3:5 \stackrel{?}{=} 8:10$ no
25. $\frac{7}{10} \stackrel{?}{=} \frac{40}{45}$ no
26. $\frac{3}{4} \stackrel{?}{=} \frac{15}{20}$ yes

430

PROPORTIONS AND SOLVING PROPORTIONS

Skill 3: Solve proportions with whole numbers

To solve a proportion, use the fact that in a proportion the cross products are equal. Remember to write a proportion using fractions first. Use cross products to write an equation. Then solve and check.

Solve: $\frac{n}{7} = \frac{3}{21}$.

$$\frac{n}{7} \bowtie \frac{3}{21}$$

$$n \times 21 = 3 \times 7$$

$$n \times 21 = 21$$

$$n = 21 \div 21$$

$$n = 1$$

Solve the proportion.

27. $\frac{n}{3} = \frac{2}{6}$ 1

28. $\frac{1}{n} = \frac{3}{12}$ 4

29. $\frac{1}{2} = \frac{n}{20}$ 10

30. $\frac{1}{5} = \frac{4}{n}$ 20

31. $\frac{1}{20} = \frac{5}{n}$ 100

32. $\frac{n}{5} = \frac{10}{25}$ 2

33. $\frac{n}{7} = \frac{12}{42}$ 2

34. $\frac{n}{9} = \frac{49}{63}$ 7

35. $\frac{8}{9} = \frac{64}{n}$ 72

36. $\frac{2}{11} = \frac{8}{n}$ 44

37. $\frac{3}{n} = \frac{30}{100}$ 10

38. $\frac{7}{n} = \frac{63}{72}$ 8

39. $\frac{9}{n} = \frac{99}{121}$ 11

40. $\frac{3}{7} = \frac{n}{91}$ 39

41. $\frac{12}{17} = \frac{n}{34}$ 24

42. $\frac{6}{10} = \frac{n}{100}$ 60

43. $\frac{7}{12} = \frac{21}{n}$ 36

44. $\frac{9}{15} = \frac{n}{30}$ 18

45. $\frac{19}{20} = \frac{57}{n}$ 60

46. $\frac{5}{6} = \frac{n}{48}$ 40

Skill 4: Solve proportions with fractions or decimals

To solve a proportion with fractions or decimals, use the same cross-product method you used with whole numbers.

Solve: $\frac{4}{0.7} = \frac{n}{7}$.

$$\frac{4}{0.7} \bowtie \frac{n}{7}$$

$$4 \times 7 = n \times 0.7$$

$$28 = n \times 0.7$$

$$28 \div 0.7 = n$$

$$40 = n$$

Solve the proportion.

47. $\frac{2}{4} = \frac{n}{3}$ 1.5 or $1\frac{1}{2}$

48. $\frac{3}{4} = \frac{n}{22}$ 16.5 or $16\frac{1}{2}$

49. $\frac{1}{4} = \frac{n}{31}$ 7.75 or $7\frac{3}{4}$

50. $\frac{n}{2} = \frac{48}{10}$ 9.6 or $9\frac{3}{5}$

51. $\frac{n}{15} = \frac{30}{60}$ 7.5 or $7\frac{1}{2}$

52. $\frac{1}{4} = \frac{0.2}{n}$ 0.8

53. $\frac{0.4}{1} = \frac{n}{2}$ 0.8

54. $\frac{n}{4.2} = \frac{1}{6}$ 0.7

55. $\frac{1}{0.5} = \frac{5}{n}$ 2.5

56. $\frac{n}{2} = \frac{1.6}{4}$ 0.8

57. $\frac{1}{3.2} = \frac{5}{n}$ 16

58. $\frac{1}{0.7} = \frac{n}{2.8}$ 4

59. $\frac{n}{4.9} = \frac{1}{7}$ 0.7

60. $\frac{2.4}{n} = \frac{3}{5}$ 4

61. $\frac{n}{6.5} = \frac{2}{26}$ 0.5

62. $\frac{\frac{1}{2}}{2} = \frac{n}{4}$ 1

63. $\frac{\frac{2}{3}}{3} = \frac{2}{n}$ 9

64. $\frac{n}{49} = \frac{\frac{4}{7}}{7}$ 4

65. $\frac{n}{25} = \frac{\frac{2}{5}}{10}$ 1

66. $\frac{3}{n} = \frac{\frac{3}{4}}{3}$ 12

EQUIVALENT FRACTIONS, DECIMALS, AND PERCENTS

Skill 1: Rename fractions with denominators of 100 as percents

Percent means **per hundred** or **hundredths.** The symbol % is used for percent. Remember that 1 (whole) = $\frac{100}{100}$ = 100%.

Rename $\frac{46}{100}$ and $\frac{138}{100}$ as percents.

$\frac{46}{100}$ = 46 hundredths = 46%

$\frac{138}{100}$ = 138 hundredths = 138%

Rename as a percent.

1. $\frac{7}{100}$ 7% **2.** $\frac{11}{100}$ 11% **3.** $\frac{21}{100}$ 21% **4.** $\frac{30}{100}$ 30% **5.** $\frac{47}{100}$ 47% **6.** $\frac{99}{100}$ 99%

7. $\frac{109}{100}$ 109% **8.** $\frac{113}{100}$ 113% **9.** $\frac{137}{100}$ 137% **10.** $\frac{151}{100}$ 151% **11.** $\frac{169}{100}$ 169% **12.** $\frac{183}{100}$ 183%

Skill 2: Rename percents as fractions in lowest terms

When you rename a percent as a fraction, first write it as a fraction with a denominator of 100. Then write the fraction in lowest terms.

Rename 16% and $33\frac{1}{3}\%$ as fractions or mixed numbers in lowest terms.

16% = 16 hundredths = $\frac{16}{100} = \frac{4}{25}$

$33\frac{1}{3}\% = 33\frac{1}{3}$ hundredths = $\frac{33\frac{1}{3}}{100}$

$= 33\frac{1}{3} \div 100 = \frac{\overset{1}{\cancel{100}}}{3} \times \frac{1}{\underset{1}{\cancel{100}}} = \frac{1}{3}$

Rename as a fraction or a mixed number in lowest terms.

13. 13% $\frac{13}{100}$ **14.** 21% $\frac{21}{100}$ **15.** 53% $\frac{53}{100}$ **16.** 67% $\frac{67}{100}$ **17.** 71% $\frac{71}{100}$ **18.** 99% $\frac{99}{100}$

19. 15% $\frac{3}{20}$ **20.** 26% $\frac{13}{50}$ **21.** 40% $\frac{2}{5}$ **22.** 75% $\frac{3}{4}$ **23.** 82% $\frac{41}{50}$ **24.** 94% $\frac{47}{50}$

25. 101% $1\frac{1}{100}$ **26.** 117% $1\frac{17}{100}$ **27.** 123% $1\frac{23}{100}$ **28.** 145% $1\frac{9}{20}$ **29.** 150% $1\frac{1}{2}$ **30.** 175% $1\frac{3}{4}$

31. $\frac{1}{4}\%$ $\frac{1}{400}$ **32.** $\frac{1}{2}\%$ $\frac{1}{200}$ **33.** $\frac{9}{10}\%$ $\frac{9}{1,000}$ **34.** $12\frac{1}{2}\%$ $\frac{1}{8}$ **35.** $66\frac{2}{3}\%$ $\frac{2}{3}$ **36.** $83\frac{1}{3}\%$ $\frac{5}{6}$

Skill 3: Rename decimals as percents

When you rename a decimal as a percent, move the decimal point 2 places to the right and add a percent sign.

Sometimes it is necessary to write additional zeros before you move the decimal.

Rename 0.21 as a percent.

0.21 → 0.21. → 21%

Rename 0.8 as a percent.

0.8 → 0.80. → 80%

EQUIVALENT FRACTIONS, DECIMALS, AND PERCENTS

Rename as a percent.

37. 0.17 17%	**38.** 0.26 26%	**39.** 0.31 31%	**40.** 0.67 67%	**41.** 0.72 72%	**42.** 0.91 91%
43. 0.432 43.2%	**44.** 0.587 58.7%	**45.** 0.621 62.1%	**46.** 0.789 78.9%	**47.** 0.801 80.1%	**48.** 0.907 90.7%
49. 0.1 10%	**50.** 0.3 30%	**51.** 0.4 40%	**52.** 0.5 50%	**53.** 0.6 60%	**54.** 0.9 90%
55. 1 100%	**56.** 1.5 150%	**57.** 2 200%	**58.** 2.25 225%	**59.** 2.5 250%	**60.** 3 300%

Skill 4: Rename percents as decimals

To rename a percent as a decimal, move the decimal point 2 places to the left and drop the percent sign.

Rename 32% as a decimal.

32% → 32 → 0.32

Sometimes it is necessary to write additional zeros before you move the decimal point.

Rename 7% as a decimal.

7% → 07 → 0.07

Rename as a decimal.

61. 11% 0.11	**62.** 18% 0.18	**63.** 37% 0.37	**64.** 42% 0.42	**65.** 79% 0.79	**66.** 83% 0.83
67. 24.3% 0.243	**68.** 36.8% 0.368	**69.** 42.7% 0.427	**70.** 59.3% 0.593	**71.** 67.2% 0.672	**72.** 81.4% 0.814
73. 1% 0.01	**74.** 2% 0.02	**75.** 5% 0.05	**76.** 6% 0.06	**77.** 8% 0.08	**78.** 9% 0.09
79. 100% 1	**80.** 125% 1.25	**81.** 150% 1.5	**82.** 175% 1.75	**83.** 200% 2	**84.** 300% 3

Skill 5: Rename fractions as percents using division

You can rename a fraction as a percent. First rename the fraction as a decimal using division. Then rename the decimal as a percent.

Rename $\frac{1}{4}$ and $\frac{4}{3}$ as percents.

$\frac{1}{4} \rightarrow 4\overline{)1.00}$ (quotient 0.25) $\rightarrow 0.25 \rightarrow 25\%$

$\frac{4}{3} \rightarrow 3\overline{)4.00}$ (quotient $1.33\frac{1}{3}$) $\rightarrow 1.33\frac{1}{3} \rightarrow 133\frac{1}{3}\%$

Rename as a percent using division.

85. $\frac{1}{10}$ 10%	**86.** $\frac{1}{5}$ 20%	**87.** $\frac{1}{4}$ 25%	**88.** $\frac{3}{10}$ 30%	**89.** $\frac{1}{2}$ 50%	**90.** $\frac{3}{4}$ 75%
91. $\frac{11}{10}$ 110%	**92.** $\frac{6}{5}$ 120%	**93.** $\frac{5}{4}$ 125%	**94.** $\frac{14}{10}$ 140%	**95.** $\frac{3}{2}$ 150%	**96.** $\frac{7}{4}$ 175%
97. $\frac{1}{3}$ $33\frac{1}{3}\%$	**98.** $\frac{1}{6}$ $16\frac{2}{3}\%$	**99.** $\frac{1}{8}$ $12\frac{1}{2}\%$	**100.** $\frac{2}{9}$ $22\frac{2}{9}\%$	**101.** $\frac{4}{7}$ $57\frac{1}{7}\%$	**102.** $\frac{7}{8}$ $87\frac{1}{2}\%$
103. $\frac{5}{3}$ $166\frac{2}{3}\%$	**104.** $\frac{7}{6}$ $116\frac{2}{3}\%$	**105.** $\frac{9}{8}$ $112\frac{1}{2}\%$	**106.** $\frac{11}{9}$ $122\frac{2}{9}\%$	**107.** $\frac{11}{7}$ $157\frac{1}{7}\%$	**108.** $\frac{13}{8}$ $162\frac{1}{2}\%$

FIND THE PERCENT OF A NUMBER

Skill 1: Find the percent of a number using a fraction

To find the percent of a number, you solve an equation of the form $a\%$ of $b = c$. First substitute the values you know. Then solve for the unknown value. Rename the percent as a fraction and then multiply.

Find $5\frac{1}{2}\%$ of 50.

$$a\% \text{ of } b = c$$
$$\downarrow \quad \downarrow \quad \downarrow \qquad \downarrow$$
$$5\frac{1}{2}\% \times 50 = c$$

THINK: $5\frac{1}{2}\% = \dfrac{5\frac{1}{2}}{100}$

$$= 5\frac{1}{2} \div 100 = \frac{11}{2} \times \frac{1}{100} = \frac{11}{200}$$

$$\frac{11}{200} \times 50 = \frac{11}{\cancel{200}_{4}} \times \frac{\cancel{50}^{1}}{1} = \frac{11}{4} = 2\frac{3}{4}$$

Find the answer using a fraction.

1. 11% of 25 $2\frac{3}{4}$	**2.** 17% of 32 $5\frac{22}{50}$	**3.** 21% of 58 $12\frac{9}{50}$	**4.** 47% of 89 $41\frac{83}{100}$
5. 10% of 60 6	**6.** 44% of 100 44	**7.** 40% of 125 50	**8.** 95% of 200 190
9. 15% of 50 $7\frac{1}{2}$	**10.** 25% of 110 $27\frac{1}{2}$	**11.** 90% of 175 $157\frac{1}{2}$	**12.** 75% of 150 $112\frac{1}{2}$
13. 8% of 20 $1\frac{3}{5}$	**14.** 13% of 60 $7\frac{4}{5}$	**15.** 21% of 44 $9\frac{6}{25}$	**16.** 35% of 75 $26\frac{1}{4}$
17. 51% of 90 $45\frac{9}{10}$	**18.** 69% of 200 138	**19.** 75% of 700 525	**20.** 80% of 800 640
21. $\frac{1}{4}\%$ of 10 $\frac{1}{40}$	**22.** $\frac{1}{2}\%$ of 25 $\frac{1}{8}$	**23.** $6\frac{1}{2}\%$ of 100 $6\frac{1}{2}$	**24.** $33\frac{1}{3}\%$ of 150 50
25. $\frac{1}{2}\%$ of 50 $\frac{1}{4}$	**26.** $\frac{1}{5}\%$ of 20 $\frac{1}{25}$	**27.** $4\frac{1}{4}\%$ of 200 $8\frac{1}{2}$	**28.** $6\frac{1}{3}\%$ of 300 19
29. $\frac{1}{8}\%$ of 40 $\frac{1}{20}$	**30.** $\frac{1}{10}\%$ of 300 $\frac{3}{10}$	**31.** $17\frac{1}{2}\%$ of 600 105	**32.** $66\frac{2}{3}\%$ of 900 600
33. $\frac{1}{2}\%$ of 60 $\frac{3}{10}$	**34.** $\frac{1}{4}\%$ of 25 $\frac{1}{16}$	**35.** $7\frac{1}{3}\%$ of 90 $6\frac{3}{5}$	**36.** $19\frac{3}{5}\%$ of 950 $186\frac{1}{5}$
37. $\frac{1}{5}\%$ of 125 $\frac{1}{4}$	**38.** $\frac{1}{3}\%$ of 150 $\frac{1}{2}$	**39.** $24\frac{2}{5}\%$ of 50 $12\frac{1}{5}$	**40.** $17\frac{2}{3}\%$ of 150 $26\frac{1}{2}$

Skill 2: Find the percent of a number using a decimal

To find the percent of a number, you can first rename the percent as a decimal and then multiply.

Find 6.5% of 100.

$$a\% \text{ of } b = c$$
$$\downarrow \quad \downarrow \quad \downarrow \qquad \downarrow$$
$$6.5\% \times 100 = c$$

THINK: $6.5\% \rightarrow 06.5 \rightarrow 0.065$

$$0.065 \times 100 = 6.5$$

FIND THE PERCENT OF A NUMBER

Find the answer using a decimal.

41. 10% of 30 3	**42.** 25% of 50 12.5	**43.** 36% of 25 9	**44.** 75% of 63 47.25
45. 10% of 110 11	**46.** 25% of 160 40	**47.** 45% of 300 135	**48.** 75% of 500 375
49. 15% of 90 13.5	**50.** 21% of 85 17.85	**51.** 52% of 98 50.96	**52.** 84% of 156 131.04
53. 10% of 70 7	**54.** 75% of 100 75	**55.** 30% of 40 12	**56.** 27% of 80 21.6
57. 11% of 56 6.16	**58.** 19% of 28 5.32	**59.** 45% of 150 67.5	**60.** 88% of 230 202.4
61. 7.5% of 30 2.25	**62.** 15.5% of 45 6.975	**63.** 25.5% of 86 21.93	**64.** 80.5% of 91 73.255
65. 13.4% of 94 12.596	**66.** 26.8% of 72 19.296	**67.** 16.45% of 40 6.58	**68.** 24.33% of 69 16.7877
69. 12.5% of 96 12	**70.** 15.5% of 50 7.75	**71.** 2.5% of 120 3	**72.** 5.5% of 24 1.32
73. 9.2% of 30 2.76	**74.** 6.4% of 25 1.6	**75.** 15.8% of 75 11.85	**76.** 12.6% of 50 6.3
77. 15.75% of 100 15.75	**78.** 2.25% of 80 1.8	**79.** 11.33% of 20 2.266	**80.** 14.125% of 40 5.65

Skill 3: Find the percent of a number when the percent is greater than 100%

When you find the percent of a number when the percent is greater than 100%, the answer will be greater than the original number. For example, 150% of 300 = 450. Notice that the answer, 450, is greater than the original number, 300.

Find 150% of 300.

$$\begin{array}{cccc} a\% & \text{of} \; b & = & c \\ \downarrow & \downarrow \;\; \downarrow & & \downarrow \\ 150\% & \times \; 300 & = & c \end{array}$$

THINK: 150% → 150 → 1.5

$1.5 \times 300 = 450$

Find the percent of the number.

81. 111% of 200 222	**82.** 150% of 400 600	**83.** 150% of 44 66	**84.** 160% of 85 136
85. 170% of 150 255	**86.** 175% of 400 700	**87.** 180% of 500 900	**88.** 200% of 225 450
89. 103% of 50 51.5	**90.** 117% of 120 140.4	**91.** 125% of 50 62.5	**92.** 165% of 222 366.3
93. 200% of 6.4 12.8	**94.** 300% of 105.5 316.5	**95.** 150% of 4.4 6.6	**96.** 175% of 55.7 97.475
97. 225% of 3.5 7.875	**98.** 450% of 2.4 10.8	**99.** 375% of 50.6 189.75	**100.** 290% of 99.9 289.71
101. 500% of 200 1,000	**102.** 250% of 400 1,000	**103.** 875% of 212 1,855	**104.** 999% of 100 999
105. 120% of 100 120	**106.** 140% of 125 175	**107.** 160% of 350 560	**108.** 180% of 36 64.8
109. 125% of 225 281.25	**110.** 132% of 70 92.4	**111.** 220% of 400 880	**112.** 275% of 500 1,375
113. 111% of 7.2 7.992	**114.** 120% of 8.9 10.68	**115.** 200% of 2.8 5.6	**116.** 360% of 9.8 35.28
117. 240% of 20.5 49.2	**118.** 350% of 50.5 176.75	**119.** 425% of 75.2 319.6	**120.** 600% of 80.5 483

FIND THE PERCENT ONE NUMBER IS OF ANOTHER

Skill 1: Find the percent using a fraction

When you solve equations of the form $a\%$ of $b = c$, substitute the values that you know. Then solve for the unknown value.

What percent of 100 is 25?

$$\begin{array}{cccc} a\% & \text{of} & b & = c \\ \downarrow & \downarrow & \downarrow & \downarrow \\ a & \times & 100 & = 25 \end{array}$$

$$a = 25 \div 100 = \frac{25}{100} = 25\%$$

Find the answer using a fraction.

1. What percent of 100 is 30? 30%
2. What percent of 100 is 45? 45%
3. What percent of 5 is 4? 80%
4. What percent of 25 is 5? 20%
5. What percent of 12 is 4? $33\frac{1}{3}\%$
6. What percent of 24 is 21? $87\frac{1}{2}\%$
7. What percent of 200 is 100? 50%
8. What percent of 90 is 15% $16\frac{2}{3}\%$
9. What percent of 100 is 10? 10%
10. What percent of 4 is 2? 50%
11. What percent of 36 is 9? 25%
12. 8 is what percent of 32? 25%
13. 50 is what percent of 500? 10%
14. 7 is what percent of 28? 25%
15. 60 is what percent of 100? 60%
16. 90 is what percent of 100? 90%
17. 16 is what percent of 64? 25%
18. 10 is what percent of 50? 20%
19. 6 is what percent of 36? $16\frac{2}{3}\%$
20. 8 is what percent of 64? $12\frac{1}{2}\%$

Skill 2: Find the percent using a decimal

When you find the percent one number is of another, you can use a decimal.

What percent of 80 is 20?

$$\begin{array}{cccc} a\% & \text{of} & b & = c \\ \downarrow & \downarrow & \downarrow & \downarrow \\ a & \times & 80 & = 20 \end{array}$$

$$a = 20 \div 80 = 0.25$$

$$a = 0.25 \rightarrow 0.25 \rightarrow 25\%$$

FIND THE PERCENT ONE NUMBER IS OF ANOTHER

Find the answer using a decimal.

21. What percent of 100 is 80? 80%

22. What percent of 10 is 6? 60%

23. What percent of 5 is 3? 60%

24. What percent of 75 is 15? 20%

25. What percent of 200 is 75? 37.5%

26. What percent of 11 is 5.5? 50%

27. What percent of 500 is 25? 5%

28. What percent of 125 is 10.5? 8.4%

29. What percent of 100 is 15? 15%

30. What percent of 9 is 3? 33.3%

31. What percent of 300 is 45? 15%

32. 70 is what percent of 100? 70%

33. 25 is what percent of 50? 50%

34. 8 is what percent of 10? 80%

35. 4 is what percent of 32? 12.5%

36. 4.2 is what percent of 42? 10%

37. 9 is what percent of 45? 20%

38. 6.7 is what percent of 20.1? 33.3%

39. 6 is what percent of 15? 40%

40. 25 is what percent of 75? 33.3%

Skill 3: Find a percent greater than 100%

Sometimes the percent can be greater than 100%. You can estimate that your answer will be greater than 100%. For example, notice that in $a \times 40 = 60$, $c > b$, or $60 > 40$. Therefore, the answer (%) is greater than 100%.

What percent of 40 is 60?

$$a\% \text{ of } b = c$$
$$\downarrow \quad \downarrow \quad \downarrow \quad \downarrow$$
$$a \times 40 = 60$$
$$a = 60 \div 40 = 1.5$$
$$a = 1.5 = 1.50 \rightarrow 150\%$$

Find the answer.

41. What percent of 100 is 150? 150%

42. What percent of 100 is 200? 200%

43. What percent of 25 is 60? 240%

44. What percent of 40 is 160? 400%

45. What percent of 10 is 55? 550%

46. What percent of 10.25 is 20.5? 200%

47. What percent of 5 is 60? 1,200%

48. What percent of 2.5 is 6.25? 250%

49. 130 is what percent of 100? 130%

50. 165 is what percent of 100? 165%

51. 150 is what percent of 50? 300%

52. 90 is what percent of 30? 300%

53. 90 is what percent of 40? 225%

54. 76.5 is what percent of 25.5? 300%

FIND A NUMBER WHEN A PERCENT OF IT IS KNOWN

Skill 1: Find the number using a fraction

To find a number when a percent of it is known, you solve an equation of the form $a\%$ of $b = c$. First substitute the values that you know. Then solve for the unknown value.

16 is 20% of what number?

$$\begin{array}{cccc} a\% & \text{of} & b & = c \\ \downarrow & \downarrow & \downarrow & \downarrow \\ 20\% & \times & b & = 16 \end{array}$$

THINK: $20\% = \frac{20}{100}$

$$\frac{20}{100} \times b = 16$$

$$b = \frac{16}{1} \div \frac{20}{100} = \frac{\overset{4}{\cancel{16}}}{1} \times \frac{100}{\underset{5}{\cancel{20}}} = \frac{400}{5} = 80$$

Find the answer using a fraction.

1. 10 is 10% of what number? 100
2. 30 is 30% of what number? 100
3. 22 is 11% of what number? 200
4. 52 is 13% of what number? 400
5. 65 is 25% of what number? 260
6. 75 is 50% of what number? 150
7. 150 is 75% of what number? 200
8. 200 is 80% of what number? 250
9. 20 is 5% of what number? 400
10. 60 is 40% of what number? 150
11. 36 is 15% of what number? 240
12. 85 is 25% of what number? 340
13. 294 is $4\frac{1}{5}\%$ of what number? 7,000
14. 150 is $7\frac{1}{2}\%$ of what number? 2,000
15. 44 is $5\frac{1}{2}\%$ of what number? 800
16. 105 is $10\frac{1}{2}\%$ of what number? 1,000
17. 45 is $2\frac{1}{4}\%$ of what number? 2,000
18. 110 is $4\frac{2}{5}\%$ of what number? 2,500

Skill 2: Find the number using a decimal

To find a number when a percent of it is known, you can use a decimal.

21 is 7% of what number?

$$\begin{array}{cccc} a\% & \text{of} & b & = c \\ \downarrow & \downarrow & \downarrow & \downarrow \\ 7\% & \times & b & = 21 \end{array}$$

THINK: 7% → .07. → 0.07

$$0.07 \times b = 21$$

$$b = 21 \div 0.07 = 300$$

FIND A NUMBER WHEN A PERCENT OF IT IS KNOWN

Find the answer using a decimal.

19. 9 is 9% of what number? 100

20. 23 is 23% of what number? 100

21. 15 is 20% of what number? 75

22. 18 is 30% of what number? 60

23. 105 is 75% of what number? 140

24. 120 is 50% of what number? 240

25. 11 is 20% of what number? 55

26. 20 is 80% of what number? 25

27. 16 is 80% of what number? 20

28. 36 is 12% of what number? 300

29. 30.5 is 50% of what number? 61

30. 7.2 is 10% of what number? 72

31. 20.25 is 50% of what number? 40.5

32. 200.4 is 80% of what number? 250.5

33. 7.5 is 20% of what number? 37.5

34. 9.75 is 75% of what number? 13

35. 130.5 is 25% of what number? 522

36. 89.3 is 38% of what number? 235

Skill 3: Find the number when the percent is greater than 100%

Remember, sometimes the percent is greater than 100%. Notice that when this happens $c > b$, for example, $84 > 70$.

84 is 120% of what number?

$$\begin{array}{cccc} a\% & \text{of } b & = & c \\ \downarrow & \downarrow\ \downarrow & & \downarrow \\ 120\% & \times\ b & = & 84 \end{array}$$

THINK: 120% → 120. → 1.2

$1.2 \times b = 84$

$b = 84 \div 1.2 = 70$

Find the answer.

37. 30 is 100% of what number? 30

38. 170 is 100% of what number? 170

39. 222 is 11% of what number? 200

40. 286 is 143% of what number? 200

41. 25 is 125% of what number? 20

42. 105 is 150% of what number? 70

43. 22 is 125% of what number? 17.6

44. 480 is 150% of what number? 320

45. 85.6 is 200% of what number? 42.8

46. 16.5 is 300% of what number? 5.5

47. 6.4 is 128% of what number? 5

48. 26.4 is 132% of what number? 20

49. 45.5 is 455% of what number? 10

50. 36.5 is 200% of what number? 18.25

PERCENT AND PROPORTION

Skill 1: Find the percent of a number

Percent problems can be solved using the percent formula, $a\%$ of $b = c$, or using the proportion, $\frac{a}{100} = \frac{c}{b}$. First substitute the values you know. Then solve for the unknown value. Remember, use cross products to solve the proportion.

Find 6% of 50.

THINK: $a = 6$, $b = 50$, $c = ?$

$$\frac{a}{100} = \frac{c}{b}$$

$$\frac{6}{100} = \frac{c}{50}$$

$$\frac{6}{100} \times \frac{c}{50}$$

$$6 \times 50 = c \times 100$$

$$300 = c \times 100$$

$$300 \div 100 = c$$

$$3 = c$$

So 6% of 50 is 3.

Find the answer using a proportion.

1. 10% of 50 5	**2.** 20% of 80 16	**3.** 25% of 60 15	**4.** 30% of 90 27
5. 10% of 160 16	**6.** 25% of 180 45	**7.** 50% of 150 75	**8.** 75% of 120 90
9. 11% of 50 5.5	**10.** 42% of 85 35.7	**11.** 41.5% of 60 24.9	**12.** 25.5% of 36 9.18
13. $\frac{1}{4}$% of 20 0.05	**14.** $\frac{1}{2}$% of 50 0.25	**15.** $3\frac{1}{2}$% of 50 1.75	**16.** $33\frac{1}{3}$% of 250 $83\frac{1}{3}$
17. 100% of 16 16	**18.** 150% of 54 81	**19.** 200% of 52 104	**20.** 300% of 75 225

Skill 2: Find the percent one number is of another

To find the percent, you can use the percent formula, $a\%$ of $b = c$, or the proportion, $\frac{a}{100} = \frac{c}{b}$. When finding the percent, remember to include the % sign in your final answer.

What percent of 16 is 4?

THINK: $a = ?$, $b = 16$, $c = 4$

$$\frac{a}{100} = \frac{c}{b}$$

$$\frac{a}{100} = \frac{4}{16}$$

$$\frac{a}{100} \times \frac{4}{16}$$

$$a \times 16 = 4 \times 100$$

$$a \times 16 = 400$$

$$a = 400 \div 16 = 25$$

So 4 is 25% of 16.

PERCENT AND PROPORTION

Find the answer using a proportion.

21. What percent of 100 is 40? 40%

22. What percent of 100 is 75? 75%

23. What percent of 10 is 8? 80%

24. What percent of 20 is 4? 20%

25. What percent of 150 is 75? 50%

26. What percent of 400 is 200? 50%

27. 4 is what percent of 12? $33\frac{1}{3}\%$

28. 150 is what percent of 400? $37\frac{1}{2}\%$ or 37.5%

29. 5.4 is what percent of 54? 10%

30. 7.8 is what percent of 39? 20%

31. 140 is what percent of 100? 140%

32. 160 is what percent of 80? 200%

Skill 3: Find a number when a percent of it is known

To find a number when a percent of it is known, you can use the percent formula, $a\%$ of $b = c$, or the proportion, $\frac{a}{100} = \frac{c}{b}$. Remember, sometimes the percent is greater than 100%. When this happens $c > b$, for example, $30 > 20$.

30 is 150% of what number?

THINK: $a = 150$, $b = ?$, c 30

$$\frac{a}{100} = \frac{c}{b}$$

$$\frac{150}{100} = \frac{30}{b}$$

$$\frac{150}{100} \bowtie \frac{30}{b}$$

$150 \times b = 30 \times 100$

$150 \times b = 3{,}000$

$b = 3{,}000 \div 150 = 20$

So 30 is 150% of 20.

Find the answer using a proportion.

33. 20 is 100% of what number? 20

34. 40 is 400% of what number? 10

35. 15 is 125% of what number? 12

36. 36 is 120% of what number? 30

37. 210 is 175% of what number? 120

38. 240 is 150% of what number? 160

39. 87.9 is 300% of what number? 29.3

40. 210 is 150% of what number? 140

41. 16.3 is 500% of what number? 3.26

42. 8.3 is 100% of what number? 8.3

43. 30 is 125% of what number? 24

44. 140 is 200% of what number? 70

45. 8.7 is 150% of what number? 5.8

46. 500 is 125% of what number? 400

47. 13.2 is 200% of what number? 6.6

48. 4.23 is 150% of what number? 2.82

PERCENT OF INCREASE AND DECREASE

Skill 1: Find percent of increase using the equation method

To find **percent of increase** using the equation method, first find the amount of change (increase). Then use this formula:

AMOUNT OF CHANGE = RATE × ORIGINAL AMOUNT

Finally, rename the answer as a percent.

Find the percent of increase.
Last week: \$20
This week: \$30

THINK: The amount of change is 30 − 20 = 10.

AMOUNT OF CHANGE = RATE × ORIGINAL AMOUNT

$$10 = r \times 20$$
$$10 \div 20 = r$$
$$0.5 = r$$

$0.5 \rightarrow 0.50 \rightarrow 50\%$

So the percent of increase is 50%.

Find the percent of increase.

1. Last week: \$50
   This week: \$60
   20%
2. Test 1: 80
   Test 2: 100
   25%
3. Game 1: 15
   Game 2: 18
   20%
4. Last year: 250
   This year: 320
   28%
5. Morning: \$500
   Afternoon: \$850
   70%
6. Bought: \$520
   Sold: \$650
   25%
7. Semester 1: 80
   Semester 2: 90 12.5% or $12\frac{1}{2}\%$
8. Last week: 600
   This week: 800 $33\frac{1}{3}\%$
9. Game 1: 30
   Game 2: 50 $66\frac{2}{3}\%$

Skill 2: Find percent of increase using the ratio method

To find percent of increase using the ratio method, first find the amount of change. Then find this ratio:

$$\frac{\text{AMOUNT OF CHANGE}}{\text{ORIGINAL AMOUNT}}$$

Finally, rename the answer as a percent.

Find the percent of increase.
Game 1: 10
Game 2: 30

THINK: The amount of change is 30 − 10 = 20.

$$\frac{\text{AMOUNT OF CHANGE}}{\text{ORIGINAL AMOUNT}} = \frac{20}{10} = 2$$

$2 \rightarrow 200 \rightarrow 200\%$

So the percent of increase is 200%.

Find the percent of increase.

10. Last year: 50
    This year: 75
    50%
11. Week 1: 25
    Week 2: 30
    20%
12. Bought: 60
    Sold: 90
    50%
13. Bought: \$40
    Sold: \$80
    100%
14. Year 1: 140
    Year 2: 175
    25%
15. Last year: \$750
    This year: \$3,000
    300%
16. Game 1: 64
    Game 2: 72
    12.5% or $12\frac{1}{2}\%$
17. Last week: 160
    This week: 180
    12.5% or $12\frac{1}{2}\%$
18. Month 1: \$120
    Month 2: \$160
    $33\frac{1}{3}\%$

442 SKILLS BANK

PERCENT OF INCREASE AND DECREASE

Skill 3: Find percent of decrease using the equation method

To find **percent of decrease** using the equation method, first find the amount of change (decrease). Then use this formula:

AMOUNT OF CHANGE = RATE × ORIGINAL AMOUNT

Finally, rename the answer as a percent.

Find the percent of decrease.
Last week: 60
This week: 48

THINK: The amount of change is 60 − 48 = 12.

AMOUNT OF CHANGE = RATE × ORIGINAL AMOUNT

$$12 = r \times 60$$
$$12 \div 60 = r$$
$$0.2 = r$$

$0.2 \rightarrow 0.20 \rightarrow 20\%$

So the percent of decrease is 20%.

Find the percent of decrease.

19. Last year: 100
This year: 60
40%

20. Morning: 80
Afternoon: 40
50%

21. Bought: 25
Sold: 5
80%

22. Last week: \$150
This week: \$90
40%

23. Year 1: 300
Year 2: 180
40%

24. Period 1: 500
Period 2: 450
10%

25. Week 1: \$90
Week 2: \$60 $33\frac{1}{3}\%$

26. Last year: 800
This year: 700 12.5% or $12\frac{1}{2}\%$

27. 10 years ago: 600
5 years ago: 420 30%

Skill 4: Find percent of decrease using the ratio method

To find the percent of decrease using the ratio method, first find the amount of change. Then find this ratio:

$$\frac{\text{AMOUNT OF CHANGE}}{\text{ORIGINAL AMOUNT}}$$

Finally, rename the answer as a percent.

Find the percent of decrease.
Week 1: 400
Week 2: 218

THINK: The amount of change is 400 − 218 = 182.

$$\frac{\text{AMOUNT OF CHANGE}}{\text{ORIGINAL AMOUNT}} = \frac{182}{400} = 0.455$$

$0.455 \rightarrow 0.455 \rightarrow 45.5\%$

So the percent of decrease is 45.5%.

Find the percent of decrease.

28. Morning: 80
Afternoon: 60
25%

29. Year 1: 50
Year 2: 10
80%

30. Last week: 25
This week: 20
20%

31. Bought: 500
Sold: 250
50%

32. Last year: \$500
This year: \$420
16%

33. Month 1: \$750
Month 2: \$525
30%

34. Session 1: 800
Session 2: 700
12.5%

35. Last week: \$600
This week: \$200
$66\frac{2}{3}\%$

36. Week 1: 1,000
Week 2: 450
55%

METRIC UNITS OF LENGTH

Skill 1: Identify the correct unit of length

kilometer (km)	centimeter (cm)
meter (m)	millimeter (mm)

Which unit of length would you use?

Distance from N.Y. to L.A.: km

Length of a car: m Height of a glass: cm

Length of a mosquito: mm

Which unit would you use to measure? Write km, m, cm, or mm.

1. Length of your foot cm
2. Height of a bus m
3. Thickness of a knife mm
4. Distance from Houston to San Diego km

Skill 2: Measure to the nearest centimeter and millimeter

A centimeter is used to measure very small lengths. On a metric ruler, each centimeter is subdivided into 10 millimeters.

Measure the line segment.

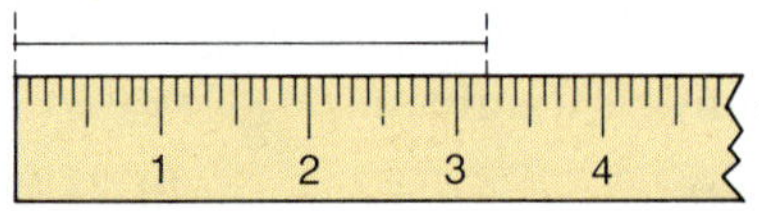

It measures 3 cm to the nearest cm.

It measures 32 mm to the nearest mm.

Measure to the nearest centimeter and to the nearest millimeter.

5. string: about ▪ cm 4
6. string: about ▪ mm 43
7. chalk: about ▪ cm 6
8. chalk: about ▪ mm 57

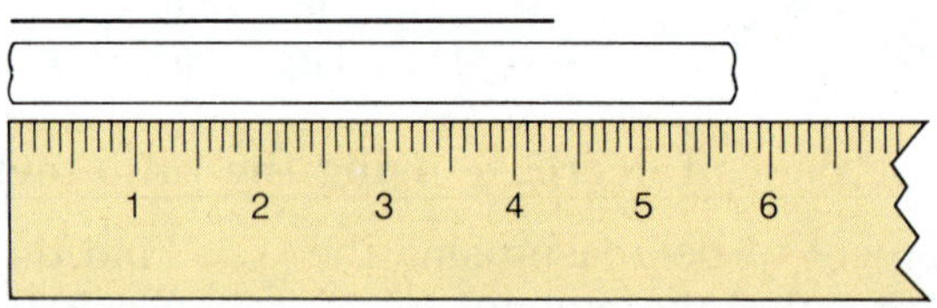

Skill 3: Convert between kilometers, meters, centimeters, and millimeters

1 centimeter = 10 millimeters
1 meter = 100 centimeters
1 kilometer = 1,000 meters

- To rename larger units to smaller units, multiply.
- To rename smaller units to larger units, divide.

Complete: 3 cm = ▪ mm.

THINK: Larger to smaller, so multiply.

1 cm = 10 mm

3 cm = ▪ mm $3 \times 10 = 30$

3 cm = 30 mm

Complete: 4,300 m = ▪ km.

THINK: Smaller to larger, so divide.

1,000 m = 1 km

4,300 m = ▪ km $4{,}300 \div 1{,}000 = 4.3$

4,300 m = 4.3 km

Complete.

9. 6 m = ▪ cm 600
10. 7 km = ▪ m 7,000
11. 1.2 cm = ▪ mm 12
12. 13 mm = ▪ cm 1.3
13. 20 mm = ▪ cm 2
14. 300 cm = ▪ m 3

METRIC UNITS OF CAPACITY AND MASS

Skill 1: Identify the correct unit of capacity or mass

UNITS OF CAPACITY
kiloliter (kL)
liter (L)
milliliter (mL)

UNITS OF MASS
kilogram (kg)
gram (g)
milligram (mg)

Which unit of capacity or mass would you use?

Capacity of a pail: L
Capacity of a dropper: mL
Mass of a person: kg
Mass of a grape: g

Which unit would you use to measure? Write kL, L, mL, kg, g, or mg.

1. Capacity of a dropper mL
2. Capacity of a water tower kL
3. Mass of a person kg
4. Capacity of a gas tank L
5. Mass of an insect mg
6. Mass of a banana g

Skill 2: Convert between kiloliters, liters, and milliliters

1 liter = 1,000 milliliters
1 kiloliter = 1,000 liters

Remember: Larger to smaller → multiply
Smaller to larger → divide

Complete: 3,000 mL = ■ L.

THINK: Smaller to larger, so divide.

1 L = 1,000 mL
3,000 mL = ■ L 3,000 ÷ 1,000 = 3
3,000 mL = 3 L

Complete.

7. 3 L = ■ mL 3,000
8. 5,000 mL = ■ L 5
9. 3 kL = ■ L 3,000
10. 6,000 L = ■ kL 6
11. 2,500 mL = ■ L 2.5
12. 3,500 L = ■ kL 3.5

Skill 3: Convert between kilograms, grams, and milligrams

1 gram = 1,000 milligrams
1 kilogram = 1,000 grams

Remember: Larger to smaller → multiply
Smaller to larger → divide

Complete: 6 kg = ■ g.

THINK: Larger to smaller, so multiply.

1 kg = 1,000 g
6 kg = ■ g 6 × 1,000 = 6,000
6 kg = 6,000 g

Complete.

13. 2 g = ■ mg 2,000
14. 4,000 g = ■ kg 4
15. 3,000 mg = ■ g 3
16. 5 kg = ■ g 5,000
17. 1,500 g = ■ kg 1.5
18. 5,500 mg = ■ g 5.5

CUSTOMARY UNITS OF LENGTH

Skill 1: Identify the correct unit of length

mile (mi)	foot (ft)
yard (yd)	inch (in.)

Which unit of length would you use?
Distance from Tampa to Dallas: mi
Length of a room: ft   Height of a cup: in.

Which unit would you use to measure? Write mi, yd, ft, or in.

1. Height of a person in. or ft
2. Length of a pencil in.
3. Height of a bookcase ft
4. Distance to a planet mi
5. Length of fabric on a bolt yd
6. Thickness of a book in.

Skill 2: Measure to the nearest inch, $\frac{1}{2}$-inch, and $\frac{1}{4}$-inch

An inch is used to measure the lengths of small objects to the nearest inch, nearest $\frac{1}{2}$-inch, and nearest $\frac{1}{4}$-inch.

Measure the line segment.

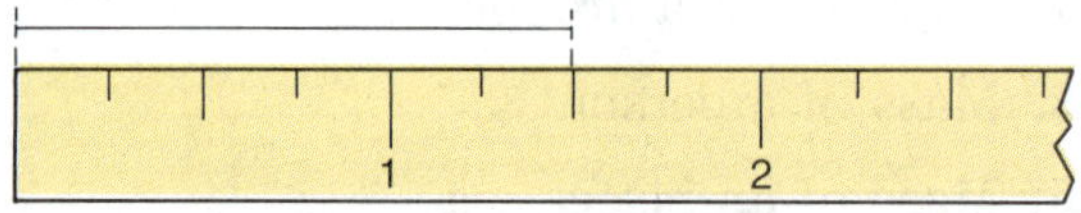

It measures 1 in. to the nearest inch.
It measures $1\frac{1}{2}$ in. to the nearest $\frac{1}{2}$-inch.
It measures $1\frac{1}{2}$ in. to the nearest $\frac{1}{4}$-inch.

Use a ruler to draw the line segment. Check students' drawings.

7. 3 in.
8. 5 in.
9. $2\frac{1}{2}$ in.
10. $4\frac{1}{2}$ in.
11. $1\frac{1}{4}$ in.
12. $3\frac{3}{4}$ in.

Skill 3: Convert between miles, yards, feet, and inches

1 foot (ft) = 12 inches (in.)
1 yard (yd) = 3 feet = 36 inches
1 mile (mi) = 1,760 yards = 5,280 feet

- To rename larger units to smaller units, multiply.
- To rename smaller units to larger units, divide.

Complete: 3 ft = ■ in.

THINK: Larger to smaller, so multiply.

1 ft = 36 in.
3 ft = ■ in. $3 \times 36 = 108$
3 ft = 108 in.

Complete: 17 ft = ■ yd.

THINK: Smaller to larger, so divide.

3 ft = 1 yd
17 ft = ■ yd $17 \div 3 = 5\frac{2}{3}$
17 ft = $5\frac{2}{3}$ yd

Complete.

13. 2 ft = ■ in. 24
14. 72 in. = ■ yd 2
15. 7 ft = ■ yd $2\frac{1}{3}$
16. 2 mi = ■ ft 10,560
17. $1\frac{1}{2}$ yd = ■ in. 54
18. 3,520 yd = ■ mi 2

CUSTOMARY UNITS OF CAPACITY AND WEIGHT

Skill 1: Identify the correct unit of capacity or weight

UNITS OF CAPACITY	
gallon (gal)	**cup (c)**
quart (qt)	**fluid ounce (fl oz)**
pint (pt)	

UNITS OF WEIGHT		
ton (T)	**pound (lb)**	**ounce (oz)**

Which unit of capacity or weight would you use?

Capacity of a gas tank: gal
Capacity of a glass: fl oz
Weight of a person: lb
Weight of a bird: oz

Which unit would you use to measure? Write gal, qt, pt, c, fl oz, T, lb, or oz.

1. Weight of an elephant T
2. Capacity of a small juice glass fl oz
3. Capacity of a pool gal
4. Weight of a bunch of bananas lb
5. Capacity of a thermos c
6. Weight of a pencil oz

Skill 2: Convert between gallons, quarts, pints, cups, and fluid ounces

1 cup = 8 fluid ounces
1 pint = 2 cups
1 quart = 2 pints
1 gallon = 4 quarts

Remember: Larger to smaller → multiply
Smaller to larger → divide

Complete: 2 qt = ■ pt.

THINK: Larger to smaller, so multiply.

1 qt = 2 pt
2 qt = ■ pt $\quad 2 \times 2 = 4$
2 qt = 4 pt

Complete.

7. 2 c = ■ fl oz 16
8. 8 c = ■ pt 4
9. 6 pt = ■ qt 3
10. 3 gal = ■ qt 12
11. 24 fl oz = ■ c 3
12. $1\frac{1}{2}$ qt = ■ pt 3

Skill 3: Convert between tons, pounds, and ounces

1 pound = 16 ounces
1 ton = 2,000 pounds

Remember: Larger to smaller → multiply
Smaller to larger → divide

Complete: 32 oz = ■ lb.

THINK: Smaller to larger, so divide.

1 lb = 16 oz
32 oz = ■ lb $\quad 32 \div 16 = 2$
32 oz = 2 lb

Complete.

13. 2 T = ■ lb 4,000
14. 48 oz = ■ lb 3
15. 6,000 lb = ■ T 3
16. 4 lb = ■ oz 64
17. $1\frac{1}{2}$ T = ■ lb 3,000
18. 3,000 lb = ■ T $1\frac{1}{2}$

PERIMETER

Skill 1: Find the perimeter of any polygon

The **perimeter** of any polygon is the distance around it. To find the perimeter of any polygon, add the lengths of all the sides.

Find the perimeter.

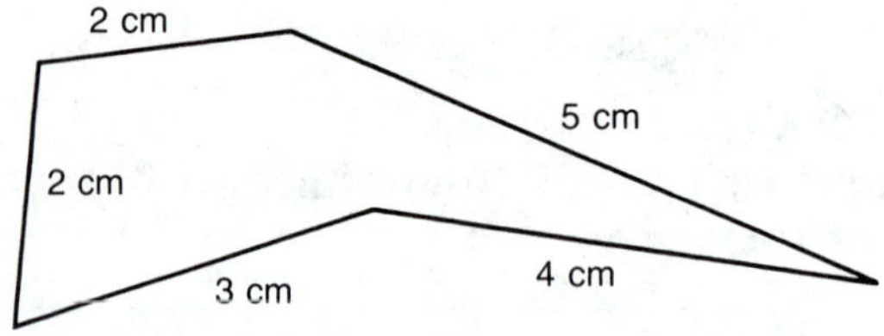

$P = 2 + 2 + 5 + 4 + 3$
$P = 16 \text{ cm}$

Find the perimeter.

1.

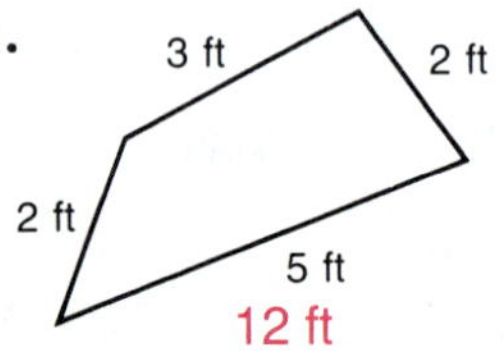

12 ft

2.

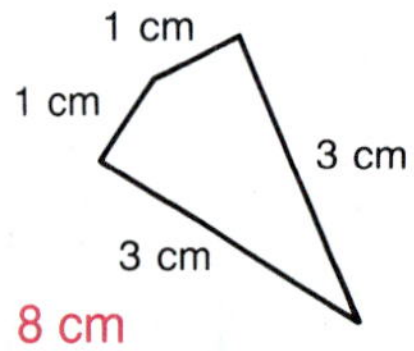

8 cm

3.

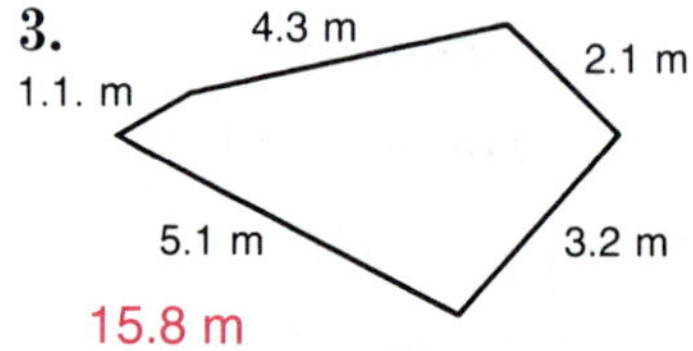

15.8 m

Skill 2: Find the perimeters of rectangles

Remember that a **rectangle** has opposite sides that are equal in length. In rectangle $ABCD$, sides BC and AD both measure 4 cm. So the length of the rectangle is 4 cm. Sides AB and CD both measure 2 cm. So the width of the rectangle is 2 cm. The perimeter of any rectangle can be found by using this formula: $P = (2 \times l) + (2 \times w)$.

Find the perimeter of rectangle $ABCD$.

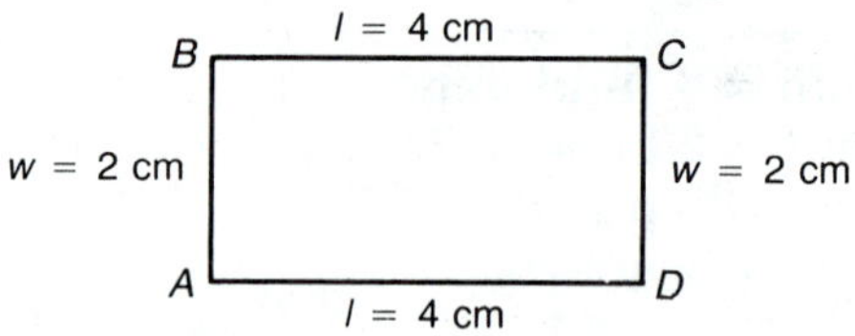

$P = (2 \times l) + (2 \times w)$
$= (2 \times 4) + (2 \times 2)$
$= 8 + 4$
$P = 12 \text{ cm}$

Find the perimeter of the rectangle.

4. $l = 4$ ft, $w = 3$ ft
14 ft

5. $l = 5.2$ m, $w = 4.1$ m
18.6 m

6. $l = 6.1$ ft, $w = 2.3$ ft
16.8 ft

Skill 3: Find the perimeters of squares

Remember that a **square** has all sides equal in length. In square $EFGH$, all sides measure 2 in. The perimeter of any square can be found by using this formula: $P = 4 \times s$.

Find the perimeter of square $EFGH$.

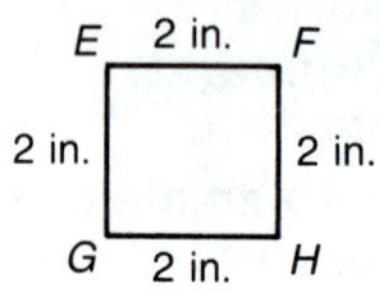

$P = 4 \times s$
$= 4 \times 2$
$P = 8 \text{ in.}$

Find the perimeter of the square.

7. $s = 3$ cm
12 cm

8. $s = 4$ ft
16 ft

9. $s = 5$ m
20 m

10. $s = 15.1$ yd
60.4 yd

CIRCUMFERENCE

Skill 1: Identify the parts of a circle

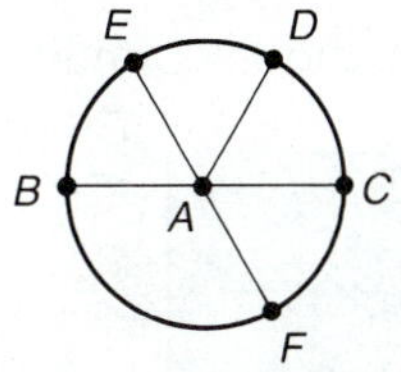

In the circle, A is the **center,** $\overline{BC}$ is a **diameter,** and $\overline{AD}$ is a **radius.** The radius of a circle is $\frac{1}{2}$ the diameter. This relationship can be written as $d = 2 \times r$ or $r = \frac{1}{2} \times d$.

Complete. Use circle A.
If $\overline{BA} = 3$ cm, then $\overline{BC} =$ ■?

THINK: $\overline{BA}$ is a radius.
$\overline{BC}$ is a diameter.

$d = 2 \times r$
$= 2 \times 3$
$d = 6$ cm

Find the diameter.

1. $r = 1$ in. 2 in. **2.** $r = 2$ cm 4 cm **3.** $r = 4$ ft 8 ft **4.** $r = 5$ m 10 m **5.** $r = 10.4$ ft 20.8 ft **6.** $r = 11.6$ yd 23.2 yd

Find the radius.

7. $d = 4$ in. 2 in. **8.** $d = 6$ cm 3 cm **9.** $d = 8$ ft 4 ft **10.** $d = 12$ m 6 m **11.** $d = 20$ yd 10 yd **12.** $d = 30.6$ km 15.3 km

Skill 2: Find the circumference

The **circumference** of a circle is the distance around it. To find the circumference use one of the following formulas:
$C = \pi \times d$ or $C = 2 \times \pi \times r$
Remember to use 3.14 for π (**pi**).

Find the circumference when $r = 3$ in.

$C = 2 \times \pi \times r$
$= 2 \times 3.14 \times 3$
$= 6.28 \times 3$
$C = 18.84$ in.

Find the circumference.

13. $d = 3$ in. 9.42 in. **14.** $d = 5$ cm 15.7 cm **15.** $d = 11$ ft 34.54 ft **16.** $r = 4$ m 25.12 m **17.** $r = 7$ yd 43.96 yd **18.** $r = 12$ km 75.36 km

Skill 3: Find the radius and diameter given the circumference

To find the radius or diameter of a circle given the circumference, use the following formulas:
$C = \pi \times d$ or $C = 2 \times \pi \times r$
Remember to use 3.14 for π.

Find the diameter of a circle when the circumference is 6.28 cm.

$C = \pi \times d$
$6.28 = 3.14 \times d$
$6.28 \div 3.14 = d$
$2 \text{ cm} = d$

Find the diameter and the radius.

19. $C = 3.14$ in. $d = 1$ in.; $r = 0.5$ in. **20.** $C = 9.42$ cm $d = 3$ cm; $r = 1.5$ cm **21.** $C = 12.56$ m $d = 4$ m; $r = 2$ m **22.** $C = 21.98$ m $d = 7$ m; $r = 3.5$ m **23.** $C = 28.26$ ft $d = 9$ ft; $r = 4.5$ ft

AREA

Skill 1: Find the areas of rectangles

The **area** of a figure is the number of square units that it contains. In **rectangle** *ABCD*, the length is 4 cm and the width is 2 cm. By counting, you can see that the area of the rectangle is 8 square cm.

You can find the area of a rectangle by multiplying. Use this formula: $A = l \times w$.

Find the area of the rectangle.

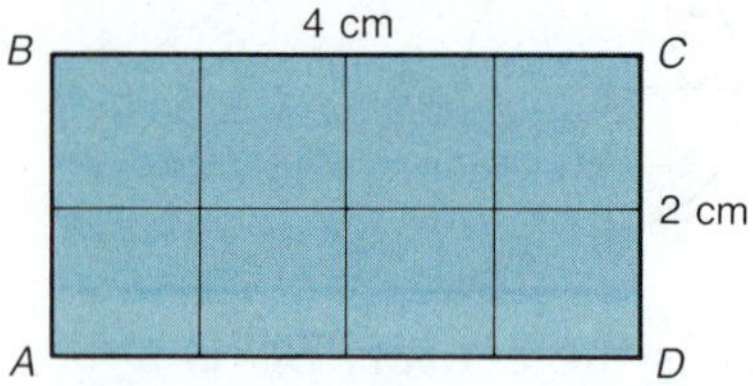

$A = l \times w$
$= 4 \times 2$
$A = 8$ square cm

Find the area of the rectangle.

1. $l = 3$ cm, $w = 1$ cm
3 square cm
2. $l = 4$ in., $w = 3$ in.
12 square in.
3. $l = 5$ ft, $w = 4$ ft
20 square ft
4. $l = 1.4$ in., $w = 1.1$ in.
1.54 square in.
5. $l = 2.1$ m, $w = 1.2$ m
2.52 square m
6. $l = 3.3$ cm, $w = 2.1$ cm
6.93 square cm

Skill 2: Find the areas of squares

Since a **square** is a rectangle with all sides equal in length, the length and the width are the same. You can find the area of a square by using this formula: $A = s \times s$ or $A = s^2$.

Find the area of a square with $s = 1.2$ cm.

$A = s \times s$
$= 1.2 \times 1.2$
$A = 1.44$ square cm

Find the area of the square.

7. $s = 3$ in.
9 square in.
8. $s = 11$ cm
121 square cm
9. $s = 1.4$ m
1.96 square m
10. $s = 10.1$ ft
102.01 square ft

Skill 3: Find the areas of triangles

In a **triangle,** the **height** is the perpendicular distance from any **vertex** to the opposite side. In triangle *DEF*, $\overline{GE}$ is a height. The height (h) = 10 in. and the **base** (b) = 32 in.

You can find the area of a triangle by using this formula: $A = \frac{1}{2} \times b \times h$.

Find the area of triangle *DEF*.

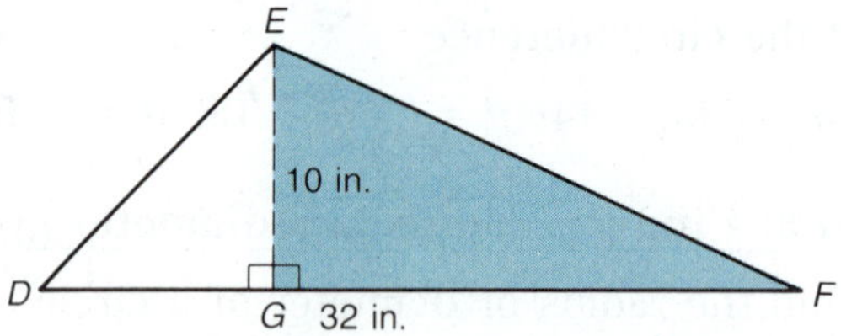

$A = \frac{1}{2} \times b \times h$
$= \frac{1}{2} \times 32 \times 10$
$= 16 \times 10$
$A = 160$ square in.

Find the area of the triangle.

11. $b = 16$ mm, $h = 8$ mm
64 square mm
12. $b = 7$ ft, $h = 5$ ft
17.5 square ft
13. $b = 4.3$ cm, $h = 2.5$ cm
5.375 square cm

AREA

Skill 4: Find the areas of parallelograms

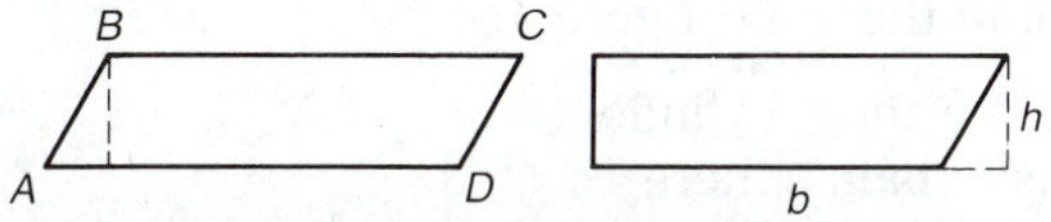

Figure *ABCD* is a **parallelogram.** A parallelogram can be changed into an equivalent rectangle. So, the area of a parallelogram is found by using this formula: $A = b \times h$.

Find the area of a parallelogram with $b = 12$ m and $h = 5$ m.

$A = b \times h$
$= 12 \times 5$
$A = 60$ square m

Find the area of the parallelogram.

14. $b = 6$ ft, $h = 4$ ft
24 square ft

15. $b = 7.1$ cm, $h = 4.2$ cm
29.82 square cm

16. $b = 9.3$ in., $h = 4.2$ in.
39.06 square in.

Skill 5: Find the areas of circles

The area of a **circle** is found by using one of these formulas: $A = \pi \times r \times r$ or $A = \pi \times r^2$. When you must find the area of a circle given the diameter, find the radius first by using the formula: $r = \frac{1}{2} \times d$. Remember to use 3.14 for π.

Find the area of a circle when $d = 4$ ft.

THINK: $r = \frac{1}{2} \times d = \frac{1}{2} \times 4 = 2$

$A = \pi \times r \times r$
$= 3.14 \times 2 \times 2$
$A = 12.56$ square ft

Find the area of the circle to the nearest hundredth.

17. $r = 3$ in.
28.26 square in.

18. $r = 5$ m
78.5 square m

19. $r = 10.2$ ft
326.69 square ft

20. $r = 15.1$ km
715.95 square km

21. $d = 6$ cm
28.26 square cm

22. $d = 10$ in.
78.5 square in.

23. $d = 12.2$ yd
116.84 square yd

24. $d = 16.8$ m
221.56 square m

Skill 6: Find the areas of combined regions

To find the area of combined regions, divide them into known polygons. Find the area of each polygon. Then add all the areas.

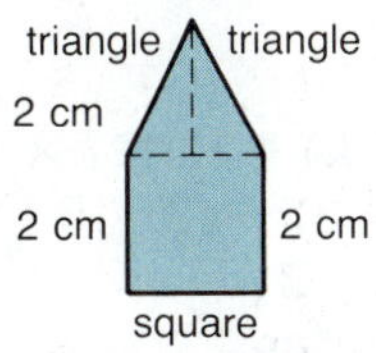

Find the area of the figure.

THINK: **Triangle with $b = 2$ cm, $h = 2$ cm**
Square with $s = 2$ cm

$A = \frac{1}{2} \times b \times h$
$= \frac{1}{2} \times 2 \times 2$
$A = 2$ square cm

$A = s \times s$
$= 2 \times 2$
$A = 4$ square cm

So the area of the combined regions is 2 square cm + 4 square cm = 6 square cm.

Find the area.

25.

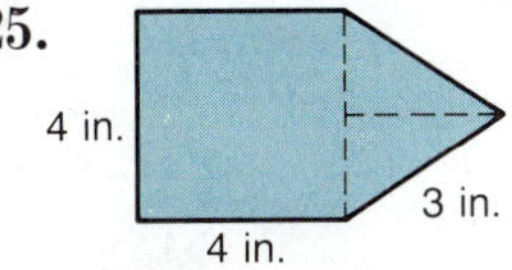

22 square in.

26.
14 square m

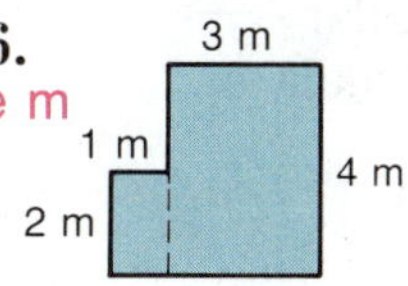

27.

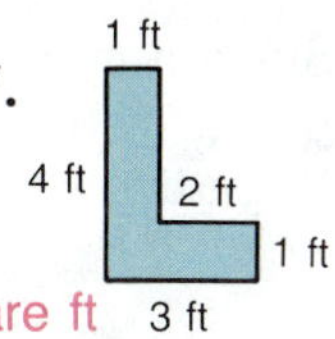

6 square ft

SURFACE AREA AND VOLUME

Skill 1: Identify solid figures

Here are some solid figures.

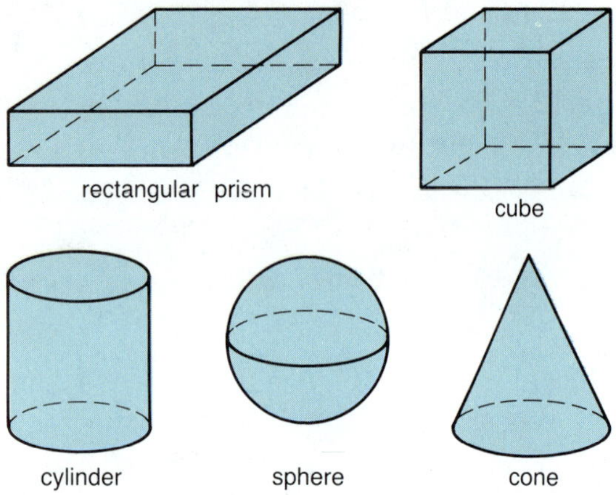

Name the solid figure.

Coffee mug: cylinder
Basketball: sphere
Party hat: cone
Shirt box: rectangular prism

Name the solid figure.

1. Cereal box
rectangular prism

2. Juice glass
cylinder

3. Funnel
cone

4. Baseball
sphere

5. Globe
sphere

6. Brick
rectangular prism

Skill 2: Find the surface areas of rectangular prisms and cubes

In a rectangular prism, each **face** (side) is a rectangle. There are 6 faces and the opposite faces have equal areas. The surface area of a rectangular prism is the sum of the area of the 6 faces.

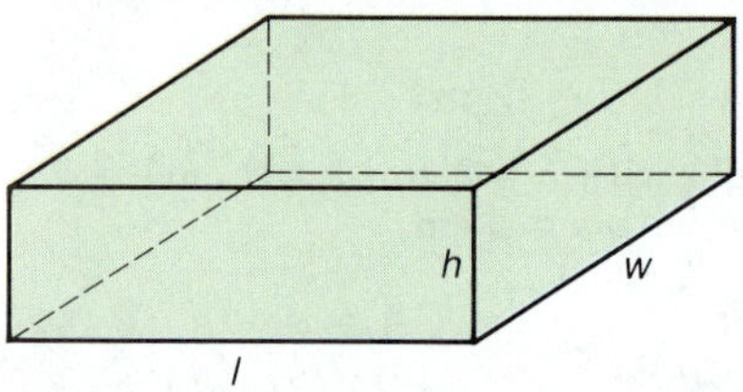

To find the surface area of a rectangular prism, use the following formula:

$$\text{SURFACE AREA} = \text{AREA OF FRONT AND BACK} + \text{AREA OF TOP AND BOTTOM} + \text{AREA OF SIDES}$$

$$SA = (2 \times l \times h) + (2 \times l \times w) + (2 \times w \times h)$$

Since a cube is a special rectangular prism, you can use the above formula or $SA = 6 \times s \times s$.

Find the surface area of the rectangular prism below.

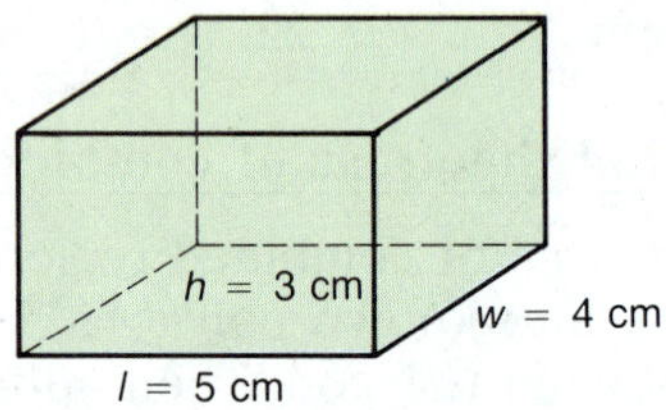

$$\begin{aligned} SA &= (2 \times l \times h) + (2 \times l \times w) + (2 \times w \times h) \\ &= (2 \times 5 \times 3) + (2 \times 5 \times 4) + (2 \times 4 \times 3) \\ &= 30 + 40 + 24 \\ SA &= 94 \text{ square cm} \end{aligned}$$

SURFACE AREA AND VOLUME

Find the surface area.

7. rectangular prism:
$l = 4$ cm, $w = 3$ cm, $h = 1$ cm
38 square cm

8. rectangular prism:
$l = 16$ in., $w = 14$ in., $h = 2$ in.
568 square in.

9. rectangular prism:
$l = 5$ ft, $w = 4$ ft, $h = 6$ ft
148 square ft

10. rectangular prism:
$l = 8$ m, $w = 2$ m, $h = 3$ m
92 square m

11. cube: $s = 8$ cm
384 square cm

12. cube: $s = 1$ in.
6 square in.

13. cube: $s = 20$ m
2,400 square m

14. cube: $s = 3$ yd
54 square yd

15. cube: $s = 4$ ft
96 square ft

16. cube: $s = 7$ m
294 square m

17. cube: $s = 11$ yd
726 square yd

18. cube: $s = 15$ cm
1,350 square cm

Skill 3: Find the volumes of rectangular prisms and cubes

The **volume** of a figure is the number of cubic units it contains. In the rectangular prism, the length is 4 in., the width is 3 in., and the height is 2 in. By counting, you can see the volume is 24 cubic in.

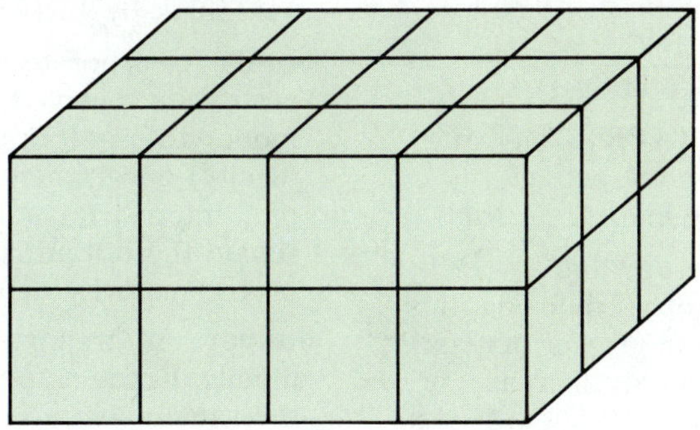

You can also find the volume of a rectangular prism by multiplying. Use the formula: $V = l \times w \times h$.

Since a cube is a special rectangular prism, you can use the above formula or $V = s \times s \times s$ or $V = s^3$.

Find the volume of a rectangular prism with 1 = 4 in., $w = 3$ in., and $h = 2$ in.

$V = l \times w \times h$
$= 4 \times 3 \times 2$
$V = 24$ cubic in.

Find the volume.

19. rectangular prism:
$l = 3$ m, $w = 2$ m, $h = 4$ m 24 cubic m

20. rectangular prism:
$l = 14$ in., $w = 13$ in., $h = 5$ in. 910 cubic in.

21. rectangular prism:
$l = 6$ m, $w = 5$ m, $h = 4$ m 120 cubic m

22. rectangular prism: 1,600 cubic cm
$l = 20$ cm, $w = 10$ cm, $h = 8$ cm

23. cube: $s = 3$ ft
27 cubic ft

24. cube: $s = 7$ m
343 cubic m

25. cube: $s = 11$ in.
1,331 cubic in.

26. cube: $s = 15$ yd
3,375 cubic yd

27. cube: $s = 14$ m
2,744 cubic m

28. cube: $s = 2$ in.
8 cubic in.

29. cube: $s = 21$ cm
9,261 cubic cm

30. cube: $s = 16$ m
4,096 cubic m

GLOSSARY

addends Numbers to be added. [2]

additive inverses Two numbers are additive inverses of each other if their sum is 0.

Example: 6 and -6 are additive inverses, since $6 + (-6) = 0$.
[371]

adjustable rate mortgage (ARM) A mortgage on which the interest rate may go up or down as interest rates in general fluctuate. [241]

all-day pass A general admission ticket that allows unlimited use of activities without paying separate admissions. [86]

algebraic expression An expression that contains one or more **variables,** with or without operational signs. [366]

annual expense [304] (See budget)

Annual Percentage Rate (APR) The true rate of interest charged on the actual amount of the outstanding loan. [145; 151]

appreciation An increase in value as general market values go up. In housing, this may be because of inflation, greater demand, rising building costs, population increase, or scarcity of vacant land. [236; 237]

area The number of square units needed to cover the surface bounded by a 2-dimensional figure such as a rectangle or a circle. [249; 262; 268]

assessed valuation The value of a property, as officially determined by a local government for taxation purposes.

$$\text{ASSESSED VALUATION} = \text{ASSESSMENT RATE} \times \text{MARKET VALUE}$$

[242]

asset [310] (See net worth)

automobile loan Money lent to help the borrower buy a car and have the use of it while paying for it in periodic installments.

The car is the security for the loan, since the lender holds the title (a registered paper that shows legal ownership) until the loan is repaid. [170]

automobile purchase price The amount paid for a car. Beginning with the sticker (offering) price, a trade-in allowance may be subtracted if the buyer owns a car to turn over to the dealer.

Other possible price adjustments are: special discounts, down payment, the cost of extra items (options), finance charges, vehicle registration fee, and taxes. [168]

average daily balance [150] (See Money Market checking account)

beneficiary The person named in an insurance policy to receive money from the insurance company should the person die who is protected under the policy. [72]

board foot The amount of wood in a board 1 in. thick by 1 ft wide by 1 ft long. [262]

bond A corporate or municipal bond is a certificate received by a person who loans money to a corporation or to a local (state or city) government. The holder of a bond is paid interest for a set period of time and is repaid the full amount (face, or par, value) on the maturity date. [327]

budget An organized plan for spending money. **Fixed expenses,** such as rent, are the same each month and are not easily changed or controlled. **Variable expenses,** such as clothing, food, and entertainment, can be changed to some extent according to necessity or choice. **Annual expenses,** such as vacation and insurance premiums, may actually be paid only once a year, but should be shown in the budget in monthly shares. [304; 306]

buying in bulk Buying in a quantity larger than is usual for normal individual use. It may also mean buying loose goods—not in packages, boxes, or bags. [275]

canceled checks [127] (See checking account statement)

cash value For insurance, the amount you would receive if you canceled a straight life policy or borrowed money with the policy as security. The cash value depends upon how much has been paid in premiums. [74]

catalog A publication that pictures and describes merchandise that can be ordered at home and received by mail. [110]

certificate of deposit (CD) A paper that is evidence of money deposited for investment. It tells the set period of time the money must be left on deposit to avoid a penalty charge for withdrawal. [322]

checking account An agreement with a bank that allows the customer to deposit money and write checks against it. [124, 135]

checking account statement A monthly statement from the bank that lists all changes in a checking account such as deposits, amounts paid by check, service charges, and interest. Checks written by the customer and stamped by the bank, to verify payment, are called **canceled** checks and are usually returned to the customer along with the monthly statement. [127]

closing costs Fees related to buying a house or condominium, and paid by the buyer before moving in. Closing costs vary, but may include lawyer's services, title search, and inspection of the property. Bank service charges, referred to as **points,** may be included. Each point is 1% of the mortgage amount. [240]

collectible Something such as a piece of fine art, stamps, or coins, bought as an investment with the expectation that its value will increase. [316]

collision insurance Insurance that pays for damage to the insured car. [175]

column-inch A measure of newspaper space, computed by multiplying the number of columns wide the space is by the number of inches long. [282]

combination A selection of objects without regard to their order. [357]

commission Money paid to an agent or salesperson, often a percentage of the gross amount of business that person generates. A **straight commission** plan pays only commission. A **graduated commission** plan pays a different percent (**rate of commission**) for different levels of sales. [66]

compatible numbers Numbers used to estimate answers in division. They are close in value to the given divisor and dividend, and divide evenly. [257; 381]

compound interest Interest calculated by periodically (quarterly, or even daily) adding earned interest to an account. The principal then increases more rapidly than with simple interest. [132]

comprehensive insurance Insurance that gives financial protection in case of fire, theft, vandalism, and weather damage that involves an insured car. [175]

computer simulation For a product, imitation by a computer of how the product will react under certain conditions. [363]

condominium An apartment that is owned, rather than rented. Each owner pays a monthly **maintenance fee** for the care of the building and grounds. [238]

Consumer Price Index (CPI) A number that represents the value of a "market basket" of goods and services bought by a typical city family. Social Security benefits and the wages of many union workers change according to how much a dollar will buy (**purchasing power** of a dollar), as indicated by the CPI. [302]

contractor A person or company who agrees to provide materials and perform work, usually for a certain price and in a given length of time. [249]

corporate stock An investment that represents shares in the ownership of a company that is organized as a corporation so that its stock can be traded by stockbrokers on a stock exchange. Owners of stock are paid **dividends** when the corporation makes a profit. [319; 324; 330; 336]

credit card A card that allows the person to whom it is issued to receive purchases and services, but pay for them at a later time. If the entire balance shown on a **monthly statement** is paid by the date due, there is no extra charge. A **minimum payment** is usually required. A **finance charge** is calculated on any **unpaid balance.** [144; 147]

credit union A cooperative group that makes loans to its members (sometimes a company's employees) at a low rate. [151]

daily interest rate [149] (See overdraft)

456

decimal A base-ten numeral that uses a decimal point and place value. [2]

deductible clause The section of an insurance policy (health, auto) that specifies how much of any incurred expense the insured person must pay. The insurance company then agrees to pay all or part of the remaining expense. [70; 175]

deflation The opposite of inflation. During a period of deflation, money is relatively scarce and of greater value. Prices fall, business activity is slowed, and unemployment is high. [303]

dependent events [352] (See event)

depreciation For automobiles, the regular decrease in the value of a car. This decrease occurs even if the car is not used heavily. Such depreciation affects the trade-in value of a car. [174]

diameter of a circle Any chord of a circle that contains the center of the circle. [264]

difference The number found by subtracting. [4]

discount The amount that is subtracted from the regular price of an item. Also called a **markdown**. [104]

dividend An amount of money paid to the shareholders of a corporation out of its earnings. [324; 330; 336]

dividend in division [8] (See divisor)

divisor The number by which another number, the **dividend,** is divided.

Example: In $6\overline{)24}$, or $24 \div 6$, 6 is the divisor and 24 is the dividend. [8]

down payment [154; 170; 240] (See automobile purchase price; installment buying; mortgage)

EPA Fuel Economy Rating (mpg) The result of tests done by the Federal Environmental Protection Agency (EPA) to find how many miles a car can go on a gallon of gasoline (miles per gallon, or mpg).

$$\text{mpg} = \frac{\text{MILES}}{\text{DRIVEN}} \div \frac{\text{GALLONS OF}}{\text{GASOLINE USED}}$$

[168; 172]

equation A mathematical sentence that contains an is equal to sign, indicating that 2 expressions are equal. [366]

estimate To calculate roughly, when an approximation is sufficient. Used also to check the accuracy of computation. [26]

evaluate an algebraic expression To replace each variable by a given value; then, simplify the resulting expression. [368]

event A subset of all possible outcomes of an experiment. Events are independent of one another if the probability of one is not affected by the probability of another.
The probability of a series of **independent events** occurring together is found by multiplying the probabilities of the individual events.
Events are **dependent** upon one another if the probability of one affects the probability of another. [346; 350; 352]

exchange rate The fluctuating ratio of the values of the money of 2 countries (as the ratio of the U.S. dollar to the English pound). [286; 287]

face value For an insurance policy, the amount printed on the policy. That is, the amount the insurance company agrees to pay. [72; 320; 327]

factors Numbers to be multiplied. [6]

Federal withholding tax A percentage of an employee's taxable pay, deducted and paid directly by the employer to the Internal Revenue Service of the Federal government. State and city income taxes are collected in a similar way. [68; 214]

finance charge [144; 154] (See credit card; installment buying)

fixed expense [304] (See budget)

fixed rate mortgage A mortgage on which the interest rate remains the same throughout the term of the mortgage. [252]

formula An equation that expresses the rule for a frequently used relationship, often among quantities such as measurements. [368]

fraction The division of 2 numbers written in the form $\frac{a}{b}$. The denominator (divisor), cannot be 0.

Examples: $\frac{1}{3}$; $\frac{7}{8}$; $\frac{12}{5}$

[10]

fringe benefit A benefit, such as free insurance, a discount on purchases, or child care, that is received by an employee in addition to regular pay. [70]

front-end estimation A procedure for estimating answers.
1. Find the greatest number.
2. Identify the place of its leading nonzero digit.
3. Compute only with the digits in that place.

[79; 121; 163, 361]

graduated commission [66] (See commission)

gross pay [68; 212] (See payroll deductions)

group rate A special reduced price for admission when tickets for a group of people are ordered at the same time. [87]

health insurance A contract by which a person regularly pays a **premium** to an insurance company. In return, the company promises to help pay medical expenses, as specified in a written policy, in case the person becomes ill or is injured. [70]

home equity loan Money borrowed to buy a house. The house is the security for the loan. That is, the lender has the right to take over the house if the loan is not repaid. [152]

homeowner's insurance A contract by which a person pays a **premium** to an insurance company. In return, the company agrees to pay money, as specified in a written policy, to cover loss or damage to the property (by fire, for example). Such insurance usually excludes causes of loss such as flood, earthquake, and nuclear accident.

Most owners insure a property for its full **replacement value,** which would cover reconstruction if it were destroyed. [244]

idle time In taxicab fares, time during which a customer is in the cab, but the cab is not moving. [201]

income tax [68] (See Federal withholding tax)

independent events [350] (See event)

Individual Retirement Account (IRA) [333] (See pension)

inflation A continuous and relatively large rise in prices caused by expansion in paper money, or bank credit, which decreases the value of money. The **inflation** rate is the percent prices increase, by comparison with a previous month or year. [302]

installment buying plan A plan by which the buyer begins using the item purchased at once, but pays for it in periodic **installments.** A **finance charge,** or interest, is added to the cash price. If there is a **down payment,** that amount is deducted before the finance charge is figured. The installment price is the total of the installment payments. [154]

insurance [70; 72; 175; 244] (See automobile insurance; health insurance; homeowner's insurance; life insurance)

interest rate [132] (See compound interest; simple interest)

inverse operations Operations that "undo" each other.

Examples: Adding and subtracting the same number
Multiplying and dividing by the same number

[370; 372]

kilowatt-hour A unit of energy, equivalent to 1,000 watts of electricity used for 1 hour. [246]

liability [310] (See net worth)

liability coverage For automobiles, insurance that gives financial protection for injury to others and damage to their property.

For homes, liability coverage provides financial protection in case someone is injured on the insured property. [175; 244]

life insurance A contract by which a person regularly pays a **premium** to an insurance company. In return, the company agrees to pay money, as specified in a written **policy,** to a **beneficiary** in case the insured person dies. A **straight life** insurance policy provides protection for as long as premiums are paid. [72]

loan Money lent at interest. The interest depends upon the amount borrowed, the interest rate (Annual Percentage Rate), and the time it takes to repay the loan. [151]

457

maintenance fee [238; 239] (See condominium)

map scale A rate (ratio) between two sets of measurements. It tells how many miles are represented by a unit (usually an inch) on a map. [190]

markdown [104] (See discount)

market research Gathering and analyzing data about consumer preferences and buying power. Such information is important in planning the development and marketing of new products. [282]

market value An approximation of the amount a property could be sold for. [242]

markup The difference between the retail and the wholesale prices of an item. The amount of the markup is found by multiplying the wholesale price by an established percent, which is the **markup rate.** [104]

mean (average) A single number used to represent a set of numbers; found by dividing the sum of the numbers by the number of numbers.

Example:

$$2 + 3 + 4 + 5 + 6 = 20 \text{ and } 20 \div 5 = 4.$$

4 is the mean, since $4 + 4 + 4 + 4 + 4 = 20$. [16]

median The middle value when a set of numbers are listed in order.

Example: 6 is the median of 2, 4, 6, 8, 9. 6 is also the median of 2, 3, 4, 8, 9, 10, since $\frac{4+8}{2} = 6$. [16]

membership fee Money paid to allow access to the facilities of an organization such as a health or swim club. [92]

mental computation Ways devised to think about operations on numbers so that external help (paper and pencil, calculator) is not needed to arrive at exact answers. [24]

minimum wage The least amount of money a person can be paid per hour; set by federal law so that it applies in all states. [42]

mixed number A number that indicates the sum of a whole number and a fraction.

Example: $3\frac{3}{4} = 3 + \frac{3}{4}$

[10]

mode The number that occurs most often in a set of numbers.

Example:

7 is the mode of 1, 1, 4, 7, 7, 7, 8, 9.

[16]

Money Market checking account A special checking account that earns interest. To earn interest and to avoid a service charge, most such accounts require that a minimum balance be kept in the account. Interest is usually compounded daily and paid monthly on the basis of the **average daily balance.** Most Money Market accounts limit the number of checks that can be written each month. [134; 135]

mortgage Money borrowed to buy a house or a condominium. The amount is approximately the purchase price less the **down payment**. The mortgage is repaid in periodic payments that include interest, and sometimes taxes. [240]

move-in costs Costs, in addition to rent, that must be paid when a renter moves into a residence. These may include a **fee** to the rental agent and a **security deposit** to the landlord. Occasionally, utility deposits are also required. [234]

mutual fund Money from a group of investors, used to buy and sell a mixture of stocks or other investments. **Dividends** are paid to members of the fund when there are profits. The price of a **share** in the fund goes up or down depending upon the changing value of the stocks held.

A "no load" fund does not charge a seller's fee; that is, the **buy** and **sell** prices of a share are the same. [330]

negotiated price [168] (See automobile purchase price)

net pay [68] (See payroll deductions)

net worth The difference between what is owned (***assets***) and what is owed (***liabilities***). [310]

NOW checking account A special checking account that earns interest. To earn interest and to avoid a service charge, most such accounts require that a minimum balance be kept in the account. Interest is usually compounded daily and paid monthly on the basis of the **average daily balance.** [134; 135]

odometer A gauge that automatically measures distance passed over by a wheeled vehicle. [172]

options [168] (See sticker price)

order of operations The order in which an expression is simplified when there is more than 1 operation. [23; 364]

outcome [346] (See event; probability; sample space, tree diagram)

overdraft A check written for more than the balance in the account. The bank may loan money to cover overdrafts up to an agreed amount. A **daily interest rate** times the **sum of the daily balances** is the usual way to compute the interest charge on overdrafts. The daily interest rate is based on an Annual Percentage Rate (APR). [149]

overestimate The result when estimating is intentionally done to make the approximation higher than the exact answer can be, as in estimating whether an amount of money is adequate for a given purpose. [32]

overtime [58; 295] (See wages)

package deal In travel, an overall charge. It includes many individual expenses (such as transportation, hotel, and meals). [203]

payroll deductions Money subtracted from a worker's full (**gross**) pay. The deductions cover taxes, insurance, and sometimes special amounts such as contributions or investments. The amount the worker receives, after deductions, is called the **net** (or take-home) pay. [68; 214]

pension Money received from a worker's pension plan after the worker retires. The amount received (annual pension benefit) depends upon the number of years the worker has belonged to the pension plan and upon her/his average annual earnings.

People can save for their retirement on their own by investing in **Individual Retirement Accounts (IRAs)**. [332]

percent A ratio that compares a number to 100. 10% means 10 hundredths, or 10 per 100.

Example: $75\% = 0.75 = \frac{75}{100}$, or $\frac{3}{4}$ [10]

perimeter The distance around the outer boundary of a 2-dimensional figure such as a rectangle or a square. [249]

permutation Any arrangement of a number of objects in a definite order. [356]

piecework A method of paying a worker on the basis of the number of items produced or jobs completed. [64]

premium In insurance, money paid by the policyholder for coverage under a contract (policy). [70; 72; 175; 244]

preventive maintenance For automobiles, periodic servicing to prevent the need for more serious and expensive repairs. [173]

probability A measure of chance. The probability, or chance, of an event occurring is a ratio: [346; 354]

$$P(\text{Event}) = \frac{\text{no. of favorable outcomes}}{\text{no. of possible outcomes}}$$

product The number found by multiplying. [6]

promotion On a job, advancement in rank or position. [295]

proportion A statement that 2 rates, or ratios, are equal. [250; 270; 286; 288]

purchasing power [302] (See Consumer Price Index)

quality control A system of inspection for making sure that an agreed-upon standard of quality is maintained in a production process. [288]

quotient The number obtained by dividing.

Example: In $6\overset{\text{4 R3}}{\overline{)27}}$, 4 is the quotient. The **remainder** is 3.

[8]

rate of commission [66] (See commission)

real estate taxes Money collected by local governments and used to pay for municipal services and schools. The real estate tax is based on the **assessed valuation** of a property, which is a percent of the property's **market value.** [242]

reconciled balance A checking account balance that reflects any changes that are needed to make the bank's monthly statement and the customer's check register agree. [127]

remainder [8] (See quotient)

remodeling Rebuilding all or part of a structure. [249]

renewable policy [74] (See term insurance)

retail Related to the sale of goods directly to the consumer, as in retail store, retail price, or "to sell at retail." [104]

rounding Replacing a number with an approximation to a nearest given unit, such as to the nearest hundred or tenth. Also, a mixed number can be rounded to the nearest whole number.

Examples: $1{,}721 \approx 1{,}700$
$1.645 \approx 1.65$
$4\frac{3}{4} \approx 5$

[26]

salary A fixed amount of money paid periodically for work or services. Most often salaries are paid monthly, semimonthly (twice a month), biweekly (every other week), or weekly. Salaries are often quoted on a "per year" basis, as in "$20,000 per year." [62]

sales tax A percent of the total price of goods and services. The money collected from sales taxes is a major source of revenue in many states and municipalities. Sales tax rates vary from state to state and from city to city. [108]

sample space A list of all possible outcomes of an experiment. [348]

sampling In manufacturing, a system of selecting a sample for testing or analysis, for the purpose of checking quality. [288]

savings account An agreement with a bank whereby a customer deposits money and in return, the bank pays **interest** out of what it earns by investing the money.

Careful records are kept of all changes in a savings account. The customer completes **deposit** and **withdrawal** slips. The bank records all changes, including the addition of interest, in a **passbook** kept by the customer. This makes a current balance available at all times. [130]

savings bond [320] (See U.S. savings bond)

scale drawing A diagram, with dimensions that are in a fixed ratio to the object it represents.

A scale drawing may be a reduction (as the blueprints of a house) or an enlargement (as a diagram of a small insect). [250]

schedules (bus, train, flight, subway) Charts, or tables, that give information about service including routes, arrival and departure times, and frequency of service. [192; 193; 195; 198]

season ticket A ticket for a series of events such as sporting events, concerts, or plays—usually at a reduced rate. [86]

secured/unsecured loans A secured loan is money lent with the understanding that if it is not repaid, some property (house, automobile, furniture) can be taken by the person who made the loan. The property is usually whatever has been bought with the borrowed money.

An unsecured loan has the guarantee of the borrower's word that the money will be repaid, but there is no property held as security. [152]

security deposit Money (usually one or two month's rent) collected and held during the period a renter occupies a residence. All or part of the deposit can be retained by the landlord if the dwelling is damaged by the renter. [234]

service contract A written agreement by which the purchaser of an appliance or a machine agrees to pay a fee regularly (usually monthly) whether or not the equipment needs any repair. In return, the serviceperson promises to make as many service calls as are needed to keep the equipment in good repair. [278]

simple interest A payment for the use of money. The amount of interest depends on the interest **rate** (expressed as an annual percent), on the **principal** (the amount of money in the account), and on the length of **time** (in years) that the money is used.

INTEREST = PRINCIPAL × RATE × TIME,
or
$I = p \times r \times t$

[132]

Social Security (FICA) The Federal Insurance Contributions Act; an insurance and pension plan contributed to, in equal shares, by employees and employers. The plan covers the cost of medical, retirement, and disability benefits paid to employees who have contributed, or to their families. [46]

sticker price The suggested price shown on a paper (**sticker**) that is taped to the window of a car that is for sale. Other information about the car is shown such as a list of items (**options**) available, but not included in the regular price, and the **EPA Fuel Economy Rating.** [168]

stock exchange An organized market for buying and selling stocks and bonds listed by that exchange. [319; 324; 330; 336]

straight commission [66] (See commission)

sum The number found by adding. [2]

sum of daily balances [149] (See overdraft)

take-home pay [54] (See payroll deductions)

term insurance Life insurance, provided for a specified number of years only. The face value is paid to the beneficiary in case the insured person dies. However, no equity, or investment, accumulates. [72]

tip The amount of money a customer leaves as a gift for a person who has provided service. [44]

title [170] (See automobile loan)

trade-in allowance [168; 170] (See automobile purchase price)

trade industry A line of skilled manual or mechanical work such as the construction, printing, or auto repair trades. [261]

tree diagram A diagram used to list all possible outcomes (the **sample space**) of a compound experiment such as rolling a die *and* tossing a coin. [348]

twofer coupon A coupon that may be exchanged at a theater box office for tickets at a discounted price. "Twofer" is short for "two for the price of one." [85]

underestimate The result when estimating is intentionally done to make the approximation less than the exact answer, as in estimating whether a certain amount has been reached. [32]

United States savings bonds (Series EE) Bonds issued by the U.S. Federal government. They are often bought through payroll deductions and at a purchase price that is half the value printed on the face of the bond (the **face value**). Such bonds can be cashed in (redeemed) at any time after 6 mo. The variable interest rate increases the longer the bond is held. [320]

unit price The price of an item given in terms of an appropriate unit of measure such as per pound, ounce, quart, or yard. [112]

unpaid balance [144] (See credit card)

utilities Goods and services such as electricity, natural gas, water, oil, and telephone. [234; 246]

value checking A checking account that requires no minimum balance. A very limited number of checks are allowed per month without charge. [135]

variable A symbol, usually a letter, used to stand for a number. [366]

variable expense [304] (See budget)

wages (hourly) Money paid for work when the amount varies according to the number of hours worked. **Overtime** is extra time worked (often any hours over 40 per wk). The hourly rate for overtime is usually $1\frac{1}{2}$ times the regular rate. [58; 294]

whole number A number in the set 0, 1, 2, 3, and so on in the same pattern. There is no greatest whole number. [2]

wholesale Related to the sale of goods in large quantities to merchants (retailers) who, in turn, sell to consumers. [104]

withholding tax [68] (See Federal withholding tax)

SELECTED ANSWERS

CHAPTER 1

Page 3
1. 757   7. 29.8   13. 1,202   19. 15.0   37. $44.45

Page 5
1. 432   7. 1,878   10. 3.6   12. 38.92   37. 385 mi

Page 7
1. 48   13. 9.6   19. 8.64   49. 2,653 mi

Page 9
1. 42   4. 47   9. 1,465   31. 2.5   36. 1.39   41. $0.49

Page 11
1. 23%   7. 63%   13. 0.25; 25%   19. 3.5; 350%   25. 0.24   31. 0.028   37. $\frac{7}{10}$ 43. 0.25   44. $\frac{1}{20}$

Page 13
1. 4   2. 6   13. 6   14. 5   25. 21   26. 10.8   37. 36 mi   38. 288 pages

Page 15
4. 150 cars and trucks   8. 4-door sedan   12. pick-up truck

Page 17
1. 90   9. 13   13. 3   15. 75.3; 78; none

CHAPTER 2

Page 23
1. 1,263,602   2. 948,405   15. 49.1   19. 70.62   23. 324.004   27. $44.84

Page 25
1. 75   11. 42   21. 1,270   29. 32   37. $18.58   38. $491

Page 27
1. 80   6. $10   13. 500   18. $0.60   32. $3

Page 29
1. 140   5. $14   21. 30   28. 2   41. $6   45. $64

Page 31
1. 19   9. 5   17. 35   25. 8   33. 15 h   37. 32 in.

Page 33
1. b   5. c   9. 10 wk   13. yes

Page 35
1. mental computation   3. paper and pencil or calculator   9. $1.05; mental computation

CHAPTER 3

Page 41
1. \$30.50 2. \$35.15 7. \$11.50

Page 43
1. \$4.65 2. \$48.60 7. Camp Crestmont; Hamburger Barn; Ray's Dept. Store 8. \$72

Page 45
1. \$2.10 3. \$10.50 4. \$27

Page 47
1. \$15.02 2. \$52.57 11. \$26.25 12. \$135

Page 49
3. 95 h, 75 h 4. 11 quarters, 5 dimes

Page 50
1. T, W, 4–8 3. 13 h 11. Big Time Burger 12. Morse's Pharmacy

CHAPTER 4

Page 59
1. \$194 2. \$184.75 7. \$246.24 8. \$0 22. \$5.75

Page 61
1. 7:00 2. 3:15

Page 63
1. \$168.27 5. \$12,872.60 9. Job B; \$2,400 15. \$9,200 per y job; \$320

Page 65
1. \$12.25 2. \$161.20 13. \$218.75 14. \$151.20

Page 67
1. \$133.75 2. \$162.50 9. \$277.50 10. \$1,480.96

Page 69
1. \$138.60 2. \$109.98 11. \$402.38 13. \$75.40 18. \$3,920.80

Page 71
1. \$300 2. \$175 11. \$19.35 12. \$232.20 19. \$29.45

Page 73
1. \$114.45 2. \$789

Page 75
1. term 3. \$3.40 11. \$130; \$170; \$475 12. SBLI

CHAPTER 5

Page 85
1. \$22.75; \$2.25 2. \$22.00; \$8.00 7. \$56 9. \$82.50; \$17.50

Page 87
1. \$17.25; pay separately

Page 89
1. \$21.44 4. 9 packets; \$1.35

Page 91
1. \$35 2. \$103

Page 93
1. \$2.38 2. \$4.18 11. \$0.46

Page 95
3. 23 cartons 7. 3 classes

Page 97
1. indoor 4. \$19.13 16. tennis 20. tennis inside, racquetball, skiing

CHAPTER 6

Page 105
1. \$18.19 6. 18% 11. \$135.00

Page 107
1. \$28.00 2. \$26.98 18. \$34 19. \$25.98

Page 109
1. \$0.17 2. \$1.37 7. \$0.94 8. \$4.95 13. \$0.63 14. \$0.64 19. \$0.88 20. \$18.38

Page 111
4. \$58.75 7. \$58.45

Page 113
1. does make sense 2. does not make sense 7. 8.2¢ per oz; 6.9¢ per oz; 10 oz for 69¢
8. 9.9¢ per oz; 10.6¢ per oz; 9 oz for 89¢

Page 115
1. less than \$5.00 2. more than \$5.00 7. \$6.42; \$13.58 8. \$5.87; \$14.13 13. \$1.72; \$3.44

Page 117
1. \$3.60 2. \$8.00 17. home 20. \$6.31

CHAPTER 7

Page 126
6. total deposit: \$160.86 11. \$386.16

Page 128
1. \$414.09 2. \$398.99

Page 131
1. total deposit: \$128.29 9. \$188.57 10. \$41.95 15. \$128.85 16. \$83.85

Page 134
1. \$60; \$310 2. \$75; \$575 9. \$50.63; \$1,050.63 13. \$5,375.60

Page 136
1. \$1,000 2. None 13. Money Market 17. \$8.00; \$96.00

Page 137
23. \$5 24. \$1,500 29. Regular savings 30. Money Market savings

CHAPTER 8

Page 146
1. 123 789 456 2 2. 2/22/89 7. Feb. 28, 1989 9. \$158.88

Page 148
1. \$218.35; \$3.28; \$257.12 2. \$80.45; \$0.80; \$181.10 10. \$7.95 15. \$234.00; \$3.51; \$582.51

Page 150
1. \$1.53 6. \$1.90; \$3,846.90 8. \$12.35; \$807.35

Page 152
1. \$403.86; \$18.86 2. \$864.72; \$55.72

Page 153
7. \$19.87; \$86.65 8. \$56.02; \$89.56 17. \$137.28; \$2,236.80 18. \$115.29; \$5,634.80

Page 155
1. \$90; \$5 2. \$126; \$24 9. \$99; \$9.05

Page 157
3. \$2.55

Page 159
1. 19.8% 4. \$100.00 15. School's Tuition Plan 17. School's Tuition Plan

CHAPTER 9

Page 169
1. \$9,958.00 5. \$12,463 8. \$14,687.76

Page 171
1. \$11,189 2. \$13,751.04 33. \$2,292.92; \$15,168.92

Page 174
1. 199 2. 2,356 13. \$40 14. \$40 25. Van

Page 177
1. \$479.17 5. 3.40 6. 2.10 11. \$223.10 12. \$183.42

Page 179
1. \$56.82 5. \$36 6. 100 mi 29. about \$300

Page 181
1. \$14,509.86 3. \$276.50 15. Car C 21. \$67.34

CHAPTER 10

Page 189
1. 1,387 mi 5. 1,000 mi 9. 45 mi 13. 20 h 14. 18 h

Page 191
1. 4.0 mi 5. 15 9. 2.0 mi

Page 194
1. 2:35 P.M. 2. 9:45 P.M. 11. 10:25 A.M. 12. 1:23 P.M.

Page 196
1. Sierra Airlines Flight #272 2. Central Airlines Flight #244 7. 3 h and 15 min
15. \$725 19. \$130

Page 199
1. Red Line 5 stops 2. Red Line 3 stops to Metro Center, change to the Orange or Blue Line, and go 8 more stops 7. \$0.80 8. \$1.25 13. 14 min 14. 27 min

Page 201
1. \$8.10 2. \$4.40 7. \$1.20 12. \$8.05 + \$1.20 tip = \$9.25

Page 203
1. \$11.25 2. \$9.00 7. \$150 8. \$300

Page 205
1. $7\frac{1}{2}$ h 2. 42 h 10. plane 13. \$192

CHAPTER 11

Page 213
1. \$1,647.90 5. \$17,731 6. \$24,080 11. \$21,546.82 12. \$4,524.34

Page 215
4. \$20,184; \$3,393 8. owe \$296.65

Page 218
1. \$21,050.37 2. \$21,292.05 7. 11,253.82 8. 319.56

Page 221
1. \$913 2. \$75 7. \$236; standard

Page 223
1. \$137.44 2. \$355.34 10. \$628.41 11. \$69.72(R)

Page 225
1. \$2,600 6. \$212.50 7. \$45

Page 227
1. Married, joint return 3. \$46,820 17. A:1040EZ; B:1040A; C:1040EZ 18. A:1040EZ; B:1040A; C:1040EZ

CHAPTER 12

Page 235
4. \$934; \$261.52 5. \$1,245; \$348.60 10. \$371 11. \$446

Page 237
1. \$82,000; \$100,000 2. \$73,000; \$114,600 7. \$5,000 8. \$55,000

Page 239
1. \$35,600; \$2,180 4. \$2,000 9. \$8,760; \$990

Page 241
1. \$445.50 6. \$24,800 7. \$99,200 8. \$907.68 18. \$2,160

Page 243
1. \$76,000 6. \$2,391.96 11. \$721 16. \$87,500 17. \$52,250

Page 245
1. \$7,800 2. \$39,000 13. \$760 14. 1,520

Page 248
4. 2,873 5. \$141.64 12. \$82.89 16. 3.051 18. \$43.87

Page 250
1. 58 ft 5. 2 gal; \$27.90

Page 251
8. \$1,512 11. l = 10 in.; w = 8 in.

Page 253
1. 11% 3. Fixed 14. Intown Bank 18. The APR, and therefore the monthly payment, may decrease.

467

CHAPTER 13

Page 263
1. \$5,400 4. \$11,250 8. 12 9. \$4.50 18. \$36 19. \$43.20

Page 265
1. $1\frac{3}{8}$ in. 2. $2\frac{1}{2}$ in. 7. 102 ft 8. 6

Page 267
1. 3 2. \$89.85 11. \$234.00

Page 269
1. 108 square ft 2. 51 square ft 26. \$336.00 30. 500 square ft

Page 271
1. 120 2. \$72 9. 1,080 10. \$162

Page 273
3. 12 posts 4. 5 times

Page 275
1. \$107.88 3. Very good 9. Good 10. Deluxe 15. 12.25¢ 17. 400 bricks
23. Truckload 24. Loose

CHAPTER 14

Page 283
1. 3.5 2. \$59.05 9. 16 10. \$375.68 17. 3 d

Page 287
1. 4,557 francs 6. 46 pounds 11. \$20.57 12. \$58.78

Page 289
1. 37 2. 160 9. 58 10. 3

Page 291
1. 3 2. 2.5 11. 28 drops per min

Page 293
1. 22 min and 31 s 7. \$178.75 11. 15 spots

Page 294
1. \$5,590 2. \$220 21. Job 2

Page 295
26. \$11,611.60 27. \$12,900 36. Yes 37. No

CHAPTER 15

Page 303
1. $2.00 2. $16.00 15. $1,900 16. $1,800

Page 305
1. $214 2. $133 8. $75 9. $38 17. $38 18. $472

Page 307
1. $64 2. $108 9. $177 10. $14

Page 309
1. $6,000 2. $19,000 17. $315,060 18. $294,130

Page 311
1. $3,892 2. $38,433 7. $111,500 8. $15,650

Page 313
1. $911 2. $1,096 15. Plan A 19. $75

CHAPTER 16

Page 321
1. $50 2. $37.50 7. $302.90 8. $344.40 13. $946.20 14. $196.20

Page 323
1. 7.92% 6. $198 7. $79.20 12. $1,131

Page 326
1. $18 2. $$30\frac{3}{4}$ 8. $1,012.50 9. $660 14. (P) $217.50 18. $0.56 19. $5.60

Page 329
1. Chapman International 2. Finch Township 7. $4,200 11. 95% 12. 87.5% 17. $525 21. 9.0%

Page 331
1. $38.12 2. $1,906 13. $4,124 14. $14,340 19. 509 shares

Page 333
1. 2.0% 2. $9,600

Page 335
3. $27,000 4. $9,000

Page 336
1. $37.50 2. 0.5%

Page 337
28. Technics 29. Jetstorm

CHAPTER 17

Page 347

1. $\frac{1}{2}$ 2. $\frac{1}{3}$ 7. $\frac{1}{20}$ 8. $\frac{3}{20}$ 17. $\frac{1}{52}$ 18. $\frac{1}{26}$

Page 349

1. (H, H) (H, T) (T, H) (T, T) 2. (T, H, T) (H, T, T) (T, T, T) (H, H, H) (H, H, T) (H, T, H) (T, H, H) (T, T, H) 9. 15 outfits

Page 351

1. $\frac{1}{36}$ 2. 0 7. $\frac{1}{4}$ 8. $\frac{3}{20}$ 13. $\frac{1}{4}$ 14. $\frac{1}{4}$ 19. $\frac{3}{10}$ 20. $\frac{1}{5}$

Page 353

1. independent 6. $\frac{2}{9}$ 7. $\frac{1}{6}$ 12. $\frac{3}{14}$ 13. $\frac{1}{14}$ 18. $\frac{25}{102}$ 19. $\frac{13}{51}$

Page 355

1. 8 d 2. 6 d 7. 48 doubles 8. 22 home runs 13. 42 games

Page 357

1. Martin, Jane, Eli; Martin, Eli, Jane; Jane, Martin, Eli; Jane, Eli, Martin; Eli, Martin, Jane; Eli, Jane, Martin 2. 48, 49, 84, 89, 94, 98

CHAPTER 18

Page 365
1. 5 2. 21 10. 11 16. 42

Page 367
1. $x + 9$ 2. $15 \div x$ or $\frac{15}{x}$ 11. $10 + 6 = y$ 12. $6(3) = x$ 22. $p = \$40 \div 5$ or $p = \frac{\$40}{5}$

Page 369
1. 11 2. 2 9. 6 10. 2 17. 9 18. 2 27. 86° F

Page 371
1. 60 2. 4.4 25. $p - \$6 = \$38$; \$44 26. $t + 8 = 89$; 81 points

Page 373
1. 32 2. 48 25. $\frac{n}{5} = 9$; 45 26. $7n = 56$; 8

Page 375
4. 7 5. 12 22. $3n + 5 = 32$; 9 23. $\frac{2}{5}n - 6 = 49$; 137.5

Page 377
3. \$24.75

TABLES

UNITED STATES CUSTOMARY SYSTEM OF MEASUREMENT

Length	12 inches (in.) = 1 foot (ft)
	3 feet, 36 inches = 1 yard (yd)
	1,760 yards, 5,280 feet = 1 mile (mi)
	6,076 feet = 1 nautical mile
Area	144 square inches (square in.) = 1 square foot (square ft)
	9 square feet = 1 square yard (square yd)
	4,840 square yards = 1 acre (A)
Volume	1,728 cubic inches (cubic in.) = 1 cubic foot (cubic ft)
	27 cubic feet = 1 cubic yard (cubic yd)
Weight	16 ounces (oz) = 1 pound (lb)
	2,000 pounds = 1 ton (T)
Capacity	8 fluid ounces (fl. oz) = 1 cup (c)
	2 cups = 1 pint (pt)
	2 pints = 1 quart (qt)
	4 quarts = 1 gallon (gal)

SYMBOLS

=	is equal to	>	is greater than
≠	is not equal to	<	is less than
$a \stackrel{?}{=} b$	is a equal to b?	%	percent
≈	is approximately equal to	π	pi
		⊥	right angles

METRIC SYSTEM OF MEASUREMENT

Length	10 millimeters (mm) = 1 centimeter (cm)
	10 centimeters, 100 millimeters = 1 decimeter (dm)
	10 decimeters, 100 centimeters = 1 meter (m)
	1,000 meters = 1 kilometer (km)
Area	100 square millimeters (square mm) = 1 square centimeter (square cm)
	10,000 square centimeters = 1 square meter (square m)
	100 square meters = 1 are (a)
	10,000 square meters = 1 hectare (ha)
Volume	1,000 cubic millimeters (cubic mm) = 1 cubic centimeter (cubic cm)
	1,000 cubic centimeters = 1 cubic decimeter (cubic dm)
	1,000,000 cubic centimeters = 1 cubic meter (cubic m)
Mass	1,000 milligrams (mg) = 1 gram (g)
	1,000 grams = 1 kilogram (kg)
	1,000 kilograms = 1 metric ton (t)
Capacity	100 milliliters (mL) = 1 liter (L)
	1,000 liters = 1 kiloliter (kL)

GEOMETRIC FORMULAS

Perimeter		**Surface Area**	
rectangle	$P = 2l + 2w$	rectangular prism	$SA = 2lw + 2lh + 2wh$
square	$P = 4s$	cube	$SA = 6s^2$
Circumference		cylinder	$SA = 2\pi rh + 2\pi r^2$
circle	$C = \pi d$ or $C = 2\pi r$	sphere	$SA = 4\pi r^2$
Area		**Volume**	
rectangle	$A = lw$	rectangular prism	$V = lwh$
square	$A = s^2$	cube	$V = s^3$
parallelogram	$A = bh$	cylinder	$V = \pi r^2 h$
triangle	$A = \frac{1}{2}bh$	rectangular pyramid	$V = \frac{1}{3}lwh$
trapezoid	$A = \frac{1}{2}h(b_1 + b_2)$	cone	$V = \frac{1}{3}\pi r^2 h$
circle	$A = \pi r^2$	sphere	$V = \frac{4}{3}\pi r^3$

FRACTION, DECIMAL, AND PERCENT EQUIVALENCIES

Fraction	Decimal	Percent	Fraction	Decimal	Percent
$\frac{1}{10}$	0.1	10%	$\frac{1}{6}$	$0.16\overline{6}$	$16\frac{2}{3}$%
$\frac{1}{5}$	0.2	20%	$\frac{1}{3}$	$0.33\overline{3}$	$33\frac{1}{3}$%
$\frac{1}{4}$	0.25	25%	$\frac{2}{3}$	$0.66\overline{6}$	$66\frac{2}{3}$%
$\frac{2}{5}$	0.4	40%	$\frac{5}{6}$	$0.83\overline{3}$	$83\frac{1}{3}$%
$\frac{1}{2}$	0.5	50%	$\frac{1}{8}$	0.125	$12\frac{1}{2}$%
$\frac{3}{5}$	0.6	60%	$\frac{3}{8}$	0.375	$37\frac{1}{2}$%
$\frac{3}{4}$	0.75	75%	$\frac{5}{8}$	0.625	$62\frac{1}{2}$%
$\frac{4}{5}$	0.8	80%	$\frac{7}{8}$	0.875	$87\frac{1}{2}$%

COMPOUND INTEREST

No. of Periods	1.5%	2%	2.5%	3%	3.5%	4%	5%	6%	7%	8%
1	1.0150	1.0200	1.0250	1.0300	1.0350	1.0400	1.0500	1.0600	1.0700	1.0800
2	1.0302	1.0404	1.0506	1.0609	1.0712	1.0816	1.1025	1.1236	1.1449	1.1664
3	1.0457	1.0612	1.0769	1.0927	1.1087	1.1248	1.1576	1.1910	1.2250	1.2597
4	1.0614	1.0824	1.1038	1.1255	1.1475	1.1699	1.2155	1.2625	1.3108	1.3605
5	1.0773	1.1041	1.1314	1.1593	1.1877	1.2167	1.2763	1.3382	1.4026	1.4693
6	1.0934	1.1262	1.1597	1.1941	1.2293	1.2653	1.3401	1.4186	1.5007	1.5869
7	1.1098	1.1487	1.1887	1.2299	1.2723	1.3159	1.4071	1.5036	1.6058	1.7138
8	1.1265	1.1717	1.2184	1.2668	1.3168	1.3686	1.4775	1.5938	1.7182	1.8059
9	1.1434	1.1951	1.2489	1.3048	1.3629	1.4233	1.5513	1.6895	1.8385	1.9990
10	1.1605	1.2190	1.2801	1.3439	1.4106	1.4802	1.6289	1.7908	1.9672	2.1589
11	1.1779	1.2434	1.3121	1.3842	1.4600	1.5395	1.7103	1.8983	2.1049	2.3316
12	1.1956	1.2682	1.3449	1.4258	1.5111	1.6010	1.7959	2.0122	2.2522	2.5182
13	1.2136	1.2936	1.3785	1.4685	1.5640	1.6651	1.8856	2.1329	2.4098	2.7196
14	1.2318	1.3195	1.4130	1.5126	1.6187	1.7317	1.9799	2.2609	2.5785	2.9372
15	1.2502	1.3459	1.4483	1.5580	1.6753	1.8009	2.0789	2.3966	2.7590	3.1722
16	1.2690	1.3728	1.4845	1.6047	1.7340	1.8730	2.1829	2.5404	2.9522	3.4259
17	1.2880	1.4002	1.5216	1.6528	1.7947	1.9479	2.2920	2.6928	3.1588	3.7000
18	1.3073	1.4282	1.5597	1.7024	1.8575	2.0258	2.4066	2.8543	3.3799	3.9960
19	1.3270	1.4568	1.5987	1.7535	1.9225	2.1068	2.5270	3.0256	3.6165	4.3157
20	1.3469	1.4859	1.6386	1.8061	1.9898	2.1911	2.6533	3.2071	3.8697	4.6610
21	1.3671	1.5157	1.6796	1.8603	2.0594	2.2788	2.7860	3.3996	4.1406	5.0338
22	1.3876	1.5460	1.7216	1.9161	2.1315	2.3699	2.9253	3.6035	4.4304	5.4365
23	1.4084	1.5769	1.7646	1.9736	2.2061	2.4647	3.0715	3.8198	4.7405	5.8715
24	1.4295	1.6084	1.8087	2.0328	2.2833	2.5633	3.2251	4.0489	5.0724	6.3412
25	1.4509	1.6407	1.8539	2.0938	2.3673	2.6658	3.3864	4.2919	5.4274	6.8485

FEDERAL INCOME TAX

1987 Tax Table

Based on Taxable Income

For persons with taxable incomes of less than $50,000.

Example: Mr. and Mrs. Brown are filing a joint return. Their taxable income on line 36 of Form 1040 is $25,325. First, they find the $25,300-25,350 income line. Next, they find the column for married filing jointly and read down the column. The amount shown where the income line and filing status column meet is $3,679. This is the tax amount they must write on line 37 of their return.

At least	But less than	Single	Married filing jointly *	Married filing separately	Head of a household
		Your tax is—			
25,200	25,250	4,807	3,664	5,374	3,973
25,250	25,300	4,821	3,671	5,391	3,987
→ 25,300	25,350	4,835	3,679	5,409	4,001
25,350	25,400	4,849	3,686	5,426	4,015

If line 36 (taxable income) is—		And you are—			
At least	But less than	Single	Married filing jointly *	Married filing separately	Head of a household
		Your tax is—			
0	5	0	0	0	0
5	15	1	1	1	1
15	25	2	2	2	2
25	50	4	4	4	4
50	75	7	7	7	7
75	100	10	10	10	10
100	125	12	12	12	12
125	150	15	15	15	15
150	175	18	18	18	18
175	200	21	21	21	21
200	225	23	23	23	23
225	250	26	26	26	26
250	275	29	29	29	29
275	300	32	32	32	32
300	325	34	34	34	34
325	350	37	37	37	37
350	375	40	40	40	40
375	400	43	43	43	43
400	425	45	45	45	45
425	450	48	48	48	48
450	475	51	51	51	51
475	500	54	54	54	54
500	525	56	56	56	56
525	550	59	59	59	59
550	575	62	62	62	62
575	600	65	65	65	65
600	625	67	67	67	67
625	650	70	70	70	70
650	675	73	73	73	73
675	700	76	76	76	76
700	725	78	78	78	78
725	750	81	81	81	81
750	775	84	84	84	84
775	800	87	87	87	87
800	825	89	89	89	89
825	850	92	92	92	92
850	875	95	95	95	95
875	900	98	98	98	98
900	925	100	100	100	100
925	950	103	103	103	103
950	975	106	106	106	106
975	1,000	109	109	109	109
1,000					
1,000	1,025	111	111	111	111
1,025	1,050	114	114	114	114
1,050	1,075	117	117	117	117
1,075	1,100	120	120	120	120
1,100	1,125	122	122	122	122
1,125	1,150	125	125	125	125
1,150	1,175	128	128	128	128
1,175	1,200	131	131	131	131
1,200	1,225	133	133	133	133
1,225	1,250	136	136	136	136
1,250	1,275	139	139	139	139
1,275	1,300	142	142	142	142
1,300	1,325	144	144	144	144
1,325	1,350	147	147	147	147
1,350	1,375	150	150	150	150
1,375	1,400	153	153	153	153

If line 36 (taxable income) is—		And you are—			
At least	But less than	Single	Married filing jointly *	Married filing separately	Head of a household
		Your tax is—			
1,400	1,425	155	155	155	155
1,425	1,450	158	158	158	158
1,450	1,475	161	161	161	161
1,475	1,500	164	164	164	164
1,500	1,525	166	166	167	166
1,525	1,550	169	169	171	169
1,550	1,575	172	172	174	172
1,575	1,600	175	175	178	175
1,600	1,625	177	177	182	177
1,625	1,650	180	180	186	180
1,650	1,675	183	183	189	183
1,675	1,700	186	186	193	186
1,700	1,725	188	188	197	188
1,725	1,750	191	191	201	191
1,750	1,775	194	194	204	194
1,775	1,800	197	197	208	197
1,800	1,825	200	199	212	199
1,825	1,850	204	202	216	202
1,850	1,875	207	205	219	205
1,875	1,900	211	208	223	208
1,900	1,925	215	210	227	210
1,925	1,950	219	213	231	213
1,950	1,975	222	216	234	216
1,975	2,000	226	219	238	219
2,000					
2,000	2,025	230	221	242	221
2,025	2,050	234	224	246	224
2,050	2,075	237	227	249	227
2,075	2,100	241	230	253	230
2,100	2,125	245	232	257	232
2,125	2,150	249	235	261	235
2,150	2,175	252	238	264	238
2,175	2,200	256	241	268	241
2,200	2,225	260	243	272	243
2,225	2,250	264	246	276	246
2,250	2,275	267	249	279	249
2,275	2,300	271	252	283	252
2,300	2,325	275	254	287	254
2,325	2,350	279	257	291	257
2,350	2,375	282	260	294	260
2,375	2,400	286	263	298	263
2,400	2,425	290	265	302	265
2,425	2,450	294	268	306	268
2,450	2,475	297	271	309	271
2,475	2,500	301	274	313	274
2,500	2,525	305	276	317	277
2,525	2,550	309	279	321	281
2,550	2,575	312	282	324	284
2,575	2,600	316	285	328	288
2,600	2,625	320	287	332	292
2,625	2,650	324	290	336	296
2,650	2,675	327	293	339	299
2,675	2,700	331	296	343	303

If line 36 (taxable income) is—		And you are—			
At least	But less than	Single	Married filing jointly *	Married filing separately	Head of a household
		Your tax is—			
2,700	2,725	335	298	347	307
2,725	2,750	339	301	351	311
2,750	2,775	342	304	354	314
2,775	2,800	346	307	358	318
2,800	2,825	350	309	362	322
2,825	2,850	354	312	366	326
2,850	2,875	357	315	369	329
2,875	2,900	361	318	373	333
2,900	2,925	365	320	377	337
2,925	2,950	369	323	381	341
2,950	2,975	372	326	384	344
2,975	3,000	376	329	388	348
3,000					
3,000	3,050	382	334	394	354
3,050	3,100	389	341	401	361
3,100	3,150	397	349	409	369
3,150	3,200	404	356	416	376
3,200	3,250	412	364	424	384
3,250	3,300	419	371	431	391
3,300	3,350	427	379	439	399
3,350	3,400	434	386	446	406
3,400	3,450	442	394	454	414
3,450	3,500	449	401	461	421
3,500	3,550	457	409	469	429
3,550	3,600	464	416	476	436
3,600	3,650	472	424	484	444
3,650	3,700	479	431	491	451
3,700	3,750	487	439	499	459
3,750	3,800	494	446	506	466
3,800	3,850	502	454	514	474
3,850	3,900	509	461	521	481
3,900	3,950	517	469	529	489
3,950	4,000	524	476	536	496
4,000					
4,000	4,050	532	484	544	504
4,050	4,100	539	491	551	511
4,100	4,150	547	499	559	519
4,150	4,200	554	506	566	526
4,200	4,250	562	514	574	534
4,250	4,300	569	521	581	541
4,300	4,350	577	529	589	549
4,350	4,400	584	536	596	556
4,400	4,450	592	544	604	564
4,450	4,500	599	551	611	571
4,500	4,550	607	559	619	579
4,550	4,600	614	566	626	586
4,600	4,650	622	574	634	594
4,650	4,700	629	581	641	601
4,700	4,750	637	589	649	609
4,750	4,800	644	596	656	616
4,800	4,850	652	604	664	624
4,850	4,900	659	611	671	631
4,900	4,950	667	619	679	639
4,950	5,000	674	626	686	646

* This column must also be used by a qualifying widow(er).

Continued on next page

Page 37

1987 Tax Table — *Continued*

If line 36 (taxable income) is—		And you are—			
At least	But less than	Single	Married filing jointly *	Married filing separately	Head of a household
		Your tax is—			
5,000					
5,000	5,050	682	634	694	654
5,050	5,100	689	641	701	661
5,100	5,150	697	649	709	669
5,150	5,200	704	656	716	676
5,200	5,250	712	664	724	684
5,250	5,300	719	671	731	691
5,300	5,350	727	679	739	699
5,350	5,400	734	686	746	706
5,400	5,450	742	694	754	714
5,450	5,500	749	701	761	721
5,500	5,550	757	709	769	729
5,550	5,600	764	716	776	736
5,600	5,650	772	724	784	744
5,650	5,700	779	731	791	751
5,700	5,750	787	739	799	759
5,750	5,800	794	746	806	766
5,800	5,850	802	754	814	774
5,850	5,900	809	761	821	781
5,900	5,950	817	769	829	789
5,950	6,000	824	776	836	796
6,000					
6,000	6,050	832	784	844	804
6,050	6,100	839	791	851	811
6,100	6,150	847	799	859	819
6,150	6,200	854	806	866	826
6,200	6,250	862	814	874	834
6,250	6,300	869	821	881	841
6,300	6,350	877	829	889	849
6,350	6,400	884	836	896	856
6,400	6,450	892	844	904	864
6,450	6,500	899	851	911	871
6,500	6,550	907	859	919	879
6,550	6,600	914	866	926	886
6,600	6,650	922	874	934	894
6,650	6,700	929	881	941	901
6,700	6,750	937	889	949	909
6,750	6,800	944	896	956	916
6,800	6,850	952	904	964	924
6,850	6,900	959	911	971	931
6,900	6,950	967	919	979	939
6,950	7,000	974	926	986	946
7,000					
7,000	7,050	982	934	994	954
7,050	7,100	989	941	1,001	961
7,100	7,150	997	949	1,009	969
7,150	7,200	1,004	956	1,016	976
7,200	7,250	1,012	964	1,024	984
7,250	7,300	1,019	971	1,031	991
7,300	7,350	1,027	979	1,039	999
7,350	7,400	1,034	986	1,046	1,006
7,400	7,450	1,042	994	1,054	1,014
7,450	7,500	1,049	1,001	1,061	1,021
7,500	7,550	1,057	1,009	1,069	1,029
7,550	7,600	1,064	1,016	1,076	1,036
7,600	7,650	1,072	1,024	1,084	1,044
7,650	7,700	1,079	1,031	1,091	1,051
7,700	7,750	1,087	1,039	1,099	1,059
7,750	7,800	1,094	1,046	1,106	1,066
7,800	7,850	1,102	1,054	1,114	1,074
7,850	7,900	1,109	1,061	1,121	1,081
7,900	7,950	1,117	1,069	1,129	1,089
7,950	8,000	1,124	1,076	1,136	1,096
8,000					
8,000	8,050	1,132	1,084	1,144	1,104
8,050	8,100	1,139	1,091	1,151	1,111
8,100	8,150	1,147	1,099	1,159	1,119
8,150	8,200	1,154	1,106	1,166	1,126
8,200	8,250	1,162	1,114	1,174	1,134
8,250	8,300	1,169	1,121	1,181	1,141
8,300	8,350	1,177	1,129	1,189	1,149
8,350	8,400	1,184	1,136	1,196	1,156
8,400	8,450	1,192	1,144	1,204	1,164
8,450	8,500	1,199	1,151	1,211	1,171
8,500	8,550	1,207	1,159	1,219	1,179
8,550	8,600	1,214	1,166	1,226	1,186
8,600	8,650	1,222	1,174	1,234	1,194
8,650	8,700	1,229	1,181	1,241	1,201
8,700	8,750	1,237	1,189	1,249	1,209
8,750	8,800	1,244	1,196	1,256	1,216
8,800	8,850	1,252	1,204	1,264	1,224
8,850	8,900	1,259	1,211	1,271	1,231
8,900	8,950	1,267	1,219	1,279	1,239
8,950	9,000	1,274	1,226	1,286	1,246
9,000					
9,000	9,050	1,282	1,234	1,294	1,254
9,050	9,100	1,289	1,241	1,301	1,261
9,100	9,150	1,297	1,249	1,309	1,269
9,150	9,200	1,304	1,256	1,316	1,276
9,200	9,250	1,312	1,264	1,324	1,284
9,250	9,300	1,319	1,271	1,331	1,291
9,300	9,350	1,327	1,279	1,339	1,299
9,350	9,400	1,334	1,286	1,346	1,306
9,400	9,450	1,342	1,294	1,354	1,314
9,450	9,500	1,349	1,301	1,361	1,321
9,500	9,550	1,357	1,309	1,369	1,329
9,550	9,600	1,364	1,316	1,376	1,336
9,600	9,650	1,372	1,324	1,384	1,344
9,650	9,700	1,379	1,331	1,391	1,351
9,700	9,750	1,387	1,339	1,399	1,359
9,750	9,800	1,394	1,346	1,406	1,366
9,800	9,850	1,402	1,354	1,414	1,374
9,850	9,900	1,409	1,361	1,421	1,381
9,900	9,950	1,417	1,369	1,429	1,389
9,950	10,000	1,424	1,376	1,436	1,396
10,000					
10,000	10,050	1,432	1,384	1,444	1,404
10,050	10,100	1,439	1,391	1,451	1,411
10,100	10,150	1,447	1,399	1,459	1,419
10,150	10,200	1,454	1,406	1,466	1,426
10,200	10,250	1,462	1,414	1,474	1,434
10,250	10,300	1,469	1,421	1,481	1,441
10,300	10,350	1,477	1,429	1,489	1,449
10,350	10,400	1,484	1,436	1,496	1,456
10,400	10,450	1,492	1,444	1,504	1,464
10,450	10,500	1,499	1,451	1,511	1,471
10,500	10,550	1,507	1,459	1,519	1,479
10,550	10,600	1,514	1,466	1,526	1,486
10,600	10,650	1,522	1,474	1,534	1,494
10,650	10,700	1,529	1,481	1,541	1,501
10,700	10,750	1,537	1,489	1,549	1,509
10,750	10,800	1,544	1,496	1,556	1,516
10,800	10,850	1,552	1,504	1,564	1,524
10,850	10,900	1,559	1,511	1,571	1,531
10,900	10,950	1,567	1,519	1,579	1,539
10,950	11,000	1,574	1,526	1,586	1,546
11,000					
11,000	11,050	1,582	1,534	1,594	1,554
11,050	11,100	1,589	1,541	1,601	1,561
11,100	11,150	1,597	1,549	1,609	1,569
11,150	11,200	1,604	1,556	1,616	1,576
11,200	11,250	1,612	1,564	1,624	1,584
11,250	11,300	1,619	1,571	1,631	1,591
11,300	11,350	1,627	1,579	1,639	1,599
11,350	11,400	1,634	1,586	1,646	1,606
11,400	11,450	1,642	1,594	1,654	1,614
11,450	11,500	1,649	1,601	1,661	1,621
11,500	11,550	1,657	1,609	1,669	1,629
11,550	11,600	1,664	1,616	1,676	1,636
11,600	11,650	1,672	1,624	1,684	1,644
11,650	11,700	1,679	1,631	1,691	1,651
11,700	11,750	1,687	1,639	1,699	1,659
11,750	11,800	1,694	1,646	1,706	1,666
11,800	11,850	1,702	1,654	1,714	1,674
11,850	11,900	1,709	1,661	1,721	1,681
11,900	11,950	1,717	1,669	1,729	1,689
11,950	12,000	1,724	1,676	1,736	1,696
12,000					
12,000	12,050	1,732	1,684	1,744	1,704
12,050	12,100	1,739	1,691	1,751	1,711
12,100	12,150	1,747	1,699	1,759	1,719
12,150	12,200	1,754	1,706	1,766	1,726
12,200	12,250	1,762	1,714	1,774	1,734
12,250	12,300	1,769	1,721	1,781	1,741
12,300	12,350	1,777	1,729	1,789	1,749
12,350	12,400	1,784	1,736	1,796	1,756
12,400	12,450	1,792	1,744	1,804	1,764
12,450	12,500	1,799	1,751	1,811	1,771
12,500	12,550	1,807	1,759	1,819	1,779
12,550	12,600	1,814	1,766	1,826	1,786
12,600	12,650	1,822	1,774	1,834	1,794
12,650	12,700	1,829	1,781	1,841	1,801
12,700	12,750	1,837	1,789	1,849	1,809
12,750	12,800	1,844	1,796	1,856	1,816
12,800	12,850	1,852	1,804	1,864	1,824
12,850	12,900	1,859	1,811	1,871	1,831
12,900	12,950	1,867	1,819	1,879	1,839
12,950	13,000	1,874	1,826	1,886	1,846
13,000					
13,000	13,050	1,882	1,834	1,894	1,854
13,050	13,100	1,889	1,841	1,901	1,861
13,100	13,150	1,897	1,849	1,909	1,869
13,150	13,200	1,904	1,856	1,916	1,876
13,200	13,250	1,912	1,864	1,924	1,884
13,250	13,300	1,919	1,871	1,931	1,891
13,300	13,350	1,927	1,879	1,939	1,899
13,350	13,400	1,934	1,886	1,946	1,906
13,400	13,450	1,942	1,894	1,954	1,914
13,450	13,500	1,949	1,901	1,961	1,921
13,500	13,550	1,957	1,909	1,969	1,929
13,550	13,600	1,964	1,916	1,976	1,936
13,600	13,650	1,972	1,924	1,984	1,944
13,650	13,700	1,979	1,931	1,991	1,951
13,700	13,750	1,987	1,939	1,999	1,959
13,750	13,800	1,994	1,946	2,006	1,966
13,800	13,850	2,002	1,954	2,014	1,974
13,850	13,900	2,009	1,961	2,021	1,981
13,900	13,950	2,017	1,969	2,029	1,989
13,950	14,000	2,024	1,976	2,036	1,996

* This column must also be used by a qualifying widow(er).

Continued on next page

1987 Tax Table — *Continued*

If line 36 (taxable income) is—		And you are—			
At least	But less than	Single	Married filing jointly *	Married filing separately	Head of a household
		Your tax is—			
14,000					
14,000	14,050	2,032	1,984	2,047	2,004
14,050	14,100	2,039	1,991	2,061	2,011
14,100	14,150	2,047	1,999	2,075	2,019
14,150	14,200	2,054	2,006	2,089	2,026
14,200	14,250	2,062	2,014	2,103	2,034
14,250	14,300	2,069	2,021	2,117	2,041
14,300	14,350	2,077	2,029	2,131	2,049
14,350	14,400	2,084	2,036	2,145	2,056
14,400	14,450	2,092	2,044	2,159	2,064
14,450	14,500	2,099	2,051	2,173	2,071
14,500	14,550	2,107	2,059	2,187	2,079
14,550	14,600	2,114	2,066	2,201	2,086
14,600	14,650	2,122	2,074	2,215	2,094
14,650	14,700	2,129	2,081	2,229	2,101
14,700	14,750	2,137	2,089	2,243	2,109
14,750	14,800	2,144	2,096	2,257	2,116
14,800	14,850	2,152	2,104	2,271	2,124
14,850	14,900	2,159	2,111	2,285	2,131
14,900	14,950	2,167	2,119	2,299	2,139
14,950	15,000	2,174	2,126	2,313	2,146
15,000					
15,000	15,050	2,182	2,134	2,327	2,154
15,050	15,100	2,189	2,141	2,341	2,161
15,100	15,150	2,197	2,149	2,355	2,169
15,150	15,200	2,204	2,156	2,369	2,176
15,200	15,250	2,212	2,164	2,383	2,184
15,250	15,300	2,219	2,171	2,397	2,191
15,300	15,350	2,227	2,179	2,411	2,199
15,350	15,400	2,234	2,186	2,425	2,206
15,400	15,450	2,242	2,194	2,439	2,214
15,450	15,500	2,249	2,201	2,453	2,221
15,500	15,550	2,257	2,209	2,467	2,229
15,550	15,600	2,264	2,216	2,481	2,236
15,600	15,650	2,272	2,224	2,495	2,244
15,650	15,700	2,279	2,231	2,509	2,251
15,700	15,750	2,287	2,239	2,523	2,259
15,750	15,800	2,294	2,246	2,537	2,266
15,800	15,850	2,302	2,254	2,551	2,274
15,850	15,900	2,309	2,261	2,565	2,281
15,900	15,950	2,317	2,269	2,579	2,289
15,950	16,000	2,324	2,276	2,593	2,296
16,000					
16,000	16,050	2,332	2,284	2,607	2,304
16,050	16,100	2,339	2,291	2,621	2,311
16,100	16,150	2,347	2,299	2,635	2,319
16,150	16,200	2,354	2,306	2,649	2,326
16,200	16,250	2,362	2,314	2,663	2,334
16,250	16,300	2,369	2,321	2,677	2,341
16,300	16,350	2,377	2,329	2,691	2,349
16,350	16,400	2,384	2,336	2,705	2,356
16,400	16,450	2,392	2,344	2,719	2,364
16,450	16,500	2,399	2,351	2,733	2,371
16,500	16,550	2,407	2,359	2,747	2,379
16,550	16,600	2,414	2,366	2,761	2,386
16,600	16,650	2,422	2,374	2,775	2,394
16,650	16,700	2,429	2,381	2,789	2,401
16,700	16,750	2,437	2,389	2,803	2,409
16,750	16,800	2,444	2,396	2,817	2,416
16,800	16,850	2,455	2,404	2,831	2,424
16,850	16,900	2,469	2,411	2,845	2,431
16,900	16,950	2,483	2,419	2,859	2,439
16,950	17,000	2,497	2,426	2,873	2,446
17,000					
17,000	17,050	2,511	2,434	2,887	2,454
17,050	17,100	2,525	2,441	2,901	2,461
17,100	17,150	2,539	2,449	2,915	2,469
17,150	17,200	2,553	2,456	2,929	2,476
17,200	17,250	2,567	2,464	2,943	2,484
17,250	17,300	2,581	2,471	2,957	2,491
17,300	17,350	2,595	2,479	2,971	2,499
17,350	17,400	2,609	2,486	2,985	2,506
17,400	17,450	2,623	2,494	2,999	2,514
17,450	17,500	2,637	2,501	3,013	2,521
17,500	17,550	2,651	2,509	3,027	2,529
17,550	17,600	2,665	2,516	3,041	2,536
17,600	17,650	2,679	2,524	3,055	2,544
17,650	17,700	2,693	2,531	3,069	2,551
17,700	17,750	2,707	2,539	3,083	2,559
17,750	17,800	2,721	2,546	3,097	2,566
17,800	17,850	2,735	2,554	3,111	2,574
17,850	17,900	2,749	2,561	3,125	2,581
17,900	17,950	2,763	2,569	3,139	2,589
17,950	18,000	2,777	2,576	3,153	2,596
18,000					
18,000	18,050	2,791	2,584	3,167	2,604
18,050	18,100	2,805	2,591	3,181	2,611
18,100	18,150	2,819	2,599	3,195	2,619
18,150	18,200	2,833	2,606	3,209	2,626
18,200	18,250	2,847	2,614	3,223	2,634
18,250	18,300	2,861	2,621	3,237	2,641
18,300	18,350	2,875	2,629	3,251	2,649
18,350	18,400	2,889	2,636	3,265	2,656
18,400	18,450	2,903	2,644	3,279	2,664
18,450	18,500	2,917	2,651	3,293	2,671
18,500	18,550	2,931	2,659	3,307	2,679
18,550	18,600	2,945	2,666	3,321	2,686
18,600	18,650	2,959	2,674	3,335	2,694
18,650	18,700	2,973	2,681	3,349	2,701
18,700	18,750	2,987	2,689	3,363	2,709
18,750	18,800	3,001	2,696	3,377	2,716
18,800	18,850	3,015	2,704	3,391	2,724
18,850	18,900	3,029	2,711	3,405	2,731
18,900	18,950	3,043	2,719	3,419	2,739
18,950	19,000	3,057	2,726	3,433	2,746
19,000					
19,000	19,050	3,071	2,734	3,447	2,754
19,050	19,100	3,085	2,741	3,461	2,761
19,100	19,150	3,099	2,749	3,475	2,769
19,150	19,200	3,113	2,756	3.489	2,776
19,200	19,250	3,127	2,764	3,503	2,784
19,250	19,300	3,141	2,771	3,517	2,791
19,300	19,350	3,155	2,779	3,531	2,799
19,350	19,400	3,169	2,786	3,545	2,806
19,400	19,450	3,183	2,794	3,559	2,814
19,450	19,500	3,197	2,801	3,573	2,821
19,500	19,550	3,211	2,809	3,587	2,829
19,550	19,600	3,225	2,816	3,601	2,836
19,600	19,650	3,239	2,824	3,615	2,844
19,650	19,700	3,253	2,831	3,629	2,851
19,700	19,750	3,267	2,839	3,643	2,859
19,750	19,800	3,281	2,846	3,657	2,866
19,800	19,850	3,295	2,854	3,671	2,874
19,850	19,900	3,309	2,861	3,685	2,881
19,900	19,950	3,323	2,869	3,699	2,889
19,950	20,000	3,337	2,876	3,713	2,896
20,000					
20,000	20,050	3,351	2,884	3,727	2,904
20,050	20,100	3,365	2,891	3,741	2,911
20,100	20,150	3,379	2,899	3,755	2,919
20,150	20,200	3,393	2,906	3,769	2,926
20,200	20,250	3,407	2,914	3,783	2,934
20,250	20,300	3,421	2,921	3,797	2,941
20,300	20,350	3,435	2,929	3,811	2,949
20,350	20,400	3,449	2,936	3,825	2,956
20,400	20,450	3,463	2,944	3,839	2,964
20,450	20,500	3,477	2,951	3,853	2,971
20,500	20,550	3,491	2,959	3,867	2,979
20,550	20,600	3,505	2,966	3,881	2,986
20,600	20,650	3,519	2,974	3,895	2,994
20,650	20,700	3,533	2,981	3,909	3,001
20,700	20,750	3,547	2,989	3,923	3,009
20,750	20,800	3,561	2,996	3,937	3,016
20,800	20,850	3,575	3,004	3,951	3,024
20,850	20,900	3,589	3,011	3,965	3,031
20,900	20,950	3,603	3,019	3,979	3,039
20,950	21,000	3,617	3,026	3,993	3,046
21,000					
21,000	21,050	3,631	3,034	4,007	3,054
21,050	21,100	3,645	3,041	4,021	3,061
21,100	21,150	3,659	3,049	4,035	3,069
21,150	21,200	3,673	3,056	4,049	3,076
21,200	21,250	3,687	3,064	4,063	3,084
21,250	21,300	3,701	3,071	4,077	3,091
21,300	21,350	3,715	3,079	4,091	3,099
21,350	21,400	3,729	3,086	4,105	3,106
21,400	21,450	3,743	3,094	4,119	3,114
21,450	21,500	3,757	3,101	4,133	3,121
21,500	21,550	3,771	3,109	4,147	3,129
21,550	21,600	3,785	3,116	4,161	3,136
21,600	21,650	3,799	3,124	4,175	3,144
21,650	21,700	3,813	3,131	4,189	3,151
21,700	21,750	3,827	3,139	4,203	3,159
21,750	21,800	3,841	3,146	4,217	3,166
21,800	21,850	3,855	3,154	4,231	3,174
21,850	21,900	3,869	3,161	4,245	3,181
21,900	21,950	3,883	3,169	4,259	3,189
21,950	22,000	3,897	3,176	4,273	3,196
22,000					
22,000	22,050	3,911	3,184	4,287	3,204
22,050	22,100	3,925	3,191	4,301	3,211
22,100	22,150	3,939	3,199	4,315	3,219
22,150	22,200	3,953	3,206	4,329	3,226
22,200	22,250	3,967	3,214	4,343	3,234
22,250	22,300	3,981	3,221	4,357	3,241
22,300	22,350	3,995	3,229	4,371	3,249
22,350	22,400	4,009	3,236	4,385	3,256
22,400	22,450	4,023	3,244	4,399	3,264
22,450	22,500	4,037	3,251	4,413	3,271
22,500	22,550	4,051	3,259	4,429	3,279
22,550	22,600	4,065	3,266	4,446	3,286
22,600	22,650	4,079	3,274	4,464	3,294
22,650	22,700	4,093	3,281	4,481	3,301
22,700	22,750	4,107	3,289	4,499	3,309
22,750	22,800	4,121	3,296	4,516	3,316
22,800	22,850	4,135	3,304	4,534	3,324
22,850	22,900	4,149	3,311	4,551	3,331
22,900	22,950	4,163	3,319	4,569	3,339
22,950	23,000	4,177	3,326	4,586	3,346

* This column must also be used by a qualifying widow(er).

Continued on next page

1987 Tax Table — *Continued*

If line 36 (taxable income) is— At least	But less than	And you are— Single	Married filing jointly *	Married filing separately	Head of a household
		Your tax is—			
23,000					
23,000	23,050	4,191	3,334	4,604	3,357
23,050	23,100	4,205	3,341	4,621	3,371
23,100	23,150	4,219	3,349	4,639	3,385
23,150	23,200	4,233	3,356	4,656	3,399
23,200	20,250	4,247	3,364	4,674	3,413
23,250	23,300	4,261	3,371	4,691	3,427
20,300	23,350	4,275	3,379	4,709	3,441
23,350	23,400	4,289	3,386	4,726	3,455
23,400	23,450	4,303	3,394	4,744	3,469
23,450	23,500	4,317	3,401	4,761	3,483
23,500	23,550	4,331	3,409	4,779	3,497
23,550	23,600	4,345	3,416	4,796	3,511
23,600	23,650	4,359	3,424	4,814	3,525
23,650	23,700	4,373	3,431	4,831	3,539
23,700	23,750	4,387	3,439	4,849	3,553
23,750	23,800	4,401	3,446	4,866	3,567
23,800	23,850	4,415	3,454	4,884	3,581
23,850	23,900	4,429	3,461	4,901	3,595
23,900	23,950	4,443	3,469	4,919	3,609
23,950	24,000	4,457	3,476	4,936	3,623
24,000					
24,000	24,050	4,471	3,484	4,954	3,637
24,050	24,100	4,485	3,491	4,971	3,651
24,100	24,150	4,499	3,499	4,989	3,665
24,150	24,200	4,513	3,506	5,006	3,679
24,200	24,250	4,527	3,514	5,024	3,693
24,250	24,300	4,541	3,521	5,041	3,707
24,300	24,350	4,555	3,529	5,059	3,721
24,350	24,400	4,569	3,536	5,076	3,735
24,400	24,450	4,583	3,544	5,094	3,749
24,450	24,500	4,597	3,551	5,111	3,763
24,500	24,550	4,611	3,559	5,129	3,777
24,550	24,600	4,625	3,566	5,146	3,791
24,600	24,650	4,639	3,574	5,164	3,805
24,650	24,700	4,653	3,581	5,181	3,819
24,700	24,750	4,667	3,589	5,199	3,833
24,750	24,800	4,681	3,596	5,216	3,847
24,800	24,850	4,695	3,604	5,234	3,861
24,850	24,900	4,709	3,611	5,251	3,875
24,900	24,950	4,723	3,619	5,269	3,889
24,950	25,000	4,737	3,626	5,286	3,903
25,000					
25,000	25,050	4,751	3,634	5,304	3,917
25,050	25,100	4,765	3,641	5,321	3,931
25,100	25,150	4,779	3,649	5,339	3,945
25,150	25,200	4,793	3,656	5,356	3,959
25,200	25,250	4,807	3,664	5,374	3,973
25,250	25,300	4,821	3,671	5,391	3,987
25,300	25,350	4,835	3,679	5,409	4,001
25,350	25,400	4,849	3,686	5,426	4,015
25,400	25,450	4,863	3,694	5,444	4,029
25,450	25,500	4,877	3,701	5,461	4,043
25,500	25,550	4,891	3,709	5,479	4,057
25,550	25,600	4,905	3,716	5,496	4,071
25,600	25,650	4,919	3,724	5,514	4,085
25,650	25,700	4,933	3,731	5,531	4,099
25,700	25,750	4,947	3,739	5,549	4,113
25,750	25,800	4,961	3,746	5,566	4,127
25,800	25,850	4,975	3,754	5,584	4,141
25,850	25,900	4,989	3,761	5,601	4,155
25,900	25,950	5,003	3,769	5,619	4,169
25,950	26,000	5,017	3,776	5,636	4,183
26,000					
26,000	26,050	5,031	3,784	5,654	4,197
26,050	26,100	5,045	3,791	5,671	4,211
26,100	26,150	5,059	3,799	5,689	4,225
26,150	26,200	5,073	3,806	5,706	4,239
26,200	26,250	5,087	3,814	5,724	4,253
26,250	26,300	5,101	3,821	5,741	4,267
26,300	26,350	5,115	3,829	5,759	4,281
26,350	26,400	5,129	3,836	5,776	4,295
26,400	26,450	5,143	3,844	5,794	4,309
26,450	26,500	5,157	3,851	5,811	4,323
26,500	26,550	5,171	3,859	5,829	4,337
26,550	26,600	5,185	3,866	5,846	4,351
26,600	26,650	5,199	3,874	5,864	4,365
26,650	26,700	5,213	3,881	5,881	4,379
26,700	26,750	5,227	3,889	5,899	4,393
26,750	26,800	5,241	3,896	5,916	4,407
26,800	26,850	5,255	3,904	5,934	4,421
26,850	26,900	5,269	3,911	5,951	4,435
26,900	26,950	5,283	3,919	5,969	4,449
26,950	27,000	5,297	3,926	5,986	4,463
27,000					
27,000	27,050	5,313	3,934	6,004	4,477
27,050	27,100	5,330	3,941	6,021	4,491
27,100	27,150	5,348	3,949	6,039	4,505
27,150	27,200	5,365	3,956	6,056	4,519
27,200	27,250	5,383	3,964	6,074	4,533
27,250	27,300	5,400	3,971	6,091	4,547
27,300	27,350	5,418	3,979	6,109	4,561
27,350	27,400	5,435	3,986	6,126	4,575
27,400	27,450	5,453	3,994	6,144	4,589
27,450	27,500	5,470	4,001	6,161	4,603
27,500	27,550	5,488	4,009	6,179	4,617
27,550	27,600	5,505	4,016	6,196	4,631
27,600	27,650	5,523	4,024	6,214	4,645
27,650	27,700	5,540	4,031	6,231	4,659
27,700	27,750	5,558	4,039	6,249	4,673
27,750	27,800	5,575	4,046	6,266	4,687
27,800	27,850	5,593	4,054	6,284	4,701
27,850	27,900	5,610	4,061	6,301	4,715
27,900	27,950	5,628	4,069	6,319	4,729
27,950	28,000	5,645	4,076	6,336	4,743
28,000					
28,000	28,050	5,663	4,087	6,354	4,757
28,050	28,100	5,680	4,101	6,371	4,771
28,100	28,150	5,698	4,115	6,389	4,785
28,150	28,200	5,715	4,129	6,406	4,799
28,200	28,250	5,733	4,143	6,424	4,813
28,250	28,300	5,750	4,157	6,441	4,827
28,300	28,350	5,768	4,171	6,459	4,841
28,350	28,400	5,785	4,185	6,476	4,855
28,400	28,450	5,803	4,199	6,494	4,869
28,450	28,500	5,820	4,213	6,511	4,883
28,500	28,550	5,838	4,227	6,529	4,897
28,550	28,600	5,855	4,241	6,546	4,911
28,600	28,650	5,873	4,255	6,564	4,925
28,650	28,700	5,890	4,269	6,581	4,939
28,700	28,750	5,908	4,283	6,599	4,953
28,750	28,800	5,925	4,297	6,616	4,967
28,800	28,850	5,943	4,311	6,634	4,981
28,850	28,900	5,960	4,325	6,651	4,995
28,900	28,950	5,978	4,339	6,669	5,009
28,950	29,000	5,995	4,353	6,686	5,023
29,000					
29,000	29,050	6,013	4,367	6,704	5,037
29,050	29,100	6,030	4,381	6,721	5,051
29,100	29,150	6,048	4,395	6,739	5,065
29,150	29,200	6,065	4,409	6,756	5,079
29,200	29,250	6,083	4,423	6,774	5,093
29,250	29,300	6,100	4,437	6,791	5,107
29,300	29,350	6,118	4,451	6,809	5,121
29,350	29,400	6,135	4,465	6,826	5,135
29,400	29,450	6,153	4,479	6,844	5,149
29,450	29,500	6,170	4,493	6,861	5,163
29,500	29,550	6,188	4,507	6,879	5,177
29,550	29,600	6,205	4,521	6,896	5,191
29,600	29,650	6,223	4,535	6,914	5,205
29,650	29,700	6,240	4,549	6,931	5,219
29,700	29,750	6,258	4,563	6,949	5,233
29,750	29,800	6,275	4,577	6,966	5,247
29,800	29,850	6,293	4,591	6,984	5,261
29,850	29,900	6,310	4,605	7,001	5,275
29,900	29,950	6,328	4,619	7,019	5,289
29,950	30,000	6,345	4,633	7,036	5,303
30,000					
30,000	30,050	6,363	4,647	7,054	5,317
30,050	30,100	6,380	4,661	7,071	5,331
30,100	30,150	6,398	4,675	7,089	5,345
30,150	30,200	6,415	4,689	7,106	5,359
30,200	30,250	6,433	4,703	7,124	5,373
30,250	30,300	6,450	4,717	7,141	5,387
30,300	30,350	6,468	4,731	7,159	5,401
30,350	30,400	6,485	4,745	7,176	5,415
30,400	30,450	6,503	4,759	7,194	5,429
30,450	30,500	6,520	4,773	7,211	5,443
30,500	30,550	6,538	4,787	7,229	5,457
30,550	30,600	6,555	4,801	7,246	5,471
30,600	30,650	6,573	4,815	7,264	5,485
30,650	30,700	6,590	4,829	7,281	5,499
30,700	30,750	6,608	4,843	7,299	5,513
30,750	30,800	6,625	4,857	7,316	5,527
30,800	30,850	6,643	4,871	7,334	5,541
30,850	30,900	6,660	4,885	7,351	5,555
30,900	30,950	6,678	4,899	7,369	5,569
30,950	31,000	6,695	4,913	7,386	5,583
31,000					
31,000	31,050	6,713	4,927	7,404	5,597
31,050	31,100	6,730	4,941	7,421	5,611
31,100	31,150	6,748	4,955	7,439	5,625
31,150	31,200	6,765	4,969	7,456	5,639
31,200	31,250	6,783	4,983	7,474	5,653
31,250	31,300	6,800	4,997	7,491	5,667
31,300	31,350	6,818	5,011	7,509	5,681
31,350	31,400	6,835	5,025	7,526	5,695
31,400	31,450	6,853	5,039	7,544	5,709
31,450	31,500	6,870	5,053	7,561	5,723
31,500	31,550	6,888	5,067	7,579	5,737
31,550	31,600	6,905	5,081	7,596	5,751
31,600	31,650	6,923	5,095	7,614	5,765
31,650	31,700	6,940	5,109	7,631	5,779
31,700	31,750	6,958	5,123	7,649	5,793
31,750	31,800	6,975	5,137	7,666	5,807
31,800	31,850	6,993	5,151	7,684	5,821
31,850	31,900	7,010	5,165	7,701	5,835
31,900	31,950	7,028	5,179	7,719	5,849
31,950	32,000	7,045	5,193	7,736	5,863

*This column must also be used by a qualifying widow(er).

Continued on next page

Page 40

1987 Tax Table — *Continued*

If line 36 (taxable income) is— At least	But less than	And you are— Single	Married filing jointly *	Married filing separately	Head of a household
		Your tax is—			
32,000					
32,000	32,050	7,063	5,207	7,754	5,877
32,050	32,100	7,080	5,221	7,771	5,891
32,100	32,150	7,098	5,235	7,789	5,905
32,150	32,200	7,115	5,249	7,806	5,919
32,200	32,250	7,133	5,263	7,824	5,933
32,250	32,300	7,150	5,277	7,841	5,947
32,300	32,350	7,168	5,291	7,859	5,961
32,350	32,400	7,185	5,305	7,876	5,975
32,400	32,450	7,203	5,319	7,894	5,989
32,450	32,500	7,220	5,333	7,911	6,003
32,500	32,550	7,238	5,347	7,929	6,017
32,550	32,600	7,255	5,361	7,946	6,031
32,600	32,650	7,273	5,375	7,964	6,045
32,650	32,700	7,290	5,389	7,981	6,059
32,700	32,750	7,308	5,403	7,999	6,073
32,750	32,800	7,325	5,417	8,016	6,087
32,800	32,850	7,343	5,431	8,034	6,101
32,850	32,900	7,360	5,445	8,051	6,115
32,900	32,950	7,378	5,459	8,069	6,129
32,950	33,000	7,395	5,473	8,086	6,143
33,000					
33,000	33,050	7,413	5,487	8,104	6,157
33,050	33,100	7,430	5,501	8,121	6,171
33,100	33,150	7,448	5,515	8,139	6,185
33,150	33,200	7,465	5,529	8,156	6,199
33,200	33,250	7,483	5,543	8,174	6,213
33,250	33,300	7,500	5,557	8,191	6,227
33,300	33,350	7,518	5,571	8,209	6,241
33,350	33,400	7,535	5,585	8,226	6,255
33,400	33,450	7,553	5,599	8,244	6,269
33,450	33,500	7,570	5,613	8,261	6,283
33,500	33,550	7,588	5,627	8,279	6,297
33,550	33,600	7,605	5,641	8,296	6,311
33,600	33,650	7,623	5,655	8,314	6,325
33,650	33,700	7,640	5,669	8,331	6,339
33,700	33,750	7,658	5,683	8,349	6,353
33,750	33,800	7,675	5,697	8,356	6,367
33,800	33,850	7,693	5,711	8,384	6,381
33,850	33,900	7,710	5,725	8,401	6,395
33,900	33,950	7,728	5,739	8,419	6,409
33,950	34,000	7,745	5,753	8,436	6,423
34,000					
34,000	34,050	7,763	5,767	8,454	6,437
34,050	34,100	7,780	5,781	8,471	6,451
34,100	34,150	7,798	5,795	8,489	6,465
34,150	34,200	7,815	5,809	8,506	6,479
34,200	34,250	7,833	5,823	8,524	6,493
34,250	34,300	7,850	5,837	8,541	6,507
34,300	34,350	7,868	5,851	8,559	6,521
34,350	34,400	7,885	5,865	8,576	6,535
34,400	34,450	7,903	5,879	8,594	6,549
34,450	34,500	7,920	5,893	8,611	6,563
34,500	34,550	7,938	5,907	8,629	6,577
34,550	34,600	7,955	5,921	8,646	6,591
34,600	34,650	7,973	5,935	8,664	6,605
34,650	34,700	7,990	5,949	8,681	6,619
34,700	34,750	8,008	5,963	8,699	6,633
34,750	34,800	8,025	5,977	8,716	6,647
34,800	34,850	8,043	5,991	8,734	6,661
34,850	34,900	8,060	6,005	8,751	6,675
34,900	34,950	8,078	6,019	8,769	6,689
34,950	35,000	8,095	6,033	8,786	6,703
35,000					
35,000	35,050	8,113	6,047	8,804	6,717
35,050	35,100	8,130	6,061	8,821	6,731
35,100	35,150	8,148	6,075	8,839	6,745
35,150	35,200	8,165	6,089	8,856	6,759
35,200	35,250	8,183	6,103	8,874	6,773
35,250	35,300	8,200	6,117	8,891	6,784
35,300	35,350	8,218	6,131	8,909	6,801
35,350	35,400	8,235	6,145	8,926	6,815
35,400	35,450	8,253	6,159	8,944	6,829
35,450	35,500	8,270	6,173	8,961	6,843
35,500	35,550	8,288	6,187	8,979	6,857
35,550	35,600	8,305	6,201	8,996	6,871
35,600	35,650	8,323	6,215	9,014	6,885
35,650	35,700	8,340	6,229	9,031	6,899
35,700	35,750	8,358	6,243	9,049	6,913
35,750	35,800	8,375	6,257	9,066	6,927
35,800	35,850	8,393	6,271	9,084	6,941
35,850	35,900	8,410	6,285	9,101	6,955
35,900	35,950	8,428	6,299	9,119	6,969
35,950	36,000	8,445	6,313	9,136	6,983
36,000					
36,000	36,050	8,463	6,327	9,154	6,997
36,050	36,100	8,480	6,341	9,171	7,011
36,100	36,150	8,498	6,355	9,189	7,025
36,150	36,200	8,515	6,369	9,206	7,039
36,200	36,250	8,533	6,383	9,224	7,053
36,250	36,300	8,550	6,397	9,241	7,067
36,300	36,350	8,568	6,411	9,259	7,081
36,350	36,400	8,585	6,425	9,276	7,095
36,400	36,450	8,603	6,439	9,294	7,109
36,450	36,500	8,620	6,453	9,311	7,123
36,500	36,550	8,638	6,467	9,329	7,137
36,550	36,600	8,655	6,481	9,346	7,151
36,600	36,650	8,673	6,495	9,364	7,165
36,650	36,700	8,690	6,509	9,381	7,179
36,700	36,750	8,708	6,523	9,399	7,193
36,750	36,800	8,725	6,537	9,416	7,207
36,800	36,850	8,743	6,551	9,434	7,221
36,850	36,900	8,760	6,565	9,451	7,235
36,900	36,950	8,778	6,579	9,469	7,249
36,950	37,000	8,795	6,593	9,486	7,263
37,000					
37,000	37,050	8,813	6,607	9,504	7,277
37,050	37,100	8,830	6,621	9,521	7,291
37,100	37,150	8,848	6,635	9,539	7,305
37,150	37,200	8,865	6,649	9,556	7,319
37,200	37,250	8,883	6,663	9,574	7,333
37,250	37,300	8,900	6,677	9,591	7,347
37,300	37,350	8,918	6,691	9,609	7,361
37,350	37,400	8,935	6,705	9,626	7,375
37,400	37,450	8,953	6,719	9,644	7,389
37,450	37,500	8,970	6,733	9,661	7,403
37,500	37,550	8,988	6,747	9,679	7,417
37,550	37,600	9,005	6,761	9,696	7,431
37,600	37,650	9,023	6,775	9,714	7,445
37,650	37,700	9,040	6,789	9,731	7,459
37,700	37,750	9,058	6,803	9,749	7,473
37,750	37,800	9,075	6,817	9,766	7,487
37,800	37,850	9,093	6,831	9,784	7,501
37,850	37,900	9,110	6,845	9,801	7,515
37,900	37,950	9,128	6,859	9,819	7,529
37,950	38,000	9,145	6,873	9,836	7,543
38,000					
38,000	38,050	9,163	6,887	9,854	7,559
38,050	38,100	9,180	6,901	9,871	7,576
38,100	38,150	9,198	6,915	9,889	7,594
38,150	38,200	9,215	6,929	9,906	7,611
38,200	38,250	9,233	6,943	9,924	7,629
38,250	38,300	9,250	6,957	9,941	7,646
38,300	38,350	9,268	6,971	9,959	7,664
38,350	38,400	9,285	6,985	9,976	7,681
38,400	38,450	9,303	6,999	9,994	7,699
38,450	38,500	9,320	7,013	10,011	7,716
38,500	38,550	9,338	7,027	10,029	7,734
38,550	38,600	9,355	7,041	10,046	7,751
38,600	38,650	9,373	7,055	10,064	7,769
38,650	38,700	9,390	7,069	10,081	7,786
38,700	38,750	9,408	7,083	10,099	7,804
38,750	38,800	9,425	7,097	10,116	7,821
38,800	38,850	9,443	7,111	10,134	7,839
38,850	38,900	9,460	7,125	10,151	7,856
38,900	38,950	9,478	7,139	10,169	7,874
38,950	39,000	9,495	7,153	10,186	7,891
39,000					
39,000	39,050	9,513	7,167	10,204	7,909
39,050	39,100	9,530	7,181	10,221	7,926
39,100	39,150	9,548	7,195	10,239	7,944
39,150	39,200	9,565	7,209	10,256	7,961
39,200	39,250	9,583	7,223	10,274	7,979
39,250	39,300	9,600	7,237	10,291	7,996
39,300	39,350	9,618	7,251	10,309	8,014
39,350	39,400	9,635	7,265	10,326	8,031
39,400	39,450	9,563	7,279	10,344	8,049
39,450	39,500	9,670	7,293	10,361	8,066
39,500	39,550	9,688	7,307	10,379	8,084
39,550	39,600	9,705	7,321	10,396	8,101
39,600	39,650	9,723	7,335	10,414	8,119
39,650	39,700	9,740	7,349	10,431	8,136
39,700	39,750	9,758	7,363	10,449	8,154
39,750	39,800	9,775	7,377	10,466	8,171
39,800	39,850	9,793	7,391	10,484	8,189
39,850	39,900	9,810	7,405	10,501	8,206
39,900	39,950	9,828	7,419	10,519	8,224
39,950	40,000	9,845	7,433	10,536	8,241
40,000					
40,000	40,050	9,863	7,447	10,554	8,259
40,050	40,100	9,880	7,461	10,571	8,276
40,100	40,150	9,898	7,475	10,589	8,294
40,150	40,200	9,915	7,489	10,606	8,311
40,200	40,250	9,933	7,503	10,624	8,329
40,250	40,300	9,950	7,517	10,641	8,346
40,300	40,350	9,968	7,531	10,659	8,364
40,350	40,400	9,985	7,545	10,676	8,381
40,400	40,450	10,003	7,559	10,694	8,399
40,450	40,500	10,020	7,573	10,711	8,416
40,500	40,550	10,038	7,587	10,729	8,434
40,550	40,600	10,055	7,601	10,746	8,451
40,600	40,650	10,073	7,615	10,764	8,469
40,650	40,700	10,090	7,629	10,781	8,486
40,700	40,750	10,108	7,643	10,799	8,504
40,750	40,800	10,125	7,657	10,816	8,521
40,800	40,850	10,143	7,671	10,834	8,539
40,850	40,900	10,900	7,685	10,851	8,556
40,900	40,950	10,178	7,699	10,869	8,574
40,950	41,000	10,195	7,713	10,886	8,591

*This column must also be used by a qualifying widow(er).

Continued on next page

Page 41

ADDITIONAL ANSWERS

CHAPTER 3
Page 47

11. $26.25	**12.** $135	**13.** $161.25	**14.** $12.11
15. $17.50	**16.** $112.50	**17.** $130	**18.** $9.76
19. $21	**20.** $120	**21.** $141	**22.** $10.59
23. $39	**24.** $165	**25.** $204	**26.** $15.32
27. $42.03	**28.** $198	**29.** $240.03	**30.** $18.03
31. $32.25	**32.** $276.89	**33.** $309.14	**34.** $23.22

CHAPTER 6
Page 113

9. 15¢ per oz; 12.5¢ per oz; 10 oz for $1.25
12. 28¢ per oz; 26.1¢ per oz; 18 oz for $4.70
15. 90¢ per lb; 87.5¢ per lb; 10 lb for $8.75
18. $1.58 per L; $1.40 per L; 2 L for $2.79

CHAPTER 9
Page 171

1. $11,189	**2.** $13,751.04	**3.** $2,562.04	**4.** $16,351.04
5. $5,258	**6.** $6,179.04	**7.** $921.04	**8.** $11,829.04
9. $7,408	**10.** $9,155.52	**11.** $1,747.52	**12.** $13,425.52
13. $8,924	**14.** $11,630.40	**15.** $2,706.40	**16.** $12,330.40
17. $9,401	**18.** $11,618.88	**19.** $2,217.88	**20.** $14,693.88
21. $9,290	**22.** $10,917.36	**23.** $1,627.36	**24.** $16,607.36
25. $8,803	**26.** $10,879.68	**27.** $2,076.68	**28.** $15,754.68
29. $8,873	**30.** $17,379.00	**31.** $8,506.00	**32.** $24,329.00

Page 177

11. $223.10	**12.** $183.42	**13.** $72.65	**14.** $479.17	**15.** 3.40	**16.** $1,629.18
17. $280.25	**18.** $197.38	**19.** $56.90	**20.** $534.53	**21.** 1.85	**22.** $988.88
23. $251.15	**24.** $141.71	**25.** $72.65	**26.** $465.51	**27.** 2.10	**28.** $977.57
29. $234.40	**30.** $197.38	**31.** $72.65	**32.** $504.43	**33.** 1.65	**34.** $832.31
35. $266.70	**36.** $141.71	**37.** $56.90	**38.** $465.31	**39.** 2.35	**40.** $1,093.48
41. $234.40	**42.** $183.42	**43.** $72.65	**44.** $490.47	**45.** 1.30	**46.** $637.61

Page 179

5. $36	**6.** 100 mi	**7.** $31	**8.** $67
9. $29	**10.** 532 mi	**11.** $138.32	**12.** $254.32
13. $48	**14.** 1,632 mi	**15.** $669.12	**16.** $765.12
17. $41	**18.** 0 mi	**19.** $0	**20.** $287
21. $48	**22.** 698 mi	**23.** $286.18	**24.** $526.18
25. $36	**26.** 263 mi	**27.** $81.53	**28.** $189.53

ADDITIONAL ANSWERS

CHAPTER 10
Page 199

2. Red Line 3 stops to Metro Center, change to the Orange or Blue Line, and go 8 more stops

3. Red Line 3 stops to Metro Center, change to the Blue Line, and go 5 more stops

6. Yellow Line 2 stops to Gallery Place, change to the Red Line, and go 1 more stop

CHAPTER 11
Page 227

21. A: Form 1040 to allow itemized deductions
B: Form 1040A because a standard deduction will be taken
C: Form 1040EZ because a standard deduction will be taken, interest income is less than $400, there are no adjustments or dependents, and the person is single

CHAPTER 13
Page 269

1. 108 square ft	**2.** 51 square ft	**3.** 57 square ft	**4.** 1 gal
5. $16.50	**6.** 375 square ft	**7.** 51 square ft	**8.** 324 square ft
9. 1 gal	**10.** $16.50	**11.** 128 square ft	**12.** 51 square ft
13. 77 square ft	**14.** 1 gal	**15.** $16.50	
16. 1,920 square ft	**17.** 102 square ft	**18.** 1,818 square ft	**19.** 10 gal
20. $165			
21. 5,600 square ft	**22.** 102 square ft	**23.** 5,498 square ft	**24.** 28 gal
25. $462			

CHAPTER 15
Page 307

12. Living expenses: $937; Fixed expenses: $616; Annual expenses: $3,223; Monthly share of annual expenses: $269; Total expenses: $1,822; Extra income needed: $119

CHAPTER 17
Page 349

1.	(H, H)	**2.** (T, H, T)	(H, H, T)
	(H, T)	(H, T, T)	(H, T, H)
	(T, H)	(T, T, T)	(T, H, H)
	(T, T)	(H, H, H)	(T, T, H)

3. (1, 1) (1, 2) (1, 3) (1, 4) (1, 5) (1, 6)
(2, 1) (2, 2) (2, 3) (2, 4) (2, 5) (2, 6)
(3, 1) (3, 2) (3, 3) (3, 4) (3, 5) (3, 6)
(4, 1) (4, 2) (4, 3) (4, 4) (4, 5) (4, 6)
(5, 1) (5, 2) (5, 3) (5, 4) (5, 5) (5, 6)
(6, 1) (6, 2) (6, 3) (6, 4) (6, 5) (6, 6)

ADDITIONAL ANSWERS

CHAPTER 17
Page 349

4. (1, 1) (1, 2) (1, 3) (1, 4) (1, 5) (1, 6) (1, 7) (1, 8)
(2, 1) (2, 2) (2, 3) (2, 4) (2, 5) (2, 6) (2, 7) (2, 8)
(3, 1) (3, 2) (3, 3) (3, 4) (3, 5) (3, 6) (3, 7) (3, 8)
(4, 1) (4, 2) (4, 3) (4, 4) (4, 5) (4, 6) (4, 7) (4, 8)
(5, 1) (5, 2) (5, 3) (5, 4) (5, 5) (5, 6) (5, 7) (5, 8)
(6, 1) (6, 2) (6, 3) (6, 4) (6, 5) (6, 6) (6, 7) (6, 8)
(7, 1) (7, 2) (7, 3) (7, 4) (7, 5) (7, 6) (7, 7) (7, 8)
(8, 1) (8, 2) (8, 3) (8, 4) (8, 5) (8, 6) (8, 7) (8, 8)

5. (R, R) (B, R) (G, R)
(R, B) (B, R) (G, R)
(R, B) (B, B) (G, B)
(R, G) (B, G) (G, B)

6. (2, 10) (4, 10) (6, 10)
(2, 20) (4, 20) (6, 20)
(2, 30) (4, 30) (6, 30)

7. (Up, 1) (Down, 1)
(Up, 2) (Down, 2)
(Up, 3) (Down, 3)

8. (A, 1) (B, 1) (C, 1) (D, 1)
(A, 2) (B, 2) (C, 2) (D, 2)
(A, 3) (B, 3) (C, 3) (D, 3)
(A, 4) (B, 4) (C, 4) (D, 4)
(A, 5) (B, 5) (C, 5) (D, 5)
(A, 6) (B, 6) (C, 6) (D, 6)

Page 357

1. Martin, Jane, Eli
Martin, Eli, Jane
Jane, Martin, Eli
Jane, Eli, Martin
Eli, Martin, Jane
Eli, Jane, Martin

3. Jason, Meg, Ben, Doris
Jason, Meg, Doris, Ben
Jason, Ben, Meg, Doris
Jason, Ben, Doris, Meg
Jason, Doris, Meg, Ben
Jason, Doris, Ben, Meg
Ben, Jason, Meg, Doris
Ben, Jason, Doris, Meg
Ben, Meg, Jason, Doris
Ben, Meg, Doris, Jason
Ben, Doris, Jason, Meg
Ben, Doris, Meg, Jason
Meg, Jason, Ben, Doris
Meg, Jason, Doris, Ben
Meg, Ben, Jason, Doris
Meg, Ben, Doris, Jason
Meg, Doris, Jason, Ben
Meg, Doris, Ben, Jason
Doris, Jason, Meg, Ben
Doris, Jason, Ben, Meg
Doris, Meg, Jason, Ben
Doris, Meg, Ben, Jason
Doris, Ben, Jason, Meg
Doris, Ben, Meg, Jason
24 arrangements

4. 357 537 735 935
375 573 753 953
359 539 759 957
395 593 795 975
379 579 739 937
397 597 793 973
24 arrangements

5. Sally and Kim
Sally and Dave
Sally and Shaun
Kim and Dave
Kim and Shaun
Dave and Shaun

ADDITIONAL ANSWERS

CHAPTER 17

Page 357

6. 1 and 2, 1 and 3, 1 and 4, 1 and 5
2 and 3, 2 and 4, 2 and 5
3 and 4, 3 and 5
4 and 5

7. 1, 2, 3, 4
1, 2, 3, 5
1, 2, 3, 6
1, 2, 4, 5
1, 2, 4, 6
1, 2, 5, 6
1, 3, 4, 5
1, 3, 4, 6
1, 3, 5, 6
1, 4, 5, 6
2, 3, 4, 5
2, 3, 4, 6
2, 4, 5, 6
2, 3, 5, 6
3, 4, 5, 6
15 combinations

8. 1st, 1st, 2nd, 3rd
1st, 1st, 1st, 2nd
1st, 1st, 1st, 3rd
1st, 1st, 3rd, 3rd
1st, 1st, 2nd, 2nd
1st, 1st, 1st, 1st

Page 358

8. **9.**

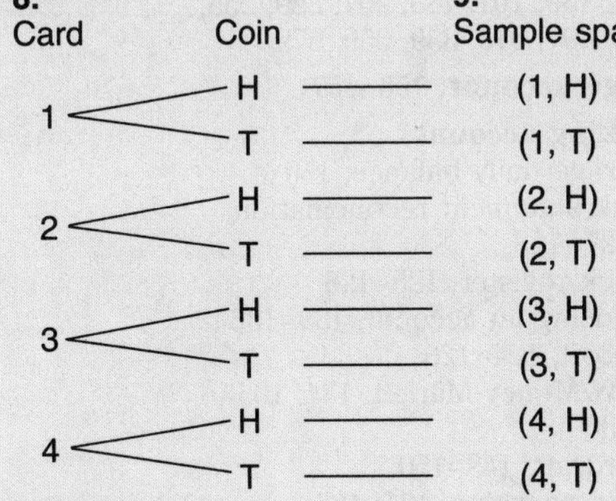

Page 359

13.

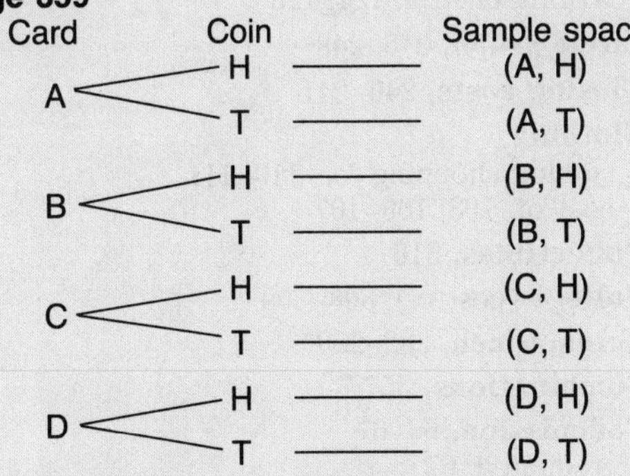

14. (1, 1) (2, 1) (3, 1) (4, 1) (5, 1) (6, 1)
(1, 2) (2, 2) (3, 2) (4, 2) (5, 2) (6, 2)
(1, 3) (2, 3) (3, 3) (4, 3) (5, 3) (6, 3)

INDEX

D

E

M

N

O

P

Q

R

S

T

U

V

W

Y

Credits

Book Design and Production: Textart Inc.
Cover Design: Textart Inc.
Cover Photo: Robert Fishman
Technical Art: Network Graphics

Illustrations: *Eva Vagreti Cockrille* 30, 31, 37, 62, 84, 86, 87, 88, 253; *Mark Giglio* 29, 33, 81, 114, 289, 291; *Steve Marchesi* 28, 92, 119; *Samantha Smith* 106, 107, 148, 154, 197, 302; 197 Subway Map "Used by permission of the Washington Metropolitan Area Transit Authority. Designed by Michael Hertz Associates, New York, for the WMATA Office of Marketing. © 1984 WMATA."

Picture Credits:
Contents Pages: iii, iv–v, vi–vii, viii–ix
HRW photo Ken Karp

Chapter 1 Opener: Page 1 Janet Pinneau/Allsport Photographic/Woodfin Camp Assoc.; **15** Stan Tess/Stock Market

Chapter 2 Opener: Page 21 HRW Photo Michal Heron; **27** Simon Nathan/The Stock Market; **29** Craig Hammell/The Stock Market; **32** Lew Merrim/Monkmeyer Press; **34–35** HRW Photo Ken Karp

Chapter 3 Opener: Page 39 HRW Photo Michal Heron; **44** Alvis Upitis/Image Bank; **48** HRW Photo Ken Karp

Chapter 4 Opener: Page 57 Palmer Kane/The Stock Market; 58 Cliff Feulner/Image Bank, **64** HRW Photo Ken Lax; **75** HRW Photo Ken Lax

Chapter 5 Opener: Page 83 Co Rentmeester; **88** National Philatelic Collection, Smithsonian Institution; **90** Kirk Schlea/Focus West; **94–95** HRW Photo Ken Karp; **97** HRW Photo Ken Lax

Chapter 6 Opener: Page 103 HRW Photo Richard Haynes; **115** HRW Photo Richard Haynes

Chapter 7 Opener: Page 123 Chris Jones/The Stock Market; **130** Jon Ferngersh/The Stock Market

Chapter 8 Opener: Page 143 HRW Photo Richard Haynes; **150** HRW Photo Ken Lax; **157** Dan McCoy/Rainbow

Chapter 9 Opener: Page 167 HRW Photo Michal Heron; **171** Gabe Palmer/The Stock Market; **173** Jay Freis/Image Bank; **179** HRW Photo Richard Haynes

Chapter 10 Opener: Page 187 Tom Bean/The Stock Market; **200** Lew Merrim/Monkmeyer Press; **202** James Cook/Folio Inc.; **204–205** HRW Photo Ken Karp

Chapter 11 Opener: Page 211 John Ficara/Woodfin Camp Assoc.; **217** HRW Photo Ken Karp; **223** Michael Quackenbush/ Image Bank; **224** HRW Photo Ken Karp

Chapter 12 Opener: Page 233 HRW Photo Richard Haynes; **236** Craig Hammell/The Stock Market; **242** Richard Hutchings/ Photo Researchers; **247** HRW Photo Ken Karp; **250** John Coletti/Stock, Boston

Chapter 13 Opener: Page 261 Gabe Palmer/The Stock Market; **268** HRW Photo Ken Lax; **273** Michael Pasdzior/Image Bank; **275** HRW Photo Ken Lax

Chapter 14 Opener: Page 281 Enrico Ferorelli/Dot; **284** Hank Morgan/Rainbow; **286** HRW Photo Ken Lax; **290** Alvis Upitis/ Image Bank; **292** Gabe Palmer/The Stock Market; **295** HRW Photo Ken Karp

Chapter 15 Opener: Page 301 Bettmann Archive; **304** HRW Photo Ken Lax; **307** Randy Duchaine/The Stock Market; **308** Roger Miller/Image Bank

Chapter 16 Opener: Page 319 Ted Russell/Image Bank; **323** Dan McCoy/Rainbow; **327** Charles Feil/Stock Boston; **331** Jay Freis/Image Bank; **334** Harvey Lloyd/The Stock Market; **337** HRW Photo Ken Karp

Chapter 17 Opener: Page 345 Hank Morgan/Photo Researchers; **349** HRW Photo Ken Karp; **351** HRW Photo Ken Karp; **355** J. Rettaliata/Focus West; **356** Alan Choisnet/Image Bank; **357** Ben Rose/Image Bank

Chapter 18 Opener: Page 363 Hank Morgan/Rainbow; **369** HRW Photo Richard Haynes; **371** Sepp Seitz/Woodfin Camp Assoc.; **373** HRW Photo Ken Lax; **376** HRW Photo Ken Karp

486